# Studies in Computational Intelligence

Volume 1010

**Series Editor**

Janusz Kacprzyk, Polish Academy of Sciences, Warsaw, Poland

The series "Studies in Computational Intelligence" (SCI) publishes new developments and advances in the various areas of computational intelligence—quickly and with a high quality. The intent is to cover the theory, applications, and design methods of computational intelligence, as embedded in the fields of engineering, computer science, physics and life sciences, as well as the methodologies behind them. The series contains monographs, lecture notes and edited volumes in computational intelligence spanning the areas of neural networks, connectionist systems, genetic algorithms, evolutionary computation, artificial intelligence, cellular automata, self-organizing systems, soft computing, fuzzy systems, and hybrid intelligent systems. Of particular value to both the contributors and the readership are the short publication timeframe and the world-wide distribution, which enable both wide and rapid dissemination of research output.

This series also publishes Open Access books. A recent example is the bookSwan, Nivel, Kant, Hedges, Atkinson, Steunebrink: The Road to General Intelligence https://link.springer.com/book/10.1007/978-3-031-08020-3.

Indexed by SCOPUS, DBLP, WTI Frankfurt eG, zbMATH, SCImago.

All books published in the series are submitted for consideration in Web of Science.

Saad G. Yaseen
Editor

# Digital Economy, Business Analytics, and Big Data Analytics Applications

 Springer

*Editor*
Saad G. Yaseen
Business Analytics and Information Systems
Faculty of Business
Al-Zaytoonah University of Jordan
Amman, Jordan

ISSN 1860-949X          ISSN 1860-9503  (electronic)
Studies in Computational Intelligence
ISBN 978-3-031-05260-6       ISBN 978-3-031-05258-3  (eBook)
https://doi.org/10.1007/978-3-031-05258-3

# Preface

As the digital economy infiltrates almost every aspect of our lives and is moving at a breakneck speed, smart technologies including computational intelligence systems are changing our global business world. The added value of smart technologies is only realized when businesses apply big data analytics to uncover insights that can improve business optimization, competitiveness, and innovation performance. A thorough examination of the antecedents, consequences, and diffusion of the business analytics and big data analytics application shows that computational intelligence systems play a crucial role in business. Thus, the current book highlights cutting-edge applied research in the fields of business analytics, big data analytics, and business intelligence paradigms, models, techniques, tools, and applications. It presents revised and extended of the best research papers submitted at the 17th SICB 2021 international conference on "Digital Economy and Business Analytics", held in Amman, Al-Zaytoonah University of Jordan, Faculty of Business from October 25–27, 2021.

The carefully selected blind peer-reviewed research papers cover a wide range of theories, technologies, methodologies, and applications in fields of the digital economy and business analytics applications.

The research papers reflect a wide and diverse team of scholars and researchers, including authors from the USA, UK, Saudi Arabia, France, Algeria, Ukraine, Tunisia, United Arab Emirates, Canada, Iraq, Bangladesh, Malaysia, India, Bulgaria, Kosovo, Morocco, Qatar, Nigeria, Palestine, Denmark, Oman, Portugal, Vietnam, Romania, Pakistan, Egypt, Turkey, and Jordan.

Though this book is a new contribution to the study of business analytics, big data, and business technology management, we hope that some essential topics that are not covered in the current book will be addressed soon.

We believe that this book will serve as a knowledge source for professors, students, decision-makers, and business leaders. Finally, we would like to thank all the authors for their added value contributions to this book.

Amman, Jordan                                                                                     Prof. Saad G. Yaseen

# Contents

# Digital Image Processing for Remote Sensing of the Earth's Surface

**Vladimir Vychuzhanin, Nickolay Rudnichenko, Yurii Bercov, Andrii Levchenko, and Alexey Vychuzhanin**

**Abstract** The article presents a theoretical generalization and development of one of the possible solutions to the problem of developing a method for adaptive image compression, which ensures the fulfillment of the requirement for the efficiency of processing and transmission of information in real time at an acceptable level of distortion of the reconstructed image. A method for assessing the degree of saturation of images has been developed, based on taking into account the global and local sensitivity of the basic functions of the Haar transformation using the energy index when analyzing the image transformant, which makes it possible to estimate the degree of saturation of the image of the earth's surface with small details from the energy distribution for different images by zonal sequences.

**Keywords** Remote sensing of the earth's surface · Compression of images · Energy indicator · Saturation of images

## 1 Introduction

Remote sensing of the earth's surface using unmanned aerial systems (UAC) is widely used, for example, to monitor vegetation and environmental parameters, aimed in particular at optimizing activities in agriculture and forestry. In this context, UACs have become useful for assessing crop health by acquiring large amounts of raw data that require processing to further support applications such as moisture, biomass, and others.

The analysis of known works indicates a constant increase in the amount of information received from the UAC during remote sensing of the earth's surface,

V. Vychuzhanin (✉) · N. Rudnichenko · A. Vychuzhanin
Odessa National Polytechnic University, Shevchenko Avenue 1, Odessa 65001, Ukraine
e-mail: vint532@yandex.ua

Y. Bercov · A. Levchenko
I. I. Mechnikov Odessa National University, Nevskogo street 36, Odessa 65016, Ukraine

S. G. Yaseen (ed.), *Digital Economy, Business Analytics, and Big Data Analytics Applications*, Studies in Computational Intelligence 1010,
https://doi.org/10.1007/978-3-031-05258-3_1

which outstrips the growth rate of the capabilities of technical means for its processing [1].

In the process of conducting remote sensing of the earth's surface, one of the ways to solve the problem of operational processing and transmission of images in real time is to reduce the amount of data by compressing images. The image compression method is understood as a set of actions that allows you to unambiguously match a set of compressed data to the original dataset [2, 3].

Analysis of existing varieties of image compression methods shows that most modern compression methods consist of several stages.

The first step in most compression methods involves changing the color model. Currently, various color models are used, which are described in sufficient detail, for example, in publications [3–6]. The high efficiency of using a color model change as a stage of image compression is confirmed by the experience of operating existing compression methods [7]. However, the use of this stage leads to a significant increase in the computational complexity of the compression algorithm (by 20/40%), which is associated with the need to carry out about 9 × n real multiplication operations and 6 × n real addition operations (where n is the number of samples in the image).

The second stage—transformation of the image (for example, orthogonal or wavelet) allows you to translate the original signal from the space–time domain into the spectral-frequency domain. In this case, the original signal (function of time) can be represented through an orthonormal set of spectral functions in the form of a series. Representation of signals in the form of a set of spectral functions allows their spectral analysis, convolution of complex signals, processing in the spectral domain with decorrelated spectral signal elements, etc. In addition, the use of transforming the incoming sequence of image samples allows redistributing the image energy. Using the transform of the processed image provides further compression procedures with decorrelated transform coefficients.

To assess the effectiveness of transformations in [8], the following particular indicators are used: $K_{sum/sub}$, $K^{(-1)}_{sum/sub}$—the number of addition/subtraction operations in direct and inverse transformations, respectively; $K_{mul/div}$, $K^{(-1)}_{mul/div}$—the number of multiplication/division operations for direct and inverse transformations, respectively; $T_{op}$—the type of operations (operands) used in the conversion procedure; $S_{coef}$—sensitivity of conversion factors; $D_{tr}$—dynamic range of transformation ratio values; $\sigma$—root-mean-square deviation during conversion.

The standard deviation (SD) is a quantitative indicator and characterizes the error of the reconstructed image relative to the original one. SD is determined by the following expression

$$\sigma = \sqrt{\frac{\sum_{i=1}^{N} \sum_{j=1}^{M} \left(x_{i,j}^{(\beta)} - x_{i,j}^{(\mu)}\right)^2}{N \cdot M}}, \tag{1}$$

where $x_{i,j}^{(\beta)}$, $x_{i,j}^{(\mu)}$—i, j-th sample of the reconstructed and original images, respectively; $N$, $M$—the dimension of the image horizontally and vertically.

**Table 1** Two-dimensional transform efficiency parameters: DCT, DWT, HT and WT in the Haar basis

|  |  | Conversion | | | |
|---|---|---|---|---|---|
|  |  | DCT | DWT | HT | WT (L steps) |
| Performance indicator | $K_{sum/sub}/K_{sum/sub}^{(-1)}$ | $4N^2 \log_2 N$ | $2N^2 \log_2 N$ | $2(N-1)$ | $N^2 \sum_{k=0}^{L-1} \left( \frac{N}{2^k} - 1 \right)$ |
|  | $K_{mul/div}/K_{mul/div}^{(-1)}$ | $4N^2 \log_2 N$ | $N$ | $N$ | $N^2 \sum_{k=0}^{L-1} \left( \frac{N}{2^k} - 1 \right)$ |
|  | $T_{op}$ | Real | Integer | Partially real | Real |
|  | $D_{tr}$ | Increases by 2–3 times | Not increasing | Not increasing | Increases by 1,5 times |
|  | $S_{coef}$ | Global | Global | Global and local | Local (with the Haar basis) |

Conversion coefficient sensitivity $S_{coef}$ divided into local and global [9]. With global sensitivity, the transformation coefficient is a function of all coordinates of the input sequence space, and for local sensitivity, only a certain part of the coordinates. Accordingly, to calculate the conversion factor with global sensitivity, it is necessary to perform more arithmetic operations than for the coefficient with local sensitivity. The sensitivity of the conversion coefficients is determined by the used orthogonal basis.

Currently, the following bases are widely used: discrete cosine transform (DCT), discrete Walsh transform (DWT), Haar transform (HT) [10] and various wavelet transforms (WT). Table 1 shows the values of the proposed performance indicators for the considered transformations [9, 11–13]. The use of WT according to the Haar basis provides the smallest increase in the dynamic range (1.5 times), when using other bases, the increase will be greater. The advantages of the orthogonal transformation of the original image according to the Haar basis include low computational complexity with a relatively low SD. Analysis of the efficiency of two-dimensional transformations: DCT, DWT, HT and WT in the Haar basis (Table 1) allows us to single out the HT and DWT transformations as the most promising in terms of the presented indicators, provided that the requirements for the efficiency of image processing are met. Despite the small SDs that DCT and WT provide, their significant drawback is the increased dynamic range, which leads to the need to perform quantization and, as a result, to an increase in SD. Analysis of the Table. 1 shows that the Haar transforms have the best values of efficiency indicators.

Existing lossy compression techniques such as JPEG, JPEG-2000, TIFF, WI and others use conversion as a preparatory step for quantization or selection. At this stage, the search and elimination of transformation coefficients is performed, the contribution of which to the formation of the restored image is minimal. In this case, it is taken into account that to obtain an accurate approximation of the initial data, it is not required to use all the transformation coefficients. To do this, the JPEG method uses a procedure called quantization. To quantize, simply divide the conversion factors by another number and round to the nearest integer. To determine the quantum value in JPEG, an array of 64 elements is used—a quantization table, in which values for each coefficient of the processed $8 \times 8$ image block are defined.

Another approach to finding and eliminating transform coefficients can be the selection of transform coefficients obtained as a result of the transformation transform. Selection is based on the zonal, threshold and zonal-threshold principle [3, 14, 15]. In zonal selection, only those coefficients are selected that are in a predetermined zone (usually in the lower spatial frequency region). In the case of threshold selection, coefficients are selected that exceed a certain threshold level [15]. Analysis of modern works [16, 17] shows that zonal-threshold selection is a combination of zonal and threshold selection and allows you to effectively use the coding procedure in the future.

The final step in most compression methods is encoding, which results in a compressed sequence. It is known that character sets in digital image processing are almost always natively represented by the use of fixed-length encoding methods such as ASCII or EBCDIC, which do not take into account the frequency of characters. Moreover, each symbol is encoded with a constant number of bits, which ensures a high computation speed. However, in applications such as image compression, where the amount of information is important, variable length codes are used.

The analysis of works [3, 18] showed that the development and widespread use of existing compression methods received statistical coding algorithms. The procedure for generating such codes for a set of values, based on their frequencies, is quite simple and may includes two stages: modeling and coding. If the elements probability distribution $p(s)$ generated by the source is known, then the length of the character codes of the alphabet is proportional $- \log p(s)$. However, in most cases, the statistical source model is not known, therefore, it is necessary to build a source model that would allow us to estimate the probability of each element occurrence at each input sequence position.

There are some algorithms [19], which form the source model as the data stream is processed or use a fixed model based on a priori ideas about the data nature. The encoding procedure effectiveness depends on the following parameters: entropy $H(S)$ coding source $S$ (for example, according to the Bernoulli scheme $H(S) = \sum_{i=1}^{n} p(s_i) \log_2 p(s_i)$); cost $C(f, S)$ source $S$ coding $f$; coding redundancy $R(f, S)$. A hallmark of optimal codes $f_0$ is that they do not expand the compressed data in the worst case and provide such an encoding cost $C(f_0, S)$ at which the inequality $R(f_0, S) \leq R(f, S)$ holds for any coding $f$. By Shannon's theorem, the best compression when represented in binary form can be obtained by encoding characters with a relative frequency $p(s)$ using the—$\log_2 p(s)$ bit.

The redundancy of Huffman coding approaches optimal only when the relative frequencies are multiples of two. It was shown in [18, 19] that the use of arithmetic coding gives results close to optimal results. Thus, the existing compression methods are a set of procedures, the implementation of which is aimed at reducing various types of redundancy in the representation of images. The analysis of the considered compression methods showed that ensuring the efficiency of processing and transmission of images of the earth's surface with UAC at an acceptable level of their distortion is an urgent task. To solve it, it is necessary to carry out: orthogonal transformation of the original image according to the Haar basis, the advantage of which is low computational complexity with a relatively low SD; carrying out zonal or zonal-threshold selection of conversion coefficients, which allows satisfying the requirements for adaptability and interactivity of the developed image compression method; coding of the selected sequence using optimal codes.

## 2  Method for Assessing the Degree of Images Saturation

To achieve this goal, we will use a direct two-dimensional transformation of images on the Haar basis.

Direct and inverse two-dimensional transformation of images on the Haar basis:

$$y_{k,p} = K_{norm} \sum_{i=0}^{N-1} \sum_{j=0}^{N-1} x_{i,j} h_{k,p}^{(1)}(i, j) \tag{2}$$

$$x_{k,p} = \sum_{i=0}^{N-1} \sum_{j=0}^{N-1} y_{i,j} h_{k,p}^{(-1)}(i, j), \tag{3}$$

where $i, j, k, p = \overline{0, (N-1)}$—the coordinates of the image reference location or its transform coefficient in the block N × N; $x_{k,p}$—image block count $X(n)$; $K_{norm} = 1/N^2$—normalization coefficient; $y_{i,j}$—Haar transform coefficient (transform element $Y(n)$); $h_{k,p}^{(1)}(i, j)$, $h_{k,p}^{(-1)}(i, j)$—$i, j$-th element of $k, p$-th submatrix $H_{i,j}^{(1)}$ direct transformation matrix $H_{np}^{(2)}(n)$ and submatrix $H_{i,j}^{(-1)}$ inverse transformation matrix $H_{op}^2(n)$ respectively.

The performed analysis of the transformation transformants showed that to reduce the dynamic range of the transformant, it is advisable to use subquantization of the coefficients in the direct transformation. Conversion factors should be grouped into selection zones that include factors with the same sensitivity. According to the selection rule, the configuration of the selection zones of the Haar transformants has the form shown in Fig. 1. (the solid line shows the boundaries of the transformant zones, and the dashed line shows the transformation coefficients in the zones).

The coefficients of the s-th (s∈a, j) selection zone of all transformants of N blocks form the s-th zonal sequence (ZP) (ZQ). So, for example, f zone ZQ consists of the coefficients f of the selection zones of transformants of all image blocks.

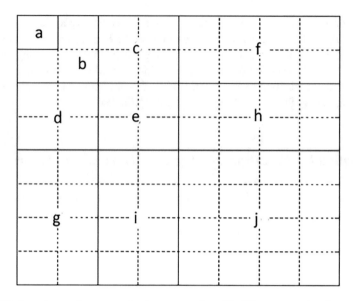

**Fig. 1** SD dependence when converting images samples with different saturation degrees

## 3 Research Original Image Orthogonal Transformation

The significant differences in the ZQ histograms, which are composed of the all blocks selection zones, for different images suggest the possibility of classifying images according to the energy distributed in them.

$$E = \sum_{s=1}^{Z} E_s = \sum_{s=1}^{Z} \sum_{k=1}^{Ks} C_{k,S}^2, \tag{4}$$

where $E_s$—energy $s$-th transform selection zones; $z$—number of breeding zones; $C_{k,s}$—$\varkappa$-th coefficient $s$-th transform zone; $K_S$—coefficients number $C_{k,s}$ at $s$-th zone, nonzero. It is proposed to use the weighted energy parameter $s$-th of the zonal sequence as an indicator that allows us to quantify the image energy distribution over ZQ.

$$Q_S = E_S/K_S \tag{5}$$

The analysis of ZQ properties showed that their energy depends on the number of nonzero samples of the Haar basis function, which take part in the ZQ coefficients formation, and on the correlation value of the original image samples. Therefore, with an increase in the zone number, zonal energy decreases, and the magnitude of this energy reflects which particular details prevail in the image. An experiment was conducted to accumulate statistics on the energy distribution over zonal sequences

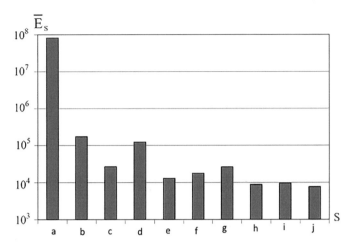

**Fig. 2** Histogram of arithmetic mean values of energy $\overline{E}_s$ s-th areas of analyzed images

for different images. The original 300 images were used in BMP (bitmap) format with a visualization parameter of 24 bits per pixel. The results of the zonal sequences average energy estimation $\overline{E} = \sum_{i=1}^{300} E_{s,i} / 300$ are presented in Fig. 2.

It is proposed to exclude zonal sequences according to the following rules: at $E_s < \overline{E}_s$ s-th ZQ considered priority for exclusion. For example, when processing images with blocks $8 \times 8$, W ZQ exceptions are proposed (W is the number of ZQ that must be excluded in order to obtain the necessary compression ratio $K_{com}$, starting from the z-th) in the following order: 10-th, 9-th and 8-th, 5-th, 7-th and 6-th, 4-th and 3-rd. The choice of this order is due to the zonal sequence influence, which is excluded, on the change $\Delta\sigma$ restored image; if necessary, eliminate ZQ, whose energy $E_s > \overline{E}_s$, the exception is in the same order.

Based on the fact that the values of the coefficients of the selection zone reflect the presence of details in the image of the corresponding size, then, for example, according to the histogram of the 10th ZQ of the third test image, it can be argued that there are more small details in comparison with other test images. According to the histograms of the first image, it can be assumed that there are large areas in the image with a smooth color change. It can be seen that with an increase in the ZQ number, the dynamic range decreases, and the percentage of coefficients with a zero value reaches 40–95%, depending on the processed digital image.

Significant differences in the ZQ histograms, compiled from the selection zones of all blocks, for different images allow for the classification of images.

Figure 3 shows the compression ratio dependence $K_{com}$ from the number of excluded selection zones W. It can be seen that when "senior" zones are excluded from processing (at $W = 1$, the tenth selection zone, at $W = 2$, the tenth and ninth selection zones, etc.). the compression ratio increases slightly (from 1.4 to 8.8 for $W = 7$). A significant increase in the compression ratio becomes possible with the "junior" zones (second and third) exclusion. It is not advisable to exclude the first

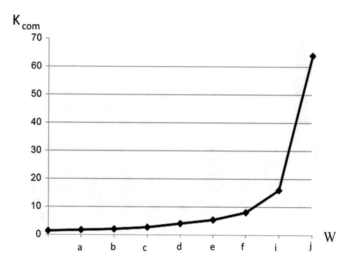

**Fig. 3** Dependence of the compression coefficients $K_{com}$ on the number of excluded selection zones W, starting from the f zone

ZQ, because at least 90% of the image energy is distributed in it, and the coefficients have a global sensitivity, therefore, its exclusion will introduce the greatest distortions. From the analysis of the graph in Fig. 3, it follows that during the execution of the lossy image compression algorithm: to minimize distortions of the reconstructed image, it should be excluded from the processing of the selection zone with the lowest energy; to obtain significant compression ratios, it is necessary to "discard" ZQ with a large number of conversion ratios.

It can be seen from the Figs. 2 and 3 that the most appropriate seems to be an exception from processing j, i. h and e zones, since they have the lowest average energy, and their "dropping" provides a compression ratio $K_{com} = 5$. However, it should be noted that the root-mean-square deviation for high- and medium-resharpened images exceeds 1.5% already with the exclusion of j, i and f ZQ, which are responsible for fine details. Wherein $K_{com}$ does not exceed more than 4–5 times, and distortions occur throughout the image, which does not allow us to choose the level of detail that can be neglected in the given processing conditions.

Methods of zonal and zonal-boundary selection of orthogonal transformation coefficients have been obtained, which differ from the known ones in that: in the case of zonal selection, the features of the energy distribution over the processed image zonal sequences are taken into account and only those that introduce minimal distortions into the reconstructed image are excluded, which makes it possible to control the ratio of the compression ratio and the standard deviation; with zonal-boundary selection, the uneven redistribution of energy over zonal sequences of each processed image block is taken into account, which makes it possible to reduce the entropy relative to the original image by an average of 4/5, increasing the efficiency of using statistical coding.

# 4 Conclusion

A method for assessing the degree of saturation of images has been developed, based on taking into account the global and local sensitivity of the basic functions of the Haar transformation using the energy index when analyzing the image transformant, which makes it possible to estimate the degree of saturation of the image of the earth's surface with small details from the energy distribution for different images by zonal sequences. A total weighted energy index is proposed, which depends on the energy distribution and the number of orthogonal transformation coefficients, forming a zonal sequence. Using the technique for assessing the degree of saturation of images will allow: more accurately, in comparison with expert judgment, to classify images; select the number of saturation classes required in each specific case; automate the classification process when evaluating an image; use the adaptive compression method as a procedure.

# References

1. Su B, Zhang H, Meng H (2017) Development and implementation of a robotic inspection system for power substations. Ind Robot 44:333–342
2. The use of a complex of operational-tactical unmanned aerial reconnaissance (1986)
3. Bondarev VN, Trester G, Chernega VS (2001) Digital signal processing: methods and means
4. Jain AK (1981) Video compression. TIIER. 3:71–117
5. Butakov EA, Ostrovsky VI, Fadeev IL (1987) Computer-aided image processing
6. Krasilnikov NN (1986) Theory of transmission and perception of images. The theory of image transfer and its applications
7. Stryuk AYu, Bohan KA (2002) Color models in video compression systems. Radio Electron Inf 1:23–25
8. Huang TS (1979) Image processing and digital filtering
9. Smirnov AV (2001) The basics of digital television: Textbook
10. Anuj B, Rashid A (2009) Image compression using modified fast Haar wavelet transform. World Appl Sci J 5:647–653
11. Haar A (1910) Theorie der ortogonalen Funktion systeme. Math Ann 69:331–371
12. Ahmed N, Rao KR (1980) Orthogonal transformations in digital signal processing
13. Tian X, Jiao L, Duan Y (2012) Video denoising via spatially adaptive coefficient shrinkage and threshold adjustment in surfacelet transform domain. Springer
14. Avduevsky VS et al (1986) Reliability and efficiency in technology. Handbook. Methodology. Organization. Terminology
15. Bohan KA (2003) Ways and means of two-dimensional image conversion in the basis of Haar
16. Chernega VS (1997) Compression of information in computer networks
17. Pratt WK (1982) Digital image processing
18. Korolyov AV, Ruban IV (1997) Intraframe and interframe coding of digital color images. Eng Simul 14:449–457
19. Vatolin D, Ratushnyak A, Smirnov M, Yukin V (2002) Data compression methods. The device archivers, image and video compression

# Image Segmentation Techniques to Support Manual Chest X-Ray Interpretation

**Norkhairani Abdul Rawi, Norhasiza Mat Jusoh,**
**Mohd Nordin Abdul Rahman, Abd Rasid Mamat, and Mokhairi Makhtar**

**Abstract** Analyzing Chest X-ray (CXR) images need to be done carefully and fast. CXR is the most affordable method of diagnosing diseases related to chest lung and related organs in that area. Radiographer analyses CXR using systematic approach and divide the image into segments to be easily examined. It needs a highly skilled radiographer to determine any abnormalities. Thus, a study on how manual interpretation can be matched with computer assisted algorithm to fasten the process has been carried out. Segmentation has been identified as one method in image processing that can be used to assist the interpretation. Few types of segmentation discussed and suggested to be used. In the next stage of research, it can be tested towards CXR datasets available to identify the best algorithm to be used.

**Keywords** CXR images · Machine learning · Image processing · Segmentation

---

Please note that the LNCS Editorial assumes that all authors have used the western naming convention, with given names preceding surnames (first name then last name). This determines the structure of the names in the running heads and the author index. No academic titles or descriptions of academic positions should be included in the addresses. The affiliations should consist of the author's institution, town, and country.

---

N. A. Rawi (✉) · M. N. A. Rahman · A. R. Mamat · M. Makhtar
Faculty of Computing and Informatics, University Sultan Zainal Abidin, 22200 Tembila Campus, Terengganu, Malaysia
e-mail: khairani@unisza.edu.my

M. N. A. Rahman
e-mail: mohdnabd@unisza.edu.my

A. R. Mamat
e-mail: arm@unisza.edu.my

N. M. Jusoh
Faculty of Medical, University Sultan Zainal Abidin, 21200 Medical Campus, Terengganu, Malaysia
e-mail: hasizamj@unisza.edu.my

S. G. Yaseen (ed.), *Digital Economy, Business Analytics, and Big Data Analytics Applications*, Studies in Computational Intelligence 1010,
https://doi.org/10.1007/978-3-031-05258-3_2

# 1  Introduction

Improving diagnosing in medical treatment is a major focus for each country. Together with the availability of technologies that become more affordable and growing fast, the need for automatically detect diseases become more crucial. Researchers keep on doing research to improve the way diseases being diagnosed and treated. Data collected during the current method of diagnosing system used being analyzed to identify the trends and methods so the medical experts can have a better guide in diagnosing diseases. One of the most collected data in medical field are radiological images. Radiological images obtained from various type of devices in medical such as plain radiograph, MRI, CT scan and many more [2]. Among all devices, X-ray machines are the most commonly available devices at medical center throughout the world. This is because of the machine is the cheapest and fastest method in diagnosing any abnormality in human. One of the most frequently performed imaging studies of all radiology is Chest radiographs (CXR) [1].

CXR normally done in two position, depending on the condition of the patient. If the patient able to stand, then the posteroanterior (PA) view will be taken, otherwise anteroposterior (AP) view will be captured through X-ray machine. In this research, the images analysed only those that has been captured through PA position.

# 2  Related Work

Analysing CXR in most hospitals is done manually. Thus, it needs a high skilled medical doctor to do that. The image of CXR obtained through few methods of image acquisition and analyzed using systematic approach [1] as discussed detail in the next part.

## 2.1  Techniques of Obtaining CXR Images

Chest X-ray can be in posteroanterior (PA), anteroposterior (AP) and lateral view. It depends on the conditions of the patient. It shown in Figs. 1, 2 and 3.

Figure 1 shows the first technique in obtaining CXR image. This technique normally used for patients that still able to stand up. This technique is called PA. This technique is where the patient is erect position and facing the upright image receptor. The superior part of the receptor must be 5 cm above the shoulder joints. Patient's chin should rise as to be out of the image area. The shoulders are switched anteriorly to let the scapulae to change laterally off the lung arenas. This can be attained by either hands located on the posterior part of the hips, elbows partially bent rolling anterior or hands are located around the image receptor in a enfolding motion with

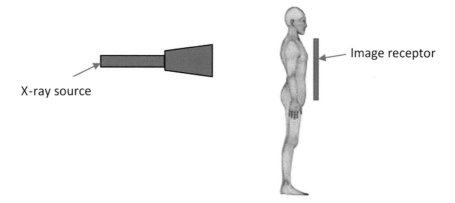

**Fig. 1** Posteroanterior (PA) position in CXR images capture

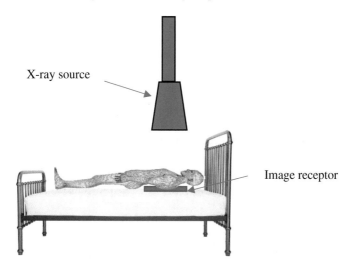

**Fig. 2** Anteroposterior (AP) position in CXR images capture

**Fig. 3** Lateral view of CXR

a focus on the lateral undertaking of the scapulae where shoulders are depressed to move the clavicles below the lung apices.

For the patients that unable to stand up, another technique named anteroposterior (AP) will be used to obtain CXR images. The AP chest view is made with the X-ray tube placed anteriorly. It then firing photons over the patient to extract the image on a sensor placed behind the patient. A detector can be located behind a relatively immovable patient. It is therefore an substitute to the PA view when the patient is too sick to abide stand-up or parting the bed. The patient can be either in sitting or flat situations as in Fig. 2.

Lateral view is a method whereby patient will be standing upright position. The left side of the patients' chest in line to the image receptor. The left shoulder positioned definitely against the image receptor. Both arms raised above the head in order to avoid superimposition over the chest. If possible, patients' arms can be positioned on the head. It also can hold onto handles where chin raised out of the image field. The position in Fig. 3 shown that midsagittal even must be vertical to the opposing X-ray beam.

## 2.2  Manual CXR Interpretation

Radiologist normally used systematic approach in reviewing the CXR images [4] as depicted in Fig. 4. This approach normally used by radiologist to identify the abnormality of CXR. By examining the CXR, the abnormalities found then will be compared to pattern before it can be classified into diagnosis. However, for this project, the abnormalities will only matched to the category of opacity and

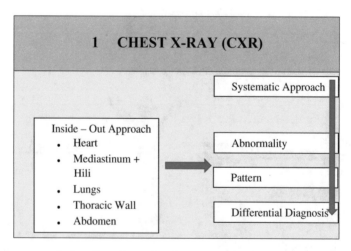

**Fig. 4** Common method used by radiographers in analysing CXR images. Adopted from Smithuis and Delden [4]

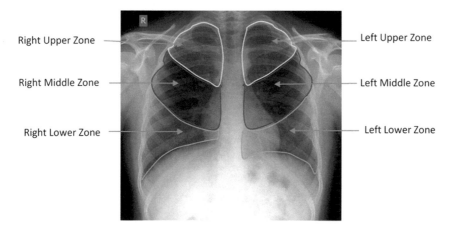

**Fig. 5** Normal CXR images adapted from Aneta Kecler-Pietryzk [3] divided into zone

LUCENCY [7] in the lung field. According to [1] the CXR images can be interpreted systematically.

Once abnormalities detected, radiologist will localize the area according to zone in order to convey the information to other medical doctors in their radiological reports. The zone mentioned earlier as in the Fig. 5. The description of the zone are as follows:

- Upper zone—Area of the lung bounded by the first two(2) anterior ribs.
- Mid zone—Area of the lung bounded between $2^{nd}$ and 4th anterior rib.
- Lower zone—Area of the lung from 4th anterior rib to diaphragm.

All medical doctors should be able to interpret CXR and detect abnormality. However, this abnormality could be missed by inexperience medical doctors. With increasing number of patients all over the world, radiologist unable to report all CXR, this may lead to incorrect interpretation by in charge doctors which may put patients at risk. Therefore, it is very beneficial if an algorithm can be developed to provide alerts to medical doctors when any abnormalities detected. This can be support by some computer aided programs. Basically, the category for the opacity and LUCENCY can be further subcategorized into multiple patterns as shown in Fig. 6.

This study focuses on identification of abnormality which then categorized into opacity and lucency. Opacity is often used interchangeably with density. Opacity refers to an area on the CXR that is brighter (white) than expected. Opacity occurs when X-rays beam are absorbed or blocked by pathology within the lung. It demonstrates as brighter or whiter area on the CXR. Opacity is very non-specific terms, and can represent a variety of lung pathologies. Lucency is the exact contrary of opacity. Lucency refers to an area on the CXR that is darker (black) than expected. It occurs when more X-rays pass through less dense regions such as air-filled lungs. This shows up as a darker or blacker area on the CXR.

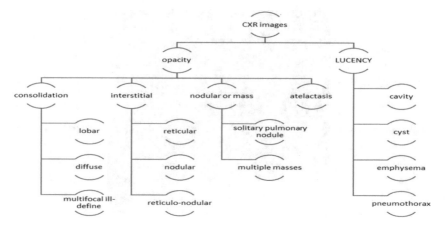

**Fig. 6** Sub categories for opacity and LUCENCY

Since CXR images are in black and white, image processing techniques can be used to identify differences/similarities between images. Among common techniques in processing images are region based, marker based and segmentation. The approach has its own benefits and flaws and depending on images type, colors, shape and density of the image.

Computer aided analysis and interpretation of CXR can provide several useful clues for diagnosis, treatment planning and medical research.

## 3 Methods of CXR Image Classification

Previously, we did discuss on the framework [11] how the semantic meaning of CXR image can be obtained to help the medical doctors and students in identifying related categories for the CXR images. Since the focus of the research only to identify two main categories of CXR images at the first stage, so this paper will discuss in detail how the two main categories which are opacity and LUCENCY will be identified.

As depicted in Fig. 5, it clearly shown how manual interpretation been done towards CXR image. As we can see, the image analyzed using part, or known as segment. Therefore, image segmentation techniques can be implemented to examine the image and identify the classification of the image [5, 6]. But the question is, which segmentation techniques is the most suitable one? Or, it might need a combination or so-called hybrid techniques in order to produce more accurate results. Let's have a look on segmentation techniques available in the next section.

## 3.1  Image Segmentation

Image can consist of single object or multiple objects. If computer need to identify the object, it can be:

(i)   Localization of the objects—when it only has single object
(ii)  Object detection—when it has more than an object

Before able to classify an image, there is a need to identify the image consist of what—which also understood consist of what segment [5, 6]. It is not appropriate to process the whole image compared to partition of images or segment of images. By dividing image into segments, we can make use important segments to process the image. Since image is a collection of pixels, we can group together pixels that have similarities using segmentation. Object detection only build the box around object while segmentation provide a wise mask of object where we get more granular understanding about object in the image. This ability can help in categorize images using segmentation techniques for fast detect especially in healthcare, road traffic, self-driving car and many more. Many algorithms can be implemented for segmentation and some of them are discussed in the next section.

## 3.2  Region-Based Segmentation

Region-based Segmentation is a technique that separate objects into different regions based on some threshold values. Thresholding is the simplest way in segmenting images. By using thresholding, the grayscale image can be an image in binary form. As discussed earlier that manual interpretation of CXR used segmentation to identify whether the CXR classification is opacity or lucency, so region-based can be used to automatically identify both categories.

Dividing an image into regions through partition is a goal of segmentation. The boundaries among regions is classified based on gaps in grayscale or color properties and known as thresholding method. Region-based segmentation is a practice for defining the region directly. Two main methods that commonly used are as follows:

(a)   Region Growing: Partitioning images into different parts according to the growth of initial pixels is the way it has been doing in region growth-based segmentation. These pixels can be nominated manually normally using prior knowledge. It also can be done automatically based on a specific application. Then the growth of the pixels is measured by the linking between the pixels. It can be stopped with the help of the prior knowledge.
(b)   Merging and Splitting Algorithms: in which end-to-end areas are equated and compound if they are close enough in some properties. Also in large non-uniform parts are broken up into smaller zones which may be uniform.

## 3.3   Edge Detection Segmentation

Edge recognition is an image processing exercise for discover the borders of objects in images. It works by spotting interruptions in intensity. Image processing, machine vision and computer are the fields that practiced edge recognition for their image segmentation and data mining [8]. There are a few common algorithms can be used in edge detection. Sobel, Canny and Prewitt are some examples of edge detection algorithms. Besides that, Roberts and fuzzy logic methods algorithm also frequently used for edge detection.

In an image, an edge is a curvature that follows a path of rapid modification in image concentration. Edges are often related with the margins of objects in a section. Edge detection is used to recognize the boundaries in an image. In Mathlab, utilizing the edge function make it easier in finding edges. Two criteria used in the function in order to get the rapid intensity changes in an image. The two criteria mentioned above are:

(i)    Spaces where the initial derivative of the intensity is greater in magnitude compared to some threshold
(ii)   Spaces where the second derivative of the intensity has a nil passage

Edge delivers some derivative estimators, each of which gears one of these descriptions. For some of these estimators, we can specify whether the operation should be sensitive to horizontal edges, vertical edges, or both edge returns a binary image containing 1's where edges are found and 0's elsewhere.

Canny method in Mathlab is the most powerful edge-detection technique for edge function [9]. This technique uses two different thresholds that make it differs from the other edge-detection methods. It uses strong and weak edges for the edge detection process. It also includes the weak edges in the output only if they are associated to strong edges. Canny method less likely than the other method to be affected by noise. It also more likely to spot true weak edges. Canny method composed of five (5) steps as follows [9]:

i.     Noise reduction;
ii.    Gradient calculation;
iii.   Non-maximum suppression;
iv.    Double threshold;
v.     Edge tracking by hysteresis.

This method is based on a grayscale image therefore there is a need to convert images into grayscale before the method can be applied.

## 3.4   Segmentation Based on Clustering

Clustering algorithms are used to group closer the data points that are more similar to each other, from other group data points. Clustering is about defining similarities and differences. Clustering algorithms aim to group the fingerprints in classes of similar elements. The clustering requires the concept of a metric. These algorithms implement the straightforward assumption that similar data belongs to the same class.

There are different types of clustering as follows [10]:

(i)   K-means clustering—group similar data points together and discover underlying patterns

(ii)  FuzzyC-means clustering—assigning membership to each data point corresponding to each cluster center on the basis of distance between the cluster center and the data point.

(iii) Mountain clustering method—based on density estimation in feature space with the highest peak extracted as a cluster center and a new density estimation created for extraction of the next cluster center.

(iv)  Subtractive clustering method—generate the tuned membership functions automatically in accordance to the domain knowledge.

From the discussion above, it is clearly shown how segmentation method can be used to be mapped with manual interpretation of CXR images. Even though the first phase of this research only focuses on two major class of CXR images, it can be further expanded to analyze the sub class of CXR images.

## 4 Conclusion

By using segmentation techniques in images processing, the CXR obtained can be analyzed and classified according to parameters selected. The main two classification of CXR images, which are opacity and LUCENCY can be easily determined. It will shorten the process in defining abnormalities in CXR images for further action to be taken. It also helps the medical team to get alert for any CXR that indicates abnormality. This method also can be used as tool in medical student learning and training.

## References

1. De Lacey G, Morley S, & Berman L (2012) The chest X-ray: a survival guide. Elsevier Health Sciences
2. Mehta IC, Khan ZJ, Khotpal RR (2006) Analysis and review of chest radiograph enhancement techniques. Inf Technol J 5:577–582
3. Kecler-Pietryzk A (n.d.) Normal chest X-ray (12-year-old, male). https://radiopaedia.org/cases/normal-chest-x-ray-12-year-old-male-3?lang=us. Accessed 21 Mar 2019
4. Smithuis R, Delden OV (2013) Chest X-ray—Basic interpretation. http://www.radiologyassistant.nl/en/p497b2a265d96d/chest-x-ray-basic-interpretation.html. Accessed 21 Mar 2019
5. Pal NR, Pal SK (1993) A review on image segmentation techniques. Pattern Recogn 26(9):1277–1294

6. Wang X, Peng Y, Lu L, Lu Z, Bagheri M, Summers RM (2017) Chest X-ray8: hospital-scale chest X-ray database and benchmarks on weakly-supervised classification and localization of common thorax diseases. In: Proceedings of the IEEE conference on computer vision and pattern recognition pp 2097–2106

7. Adam A, Dixon A, Gillard J, Schaefer-Prokop C, Grainger R (2014) Grainger & Allison's Diagnostic Radiology 6th Edition—reference for opacity and lucency. Churchill Livingstone

8. Muthukrishnan R, Radha M (2011) Edge detection techniques for image segmentation. Int J Comput Sci Inf Technol 3(6):259

9. Sahir S (2019) Canny edge detection step by step in python—computer vision. https://towardsda tascience.com/canny-edge-detection-step-by-step-in-python-computer-vision-b49c3a2d8123

10. Seif G (2018) The 5 clustering algorithms data scientists need to know. https://towardsdatascie nce.com/the-5-clustering-algorithms-data-scientists-need-to-know-a36d136ef68. Accessed 10 Jul 2021

11. Rawi NA, Rahaman MNA, Makhtar M, Jusoh NM, Mamat AR (2017) Architecture model for semantic classification of chest X-ray images using machine learning algorithm. World Appl Sci J 35(10):2124–2128

# Determinants of Brand Switching in Cellular Networks in Pakistan

**Waseem Ahmad, Noor Fatima Raza, Tanvir Ahmed, and Fahad Gill**

**Abstract** Customer switching behavior has a straight effect on the company's productivity, sustenance and market shares. Getting a new customer requires a significantly more amount of cost relative to retaining an existing customer. The primary aim of the current study is to identify the factors influencing consumers switching behavior for the cellular service providers in two major metropolitan cities of Pakistan, i.e., Faisalabad and Lahore. For this purpose, the data were collected from 384 users of cellular networks, and a logistic regression model was applied for data analysis. The results revealed positive and significant influence of high internet charges, signal dropping, low internet speed, competitor ads, competitor's better technology, and influence of family and friends on brand switching behavior. This study is beneficial for the service providers to understand the customer's preference for a particular service attribute.

**Keywords** Switching behavior · Internet charges · Signal dropping · Internet speed · Competitor ads · Better technology · Location

## 1 Introduction

Telecommunication includes a wide range of technologies for data transmitting such as tv, phones, internet, fiber optics, etc. Wireless and cellular systems are rapidly becoming the most critical network access types for media transmission services

W. Ahmad (✉) · N. F. Raza
Institute of Business Management Sciences, Faculty of Social Sciences, University of Agriculture, Faisalabad, Pakistan
e-mail: waseem@uaf.edu.pk

T. Ahmed
Department of Economics, Forman Christian College (A Chartered University), Lahore, Pakistan
e-mail: tanvirahmed@fccollege.edu.pk

F. Gill
Senior Quant, NextEra Energy Inc., Florida, USA
e-mail: Fahad.gill@nexteraenergy.com

© The Author(s), under exclusive license to Springer Nature Switzerland AG 2022
S. G. Yaseen (ed.), *Digital Economy, Business Analytics, and Big Data Analytics Applications*, Studies in Computational Intelligence 1010,
https://doi.org/10.1007/978-3-031-05258-3_3

[16]. The telecommunication industry in today's modern and dynamic world plays a vital part by offering communication services that are upheld by new technologies and advancements [17].

Competition between wireless telecommunication industries is growing globally during the last decade [1]. Pakistan's cellular service industry shows significant growth. The number of customers is growing day by day; service network companies are doing their best job to fascinate, facilitate and hold them [1]. The total number of cellular subscribers in Pakistan in June 2021 was 184.25 million [13]. Domestic and International companies are setting up in Pakistan's market, and competition within cellular networks has become more intense [1]. As of now, four cellular service providers named Jazz, Telenor, Zong, and Ufone are working in Pakistan. These four companies work in a very competitive environment for survival across all regions of Pakistan.

Customers are more advantageous for the existing company than the new ones. So, creating new business practices is crucial to understand the components that influence brand switching [5]. The leading issue that a service provider confronts is to provide proficient services to fulfill the necessities/requirements of customers [7]. Ordinarily, the importance given by the service providers is cost and benefits rather than superior service quality.

## 2 Literature Review

According to Anderson and Sullivan [2], switching over to another brand means consumers felt no more motivation in the existing strategies of a brand. Afzal [1] stated that customer retention is possible when companies develop distinguishing competitive advantages after considering different factors. Keaveney [8] proposed the earliest theoretical framework of consumer switching in the 45 different service types, based on eight components i.e. price, service encounter failure, inconvenience, employee response to service failure, attraction by competitors, and core service failure. Behind customer's switching of a brand, core service failure was the top common reason. Masud-Ul-Hasan [12] pointed out the variables influencing brand switching behavior. Results revealed that the foremost reason for brand switching is service failure [10]. Further, Lim [10] concluded that price, competition, and customer service had no significant relationship with customer switching behavior. Kouser [9] used binary logistic regression to study customer satisfaction and concluded that customer satisfaction was more affected by call charges [15], SMS charges, GPRS service, and network services. Fatima [4] evaluated the impact of advertisement of cellular service providers. The study results showed a significant effect of advertising, quality, signal strength, customer service, sales promotion, and coverage on customer perception. Similarly, Manzoor [11] concluded that network quality, brand image, price, value-added services, and promotional activities significantly impact customer switching behavior. Cellular network providers use compelling ways to fulfill their customers' satisfaction because customers' dissatisfaction could become a reason for their switching to some other network. For this reason, companies are examining the market to understand customer's requirements and fulfilling them by offering value-added service. Based on the literature review, the present study has identified various factors that can have an impact on customer brand switching such as SMS charges, call rates, internet charges, signal dropping, internet speed, network coverage, employee response during an issue with the service, competitor's as, competitor's technology, brand image, recommendation from family and friend, hidden charges and location (as shown in Fig. 1). Therefore, the following hypotheses are tested in the present study:

H1: High SMS charges has no influence on consumer's switching behavior.
H2: High call rates has no effect on consumer's switching behavior.
H3: High internet charges has no significant influence on consumer's switching behavior.
H4: Signal dropping has no effect on consumer's switching behavior.
H5: Low internet speed has no significant influence on consumer's switching behavior.
H6: Poor network coverage has no effect on consumer's switching behavior.

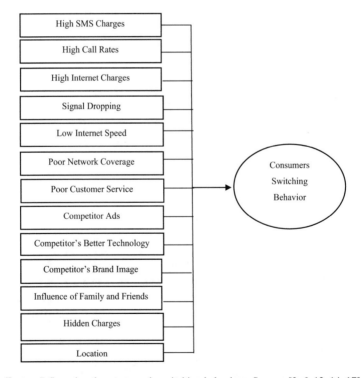

**Fig. 1** Factors Influencing the consumer's switching behaviour. *Sources* [3, 6, 12, 14, 17]

H7: Poor customer service has no significant influence on consumer's switching behavior.

H8: Competitor ads has no influence on consumer's switching behavior.

H9: Competitor's better technology has no significant influence on consumer's switching behavior.

H10: Competitor's brand image has no effect on consumer's switching behavior.

H11: Influence of family and friends has no significant influence on consumer's switching behavior.

H12: Hidden charges has no effect on consumer's switching behavior.

H13: Location has no significant influence on consumer's switching behavior.

## 3 Methodology

For this study, cross-sectional data are collected from two major metropolitan cities. Data were gathered equally from both metropolitan cities i.e., Faisalabad and Lahore, Pakistan. A structured questionnaire is used to collect information from 384 customers relating to internet charges, SMS charges, call rates, hidden charges, internet speed, network coverage, signal dropping, customer service, competitor's technology, competitor's brand image, competitor's ads, influence of family and friends and location. Further, the data are collected whether the customer switched cellular network services or not. Since the response variable is dichotomous, so the probit and logit models are the commonly used approaches. Both the models have many common properties, but they differ in their distribution. The logit specification is more appealing as the properties of the estimation procedure of logistic function are more desirable then associated with the probit model. Therefore, a binary logistic regression model is used. The marginal effect for each predicted coefficient is estimated to rank the influencing factors for customers switching over another cellular network.

## 4 Data Analysis and Results

The outcomes of descriptive statistics, logistic regression analysis, and marginal effect are discussed in this section. Table 1 shows the socio-demographic attributes of respondents. According to the results, 66% respondents were male and the remaining 34% were female. The maximum number of respondents falls in the age bracket of 20–30 years i.e., 70%. The number of respondents with the proportion of 62% are having education from undergraduate to postgraduate level. The brand share of different cellular networks varied from 11 to 59% in the sample.

Table 2 defines variables and their descriptive statistics. It shows that 38%, 46%, and 79% of respondents said that SMS charges, call rates, and internet charges are high, respectively. In addition, 58%, 32%, 70%, and 66% of respondents

**Table 1** Distribution of respondents according to socio-demographic attributes

| Variables | Levels | Frequency | Percentage |
|---|---|---|---|
| Gender | Male | 254 | 66.1 |
| | Female | 130 | 33.9 |
| Location | Faisalabad | 192 | 50.0 |
| | Lahore | 192 | 50.0 |
| Age | 19 or below | 66 | 17.2 |
| | 20–30 | 271 | 70.6 |
| | 31–40 | 38 | 9.9 |
| | Above 40 | 9 | 2.3 |
| Education | Up to Intermediate | 34 | 8.9 |
| | Intermediate to undergraduate | 100 | 26.0 |
| | Undergraduate to postgraduate | 241 | 62.8 |
| | More than postgraduate | 9 | 2.3 |
| Current Cellular Network (whether switched or not) | Jazz | 226 | 58.9 |
| | Zong | 56 | 14.6 |
| | Ufone | 58 | 15.1 |
| | Telenor | 44 | 11.4 |

believe that signal dropping, low internet speed, poor network coverage, and poor customer service, respectively, are issues they are facing. Competitor's position in the market also plays an important role in customer switching. Attractive competitor's ads, competitor's better technology, and competitor's brand image play an important role in the switching between different cellular service network as reported by 56%, 72%, and 72% respondents respectively.

The results of the binary logistic regression are given in Table 3. The value of the likelihood ratio test shows that the overall model is significant, and the value of Pseudo $R^2$ indicates that different independent variables explain 10% variation in the switching behaviour. The study results show that internet charges, signal dropping, low internet speed, attractive competitor's ad, competitor's better technology, and influence of family and friend have a significant impact on the customer's switching behaviour. The marginal value of high internet charges indicates that customers are more likely to switch cellular network services due to increased internet charges. This finding is supported by the previous research of [12]. The possible reason for this behaviour may be that the customer from these metropolitan cities is educated and uses internet services to access social media, web browsing, etc. Results further indicate that signal dropping also results in dissatisfaction in customer and as a result to this behaviour, the customer switch to other networks. There is a significant and positive impact of low internet speed on consumers switching behaviour. Similar findings are supported by Masud-Ul-Hasan [12] and Kouser [9] research. The possible explanation for this behaviour is that youth want internet facilities with high browsing speed. Competitor's behaviour also plays a significant role in the attraction

**Table 2** Definition of variables and descriptive statistics

| Variables | Definition | Mean | S.D |
|---|---|---|---|
| High SMS charges | High SMS charges = 1, if respondent's says high SMS charges, 0 otherwise | 0.3880 | 0.4879 |
| High call rates | High call rates = 1, if respondent's says high call rates, 0 otherwise | 0.4635 | 0.4993 |
| High internet charges | High internet charges = 1, if respondent's says high internet charges, 0 otherwise | 0.7943 | 0.4047 |
| Signal dropping | Signal dropping = 1, if respondent's says frequent signal drops of cellular network, 0 otherwise | 0.5833 | 0.4936 |
| Low internet speed | Low internet speed = 1, if respondent's says low internet speed, 0 otherwise | 0.3281 | 0.4701 |
| Poor network coverage | Poor network coverage = 1, if respondent's says poor network coverage, 0 otherwise | 0.7031 | 0.4574 |
| Poor customer service | Poor customer service = 1, if respondent's says poor customer service, 0 otherwise | 0.6693 | 0.4710 |
| Competitor ads | Competitor ads = 1, if respondent's says competitor ads is attractive, 0 otherwise | 0.5625 | 0.4967 |
| Competitor's better technology | Competitor's better technology = 1, if respondent's says competitor is using better technology, 0 otherwise | 0.7240 | 0.4476 |
| Competitor's brand image | Competitor's brand image = 1, if respondent's says competitor's brand image is good, 0 otherwise | 0.7266 | 0.4463 |
| Influence of family and friends | Influence of family and friends = 1, if respondent's says family and friends influence the behaviour, 0 otherwise | 0.7135 | 0.4527 |
| Hidden charges | Hidden charges = 1, if respondent's says cellular network has hidden charges, 0 otherwise | 0.6667 | 0.4720 |
| Location (Faisalabad) | Faisalabad = 1, if respondent's belongs to Faisalabad city, 0 otherwise | 0.5000 | 0.5006 |
| Location (Lahore) | Lahore = 1, if respondent's belongs to Lahore city, 0 otherwise | 0.5000 | 0.5006 |
| Consumers switching behavior | Consumers switching behaviour = 1, if respondent's switched cellular network in past six years, 0 otherwise | 0.4557 | 0.4986 |

**Table 3** Estimated coefficients of logistic regression model

| Variables | Coefficient | Standard error | P value | Marginal effect |
|---|---|---|---|---|
| High SMS charges | 0.293 | 0.248 | 0.237 | 0.072 |
| High call rates | 0.098 | 0.234 | 0.676 | 0.024 |
| High internet charges | 0.650* | 0.313 | 0.038 | 0.155 |
| Signal dropping | 0.409** | 0.248 | 0.099 | 0.100 |
| Low internet speed | 0.738* | 0.251 | 0.003 | 0.182 |
| Poor network coverage | −1.312 | 0.280 | 0.264 | –0.077 |
| Poor customer service | −0.376 | 0.254 | 0.139 | −0.093 |
| Competitor ads | 0.418** | 0.230 | 0.069 | 0.102 |
| Competitor's better technology | 0.618* | 0.264 | 0.019 | 0.148 |
| Competitor's brand image | 0.256 | 0.268 | 0.339 | 0.062 |
| Influence of family and friends | 0.488** | 0.257 | 0.057 | 0.118 |
| Hidden charges | −0.023 | 0.250 | 0.928 | −0.005 |
| Location (Faisalabad) | 0.297 | 0.226 | 0.189 | 0.073 |
| Constant | −2.2435 | 0.441 | 0.000 | |
| | $LR\chi^2 = 52.44, 0.000*$ | | Pseudo $R^2 = 0.099$ | |

*, ** represent statistical significance at 5 and 10 % level of significance respectively

and retention of customers. For this purpose, the competitors are using different kinds of advertising techniques to attract. The results indicate that competitor's attractive ad also significantly impacts the customer's switching between different cellular service networks. Competitor's better technology is also another factor of customer switching between different cellular service networks. Similar results were observed by Rao [14], who found that consumers prefer to switch their existing brand when they found no product improvements in the existing brand or competitor offers additional features in services. But this finding is in contrast with [3], who found little differentiation in service providers' advanced technology and found no impact of change in advanced technology on customer switching behaviour. There is a significant impact of factor named as influence of family and friends on consumers switching behaviour. However, the high SMS charges, call rates, poor network coverage, poor customer service, hidden charges, and location of the respondent has no significant impact on the customer switching behaviour between different cellular networks. The possible reason for the non-significance of location and other variables could be that the respondents belong to major metropolitan cities. So, there is no problem of poor network and poor customer service.

## 5   Conclusion and Managerial Implications

The study's main aim is to identify the factors that affect the brand switching behavior of consumers towards cellular networks. For this purpose, 13 hypotheses are tested after collecting information from 384 respondents from two major metropolitan cities. The binary logistic regression model is applied to test these hypotheses. The study results indicate that high internet charges, low internet speed, signal dropping, attractive competitor ads, competitor's better technology, and influence of family and friends were positive and significant factors that affected brand switching behavior of customers towards cellular networks. At the same time, high SMS charges, high call rates, poor network coverage, poor customer service, hidden charges, competitor's brand image and location show the non-significant impact on consumers switching behavior towards cellular networks. Therefore, the present study suggests that service providers should focus on quality of service, internet charges, internet speed, attractive and informative advertisements. However, the companies do not need to focus on each metropolitan region separately as different regions have no differential impact on customer switching between different cellular service networks.

## References

1. Afzal S, Chandio AK, Shaikh S, Bhand M, Ghumro BA (2013) Factors behind brand switching in cellular networks. Int J Asian Soc Sci 3(2):299–307
2. Anderson EW, Sullivan M (1993) The antecedents and consequences of customer satisfaction for firms. Market Sci 12(2):125–143
3. Awwad MS, Neimat BA (2010) Factors affecting switching behavior of mobile service users: the case of Jordan. J Econ Admin Sci 26(1):27–51
4. Fatima SR, Aslam N, Azeem M, Tufail HS, Humayon AA, Luqman R (2019) The impact of advertisement on customer perception: a case of telecom sector of Pakistan. Euro Online J Nat Soc Sci 114–120
5. Fintikasari I, Ardyan E (2018) Brand switching behaviour in the generation Y: Empirical studies on smartphone users. Jurnal Manajemen Dan Kewirausahaan 20(1):23–30
6. Ghouri AM, Khan NU, Siddqui UA, Shaikh A, Alam I (2010) Determinants analysis of customer switching behavior in private banking sector of Pakistan. Interdisc J Contemp Res Bus 2(7):96–110
7. Hsieh YH, Yuan ST, Liu HC (2014) Service interaction design: a Hawk-Dove game based approach to managing customer expectations for oligopoly service providers. Infor Sys Fron 16(4):697–713
8. Keaveney SM (1995) Customer switching behavior in service industries: an exploratory study. J Market 59(2):71–82
9. Kouser R, Qureshi S, Shahzad FA, Hasan H (2012) Factors influencing the customer's satisfaction and switching behavior in cellular services of Pakistan. Inter J Res Bus 2(1):15–25
10. Lim K, Yeo S, Goh M, Koh W (2018) A study on consumer switching behaviour in telecommunication industry. J Fund App Sci 10:1143–1153
11. Manzoor U, Baig SA, Usman M, Shahid MI (2020) Factors affecting brand switching behavior in telecommunication: a quantitative investigation in Faisalabad region. J Mark Inf Sys 3(1):63–82

12. Masud-Ul-Hasan M (2016) Factors affecting customers brand switching behavior in mobile telecommunication industry: a study of Pabna district in Bangladesh. Asian Bus Rev 6(3):125–130
13. Pakistan Telecommunication Authority (2021) Monthly cellular subscribers. https://www.pta.gov.pk/en/telecom-indicators/1. Accessed 15 Jul 2021
14. Rao UV, VCSMR, P., & Gundala, R. R. (2016) Brand switching behavior in Indian wireless telecom service market. J Mark Manage 4(2):100–109
15. Sathish M, Kumar KS, Naveen KJ, Jeevanantham V (2011) A study on consumer switching behaviour in cellular service provider: a study with reference to Chennai. Far East J Psycho Bus 2(2):71–81
16. Shah AA, Memon H, Noor A, Sidra S, Bhutto A, Khan A (2020) The impact of sponsorship on brand equity of cellular networks in Hyderabad Pakistan. Asian J Econ Bus Acc 13(1):1–12
17. Shah MAR, Husnain M, Zubairshah A (2018) Factors affecting brand switching behavior in telecommunication industry of Pakistan: a qualitative investigation. Am J Ind Bus Manage 8(2):359–372

# User Interface Development Tools and Software for Arduino "A Comparative Study"

Ali Al-Dahoud, Mohamed Fezari, and Ahmad Al Dahoud

**Abstract** In this paper, we have introduced some of the most interesting user interface software for developing applications with Arduino modules. A brief comparison of the most used Graphical interfaces is done based on student's feedback and experiment. The first scenario was a questionnaire presented to License L3 students and Master students in their final project work. The second scenario is taken from research's documentation from the web. We concluded that students like to have GUI in their applications however they prefer simplicity, clearness in the GUI and real-time interaction.

**Keyword** Development tools · Simulation · Graphic interface · Arduino modules · E learning

## 1 Introduction

One of the most popular hardware platforms these days is Arduino module. We have come to the place to learn about connecting an Arduino to other programing devices, whatever software is running on those other devices (PC, laptop, tablet, smartphone). The Arduino can "talk", (transmit or receive data) via a serial channel using wired or wireless connection, so any other device with serial capabilities can communicate with an Arduino. It does not matter what program/programming language is driving the other device.

A. Al-Dahoud (✉)
Faculty of IT, Al-Zaytoonah University of Jordan, Amman, Jordan
e-mail: aldahoud@zuj.edu.jo

M. Fezari
Department of Computer Science, Faculty of Engineering, Badji Mokhtar Annaba University, Annaba, Algeria

A. Al Dahoud
Faculty of Architecture and Design, Al-Zaytoonah University of Jordan, Amman, Jordan

© The Author(s), under exclusive license to Springer Nature Switzerland AG 2022
S. G. Yaseen (ed.), *Digital Economy, Business Analytics, and Big Data Analytics Applications*, Studies in Computational Intelligence 1010,
https://doi.org/10.1007/978-3-031-05258-3_4

31

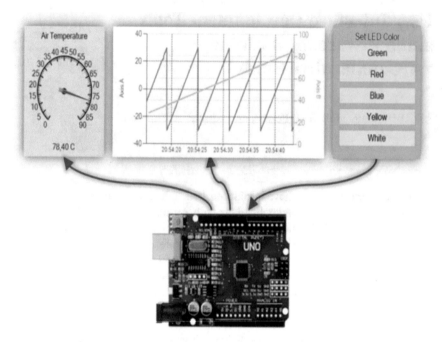

**Fig. 1** Azande GUI presentation

We can either use the Arduino's "main" serial port, the one it uses when we transmit program into it, or we can leave that channel dedicated to programming (and the development environment's serial monitor), and use two other pins for an extra serial link dedicated to the external device by adding a serial communication library.

Some programs (like Flash) do not have native serial capabilities. They can still communicate with Arduino through an intermediary, which, like a "translator", enables them to talk to each other (Fig. 1).

## 2 Main Development Tools for Arduino

We present some interesting development tools for Arduino modules [1]:

### 2.1 Azande

With the development tool Azande [2], we can control and monitor Arduino from a Windows PC [3].

Azande Feature List are:

- Control the Arduino module by sending Commands to it.
- Live-Monitor of internal data. Example: view the temperature of a sensor.
- Log and view our data in a Live-Chart
- Supports USB/Virtual-COM, Ethernet/Wi-Fi and Bluetooth
- Single-Source-Configuration: the Arduino Sketch holds all configurations.
- Supplied. Components:
- Azande Studio is a Windows application and is the hub in Azande.
- An Open-Source library for Arduino with a small footprint and is easy to use more details can be found in the link [3].

## 2.2  Jubito

Jubito is a front-end implementation of the free and open source jaNET Framework [4]. jaNET Framework is a set of built-in functions and a native API where Jubito can utilize in order to interact with multiple vendor hardware (especially open hardware, e.g. Arduino).

It is designed for interoperability. It can operate on any device that is capable of running .NET Framework or mono-runtime (Linux, Windows, Mac, including single-board computers, such as Raspberry Pi and Banana Pi).

Using Jubito we can take control of our developed system via a mobile phone, tablet, laptop or anything can connect to the internet/intranet. It provides an intuitive user interface to control and manage an abundance of different technologies via a centralized system. The advantage of the tool lies in ease of use and the usability that is offered by its ecosystem, so anyone can very quickly setup his controlled environment. No additional programming is required, but someone with that skill can easily adapt its own modules. Figure 2 illustrates an example of the GUI using Jubito. More sources of information can be found in these links: [5, 6].

## 2.3  Ardulink

Ardulink is a complete, open source, java solution for the control and coordination of Arduino boards. It defines a communication protocol and a communication interface allowing several protocol implementations.

It is composed by several java libraries/applications:

- Ardulink Core
- Ardulink SWING that is a ready java SWING components collection able to communicate with Arduino.
- Ardulink Console that is a SWING application that can be used to control an Arduino board without programmer skill.

**Fig. 2** Jubito GUI interface features

- Ardulink Network Proxy Server that is a command line application. It is a network client/server technology for remote control purpose.
- Ardulink Mail that is a command line application. With this application, a user can control several Arduino boards sending just e-mails.
- Ardulink MQTT that is a command line application able to connect Arduino boards with an MQTT broker.

  You can read more detail on Ardulink official site or on GitHub.

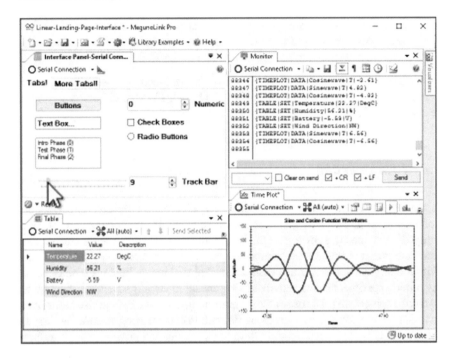

**Fig. 3** Meguno interface and programming

## 2.4  MegunoLink

MegunoLink is a user interface development tool for your Arduino. It provides data-plotting, monitoring and user interface construction to kick start a new project. In Fig. 3, we present main windows of Meguno.

More documentation can be found in the following links: [7, 8]. Plotting Arduino data, Arduino library for processing serial commands and Building user interface panels for an Arduino.

## 2.5  CmdMessenger Messaging Library

A Messaging library for both Arduino and C# & VB .NET/Mono. It implements [9].

(1)  Commands that can be sent or received.
(2)  Multiple arguments can be appended to commands.
(3)  Callback functions can be triggered on received commands.
(4)  All basic data-types (char arrays, floats, int, bytes), both for sending and receiving.
(5)  Optional waiting for a acknowledge commands.

(6)  Escaping data. The special characters that would be interpreted to be field separators or command separators, can be escaped and used in arguments.
(7)  Sending and receiving both plain text and binary data.

Simple software control of an Arduino Uno or Leonardo using the standard firmata. Allows you to control digital Input, Output, PWM, and Servos as well as read analog values.

## 2.6  Instrumentino

Instrumentino is an open-source modular graphical user interface framework for controlling Arduino based experimental instruments. It expands the control capability of Arduino by allowing instruments builders to easily create a custom user interface program running on an attached personal computer. It enables the definition of operation sequences and their automated running without user intervention. Acquired experimental data and a usage log are automatically saved on the computer for further processing. The use of the programming language Python also allows easy extension. Complex devices, which are difficult to control using an Arduino, may be integrated as well by incorporating third party application programming interfaces into the Instrumentino framework [10, 11].

## 2.7  MakerPlot

Maker Plot is Windows software for plotting analog and digital data generated by your microcontroller and other devices with ASCII serial outputs. No proprietary hardware is required—just a serial connection from your microcontroller or other device to your PC—that's it! MakerPlot is software that allows us to build custom interfaces to measure and control the analog and digital data from our microcontrollers.

With MakerPlot, the PC now becomes a laboratory instrument [12]. MakerPlot interface is well presented on Fig. 4, with board design and signal evolution on time.

Each screen can be created with dials, meters, buttons, switches and message areas that display and control the micro's data and internal functions. Meters can be configured with alarm settings (both high and low) then made to sound audio tones using any WAV file we select from the library. Use any of our interfaces that come standard with the software or create our own...that's the power of MakerPlot. For more details Visit http://www.makerplot.com.

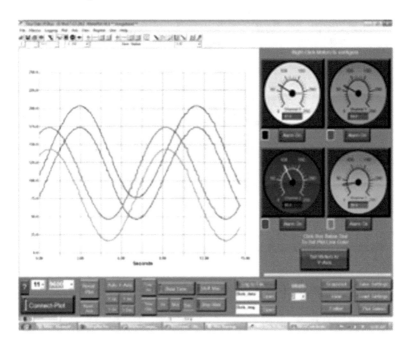

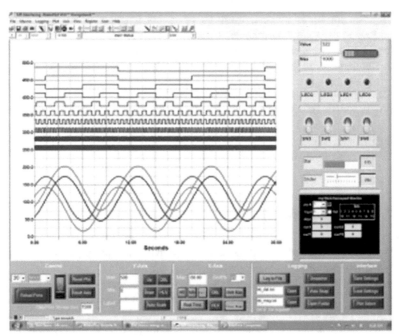

**Fig. 4** Maker plot interface as a lab instrument

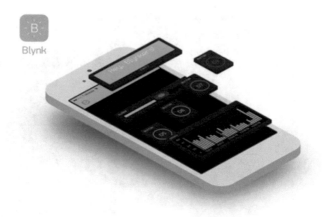

**Fig. 5** Smartphone app to control Arduino i.e. App Inventor from MIT

# 3 Smart Phone Apps to Control Arduino

## 3.1 Blynk

**Blynk** is an app for all makers, inventors, designers, teachers, nerds and geeks who would love to use their smartphones to control Arduino. All the hard work of establishing Internet connection, building an app and writing hardware code is made by Blynk Team, you can just build on top of it (Fig. 5).

It works over the Internet, but Bluetooth is on the way [13].

Simply snap together a visual interface from various widgets like buttons, sliders, Graphs, joysticks, etc., upload the example code to the hardware and see first results in under 5 min.

We can download Blynk for iOS and Android from [14].

## 3.2 Phiro—Pocket

Phiro—Pocket Code for Phiro is a smartphone app that can program & control both. Arduino and Phiro simultaneously via Bluetooth [15].

Pocket Code for Phiro is a free open-source visual programming mobile application. Special "Scratch"-like blocks have been created to access Phiro + Smartphone + Arduino sensors and output devices.

Phiro's smartphone app opens up endless possibilities for the Arduino tinkering community and the Maker movement world-wide to extend the capabilities for Phiro and Arduino with Pocket Code.

Check out these 2 demo videos:

Program & Control an Arduino board + Phiro robot simultaneously with smart phone pocket code apps.

Pocket Code demo with Phiro and Arduino at the same time! Phiro works with open-source languages viz (1) Scratch 2.0 from MIT-USA, (2). Swish Cards from Robotix, (3) Sequential Keys from Robotix, (4). Snap4Arduino from UC Berkeley-USA/Citilab-Spain (5) Pocket Code mobile app's from Graz interface.

### 3.3 iArduino App for iPhone and iPad

This App let you control your Arduino Board Wirelessly with your iPhone and iPad. It provides interactive features like GPIO control of Arduino board, LCD display control, Robot Control, Universal Remote, Wireless Servo Control and much more. We can see iArduino App in reference [16]. Figure 6 illustrates iArduino simulation interface.

**Fig. 6** Iarduino simulator interface

# 4 Other Software for Arduino

A just description on other software for Arduino in different applications: Industrial, health, automotive, Smart home, precision Agriculture, Smart building and cities. Among these software's we can find:

- Interactive Arduino BASIC interpreter:—An Interactive Arduino BASIC interpreter that runs on the arduino.
- **Scada for Arduino**—SCADA Acimut Monitoriza for Arduino [17].
- **Processing Modbus** Master for Arduino—ModbusMaster class for the Processing environment. ModbusMaster is a class developed in and for the Processing environment. It allows you to communicate with modbus slave devices from within your Processing sketch. It works with any hardware serial port.[18].
- WhiteCat Lighting Board—A lighting application dedicated to theatre and dance embeds communication with the Arduino (use of sliders and buttons, dimmers and motors).
- **Firmata**—a standard firmware for communication with a variety of software on the computer [19].
- **GoBetwino**—a generic proxy that runs on your PC and can do a lot of things that Arduino cannot do alone, like starting programs sending e-mails, and a lot more [20].
- Bitlash—A command shell that interprets commands you type or send programmatically over the serial port.
- Avros:—Yet another small human-readable-writable serial protocol.
- Serial-to-network proxies—programs that allow communication with an Arduino via a network connection.
- Generic case: An example of connecting an Arduino which has an LED and a switch to an external device via a serial link. The external device is the master; the Arduino the slave. While the example uses Delphi in the master, what is in the Arduino could be used unchanged for interfacing to any external master. (Master can turn Arduino LED on or off, and it displays the state of the switch.)
- Arduino Manager for iPhone—iPad - Mac OSX

  Integration between iOS devices or Mac and Arduino has never been so easy!!!
  Arduino Manager is an app to control your Arduino board and receive information from it through the new official WiFi Shield or the Ethernet Shield.
  The app shows a grid and tapping on it you can insert specialized widgets to send and receive information from Arduino.
  More Details on Video Tutorial, For iPhone & iPad, For Mac can be found in [21].

- Arduino + iOS with ArduIP & ArduIP-HD

ArduIP is an app to control your Arduino over a WiFi Network or an Ethernet Network. The configuration is full changeable in the necessary ion-Project File. You can control to switch Pins, read Sensor values and act with servos.

ArduIP HD (iPad) AppStore-Link: https://itunes.apple.com/de/app/arduip hd/id581731630?mt = 8.

Link to the Project-File: http://rkmobilearts.de/index.php/arduip/11-arduip-ino-projekt

**Device Druid GUI**

The successor to druid4arduino, Device Druid is an automatic, configuration-free, GUI for any Arduino project using SerialUI (Terminal based UI) [22].

Automatically reproduces the commands, sub-menus and input requests you have configured for your device. Provides access to commands and sub-menus to any depth, and handles user input, error reporting and more. Desktop versions for Windows and Linux available, with a Druid Builder code generation wizard available, too (Fig. 7).

Device Druid and Builder intros 3 min to an Arduino GUI Walk-through and tutorials, Screenshots:

**Arduino + iOS (iPhone + iPad** : Ardumote An app for iOS that lets the user design an interface on the iPhone/iPad and send messages to the Arduino wirelessly over WiFi as UDP packets. Supports control of all pins, PWM, and receiving messages for reading sensors.

- Arduino + iPhone/iPad: iArd is an app for iPhone that allows you to communicate with Arduino (and compatible) via the Ethernet shield and a special sketch.
- Arduino & iOS: Arduino Control You can use this App to control an Arduino using your iPhone, iPad or iPod Touch. It uses JSON for communication. You can change the sketch to make it do or return what you want.

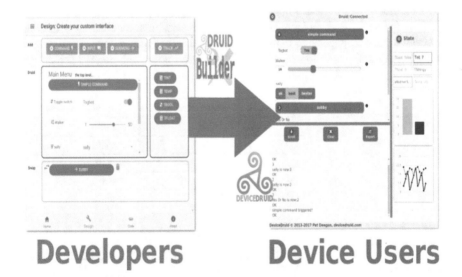

**Fig. 7** Device druid interface

# 5   Experiments

We have gone on two scenarios, the first one we conducted a questionnaire and presented in it 6 main questions related to the choice of a good GUI software to resent an interactive Human machine Application, where the machine is Arduino module, we prospected 150 License level students in third year and 50 Master students in their 2 year and doing their final year project.

The main questions selected are:

(1)   Clear: Is the GUI clear, that it prevents user from errors?
(2)   Consistent: is it Consistent that allows user to apply previous learned knowledge to new task?
(3)   Simplicity: It should be Simple, so it is easy to learn and to use then give the interface consistent look.
(4)   Control: The user must control all actions, not the machine.
(5)   Corrections: Users make mistakes. User actions should be reversible. A good interface facilitates exploration and trial and error learning.

Based on these questions, we have the following Table 1

Based on the results it is clear that Master students do have more experiments with GUI than License students do.

Concerning question (6) the response is on next Table2, the question is related to the choice between PC and Smartphone to develop GUI.

We notice the interest more on Smartphones and developing apps, maybe because all students do have smartphones, moreover, they are always using this device'Smartphone' and less PC.

Second scenario: based on web searches, most of authors would prefer to develop GUI interface for different microcontrollers using Visual language such a Visual C, C#, others prefer using language that facilitate serial communication with other processor via web pages and in this case, they subject to use: python, Java, or Java script. There software allows design good GUI and good web pages interactions.

**Table 1** Number of students do agree on 5 main questions

| Question | License students (L3) | Master 2 students |
|---|---|---|
| Q1 | 120/150 | 43/50 |
| Q2 | 100/150 | 45/50 |
| Q3 | 140/150 | 45/50 |
| Q4 | 115/150 | 42/50 |
| Q5 | 145/150 | 45/50 |

**Table 2** Smartphone or personal computer to developed GUI

| Students | For smartphone | PC | Don't care |
|---|---|---|---|
| License L3 | 35 | 102 | 13 |
| Master | 27 | 16 | 07 |

Others prefer high-level interaction languages rich with library example: LabVIEW and Mat lab, however these students are most of them Master level with high level in programming. We also noticed that students want more interaction with GUI, some would like to get sensor data and control the actuators semi-automatically.

# 6  Conclusion

We presented on this paper some interesting graphic user interface software, and we have conducted statistical research among our students to select a good GUI software that can be taught in future, the questions were based on clearness, consistency, simplicity, control and corrections.

We noticed, working on achieving some of these characteristics of GUI might clash with working on others. For example, by trying to make an interface clear, you may be adding too many descriptions and explanations that end up making the whole thing big and bulky. Cutting stuff out to make things concise may have the opposite effect of making things ambiguous. Achieving a perfect balance takes skill and time, and each solution will depend on a case-by-case basis. That is what we can conclude from students reply to the questionnaire. We can assure, now, UI design is poised for a radical change, primarily Brought on by the rise of the World Wide Web, ubiquitous computing, recognition-based user interfaces, handheld devices, wireless communication, and other technologies. Therefore, we expect to see a resurgence of interest in and research on GUI software tools to support the new user interface styles. This is not concerning PC but also Smartphones and tablets.

# References

1. http://jubitoblog.blogspot.com/search/label/arduino
2. https://zeijlonsystems.se/products/azande/index.html
3. https://playground.arduino.cc/Main/InterfacingWithSoftware
4. http://www.jubito.org
5. https://github.com/jambelnet/janet-framwork
6. https://360.eai.eu/projects/
7. https://www.megunolink.com/documentation/ploting-data/
8. https://www.megunolink.com/documentation/build-Arduino-interface/
9. https://playground.adruino.cc/Code/Cmdmessenger
10. http://www.chemie.unibas.ch/hauser/open-source-lab/instrumentino/index.html
11. https://github.com/yoek/instrumentino
12. Visit http://www.makerplot.com for complete details.
13. http://blynk.cc
14. http://community.blynk.cc
15. Petri A, Schindler C, Slany W, Spieler B (2015) Pocket code game jams: a constructionist approach at schools. In: proceedings of the 17th international conference on human computer interaction with mobile devices and services adjunct, p 1207–1211
16. https://duino4projects.com/android-based-arduino-projects-list/

17. Creery A, Byres EJ (2005) Industrial cyber security for power system and SCADA networks‖. IEEE Paper No. PCIC-2005-DV45
18. http://code.google.com/p/processing-modbus
19. http://playground.arduino.cc/interfacing/firmata
20. http://playground.arduino.cc/interfacing/GoBitwino
21. https://sites.google.com/site/lurvill/arduinomanager_1-6
22. https://devicedruid.com

# Adaptive Digital Image Compression

**Nickolay Rudnichenko, Vladimir Vychuzhanin, Alexey Vychuzhanin, Yurii Bercov, Andrii Levchenko, and Tetiana Otradskya**

**Abstract** The article is devoted to the transformation of digital images based on their adaptive compression during processing and transmission in real time. To meet the requirements for the efficiency of processing and transmission of digital images when implementing compression methods, it is proposed to carry out an orthogonal transformation of the original image. A technique is proposed for compressing digital images in real time based on their adaptive compression, based on taking into account the sensitivity of the basic functions of the Haar transform. The development and application of the method of adaptive compression of digital images will ensure the fulfillment of the requirements for high-quality processing and transmission of digital video information in real time.

**Keywords** Digital image · Image conversion · Adaptive image compression · Haar transform

## 1 Introduction

Meeting the performance requirement for processing and transmitting digital images in real time is based on reducing the amount of data associated with images by compressing them. However, the efficiency of image compression significantly affects the quality of the reproduced video information.

The use of currently known digital image compression methods allows:

N. Rudnichenko (✉) · V. Vychuzhanin · A. Vychuzhanin · T. Otradskya
Odessa National Polytechnic University, Shevchenko Avenue 1, Odessa 65001, Ukraine
e-mail: nickolay.rud@gmail.com

V. Vychuzhanin
e-mail: vint532@yandex.ua

T. Otradskya
e-mail: tv_61@ukr.net

Y. Bercov · A. Levchenko
I. I. Mechnikov Odessa National University, Nevskogo street 36, Odessa 65016, Ukraine

© The Author(s), under exclusive license to Springer Nature Switzerland AG 2022
S. G. Yaseen (ed.), *Digital Economy, Business Analytics, and Big Data Analytics Applications*, Studies in Computational Intelligence 1010,
https://doi.org/10.1007/978-3-031-05258-3_5

- reduce the time required to transfer images;
- to increase the number of images transmitted over the communication channel per unit of time;
- to reduce the requirements for the speed of information transmission devices;
- reduce the storage capacity of images.

As stated in [1–4], the representation of digital images is inherent in redundancy. There are types of redundancy: semantic: statistical; structural; psycho-visual. The methods used to compress digital images are a set of procedures, the implementation of which makes it possible to reduce various types of redundancy in the processing and transmission of images.

Currently, many methods have been developed for compressing digital images [5–7], which can reduce this or that redundancy. Nevertheless, for this purpose, intensive development of new methods of compression of digital images continues [8–10].

Analysis of digital image compression methods shows that most of them are complex, consisting of stages:

- change of color model [11, 12];
- image conversion;
- image encoding.

An effective and simple way to increase the compression ratio provided by image compression algorithms is to select the optimal color model in which the image data is represented. Compression algorithms occupy an essential place in the theory of digital image processing.

This is due to the fact that images presented in digital form require a large amount of memory to store, and when transferring images through communication channels, it takes a long time.

To transform a digital image, an orthogonal or wavelet transform is used.

Orthogonal transformations consist of direct and inverse transformations [13]. Discrete forward and backward transformations in general:

$$Y(k) = < \frac{1}{N} > \cdot \sum_{m=0}^{N-1} X(m) \cdot W(k, m), \quad k = \overline{0, N-1} \tag{1}$$

$$X(m) = \sum_{k=0}^{N-1} Y(k) \cdot W(k, m), \quad m = \overline{0, N-1} \tag{2}$$

where $X(m)$—$m$-th sample of the original discrete signal; $Y(k)$—$k$-th conversion coefficient; $W(k, m)$—$m$-th sample of the $k$-th basis function of the orthogonal transformation; $N$—the number of samples in the original signal or in the block of the original signal during block processing; $\langle 1/N \rangle$—normalization coefficient, which may be absent in some orthogonal transformations (for example, a discrete-cosine transform).

The entire image or its fragments is subjected to conversion. A typical image fragment consists of 8 × 8, 16 × 16 or 32 × 32 samples. A fragment size selection is due to the fact that the correlation interval for images does not exceed 8 … 32 samples.

To assess the efficiency of transformations in works [5, 13–15] are used, for example, as a private indicator of the sensitivity of the transformation coefficients.

The orthogonal basis determines the sensitivity of the transformation coefficients (discrete cosine transform (DCT), discrete Walsh transform (DWT), Haar transform (HT), wavelet transform (WT)) [16].

The complexity of transformations is greatly reduced when using separable orthogonal transformations having fast computational algorithms. These include Fourier transforms, Haar transforms, and cosine transforms, which are most useful for compressing image data.

The main property of the discrete cosine transform is that its basis vectors approximate very well the eigenvectors of Telltz matrices.

In its decorrelating properties, it approaches the Karunen-Loeve transform, while retaining the capabilities of fast Fourier algorithms. This explains the use of discrete cosine transform in JPEG-formats of representation and transmission of images.

The most efficient, from the point of view of computational costs per one transformation ordinate, is the orthonormal system of piecewise constant functions of the Haar basis. The Haar basis possesses the locality property, which underlies the modern theory of wavelet transforms.

When converting an image, the standard deviation (SD) is used—a quantitative indicator (σ) corresponding to the errors of the reconstructed image in relation to the original image. In Fig. 1 shows the corresponding values of the transformation efficiency indicators for the orthogonal basis [5, 13–15].

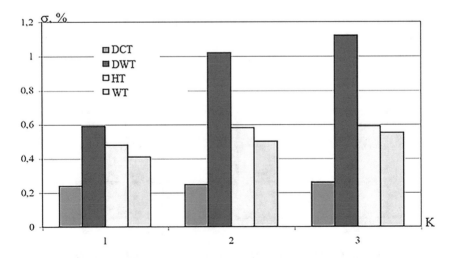

**Fig. 1** SD dependence when converting images samples with different saturation degrees

Thus, Fig. 1 illustrates the dependence of the root-mean-square deviation σ when using the analyzed transformations for digital images of three classes: k = 1—monotonic, k = 2—of the "portrait" type; k = 3—with a large number of contours.

As follows from Fig. 1, the Haar transforms have the best values of efficiency indicators.

Despite the small SD values provided by DCT and WT, their significant disadvantage is the increased dynamic range.

This leads to the need to perform quantization, as a result of which the SD is increased.

One of the stages of digital image compression is image encoding. As a result of encoding a digital image, a compressed sequence is formed using fixed—length encoding methods.

When compressing digital images, for which the amount of information plays an important role, variable length codes are used.

A distinctive feature of optimal codes is that they do not lead to expansion of the compressible data.

According to the data in Fig. 2, the results of optimal coding (shown by the dotted line) and Huffman coding (shown by the solid line) can be compared. $C(f, S)$ is the source $S$ coding cost $f$, $p(s)$ is the relative symbol coding frequency.

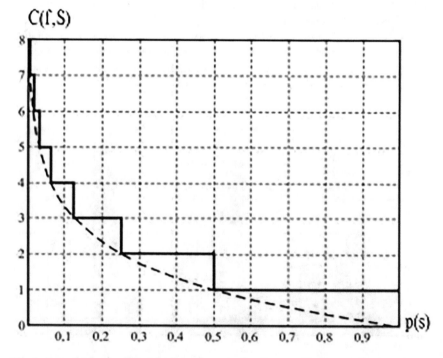

**Fig. 2** Dependence of coding redundancy for optimal coding—and huffman coding

It can be seen from the graphs that the Huffman coding redundancy approaches optimal only when the relative symbol coding rates are multiples of two. Thus, from the analysis of digital image compression procedures, it follows that in order to meet the requirements for the efficiency of image processing and transmission when implementing methods for compressing them with an acceptable level of distortion, it is necessary to carry out an orthogonal transformation of the original image.

## 2   A Technique for Adaptive Compression of Digital Images

The real-time digital image compression technique is based on the use of adaptive compression for forward and inverse two-dimensional transformations of images on the Haar basis.

To reduce the dynamic range of the transform, it is advisable to use subquantization of the Haar transform coefficients in the direct transformation.

The purpose of subquantization is to reduce the dynamic range of the transform with a controlled decrease in the quality of the reconstructed image by reducing the normalizing factor by a factor of V. According to the selection rule, the configuration of the selection zones of the Haar transformant, for example, with a dimension of 8 × 8, has the form shown in Fig. 3 (the solid line shows the boundaries of the transformant zones, and the dashed line shows the transformation coefficients in the zones). Conversion coefficients are grouped into selection zones containing coefficients with the same sensitivity. During subquantization, the value of the transformation coefficients is quantized with less accuracy in relation to the samples of the original digital image.

The coefficients of the s-th ($s \in 1,10$) selection zone of all N × N blocks transforms form the s-th zonal sequence (ZQ). So, for example, the sixth ZQ consists of the sixth zones selection coefficients transforms of all image blocks.

## 3   Research on the Method of Adaptive Compression of Digital Images

When conducting a study using the method of adaptive compression of digital images in real time, the distribution of the number of repetitions of the values of the coefficients ZQ N ($y_i$) was considered using the example of histograms of 3, 5 and 10 zones for test images shown in Figs. 4 and 5.

Based on the fact that the values of the coefficients of the selection zone reflect the presence of details in the image of the corresponding size, then, for example, according to the histogram of the 10th ZQ of the third test image, it can be argued that there are more small details in comparison with other test images. According to the histograms of the first image (Fig. 4a), it can be assumed that there are large

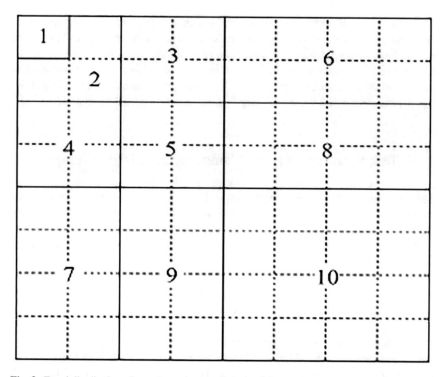

**Fig. 3** Zonal distribution of transformation coefficients of the two-dimensional Haar transform

areas in the image with a smooth color change. It can be seen that with an increase in the ZQ number, the dynamic range decreases, and the percentage of coefficients with a zero value reaches 40–95%, depending on the processed digital image.

Significant differences in the ZQ histograms, compiled from the selection zones of all blocks, for different images allow for the classification of images.

# 4 Conclusion

A technique is proposed for compressing digital images in real time based on their adaptive compression, based on taking into account the sensitivity of the basic functions of the Haar transform. The development and application of the method of adaptive compression of digital images will ensure the fulfillment of the requirement for the efficiency of processing and transmission of digital video information in real time at an acceptable level of distortion of the reconstructed image.

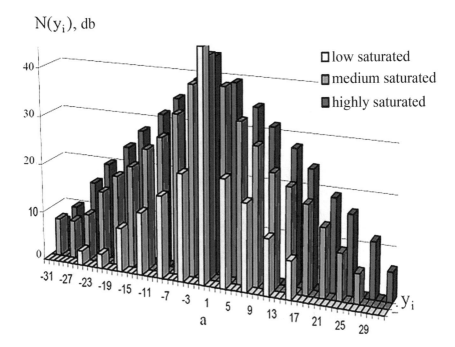

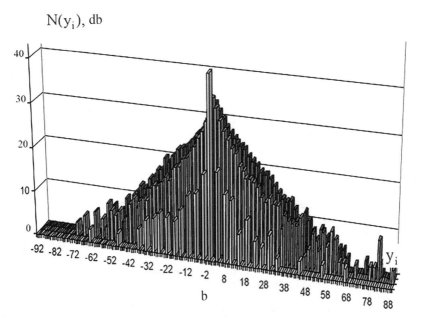

**Fig. 4** Histograms of the two-dimensional Haar transformation coefficients values distribution ZQ: 10th ZQ (**a**); 5th ZQ (**b**)

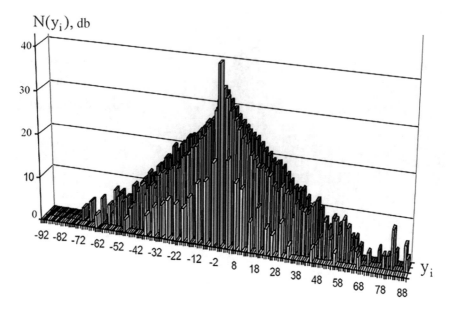

**Fig. 5** The two-dimensional Haar transformation coefficients val-ues 3rd ZQ distribution histogram

# References

1. Rahman Md (2019) Lossless image compression techniques: a state-of-the-art survey. Symmetry 11:1274. https://doi.org/10.3390/sym11101274
2. Annalakshmi N (2021) Lossy image compression techniques. Int J Comput Appl 183:30–34. https://doi.org/10.5120/ijca2021921558
3. Tyagi V (2018). Image Compression. https://doi.org/10.1201/9781315123905-10
4. Dahiwal P, Kulkarni A (2020) An analytical survey on image compression. pp 656–661 https://doi.org/10.1109/WorldS450073.2020.9210364
5. Tiwari D, Tyagi V (2017) Digital image compression. Indian Sci Cruiser 31:44. https://doi.org/10.24906/isc/2017/v31/i6/166460
6. Mohan P (2019) Enhanced image compression system. Int J Mob Comput Appl 6:1–7. https://doi.org/10.14445/23939141/IJMCA-V6I3P101
7. Zhou S, Zhang Q, Wei X, Zhou C (2010) A summarization on image encryption. IETE Tech Rev 27:503–510. https://doi.org/10.4103/02564602.2010.10876783
8. Gong Q, Wang H, Qin Yi, Wang Z (2019) Modified diffractive-imaging-based image encryption. Opt Lasers Eng 121:66–73. https://doi.org/10.1016/j.optlaseng.2019.03.013
9. Kim J, Lee K (2015) Digital television signal, digital television receiver, and method of processing digital television signal
10. Seel P (2020) Digital Television and Video. https://doi.org/10.4324/9780367817398-8
11. Sinclair I (2011). Digital Television and Radio. https://doi.org/10.1016/B978-0-08-097063-9.10016-0
12. Lee S (2017) A study on the digital television loudness analysis before and after introducing the digital television loudness legislation. J Broadcast Eng 22:128–135. https://doi.org/10.5909/JBE.2017.22.1.128
13. Misawa R (2015) Image compression apparatus, image compression method, and storage medium

14. Mohan P (2019) Enhanced image compression system. Int J Mobile Comput Appl 6:1–7. https://doi.org/10.14445/23939141/IJMCA-V6I3P101
15. Aly M Mahmood M (2018) 3D medical images compression. https://doi.org/10.4018/978-1-5225-5246-8.ch010
16. Katharotiya A, Patel S, Goyani M (2011) Comparative analysis between DCT & DWT techniques of image compression. J Inf Eng Appl 1(2):334–348

# Machine Learning and Deep Learning Applications for Solar Radiation Predictions Review: Morocco as a Case of Study

**Mohamed Khalifa Boutahir, Yousef Farhaoui, and Mourade Azrour**

**Abstract** With the increased number of solar power plants, the variation in solar resources causes many problems with grid management and energy systems in general. This is why artificial intelligence (AI) technologies have become useful. AI abilities have been set to use in a variety of situations to handle difficult challenges. The forecast of solar radiation using AI approaches provides a good picture of the integrity of the solar system. This process is simplified by the availability and easy use of different data sources. In reality, there are two famous methods for solar radiation predictions. The first was to use historical solar data, while the second was to integrate other weather parameters. This paper describes the solutions to solar systems and grid management issues using artificial intelligence approaches. It also outlines algorithms for solar radiation prediction by various Machine Learning and Deep Learning techniques, such as ANN, MLP, BPNN, DNN, and LSTM, utilized in various Morocco regions.

**Keywords** Machine learning · Deep learning · Artificial intelligence · Renewable energies · Solar radiation

## 1 Introduction

Solar radiation is the energy we receive from the sun. It is a permanent source of natural and renewable energy. For energy systems, a thorough understanding of the availability and variability of solar radiation intensity is fundamental and crucial.

M. K. Boutahir (✉) · Y. Farhaoui · M. Azrour
Engineering Science and Technology Laboratory, IDMS Team, Faculty of Sciences and Techniques, Moulay Ismail University of Meknes, Errachidia, Morocco
e-mail: moha.boutahir@edu.umi.ac.ma

Y. Farhaoui
e-mail: y.farhaoui@fste.umi.ac.ma

M. Azrour
e-mail: mo.azrour@umi.ac.ma

© The Author(s), under exclusive license to Springer Nature Switzerland AG 2022
S. G. Yaseen (ed.), *Digital Economy, Business Analytics, and Big Data Analytics Applications*, Studies in Computational Intelligence 1010,
https://doi.org/10.1007/978-3-031-05258-3_6

It was known as the fuel of any solar energy system. Due to the growing electricity demand, several countries have been targeting the renewable energy production market, Morocco is one of them. Morocco's government launched a range of projects that will hit about 2000 MW of electricity and make 42% of its energy renewable by 2020 and bring it to 52% by 2030 [1]. The kingdom has a high sunshine rate: around 3000 h of sunshine per year. All projects implemented or under construction offer the country the chance to be the leader in the MENA region in this field [1].

One of the most well-known projects in Morocco is Noor, Ouarzazate currently has four factories at different stages of development: Noor I is a 160 MW cylindrical-parabolic mirror with 3 h of thermal storage and an annual production of 520 GWh; Noor II is a 200 MW cylindrical-parabolic mirror with 7 h of thermal storage and an annual production of 699 GWh; Noor III is a 150 MW tower with 8 h of thermal storage with an annual production of 515 GWh and Noor IV of 72 MW with an annual production of 125 GWh [17]. Noor has other proposed projects such as Noor Midelt, Noor Tafilalet (Zagora, Erfoud and Missour), Noor PV II (Taroudant, Kalâat Seraghna, Bejaâd, Guercif and El Hajeb), Noor Atlas (Tata, Tahla, Tin Tan, Outat El Haj, Ain Beni Mathar, Boudnib, Bouanane and Boulemane) and Noor Argana (Boumalne, Tinghir, Errhamna and Essaouira) [18].

An analysis of previous solar radiation studies in Morocco are presented in this article. Section 2 presents the techniques for the estimation of solar radiation. Section 3 presents some methods and sources for obtaining the required data to train the AI models; Sect. 4 discusses Artificial Intelligence and its axes; Sect. 5 discusses several ways for evaluating models; Sect. 6 discusses the solar radiation Machine Learning and Deep Learning approaches used in Moroccan researches.

## 2 Solar Radiation Forecasting Methods

Due to the rapid rise in implementation and high penetration of solar power in electricity grids worldwide, forecasting of solar radiation production has become a crucial need, for this purpose (Jai Singh Arya et al. [2]) presented the different types of solar radiation forecasting methods available:

- **Stochastic Learning techniques**: These approaches use current data from photovoltaic power plants or radiometer outputs to forecast shifts in sun angles.
- **Artificial Neural Network**: ANN deals with meteorological variables taken as inputs to predict various time scales (see Fig. 1) of solar radiation.
- **Numerical Weather Prediction Method**: To forecast solar radiations, this method combined the Autoregressive Moving Average (ARMA) and Autoregressive Integrated Moving Average (ARIMA) methodologies with numerical weather data.
- **Satellite Image techniques**: This method is based on cloud light measurements taken by satellites.

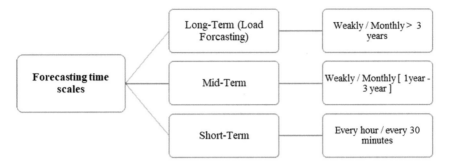

**Fig. 1** Different forecasting time scales for solar radiation

- **Ground-based image techniques**: The strategy used the total sky images (TSI) to present a clear view of cloud shadows for forecasting solar radiation.

## 3  Data Sources

There are numerous approaches for source data methods that can be used to forecast solar radiation, some of which are:

- **The Prediction of Worldwide Energy Resources (POWER)** is a NASA project, it has a goal to observe, understand, and model the Earth system to discover how it is changing, to better predict change, and to understand the consequences for life on Earth. The project was initiated to improve upon the current renewable energy data set and to create new data sets from new satellite systems. The POWER project targets three user communities: (Renewable Energy, Sustainable Buildings, and Agroclimatology. They provide two different datasets, the Meteorology (starting from January 1, 1981, to Now) and the Solar Radiation data (from July 1, 1983, to Now).
- **The Copernicus Atmosphere Monitoring Service (CAMS)** provides consistent and quality-controlled information related to air pollution and health, solar energy, greenhouse gases, and climate forcing, everywhere in the world. CAMS offers information services based on satellite Earth observation, in situ (non-satellite) data, and modeling.
- **METEONORM software** is a unique combination of reliable data sources and sophisticated calculation tools. It provides access to typical years and historical time series for more than 30 different weather parameters. The database consists of more than 8 000 weather stations, five geostationary satellites, and a globally calibrated aerosol climatology. On this basis, sophisticated interpolation models, based on more than 30 years of experience, provide results with high accuracy worldwide.

- **Local laboratory:** Many universities around the world have already established local laboratories where various sensors (pyrometers, anemometers, pluviometers, radiometers, and thermo-hygrometers) are mounted to capture different meteorological and solar parameters.

## 4 Artificial Intelligence Methods

From the solar radiation methods which we already discussed on the (Sect. 2), we would focus our research on Artificial Neural Network Forecasting, which comes under the umbrella of Artificial Intelligence methodologies, but what are those methods? What impact does it have on solar radiation forecasting?

**Artificial Intelligence (AI)** is an area of computer science dealing with the creation of intelligent machines capable of doing activities that would normally necessitate human Intelligence. With strong prediction and automation capabilities, AI can drive with excellence several areas [4].

**Machine Learning (ML)** Artificial Intelligence strives to mimic human cognition, thus Machine Learning grows deeper. Machine Learning is a set of algorithms that allows machines to learn without the need for human intervention. In fact, to solve problems, the "machine" is an algorithm that analyzes a huge volume of data that would be unmanageable for a human being. Machine learning, in other words, allows the machine to be educated to automate tasks that are difficult for a human being, and it can make predictions with this learning.

**Deep Learning (DL)** may be viewed as a form but more complex of Machine Learning (see Fig. 2). Deep Learning is a series of algorithms that simulate the human brain's neural networks. The computer learns by itself, but in phases or layers, in this technology. The model's depth would depend on the number of layers in the model.

## 5 Model Accuracy Evaluation

Typically, or always, most research papers using Artificial Intelligence methods use an evaluation metric to measure how effective the model is. This section will introduce some of the most used evaluation model accuracies:

**Root Mean Square Error (RMSE)** is one of the most used model accuracy metric, especially for regression problems. It is based on the standard deviation of residuals. It is widely used to validate experimental findings in climatology, forecasting, and regression analysis. It detects how oriented the data is around the best fit line.

$$RMSE = \sqrt{\frac{\sum_{i=1}^{n}(X_{obs\rightleftarrows,i} - X_{model,i})^2}{n}}$$

**Fig. 2** The difference
between AI, ML, and DL

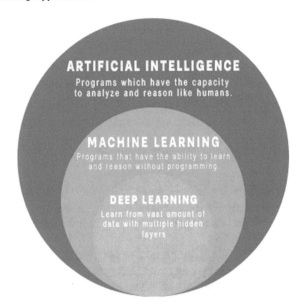

*where Xobs has observed values and Xmodel has modelled values at time/place i.*

**R-Squared** (R2 or the decision coefficient) is a statistical metric that measures how much the independent variable explains the variation in the dependent variables. In other words, the R2 coefficient detects how well the data matches the regression model.

$$R^2 = 1 - \frac{\sum_{i=1}^{n} (y_i - \hat{y}_i)^2}{\sum_{i=1}^{n} (y_i - \overline{y})^2}$$

*where n is the number of measurements, yi is the value of the measurement, ŷi is the corresponding predicted value, and ȳ is the mean of the measurements.*

**Mean Absolute Error (MAE)** is a statistical measure that determines the average size of errors in a group of forecasts without taking into account the direction of the errors. It checks the precision of continuous variables.

$$MAE = \frac{\sum_{i=1}^{n} |y_i - x_i|}{n} = \frac{\sum_{i=1}^{n} |e_i|}{n}.$$

*where yi is the prediction and xi is the true value.*

**The mean square error (MSE)** uses the squared difference between measured and forecasted values.

$$MSE = \frac{1}{n} \sum_{i=1}^{n} (Y_i - \hat{Y}_i)^2$$

# 6    Solar Radiation Researches in Morocco

In this section, we will examine a few papers that deal with solar radiation forecasting. In addition, we will discuss and explain the most often used machine learning and deep learning algorithms.

## 6.1    Artificial Neural Networks

Because of its noise-immune and fault-tolerant properties, the Artificial Neural Network (ANN) as a Machine Learning method may be the best choice for processing complicated data inputs. Forecasting, regression, and curve fitting may all be done with ANNs. [5].

Jallal et al. [6], proposed a data-driven ANN for a half-hour global solar radiation prediction by using only historical measurements of global solar radiation data as inputs, without the need for additional meteorological data for the Agdal site, Marrakesh, Morocco. The authors divided the dataset into two subsets, the first one is for detecting the relationship between the endogenous inputs and the next global solar radiation values for the year 2008, and the second one is used to produce precise predictions for the next six years (from 2009 to 2014). For the year 2008, the built data-driven model achieved an accuracy of 99.12% and 99.13% for the next six years (from 2009 to 2014).

Ettaybi et al. [7], use the hourly global horizontal radiation dataset to analyse the hourly global solar radiation in the Rabat region. The data was collected using the Copernicus Atmosphere Monitoring Service (CAMS) with hourly meteorological data (pressure, humidity, temperature, wind speed, wind direction, and rainfall) from the modern-era retrospective analysis for research and applications version 2 (MERRA-2), for the period of (2012–2016). The authors compared two methodologies, the first used the SARIMA model as a Univariate Linear Model (ULM), and the second was the machine learning MultiLayer Perceptron (MLP) of the Artificial Neural Networks (ANN). The results showed that ANN had better performance than the SARIMA models.

Gaizen et al. [8], tried to build a model that can reduce the uncertainty in forecasting the near future solar power generation under various weather conditions by using the ANN algorithm associated with the transform wavelet (WT) and the SARIMA models. The findings show a contrast between the various cities of the world, like Rabat, Morocco. The results show that the combination of WT with SARIMA and ANN improves the precision and efficiency of the forecast.

## 6.2 Multilayer Perceptron and Nonlinear Autoregressive Exogenous Model

**Multilayer perceptron (MLP)** is an artificial neural network with a multilayer feedforward network that uses supervised learning to process mapping relationships between inputs and outputs. The algorithm consists of three layers. [9] The first layer is the input layer, which transmits external data. [9] The second and third layers are respectively the hidden layer and the output layer, which perform the default neural networks steps (weights, biases, and the activation function) [9].

**The nonlinear autoregressive exogenous model (NARX)** is a strong predictor of time series. Since solar radiation is a time series, NARX has become one of the best options for solar radiation predictions. NARX is Auto-Regressive as its output is the regression of its previous values. [10, 22]

The two neural networks were compared by Loutfi et al. [11] to predict hourly global solar irradiation on a horizontal plane using data collected for the duration (From 1 January 2010 to 31 December 2014) from a local station in the laboratory of the Faculty of Science Fes, Morocco. The authors used various instruments and sensors to gather meteorological parameters (Humidity, Temperature, and Wind Speed). As a result, the authors have been showed that the NARX model with five inputs provides the best efficiency after ten tested models.

In the region of Benguerir, El Alani et al. [20] used multilayer perception (MLP) to forecast global horizontal irradiance for a hot semi-arid atmosphere. For this reason, the authors used data obtained from the weather station located in Green Energy Park, Benguerir, Morocco, formed by different meteorological data (Temperature, Relative Humidity, Barometric Pressure, Wind Speed, Wind Direction, precipitation, …). Using the correlation coefficient, they find that the global irradiance at the top of the atmosphere and the solar zenithal angle are the most correlated astronomy parameters, and the temperature is the most correlated meteorological parameter with solar irradiance.

Nait Mensour et al. [21] developed an artificial neural network (ANN) model with a multilayer perceptron (MLP) technique to measure the monthly average global solar irradiation on the horizontal surfaces of the Souss-Massa region in Morocco. To train the model, the authors used data from 175 locations spread across the Souss-Massa region for 10 years (1996–2005) provided by the NASA geo-satellite database and Google Maps. The authors have chosen a set of 24 different climate locations to achieve a stable design of the ANN model and validate it for the remaining 151 sites. As a result, the optimal model has 25 nodes in the hidden layer with an RMSE of 0.234 and a correlation coefficient R 0.988. The authors tested the models for clear and unclear days, the results are very acceptable for clear days with NMBE: 0.015%, NRMSE: 0.10%, and a correlation coefficient of 0.99, for unclear days the accuracy was NMBE: 0.14%, NRMSE: 0.39% and a correlation coefficient of 0.96.

## 6.3   Random Forest

Random Forest (RF) is a machine learning method and a tweaked algorithm based on decision trees, including a variety of decision tree fittings for various sub-samples of the initial data set at the training level to produce decision trees for computation and to arrange trees.

Bounoua et al. [19] tested 22 empirical and 4 machine learning models to measure Global Solar Radiation at five locations in Morocco (Oujda, Missour, Erfoud, Zagora, and Tan-Tan). The models tested were the Multilayer Perceptron Artificial Neural Network Model (MLP) and three-set methods (Boosting, Bagging, and Random Forest) with some measured meteorological variables and some astronomical parameters (ambient air temperature, relative humidity, wind speed, etc.) were used to train these models. The findings achieved suggest that the Temperature and Geographic factors model were the more accurate with R: 72.38–93.46%; nMAE: 6.96–17.94%; nRMSE: 9.89–22.39%. The Random Forest (RF) has also proven to be the highest performing in all stations between the four machine-learning methods.

## 6.4   Back Propagation Neural Networks

Back-propagation Neural Networks (BPNN) are an Artificial Neural Network aspect. It tries to adjust the network weights using the error rate of the past epoch. Proper tuning of the weights helps to reduce the error rate and makes the model accurate by increasing its generalization.

Aghmadi et al. [16] used BPNN with the Empirical Mode Decomposition (EMD) to improve the accuracy of solar radiation estimation and simplify the energy management system. A one-hour data for the year 2018 of Measured Direct Normal Irradiance (DNI) collected from a meteorological ground station located in Rabat, Morocco, is used. Three concepts of reliability and performance quality criteria are used: MAE, MAPE, and RMSE. The tests of the EMD-BPNN hybrid approach revealed a RMSE of 28.13% and a MAE of 20.99%, much less than other traditional approaches such as the conventional neural network or the ARIMA time series.

## 6.5   Deep Neural Networks

A deep neural network (DNN) is a sophisticated neural network that uses advanced mathematical technics to process data in complex ways. It is designed to imitate human behavior, particularly the identification of patterns and the passing of inputs through different layers to grow the forecasting performance [12].

Jallal et al. [13] proposed the use of a Deep Neural Network (DNN) to handle the dynamic behaviour of meteorological data and providing an accurate hourly global solar radiation prediction. The neural network used hourly data on global solar radiation and meteorological parameters based on the METEONORM datasets of El Kelaa des Sraghna, Morocco. The authors used the Elman neural network (ENN) with the Levenberg–Marquardt Optimizer. With a 99.38% in a correlation coefficient ®, the deep neural network implemented proves to be very effective and accurate.

## 6.6 Long Short-Term Memory

Long Short-Term Memory (LSTM) enables the simulation of very long-term dependencies. It is based on a memory cell and three gates (Forgotten Gate—Input Gate—Output Gate). The complete process of the LSTM can be outlined in three steps:

1. Detect knowledge from the past, drawn from the memory cell via the forgotten gate;
2. Choose, from the current entrance, the ones that will be useful in the long term, via the input gate. These would be applied to the memory cell;
3. Collect valuable short-term information from the current cell state to produce the next hidden state via the output gate [14].

Benamrou et al. [3] suggested a very short-term prediction of horizontal global solar irradiation using the LSTM model based on two separate data sources. The first was obtained from the Al-Hoceima, Morocco Meteorological Station for the duration (2015–2017), and the second was a satellite-derived data retrieved from the CAMS dataset around the Al-Hoceima Meteorological Station for the same period. The authors used the Recursive Function Elimination (RFE) approach to find the desired features of the model. Three scenarios were suggested to model solar irradiation using various algorithms (XGBoost, Random Forest, and SVR). As a result, XGBoost provided the best output model with an R2 coefficient of 0.916.

Bendali et al. [15] proposed a hybrid approach to refine the forecasting of solar irradiance using a combination of Deep Neural Network (DNN) and genetic algorithms. For this purpose, the authors evaluated Long-Term Memory (LSTM), Gate Recurrent Unit (GRU), and Recurrent Neural Network (RNN) models. The Genetic algorithm was used to find the most suitable number of window sizes and the number of neurons in each layer. For this work, the Global Horizontal Irradiance (GHI) time series of Fes, Morocco, was used from 2016 to 2019 as a dataset derived from the METEONORM Platform. As a result, the MSE and MAE metrics indicated that the combination of the genetic algorithm and the LSTM model had the best accuracy for the four seasons of the year.

# 7   Results

This paper provided an analysis of forecasting solar radiation using artificial intelligence techniques in Morocco. As seen in Table 1, several Machine Learning and Deep Learning methods have been used, the most widely used are Machine Learning algorithms and, in particular, ANNs. Its methods used various model precision performances for different data sources, we can detect that the best performance is for Jallal et al. [6] with a decision coefficient R2 of 99.12%, using data obtained from a local laboratory in the region of Marrakech.

We also note a rise in the use of Deep Learning approaches in recent years, as can be seen in Table 1, in particular Deep Neural Networks and Long-Term Short-Term Memory. As well as the ANN case, Jalal et al. [13] made the best performance with a decision coefficient $R^2$ of 99.38% using the DNN model and the METEONORM datasets for the Elkelaa des Sraghna region.

From our study, we have seen different choices of geographical, meteorological, and solar input parameters. This choice is the most critical consideration for the reliable and accurate estimation of solar radiation. Unless few studies are working on this problem, we may take, for example, the cross-correlation function CCF is used by Ettaybi et al. [7], to calculate the correlation between the clear-sky index and each meteorological parameter to determine which one will be used to train the model. On the other hand, Benamrou et al. [3], use the Recursive Feature Elimination (RFE) approach with the XGBoost algorithm to find the best features to be used for model learning.

# 8   Conclusion

Renewable energy has been highlighted as a crucial strategic source for green development in the world. Morocco has an immense solar energy capacity; the Kingdom is implementing many policies and initiatives to meet the optimistic goal of 2030 by achieving 52% of overall electricity generation.

To encourage potential studies in this area, our paper provides an updated summary of predicting solar radiation researches in Morocco. Indeed, due to advances in the AI methods, the efficiency and availability of daily data, and the development of actual Solar energy projects, the example of Noor projects as we've seen early involve more and more studies and applications for solar radiation and energy systems in general.

**Table 1** List of reference papers on global forecasts of solar radiation in Morocco laboratories

| References | Models used | Year | Performance indicator | Horizon | Data source | Location |
|---|---|---|---|---|---|---|
| [6] | ANN | 2020 | $R^2$: 99.12% | Half hour | Local station | Marrakech, Morocco |
| [7] | ANN | 2018 | MAPE: 8.95% nRMSE: 5.56% RMSE: 17.17 Kwh/m$^2$ | 24 h | CAMS + MERRA2 | Rabat, Morocco |
| [8] | ANN | 2020 | MAPE:8% | _ | _ | Rabat, Morocco |
| [11] | MLP & NARX | 2017 | $R^2$: 95% | 4 h | Local station | Fez, Morocco |
| [20] | MLP | 2019 | NMBE: 0.015%, NRMSE: 0.10% Correlation coefficient: 0.99 | Hourly | Local station | Benguerir, Morocco |
| [21] | MLP | 2017 | RMSE: 0.234 R: 0.988 | Daily | NASA + Google Maps | Souss-Massa region, Morocco |
| [18] | RF | 2020 | R: 72.38–93.46% nMAE: 6.96–17.94%; nRMSE: 9.89–22.39% | Daily | Local stations | Oujda, Missour, Erfoud, Zagora, and Tan-Tan |
| [16] | BPNN | 2020 | RMSE: 28.113 MAE: 20.99% | Hourly | Local station | Rabat, Morocco |
| [13] | DNN | 2020 | R: 99.38% | Hourly | METEONORM | Elkelaa des Sraghna, Morocco |
| [3] | LSTM | 2020 | $R^2$: 91.6% | Hourly | Local station + CAMS | Al-Hoceima, Morocco |
| [15] | LSTM | 2020 | MSE/Autumn: 0.0019 MSE/Winter: 0.00301 MSE/Spring: 0.00322 MSE/Summer: 0.0015 | Hourly | METEONORM | Fez, Morocco |

# References

1. Azeroual M, El Makrini A, El Moussaoui H, El Markhi H (2018) Renewable energy potential and available capacity for wind and solar power in morocco towards 2030. J Eng Sci Technol Rev
2. Arya JS, Singh O (2015) Solar energy radiation forecasting methods in the current scenario. In: Sixth inernational conference on advances in engineering and technology, AET
3. Benamrou B, Ouardouz M, Allaouzi I, Ben Ahmed M. A proposed model to forecast hourly global solar irradiation based on satellite derived data, deep learning and machine learning approaches. J Ecol Eng 21(4):26–38
4. Mortier T (2020) Why artificial intelligence is a game-changer for renewable energy. Ernst & Young Global Limited
5. Mosavi A, Salimi M, Faizollahzadeh Ardabili S, Rabczuk T, Shamshirband S, Varkonyi-Koczy AR (2019) State of the art of machine learning models in energy systems, a systematic review. Energies
6. Jallal MA, El Yassini A, Chabaa S, Zeroual A, Ibnyaich S (2020) AI data driven approach-based endogenous inputs for global solar radiation forecasting. Ingénierie des systems d'Information 25(1):27–34. https://doi.org/10.18280/isi.250104
7. Ettaybi H, HimdiKE (2018) Artificial neural networks for forecasting the 24 hours ahead of global solar irradiance. AIP conference proceedings 2056:020010. https://doi.org/10.1063/1.5084983
8. Gaizen S, Fadi O, Abbou A (2020) Solar power time series prediction using wavelet analysis. Int J Renew Energy Res IJRER 10(4)
9. Wei CC (2017) Predictions of surface solar radiation on tilted solar panels using machine learning models: case study of Tainan City, Taiwan. Energies 10:1660
10. Boussaada Z, Curea O, Remaci A, Camblong H, Bellaaj NM (2018) A nonlinear autoregressive exogenous (NARX) neural network model for the prediction of the daily direct solar radiation. Energies 620(11):2–21
11. Loutfi H, Bernatchou A, Tadili R (2017) Generation of horizontal hourly global solar radiation from exogenous variables using an artificial neural network in fes (Morocco). Int J Renew Energy Res 7(3)
12. Deep Neural Network. Techopedia (2018, Apr 13). https://www.techopedia.com/definition/32902/deep-neural-network
13. Jallal MA, El Yassini A, Chabaa S, Zeroual A, Ibnyaich S (2020) A deep learning algorithm for solar radiation time series forecasting: a case study of El Kelaa Des Sraghna City. Revue d'Intelligence Artificielle 34(5):563–569
14. Alouini S, Calcagno S (2019) Les réseaux de neurones récurrents : des RNN simples aux LSTM. Octo Blog. https://blog.octo.com/les-reseaux-de-neurones-recurrents-des-rnn-simples-aux-lstm/
15. Bendali W, Mourad Y, Saber I, Boussetta M (2020 Dec) Deep learning using genetic algorithm optimization for short term solar irradiance forecasting. Conference: international conference on intelligent computing in data sciences (ICDS) At: Fes
16. Aghmadi A, El Hani S, Mediouni H, Naseri N, El Issaoui F (2020) Hybrid solar forecasting method based on empirical mode decomposition and back propagation neural network. E3S web of conferences 2020 2nd international conference on power, energy and electrical engineering, vol 231. PEEE, p 02001
17. Stitou (2017) Case study: Masen NOOR ouarzazate solar complex. The Center for Mediterranean Integration. https://www.cmimarseille.org/menacspkip/wp-content/uploads/2017/08/Youssef_Stitou_MENA_CSP_KIP_Jordan_Workshop_25_July_2017.pdf
18. "Solar power in Morocco" From Wikipedia. https://en.wikipedia.org/wiki/Solar_power_in_Morocco
19. Bounoua Z, Chahidi LO, Mechaqrane A (2021 Jul) Estimation of daily global solar radiation using empirical and machine-learning methods: a case study of five Moroccan locations. Sustain Mater, Technol 28:e00261

20. El Alani O, Ghennioui H, Ghennioui A (2019) Short term solar irradiance forecasting using artificial neural network for a semi-arid climate in Morocco. In: 2019 International Conference on Wireless Networks and Mobile Communications (WINCOM)
21. Nait Mensour O, Bouaddi S, Abnay B, Hlimi B, Ihlal A (2017) Mapping and estimation of monthly global solar irradiation in different zones in Souss-Massa area, Morocco, using artificial neural networks. Int J Photoenergy 2017(8547437):19 p
22. Boussaada Z, Curea O, Remaci A, Camblong H, Mrabet Bellaaj N (2018) A nonlinear autoregressive exogenous (NARX) neural network model for the prediction of the daily direct solar radiation. Energies 11:620. https://doi.org/10.3390/en11030620

# Applicability of CAD/CAM Technology in Orthodontics

Sara Jasen

**Abstract** The number of studies regarding the applicability of CAD/CAM technology in orthodontics is steadily increasing due to the increasing demand. The purpose of this study is to present a comprehensive review of recent literature investigating the practical uses of CAD/CAM technology in orthodontics, highlight the areas for improvement and identify possible future applicability. After an intensive review of previous publications, 71 recent research papers were identified and selected. Analysis of the literature revealed that the vast majority of the researchers consider CAD/CAM technology a helpful method of enhancing almost all aspects of orthodontic treatment. The authors describe its benefits in almost all stages of orthodontic treatment, from oral scanning with digital casts, treatment planning to the production of individualized orthodontic appliances with improved physical properties. Although the technology is relatively new in orthodontic offices its applicability seems unlimited. Nevertheless, authors highlight the need for improvements in certain areas such as improving orthodontic software to facilitate its usage.

**Keywords** CAD/CAM · Orthodontic appliances · Lingual archwires · Herbst appliance · Miniscrews · Piezocision · Sleep apnea devices

## 1 Introduction

The last 25 years have shown a steady increase in the number of orthodontists that use Computer aided-design (CAD) and Computer aided-manufacturing (CAM) to facilitate their work [1]. Orthodontists are known to rapidly embrace new technology with the aim of enhancing their clinical efficiency and practical workflow. The CAD/CAM technology is primarily used for orthodontic diagnosis, treatment planning and the production of individualized orthodontic appliances. Modern technologies have introduced new treatment methods to modern orthodontic offices

S. Jasen (✉)
Section of Orthodontics, Department of Dentistry And Oral Health, Aarhus University, Vennelyst Boulevard 9, 8000 Aarhus C, Denmark
e-mail: Au705489@uni.au.dk

© The Author(s), under exclusive license to Springer Nature Switzerland AG 2022
S. G. Yaseen (ed.), *Digital Economy, Business Analytics, and Big Data Analytics Applications*, Studies in Computational Intelligence 1010,
https://doi.org/10.1007/978-3-031-05258-3_7

such as: oral scanning with digital casts, digital analysis of orthodontic treatment needs, aesthetic clear aligners, indirect bonding systems, individualized designs of orthodontic appliances and computer designed guides for surgical orthodontic procedures [2–4]. CAD/CAM technology enhances the precision of orthodontic treatment as it introduces a more individualized approach. It increases the predictability of the treatment meanwhile reducing the chair time required for performing orthodontic treatment. Also, it minimalizes the risk linked to performing surgical procedures or the risk linked to low clinician experience. Although digitalisation has its disadvantages as there is still need for orthodontic software innovations, need for more powerful computing requirements and more affordable systems it is undoubtedly the future of orthodontics.

The aim of this paper is to identify practical uses of CAD/CAM technology in orthodontics, stressing the advantages over traditional orthodontics but also underlying the areas for improvement and possible future applicability.

## 2   Literature Review

For orthodontic planning and diagnosis orthodontists traditionally use plaster models with all their disadvantages. Plaster models are prone to breakage, degradation issues and they also require huge storage space [5]. The era of digital technology has introduced intraoral cameras with built-in software to capture digital impressions. This software is essential as it controls the 3D reconstruction's resolution, extrapolates the non-registered zones that were not scanned by the orthodontist and assembles the 3D images according to a precise algorithm. The study models acquired using this technology are easy to store and access, have long durability, and are easy to transfer thus enhance the communication between the dental office and dental laboratories [6]. In addition, most patients find intraoral scanning a much more comfortable procedure than traditional impression taking; especially patients with a gag-reflex.

Most of the Software systems allow the orthodontists to view the digital cast in different planes, perform orthodontic measurements such as Bolton analysis, space analysis, arch length and tooth size. Moreover, it allows to measure tooth movement in different planes [6]. Recent studies have shown that intraoral scans are a valid way of performing orthodontic measurements as they are accurate and clinically applicable [7–14].

After treatment planning comes the treatment itself. The era of CAD/CAM has introduced new orthodontic treatment methods such as clear aligner therapy (e.g., Invisalign), custom made brackets and robotically formed archwires (e.g., Suresmile) [15, 16]. This enables the orthodontist to reduce treatment time which is of upmost importance. It's been documented that lengthy orthodontic treatment correlates with external apical root resorption and white spot lesions; both of which are important considerations [17–21].

Furthermore, CAD/CAM is used in indirect bonding system which in comparison to the traditional direct bonding system not only reduces treatment time but also allows for a more accurate bracket positioning, facilitates teeth movement control and reduces the staff required during treatment [22]. In order to benefit from the indirect bonding technique first the CAM/CAM program designs a virtual model of the teeth and supporting tissues, then the program designs a customized transfer jig that is then printed by a 3D printer (e.g., Project HD 3000 Plus).

Kim et al. [23] used The Geomagic Verify program (3D Systems) to evaluate the precision of bracket positioning after indirect bonding technique and the intended bracket position according to the software's design. Their results showed no significant differences although the authors underline the importance of specific clinical situations that may hinder the results.

As the demand on decreasing the duration of orthodontic treatment increases, CAD/CAM technology has found its applicability in computer-assisted piezocision guide (CAPG) for surgically facilitated orthodontics.

Piezocision is characterized by performing minimal piezoelectric osseous cuts using a piezoelectric knife in the bonce cortex [24, 25]. Piezocision procedure facilitates tooth movement thus decreases orthodontic treatment time [25]. The procedure has its risks related to clinician's experience. The risks include an incidental contact of tooth root or critical anatomical structures [26]. The CAPG acts as a translucent guide, digitally designed with slots that guide the clinician where to perform piezocision in the right position, depth as well as angulation therefore minimizing risks associated the surgical procedure [27].

Although duration of orthodontic treatment time is of great importance for the patient, the aesthetics aspects are just as valid. The demand for aesthetics has pushed orthodontists into treating malocclusions using lingual orthodontics where the archwires are bonded to the lingual surface of teeth and aren't visible facially [28, 29].

Lingual orthodontics has its difficulties as the lingual teeth surfaces vary much more that the facial teeth surfaces between individuals. This intensifies the need for individualized archwires and brackets.

The Orthomate system (Syrinx Medical Technologies GmbH, Berlin, Germany) allows automatic individualized archwire bending using a bending robot. In addition, optical 3D scanners (GOM, Braunschweig, Germany) with Incognito® bracket system and ProLingual (TOP-Service für Lingualtechnik, 3M Unitek, Bad Essen, Germany) program simplifies the production of individualized brackets; designed according to teeth surface anatomic structure [30–33]. The system also allows to reduce bracket thickness enhancing patient's comfort and decreasing debonding incidences [34, 35].

Wiechmann et al. [36] described a method of using customized lingual orthodontic appliances with Herbst telescopes. Herbst appliances are used to treat distal occlusion by keeping the mandible in a protruded position. The author used CAD/CAM with rapid prototyping techniques (iBraces/Incognito, Lingualcare, Dallas, Tex) to construct brackets made of high gold content alloy with the desired slot plane, angulations and torque.

The Herbst appliance is made up of a tube, plunger and 2 pivots and 2 screws. CAD ensures the opportunity to enhance the design by improving pivot parallelism in the mandible and maxilla, increase the interpivot distance so that the plunger doesn't slip out of the tube during function. In addition, it helps design individualized pivots that have the same curvature as the tooth surfaces and provide custom-made screw threads are important during heavy mastication.

However, the most important part of the construction is the interface; where the bracket and pivot attach, the author could virtually design them as a single unit reducing the probability of the interface breakage [36].

Simon Graf et al. [37] published a pilot study where he used Computer-aided design to design then fabricate a Hyrax device that is used for palate expansion. The appliance was modelled using 3Shape Appliance Designer Software and was fully individualised to the malocclusion needs. The appliance bands were modelled so that they don't extend into the interdental space as they are ideally modelled according to tooth surface anatomy; this eliminates the need for the use of separators decreasing chair-side time. In addition, the appliance design was controlled in all three dimensions, an ideal bonding gap could be designed decreasing to a minimum debonding incidences. The author stresses how important it is to alter force distribution in accordance to dental needs and CAD/CAM renders it possible to have an individualized approach to each and every patient instead of using prefabricated appliances.

Wilmes et al. [38] describes the applicability of CAD/CAM in mini-implant anchorage devises used in orthodontics. Many practitioners don't feel confident performing surgical procedures, a CAD-CAM produced insertion guides for mini-implants could change that. The computer aided design helps with locating the optimal location, length and angulation that is planned beforehand using a virtual planning software for a specific patient. The predetermined design is then fabricated as an acrylic cast. This approach decreases the failure risk of mini-implant insertion by a less experienced clinician. Also, the guide allows the attachment of orthodontic appliance as well as implant insertion on a single visit.

Miniscrews have been used as anchorage devises for years [7]. However, their insertion is not an easy procedure and according to previews literature their failure rate is high; ranging from 11 to 30.4% [7–9]. The insertion of a miniscrew requires precise and safe placement as there's risk of injury to important anatomic structures as well as they need an ideal trajectory to be useful in orthodontic treatment during tooth movement.

Liu et al. used MIMICS (medical image control system) and Magic (both, Materialise, Leuven, Belgium) software to visualize a 3D image. The author was able virtually assess the ideal location of miniscrew in relation to teeth roots then fabricate a surgical template for the miniscrews. The miniscrews were then successfully and accurately inserted with limited deviation [39].

After the completion of orthodontic treatment, retention is of upmost importance. Retention aims to maintain the teeth in their correct position after the end of orthodontic treatment. Past studies indicate that approximately half of maxillary and one fifth of mandibular lingual retainers fail [10–14, 40–43]. Traditionally, retention is made by bending the wire by the clinician or the dental technician by hand which increases the risk of wire fracture.

**Table 1** Advantages of CAD/CAM technology and its limitations in relation to the orthodontic treatment

| Advantages of CAD/CAM | Limitations of current CAD/CAM capabilities |
|---|---|
| • Increases the predictability of orthodontic treatment<br>• Orthodontic treatment more efficient; less time consuming<br>• More individualized approach to orthodontic treatment<br>• Side effects of orthodontic treatment<br>• Patient friendly work<br>• Easy flow work between dental practice and dental LAB<br>• Reduce physical storage needs<br>• Accurate control of material thickness, design and position within the oral cavity | • Expensive<br>• Not all are user friendly, more training and customer support is needed<br>• Many clinicians have concerns about the future, concerns include self-treatment trends and direct-to-consumer products |

Kravitz et al. [44] describes the use of Memotain; a CAD/CAM lingual retainer that is not bent at all but is a custom-cut wire with smooth curvature. After intraoral scanning of patient's teeth, Memotain is digitally positioned where it won't interfere with patient's occlusion, is digitally shaped to an accurate fit with ideal interproximal adaptation. It is also electropolished providing a smooth surface finish thus decreases the incidence of plaque accumulation of all its negative clinical effects (Table 1).

Additive manufacturing along with computer-aided manufacturing can also aid in designing sleep-apnea devices. Mortadi et al. had used a FreeForm software (version 11; Geo Magics SensAble Group, Wilmington, Mass) with a Phantom Desktop (Geo Magics SensAble Group) arm to design a sleep-apnea device virtually. Next, the design was exported to jet 3000 plus machine to manufacture the device. The author states that the device displayed an ideal fit [45].

# 3 Conclusion

The main objective of an orthodontic treatment is to accomplish exemplary treatment results in a reasonable period of time. The treatment should be effective as well as efficient in regards to the number of appointments. In order to achieve these goals, it is beneficial to use digitally driven orthodontic appliance manufacture and positioning within the oral cavity such as customized brackets with patient-specific torque.

Although the areas for CAD/CAM applicability are vast, there are still areas for improvement, more research with precise evaluation of the technology in comparison to traditional orthodontic treatment are required in order to assess and the system and identify its weaknesses for future improvements. Future research should include prospective randomized controlled trials with larger sample sizes to minimalize result bias due to clinician's clinical judgment and individual biologic response to orthodontic treatment.

Furthermore, there is a consistent need for orthodontic software innovations, need for more powerful computing requirements as we haven't yet reached the full potential of CAD/CAM revolution in modern orthodontic offices.

# References

1. Duret F, Blouin JL, Duret B (1988) CAD-CAM in dentistry. J Am Dent Assoc 117(6):715–720
2. Brown MW, Koroluk L, Ko C-C, Zhang K, Chen M, Nguyen T (2015) Effectiveness and efficiency of a CAD/CAM orthodontic bracket system. Am J Orthod Dentofac Orthop 148(6):1067–1074
3. Müller-Hartwich R, Jost-Brinkmann PG, Schubert K (2016) Precision of implementing virtual setups for orthodontic treatment using CAD/CAM-fabricated custom archwires. J Orofac Orthop 77:1–8
4. Davidowitz G, Kotick PG (2011) The use of CAD/CAM in dentistry. Dent Clin North Am 55(3):559–570
5. Westerlund A et al (2015) Digital casts in orthodontics: a comparison of 4 software systems. Am J Orthod Dentofacial Orthop 147:509–516
6. Alford TJ, Roberts WE, Hartsfield JK Jr, Eckert GJ, Snyder RJ (2011) Clinical outcomes for patients finished with the Suresmile method compared with conventional fixed orthodontic therapy. Angle Orthod 81:383–388
7. Cheng SJ, Tseng IY, Lee JJ, Kok SH (2004) A prospective study of the risk factors associated with failure of mini-implants used for orthodontic anchorage. Int J Oral Maxillofac Implants 19:100–106
8. Kravitz ND, Kusnoto B (2007) Risks and complications of orthodontic miniscrews. Am J Orthod Dentofacial Orthop 131:43–51
9. Wiechmann D, Meyer U, Büchter A (2007) Success rate of mini- and micro-implants used for orthodontic anchorage: a prospective clinical study. Clin Oral Implants Res 18:263–267
10. Schneider E, Ruf S (2011) Upper bonded retainers. Angle Orthod 8:1050–1056
11. Becker A, Goultschin J (1984) The multi-stranded retainer and splint. Am J Orthod Dentofacial Orthop 85:470–474
12. Zachrisson BU (2007) Long-term experience with direct-bonded retainers: update and clinical advice. J Clin Orthod 41:728–737
13. Dahl EH, Zachrisson BU (1991) Long-term experience with direct-bonded lingual retainers. J Clin Orthod 25:619–630
14. Artun J, Spadafora AT, Shapiro PA (1997) A 3-year follow-up study of various types of orthodontic canine-to-canine retainers. Eur J Orthod 19:501–509
15. Brown MW, Koroluk L, Ko CC, Zhang K, Chen M, Nguyen T (2015) Effectiveness and efficiency of a CAD/CAM orthodontic bracket system. Am J Orthod Dentofacial Orthop 148:1067–1074
16. Jacox LA, Mihas P, Cho C, Lin F-C, Ko C-C (2019) Understanding technology adoption by orthodontists: a qualitative study. Am J Orthod Dentofac Orthop 155(3):432–442
17. Khalaf K (2014) Factors affecting the formation, severity and location of white spot lesions during orthodontic treatment with fixed appliances. J Oral Maxillofac Res 5:e4
18. Sameshima GT, Sinclair PM (2001) Predicting and preventing root resorption: Part II. Treatment factors. Am J Orthod Dentofacial Orthop 119:511–515
19. Segal GR, Schiffman PH, Tuncay OC (2004) Meta analysis of the treatment-related factors of external apical root resorption. Orthod Craniofac Res 7:71–78

20. Hurt AJ (2012) Digital technology in the orthodontic laboratory. Am J Orthod Dentofac Orthop 141(2):245–247
21. Fox N (2005) Longer orthodontic treatment may result in greater external apical root resorption. Evid Based Dent 6:21
22. Kalange JT (2004) Indirect bonding: a comprehensive review of the advantages. World J Orthod 5:301
23. Kim J, Chun Y-S, Kim M (2018) Accuracy of bracket positions with a CAD/CAM indirect bonding system in posterior teeth with different cusp heights. Am J Orthod Dentofac Orthop 153(2):298–307
24. Yi J, Xiao J, Li Y, Li X, Zhao Z (2017) Efficacy of piezocision on accelerating orthodontic tooth movement: a systematic review. Angle Orthod 87:491–498
25. Dibart S, Surmenian J, Sebaoun JD, Montesani L (2010) Rapid treatment of Class II malocclusion with piezocision: two case reports. Int J Periodontics Restorative Dent 30:487–493
26. Cassetta M, Ivani M (2017) The accuracy of computer-guided piezocision: ma prospective clinical pilot study. Int J Oral Maxillofac Surg 46:756–765
27. Hou H-Y, Li C-H, Chen M-C, Lin P-Y, Liu W-C, Cathy Tsai Y-W, Huang R-Y (2019) A novel 3D-printed computer-assisted piezocision guide for surgically facilitated orthodontics. Am J Orthod Dentofac Orthop 155(4):584–591
28. Hiro T, Takemoto K (1998) Resin core indirect bonding system—improvement of lingual orthodontic treatment. J Jpn Orthod Soc 57:83–91
29. Huge SA(1998) The customised lingual appliance set-up service (CLASS) system. In: Romano R (ed) Lingual orthodontics. Decker, Hamilton-London, pp 163–173
30. Wiechmann D (2000) La therapeutique eco-linguale. Premiere partie: Unetheorie pour un concept moderne de traitement lingual. J Edge 42:53–69
31. Wiechmann D (1999) Lingual orthodontics. Part 2: archwire fabrication. J Orofac Orthop 60:416–426
32. Wiechmann D (2003) A new bracket system for lingual orthodontic treatment. Part 2: first clinical experiences and further development. J Orofac Orthop 64:372–388
33. Wiechmann D (2002) A new bracket system for lingual orthodontic treatment. Part 1: theoretical background and development. J Orofac Orthop 63:234–245
34. Hohoff A, Stamm T, Ehmer U (2003) Comparison of the effect on oral discomfort of two positioning techniques with lingual brackets. Angle Orthod 73:25–32
35. Hohoff A, Stamm T, Goder G et al (2003) Comparison of three bonded lingual appliances by auditive and subjective assessment. Am J Orthod Dentofacial Orthop 124:737–745
36. Wiechmann D, Schwestka-Polly R, Hohoff A (2008) Herbst appliance in lingual orthodontics. Am J Orthod Dentofac Orthop 134(3):439–44642
37. Graf S, Cornelis MA, Hauber Gameiro G, Cattaneo PM (2017) Computer-aided design and manufacture of hyrax devices: can we really go digital? Am J Orthod Dentofac Orthop 152(6):870–874
38. Wilmes B, Vasudavan S, Drescher D (2019) CAD-CAM–fabricated mini-implant insertion guides for the delivery of a distalization appliance in a single appointment. Am J Orthod Dentofac Orthop 156(1):148–156
39. Liu H, Liu D, Wang G, Wang C, Zhao Z (2010) Accuracy of surgical positioning of orthodontic miniscrews with a computer-aided design and manufacturing template. Am J Orthod Dentofac Orthop 137(6):728.e1-728.e10
40. Renkema AM, Renkema A, Bronkhorst E, Katsaros C (2011) Long-term effectiveness of canine-to-canine bonded flexible spiral wire lingual retainers. Am J Orthod Dentofacial Orthop 139:614–621
41. Lie Sam Foek DJ, Ozcan M, Verkerke GJ, Sandham A, Dijkstra PU (2008) Survival of flexible, braided, bonded stainless steel lingual retainers: a historic cohort study. Eur J Orthod 30:199–204
42. Aldrees AM, Al-Mutairi TK, Hakami ZW, Al-Malki MM (2010) Bonded orthodontic retainers: a comparison of initial bond strength of different wire-and-composite combinations. J Orofac Orthop 71:290

43. Taner T, Aksu M (2012) A prospective clinical evaluation of mandibular lingual retainer survival. Eur J Orthod 34:470–474
44. Kravitz ND, Grauer D, Schumacher P, Jo Y (2017) Memotain: a CAD/CAM Nickel-Titanium lingual retainer. Am J Orthod Dentofac Orthop 151(4):812–815
45. Al Mortadi N, Eggbeer D, Lewis J, Williams RJ (2012) CAD/CAM/AM applications in the manufacture of dental appliances. Am J Orthod Dentofac Orthop 142(5):727–733

# The Context of "Globalization Versus Localization" After the World Pandemic and Quarantine

Oleh Sokil, Svitlana Kucherkova, Anna Kostyakova, Nazar Podolchak, Yana Sokil, and Natalia Shkvyria

**Abstract** The consequences of the popular trend of globalization, a certain closeness of the principles of localization, and, of course, the impact of the global crisis caused by COVID-19 were investigated in the scientific work. An index analysis of the main indicators of sustainable development is provided in the paper. On the basis of the integrated index of sustainable development, a trend line was built and a forecast of the vector of globalization/localization was made. An upward forecast of the integrated index of sustainable development, which indicates a continuous process of globalization was proved in the paper. The calculation and forecast of the transition from global globalization to national localization as a result of the impact of the pandemic and quarantine were presented in the scientific work too.

**Keywords** Sustainable development · Globalization · Localization · Global pandemic · COVID-19

O. Sokil (✉) · S. Kucherkova · A. Kostyakova · Y. Sokil · N. Shkvyria
Dmytro Motornyi, Tavria State Agrotechnological University, 8 B. Khmelnytsky Ave, Melitopol, Zaporizhia Oblast 72310, Ukraine
e-mail: oleh.sokil@Tsatu.edu.ua

S. Kucherkova
e-mail: svitlana.kucherkova@Tsatu.edu.ua

A. Kostyakova
e-mail: anna.kostyakova@Tsatu.edu.ua

Y. Sokil
e-mail: yana.sokil@Tsatu.edu.ua

N. Shkvyria
e-mail: natalia.shkvyria@Tsatu.edu.ua

N. Podolchak
Lviv Polytechnic National University, Lviv, Ukraine
e-mail: nazarpodolchak@gmail.com

S. G. Yaseen (ed.), *Digital Economy, Business Analytics, and Big Data Analytics Applications*, Studies in Computational Intelligence 1010,
https://doi.org/10.1007/978-3-031-05258-3_8

# 1 Introduction

Globalization is often spoken of today. Globalization enhances the interconnectedness and uniqueness of people and civilizations. However, in addition to the positive aspects, some factors are alarming. Some scholars speculate that the role of transnational organizations will soon be so great that they can question the existence of nation-countries. Many experts have high hopes for globalization in terms of solving economic problems.

The anti-globalists who criticize the processes of globalization have the opposite point of view. However, everyone agrees that globalization has opened a new stage in world economic relations. The world development in current conditions of pandemic determines the advisability of the participation of each country in the processes of globalization and self-localization into the world economy.

# 2 Analysis of Recent Studies and Publications

Before starting the development of this scientific article, we researched a large amount of scientific, statistical and analytical literature. The following sources have made the greatest impact on our research in the study of the principles of globalization [1–5]. Also, a special influence on this scientific article was made by such sources as [6–9]. And of course, the main reason for comparing the principles of globalization and localization was the sudden global crisis in the sustainable development of the world caused by the COVID-19 virus [10, 11]. In this regard, we decided to do some research on the confrontation between globalization and localization as a result of recent events.

# 3 The Aim

The statement of the objective and tasks of the study is to confirm the transformation of sustainable development trends and the transition from worldwide globalization to national localization in the modern conditions of the global crisis of COVID-19.

# 4 Methods

The goal will be achieved through accounting and analytical support and mathematical analysis of the quadratic correlation and regression dependence of the integrated indicator of sustainable development. The research methodology is shown in Fig. 1.

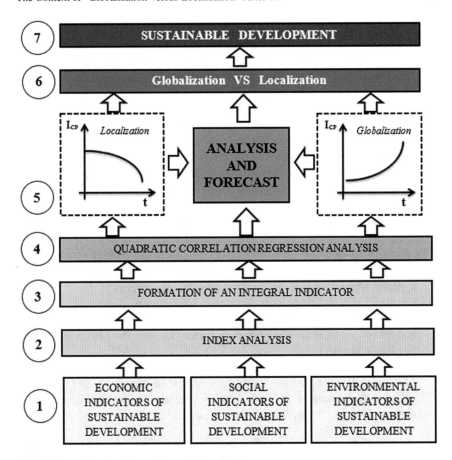

**Fig. 1** Research methodology. *Source* Author findings

① Analysis and collection of economic, environmental, and social data. At this stage, the most significant indicators of five countries were selected: Ukraine, Germany, Moldova, Romania, Belarus and The Sultanate of Oman. Such indicators that had a significant impact as a result of the pandemic crisis, lockdowns, and quarantine measures in 2020 include GDP, export and import volumes, mortality population, unemployment rate, $CO_2$ emissions, volumes of household waste. The analysis period covers the period from 2009 to 2020. The ecological and social groups of indicators included 2 indicators each, and the economic - 3, as the most significant and most mutual influence on the previous two. Data were obtained from official and verified sources of the international statistical base Knoema [12].

② Index analysis—the use of accounting and analytical procedures for generalization and primary processing of data. To unify all indicators for one absolute format, the procedure for calculating indices was carried out using the formula:

$$I_n = P_n/P_{n-1}; \tag{1}$$

where $n$ is the period (year) of the study; $I_n$—indicator index for the period of analysis; $P_n$—indicator for the period of analysis; $P_{n-1}$—indicator for the previous period.

To calculate the indices of economic, social, and environmental indicators for 2010–2020, the period from 2009 to 2020 was analyzed.

③ Formation of an integral indicator—a sustainable development index. At this stage, it was decided to evenly distribute the share and level of significance of each indicator of the country's sustainable development. The sustainable development index for each period is calculated by the formula:

$$I_{CPn} = (I_1 + I_2 + I_3 + \ldots + I_n)/n = \sum I_{n(i=1)}/n. \tag{2}$$

where, $I_{CPn}$—index of sustainable development of the period; $n$—the period (year) of the study; $I_n$—indicator index for the period of analysis; (Providing the results of calculations by personal appeal to the authors.)

④ Carrying out analytical procedures using technical means of information support Microsoft Excel, namely, correlation and regression analysis followed by the formation of a trend line of the fourth degree [13, 14]. The calculation results are shown in Fig. 2.

⑤ Analysis and forecast of the results obtained. After visualizing the trend line of the integral index of sustainable development, the trajectory and rhythm of globalization processes within the country becomes obvious. When predicting this trend for 2–3 years, 2 scenarios are possible:

a. If $y = ax^4 + bx^3 - cx^2 + dx - e \to \infty$, then the processes of globalization have a positive trend and the country as a whole is pursuing such a policy.
b. If $y = ax^4 + bx^3 - cx^2 + dx - e \to 0$, then the processes of globalization have a negative tendency and the country as a whole is pursuing a policy of localization.

In this case, the value of the reliability of the approximation R2 is important, which can be within the following limits:

- more than 0.8—the constructed trend line forecast has greater reliability;
- from 0.5 to 0.8—the constructed trend line forecast has average reliability;
- up to 0.5—the constructed trend line forecast has low reliability;

⑥ Preliminary analysis of globalization and localization trends. Assessment of the relationship between the sustainable development trend and the policies of globalization and localization. Formation of the conclusion on the level of sustainable development and its dynamics. Confirmation of the theory about: direct and inverse dependence of the trend of globalization and localization on the integral index of sustainable development.

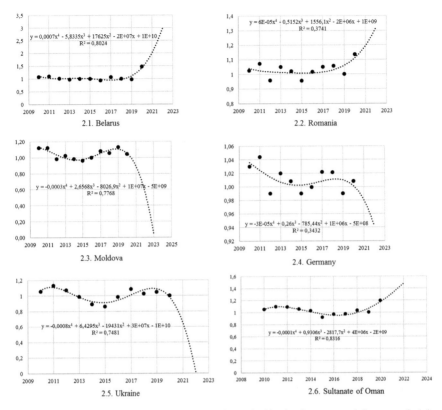

**Fig. 2** Trend line of the integrated indicator of sustainable development and forecast of globalization/localization of five European countries and The Sultanate of Oman. *Source* Author calculations

⑦   Formation of a research hypothesis: is there a theoretical, methodological, and methodological ability of the integral index of sustainable development to predict and solve the problems of globalization and localization policy?

## 5   Results

The analysis of the trend and forecast of the integrated indicator of sustainable development of five countries is presented in Fig. 2.

When analyzing identical drawings Fig. 2c Moldova, Fig. 2d Germany, and Fig. 2e Ukraine's trend line has a cyclical nature of changes in the integral indicator of sustainable development until 2020. A particularly tangible indicator of the world turning point was the geopolitical conflict between Ukraine and Russia in 2014 [15] when the trend line of the integrated indicator of sustainable development began to move towards globalization and it sought to increase until 2019. In the period

2014–2019. The globalization directions of the state policy of Ukraine were felt. During this period, Ukraine received significant support in the world arena and the European integration processes intensified. But in Fig. 2c Moldova, 2d Germany, and 2e Ukraine in 2020 began the crisis caused by the global pandemic Covid-19. Through a significant impact on individual economic, social and environmental indicators of sustainable development (their indices began to decline), these countries began to change the globalization policy to localize their own resources, accumulate internal potential necessary for a quick exit from the crisis with minimal losses and negative changes in indicators (sustainable development indices). Figure 2c, 2d, 2e, it becomes obvious that until 2023 Moldova will finally change the policy of globalization to localization, Germany—until 2028, and Ukraine - until the end of 2022.

Analyzing the data in Fig. 2a Belarus, 2b Romania and 2f The Sultanate of Oman had the largest negative dynamics of sustainable development indices among the analyzed countries, where the increase in positive economic indices was several times less than the increase in negative indices caused by the consequences of the pandemic. Because of this, the forecast for the trend of the indices of sustainable development and globalization tends to increase in the future. Therefore, for these countries, it is especially important to analyze the annual indicators of sustainable development and build on this basis a new forecast to track the pace of the ratio of globalization and localization.

The value of the reliability of approximation R2 for calculating the trend of the integral index of sustainable development and the forecast of the globalization/localization policy for Ukraine, Belarus, and Moldova (over 0.7) are quite high and their forecasts are the most reliable. For Romania and Germany, the reliability is on the verge of 0.34–0.37. These data may indicate the inaccuracy of the onset of the final transition from globalization to localization, but these values are enough to assert the direction of increase or decrease.

## 6    Conclusions

As mentioned at the beginning of the work, to identify the impact on the behavior model of globalization/localization, 7 factors-indices of sustainable development were taken, including 3 economic and 2 indices of social and environmental nature. All factors are reduced to an index value and an integrated indicator is determined. This made it possible to unify all indices and indicators for comparison and analysis, not only for aggregation into an integrated index but also for comparing it between countries.

Now we will consider each component of sustainable development, the change in its index, and the principle of referring to the signs of globalization or localization.

1.   The economic component includes GDP indices, export, and import volumes. If these indices have annual positive dynamics, this indicates the openness of the economy, the growth of external economic relations, and, as a result, the

strengthening of globalization processes in the economy. And if this aggregate index goes down it is a sign of internal localization.

2. The social component includes mortality and unemployment rates. The positive dynamics of these indices testifies to the negative consequences of demographic and social crises and pandemics. The increase in mortality indicates open borders, ineffective quarantine measures, and lockdowns, etc. Accordingly, the positive dynamics of social indices of sustainable development indicates the presence and predominance of globalization over localization.

3. The environmental component includes indices of CO2 emissions and volumes of household waste. The positive dynamics of these indicators, even under the conditions of world quarantine, testifies to globalization sentiments, the expansion of production, or the reduction of measures to reduce emissions and household waste.

So, we can visualize two options for the flow of events in which the following scenario development is possible. The first scenario is when we have a gradual decrease in the predicted integral index of sustainable development, which after some time reaches zero (point n Fig. 3). This point in time is the end of the dominant globalization and the beginning of an expanded localization of resources and means of preserving the sustainable development of the country as a whole and overcoming the consequences of the pandemic crisis with minimal losses. (Fig. 3).

Model Fig. 3 typical for Ukraine, Moldova, and Germany, which were calculated in the previous paragraph. These experimental calculations indicate that the lower the value of the approximation reliability R2, the later the maximum localization occurs (line y' in Fig. 3) and, accordingly, the transformation point n' shifts to the right.

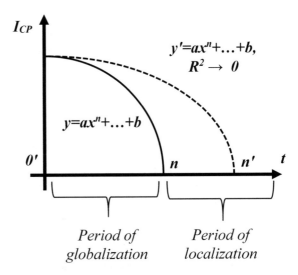

**Fig. 3** Model of the transition of national politics from globalization to localization. *Source* Author findings

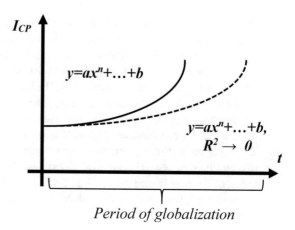

**Fig. 4** National policy model of globalization. *Source* Author findings

*Period of globalization*

At the same time, analyzing the calculations and graphical representation of the integral index of sustainable development of Belarus, Romania and The Sultanate of Oman, it becomes possible to present the following theory (Fig. 4).

In this theory, the level of reliability of approximation R2 also has a significant impact on the elasticity of the integrated index and, accordingly, on the policy of globalization. If R2 tends to zero, then the globalization process has features of its own experience and careful adaptation to external factors. But in the end, this model is devoid of localization principles. Although, as we have already noted, annual analysis and recalculation of the forecast will be the best method to reduce risks and increase the accuracy of forecasts.

The pandemic of acute respiratory disease COVID-19 caused by the SARS-CoV-2 coronavirus has become the factor that forced the whole world to revise not only its development forecasts, its short-term economic and social policy, but also to approach the formation of its own priorities for the long-term in a different way. The lessons learned by humankind from this pandemic can change not only governance models in the future but also the social behavior of humankind from globalization to localization.

With our calculations, we tried to prove that the global trend of globalization is losing its positions among countries and especially those that are actively developing and/or are at an active stage of internal changes and reforms.

Summing up, it becomes possible to summarize the following:

- The Institute of Accounting with the help of accounting and analytical support tools is an effective forecasting tool and, accordingly, a means of building an effective model of sustainable development of society.
- An integral indicator of sustainable development, which combines economic and eco-social indicators, is a useful source of information on the dynamic development and direction of national policy.
- Correlation-regression analysis, trend line construction, and forecasting can be a confirmation of the concentration of globalization or the moment of localization.

- The 2020 crisis caused by the COVID-19 pandemic has become a powerful catalyst for change and the transition from globalization to resource localization, effective policies, and a focus on self-interest within the country.

# References

1. Fayol A (1992) General and Industrial Management. Moscow Controlling 2:151–152 (in Russian)
2. Kharazyshvyly YuM (2012) Innovation as a characteristic of quality of socio-economic development. In: Problems and prospects for innovative economic development, pp 175–180 (in. Russian)
3. Prodanchuk MA (2015) Theoretical and applied aspects of the concept of accounting product formation in agricultural business management. APK Econ 9:4–5 (in Ukrainian)
4. Pshyk BI (2010) Ways to improve the depreciation policy and strengthen its role in financial relations in Ukraine. Regional Economy Part 1:115–122 (in Ukrainian)
5. Zhuk V, Trachova D, Semenyshena N, Ionin Y, Zhuk N (2020) Problems of amortization methodology in accounting policy (On the example of institutional sectors of the Ukrainian economy). Viešoji Politika Administravimas 4(19):142–154
6. Olefir VK (2016) Assessment of import dependence of the Ukrainian food market. Econ Forecast 4:91–105 (in Ukrainian)
7. Shinkaruk LV (ed) (2012) World economic disproportionality: peculiarities, trends, influence on the economy of Ukraine. National Academy of Sciences of Ukraine, Institute for Economics and Forecasting, NAS of Ukraine, Kyiv (in Ukrainian)
8. Shovkun IA (2017) Localization of production—world practice and conclusions for Ukraine. Econ Forecast 2:31–56 (in Ukrainian)
9. Legenchuk S, Pashkevych M, Usatenko O, Driha O, Ivanenko V (2020) Securitization as an innovative refinancing mechanism and an effective asset management tool in a sustainable development environment. In E3S web of conferences. The International conference on sustainable futures: environmental, technological, social and economic matters, vol. 166. EDP Sciences, p. 13029
10. Open source Google News (2021). Covid-19 (2021) https://news.google.com/covid19/map?hl=ru&gl=RU&ceid=RU%3Aru. Accessed 21 Mar 2021
11. Ukraine: influence of covid-19 on the economy and society (2021). https://www.me.gov.ua/Documents/Download?id=bc5d2c61-1a7f-4ec7-8071b996f2ad2b5a. In Ukrainian. Accessed 21 Mar 2021
12. Knoema (2021) Professional data discovery and data management tools, https://knoema.com/. Accessed 21 Mar 2021.
13. Sokil O, Zhuk V, Holub N, Levchenko O (2019) Accounting and Analytical methods for identifying risks of agricultural enterprises' sustainable development. In Nadykto V (ed) Modern development paths of agricultural production: trends and innovations. Springer, Cham, pp 561–569. https://doi.org/10.1007/978-3-030-14918-5_55
14. Sokil O, Zvezdov D, Zhuk V, Kucherkova S, Sahno L (2020) The impact of accounting and analytical support of social and environmental costs on enterprises' sustainable development. Economic Annals-XXI 181(1–2):124–150. https://doi.org/10.21003/ea.V181-11
15. Zhyhlei I, Legenchyk S, Syvak O (2020) Hybrid war as a form of modern international conflicts and its influence on accounting development. Przegląd Wschodnioeuropejski 11(1):191–205, https://doi.org/10.31648/pw.5980

# The Impact of the COVID 19 Shock on Intention to Adopt Social Commerce

Khaled Saleh Al-Omoush, Saad G. Yaseen, and Ihab Ali El Qirem

**Abstract**  This study aims to explore the impact of the COVID 19 shock on the intention to adopt social commerce. It also examines the impact of the pandemic shock on Electronic Word of Mouth (EWOM) and the perceived usefulness of social commerce. Data were collected from Facebook users in Jordan and analyzed using smart PLS software. The results show a significant impact of the COVID 19 shock on EWOM, perceived usefulness, and intention to adopt social commerce. The findings also show a significant mediating impact of EWOM and perceived usefulness on the relationship between the COVID 19 shock and intention to adopt social commerce.

**Keywords**  COVID 19 · Shock · Social commerce · EWOM · Perceived usefulness · Intention to adopt

## 1  Introduction

The global COVID-19 epidemic, countermeasures, and other procedures applied against it have prompted sharp changes in all aspects of people's lives. Beyond direct effects on people's health, jobs, and incomes, this unprecedented crisis has created waves of shock to consumers, organizations, industries, the business environment, and global trade. At the consumer level, preventive measures, including quarantine, distancing between individuals, commercial sector lockdown, travel restrictions, and urging people to stay home, have limited consumers' freedom and their ability to access the goods and services they need.

K. S. Al-Omoush (✉) · S. G. Yaseen · I. A. E. Qirem
Al-Zaytoonah Univeristy of Jordan, Amman, Jordan
e-mail: K.Alomoush@zuj.edu.jo

S. G. Yaseen
e-mail: Saad.Yaseen@zuj.edu.jo

I. A. E. Qirem
e-mail: i.elqirem@zuj.edu.jo

© The Author(s), under exclusive license to Springer Nature Switzerland AG 2022
S. G. Yaseen (ed.), *Digital Economy, Business Analytics, and Big Data Analytics Applications*, Studies in Computational Intelligence 1010,
https://doi.org/10.1007/978-3-031-05258-3_9

The outbreak of COVID-19 and counter restrictions have added new prospects for the necessity of social media during crises. With the advancements in social media platforms, people were able to find alternatives to their direct and physical interaction with their environment. At the same time, these platforms have become a feature of entrepreneurial orientation in revolutionizing the form business organizations interact with consumers in the context of applying the social commerce paradigm. The transformation toward online shopping and cashless payments has opened opportunities for social commerce platforms to attract more consumers [1].

Social commerce integrates online social networking and shopping, creating innovative virtual markets and consumer communities via social media platforms. It represents a new era of electronic commerce, applying the rule of following and consumers' trends and serving them wherever they go, adjusting business models accordingly. COVID 19 crisis has enhanced the role of social commerce to be not only a platform for initiating a purchasing intention but also to complete purchase transactions right on social media platforms themselves, without any necessity for external links. However, many scholars (e.g., [2, 3]) have emphasized the role of the Electronic Word of Mouth (EWOM) as a manifestation of social commerce that provides a large-scale consumer-to-consumer as well as consumer-to-brand interaction. Nowadays, in unstable and crisis situations, understanding EWOM is a critical area of concern for business [4]. In general, the high and extensive collaboration and social interaction in the form of WOM are regarded as significant drivers of consumers' behavior and purchasing intention [5].

The modern world, in which social media and information and knowledge societies have emerged, has not experienced such a global epidemic. Although there is a growing literature on social commerce, far less attention has been paid to studying this new chapter of electronic commerce during global pandemic crises, such as COVID 19. A review of the literature indicates a lack of empirical research on the impact of the COVID 19 shock on the consumers' decision to adopt social commerce applications. Furthermore, prior research has not addressed the relationships between the COVID 19 shock, the power of EWOM, perceived usefulness, and intention to adopt social commerce in a unified framework.

Bridging the above-mentioned gap, this study aims to examine the impact of the COVID 19 shock on the intention to adopt social commerce from a consumer perspective. Moreover, it intends to study the impact of the power of EWOM and perceived usefulness on the intention to adopt social commerce during such unprecedented crises.

## 2   Literature Review

The emergence of the COVID-19 pandemic has caused an unprecedented global shock, urging scholars from various disciplines and fields to study its effects and how to confront its threats and exploit escort opportunities, if any. Given its pivotal role during COVID 19 crisis, major research efforts have intended to investigate different issues on Information and Communication Technology (ICT), with particular attention to social media platforms.

Social computing represents one of the most powerful platforms for reforming the way to conduct marketing and shopping. More importantly, in recent years, social media has invited novel changes to e-commerce applications, empowering the rise of the social commerce paradigm. Social commerce embraces the employment of Web 2.0 in allowing consumers to contribute to the marketing, match, sell, purchase, and share information via social media platforms [6]. According to Hajli [7], social commerce applications enable online shoppers to co-create interactive virtual communities of consumers that support their members in all phases of purchase decision-making.

An examination of the literature reveals that a considerable stream of research (e.g., [8, 9]) has focused on the impact of COVID 19 on purchase behavior, customers' attitudes, and customer co-creation. A significant line of studies (e.g., [10, 11]) have been dedicated to examining the impact of uncertainty, risk perception, and panic on consumer behavior. E-commerce has received a large share of recent studies (e.g., [12–14]) that have examined the role of ICT in the COVID 19 crisis. However, understanding the impact of social media platforms on shopping, consumer behavior, and purchase intention during the pandemic crisis has also gained special attention from scholars (e.g., [15, 16]). Furthermore, the literature reveals a considerable effort (e.g., [17, 18]) devoted to the role of digital marketing strategy in increasing customer engagement in responding to the COVID-19 crisis.

Despite the increasing interest in social commerce before the emergence of COVID-19, the trend towards studying it was less than expected during the epidemic, especially at a time people turned to social media in different affairs of their lives. For example, Diwanji and Cortese [1] investigated the influence of the presentation format of consumer-generated reviews on shoppers' perception and purchasing decisions from social commerce platforms. Zhang et al. [9] examined the impact of COVID 19 on consumer social activities and purchase decisions in the context of social commerce. De Silva et al. [3] discussed the role of social commerce in overcoming the negative consequences of Covid-19. However, it is worth noting that the previously mentioned studies were analytical and did not depend on empirical research or data representing the consumers' perspective.

EWOM is considered as one of the key drivers of social commerce platforms. EWOM refers to all informal communications with other consumers via Internet-based platforms about the characteristics, ownership, or/and usage of a product, service, or vendor [5]. It contains positive, negative, or neutral comments that consumers publish for others to view on the internet and about products, vendors, or companies [19]. EWOM became an increasingly important channel for the diffusion of information and the exchange of opinions among consumers. The literature revealed that consumers tend to believe EWOM rather than business-generated advertising campaigns, where people prefer to trust recommendations and EWOM coming from the crowd than those from the messages provided by marketers [20]. However, although there is a growing interest (e.g., [4, 19]) in the impact of EWOM on the purchase decision and consumer behavior during the COVID-19 crisis, the previous studies did not investigate any relationship between the power of EWOM and intention to adopt social commerce.

## 3    Research Model and Hypotheses

As shown in Fig. 1, this study suggests a direct impact of the COVID 19 shock on EWOM and the perceived usefulness of social commerce. Furthermore, the research model proposes that the COVID 19 shock has a direct impact on the intention to adopt social commerce. Finally, this study proposes a mediating role of EWOM and perceived usefulness in the relationship between the COVID 19 shock and intention to adopt social commerce.

The proposed relationships between the research constructs are discussed below in detail.

### 3.1    The Impact of the COVID 19 Shock on the EWOM

The COVID-19 shock has led to fears, anxiety, and worries affecting consumer behavior, attitudes, and interaction with online communities. Shi and Chow [2] emphasize that social commerce is an innovative model of WOM, applied to the e-commerce paradigm. Scholars (e.g., [6, 21]) revealed that compared to positive WOM, negative WOM has a stronger impact on consumer purchase intention, where the intensity of consequences will be greater when negative WOM is disseminated through online social platforms. Recent studies (e.g. [4, 22]) emphasized that, in pandemic crises, a consumer who receives WOM in a supportive and emotional manner is likely to experience an enhanced relationship with a company. According to Sheth [23], the promotion of brands or products without responding to consumers' concerns about the COVID 19 epidemic, in an empathetic way, mostly leads to dissatisfaction and negative EWOM. The content of WOM on social media during the pandemic has been different than usual, having more effects on consumers' shopping behavior [19]. Naeem [24] indicates that social media has enabled consumers to

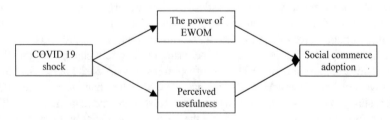

**Fig. 1**  Research model

connect and share WOM globally during the COVID-19 crisis. Therefore, this study proposes:

H1: The COVID 19 shock has a significant impact on the EWOM.

## 3.2    The Impact of the COVID 19 Shock on Social Commerce Usefulness

According to Hajli [7], the decision of online purchase is an antecedent of many functional, social, and emotional dimensions of consumer perceived usefulness. Prior research (e.g., [22, 25]) reveals that COVID-19 has enforced consumers to review their shopping attitudes and practices, exploring the benefits of online services they had never used before. Since people were confined to their homes due to the pandemic, the significance of social media platforms has increased, influencing the behavior of consumers in unusual conditions [24]. Previous studies (e.g., [4, 10]) confirmed that many consumers have switched to online purchases due to the safety and benefits of online ordering, cashless payment, and home delivery. Adopting online ordering using social commerce was an effective way to overcome the barriers of countermeasures against the pandemic [1, 9]. De Silva et al. [3] have explained the pivotal role of social commerce in preserving the lives of people while arriving at the essential and daily needs of consumers efficiently and effectively. Therefore, this study hypothesizes:

H2: The COVID 19 shock has a significant impact on the perceived usefulness of social commerce.

## 3.3    The Impact of the COVID 19 Shock on Intention to Adopt Social Commerce

The consumers' greater health concerns, combined with lower accessibility of marketplaces and stores, have created an immediate persistent need for alternative shopping channels [25]. Consumers who are deeply concerned about the impact of the pandemic and lockdowns have recognized an extreme adoption of social media [26]. One of the immediate effects of the COVID-19 crisis was the increase in online shopping [23]. Donthu and Gus-tafsson [26] claimed that the Internet became a major channel to purchase essential goods and services during the epidemic crisis. In the context of the fear-inducing phenomenon, several studies (e.g., [14] [24]) affirmed that during the COVID 19 crisis, consumers intended to purchase more online, perceiving the usefulness of e-commerce and social commerce applications. Furthermore, less digitally socio-economic people and older consumers have begun to discover and enjoy online shopping, welcoming the safety provided by social commerce [3]. According to the preceding discussion, this study proposes:

H3: The COVID 19 shock has a significant impact on the intention to adopt social commerce.

### 3.4  The Mediating Impact of EWOM and Perceived Usefulness

Recent literature (e.g., [4, 23, 24]) confirms the role of the COVID 19 shock in empowering and forming the trend of EWOM affected by the extent of corporate response to consumers' concerns in a supportive and emotional manner during the pandemic crisis. On the other hand, Ng [27] describes EWOM as the most powerful driver of social commerce evolution and adoption. Shi and Chow [2] also emphasized the importance of WOM communication as an essential aspect of social interaction in social commerce adoption and consumers' decision-making process. According to social interaction theory, the complex network of social relations built by the consumers through frequent online interaction and communication, generating EWOM, is a significant feature of social commerce adoption [5]. However, the literature (e.g., [19, 20]) emphasized that WOM has a critical effect on purchasing intention either directly or indirectly, as a mediating factor, by affecting consumers' trust. Therefore, the following hypothesis is proposed:

H4: The relationship between the COVID 19 shock and intention to adopt social commerce is mediated by EWOM.

Previous studies (e.g., [14, 24]) have implied the role of COVID 19 in perceiving social commerce' usefulness in responding to threats and related preventive measures against the pandemic. At the same time, online social networks and interactions have enabled consumers to participate actively in building a value co-creation circle in social commerce, increasing consumer intention to buy from social commerce platforms [1, 7]. Prior research (e.g., [6, 27]) analyzed the impact of social relationships on purchase intention, showing that reviews and ratings of the crowd related to the products, services, and vendors could enhance users' perceived usefulness, thereby enhancing their willingness to adopt social commerce. Furthermore, Yin et al. [20] explained that the perceived usefulness of social commerce reduces consumer suspicions about the purchase risk, promoting the adoption of purchasing from its platforms. Therefore, this study proposes:

H5: The relationship between the COVID 19 shock and intention to adopt social commerce is mediated by perceived usefulness.

**Table 1** References for measures

| Construct | Code | Items | References |
|-----------|------|-------|------------|
| COVID 19 shock | COVDS | 4 | [11, 24] |
| EWOM | EWOM | 4 | [9, 15] |
| Perceived usefulness | PU | 4 | [1, 16] |
| Intention to adopt social commerce | SCA | 4 | [10, 15] |

# 4 Research Method

## 4.1 Measurements and Instrument Development

To measure the constructs of the research model items were derived from previous studies as shown in Table 1.

To validate the survey instrument, a small-scale pre-test was used with 15 participants to detect any problems with formatting and understanding of the measurement Items. Based on this feedback the items were refined to confirm that the instrument is clear and suitably validated. However, all items (Table 2) were measured using a five-point Likertscale.

## 4.2 Sampling and Questionnaire Distribution

An online questionnaire was employed via Facebook, as a primary shopping website, to attract more participation from the online consumer population. The link to the developed online survey was shared on social commerce platforms and large Facebook Jordanian groups to attract more participation from members. The introduction of the questionnaire declared that only those with social commerce experience would be qualified to take part in the survey. However, 189 valid responses were received and included to further analysis. Table 3 presents the sample characteristics.

## 4.3 Data Analysis

Smart PLS program 2.0 was employed for data analysis and to estimate the hypothesized causal relationships based on two steps, including the measurement model analysis and structural model analysis.

**Table 2** Items of the questionnaire

| Constructs | Code | Measurement items |
|---|---|---|
| The COVID 19 shock | COVDS1 | COVID-19 has created unprecedented fear and panic |
| | COVDS2 | I have never expected that the world would be exposed to such a pandemic crisis and preventive measures |
| | COVDS3 | The emergence of this epidemic created a worldwide shock |
| | COVDS4 | COVID-19 and its consequences have reinforced radical changes in people's lives |
| EWOM | EWOM1 | I share my personal negative or positive experiences on social media platforms about vendors or brands |
| | EWOM2 | I use online discussion forums and communities for acquiring information and benefiting from others' experiences before I buy a product |
| | EWOM3 | Social media is a powerful channel that provides a resourceful environment of experts who spread their judgments and consumption-related advice |
| | EWOM4 | I am willing to recommend a product or service that is worth buying to others on social media platforms |
| Perceived usefulness | PU1 | I find social commerce very useful in the COVID-19 crisis |
| | PU2 | Using social media in the online purchasing process is consistent with the preventive measures for COVID-19 |
| | PU3 | Online ordering and e-payment reduce my anxiety and fear of contracting the coronavirus |
| | PU4 | In general, social commerce has plenty of benefits and advantages compared with traditional ways of shopping |
| Intention to adopt social commerce | SCA1 | I have an intention to continue using social commerce services |
| | SCA2 | I plan to purchase/repurchase from social commerce platforms |
| | SCA3 | I recommend my friends to purchase through social commerce platforms |
| | SCA4 | I would be likely to continue shopping online via social commerce platforms |

**Table 3** Sample demographics

| Sample information | | No | % |
|---|---|---|---|
| Gender | Female | 107 | 57 |
| | Male | 81 | 43 |
| Total | | 189 | 100 |
| Age | = < 24 | 57 | 30 |
| | 25–34 | 72 | 38 |
| | 35–44 | 36 | 19 |
| | 45 + | 24 | 13 |
| Total | | 189 | 100 |

**Table 4** The measurement model results

| No | Construct | α | ρ_A | CR | AVE | 1 | 2 | 3 | 4 |
|---|---|---|---|---|---|---|---|---|---|
| 1 | COVDS | 0.804 | 0.845 | 0.867 | 0.619 | **0.787** | | | |
| 2 | EWOM | 0.739 | 0.740 | 0.852 | 0.657 | 0.556 | **0.811** | | |
| 3 | PU | 0.855 | 0.864 | 0.901 | 0.695 | 0.593 | 0.426 | **0.834** | |
| 4 | SCA | 0.828 | 0.843 | 0.885 | 0.658 | 0.639 | 0.632 | 0.594 | **0.811** |

# 5 Results

## 5.1 Measurement Model

The outer loadings of measurement items were greater than 0.7, at level $\alpha = 0.05$, thus demonstrating validity. However, one item with a factor loading less than 0.5, was excepted from the construct of the EWOM scale (EWOM4). As shown in Table 4, the results of testing Cronbach's alpha ($\alpha$), rho_A, and Composite Reliability (CR) indicate that all constructs had values exceeding the 0.70 thresholds [28], confirming adequate internal consistency reliability.

Furthermore, all values of the Average Variance Extracted (AVE) were above 0.5, showing adequate convergent validity. Finally, drawing on Fornell and Larcker's [29] criterion, the measurement model analysis indicates adequate discriminant validity, where the square roots of AVE of all constructs were greater than the correlation with other constructs.

## 5.2 Structural Model and Hypothesis Testing

Figure 2 displays the result of the structural modelling analysis, representing the causal relationships between research constructs.

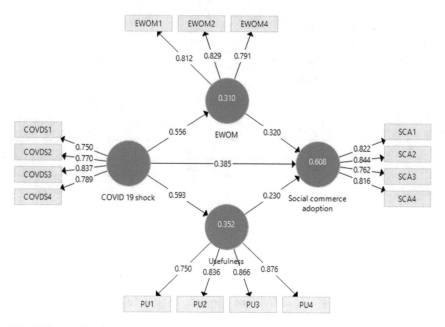

**Fig. 2** The results of the structural modeling analysis

Path coefficients (β), t values, and significance level for each relationship were used to test the hypothesized relationships. A rule of thumb is that path coefficient above 0.1 with at value greater than 1.96 is significant, at level α = 0.05 [28]. However, Table 5 shows that all hypothesized impacts of the COVID 19 shock on EWOM, perceived usefulness, and intention to adopt social commerce are significant. Accordingly, H1, H2, and H3 are supported.

The present study has conducted the Sobel test (Table 6) to evaluate the mediating role of EWOM and perceived usefulness.

**Table 5** Results of the hypotheses testing

| H | β | t value | Sig | Results |
|---|---|---------|-----|---------|
| 1 | 0.556 | 7.288 | 0.000 | Accepted |
| 2 | 0.593 | 12.395 | 0.000 | Accepted |
| 3 | 0.385 | 4.736 | 0.000 | Accepted |

**Table 6** Results of the Sobel test

| H | z test | Sig | Results |
|---|--------|-----|---------|
| 4 | 3.399 | 0.001 | Accepted |
| 5 | 2.531 | 0.011 | Accepted |

Table 6 confirms the mediating role of EWOM and perceived usefulness in the impact of the COVID 19 shock on intention to adopt social commerce, supporting the associated hypotheses (H4 and H5).

# 6   Discussion, Conclusion, and Implications

The emergence and global-scale impact of COVID 19 and the panic and anxiety that it sparked have changed dramatically people's lives leading to a shock that swept the whole world. The risk and threats of this epidemic and the preventive measures that have emerged to protect society have led to many transformations in consumer trends and behaviors, including resorting to social media platforms to interact with others and purchase online. Social commerce during this crisis has created a safe and comfortable shopping and purchasing environment in line with the commitment to preventive measures. However, although there are valuable efforts and evolving literature on social commerce, far less attention has been given to drivers of its adoption during pandemic crises. Therefore, this study has aimed to explore empirically the impact of the COVID 19 shock on EWOM, perceived usefulness, and intention to adopt social commerce. It also examined the mediating role of EWOM and perceived usefulness in the impact of the COVID 19 shock on intention to adopt social commerce.

The results of this study show a significant impact of the COVID 19 shock on EWOM. These findings are consistent with prior research (e.g., [4, 23]) that has investigated the impact of the COVID 19 on consumer participation and forming the trend of EWOM in terms of the corporate interaction with consumers' concerns in a supportive and emotional manner.

The results reveal that the COVID 19 shock has a significant impact on the perceived value of social commerce. These findings are in line with recent studies (e.g., [4, 10]), emphasizing the positive role of social commerce in overcoming the barriers of countermeasures against the pandemic, creating a safe shopping and purchasing environment through online ordering, e-payment, and home delivery.

The results indicate a significant impact of the COVID 19 shock on the intention to adopt social commerce. These findings agree with previous studies (e.g., [23, 26]) that have confirmed the effects of the COVID-19 crisis on intensifying consumers' attitudes toward online shopping. They are also consistent with recent research (e.g., [14, 24]) that has explained the consumers' intention to purchase more online during the COVID 19 crisis in the context of the fear-inducing effects.

The results indicate a significant mediating impact of EWOM and perceived usefulness on the relationship between the COVID-19 shock and on intention to adopt social commerce. These findings are in agreement with conclusions drawn from the literature (e.g., [5, 20]), emphasizing the vital role of WOM in consumers' purchase intention from social commerce platforms either directly or indirectly as a mediating factor. Furthermore, these results are consistent with other findings (e.g., [1, 10]) confirming that awareness of the risks of the COVID-19 and the perceived usefulness of social commerce in protecting consumers from the pandemic in line with preventive measures will enhance the trend towards adopting it.

Under conditions of crisis that induce public panic and anxiety and the imposition of preventive measures restricting consumer's choices, freedom of movement, and shopping decisions, the gravity of social commerce emerges. The COVID-19 shock has enriched the consumers' perception of social commerce usefulness. Social commerce provides a safe and comfortable environment for consumers with tools that enable the consumers' community to share information, reviews, referrals, ratings, and knowledge about products and vendors in the context of EWOM supporting their decision-making. Companies need to know that the consistency of their operations, marketing strategies, and promotional campaigns with the events of the epidemic and consumer concerns represents an important motivation to spread a positive EWOM.

The history of epidemics implies that COVID 19 will not be the last or the most dangerous. This study makes valuable academic and practical contributions to the field of social commerce during unprecedented global crises, such as pandemics. This study elucidates the importance of social trade during epidemics and the need for more studies to help in developing this paradigm as a part of preventive measures to protect people and secure their needs and wants at the same time during such crises. However, this study contributes to continuing discussions on the determinants of social commerce adoption, introducing the impact of unprecedented crisis shocks as a new subject matter to predict the consumer's intention to adopt social commerce. To the best of the authors' knowledge, there is no prior research that has explored the impact of the COVID 19 shock on EWOM, and perceived usefulness in enhancing the opportunities for social commerce adoption. On the other hand, this study provides businesses with a better understanding and knowledge of developing successful social commerce initiatives from the consumer viewpoint, especially during pandemic crises. Understanding consumers' needs for social commerce and the drivers of its adoption in facing such pandemics is a critical requirement for all business partners to implement adaptive strategies and operations, enabling them to survive and conduct their social responsibility in these difficult situations.

This study has some limitations that can provide an impetus for further studies in the future. Data were collected from a small sample of Facebook users in Jordan. Therefore, it can be said that this sample is not representative of all consumers who have adopted or are interested in social commerce adoption. Furthermore, the pandemic shock and social commerce usefulness may change over time depending on the progress or decline in the spread of the virus. Therefore, longitudinal studies are beneficial for improving our understanding of the dynamics of these factors and their long-term impact on social commerce adoption.

# References

1. Diwanji VS, Cortese J (2021) Comparing the impact of presentation format of consumer generated reviews on shoppers' decisions in an online social commerce environment. J Electron Commer Res 22(1):22–45
2. Shi S, Chow WS (2015) Trust development and transfer in social commerce: prior experience as moderator. Ind Manage Data Syst 115(7):1182–1203
3. De Silva PO, Perera KJT, Rajapaksha RMMC, Idroos AA, Herath HMJP, Udawaththa UDIC, Ratanayake RMCS (2020) The future of s-commerce in the age of COVID-19-review paper. J Manage Tourism Res (JMTR) 38:38–52
4. Wang Y, Zhang M, Li S, McLeay F, Gupta S (2021) Corporate responses to the coronavirus crisis and their impact on electronic-word-of-mouth and trust recovery: evidence from social media. Br J Manage 1–19. https://doi.org/10.1111/1467-8551.12497
5. See-To EW, Ho KK (2014) Value co-creation and purchase intention in social network sites: the role of electronic word-of-mouth and trust–a theoretical analysis. Comput Hum Behav 31:182–189
6. Kim Y, Chang Y, Wong SF, Park MC (2014) Customer attribution of service failure and its impact in social commerce environment. Int J Electron Custom Relat Manage 8(1):136–158
7. Hajli N (2015) Social commerce constructs and consumer's intention to buy. Int J Inf Manage 35(2):183–191
8. Larios-Gómez E, Fischer L, Peñalosa M, Ortega-Vivanco M (2021) Purchase behavior in COVID-19: a cross study in Mexico, Colombia, and Ecuador. Heliyon 7(3):e06468
9. Zhang YL, Kwark Y, Wang Y, Shin D (2020) Impact of COVID-19 crisis on social commerce: an empirical analysis of e-commerce social activities during the pandemic (September 21, 2020). Available at SSRN: https://ssrn.com/abstract=(2020)
10. Byrd K, Her E, Fan A, Almanza B, Liu Y, Leitch S (2021) Restaurants and COVID-19: what are consumers' risk perceptions about restaurant food and its packaging during the pandemic? Int J Hosp Manage 94:102821
11. Sharma P (2021) Customer co-creation, COVID-19 and sustainable service outcomes. Benchmarking Int J 1–27. https://doi.org/10.1108/BIJ-10-2020-0541
12. Bhatti A, Akram H, Basit HM, Khan AU, Raza SM, Naqvi MB (2020) E-commerce trends during COVID-19 pandemic. Int J Future Gener Commun Netw 13(2):1449–1452
13. Shahzad A, Hassan R, Abdullah NI, Hussain A, Fareed M (2020) COVID-19 impact on e-commerce usage: an empirical evidence from Malaysian healthcare industry. Humanit Soc Sci Rev 8(3):599–609
14. Tran LTT (2021) Managing the effectiveness of e-commerce platforms in a pandemic. J Retail Consum Serv 58:102287
15. Alhubaishy A, Aljuhani A (2021) The influence of information sharing through social network sites on customers' attitudes during the epidemic crisis of COVID-19. J Theor Appl Electron Commer Res 16(5):1390–1403
16. Ali Taha V, Pencarelli T, Škerháková V, Fedorko R, Košíková M (2021) The use of social media and its impact on shopping behavior of Slovak and Italian consumers during COVID-19 pandemic. Sustainability 13(4):1710
17. Habes M, Alghizzawi M, Ali S, SalihAlnaser A, Salloum SA (2020) The relation among marketing ads, via digital media and mitigate (COVID-19) pandemic in Jordan. Int J Adv Sci Technol 29(7):12326–12348
18. Febrian A, Bangsawan S, Ahadiat A (2021) Digital content marketing strategy in increasing customer engagement in Covid-19 situation. Int J Pharm Res 13(1):1–9
19. Yasir A, Hu X, Ahmad M, Rauf A, Shi J, Ali Nasir S (2020) Modeling impact of word of mouth and e-government on online social presence during COVID-19 outbreak: a multi-mediation approach. Int J Environ Res Public Health 17(8):1–21
20. Yin X, Wang H, Xia Q, Gu Q (2019) How social interaction affects purchase intention in social commerce: a cultural perspective. Sustainability 11(8):1–18

21. Mikalef P, Giannakos MN, Pappas IO (2017) Designing social commerce platforms based on consumers' intentions. Behav Inf Technol 36(12):1308–1327
22. Odekerken-Schröder G, Mele C, Russo-Spena T, Mahr D, Ruggiero A (2020) Mitigating loneliness with companion robots in the COVID-19 pandemic and beyond: an integrative framework and research agenda. J Serv Manage 32(6):1149–1162
23. Sheth J (2020) Impact of Covid-19 on consumer behavior: will the old habits return or die? J Bus Res 117:280–283
24. Naeem M (2021) Do social media platforms develop consumer panic buying during the fear of COVID-19 pandemic. J Retail Consum Serv 58:1–10
25. Eger L, Komárková L, Egerová D, Mičík M (2021) The effect of COVID-19 on consumer shopping behaviour: generational cohort perspective. J Retail Consum Serv 61:1–11
26. Donthu N, Gustafsson A (2020) Effects of COVID-19 on business and research. J Bus Res 117:284–289
27. Ng CSP (2013) Intention to purchase on social commerce websites across cultures: a cross-regional study. Inf Manage 50(8):609–620
28. Hair JF, Hult GT, Ringle CM, Sarstedt M (2013) A primer on partial least squares structural equation modelling (PLS-SEM). Sage, Thousand Oaks
29. Fornell C, Larcker D (1981) Evaluating structural equation models with unobservable variables and measurement error. J Mark Res 18(1):39–50

# Exploring the Understanding and Criticality of Digital Literacy in MENA Digital Start-Ups

Manal Belghali and Salwa Bahyaoui

**Abstract** This paper aims primarily to explore how the concept of digital literacy fits in start-ups operating in a digital setting, with a focus on the MENA region. It explores the perceived criticality of digital literacy in this context, highlighting challenges relative to the emerging and growing need for digitally literate, and at least digitally competent employees. The data is analysed within what is pertinent to the available definitions of digital literacy in literature, highlighting a focus on skills pertaining to basic digital competence and digital usage in a professional context. Findings also highlight that digital literacy is perceived as critical in light of digital start-up's challenges.

**Keywords** Digital literacy · Digital skills and competences · Digital Start-ups

## 1 Introduction

The emergence of new technologies and the continuous incoming of digital technologies in the realm of entrepreneurship and within the start-up ecosystem yields new challenges. There is a growing focus on how digital literacy can be approached to channel the benefits of digitalisation and tackle its challenges so that the digital era becomes auspicious and yielding of opportunities, rather than letting digital advancements exceed and overtake people's abilities to make use of them.

These changes connected to creating and operating businesses that are digital or digitally-enabled also yield changes at the level of the individuals. With the rhythm at which digital acceleration is happening and with the growing demand for digital talents, attention should be given to emerging skills and potential skill gaps.

M. Belghali (✉) · S. Bahyaoui
Laboratory for Research on Economic Competitiveness and Managerial Performance (LARCEPEM), Mohamed V University of Rabat, Al Irfane, Rabat, Morocco
e-mail: manal.belghali@um5r.ac.ma

Consistent with the specific features of the digital start-up ecosystem, supporting digital start-ups becomes a critical task that needs to be carried out by specialized employees that have access to relevant skill sets. This entails developing and embedding digital skills within roles across many, if not all disciplines.

The focus on digital start-ups is self-explanatory, as while other organisations face an inevitable digital transformation in some areas, digital start-ups have built their business models entirely either on contributing to or making use of technologies. Their main assets are linked to technological investments, which calls for the question: how critical does it become for these start-ups to have access to the right people to pilot a business that is digitally enabled?

## 2 Defining Digital Literacy

The notion of 'literacy' has been enriched by skill-based literacies relative to the growing complexity of information and the emergence of new technologies [1]. Its semantic reach extended to "the ability to understand information however presented" [2].

The concept of digital literacy seems to find its origin in this notion, as it was defined by Paul Gilster as "the ability to understand and use information in multiple formats from a wide variety of sources when it is presented via computers" [3].

In the relevant literature, this concept is relatively broad and approached as a spectrum going from general awareness and perspective to specific skills and competencies. An occurrence analysis highlighted the following associated terminology: computer literacy, Information Technology (IT) literacy, electronic (information) literacy, media literacy, network literacy, Internet literacy, e-literacy, digital (information) literacy.

The concept and term of digital literacy are widely used according to Gilster's broad definition [3], but other articles have assimilated it with computer literacy with a focus on IT skills contained within a broader information literacy [4].

A Norwegian whitepaper presented a conceptual definition for it as a broad competency that links traditional skills with the capacity to use ICT critically and creatively [5, 6].

The term "e-literacy" (electronic literacy) is less frequently used as a synonym for digital literacy. It appears in the glossary of teaching technology by the Leeds University as a term that synthesizes traditional computer literacy skills, some aspects of information literacy (handling digital information, including finding, organising, and using it), in addition to interpretation and knowledge construction and expression [7]. Digital literacy was also centred around the acquisition and use of knowledge, techniques, attitudes, and personal qualities with the ability to plan, execute and evaluate digital actions towards problem-solving in day-to-day tasks [8].

The concept has been incorporated in the DigEuLit project as "the awareness, attitude and ability" to properly use digital tools and digital facilities to "identify, access, manage, integrate, evaluate, analyse and synthesise digital resources, construct new knowledge, create media expressions, and communicate in the context of specific life situations", to enable a constructive social action and reflect upon the process [8].

## 2.1	Digital Literacy Skills and Competences

No exhaustive list of digital literacy skills was directly presented by Paul Gilster, but one was rather derived [1]. It includes: "knowledge assembly" skills, retrieval skills, the ability to exercise critical thinking to make informed judgments on the validity and completeness of retrieved information, the ability to reliably construct an information hoard from various sources, reading and understanding skills in a dynamic non-sequential hypertext context, the distinction between fact and opinion; (Internet) searching skills; managing multimedia flow, utilising information filters and agents, personal information strategy creation; awareness of one's expanded ability and that of others via networks through discussion and collaboration on issues; ability to understand a problem through a set of questions pertaining to information needed; ability to back-up traditional forms of content with networked tools; caution and alertness when judging referenced material's validity and completeness [3, 9].

This exploratory study starts from the definition that considers digital literacy as a broader concept that incorporates elements of associated literacies mentioned above [10]. It highlights skills and competencies relative to the ability to read and understand the information presented in digital and non-digital format; its evaluation; creation and communication; knowledge assembly; information and media literacy. A three-level approach was incorporated into this definition considering anyone at or above level II as digitally literate [10]. A synthesis of this approach is modelled in Table 1.

## 2.2	Digital Literacy in the Context of Digital Start-Ups and Digital Entrepreneurship

The concept of digital literacy has been approached in the light of the digital start-up ecosystem, and more specifically digital entrepreneurship. Digital entrepreneurship refers to the act of creating ventures or transforming existing businesses through innovative technologies or innovative uses of existing technologies [11].

Digital technologies have found their way into business, leading to deep remodelling and transformation of entrepreneurial processes and outcomes [11], including ways of creating and doing business, prompting a new specific type of entrepreneurship [12]. In this context, the validity of the business idea and its consistency with the existing skills become critical elements. Several aspects could become dependent on the employees and the entrepreneur, assuming these have the digital skills enabling

**Table 1** Levels and respective skills of digital literacy. Based on Allan Martin's work

| Digital literacy | Level III: Digital transformation |
|---|---|
| | Innovation and Creativity<br>Developing digital usages to the point where they stimulate and initiate change within the professional context or knowledge domain, at the individual, group, or organisational level |
| | Level II: Digital usage |
| | Professional/Discipline Application<br>Informed application of digital competence within specific professional contexts, specific to an individual, a group, or an organisation<br>Drawing on relevant competences and skills for specific professions, domains, life contexts, based on the situation requirements<br>Involving digital tools to find and process needed information, to construct the output for task achievement or problem-solving |
| Digital competence | Level I: Digital competence |
| | Skills and Attitudes for Confident and Successful Use of Technologies<br>*Basic visual recognition and manual action skills* (finding information, word processing, electronic communication, use of spreadsheets, presentation creation, web and desktop publishing, databases creation and manipulation, modelling and simulations…)<br>*Critical, evaluative, and conceptual skills* (Stating the problem to solve, identifying and accessing digital resources, evaluating their objectivity and reliability, interpreting meanings, relevantly integrating and combining resources to create new knowledge…)<br>*Attitudes and awareness* (Communicating to interact with relevant actors while working on the problem or task, present outputs, and solutions for knowledge dissemination, and reflecting on the success of the problem-solving or task achievement as well as on self-development pertaining to digital literacy) |

them to collect, market, and transform business opportunities in this specific setting [13].

To achieve growth and sustainability, these digital start-ups need to incorporate new digital competences and integrate them into existing and new management functions from operations to marketing. These competences pertain to the ability to possess, manage, dominate, and develop digital technologies [14]. Digital competences are broad and they range from hard skills relative to using computer programmes, packages, tools, social and mobile analytics, cloud, robotics, Internet… to soft skills associated with the more critical skills, behaviours, and attitudes that allow individuals to make effective use of digital tools for problem-solving, knowledge networking… [15].

Digital literacy of entrepreneurs was defined as the ability to adapt to the development of technology and make use of media as communication, to analyse trends… It was associated with the use of applications that enable business management [16]. Within the perspective of fuelling and enabling digital start-ups and digital entrepreneurship as such, the European Commission proposes, among other initiatives, the introduction of digital knowledge and diffusion of digital skills, along with nurturing e-leadership [13].

Regarding digital literacy in the MENA region, and in the Arab business setting in general, studies have explored the factors affecting IT adoption with a focus on cultural dimensions [17], fewer studies looked at the competence dimension for IT adoption in the Arab business context. A study highlighted the criticality of digital literacy and related digital competency to enable the country's growth through digitally literate local populations [18]. Another exploration of the role of education in building capacity for digital development in the Arab world highlighted that new technologies require communication skills, critical thinking and problem solving, creative thinking, leadership skills, technical skills, and collaboration/teamwork skills and concluded that a digital curriculum involving digital literacy, internet connectivity, life-long learning, and computer-based applications, must be incorporated into the educational system to increase readiness for new technologies [19], especially given the growing disconnect between the content-driven education model inherited from the nineteenth century and the rapidly evolving skills-based work setting [20]. Another study conducted the first exploration of digital literacy in the context of the Palestinian refugee community in the Middle East and again identified a deficit in access to digital literacy education arguing that digital literacy has not been developed or discussed sufficiently in middle eastern education systems and outlining objectives and learning strategies to bridge the gap between what is provided by the educational systems and what is needed in a digital environment that is not homogeneous [21].

Overall, there is a growing interest in exploring digital literacy in the Arab context and from a competency and skill perspective, along with a focus on education as a driving pillar for the digital socio-economic transformation.

## 3 Exploring the Understanding and Criticality of Digital Literacy in Digital Start-Ups from the MENA Region

### 3.1 Methodology

This research utilised a qualitative descriptive approach for a qualitative exploration of the understanding and the perceived criticality of digital literacy as described by respondents based on their experiences. It utilised two qualitative data collection methods: semi-structured interviews with three participants, and 12 fully qualitative surveys. Both were directed at experts and stakeholders from the digital start-up ecosystem in North Africa and the Middle East. The subjects of this study consisted of start-up founders, executives, mentors, and consultants from Morocco, Tunisia, Algeria, Egypt, Sudan, Jordan, Kuwait, Saudi Arabia, United Arab Emirates, and Oman.

Surveys were self-administered online, and semi-structured interviews were conducted in virtual calls and were transcribed to supplement data from the qualitative surveys.

The exploration followed opportunity and snowball sampling. Participants were contacted via email and LinkedIn, depending on their willingness and availability, and were asked to recommend individuals within the scope of research.

The surveys and interviews majorly involved topic-based questions. No questions pertaining to roles or industry of respondents were included, as the identity of respondents and therefore their relevance to the exploration was known at the time of initial contact. The qualitative surveys consisted of open-ended questions, centred on the topic of digital literacy skills and perceptions of how digitally enabled employees are.

For the data, a thematic analysis was conducted, partially using qualitative data analysis software to detect dominant themes which were discussed considering the key points from the review of literature. Analysis involved frequency count of words representing key concepts to identify key points that were clustered together into descriptive categories, grouped into final themes that emerged inductively and were not predetermined.

## 3.2 Findings and Discussion

This study explored the concept of digital literacy within start-ups operating in a digital setting that are based or have operations in the MENA region. The exploration was conducted from two standpoints: the understanding and the perceived criticality of digital literacy in that specific context.

Five themes emerged from the data: 'Digital skills', 'Digital aspects of work', 'Digital start-up challenges', 'Employee background', and 'Development opportunities'. The three first were dominant and closely linked reflecting the study aims. The themes associated with 'Digital literacy development opportunities' and 'Employee background' emerged as the respondents' views on relevant channels and responsible parties to develop digital literacy. These two themes seem relevant to explore, although they are outside the specific scope of this paper, they can fit into a broader discussion.

Before integrating the key findings in the discussion, the themes are outlined separately with illustrations from the data below.

Digital Skills: Respondents' understanding of digital literacy was approached through questions pertaining to how they defined digital literacy in terms of skills, how they described a digitally literate employee, with probes relative to roles these employees hold to make sure their understanding is not limited to roles usually associated with technology. The skills highlighted in data analysis ranged from basic skills pertaining to using computers, Internet, and various online and offline tools for work purposes, to specific skills associated with design, web creation, computerised data analysis…

Good literacy of digital products and technologies at least in concept. Ability to work with online applications with ease [Digital literacy skills].

Adaptable, agile, and open to new ways of working. Able to set goals and track key performance indicators [Describing the ideal digitally literate employee].

The ability to learn is mentioned by several respondents highlighting a progressive aspect of digital literacy. Employees need to have the fundamentals to be able to learn and use new technologies but also know when new learnings are needed to deliver the needed results. The term 'tech-savvy' is mentioned eight times, referring to employees that can navigate their ways around new or existing technologies.

… must be able to use mainstream apps and software in their specific field and have the ability to quickly learn how to use any new ones.

Ideally the candidate should be able to understand what tool to use, how much time the activity would need, and if he/she needs to learn anything more to deliver the right results in the expected time.

When asked about digital skills perceived as lacking, 'ability to think algorithmically', 'the ability to use cloud applications', 'basic programming knowledge/skills" and 'the ability to use data and provide insights using computerised tools' were highlighted.

**Digital Aspects of Work**: Respondents highlighted areas of work where being digitally literate (or digitally competent at the very least) was critical. The questions pertained to both the perceived criticality and the areas of work concerned with this criticality.

Disciplines that are more exposed to digital platforms like marketing (including design, social media… are naturally more digitally critical.

You make the conscious decision to adopt a new piece of technology, or you have seen a problem that can be solved by utilising it, that mindset is important.

**Digital Start-Up Challenges**: When approaching the criticality of being digitally literate in the context of digital start-ups, respondents highlighted how digital skills allow start-ups to fulfil their goals.

Understanding how technology can help accelerate, automate, and augment desired outcomes is not only crucial for technology adoption, but for embracing technology as an enabler of growth.

…to be successful, we have to adapt, and technology allowed us to do that quickly.

Overall, perceptions and instantiations of digital literacy vary from one individual to another. The reason for this could be linked to the fact that the settings, contexts, and situations in which respondents operate vary, as well as the nature of their work, their background, and the tools and facilities they use or develop also differ. Although common themes are detected in terms of the understanding of digital literacy in the context of a digital start-up, and despite the unanimous agreement that digital literacy is critical across functions in digital start-ups, respondents draw upon digital literacy and the opportunities and challenges it poses as appropriate and relevant to their specific settings.

Analysis of participants' responses highlighted whom they perceived as responsible for building a digitally literate workforce.

"The most senior leaders of the business,"; "Employee onboarding and building an internal knowledge management solution accessible to all employees"; "The fundamentals of digital skills seem to be acquired through the natural process of living in the digital age"; "Education for the basics, and more if possible. But employees can always learn onsite for the specifics".

Respondents mentioned the employee's background several times, making the distinction between expectations from employees with a tech background and those with a business educational background. This shows that education (mentioned five times) is viewed as critical to have basic digital skills. Other researchers have argued that developing the skills and competences is not sufficient, that these must be grounded in an encompassing moral framework of being an educated individual [22].

Based on the analysed data, examined in light of the three-level digital literacy model discussed above in this paper, it appears that the digitally enabled employees and the skills described by respondents as pertaining to being digitally literate mostly fall under the category of digital competence.

However, when describing how critical digital literacy is for different functions including general business, respondents refer to the second level: 'digital usage', which places the skills and competences in a specific work context or discipline. This is associated with having the skills and competences first, then making informed usage of their skills and competences to achieve tasks in their respective roles. When speaking about the 'mindset' relative to adopting technology as an enabler of growth, and using digital skills and technology to adapt, and therefore succeed, respondents referred to a higher level of digital literacy (Level III). In respect to the three-level model of digital literacy, digital skills, and their usage as described by respondents oscillate mainly between the level of digital competence (basic skills for confident and successful use of Information Society Technologies) and the level of digital usage considered as the first level of digital literacy (professional application of skills). These are the skills that are critically expected by respondents from digitally literate employees in the digital start-up context.

## 3.3   Research Implications and Limitations

This research provides a primary conceptual exploration of digital literacy in countries from the MENA region, adding to findings from the few studies that explored this concept in the Arab business setting. Regarding digital literacy, updating the understanding and bringing in new competences is highlighted as a necessary matter, particularly in a context marked by change in terms of the digital information environment [23].

Implications that should be highlighted are relative to setting a foundation to bridge the digital skill gaps in the start-up ecosystem. It will be relevant to conduct a deeper and larger exploration of digital literacy in the context of start-ups, particularly those operating in a digital context. Also, as this primary exploration reported some skills that are perceived as both critical and lacking, a relevant track could be that of exploring these skills to identify any potential digital skill gap. Another relevant track can be exploring how digital start-ups contribute to building a digitally literate workforce internally.

The findings from this exploration are limited by the small sample size and the limited representativity of the MENA region and Arab start-up eco-system.

# References

1. Bawden D (2001) Information and digital literacies: a review of concepts. J Doc 57(2):218–259
2. Lanham R (1995) Digital literacy. Sci Am 273(3):160–161
3. Gilster P (1997) Digital literacy. Wiley, New York
4. Williams P, Minnion A (2007) Exploring the challenges of developing digital literacy in the context of special educational needs. In: Andretta S (ed) Change and challenge: information literacy for the 21st century. Auslib Press, Adelaide, pp 115–144
5. Søby M (2003) Digital Kompetanse: Fra 4. Basisferdighet til Digital Dannelse [Digital competence: from the fourth basic skills to digital education]. Oslo
6. Knobel M, Lankshear C (2006) Digital literacy and digital literacies: policy, pedagogy and research considerations for education. Nordic J Digit Literacy 1
7. Leeds University (UK) Glossary of teaching technology. Leeds
8. Martin A (2005) DigEuLit—A European framework for digital literacy: a progress report. University of Glasgow, Glasgow
9. Nicholas D, Williams P (1998) Review of Paul Gilster digital literacy book. J Doc 54(3):360–362
10. Martin A (2008) Digital literacy and the "digital society". In: Lankshear C, Knobel M (eds) Digital literacies: concepts, policies, and practices. New York, pp 151–176
11. Nambisan S, Baron RA (2013) Entrepreneurship in innovation ecosystems: entrepreneurs' self-regulatory processes and their implications for new venture success. Entrep Theory Pract 37(5):1071–1097
12. Nambisan S (2017) Digital entrepreneurship: toward a digital technology perspective of entrepreneurship. Entrep Theory Pract 41(6):1029–1055
13. European Commission (EC) (2014) Fuelling digital entrepreneurship in Europe, Background paper
14. Thomas A, Passaro R, Quinto I (2020) Developing entrepreneurship in digital economy: the ecosystem strategy for startups growth. In: Strategy and behaviors in the digital economy. Books on Demand
15. Iordache C, Mariën I, Baelden D (2017) Developing digital skills and competences: a QuickScan analysis of 13 digital literacy models. Ital J Sociol Educ 9(1):6–30
16. Sariwulan T, Suparno S, Disman D, Ahman E, Suwatno S (2020) Entrepreneurial performance: the role of literacy and skills. J Asian Financ Econ Bus 7(11):269–280
17. Dajani D, Yaseen SG (2016) Applicability of technology acceptance models in the Arab business setting. J Bus Retail Manage Res 10
18. Jewels T, Albon R (2011) Reconciling culture and digital literacy in the United Arab Emirates. Int J Digital Literacy Digital Competence 2(2):27–39
19. Al-Roubaie A (2019) Building capacity for digital development in the Arab world: the role of education. Int J Eng Adv Technol 8(5):1530–1537
20. Deloitte, Global Business Coalition for Education (2018) Preparing tomorrow's workforce for the fourth industrial revolution

21. Traxler J (1983) Digital literacy: a Palestinian refugee perspective. Res Learn Technol 26:2018
22. Bawden D (2008) Origins and concepts of digital literacy: concepts, policies and practices. In: Lankshear C, Knobel M (eds) Digital literacies. Peter Lang Publishing, New York, pp 17–32
23. Wilson MC (1998) To dissect a frog or design an elephant: teaching digital information literacy through the library gateway. Inspel 32(3):189–195

# Knowledge Management by Firms: A Systematic Review

Enas Al-lozi and Ra'ed Masa'deh

**Abstract** Knowledge management is well researched topic in the empirical as well as theoretical literature. The purpose of this study was to understand the literature that has already been published in the field of knowledge management especially related to firms. The literature mainly focused on knowledge management and organizational performance. While reviewing the literature it has been observed that leadership play a significant role in developing the culture for learning and knowledge management. The systematic review identified several research streams and has identified several inconsistencies which raise certain ambiguities and open the doors for research. The purpose of this study is to understand the new directions that can be set through the review of existing literature in the field of knowledge management. For the said purpose a systematic review was conducted. The articles were initially collected based on key words knowledge management and firm performance. While filtering the articles all the articles related to other fields were removed and finally 75 chosen articles were kept on the basis of which the analysis has been made. The study opened the horizons for future research and concluded that knowledge management in becoming more and more important for the sustainability, survival, and performance of businesses.

**Keywords** Knowledge management · Knowledge sharing · Firm performance · Innovation

## 1 Introduction

In the current dynamic and competitive environment, knowledge is considered as the key to element that plays a key role in the survival and sustainability of firms [34].

E. Al-lozi (✉)
Faculty of Business, Al-Zaytoonah University of Jordan, Amman, Jordan
e-mail: Enas.al-lozi@zuj.edu.jo

R. Masa'deh
Faculty of Business, The University of Jordan, Amman, Jordan
e-mail: r.masadeh@ju.edu.jo

© The Author(s), under exclusive license to Springer Nature Switzerland AG 2022
S. G. Yaseen (ed.), *Digital Economy, Business Analytics, and Big Data Analytics Applications*, Studies in Computational Intelligence 1010,
https://doi.org/10.1007/978-3-031-05258-3_11

Firms with strong knowledge performs better than others [79]. Thus, knowledge management has a significant importance for leveraging performance and gaining competitive advantage [18].

The need for big data is the base line for knowledge management as data analytics is one of the major sources of knowledge seeking leading to a successful knowledge management process [52, 88]. Previous literature of different fields highlights the need for knowledge management for gaining competitive advantage through innovation and continuous improvement [1, 39, 40]. Not only gaining competitive advantage, but firms seek to developing an innovative knowledge base seeking to improve productivity [19, 20], performance, and strategic capabilities [36].

Lack of knowledge management and sharing may cause harmful consequences. Implicit knowledge must be shared as it serves as building blocks for firms' success [7]. In the globalized world knowledge is considered as a key and strategic resource and its management is now not limited to local boundaries only but exceeds that [27]. Knowledge management covers transference of knowledge as well as knowledge seeking from subsidiaries or parent companies. In the current scenario many international collaborations are taking place for knowledge sharing and knowledge management [36, 76].

The purpose of this study is to unveil the issues and perspectives of knowledge management with regards to businesses. For the said purpose authors has scrutinized several studies conducted by different researchers in the field of knowledge management and its impact on different dimensions of a firm's performance. Furthermore, several studies where knowledge management has been taken as a mediating variable have also been studied. The author has reviewed the literature of last ten years and have tried to cover major parts of the unexplored. By the process of systematic review issues, antecedents, trends, future directions and challenges related to knowledge management have been identified. The current study followed a standardized approach for understanding knowledge management from different aspects, which will help practitioners, academicians and policy makers to understand the common issues being faced by the firms in managing knowledge and what advantages can be gained by managing knowledge properly. Furthermore, proper understanding about the main advantages can be understood to optimally utilize the benefit of knowledge management.

## 2 Methodology

For the said purpose a systematic meta-analysis has been conducted because meta-analytics approach helps to establish propositions for generalizations. The focus of this approach is to understand the objective knowledge through a scientific and systematic way. Likewise in this study observations have been made based on the large number of prior studies, previously conducted with different methods with certain commonalities. For the said purpose key terms used are knowledge management and firm performance has been used and papers that have been published mainly in the last ten years have been reviewed. Only those papers published in the peer-reviewed journals have been considered. Regardless of the type of the paper, the reviewed literature contains the term knowledge management with regard to any of effect over the firm.

Initially, articles were screened out and included in the study based on the following selecting condition i.e., the article title should contain the key word knowledge management and any impact or relationship with influence over the firm. The impact or influence may be over innovation [73], strategic performance, competitive advantage, leadership role, human resource practices, or quality management. Only those articles were included in the study that have a direct link with the firms and were published in peer reviewed papers between 2010 to 2021. Thus, 75 articles met the inclusion criterion for this study. All the selected articles were structured in matrix form that have commonalities. The summaries of the articles can be seen in the tables that have been mentioned in the discussion section.

## 3   Discussion

Importance of knowledge management is growing and businesses are becoming more concerned about managing this valuable resource [51]. However, knowledge management challenging as it includes unstructured data due to the nature of tacit knowledge to which several barriers hinder the successful management of knowledge [28, 83]. Prior studies have elaborated several factors like enablers, facilitators, motivators, barriers, and deterrents, which promote or demote knowledge management practices in the firms [47]. Based on main research objectives the contents from the reviewed articles were extracted and were grouped into different categories; knowledge management, innovation [1], organizational culture [6, 67], intellectual capital [15, 54], competitive advantage [7, 49], strategy [44], total quality management [3, 69], entrepreneurship [23], leadership, human resource, and knowledge management.

## 4   Knowledge Management and Organizational Performance

Several studies have been conducted with regards to knowledge management and organizational performance. It has been observed that knowledge management has been used as independent and mediating variable and it has a significant direct or indirect effect [22, 31, 33, 35, 38, 62, 64, 84]. The studies that have been conducted in the field of knowledge management and organizational performance have been identified in Table 1.

**Table 1** Knowledge management and organizational performance

| Source | Issues | Findings |
|--------|--------|----------|
| Sun [82] | Identifying knowledge acquisition, creation, utilization and sharing | Results identified five organizational themes namely strategic knowledge, strategic engagement, social networking, cultural context, and structural context |
| Madhoushi et al. [48] | Identifying the mediating role of knowledge management in entrepreneurial orientation and innovation performance | Entrepreneurial orientation directly and indirectly influence innovative performance and knowledge management holds the mediating role |
| Mills and Smith [55] | Analyzing the impact of knowledge management resources over organizational performance | Knowledge management resources directly impact organizational performance |
| Schiuma [78] | Analyzing the relationship between knowledge management and business performance improvement | Knowledge management mechanism is required for the improvement of organizational performance |
| Rašula et al. [71] | To identify the role of knowledge management in organizational performance | Knowledge management through information technology positively affect organizational performance |
| Zaied et al. [87] | Enhancing organizational performance through knowledge management capabilities in due to increasing importance of knowledge management as a weapon for sustaining competitive advantage | All elements of knowledge management capabilities have a significant positive impact over performance |
| Wu and Chen [86] | Defining a model for evaluating knowledge management | Knowledge asset and process capabilities are different but relevant drivers of value creation process and businesses processes mediate the relationship |
| Jamil and Lodhi [41] | Examining the impact of knowledge management practices in enhancing universities' performance | Knowledge management process and infrastructure have a significant impact over performance of universities |

(continued)

The stream of literature in the field of knowledge management and organizational performance is vast [2, 9, 58]. The studies conducted in the field covered knowledge-based view in collaboration with other theories [48, 74], Jianfeng, [46]. Some studies proposed future implications while other empirically tested developed models [24, 66, 78, 82]. Empirical studies in the domain mainly focused on firm performance,

**Table 1**  (continued)

| Source | Issues | Findings |
|---|---|---|
| Imran et al. [37] | Analyzing the role of knowledge management in achieving organizational performance through organizational learning | Knowledge management capabilities are of high importance, use of updated technology, supportive culture, knowledge acquisition and application process play a crucial role in organizational learning and overall performance |
| Alkaffaf et al. [10] | Identifying the mediating role of knowledge management process between knowledge management enablers and organizational creativity | Six individual elements of knowledge management enablers were identified and significantly identified that knowledge management process mediates between knowledge management enablers and organizational creativity |
| Abualoush et al. [4] | Identifying the inter connections among knowledge management infrastructure and process, intellectual capital, and firm performance | Exploiting knowledge management with the help of intellectual capital helps in achieving performance |
| Alsheikh et al. [14] | Linking knowledge management practices with human resource practices and analyzing their impact on job performance | Highlighting a significant role of human resource management practices, organizational culture, motivation and knowledge management on job performance |
| Alzghoul et al. [16] | Analyzing knowledge management as a moderator between authentic leadership, workplace harmony, worker's creativity and job performance | Identifying the positive role that authentic leadership has on workplace climate, creativity, and job performance, furthermore, workplace climate mediates and knowledge management moderates the relationship between authentic leadership and workplace climate which influence job performance |
| Iqbal et al. [39] | Identifying the impact of knowledge management enablers on knowledge management processes and ultimately on organizational performance along with mediating role of intellectual capital and innovation between knowledge management processes and performance has been researched | The findings of the study revealed that knowledge management enablers significantly impact knowledge management processes and knowledge management processes impact organizational performance directly and also through the mediating role of innovation and intellectual capital |

(continued)

**Table 1** (continued)

| Source | Issues | Findings |
|---|---|---|
| Riaz and Hassan [72] | Unpacking the link between knowledge management processes and organizational performance by inculcating the mediating role of organizational creativity | Organizational creativity mediates the relationship and invigorating organizational performance because of knowledge management |
| Butt et al. [25] | Identifying the mediating role of knowledge-worker productivity between individual knowledge management engagement and innovation | In knowledge-based organizations the aspect of knowledge worker productivity is significant between knowledge management engagement and innovation |
| Sahibzada et al. [74] | Identifying the mediating role of knowledge worker productivity between knowledge management process and organizational performance | The knowledge worker productivity hold a mediating role and knowledge management itself is important directly and indirectly for gaining organizational performance |
| Shabbir and Gardezi [79] | Identifying the mediating role of knowledge management between application of big data and performance of SMEs | The application of big data analytics has a significant positive impact and knowledge management holds partial mediating role |
| Wijaya and Suasih [85] | Identifying the impact of knowledge management on organizational performance with mediating role of competitive advantage | There is no direct impact of knowledge management but through mediating variable it has a significant impact |
| Khakpour and Hasani | Exploring the impact of knowledge management processes over performance of entrepreneurial firms | Knowledge management and knowledge creation are strong predictors of performance of entrepreneurial firms |

however, the impact is different [12, 39, 50, 55, 65, 71, 85–87]. Previous research also covered aspects of human resource and organizational culture [53], knowledge management process [8], learning, intellectual capital for understanding the impact of knowledge management in gaining performance [4, 14, 17, 37, 63, 72, 75].

# 5 Knowledge Management and Leadership

Another cluster of studies related to knowledge management covered leadership aspects highlighting the importance of leadership in different cultural as well as environmental context [11, 59, 80]. These studies particularly identified the role of leadership for knowledge management and organizational performance. The synthesis of the literature is mentioned in Table 2.

The last stream identified was related to knowledge management and leadership [30, 57, 61]. This stream covers the role and importance of leadership in capturing knowledge, managing it, and creating a learning culture which promotes innovation and helps in gaining competitive advantage [16, 21, 27, 45, 46, 77]. The knowledge-oriented leadership especially in the entrepreneurially oriented firms is required to promote a culture of managing and sharing [5, 43, 56, 70].

# 6 Future Directions

The concept of knowledge management is well researched in the past two decades. Different literature streams are available, however, in this review study, literature covered has been divided into four parts. Knowledge management and performance, knowledge management and innovation and performance, knowledge management and organizational culture and competitive advantage, and finally knowledge management and leadership. Knowledge management has been used widely as independent and mediating variable, and in some context's dependent variable as well. Majority of the literature agreed that knowledge management has a significant positive impact over performance. While analysing the varied literature significant contradictions have also been observed [13, 26]. The synthesized and systematic review of literature has opened several horizons for research. The contradiction among findings especially in the relationship between knowledge management, innovation, competitive advantage and performance calls for further in depth research in developing countries. Another important aspect observed that knowledge sharing being the dimension of knowledge management has certain ambiguities that needs to be resolved through empirical research. The study synthesised the controversial findings and opened the gate for empirical testing of the controversial literature to resolve the gap.

# 7 Conclusion

The current systematic review aims to conduct a synthesis of current literature in the field of knowledge management and proposed several opportunities for conducting research and covering the gap in the field of knowledge management. Knowledge

**Table 2** Knowledge management and leadership

| Source | Issues | Findings |
|---|---|---|
| Noruzy et al. [60] | Determining the relationship between transformational leadership, organizational learning, knowledge management, organizational innovation, and performance | Transformational leadership has a significant impact over organizational learning, organizational innovation, knowledge management, and organizational performance, whereas, organizational learning influence knowledge management which influence organizational innovation and performance |
| Chin-Fu Ho et al. [29] | Identifying the impact of knowledge management through integrated model considering knowledge enabler, knowledge circulation over performance | Organizational culture has a significant impact over knowledge circulation processes and performance |
| Edú-Valsania et al. [32] | Exploring the relationships between authentic leadership, knowledge sharing, and intervening processes | There is a direct impact of authentic relationship as well as indirect impact through group innovation climate |
| Ramezani et al. [70] | Identifying the impact of leadership on trust, knowledge management and organizational innovation performance | Knowledge-oriented leadership has an indirect impact over organizational innovation performance mediated by knowledge management |
| Naqshbandi and Jasimuddin [56] | Identifying the mediating role of knowledge management capability between knowledge-oriented leadership and open innovation | Knowledge management capability mediates between leadership and open innovation in terms of developing and maintaining competitive advantage |
| Alzghoul et al. [16] | Identifying the mediating role of workplace and moderating role of knowledge sharing between authentic leadership, workplace harmony, creativity and performance | Authentic leadership has a significant impact over workplace climate, creativity and performance. Moreover, workplace climate mediates between authentic leadership and creativity and performance and knowledge sharing behavior moderate's relationship between authentic leadership and workplace climate |

(continued)

**Table 2** (continued)

| Source | Issues | Findings |
|---|---|---|
| Abubakar et al. [5] | The aim of the qualitative review was to develop a framework for organizational performance based on knowledge creation, knowledge management enablers and knowledge creatin process | The study concluded that knowledge creation mediates and decision style moderates the relationship between knowledge management enabler factors and organizational performance |
| Pellegrini et al. [68] | Reviewing literature regarding human and relational aspects, systematic and performance aspects, contextual and contingent aspects and cultural and learning aspects with regard to knowledge management | The need for empirical testing of the well polarized clusters of knowledge management represented by; human and relational aspects, systematic and performance aspects, contextual and contingent aspects and cultural and learning aspects with regard to knowledge management |
| Latif et al. [45, 46] | Empirically testing the impact of knowledge management enablers including entrepreneurial orientation and knowledge-oriented leadership over knowledge management process and success of the projects | Entrepreneurial orientation and knowledge-oriented leadership are strong predictors of knowledge management processes, however, knowledge management processes are unable to predict project success |
| Latif et al. [45, 46] | Identifying the direct and indirect impact of entrepreneurial leadership over project success mediated by knowledge management processes | Knowledge management fully mediates entrepreneurial leadership and project success |

management is vital to study because of its varied dimensions and its role in resource-based view as well as knowledge-based view. It helps in developing a culture of innovation culture and promote organizational performance. Based on the systematic review it is observed that knowledge management is a significant area and has several dimensions which need to be further explored. It is important to mention here that literature analysis is affected and determined based on the cultural, industrial and other contextual settings.

This it would be right to say that still further in depth research is needed in the field of knowledge management. The current synthesis supports the view of knowledge base and is considered it vital in developing and under developing countries. The concept of employee learning and organizational learning promoted through knowledge management is crucial for the countries that lack financial or natural resources. Therefore, it is recommended that businesses in the current competitive and turbulent environments pay significant attention towards developing best practices in knowledge management for the purpose of gaining competitive advantage and improving overall performance.

# References

1. Abdi K, Mardani A, Senin AA, Tupenaite L, Naimaviciene J, Kanapeckiene L, Kutut V (2018) The effect of knowledge management, organizational culture and organizational learning on innovation in automotive industry. J Bus Econ Manag 19(1):1–19. https://doi.org/10.3846/jbem.2018.1477
2. Aboelmaged MG (2014) Linking operations performance to knowledge management capability: the mediating role of innovation performance. Prod Plan Control 25(1):44–58. https://doi.org/10.1080/09537287.2012.655802
3. Aboyassin NA, Alnsour M, Alkloub M (2011) Achieving total quality management using knowledge management practices: a field study at the Jordanian insurance sector. Int J Commer Manag 21(4):394–409. https://doi.org/10.1108/10569211111189383
4. Abualoush S, Masa'deh R, Bataineh K, Alrowwad A (2018) The role of knowledge management process and intellectual capital as intermediary variables between knowledge management infrastructure and organization performance. Interdisc J Inf Knowl Manage 13:279–309. https://doi.org/10.28945/4088
5. Abubakar AM, Elrehail H, Alatailat MA, Elçi A (2019) Knowledge management, decision-making style and organizational performance. J Innov Knowl 4(2):104–114. https://doi.org/10.1016/j.jik.2017.07.003
6. Adeinat IM, Abdulfatah FH (2019) Organizational culture and knowledge management processes: case study in a public university. VINE J Inf Knowl Manage Syst 49(1):35–53. https://doi.org/10.1108/VJIKMS-05-2018-0041
7. Al-Abdullat BM, Dababneh A (2018) The mediating effect of job satisfaction on the relationship between organizational culture and knowledge management in Jordanian banking sector. Benchmark: Int J 25(2):517–544. https://doi.org/10.1108/BIJ-06-2016-0081
8. Albream F, Maraqa M (2019) The impact of adopting e-collaboration tools on knowledge management processes. Manage Sci Lett 9:1009–1028. https://doi.org/10.5267/j.msl.2019.4.004
9. Alegre J, Sengupta K, Lapiedra R (2013) Knowledge management and innovation performance in a high-tech SMEs industry. Int Small Bus J 31(4):454–470. https://doi.org/10.1177/0266242611417472
10. Alkaffaf M, Muflish M, Al-Dalahmeh M (2018) An integrated model of knowledge management enablers and organizational creativity: the mediating role of knowledge management processes in social security corporation in Jordan. J Theor Appl Inf Technol 96(3):677–701
11. Alkhuzaie AS, Asad M (2018) Operating cashflow, corporate governance, and sustainable dividend payout. Int J Entrepreneur 22(4):1–9
12. Alrubaiee LS, Aladwan S, Joma MH, Idris WM, Khater S (2017) Relationship between corporate social responsibility and marketing performance: the mediating effect of customer value and corporate image. Int Bus Res 10(2). https://doi.org/10.5539/ibr.v10n2p104
13. Al-Sa'di AF, Abdallah AB, Dahiyat SE (2017) The mediating role of product and process innovations on the relationship between knowledge management and operational performance in manufacturing companies in Jordan. Bus Process Manage J 23(2):349–376. https://doi.org/10.1108/BPMJ-03-2016-0047
14. Alsheikh GA, Alnawafleh EA, Halim MS, Tambi AM (2018) The impact of human resource management practices, organizational culture, motivation and knowledge management on job performance with leadership style as moderating variable in the Jordanian commercial banks sector. J Rev Glob Econ 6:477–488

15. Altarawneh II, Altarawneh K (2017) Knowledge management practices and intellectual capital: a case from Jordan. Int J Bus 22(4):341–367
16. Alzghoul A, Elrehail H, Emeagwali OL, AlShboul MK (2018) Knowledge management, workplace climate, creativity and performance: the role of authentic leadership. J Work Learn 30(8):592–612. https://doi.org/10.1108/JWL-12-2017-0111
17. Anand A, Muskat B, Creed A, Zutshi A, Csepregi A (2021) Knowledge sharing, knowledge transfer and SMEs: evolution, antecedents, outcomes and directions. Pers Rev. https://doi.org/10.1108/PR-05-2020-0372
18. Arsenijević J, Tot V, Arsenijević D (2010) The comparison of two groups in perception of knowledge management in the environment of higher education. Afr J Bus Manage 4(9):1916–1923. https://doi.org/10.5897/AJBM.9000553
19. Asad M, Ahmad I, Haider SH, Salman R (2018) A critical review of Islamic and conventional banking in digital era: a case of Pakistan. Int J Eng Technol 7(4.7):57–59
20. Asad M, Altaf N, Israr A, Khan GU (2020) Data analytics and SME performance: a bibliometric analysis. In: 2020 International conference on data analytics for business and industry: way towards a sustainable economy (ICDABI). IEEE, Sakhir, pp 1–7. https://doi.org/10.1109/ICDABI51230.2020.9325661
21. Asad M, Haider SH, Fatima M (2018) Corporate social responsibility, business ethics, and labor laws: a qualitative analysis on SMEs in Sialkot. J Legal Ethical Regul Issues 21(3):1–7
22. Ayoub HF, Abdallah AB, Suifan TS (2017) The effect of supply chain integration on technical innovation in Jordan: the mediating role of knowledge management. Benchmark: Int J 24(3). https://doi.org/10.1108/BIJ-06-2016-0088
23. Barua B (2021) Impact of total quality management factors on knowledge creation in the organizations of Bangladesh. TQM J. https://doi.org/10.1108/TQM-06-2020-0145
24. Berraies S, Achour M, Chaher M (2015) Focusing the mediating role of knowledge management practices: how does institutional and interpersonal trust support exploitative and exploratory innovation? J Appl Bus Res 31(4):1479–1492. https://doi.org/10.19030/jabr.v31i4.9331
25. Butt MA, Nawaz F, Hussain S, Sousa MJ, Wang M, Sumbal MS, Shujahat M (2019) Individual knowledge management engagement, knowledge-worker productivity, and innovation performance in knowledge-based organizations: the implications for knowledge processes and knowledge-based systems. Comput Math Organ Theory 25(3):336–356
26. Byukusenge E, Munene JC, Ratajczak-Mrozek M (2017) Knowledge management and business performance: does innovation matter? Cogent Bus Manage 4(1). https://doi.org/10.1080/23311975.2017.1368434
27. Cabrilo S, Dahms S (2018) How strategic knowledge management drives intellectual capital to superior innovation and market performance. J Knowl Manag 22(3):621–648. https://doi.org/10.1108/JKM-07-2017-0309
28. Chikati R, Mpofu N (2013) Developing sustainable competitive advantage through knowledge management. Int J Sci Technol Res 2(10):77–81
29. Chin-Fu H, Pei-Hsuan H, Wei-Hsi H (2014) Enablers Processes Effective Knowl Manage 114(5):734–754. https://doi.org/10.1108/IMDS-08-2013-0343
30. Donate MJ, Pablo JD (2015) The role of knowledge-oriented leadership in knowledge management practices and innovation. J Bus Res 38(2):360–370. https://doi.org/10.1016/j.jbusres.2014.06.022
31. Durmuş-Özdemir E, Abdukhoshimov K (2018) Exploring the mediating role of innovation in the effect of the knowledge management process on performance. Technol Anal Stategic Manage 30(5):596–608. https://doi.org/10.1080/09537325.2017.1348495
32. Edú-Valsania S, Moriano JA, Molero F (2016) Authentic leadership and employee knowledge sharing behavior: mediation of the innovation climate and workgroup identification. Leadersh Org Dev J 37(4):487–506. https://doi.org/10.1108/LODJ-08-2014-0149
33. Esterhuizen D, Schutte C, Toit AD (2012) Knowledge creation processes as critical enablers for innovation. Int J Inf Manage 32(4):354–364. https://doi.org/10.1016/j.ijinfomgt.2011.11.013
34. Ferraris A, Mazzoleni A, Devalle A, Couturier J (2019) Big data analytics capabilities and knowledge management: Impact on firm performance. Manag Decis 57(8):1923–1936. https://doi.org/10.1108/MD-07-2018-0825

35. Giudice MD, Peruta MR (2016) The impact of IT-based knowledge management systems on internal venturing and innovation: a structural equation modeling approach to corporate performance. J Knowl Manag 20(3):484–498. https://doi.org/10.1108/JKM-07-2015-0257

36. Heisig P, Suraj OA, Kianto A, Kemboi C, Arrau GP, Easa NF (2016) Knowledge management and business performance: global experts' views on future research needs. J Knowl Manag 20(6):1169–1198. https://doi.org/10.1108/JKM-12-2015-0521

37. Imran MK, Ilyas M, Fatima T (2017) Achieving organizational performance through knowledge management capabilities: mediating role of organizational learning. Pakistan J Commerce Social Sci (PJCSS) 11(1):106–125. http://hdl.handle.net/10419/188284

38. Inkinen HT, Kianto A, Vanhala M (2015) Knowledge management practices and innovation performance in Finland. Balt J Manag 10(4):432–455. https://doi.org/10.1108/BJM-10-2014-0178

39. Iqbal A, Latif F, Marimon F, Sahibzada UF, Hussain S (2019) From knowledge management to organizational performance: modelling the mediating role of innovation and intellectual capital in higher education. J Enterp Inf Manag 32(1):36–59. https://doi.org/10.1108/JEIM-04-2018-0083

40. Iqbal S, Rasheed M, Khan H, Siddiqi A (2020) Human resource practices and organizational innovation capability: role of knowledge management. VINE J Inf Knowl Manage Syst. https://doi.org/10.1108/VJIKMS-02-2020-0033

41. Jamil RA, Lodhi MS (2015) Role of knowledge management practices for escalating universities' performance in Pakistan. Manage Sci Lett 5(10):945–960. https://doi.org/10.5267/j.msl.2015.8.002

42. Jyoti J, Rani A (2017) High performance work system and organisational performance: role of knowledge management. Pers Rev 46(8):1770–1795. https://doi.org/10.1108/PR-10-2015-0262

43. Khalil R, Asad M, Khan SN (2018. Management motives behind the revaluation of fixed assets for sustainability of entrepreneurial companies. Int J Entrepreneur 22(Special):1–9

44. Kucharska W, Bedford DA (2019) Knowledge sharing and organizational culture dimensions: does job satisfaction matter? Electron J Knowl Manag 17(1):1–18

45. Latif KF, Afzal O, Saqib A, Sahibzada UF, Alam W (2020) Direct and configurational paths of knowledge-oriented leadership, entrepreneurial orientation, and knowledge management processes to project success. J Intellect Cap 22(1):149–170. https://doi.org/10.1108/JIC-09-2019-0228

46. Latif KF, Nazeer A, Shahzad F, Ullah M, Imranullah M, Sahibzada UF (2020) Impact of entrepreneurial leadership on project success: mediating role of knowledge management processes. Leadersh Org Dev J 14(2):237–256. https://doi.org/10.1108/LODJ-07-2019-0323

47. Lee V-H, Foo AT-L, Leong L-Y, Ooi K-B (2016) Can competitive advantage be achieved through knowledge management? A case study on SMEs. Expert Syst Appl 65(15):136–151. https://doi.org/10.1016/j.eswa.2016.08.042

48. Madhoushi M, Sadati A, Delavari H, Mehdivand M, Mihandost R (2011) Entrepreneurial orientation and innovation performance: the mediating role of knowledge management. Asian J Bus Manage 3(4):310–316

49. Mahawrah F, Shehabat I (2016) The impact of knowledge management on customer relationship management: a case from the fast food industry in Jordan. Int J Electron Customer Relationship Manage 10:138–157

50. Mardani A, Nikoosokhan S, Moradi M, Doustar M (2018) The relationship between knowledge management and innovation performance. J High Technol Manage Res 29(1):12–26. https://doi.org/10.1016/j.hitech.2018.04.002

51. Marjanovic O, Freeze R (2012) Knowledge-intensive business process: deriving a sustainable competitive advantage through business process management and knowledge management integration. Knowl Process Manag 18(4):180–188

52. Migdadi MM, Zaid MK (2016) An empirical investigation of knowledge management competence for enterprise resource planning systems success: Insights from Jordan. Int J Prod Res 54(18):5480–5498. https://doi.org/10.1080/00207543.2016.1161254

53. Migdadi MM, Zaid MK, Yousif M, Almestarihi R (2018) An empirical examination of collaborative knowledge management practices and organisational performance: the mediating roles of supply chain integration and knowledge quality. Int J Bus Excell 14(2):180–211
54. Migdadi MM, Zaid MK, Yousif M, Almestarihi R, Al-Hyari K (2017) An empirical examination of knowledge management processes and market orientation, innovation capability, and organisational performance: insights from Jordan. J Inf Knowl Manage 16(1):1–32. https://doi.org/10.1142/S0219649217500022
55. Mills AM, Smith TA (2011) Knowledge management and organizational performance: a decomposed view. J Knowl Manag 15(1):156–171. https://doi.org/10.1108/13673271111108756
56. Naqshbandi MM, Jasimuddin SM (2018) Knowledge-oriented leadership and open innovation: Role of knowledge management capability in France-based multinationals. Int Bus Rev 27(3):701–713. https://doi.org/10.1016/j.ibusrev.2017.12.001
57. Nawab S, Nazir T, Zahid MM, Fawad SM (2015) Knowledge management, innovation and organizational performance. Int J Knowl Eng 1(1):43–48. https://doi.org/10.7763/IJKE.2015.V1.7
58. Nawaz MS, Shaukat S (2014) Impact of knowledge management practices on firm performance: testing the mediation role of innovation in the manufacturing sector of Pakistan. Pakistan J Commerce Soc Sci 8:99–111. http://hdl.handle.net/10419/188128
59. Neeru M, Karishma G, Renu V (2011) Achieving competitive advantage through knowledge management and innovation: empirical evidences from the Indian IT sector. IUP J Knowl Manage 9(2):7–25
60. Noruzy A, Dalfard VM, Azhdari B, Nazari-Shirkouhi S, Rezazadeh A (2013) Relations between transformational leadership, organizational learning, knowledge management, organizational innovation, and organizational performance: an empirical investigation of manufacturing firms. Int J Adv Manuf Technol 64:1073–1085. https://doi.org/10.1007/s00170-012-4038-y
61. Nowacki R, Bachnik K (2016) Innovations within knowledge management. J Bus Res 69(5):1577–1581. https://doi.org/10.1016/j.jbusres.2015.10.020
62. Obeidat BY, Al-Suradi MM, Masa'deh R, Tarhini A (2016) The impact of knowledge management on innovation: an empirical study on Jordanian consultancy firms. Manage Res Rev 39(10):1214–1238. https://doi.org/10.1108/MRR-09-2015-0214
63. Obeidat BY, Tarhini A, Masa'deh R, Aqqad NO (2017) The impact of intellectual capital on innovation via the mediating role of knowledge management: a structural equation modelling approach. Int J Knowl Manage Stud 8(3):273–298. https://doi.org/10.1504/IJKMS.2017.087071
64. Odea E, Ayavoo R (2020) The mediating role of knowledge application in the relationship between knowledge management practices and firm innovation. J Innov Knowl 5(3):210–218. https://doi.org/10.1016/j.jik.2019.08.002
65. Okour MK, Chong CW, Asmawi A (2019) Antecedents and consequences of knowledge management systems usage in Jordanian banking sector. Knowl Process Manage 26:10–26
66. Overall, Jeffrey (2015) A concept framework of innovation and performance: the importance of leadership, relationship quality, and knowledge management. Acad Enterpreneur J 21(2):41–54
67. Paliszkiewicz J, Svanadze S, Jikia M (2017) The role of knowledge management processes on organizational culture. Online J Appl Knowl Manage (OJAKM) 5(2):29–44. https://doi.org/10.36965/OJAKM.2017.5(2)29-44
68. Pellegrini MM, Ciampi F, Marzi G, Orlando B (2020) The relationship between knowledge management and leadership: mapping the field and providing future research avenues. J Knowl Manag 24(6):1445–1492. https://doi.org/10.1108/JKM-01-2020-0034
69. Qasrawi BT, Almahamid SM, Qasrawi ST (2017) The impact of TQM practices and KM processes on organisational performance: An empirical investigation. Int J Qual Reliab Manage 34(7):1034–1055. https://doi.org/10.1108/IJQRM-11-2015-0160
70. Ramezani Y, Safari Z, Hashemiamin A, Karimi Z (2017) The impact of knowledge-oriented leadership on innovative performance through considering the mediating role of knowledge management practices. Ind Eng Manage Syst 16(4):495–506. https://doi.org/10.7232/iems.2017.16.4.495

71. Rašula J, Vukšić VB, Štemberger MI (2012) The impact of knowledge management on orgaizational performance. Econ Bus Rev Central South-Eastern Europe 14(2):147–168
72. Riaz H, Hassan A (2019) Mediating role of organizational creativity between employees' intention in knowledge management process and organizational performance: an empirical study on pharmaceutical employees. Pakistan J Commerce Soc Sci 13(3):635–655. http://hdl.handle.net/10419/205271
73. Sabri MO, Odeh M (2019) The impact of knowledge management on the development of innovative business process architecture modelling: the case of banking in Jordan. Int J Technol Manag Sustain Dev 18(21):197–220
74. Sahibzada UF, Jianfeng C, Latif KF, Shah SA, Sahibzada HF (2020). Refuelling knowledge management processes towards organisational performance: mediating role of creative organisational learning. Knowl Manage Res Practice 1–13. https://doi.org/10.1080/14778238.2020.1787802
75. Salloum SA, Shaalan K (2021) The role of knowledge management processes for enhancing and supporting innovative organizations: a systematic review. Recent Adv Intell Syst Smart Appl 143–161
76. Santoro G, Mazzoleni A, Quaglia R, Solima L (2021) Does age matter? The impact of SMEs age on the relationship between knowledge sourcing strategy and internationalization. J Bus Res 128:779–787. https://doi.org/10.1016/j.jbusres.2019.05.021
77. Santoro G, Vrontis D, Thrassou A, Dezi L (2018) The Internet of Things: building a knowledge management system for open innovation and knowledge management capacity. Technol Forecast Soc Chang 136:347–354. https://doi.org/10.1016/j.techfore.2017.02.034
78. Schiuma G (2012) Managing knowledge for business performance improvement. J Knowl Manag 16(4):515–522. https://doi.org/10.1108/13673271211246103
79. Shabbir MQ, Gardezi SB (2020) Application of big data analytics and organizational performance: The mediating role of knowledge management practices. J Big Data 7(1):1–17. https://doi.org/10.1186/s40537-020-00317-6
80. Shao Z, Feng Y, Liu L (2012) The mediating effect of organizational culture and knowledge sharing on transformational leadership and Enterprise Resource Planning systems success: an empirical study in China. Comput Hum Behav 28(6):2400–2413. https://doi.org/10.1016/j.chb.2012.07.011
81. Sheikh UA, Asad M, Mukhtar U (2020) Modelling asymmetric effect of foreign direct investment inflows, carbon emission and economic growth on energy consumption of South Asian region: a symmetrical and asymmetrical panel autoregressive distributive lag model approach. Account Bus Public Interest 19:193–221
82. Sun P (2010) Five critical knowledge management organizational themes. J Knowl Manag 14(4):507–523. https://doi.org/10.1108/13673271011059491
83. Tong C, Tak WI, Wong A (2015) The impact of knowledge sharing on the relationship between organizational culture and job satisfaction: The perception of information communication and technology (ICT) practitioners in Hong Kong. Int J Human Resour Stud. https://doi.org/10.5296/ijhrs.v5i1.6895
84. Valdez-Juárez LE, Lema DG-P, Maldonado-Guzmán G (2016) Management of knowledge, Innovation and performance in SMEs. Interdisc J Inf Knowl Manage 11(4):141–176
85. Wijaya PY, Suasih NN (2020) The effect of knowledge management on competitive advantage and business performance: a study of silver craft SMEs. Entrepreneurail Bus Econ Rev 8(4):105–121. https://doi.org/10.15678/EBER.2020.080406
86. Wu I-L, Chen J-L (2014) Knowledge management driven firm performance: the roles of business process capabilities and organizational learning. J Knowl Manage 18(6):1141–1164. https://doi.org/10.1108/JKM-05-2014-0192
87. Zaied AN, Hussein GS, Hassan MM (2012) The role of knowledge management in enhancing organizational performance. Int J Inf Eng Electron Bus 5:27–35. https://doi.org/10.5815/ijieeb.2012.05.04
88. Zhuang Y-T, Wu F, Chen C, Pan Y-H (2017) Challenges and opportunities: from big data to knowledge in AI 2.0. Front Inf Technol Electron Eng 18:3–14

# Deep Neural Network to Forecast Stock Market Price

Qeethara Al-Shayea

**Abstract** The forecasting of the stock exchange asking price has been affected by a number of monetary and nonmanetary indexes that might be used as a warning rule for investors. Expecting the future trend of the stock market is a critical issue in investment sector. In this work, the forecasting of futurity open and close asking price of Dow Jones Industrial Average (DJIV) has been performed utilization deep neural network. The Long short term memory (LSTM) network was used to predict values of futurity time steps of a sequence of opening and closing into Dow Jones Industrial Average stock market. The LSTM network learns to forecast the value of the next step. By train the LSTM network, we have expect the value of future time steps of open and close of the stock market. The performance of the proposed technique is promising for DJIV stock market expectation.

**Keywords** Deep neural network · Forecasting · Stock market price · Long short term memory network · Artificial intelligence

## 1 Introduction

The stock exchange market has been an appealing domain to forecast for a considerable number of investors [1]. The expectation of the stock price can be provided investors with the right decision-making, references for higher return and cut back the risk.

Stock market price forecasting is a challenging research field where various methods have been developed to expect stock exchange price movement [2]. Artificial neural networks have been used for years as a means to expect stock exchange prices. While it is difficult to expect with classical statistical and econometric procedures because of nonlinear connections and the ability of neural networks to model nonlinear relations without a priori presumptions [3]. The neural networks used in time series forecasting have enhanced because of the rising of deep learning. Deep

Q. Al-Shayea (✉)
Faculty of Business, Al-Zaytoonah University of Jordan, Amman, Jordan
e-mail: drqeethara@zuj.edu.jo

© The Author(s), under exclusive license to Springer Nature Switzerland AG 2022
S. G. Yaseen (ed.), *Digital Economy, Business Analytics, and Big Data Analytics Applications*, Studies in Computational Intelligence 1010,
https://doi.org/10.1007/978-3-031-05258-3_12

neural network has been used to train a complex nonlinear relationship. Deep learning schemes have promising achievements in many research areas such as stock price expectation [4].

The long short term memory (LSTM) neural network is predominant among deep learning models in time series field [5, 6]. Hochreiter in [7] mentioned that LSTM solves complex tasks.

In this work, we propose an LSTM model to forecast the open and close of the stock exchange market depending on time series data for Dow Jones Industrial Average (DJIV) from January 2018 up to June 2021. We show that the application of the LSTM model gives satisfactory results and upgrade the capability of it to expect the stock exchange price.

The remainder of the work has been organized as follows. Section 2 presents the literature review. In Sect. 3, we describe the data selection and methodology used in this paper plus the performance evaluation. Lastly, Sect. 4 concludes the paper.

## 2 Literature Review

In past years, many researchers proposed and enhanced algorithms such as artificial neural networks, support vector machine and deep learning for predicting financial series based on stock market historical data.

Nikou et al. [8] proposed LSTM scheme. The results illustrated that the LSTM scheme accomplished better compared with the artificial neural network scheme, support vector regression scheme and random forest scheme in the expectation of the closing prices of Britain stock exchange. Fazeli and Houghten [9] presented LSTM network to expect stock prices. They tested the impact of the Relative Strength Indicator, and dispersal on the loss of the scheme. Lakshminarayanan [10] presented a comparative study of the performance of LSTM network models with Support Vector Machine (SVM) regression models. Nguyen et al. [11] introduced dynamic LSTM network to expect future stock exchange prices. Li et al. [12] proposed LSTM model with multi input, which accomplished better effectiveness in extracting potential information and filtering noise. Kim and Won [13] presented a combining scheme that hybrid an LSTM scheme with three of generalized autoregressive conditional heteroscedasticity schemes. Baek and Kim [14] presented two modules. The first one is an overfitting prevention LSTM module while the other one is an expectation LSTM module. Qian and Chen [15] presented an LSTM scheme and Auto Regressive Moving Average scheme. Both models give acceptable results. Chatzis et al. [16] proposed multiple machine learning algorithms. They concluded that apply Deep Neural Networks significantly increases the efficiency of stock market expectation. Lai et al. [17] presented an LSTM model that used stock market indicators to expect the next five days. Xu and Keselj [18] presented an attention-based LSTM to predict the stock market. They found that attention-based LSTM give better results than conventional LSTM. Sohangir et al. [19] implemented multiple neural network models and deep learning to enhance the performance of sentiment analysis for

StockTwits. The results illustrate that deep learning model gives better effectiveness. Zhang et al. [20] presented neural network scheme depend on a deep factorization, and attention technique. They tried to enhance the expectation precision of exchange price movement by improving feature learning. Fischer and Krauss [21] applied a random forest, a deep neural network, and a logistic regression classifier as well as LSTM network. They concluded that the LSTM network outperforms memory-free classification techniques. Weng et al. [22] presented a prediction expert system for one-day ahead 587 stock price. Eapen et al. [23] implemented a deep neural network scheme that hybrids various pipelines of convolutional neural network and bi-directional LSTM. Liu et al. [24] provided a convolutional network in addition to a LSTM network for analyzing multiple quantitative strategy in stock market exchange. The convolutional neural network serving on the stock choice strategies and features extraction according to quantitative data followed by LSTM to keep the time-series features for gains improvement. Farahani and Hajiagha [25] proposed a hybrid metaheuristic-based artificial neural network. Then, they used genetic algorithms as a heuristic algorithm for feature selection and picking the best and most concerning indicators. At the same time, they presented ARMA and ARIMA for stock price prediction. Budiharto [26] proposed a Long Short Term Memory scheme.

# 3  Methodology

## 3.1  Data Selection and Normalization

The dataset is a daily exchange for Dow Jones Industrial Average (DJIA).It has been collected from the Yahoo Finance website. Thirty outstanding companies are listed on DJIA stock exchanges in the United States. The DJIA is one of the elder and the most common followed equity indices. There are two parameters used in this paper. The first one is the asking price of the exchange at the starting of the day (open). The asking price of the exchange at the end of the day (close) is the second one. The dataset covers the time from January 2018 up to June 2021. The dataset has 859 records. Figure 1 shows the single time series with time steps versus to the days and values versus the amount of daily openings. Figure 2 shows the single time series with time steps versus to the days and values versus the amount of daily close. In this work, long short term memory (LSTM) neural network has been created using deep learning for a time series forecast from MATLAB R2019a.

## 3.2  Results and Performance Evaluation

The first step standardizes the training data for a better convenient and diverging prevention. The same step is implemented on test data. The second step is training

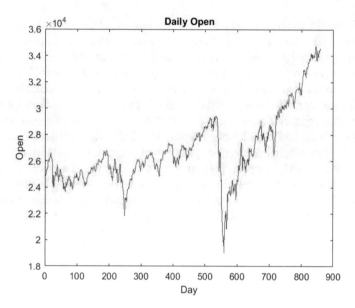

**Fig. 1** Daily open of the DJIA stock market

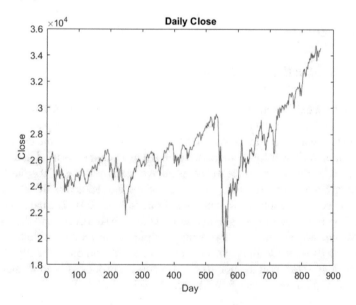

**Fig. 2** Daily close of the DJIA stock market

the LSTM network. Specific options has been used to train the proposed network. Figures 3 and 4 illustrate the training progress. The LSTM network has been applied to forecast the amount of the next time-step. Two hundred hidden units have been utilized while the initial learning rate is 0.005.

Figures 5 and 6 show the predicted values of numerous time-steps in futurity for open and close sequentially. The root mean square-error (RMSE) for open from the unstandardized predictions is 4.7710e+03. The RMSE for close from the unstandardized predictions is 8.3959e+03. The RMSE has been calculated from the standardized data as illustrated in Figs. 7 and 8. Figures 9 and 10 show the forecasted values in

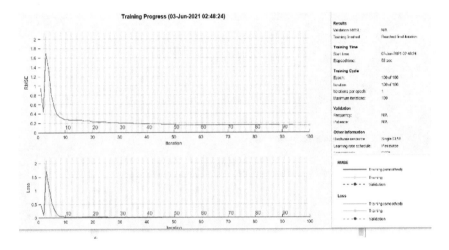

**Fig. 3** The network training progress round 1

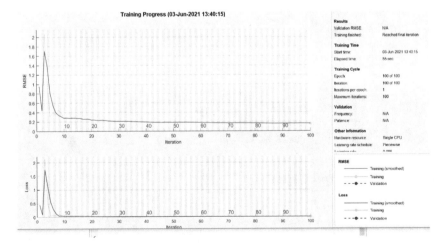

**Fig. 4** The network training progress round 2

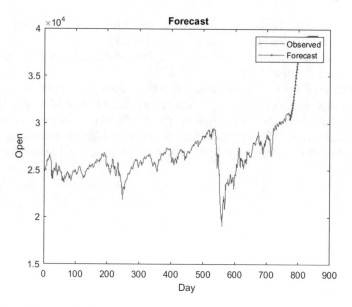

**Fig. 5** The predicted values for open

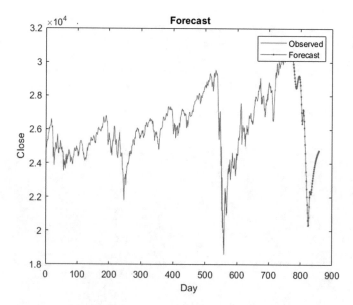

**Fig. 6** The predicted values for close

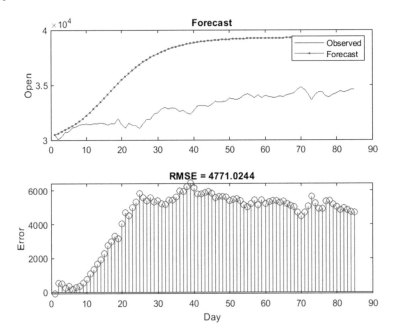

**Fig. 7** The predicted values against the train data

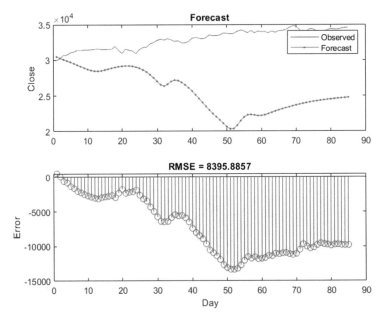

**Fig. 8** The predicted values against the test data

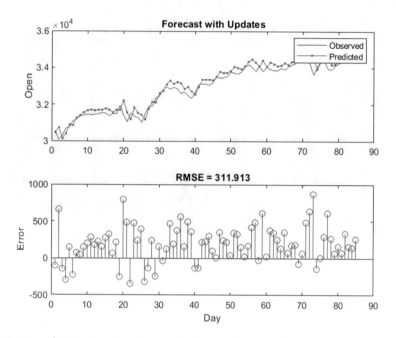

**Fig. 9** The predicted values for updates against the test data

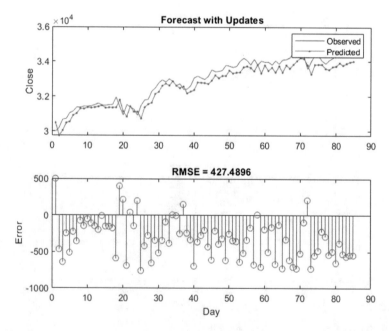

**Fig. 10** The predicted values for updates versus the test data

the test data for open and close respectively. The LSTM network has been up-to-date with the noticed values rather than the expected values. It is obvious, that we have higher accurate prediction just after updating the network.

# 4   Conclusion

A Long Short Term Memory deep neural network is proposed to use the historical information in the forecasting system. The proposed deep neural network is trained and assessed rely on the historic data of the Dow Jones Industrial Average stock market. This study leverages the power of Long Short Term Memory deep neural network and its capability to forecast the stock price. Experimental results indicate that the proposed forecasting market asking price for open and close has the best performance.

# References

1. Rosas-Romero R, Diaz-Torres A, Etcheverry G (2016) Forecasting of stock return prices with sparse representation of financial time series over redundant dictionaries. Expert Syst Appl 57:37–48
2. Thakkar A, Chaudhari K (2021) A comprehensive survey on deep neural networks for stock market: the need, challenges, and future directions. Expert Syst Appl 177
3. Al-Shayea Q (2017) Neural networks to predict stock market price. In: Proceedings of the world congress on engineering and computer science 2017 WCECS, San Francisco, USA
4. Hu Z, Zhao Y, Khushi M (2021) A survey of forex and stock price prediction using deep learning. Appl System Innov 4
5. Sezer OB, Gudelek MU, Ozbayoglu AM (2020) Financial time series forecasting with deep learning: a systematic literature review: 2005–2019. Appl Soft Comput 90:106–181
6. Torres JF, Hadjout D, Sebaa A, Martinez-Alvarez F, Troncoso A (2020) Deep learning for time series forecasting: a survey, big data
7. Hochreiter S, Schmidhuber J (1997) Long short-term memory. Neural Comput 9(8):173–1780
8. Nikou M, Mansourfar G, Bagherzadeh J (2019) Stock price prediction using DEEP learning algorithm and its comparison with machine learning algorithms. Intell Syst Acc Finance Manage 26:164–174
9. Fazeli A, Houghten S (2019) Deep learning for the prediction of stock market trends. In: Proceedings of the 2019 IEEE international conference on big data (Big Data), Los Angeles, CA, USA, 9–12 Dec 2019
10. Lakshminarayanan SK, McCrae J (2019) A comparative study of SVM and LSTM deep learning algorithms for stock market prediction. In: Proceedings of the 27th AIAI Irish conference on artificial intelligence and cognitive science (AICS 2019), Galway, Ireland
11. Nguyen D, Tran L, Nguyen V (2019) Predicting stock prices using dynamic LSTM models. In: Proceedings of the international conference on applied informatics, Madrid, Spain, 7–9 Nov 2019. Springer, Berlin, Heidelberg, Germany
12. Li H, Shen Y, Zhu Y (2018) Stock price prediction using attention-based multi-input LSTM. In: Proceedings of the Asian conference on machine learning, Beijing, China
13. Kim HY, Won CH (2018) Forecasting the volatility of stock price index: a hybrid model integrating LSTM with multiple GARCH-type models. Expert Syst Appl 103:25–37

14. Baek Y, Kim HY (2018) ModAugNet: a new forecasting framework for stock market index value with an overfitting prevention LSTM module and a prediction LSTM module. Expert Syst Appl 113:457–480

15. Qian F, Chen X (2019) Stock prediction based on LSTM under different stability. In: Proceedings of the 2019 IEEE 4th international conference on cloud computing and big data analysis (ICCCBDA), Singapore, 17–18 Apr 2019

16. Chatzis SP, Siakoulis V, Petropoulos A, Stavroulakis E, Vlachogiannakis N (2018) Forecasting stock market crisis events using deep and statistical machine learning techniques. Expert Syst Appl 112:353–371

17. Lai CY, Chen RC, Caraka RE (2019) Prediction stock price based on different index factors using LSTM. In: Proceedings of the 2019 international conference on machine learning and cybernetics (ICMLC), Kobe, Japan, 7–10 July 2019

18. Xu Y, Keselj V (2019) Stock prediction using deep learning and sentiment analysis. In: Proceedings of the 2019 IEEE international conference on big data (big data), Los Angeles, CA, USA, 9–12 Dec 2019

19. Sohangir S, Wang D, Pomeranets A, Khoshgoftaar TM (2018) Big data: deep learning for financial sentiment analysis. J Big Data 5(3)

20. Zhang X, Liu S, Zheng X (2021) Stock price movement prediction based on a deep factorization machine and the attention mechanism. Mathematics 9(15)

21. Fischer T, Krauss C (2018) Deep learning with long short-term memory networks for financial market predictions. Eur J Oper Res 270:654–669

22. Weng B, Lu L, Wang X, Megahed F, Martinez W (2018) Predicting short-term stock prices using ensemble methods and online data sources. Expert Syst Appl 112:258–273

23. Eapen J, Bein D, Verma A (2019) Novel deep learning model with CNN and bi-directional LSTM for improved stock market index prediction. In: 2019 IEEE 9th annual computing and communication workshop and conference (CCWC), pp 0264—0270

24. Liu S, Zhang C, Ma J (2017) CNN-LSTM neural network model for quantitative strategy analysis in stock markets, vol 1. Springer, pp 198—206

25. Farahani MS, Hajiagha SHR (2021) Forecasting stock price using integrated artificial neural network and metaheuristic algorithms compared to time series models. Soft Comput 25:8483–8513

26. Budiharto W (2021) Data science approach to stock prices forecasting in Indonesia during Covid-19 using Long Short-Term Memory (LSTM). J Big Data 8(47)

# A VBA-Module for Converting Integrated Variables into Negative and Positive Cumulative Partial Sums

Abdulnasser Hatemi-J and Alan Mustafa

**Abstract** The objective of this paper is to present our software component entitled TDICPS. We demonstrated how this software component can be utilized by practitioners that are interested in transforming time-series variables into positive and negative parts to account for asymmetric impacts in empirical analyses. TDICPS is created by authors in VBA (Visual Basics for Applications) and it is a module for MS-Excel, which can convert a time-series variable into partial cumulative sums of positive and negative components. It also provides the graphs of these components for a large sample size consisting of more than one million values potentially. The software has also numerous options. In addition to the stochastic trend, the variable can have deterministic fragments also, such as both the drift and the trend or only the drift. Furthermore, an application is provided for decomposing the consumer price index and the interest rates of the US economy into partial segments of positive and negative variations. Any other potential variables can also be decomposed in a similar manner. The decomposed data can then be utilized for conducting asymmetric causality tests or producing the asymmetric impulse response functions as introduced by Hatemi-J (Empir Econ 43:447–456 [5]; Econ Model 36:18–22 [6]). This code is accessible at (Appl Comput Inform. forthcoming. [20]; https://github.com/alanmustafa/Transforming_Data_Into_Cumulative_Partial_Sums [25]).

**Keywords** The VBA · Software-component · Asymmetry · Positive changes · Negative changes · CPI · The US interest rates · Dynamic models

**JEL Classifications** G15 · G11 · C32

A. Hatemi-J
College of Business and Economics, UAE University, Al Ain, United Arab Emirates
e-mail: AHatemi@uaeu.ac.ae

A. Mustafa (✉)
IEEE, Duhok, Kurdistan Region, Iraq
e-mail: Alan.Mustafa@ieee.org

# 1  Introduction

Econometrics as a scientific discipline was introduced by the prominent Norwegian economist Ragnar Frisch in the sense that it is utilized nowadays. The underlying scientific discipline had a major boost when the Econometric Society was established by launching the publication of Econometrica [1]. Since then econometrics, which verbally means measuring economics, has exponentially grown both in terms of theoretical development as well as applications. Nowadays, econometrics is a crucial tool for analyzing the potential relationship between any set of random variables not necessary only in the social sciences but also anywhere else that quantitative data is available. This data could be cross-sectional data, time-series data or panel data. It is very common nowadays to find published articles that are making use of econometrics in disciplines such as medicine, physics, chemistry, agriculture, psychology, mathematics as well as engineering in addition to its enormous applications in social sciences. A lot of theoretical questions can find numerical answers via an econometric model. The fact that we have more access to computing power for both storage of data and the estimation of complicated models has also contributed to this continuous popularity of econometrics in different scientific disciplines.

It is well known in the literature that it is important to allow for asymmetric impacts in empirical analyses because it is in harmony with the way reality functions. Several logical reasons justify this conclusion. For instance, it is commonly agreed in the finance literature that individuals respond stronger to the negative news in contrast to the positive news. Consequently, the effect of a negative condition can be different from the effect of a positive condition in absolute terms. An additional motive for potential asymmetric impacts is the point that information is imperfect in financial markets as is established by influential works of Akerlof [2], Spencer [3] and Stiglitz [4]. Furthermore, asymmetric impacts are a consequence of the natural constraint on any asset price. This implies that the price of any asset or any commodity can rise potentially beyond any limit. While there exists a natural limitation on the price decreases because the price of any normal asset or commodity cannot be a negative value in the real markets with rational actors. Hence, a given amount of price growth has a different implication than a price reduction by the same magnitude in the absolute terms for the same asset. There are also additional facts observed in markets that verify the presence of asymmetric impacts. For instance, the oil price changes can be mentioned. We usually observe that fuel prices increase directly when there is an increase in oil prices. Contrariwise, the value of fuel does not adjust directly when there is an oil price decrease. The adjustment of the fuel price to the negative oil price changes is usually partial and it takes time while the adjustment is normally incomplete, which is an asymmetric behaviour. Moral or legal restrictions are also extra reasons for asymmetric structures. The econometric practice has been attempting to advance diverse methods for taking into account the potential asymmetric impacts such as thresholds, dummy variables or Markov regime-switching approaches. Several well-known econometric packages (such as EViews, RATS, STATA, Gauss, Microfit, PcGive, etc.) make these mentioned methods accessible to the practitioners. Nonetheless, recent developments exist in the literature that is based on decomposing the underlying variables into partial cumulative sums of positive as well as negative. The converted data can then be used for conducting asymmetric causation tests and estimating impulse response functions that are asymmetric or variance decompositions. Given that these approaches are relatively recent they are generally not easily accessible to regular practitioners due to the lack of suitable software. Therefore, we have produced a consumer-friendly MS-Excel module that is implemented in Visual Basics for Applications (VBA) for transforming a variable with a unit root into cumulative partial sums for negative or positive variations. The software can also provide graphs for a large sample size consisting of more than one million observations. This chapter aims at showing how this software component can be utilized step by step by using an application from real data.

The remaining part of this chapter is organized as the following. Section 2 defines the equations that are necessary for decomposing an integrated time-series variable into cumulative sums. It also presents the vector autoregressive (VAR) model and information criterion for determining the optimal lag order. Section 3 describes how our module, the TDICPS can be applied for this purpose by a clear illustration of two real-time series variables using a long period of sample data. The last section provides conclusions.

## 2 The Mathematical Equations

Consider a variable that is integrated of the first degree, denoted by $w_t$, with constant and trend that is generated by the following process:

$$w_t = a + bt + w_{t-1} + \varepsilon_t \tag{1}$$

Here $t = 1, 2, \ldots, T$. The denotation $\varepsilon_t$ is an error term that is assumed to be white noise. The parameters $a$ and $b$ are constants that need to be calculated. Via the substitution method, the solution to Eq. (1) is Eq. (2) as presented below

$$w_t = at + \frac{t \times (t+1)}{2}b + w_0 + \sum_{i=1}^{t} \varepsilon_i \tag{2}$$

Observe that $w_0$ signifies the startup value. Both positive and negative variations for this variable are acquired as the following in Eqs. (3) and (4):

$$w_t^+ = \frac{at + \left[\frac{t(t+1)}{2}\right]b + w_0}{2} + \sum_{i=1}^{t} \varepsilon_i^+. \tag{3}$$

and

$$w_t^- = \frac{at + \left[\frac{t(t+1)}{2}\right]b + w_0}{2} + \sum_{i=1}^{t} \varepsilon_i^-$$  (4)

It should be mentioned that the positive and negative shocks are identified as follows:

$\varepsilon_i^+ = \max(\varepsilon_i, 0)$ and $\varepsilon_i^- = \min(\varepsilon_i, 0)$. For additional information and proof on these results, see [5–7].

It should be pointed out that our module estimates the parameters $a$ and $b$ via the ordinary least squares (OLS) method. Namely, the following solutions are used in order to estimate the underlying parameters in Eqs. (5) and (6):

$$a = \overline{\Delta w} - b\overline{t}$$  (5)

and

$$b = \frac{\sum_{t=1}^{T}\left(\Delta w_t - \overline{\Delta w}\right)\left(t - \overline{t}\right)}{\sum_{t=1}^{T}\left(t - \overline{t}\right)^2}$$  (6)

where $\overline{\Delta w} = \frac{\sum_{t=1}^{T} \Delta w_t}{T}$, $\overline{t} = \frac{\sum_{t=1}^{T} t}{T}$, and $\Delta$ is the first difference operator.

The dynamic interaction between these variables can be captured by making use of the vector autoregressive model (VAR). This model is introduced by Sims [8] and it is regularly used by practitioners when the variables of interest are quantities that are observed across time. The reason for the popularity of the VAR model is that it is a multivariate dynamic model that is identified unequally with very good forecasting properties. It is also operational to test for causality by estimating a VAR model. The vector error correction model, which is a representation of the VAR model, can be used in order to test for the long-run relationship (i.e. cointegration as defined by Granger [9])[1] along with short-run adjustment behaviour. This is an important issue to avoid spurious relationships between time series variables that have unit-roots. In addition, it is possible to transform the VAR into its vector moving average representation to estimate the impulse response functions on variance decompositions for capturing the impact of a shock in the system. See [8, 17] for more details.

A crucial issue within this context is to determine the optimal lag order in the VAR model. This is the case because all the conducted inference in the estimated VAR model is based on the chosen lag order. For example, assume that we are interested in the causal impacts between the positive components of $m$ variables. Then, our vector of interest is defined as $y_t^+ = \left(w_{1t}^+, \ldots, w_{mt}^+\right)$. Thus, the test for causality can be implemented by using the vector autoregression model with lag length $p$, VAR ($p$) expressed below

$$y_t^+ = C_0 + C_1 y_{t-1}^+ + \cdots + C_p y_{t-p}^+ + u_t^+$$  (7)

---

[1] For cointegration tests see [10–15], among others.

here the denotation $y_t^+$ represents a m × 1 vector of the time series variables, $C_0$ signifies the $n$ × 1 vector of intercepts, and $u_t^+$ is a $m$ × 1 vector of error terms (corresponding to each of the variables representing the cumulative sum of positive shocks). The matrix $A_j$ is a $m$ × $m$ matrix of parameters for lag $j$ ($j = 1, ..., p$). The information criterion presented below can be utilized for selecting the optimal lag length ($p$):

$$ HJC = \ln\left(\left|\widehat{\Omega}_j\right|\right) + j\left(\frac{m^2 \ln T + 2m^2 \ln(\ln T)}{2T}\right), \quad j = 1, \ldots, p \qquad (8) $$

here $\left|\widehat{\Omega}_j\right|$ is the estimated determinant of variance-covariance matrix for $u_t^+$ that is estimated with lag order $j$, $m$ is the number of variables and $T$ is the number of observations. This information criterion is suggested by Hatemi-J [18]. The simulation results provided in [19] reveals that this information criterion is robust to the autoregressive conditional heteroscedasticity (ARCH) effects and it has good forecasting properties. Mustafa and Hatemi-J [20] provide further simulations results pertinent to this information criterion.[2] It should be mentioned that other combinations can also be considered. For example, another VAR model can be estimated to capture the dynamic relationship between the negative partial components of the underlying variables.

It is important to emphasize that the error term in the VAR model (i.e. $u_t^+$) needs to fulfil certain statistical assumptions for a good regression model such as zero expected value, no autocorrelation, homoscedasticity and normality. Each of these assumptions needs to be tested for and in case any underlying assumption is not fulfilling a remedy needs to be found before any reliable empirical inference can be conducted based on the estimated VAR model. This is the case because a model is only as good as the underlying assumptions and if the assumptions are not fulfilled the conducted inference based on the model can be misleading. Thus, this crucial issue requires special attention in empirical analyses.

## 3   The TDICPS Module

To show how the TDICPS (Transforming Data Into Cumulative Partial Sums) for negative and positive components with and without deterministic trend parts module operates we use two sample datasets. In Sect. 3.1, the consumer price index for the US economy as an example was selected. The variable covers the years of January 1871–May 2018 (1769 records) on a monthly basis. The source for this dataset is

---

[2] See also [16] for the performance properties of different information criteria pertinent to the optimal lag order of the VAR model.

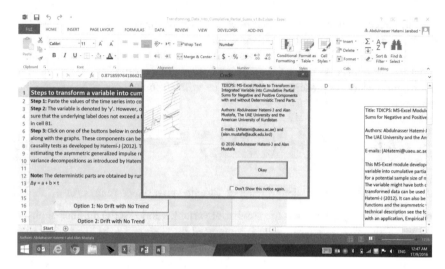

**Fig. 1** A Screenshot of entrance to the program with credit to authors

from the website of Professor Robert J Shiller.[3] In Sect. 3.2 the interest rates for the US for the same period of time has been analyzed.

After initiating the TDICPS module in MS-Excel, the following window opens (Fig. 1):

By activating the module, a dialogue box appears presenting a credit message to its developer. By confirming the message, the module is ready to function. The next step would be copying data from any dataset source and pasting it to the shown column as "B", denoted by y, which can be labelled as per its original column name. It is possible to have a dataset of just above one million records for processing by the module which is the ultimate number of rows accepted in a sheet in Microsoft Excel version 2019. There are three options available to process the data, labelled on each button with yellow background, as given in Fig. 2.

To process the data based on no-drift and no-trend, option one can be chosen (i.e. $a = 0$ and $b = 0$). The second option will process the data based on the drift and no-trend (i.e. $a \neq 0$ and $b = 0$). The third option will process the data based on the drift and the trend selected (i.e. $a \neq 0$ and $b \neq 0$). The last option will produce a processed data with all three options selected in a workbook ready for further analysis.

## 3.1 The Consumer Price Index (CPI) Dataset

In order to show how the TDICPS module can be used, we choose option three and the following results are obtained (Fig. 3).

---

[3] The data is available for download from: http://www.econ.yale.edu/~shiller/data.htm.

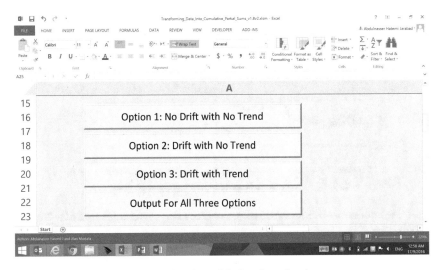

**Fig. 2** A view of availability of combinations of draft and trend options

| | A | B | C | D | E |
|---|---|---|---|---|---|
| 1 | **Data Transformed with drift and trend** | | | | |
| 2 | | | | | |
| 3 | t | CPI | CPI+ | CPI- | |
| 4 | 0 | 2.522849396 | | | |
| 5 | 1 | 2.552926706 | 1.291789889 | 1.26113677 | |
| 6 | 2 | 2.567635892 | 1.306785703 | 1.260850191 | |
| 7 | 3 | 2.530455571 | 1.306500316 | 1.223955154 | |
| 8 | 4 | 2.507467852 | 1.306216359 | 1.201251507 | |
| 9 | 5 | 2.491839312 | 1.305933595 | 1.185905695 | |
| 10 | 6 | 2.491839312 | 1.306215048 | 1.185624242 | |
| 11 | 7 | 2.475969448 | 1.305934787 | 1.170034528 | |

**Fig. 3** A view of the outcome of option 3 with both drift and trend

To clarify the detailed labelling for the processed data outcome, note that the first column (column *A*) gives the time index for the record, the second column (column *B*) is the original dataset, the third column (column *C*) is the converted outcome for positive changes of the dataset, and column *D*, the fourth column is for the negative changes of the transformed dataset. At the end of the process, the module creates a graph for all three options mentioned above for the simulation of transformed data. An example is given in Fig. 4.

Observe that *CPI* signifies the original data for the US consumer price index (Fig. 4). Figure 5 presents the partial cumulative sum of the positive changes is

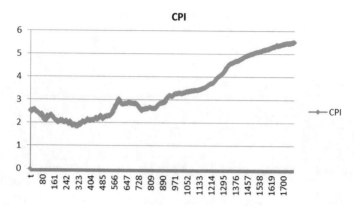

**Fig. 4** Figure of sample data for the US consumer price index, period of Jan 1871-May 2018 on a monthly basis

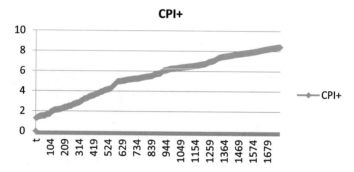

**Fig. 5** Partial cumulative sum for positive changes for consumer price index (CPI+)

represented by *CPI+*, and the partial cumulative sum of the negative changes is denoted by *CPI−* in Fig. 6.

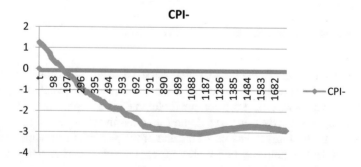

**Fig. 6** Figure of partial cumulative sum for negative changes for consumer price index (CPI−)

| | A | B | C | D |
|---|---|---|---|---|
| 1 | **Data Transformed with drift and trend** | | | |
| 2 | | | | |
| 3 | t | Interest Rates | Interest Rates+ | Interest Rates- |
| 4 | 0 | 1.782743854 | | |
| 5 | 1 | 1.689606614 | 0.890316844 | 0.799289763 |
| 6 | 2 | 1.627501675 | 0.889262557 | 0.73823905 |
| 7 | 3 | 1.715307663 | 0.978121996 | 0.737185597 |
| 8 | 4 | 1.488316392 | 0.977069378 | 0.51124692 |
| 9 | 5 | 1.396585471 | 0.976017594 | 0.42056784 |
| 10 | 6 | 1.361018187 | 0.974966586 | 0.386051506 |
| 11 | 7 | 1.492291869 | 1.107290387 | 0.385001361 |

**Fig. 7** Figure of sample data for the US interest rates for the period of Jan 1871–May 2018 on a monthly basis

It should be clarified that the converted data can be utilized for conducting the asymmetric causality tests as introduced by Hatemi-J [5]. The transformed data can likewise be used for the estimation of the asymmetric generalized impulse response functions as developed by Hatemi-J [6]. This is operational since the transformed data can be imported to any well-known econometric software package like EViews, RATS, Stata, PcGive or Microfit, inter alia.

## 3.2 The Interest Rate Dataset

In this section, we use the US interest rate dataset with the same category of 'with drifts and trend' from TDICPS simulation software (Fig. 7).

For this group of datasets we find the following two graphs of partial cumulative sums for positive and negative changes with drift and trend options selected (Figs. 8, 9 and 10).

## 3.3 The OptLag-HJC Module for Identifying the Optimal Lag Order

It is worth noting that we also estimated the VAR model presented by Eq. (7) for the original data as well as for different combinations of the transformed data with and without deterministic trend parts. For example, Fig. 11 provides a graph for identifying the optimal lag order for original sets for both *CPI* and the US interest

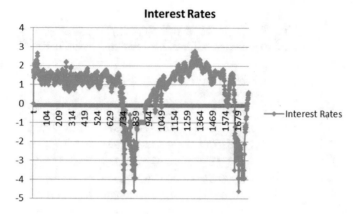

**Fig. 8** Original dataset for US interest rates for Jan 1871–May 2018 on a monthly basis

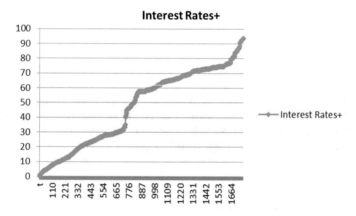

**Fig. 9** Partial cumulative sum for positive changes for interest rates

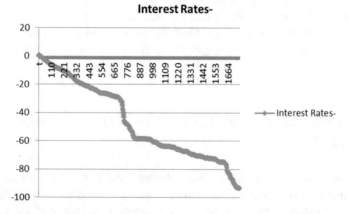

**Fig. 10** Figure of partial cumulative sum for negative changes for interest rates

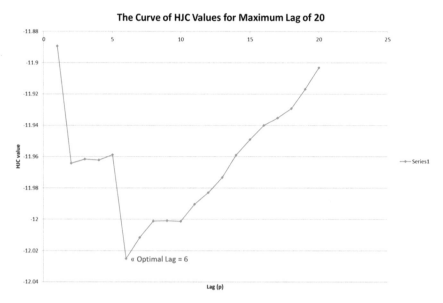

**Fig. 11** The optimal lag order of 6 was found between two datasets of the *CPI* and the US interest rate

rate, hence no option of 'with or without drift and trend' is applicable, and so, the optimal lag order of 6 is calculated.

For the rest of the categories of 'with or without drift and trends' in TDICPS options between cumulative sums of positive and negative changes of both consumer price index and interest rates, different outcomes are given, as is shown in the example provided in Fig. 12.

The results are presented in Table 1. These results show that the optimal lag order of the VAR model based on the minimization of the information criterion in Eq. (8) is six for the original data and it is five for the transformed data regardless of the properties of the deterministic trend part. This means that the optimal lag order for the VAR model is not the same for the transformed data compared to the original data. Thus, the practitioner needs to determine the optimal lag order of the VAR model for the transformed data separately from the original data before testing for causality or estimating the impulse response functions of the variance decompositions. Another point that can be made is that the existence or absence of deterministic trend parts does not impact the optimal lag order.

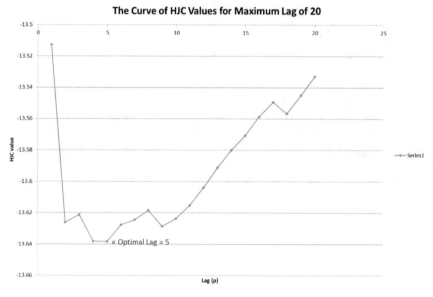

**Fig. 12** The optimal lag order of 5 was found between two sets of data of the *CPI+* with the positive component and the US interest component with the positive component

| | Original data | Positive components | Negative components |
|---|---|---|---|
| No constant or trend | 6 | 5 | 5 |
| Constant | | 5 | 5 |
| Constant and trend | | 5 | 5 |

**Table 1** The estimation results for determining the optimal lag order of the VAR model

*Note* the optimal lag order in each VAR model is determined by minimizing the information criterion presented by Eq. (8)

## 4 Concluding Remarks

It is commonly accepted in the literature that it is important to consider potential asymmetric impacts while conducting empirical studies. In this chapter, a module has been developed using Visual Basic for Applications (VBA) to strengthen and extend the capabilities of a commonly used application software known as Microsoft Excel. It is the aim of this software to be utilized as a prototype prior to the use of other programming languages such as C, C++, Python, or specialist languages such as Gauss or EViews.

To demonstrate the capabilities of this module, the consumer price index and the interest rates in the US markets for the period of January 1871–May 2018 have been used, which transforms data into the positive and the negative sets of records in the form of cumulative parts. Subsequently, the researcher can use the outcome of this transformation for any asymmetric causality tests or the process of creating the generalized impulse response functions. Further technical details on these methods can be obtained from [5–7]. Another example of using the transformed data is to test for hidden cointegration as developed by Granger and Yoon [21], as well as hidden panel cointegration suggested by Hatemi-J [22], the testing for asymmetric panel causality (see Hatemi-J [23]), and for implementing asymmetric GARCH model as recommended by Hatemi-J [24]. The TDICPS module can be accessed online at Hatemi-J and Mustafa [25]. The same module has been developed by Hatemi-J and Mustafa [26] in Octave for ease of accessibility for different practitioners. Although, the Octave module has only one option of both drift and trend for the underlying variables, unlike the VBA version that has three options. In addition, the current module is very consumer-friendly compared to the alternative ones and more accessible since the MS-Excel is the software that is widely utilized for data analysis of this kind and it is part of a common package that is accessible in most computers.

**Acknowledgements** This research is partially been funded by a grant provided by the UAE University (Grant # 31B028), which is enormously appreciated.

# References

1. Frisch R (1933) Editor's note. Econometrica 1:1–4
2. Akerlof G (1970) The market for lemons: quality uncertainty and the market mechanism. Quart J Econ 84:485–500
3. Spence M (1973) Job market signaling. Quart J Econ 87:355–374
4. Stiglitz J (1974) Incentives and risk sharing in sharecropping. Rev Econ Stud 41:219–255
5. Hatemi-J A (2012) Asymmetric causality tests with an application. Empir Econ 43(1):447–456
6. Hatemi-J A (2014) Asymmetric generalized impulse responses with an application in finance. Econ Model 36:18–22
7. Hatemi-J A, El-Khatib Y (2016) An extension of the asymmetric causality tests for dealing with deterministic trend components. Appl Econ 48:4033–4041
8. Sims C (1980) Macroeconomics and reality. Econometrica 48(1):1–48
9. Granger C (1981) Some properties of time series data and their use in econometric model specification. J Economet 16(1):121–130
10. Engle R, Granger C (1987) Co-integration and error correction: representation, estimation, and testing. Econometrica 55(2):251–276
11. Phillips PCB (1987) Time series regression with a unit root. Econometrica 55:277–301
12. Johansen S (1988) Statistical analysis of cointegration vectors. J Econ Dyn Control 12(2–3):231–254
13. Johansen S (1991) Cointegration and hypothesis testing of cointegration vectors in gaussian vector autoregressive models. Econometrica 59(6):1551–1580
14. Stock J, Watson M (1988) Testing for common trends. J Am Stat Assoc 83:1097–1107
15. Johansen S, Juselius K (1990) Maximum likelihood estimation and inference on cointegration—With applications to the demand for money. Oxford Bull Econ Stat 52(2):169–210
16. Hacker S, Hatemi-J A (2008) Optimal lag-length choice in stable and unstable VAR models under situations of homoscedasticity and ARCH. J Appl Stat 35(6):601–615

17. Pesaran H, Shin Y (1998) Generalized impulse response analysis in linear multivariate models. Econ Lett 58:17–29
18. Hatemi-J A (2003) A new method to choose optimal lag order in stable and unstable VAR models. Appl Econ Lett 10(3):135–137
19. Hatemi-J A (2008) Forecasting properties of a new method to choose optimal lag order in stable and unstable VAR models. Appl Econ Lett 15(4):239–243
20. Mustafa A, Hatemi-J A (2021) A VBA module simulation for finding optimal lag order in time series models and its use on teaching financial data computation. Appl Comput Inf forthcoming. https://doi.org/10.1016/j.aci.2019.04.003
21. Granger C, Yoon G (2002) Hidden cointegration. In: No 92, Royal Economic Society annual conference 2002. Royal Economic Society. https://econpapers.repec.org/scripts/showcites.pf?h=repec:ecj:ac2002:92
22. Hatemi-J A (2020) Hidden panel cointegration. J King Saud Univ Sci 32(1):507–510
23. Hatemi-J A (2020) Asymmetric panel causality tests with an application to the impact of fiscal policy on economic performance in Scandinavia. Int Econ 73(3):389–404
24. Hatemi-J A (2013) A new asymmetric GARCH model: testing, estimation and application. In: MPRA Paper 45170. University Library of Munich, Germany
25. Hatemi-J A, Mustafa A (2019) Transforming data into cumulative partial sums using VBA module for MS excel. Available from https://github.com/alanmustafa/Transforming_Data_Into_Cumulative_Partial_Sums
26. Hatemi-J A, Mustafa A (2016) TDICPS: OCTAVE module to transform an integrated variable into cumulative partial sums for negative and positive components with deterministic trend parts. In: Statistical software components. Nr. OCT001, Boston College Department of Economics. https://ideas.repec.org/c/boc/bocode/oct001.html

# Optimistic Retention Contract Evaluation

Belkacem Athamena, Zina Houhamdi, and Ghaleb El Refae

**Abstract**  This paper focuses on utilizing retention contracts to screen and discipline managers in a context in which the council, board of directors, possesses incomplete information about the consequences of managers' decisions. The analysis enlightens us on empire building, on the slight connection between achievement and firing, and describes concerns about the belief that low achievements result from bad managers. This paper analyzes a basic model to show the resulting dilemmas. The desire to screen managers to enhance their future well-being motivates managers to show their credentials by becoming excessively active. The council can address this bias by firing a manager whose project is proven to ruin value. Moreover, the council can replace the manager if he has implemented a project, but its outcomes remain unobservable. Both decisions decrease the attraction to develop loss-generating projects. However, the dismissing decision on either ground will affect the council deduction that the expected competence of the incoming manager is lower than that of the dismissed manager. This study shows how the selection option is preferred over the disciplining option using an optimistic retention contract.

**Keywords**  Moral hazard · Retention contract · Principal-agent problem

## 1  Introduction

Tasks' delegation is important for multiple reasons. For example, in a company, the top executive (the principal) does not have time to make decisions related to the

B. Athamena (✉) · G. E. Refae
Business Administration Department, College of Business, Al Ain University, Al Ain, UAE
e-mail: belkacem.athamena@aau.ac.ae

G. E. Refae
e-mail: ghalebelrefae@aau.ac.ae

Z. Houhamdi
Cybersecurity Department, College of Engineering, Al Ain University, Al Ain, UAE
e-mail: zina.houhamdi@aau.ac.ae

© The Author(s), under exclusive license to Springer Nature Switzerland AG 2022
S. G. Yaseen (ed.), *Digital Economy, Business Analytics, and Big Data Analytics Applications*, Studies in Computational Intelligence 1010,
https://doi.org/10.1007/978-3-031-05258-3_14

quotidian routine. Therefore, the top executive assigns these decisions to a director (the agent). In a typical democracy, the public has a frail motivation to examine the complete implications of possible policies. Thus, the public delegates policy decisions to politicians. In these two cases, it is clear that the decision-making delegation is advantageous. Nevertheless, decision delegation is problematic in case the agent has different preferences than the principal. Particularly with the existence of asymmetric information. The typical problem of principal-agent is the interest conflict between the chief executive officer and the stockholders. The council, on account of stockholders, has the authority to run the company. Directors, on the other hand, do not possess the information and the time needed for decision-making. Accordingly, the decision-making is assigned to the chief executive officer. The council's role is twofold. (1) the council should nominate and dismiss managers. (2) the council should supervise the managers' achievements and oust them whenever inevitable [1]. The challenge is that the council has to perform these two tasks based on incomplete information. To discipline executives, the council has to adhere to norms and rules to discipline managers. Often adhering to rules is the unique manner to discipline managers. Usually, human behavior is very intricate in a small band, and behavioral norms (culturally created and implanted overtime) play a considerable role in molding companies [2, 3]. This paper considers the exploitation of retention contracts intended to screen and discipline managers in a context where the council possesses incomplete information about the consequences of managers' decisions. The contract is implicit and indicates the circumstances under which a manager is kept or replaced. The contract is seen as a rule or norm shared by the council and the manager. For instance, a manager who has developed an exceptionally profitless project understands he will need to quit the company immediately after it becomes public. In case the manager is qualified, the council finds it harsh to replace the manager; however, the council must respect the norms. Accordingly, the dismissal resulting from awful achievement is considered disappointing yet unavoidable [2].

The remaining of the paper contains five sections. Section 2 describes some related literature. Section 3 describes the proposed model and the trade-off the council faces between screening and disciplining managers. Section 4 addresses how the council shapes the manager's behavior, considering the optimistic contract (in which the council keeps the manager when the value of the implemented project is not observable). Section 5 concludes the paper.

## 2 Related Literature

This study contributes to the councils' (board of directors) literature. Hermalin and Weisbach [4], In their survey paper, observed that "the experimental literature on councils in governmental organizations is relatively properly developed, whereas its theory remains in its childhood". Furthermore, in their survey on councils, Grace et al. [5] reached a similar conclusion concerning the theory dearth. Hermalin [6] modeled the council of directors chooses an applicant for a manager post, makes an idea on the manager competence, and determines to keep or dismiss him. There are two significant differences with the proposed model. (1) Hermalin perception of the manager's competence depends on his communication and presentation skills in council meetings, not on his perceived organizational performance. (2) Hermalin emphasizes the unique role of the council, screening manager competitions. Consequently, the council does not have to harmonize clashing objectives. Graziano and Luporini [7] propose an identical selection model and retention-dismissal decision. Since the council mistakenly hires an unqualified manager in the selection phase, the council can be doubtful in the appraisal phase to fire the manager because this can indicate its own deficiency of ability and probably provoke its substitution.

In this study, the council utilizes the retention contract to handle the manager moral hazard problem, similar to the electorate employing its reelection approach to control politicians. For all we know, this correspondence was never exploited in business governance literature. The considered contract is not explicit (tacit) and is not imposed by a third party, as per the literature on the political agency. It forms common expectations between the agent (manager, director, or minister) and the principal (council, electors, or congress) about the cases in which the current agent is kept or replaced [8]. However, Ferejohn [9] and Barro [10] were the leaders to claim that the power of replacing agents disciplines agent who tends to use his office to pursue his own goals. The main goal of our study is to determine the ideal or the best implicit contract. As mentioned previously, this type of implicit contract is seen as standard. IT is proven that this approach (implicit contracts) effectively determines some relationship between the council and its managers. Despite everything, just as it is difficult to measure the contribution of foreign affairs ministers to the country's well-being, it is also difficult to identify a manager's contribution to the company's continuation and profitability in the long-term. The easiest thing to perceive is the minister's or manager's activeness (such as state a re-structuring, signing an agreement, implementing a strategy, etc.). Moreover, acting as a parliament writes an implicit contract without defining when a minister will be dismissed, a classic council will not specify explicitly in the contract what causes the manager dismissal.

# 3 The Model

One of the literature backbones of corporate governance is that managers become builders of an empire if not controlled by certain serious kinds of governance. Also, it is continuously asserted that the building of such an empire indicates managers' desire for prestige, power, and status [11–13]. Therefore, building Empires originated from the divergence between the board council and the managers' preferences and visibility lacking (classical moral hazard problem). Ayllon and Nollenberger [11] added another cause for growth by saying that "decisions that lead to successful increase signifies that the manager is qualified and deserves rewards. Thus, individual competence is evaluated by accomplished expansion". Such signaling is beneficial for the council, which possesses only restricted information on a manager's competence. Therefore, there are many questions: How does the council handle a probable conflict between stopping empire building and requesting information? What kind of retention approaches are available? How do they balance between the achievements of the council's goals?

A simple two-phase model is used for answering these questions. In each phase, a manager conceives a project and then must decide to implement it or not. The project represents something that has considerable and strong effects on the organization, such as reorganizing, diversifying, and purchasing. The project quality depends on external conditions and the manager's competence. The manager knows his ability and observes the external conditions, but the council does not. The board perceives only the manager's implementation decision. Thus, the council determines, with a probability, the project quality only after its implementation. At the end of the first phase, where the manager took the implementation decision, the council chooses between retaining the manager and dismissing him. The significant property of the proposed model is that a qualified manager implements more probably a project than an unqualified one because generally, a qualified manager conceives more appropriate projects (beneficial projects even in a more hostile situation). Hence, the activism expresses aptitude and qualification. Thus, activism is used for screening. Consequently, the council desires a qualified manager to implement a project that is undesirable per se. Furthermore, the council sometimes desires an unqualified manager to desist from implementing an appropriate project. Accordingly, the relation between low performance and a poor quality manager is relapsed.

After establishing the screening function, we find out that a manager's desire to preserve his position (by the love of power, compensation, prestige, etc.) leads him to abuse this function and sometimes to deform the implementation decision. The manager partially bases his decisions on the decision effects on his career. Higher is the desire for power, prestige, and the more is the distortion of his decisions (building his empire). Therefore, the usage of the implementation decision by the council for screening causes a moral-hazard problem. The council reduces this problem by firing a manager who implemented a very low-quality project. Nevertheless, the signaling function of the implementation decision indicates that bad projects are implemented by qualified managers particularly. Consequently, the council finds it hard to intentionally fire a qualified manager and replace him with a manager with anonymous quality. To surmount this problem, a council has to adhere to norms or rules. Accordingly, dismissal from a low performance is usually considered unfortunate yet unavoidable. The council also decides what action to perform in case the project quality remains unobservable. Anew, it is on the dilemma horns.

The model analysis results are shown in Fig. 1 (which illustrates the qualified manager decision) and Fig. 2 (which describes the unqualified manager decision). Figure 1a, b presents the $V_1$ values range for which the project is implemented (not implemented) by a qualified manager in case $\beta = 0(\beta > 0)$. However, Fig. 2 describes alike information but for an unqualified manager.

Where,

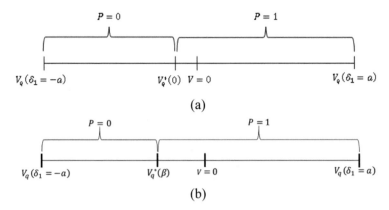

**Fig. 1** Qualified manager decision

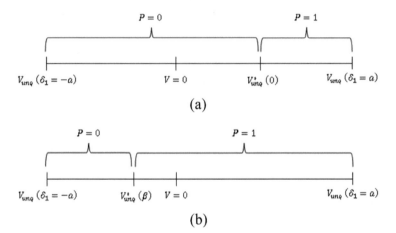

**Fig. 2** Unqualified manager decision

- $\delta_x$: the world state ('business conditions') is selected randomly from a uniform distribution on $[-a, a]$
- $V_{unq}$: the value of the project developed by an unqualified manager.

$$V_x = V_{unq}(\delta_x) = \rho + \delta_x \tag{1}$$

- $V_q$: the value of the project developed by a qualified manager.

$$V_x = V_q(\delta_x) = \rho + f + \delta_x, \; f > 0 \tag{2}$$

- $\rho$ defines the manager's performance.

- $x \in \{1, 2\}$ defines the phases; $\beta$ denotes the benefits drawn by the manager from possessing his post.

The following section discusses the optimistic retention contract known as 'No news is good news' where the council keeps the manager if the project is implemented, but its value is unknown. Contrary to the pessimistic retention contract known as 'no news is bad news', the manager is dismissed if the project value is unknown.

# 4   Optimistic Contract: 'No News is Good News'

In short, the optimistic contract accepts all information as facts until demonstrated to be false. The optimistic contract reduces the data verification cost, as the data verification is required only in case of certainty that the data is false. Under the optimistic contract, the council decides to:

- Keep the manager if he implemented a project that its value is unobservable or bigger than the threshold ($V_1 > t$).
- Dismiss the manager if he did not implement any project or implementing a project, but its value is unobservable.

The primary interest is to determine the optimal value of the threshold $t$ from the council's perspective. The optimal threshold value defines the extent to which a manager is disciplined and the possibility that a qualified manager is chosen for phase 2.

As Figs. 1 and 2, the council decision affects the manager's decision for implementing a project in the first phase. Assume that the council selects $t \in \left[V_q'(\beta), V_{unq}'(\beta)\right]$. Thus, the qualified tenured decision on $V_1$ is affected by $t$. If the council notices $V_1 \leq t$, $P_1 = 1$ causes firing. Therefore, in comparison to the previous situation, where $P_1 = 1$ leads always to keep the manager, the motivation to select $P_1 = 1$ is attenuated. If the tenured in the first phase is unqualified, $t \in \left[V_q'(\beta), V_{unq}'(\beta)\right]$ has no impact on the tenured implementation decision since $t$ is no constraining. However, if the council selects $t \geq V_{unq}'(\beta)$, $t$ is constraining for a qualified and unqualified manager. $\mu$ is the probability that the selected tenured is qualified.

$$V_q(\delta_1) \geq V_q'(\beta) := -\left(\Pi_q - \Pi_\mu\right) - \beta \tag{3}$$

$$V_{unq}(\delta_1) \geq V_{unq}'(\beta) := \left(\Pi_\mu - \Pi_{unq}\right) - \beta \tag{4}$$

According to Eqs. (3) and (4), the motivation to select $P_1 = 1$ is attenuated. Consequently, there are two options for the manager's selection of $t$ value:

1. If $t \in \left[ V'_q(\beta), V'_{unq}(\beta) \right]$ then, the council disciplines the qualified manager, believing that the unqualified manager distorts the implementation decision (Eq. 3)
2. If $a \geq V * IC(\lambda)$, then the council influences the implementation decision of both qualified and unqualified managers. Where $\lambda$ defines the probability that the manager perceives $V_1$

Now we will discuss the impacts of $t$ value on the behavior of qualified and unqualified managers separately.

### Qualified Manager Discipline

In an ideal situation, the council desires a qualified tenured to select $P_1 = 1$ if and only if $V_1 > V'_q(0)$. Nevertheless, a qualified manager selects $P_1 = 1$ if and only if $V_1 \geq V'_q(\beta)$ (Eq. 1). The usage of the threshold value $t$ in his retention contract, the council, disciplines the manager. We consider a manager "fully disciplined" if there is no distortion in the implementation decision, whereas a manager is "slightly disciplined" if the distortion is diminished. Assume that $V^*_q$ indicates the cut-off value used by a qualified tenured if the council defines a high-value threshold $t$. Note that to have an impact on a qualified manager's implementation decision, the council must define $t \geq V^*_q$.

Assume that a qualified manager notices $V_q(\delta_1) < t$. Thus, the manager implements the project if and only if

$$V_q(\delta_1) + \beta + \left[ (1 - \lambda)(\Pi_q + \beta) + \lambda \Pi_\mu \right] < \beta + \Pi_\mu \tag{5}$$

Therefore, the manager decides to implement the project if

$$V_q(\delta_1) \geq V^*_q(\beta) := (1 - \lambda)(\Pi_\mu - \Pi_q) - (1 - \lambda)\beta \tag{6}$$

We can deduce four important points:

1. If $V_1 \in \left[ V^*_q(\beta), t \right]$, the manager implements the project hoping that the council misses the project value; thus, he holds the bureau.
2. The council changes the threshold value without influencing the cut-off value set by the manager, considering that $t \geq V^*_q(\beta)$.
3. According to Eqs. (3) and (6), we deduce that $V'_q(\beta) < V^*_q(\beta)$. Therefore, the manager will be at least "slightly disciplined".
4. If $\beta < \beta' := \frac{\lambda}{1-\lambda}(\Pi_q - \Pi_\mu)$ then the cut-off value satisfies $V'_q(0) < V^*_q(\beta)$. Hence, if the qualified manager desires to hold the bureau, then the impact of fixing the $t$ value is very strong: this hampers the screening function of the implementation decision. However, it also indicates that if $\beta < \beta'$, the council induces the manager to use $V'_q(0)$ as his $t$ value by putting $t = V'_q(0)$ which prevents the manager to distort the implementation decision.

Accordingly:

$$V_q^* = \begin{cases} V_q'(0) & if\ \beta < \beta' \\ (1-\lambda)\big(\Pi_\mu - \Pi_q\big) - (1-\lambda)\beta & if\ \beta \geq \beta' \end{cases} \tag{7}$$

where $V_q^* \in \big[V_q'(\beta),\, V_q'(0)\big]$.

## Unqualified Manager Discipline

Here we discuss the case where the manager fixes the threshold value $t$ so that the behavior of an unqualified manager is impacted, $t \geq V_{unq}'(\beta)$. Thus, it is impossible to discipline an unqualified manager completely. He can be "slightly-disciplined" by fixing

$$t \geq V_{unq}^*(\beta) := (1-\lambda)\big(\Pi_\mu - \Pi_{unq}\big) - (1-\lambda)\beta \tag{8}$$

Therefore, the council provokes an unqualified manager to select $P_1 = 1$ if and only if

$$V_1 \geq V_q^*(\beta) \tag{9}$$

where

$$V_{unq}^*(\beta) \in \big[V_{unq}'(\beta),\, V_{unq}'(0)\big] \tag{10}$$

***Proof*** Assume that $t$ value for which the unqualified manager's decision is impacted $(t \geq V_{unq}'(\beta))$. Obviously, for $V_{unq}(\delta_1) \geq t$, the manager's decision will be $P_1 = 1$. If

$$V_{unq}(\delta_1) < t,\ P_1 = 1 \tag{11}$$

yield

$$V_{unq}(\delta_1) + \beta + \big[(1-\lambda)\big(\Pi_{unq} + \beta\big) + \lambda\Pi_\mu\big] \tag{12}$$

whereas $P_1 = 0$ yields $\beta + \Pi_\mu$.

Accordingly, it is clear that $P_1 = 1$ is better than $P_1 = 0$ if

$$V_{unq}(\delta_1) \geq V_{unq}^*(\beta) := (1-\lambda)\big(\Pi_\mu - \Pi_{unq}\big) - (1-\lambda)\beta \tag{13}$$

where

$$V_{unq}'(\beta) < V_{unq}^*(\beta) < V_{unq}'(0)] \tag{14}$$

Consequently, it affirms that the council will just slightly discipline an unqualified manager. To figure out why the council cannot completely discipline an unqualified manager, assume that the council sets the $t$ value $t = V'_{unq}(0)$ and the manager perceives $\delta_1$ so that

$$V_{unq}(\delta_1) = V'_{unq}(0) \ and \ \beta > 0 \tag{15}$$

Remember that $V'_{unq}(0)$ represents the cut-off used by the manager in case he is not interested in the advantages of keeping the bureau and the project value is enough for reselection. Since $\beta > 0$, actually, the manager prefers $P_1 = 1$ to $P_1 = 0$ from a benefits perspective. From the project value perspective, If the project value is satisfactory to retain the manager, an unqualified tenured must select $P_1 = 0$ in order to guarantee his replacement by a medium-qualification manager to increase the intended payoff in the next phase. Nevertheless, according to the optimistic contract, the decision $P_1 = 1$ no more ensures retention. Thus, the advantages of implementing a beneficial project in the first phase ($V'_{unq}(0) > 0$) are combined with the dismissal (increased payoff in the second phase). Consequently, in this case, the manager prefers $P_1 = 1$ to $P_1 = 0$. Since the manager prefers always to implement the project from a career perspective when

$$V_{unq}(\delta_1) = V'_{unq}(0) \tag{16}$$

He will not be complete-disciplined.

## Threshold Value t

We analyzed the impact of $t$ value on the behavior of a different type of tenured separately. In this section, we analyze the impact of the selection of $t$ value on the council's utility. For this, we examine the impact of the selection of $t$ value on the applied discipline in phase 1 and on the probability that a qualified manager will be chosen in phase 2.

There are two alternatives for the council:

- Discipline a qualified manager whenever possible (completely or slightly) by defining $t = V_q^*$
- Discipline all manager types (qualified or unqualified) by defining $t \in \left[V'_{unq}(\beta), V_{unq}(\delta_1 = a)\right]$.

Figure 3 grasps the above alternations.

Figure 3 illustrates a case where $V_q^* < V'_{unq}(\beta)$. In this situation, the council's main policy is disciplining as well unqualified manager. By defining $t = V_q^*$ also influences the unqualified manager behavior. Assume that the council defines $t < V'_{unq}(\beta)$. Accordingly, the council defines $t$ value in such a way to affect the behavior of a qualified manager, assuming that an unqualified manager's decision depends on $V'_{unq}(\beta)$. In [14], we claimed that if the council ignores the manager's competence, preferably the council desires a qualified manager to select $P_1 = 1$ if and only if $V_1 > V_q^*(0)$. Also, we state that if $\beta \geq \beta'$, the council reaches this objective by fixing $t = V'_q(0)$: a qualified tenured could be completely disciplined. If instead $\beta \geq \beta'$, the

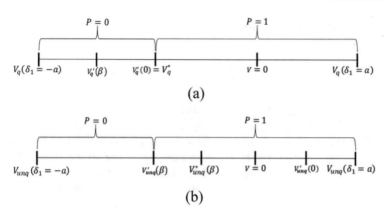

**Fig. 3** Threshold value $t$ alternatives

council will only slightly discipline a qualified manager, $V_q^*(\beta) < V_q'(0)$, by defining $t \geq V_q^*(\beta)$. For sure that the council has no incentive to define $t > V_q^*(\beta)$, instead of $t = V_q^*(\beta)$. If the council perceives that $V_1 \in \left[V_q^*(\beta), V_{unq}'(\beta)\right]$, it concludes that the manager is qualified. The tenured manager will be retained in the second phase.

Now assume that the council defines $t \geq V_q^*(\beta)$. Then, the council disciplines an unqualified manager and because of $V_q^* < V_{unq}'(\beta)$, the council disciplines a qualified manager. Therefore, by setting choosing $t \geq V_{unq}^*(\beta)$, the council profits as much as possible from applying the optimistic contract. Note that in case the council perceives $t$ value $V_1 \in \left[V_q^*, V_{unq}'(\beta)\right]$, the council knows that the manager is qualified; however, the manager is dismissed. The contract cost to discipline an unqualified manager is that the probabilities for choosing a qualified manager are not entirely exploited. So, what is the best $t$ value that the council should set? Remember that ideally, the council desires an unqualified manager to select $P_1 = 1$ if and only if $V_{unq}(\delta_1) \geq V_{unq}'(0)$. But, within the optimistic contract, the council cannot completely discipline an unqualified manager. The council can only slightly discipline an unqualified manager by fixing $t \geq V_{unq}^*(\beta)$. For sure, the council has no incentive to fix $t > V_{unq}^*(\beta)$, instead of $t = V_{unq}^*(\beta)$. For $t \in \left[V_{unq}^*(\beta), V_{unq}(\delta_1 = a)\right]$, the $t$ value does not affect the implementation decision of the two manager types or the selection. The council does not fix $t < V_{unq}(\delta_1 = a)$ since $V_1 > V_{unq}(\delta_1 = a)$ is obvious proof that the manager is qualified.

To summarize, remind that the equality of retention and the project implementation increases manager activism. The council uses information on the project value (that is not always observable) to make its decision. The optimistic contract prides two alternatives for the council to control the manager behavior:

1. The council focuses on qualified manager discipline only if $t = V_q^*$. This selection alternative maximizes the exploitation of selection opportunities. This alternative maximizes the probability that the tenured in the second phase is qualified. Figure 2 shows the length of the interval $\left[V_q^*, V_{unq}'(\beta)\right]$.

2.  The council focuses on qualified and unqualified managers' discipline if $t = V_{unq}^*(\beta)$. This is referred as disciplining alternative. This manager activism reduces the distortion of the implementation decision by an unqualified manger. This advantage is absolutely influenced by on the length of the interval $[V_{unq}'(\beta), V_{unq}^*(\beta)]$.

Now, what is the best alternative for the council?

If $V_q^* < V_{unq}'(\beta)$ then a rise in $\beta$ or $\lambda$ or a reduction in $(\Pi_q - \Pi_{unq})$ or $\mu$ $\rho$, expand the parameters interval for which the council selects the disciplining alternative. However, if $V_{unq}'(\beta) \leq V_q^*$ then the council selects the disciplining alternative by defining $t = V_{unq}^*(\beta)$. The proof is in the appendix.

Evidently, if $V_{unq}'(\beta) \leq V_q^*$, then the council applies the disciplining alternative because, in this situation, if the council was to set $t = V_q^*$, it also affects an unqualified manager's behavior. Thus, the disciple of a qualified manager only, the selection alternative, is not the right alternative. Furthermore, the advantage of the selection alternative depends directly on the interval length $[t, V_{unq}'(\beta)]$. Clearly, if $V_{unq}'(\beta) \leq t$, so there are no advantages of the selection alternative.

Assume that $V_q^* < V_{unq}'(\beta)$, then the council actually chooses between the disciplining and the selection alternatives. To figure out how the modification of the parameters influences the council's decision regarding the two alternatives, we investigate the impacts of this modification on the ranges of $[V_q^*, V_{unq}'(\beta)]$ and $[V_{unq}'(\beta), V_{unq}^*(\beta)]$ where $\beta \geq \beta'$ (and so $V_q^* = V_q^*(\beta)$). It is easy to show that

$$V_{unq}'(\beta) - V_q^* = \Pi_q - \Pi_{unq} - \lambda(\Pi_q - \Pi_\mu) - \lambda\beta$$
$$= (1 - \lambda(1 - \mu))(\Pi_q - \Pi_{unq}) - \lambda\beta \tag{17}$$

$$V_{unq}^*(\beta) - V_{unq}'(\beta) = -\lambda(\Pi_\mu - \Pi_{unq}) - \lambda\beta = -\lambda(\Pi_q - \Pi_{unq}) + \lambda\beta \tag{18}$$

Larger is the (Eq. 17) value, the selection alternative is more adequate. Conversely, larger is the (Eq. 18) value, the disciplining alternative is more adequate. According to Eqs. (17) and (18), a rise in advantages from keeping the bureau $\beta$ enhances the value the council assigns to disciplining alternative and diminishes the value of the selection alternative because the manager's desire to keep the bureau is behind the council disciplining in the first place.

Actually, the council desires a qualified, instead of an unqualified manager to implement a project. Consequently, it is scarcely strange that the rise in $(\Pi_q - \Pi_{unq})$ increases the parameters interval for which the council selects the selection alternative. It should be pointed that a raise in $(\Pi_q - \Pi_{unq})$ diminishes Eq. (18) because a raise in $(\Pi_q - \Pi_{unq})$ increases the decision distortion cost for the manager. Therefore, the necessity of manager disciplining decreases. On the other hand, the $\lambda$ increase indicates an increase in the probability that the council observes the project value. Clearly significant the impact of $\lambda$ on the alternative (selection or disciplining) choice since the observation of the project value is a prerequisite for managers disciplining. Thus, it is insightful that an increase in λ raises the attraction of the disciplining alternative. It is confirmed by Eq. (18) where the disciplining alternative value raises λ. Similarly, the selection alternative value, Eq. (17), declines since the greater possibility of the observed project value diminishes a qualified manager's enthusiasm to select $P_1 = 1$. Furthermore, the parameter μ expresses the probability that a manager is qualified. Thus, μ increase direct impact is lesser harm resulting from supporting unqualified managers. For this reason, the μ increase expands the parameters interval for which the council selects the selection alternative ((17) increases in μ, whereas (18) declines in μ).

## 5   Conclusion

The Board council has restricted information that it uses for disciplining and screening the managers of its organization. This paper analyzed a simple model showing the resulting difficulties. The desire to preserve the manager to enhance the company's future benefits incites the manager to become extremely active and demonstrates his competencies and skills. The council can address this bias by firing a manager who implements projects destroying the business value. Furthermore, the council can dismiss the manager if he implements a project, but its outcomes are not observed. Therefore, both decisions decrease the attraction for implementing projects that generate losses. However, regrettably, the dismissing decision on either ground leads to the council deduction that the expected qualification of the new manager is lower than the fired manager. This study considers an optimistic retention contract known as 'no news is good news where the council retains the manager if the manager implements a project, but the project value remains unobservable. We have demonstrated under what circumstances the selection alternative is preferred over the disciplining option.

This study considered the employment of retention contracts to discipline and screen managers in a context where the board council possesses poor information about the managers' decision consequences. The considered contracts are implicit and specify the used conditions to retain or to dismiss the manager. The proposed model can be applied in electoral competition in two-party systems or in a reelection strategy to discipline politicians. A popular justification for representative democracy is "pure democracy causes a free-rider problem concerning the information collection". Therefore, this problem takes for granted that the representative collects information. The proposed model can be used to examine if electoral competition convinces candidates or political parties to gather information regarding policy outcomes. Where the electorate desires that parties execute two activities. The first activity, assigned to both the incumbent and the opposition parties, to collect information about possible policy alternatives. After that, the second task assigned to the incumbent party is the selection of the best policy. The model identifies the cases where voters provoke political parties to gather information and then select optimal policies from the voter's perspectives.

As future work, we will investigate the pessimistic retention contract known as 'no news is bad news where the manager is fired if the council cannot perceive the project value.

## Appendix

The proposition contains two parts:

1. Definition of a parameter range $(V'_{unq}(\beta) \leq V^*_q)$ for which the council's leading approach is disciplining a manager regardless of his type and a parameter range $(V'_{unq}(\beta) > V^*_q)$ for which the council sometimes selects the disciplining alternative and sometimes selection alternative.
2. Comparative statics results for $V'_{unq}(\beta) > V^*_q$. We derived the constraints on the parameter range. Thus, now we prove the comparative statics results.

We identify two different cases:

1. if $\beta < \beta' = \frac{\lambda}{1-\lambda}(\Pi_q - \Pi_\mu)$, the council can fully discipline the qualified manager.
2. if $\beta \geq \beta'$, the council can only slightly discipline the qualified manager.

**Case 1** Comparative statics results for $\beta < \beta'$.

We start by determining the council's expected utility if it selects the selection alternative. Then, we define the council's expected utility if it selects the disciplining alternative. Assume that the council selects the selection alternative and fixes $t = V^*_q(0)$, which implies that a qualified manager decides $P_1 = 1$ if and only if $V_1 > V^*_q(0)$ and an unqualified manager chooses $P_1 = 1$ if and only if $V_1 \geq V'_{unq}(\beta)$. Then the council's expected utility is:

$$\Pi_\mu - \frac{1}{4a}\left(\mu\left(V'_q(0)\right)^2 + (1-\mu)\left(V'_{unq}(\beta)\right)^2\right)$$
$$+ \left[\Pi_\mu + \frac{\mu(1-\mu)}{2a}\left(f + V'_{unq}(\beta) - V'_q(0)\Pi_q - \Pi_{unq}\right)\right] \qquad (19)$$

Replacing $V'_{unq}(\beta)$ (from Eq. 6) and $V'_q(0)$ (from Eq. 4) with $= 0$, then the rewriting leads to:

$$\Pi_\mu - \frac{1-\mu}{4a}\left(\mu\left(\Pi_q - \Pi_{unq}\right)^2 + \beta^2 - 2\mu\left(\Pi_q - \Pi_{unq}\right)\beta^2\right)$$
$$+ \left[\Pi_\mu + \frac{\mu(1-\mu)}{2a}\left(f + \Pi_q - \Pi_{unq}\right) - \beta\left(\Pi_q - \Pi_{unq}\right)\right] \qquad (20)$$

Assume that the council selects the disciplining alternative and fixes $t \in [V'_{unq}(\beta), V_{unq}(\delta_1 = a)]$ which implies that a qualified (unqualified) manager decides to implement the project if and only if $V_1 \geq V_q(\beta)$ ($V_1 \geq V_{unq}(\beta)$). Thus the expected manager utility is:

$$\Pi_\mu - \frac{1}{4a}\left(\mu\left(V_q^*(\beta)\right)^2 - (1-\mu)\left(V_{unq}^*(\beta)\right)^2\right)$$
$$+ \left[\Pi_\mu + \frac{\mu(1-\mu)}{2a}\left(f + (1-\lambda)(V_{unq}^*(\beta) - V_q^*(\beta))\right)(\Pi_q - \Pi_{unq})\right] \quad (21)$$

Replacing $V_q^*(\beta)$ (using Eq. 17) and $V_{unq}^*(\beta)$ the rewriting leads to:

$$\Pi_\mu - \frac{(1-\lambda)^2}{4a}\left(\mu(1-\mu)\left(\Pi_q - \Pi_{unq}\right)^2\right)$$
$$+ \beta^2\left[\Pi_\mu + \frac{\mu(1-\mu)}{2a}\left(f + (1-\lambda)^2(\Pi_q - \Pi_{unq})\right)\left(\Pi_q - \Pi_{unq}\right)\right] \quad (22)$$

The choice between the two alternatives (selection and disciplining) consists of the comparison between the two Eqs. (20) and (22). The council selects the selection alternative if

$$(\lambda(2-\lambda) - \mu)\beta^2 - \mu(1-\mu)\lambda(2-\lambda)\left(\Pi_q - \Pi_{unq}\right)^2 < 0 \quad (23)$$

The inequality (23) is always valid if $(\lambda(2-\lambda)) - \mu \leq 0$. Accordingly, the council always selects the selection alternative. However, if $(\lambda(2-\lambda) - \mu) > 0$, the council selects the selection alternative if and only if

$$\beta < \overline{\beta_1} = \left(\Pi_q - \Pi_{unq}\right)\sqrt{\frac{\mu(1-\mu)\lambda(2-\lambda)}{\lambda(2-\lambda) - \mu}} \quad (24)$$

where $\overline{\beta_1}$ represents the $\beta$ value at which the council is indifferent to discipline just the qualified manager and to discipline both managers type. At this point, we can discover the impact of $\beta$, $f$, $\lambda$, and $\mu$ on choosing between the selection and disciplining alternatives:

1. A rise in $\beta$ expands the parameters interval for which the council selects to discipline both managers type (see Eq. (24)).
2. An increase in the parameter f reduces the disciplining alternative attractiveness. Because a high $f$ value impacts positively $\overline{\beta_1}$

$$\frac{d\overline{\beta_1}}{df} = \frac{d\left(\Pi_q - \Pi_{unq}\right)}{df}\sqrt{\frac{\mu(1-\mu)\lambda(2-\lambda)}{\lambda(2-\lambda) - \mu}} \quad (25)$$

where

$$\frac{d\left(\Pi_q - \Pi_{unq}\right)}{df} = 2(a + \rho + f) > 0 \tag{26}$$

3.  An increase in parameter $\mu$ also reduces the disciplining alternative attractiveness. Note that high $\lambda$ value expands the interval for which the council always chooses selection alternative $\frac{d(\lambda(2-\lambda)-\mu)}{d\mu} = -1 < 0$. Furthermore, if $\lambda(2 - \lambda) - \mu > 0$, then a raise in $\mu$ value affects positively $\overline{\beta_1}$

$$\frac{d\overline{\beta_1}}{d\mu} = \frac{\left(\Pi_q - \Pi_{unq}\right)\lambda(2 - \lambda)(1 - 2\mu)(\lambda(2 - \lambda) - \mu) + \mu(1 - \mu)}{2(\mu(1 - \mu)\lambda(2 - \lambda)^{\frac{1}{2}}(\lambda(2 - \lambda) - \mu)^{\frac{3}{2}}} > 0 \tag{27}$$

To demonstrate that $\frac{d\overline{\beta_1}}{d\mu} > 0 >$, we must prove that

$$(1 - 2\mu)(\lambda(2 - \lambda) - \mu) + \mu(1 - \mu) > 0 \tag{28}$$

Suppose that

$$Y = (1 - 2\mu)(\lambda(2 - \lambda) - \mu) + \mu(1 - \mu) \tag{29}$$

We can demonstrate that $Y > 0$ by proving that the smallest possible value of $Y > 0$. The proof consists of showing that

$$\frac{dY}{d\mu} = -(2 - \lambda)2\lambda + \mu < 0 \tag{30}$$

Remember that $\mu < (2 - \lambda)\lambda$. Thus, by choosing

$$\mu = \lim_{\varepsilon \to 0}(\lambda(2 - \lambda) - \varepsilon) = \lambda(2 - \lambda) \tag{31}$$

We derive the smallest $Y$ value, noted $Y_{min}$. Thus

$$Y_{min} = \lambda\left(-\lambda^3 + 4\lambda^2 - 9\lambda + 6\right) \tag{32}$$

The $Y_{min}$ sign is dependent on the sign of

$$Y' = \left(-\lambda^3 + 4\lambda^2 - 9\lambda + 6\right) \tag{33}$$

We take the derivative concerning $\lambda$ we obtain

$$\frac{dY'}{d\lambda} = -3\lambda^2 + 8\lambda + 9 < 0, \quad \text{for} \quad \lambda \in \,]0, 1[ \tag{34}$$

Consequently, by considering $\mu = \lim_{\varepsilon \to 0}(1 - \varepsilon) = 1$, we conclude that the minimum value of $Y_{min}$ is 0. Thus, we deduce that $\frac{d\overline{\beta_1}}{d\mu} > 0$.

4.  An Increase in the $\lambda$ parameter makes increases the disciplining alternative attractiveness. Firstly, remember that the high value of $\lambda$ reduces the parameters interval for which the council always prefers the selection alternative $\left(\frac{d(\lambda(2-\lambda)-\mu)}{d\lambda} > 0\right)$. Furthermore, if $(\lambda(2 - \lambda) - \mu) > 0$ then a raise in $\lambda$ impacts negatively $\overline{\beta_1}$.

**Case 2** Comparative statics results for $\beta \geq \beta'$.

For this purpose, we compare the council's expected utilities for the selection and the disciplining alternatives. Since we calculated already the expected utility for the disciplining alternative, now we calculate the council's expected utility for the selection alternative. Recall that the selection alternative means that.

- The council sets $t = V_q^*(\beta)$,
- A qualified manager chooses $P_1 = 1$ if $V_1 \geq V_q'(\beta)$, and
- An unqualified manager chooses $P_1 = 1$ if $V_1 \geq V_{unq}'(\beta)$.

The council's expected utility equals

$$\Pi_\mu - \frac{1}{4a}\left(\mu\left(V_q^*(\beta)\right)^2 + (1 - \mu)\left(V_{unq}'(\beta)\right)^2\right)$$
$$+ \left[\Pi_\mu + \frac{\mu(1 - \mu)}{2a}(f + V_{unq}'(\beta) - V_q^*(\beta))(\Pi_q - \Pi_{unq})\right] \quad (35)$$

Replacing $V_q^*(\beta)$ (Eq. (17)) and $V_{unq}'(\beta)$ (Eq. (7)) and then the rewriting leads to:

$$\Pi_\mu - \frac{1}{4a}\mu\left((\lambda - 1)(1 - \mu)(\Pi_q - \Pi_{unq}) - (1 - \lambda)\beta\right)^2 + (1 - \mu)\left(\mu(\Pi_q - \Pi_{unq}) - \beta^2\right)$$
$$+ \left[\Pi_\mu + \frac{\mu(1 - \mu)}{2a}(f + (1 - \lambda(1 - \mu))(\Pi_q - \Pi_{unq}) - \lambda\beta)(\Pi_q - \Pi_{unq})\right] \quad (36)$$

The comparison between Eqs. (36) and (22) informs about the choice between the two alternatives, selection and disciplining. Hence, the council chooses the selection alternatives if

$$\beta < \overline{\beta_2} = \frac{\Pi_q - \Pi_{unq}}{2 - \lambda}\left(\mu(1 - \lambda) + \sqrt{2(1 - \lambda)(2 - \lambda) + \mu}\right) \quad (37)$$

We start by determining the impact of the parameters $\beta$, $f$, $\mu$, and $\lambda$ on the decision of choosing between the two alternatives: selection and disciplining.

- A rise in $\beta$ expands the parameters interval for which the council selects the disciplining alternative for both manager types instead of only the discipline of a qualified manager (inequality 37).

- The high value of f reduces the disciplining alternative attractiveness. Thus, a rise in $f$ value impacts positively $\overline{\beta_2}$.

$$\frac{d\overline{\beta_2}}{df} = \frac{d\left(\Pi_q - \Pi_{unq}\right)}{df} * \frac{(\mu(1 - \lambda) + \sqrt{\mu(2(1 - \lambda)(2 - \lambda) + \mu)}}{2 - \lambda} \tag{38}$$

where

$$\frac{d\left(\Pi_q - \Pi_{unq}\right)}{df} = 2(a + \rho + f) > 0 \tag{39}$$

- The high value of $\mu$ parameter reduces the disciplining alternative attractiveness. A rise in $\mu$ value in ρ impacts positively $\overline{\beta_2}$.

$$\frac{d\overline{\beta_2}}{d\mu} = \frac{\left(\Pi_q - \Pi_{unq}\right)}{2 - \lambda}\left((1 - \lambda) + \frac{(\mu(1 - \lambda)(2 - \lambda)}{\sqrt{\mu(2(1 - \lambda)(2 - \lambda) + \mu)}}\right) > 0 \tag{40}$$

- The high value of is λ increases the disciplining alternative attractiveness. A rise in $\mu$ value in ρ impacts negatively $\overline{\beta_2}$.

$$\frac{d\overline{\beta_2}}{d\lambda} = \frac{\left(\Pi_q - \Pi_{unq}\right)}{2 - \lambda}\left(-\mu + \frac{-\mu(2 - \lambda - \mu)}{\sqrt{\mu(2(1 - \lambda)(2 - \lambda) + \mu)}}\right) < 0 \tag{41}$$

# References

1. Rouen E (2020) Rethinking measurement of pay disparity and its relation to firm performance. Account Rev 95:343–378. https://doi.org/10.2308/ACCR-52440
2. Campbell DE (2018) Incentives: motivation and the economics of information. Cambridge University Press
3. Ruccia N (2021) The single resolution board: salient features, peculiarities and paradoxes. In: The role of EU agencies in the Eurozone and migration crisis. Springer, pp 103–125
4. Hermalin B, Weisbach MS (2003) Boards of directors as an endogenously determined institution: a survey of the economic literature. Econo Policy Review 9:7–26. https://doi.org/10.3386/w8161
5. Grace A, Frazer L, Weaven S et al (2020) Franchisee advisory councils and justice: franchisees finding their voice. J Strateg Mark 30(2):1–20. https://doi.org/10.1080/0965254X.2020.1740767
6. Hermalin BE (2005) Trends in corporate governance. J Finance 60(5):2351–2384. https://doi.org/10.1111/j.1540-6261.2005.00801.x
7. Graziano C, Luporini A (2003) Board efficiency and internal corporate control mechanisms. J Econo Manage Strategy 12(4):495–530. https://doi.org/10.1111/j.1430-9134.2003.00495.x
8. Drago F, Galbiati R, Sobbrio F (2020) The political cost of being soft on crime: evidence from a natural experiment. J Eur Econ Assoc 18:3305–3336. https://doi.org/10.1093/JEEA/JVZ063
9. Ferejohn J (1986) Incumbent performance and electoral control. Public Choice 50(1/3):5–25. http://www.jstor.org/stable/30024650

10. Barro RJ (1973) The control of politicians: an economic model. Public Choice, 14:19–42. http://www.jstor.org/stable/30022701
11. Ayllón S, Nollenberger N (2021) The unequal opportunity for skills acquisition during the great recession in Europe. Rev Income Wealth 67:289–316. https://doi.org/10.1111/ROIW.12472
12. Yahaya A, Mahat F, Abdulkadir J (2020) Effect of corporate governance practice and bank regulatory capital on performance: evidence from deposit money banks in Nigeria. Int J Soc Polit Econ Res 7:838–862. https://doi.org/10.46291/IJOSPERvol7iss4pp838-862
13. Miglietta A, Peirone D (2021) Finance, innovation and the value of the firm. Int J Bus Econ 8:31–45
14. Athamena B, Houhamdi Z, El Refae GA (2020) Managing moral hazard impact in decision making process. In: Proceedings—2020 21st International Arab conference on information technology, ACIT 2020. Institute of Electrical and Electronics Engineers Inc., pp 1–5

# Cloud ERP Systems and Firm Performance

Mua'th J. Hamad and Mohammed M. Yassin

**Abstract** Cloud ERP systems provide solutions to all difficulties and challenges which traditional ERP systems may encounter, in addition to providing more benefits. This study aimed at providing an evidence on whether cloud ERP adoptions improved the financial performance of companies. A range of financial indicators were used to test the effect of implementing cloud ERP over a window of three years (One year before and two years after). The results showed that the return ratios had increased and the cost ratios had decreased after the cloud ERP implementation.

**Keywords** Cloud ERP · Financial performance · Amman stock exchange · Jordan

## 1 Introduction

ERP systems were qualified as "the most important development in the corporate use of information technology (IT) in the 1990s" [3]. ERP is a technological tool used to manage supply chain processes in firms [1]. ERP systems are strong business packages that facilitate complex functions, integrate departments and manage resources. Business firms experience higher profitability, superior management and an organization wide view of the business in a single platform for the business operations [2]. This technology has radically changed organizational computing through simplifying the integrated planning, production, and customer responses. ERP systems provide the means for managing and controlling data, information, and materials [12].

"Accompanied with the emergence of cloud computing technologies in the late 2000s, there is an increasing trend for companies to migrate their hitherto internal ERP applications and databases into the cloud" [17]. Despite the high implementation costs for ERP systems, many firms are enhanced to use them via cloud environment [2]. They will be committed to pay annual subscription only instead of investing in expensive IT infrastructure. This encouraged firms to use Cloud-based Enterprise Resource Planning Systems in large businesses and more recently, in medium-sized

M. J. Hamad · M. M. Yassin (✉)
Al-Zaytoonah University of Jordan, Amman, Jordan
e-mail: Mohammed.yassin@zuj.edu.jo

© The Author(s), under exclusive license to Springer Nature Switzerland AG 2022
S. G. Yaseen (ed.), *Digital Economy, Business Analytics, and Big Data Analytics Applications*, Studies in Computational Intelligence 1010,
https://doi.org/10.1007/978-3-031-05258-3_15

ones. Cloud ERPs help firms to reach greater levels of sustainable performance [6]. While some adopters have accomplished considerable competencies via cloud ERPs, others faced unsuccessful implementation, budget surpasses, and frustrating performance.

This paper is motivated by the benefits that are driven from using cloud ERP systems, especially the financial benefits that are generated from using such technology. Different results have been reached by prior studies performed in this area on traditional ERP systems. Hunton et al. [9] argued that the financial performance of ERP adopters did not change after the ERP implementation. Also, Poston and Grabski [19, 18] reported very few differences in a number of financial performance measures between firms that adopted ERP systems and those which did not. However, Nicolaou [14] reported that adopters need a lag of 2 years minimum before they would start to achieve additional benefits comparing to the non-adopters. This result may explain what Poston and Grabski [19, 18] achieved. In their study, Hult et al. [8] asserted that combining the several different modules of ERP system have stronger effect on performance than simply the direct effect of each module solely.

ERP systems are designed to increase firms' performance through the improved business processes, enhanced information quality and decision-making [16]. Accordingly, cloud ERP systems are categorized as an innovative technology due to their prospective benefits to the business organizations. Based on that, we expect to find significant positive effect of cloud ERP adoption on the firm performance. Cloud ERPs are expected to boost firms' performance and market value.

Despite the benefits of ERP systems to the business organization, implementing ERP systems is an expensive and complex process. The scale of the projects exacerbates traditional technological troubles like late delivery or cost overrun. Also, cloud ERP implementation may cause major disruptions to the operations of the adopters, which can threaten their financial viability. Based on that, our paper comes to examine the effect of cloud ERP systems implementation on the long-term financial performance of companies that have previously adopted a cloud ERP system in the industrial and service sectors' companies in Jordan.

Due to the importance of SMEs in the local and global economy, many ERP providers, like SAP, Oracle, NetSuite and Microsoft, have begun to launch cloud-based ERP systems in the markets. A number of researchers have attempted to study the factors that encourage firms to adopt cloud ERP systems. Prior studies have identified cost reductions, ease of use, and the ability to concentrate on the business core activities, as the major factors that affect the adoption of cloud ERPs [20].

Johansson et al. [10] found that a hybrid solution between traditional and cloud ERP can allow organizations, especially large ones, to settle many concerns while at the same time enable them to gain some benefits of cloud computing. Navaneethakrishnan [13] argued that cloud ERP systems provide solutions to all the difficulties encountered by traditional ERP systems.

Based on that, the study tries to answer the following question: Does the financial performance of cloud ERP adopters improve after the implementation of cloud ERP system? The main objective of this study is to provide an empirical evidence on whether cloud ERP adoptions improved the financial performance of industrial and service sectors' companies that have adopted one of the well-known cloud ERP systems. This objective will be achieved through comparing a number of financial ratios for cloud ERP adopters before and after the cloud ERP implementation.

The paper consists of four parts. First, it defines the cloud ERP. Second, it reviews the related literature and develops hypotheses. Third, the methodology is presented. Fourth, the findings are discussed. Finally, the paper concludes.

## 2  Literature Review and Hypothesis Development

A framework that links cloud ERP systems adoption with financial performance could be the agency theory for Jensen and Meckling in 1976 [22, 23]. This theory shows that the costs arise from the conflict of interests between shareholders (principal) and managers (agent). Agency costs are defined as the costs that result from the discrepancies between the principals' and the agents' objectives. Along with the costs of developing a suitable incentive contract, these costs include monitoring the work efforts of the agent, the agent's non-value-added tasks of documenting and reporting activities, and the welfare loss that can result from any miscommunications with the agent and inefficiencies.

In regards to ERP systems, monitoring costs should be minimized by implementing the software through automating processes steps and setting up an electronic trace of employees' responsibilities [7]. By giving managers an inclusive access to one database, they become able to review their employees' actions in an efficient and effective way and on timely basis. In turn, this will reduce the need for the extra monitoring tasks, lower human errors and defects, and so the dispense of the investigation and rework employees.

Many studies on ERP systems have taken a place through the last decade. Elragal et al. [4] performed a case study in which he investigated the influence of the ERP system on the business performance and found out that several operational and financial advantages have been achieved by adapting an ERP system. Galani et al. [5] studied a sample of firms that adopt ERPs and focused on factors like; cost cut, improved flow of information, connections between customers and suppliers, and the response time. They concluded that ERPs adopters have achieved extra advantages than non-adopters and cost reduction was the most significant benefit. They indicated that 40% of the companies that adopted an ERP system have minimized their production cost and enhanced their productivity. A case study on ERP systems implementation took place in Indian SME [11], ERP vendors were interviewed and identified some issues to SME that must be considered before implementing an ERP system including; cost and limited funds, awareness and perception, change management and the implementation approach. In addition to other factors like, top management support infrastructure resources, human resources and the required education and training about ERPs. Another case study by Tsai et al. [21] used a questionnaire and ANOVA analysis to examine ERPs performance and process problems. Their conclusion was that companies can achieve a better ERP system performance if they redesign their business process where business processes and systems are related to each other and must be consistent.

Poston and Grabski [18] used a sample of 50 ERP-adopting firms to examine their post-implementation performance over a 3-years period after controlling for the pre-implementation performance. The results did not show an increase in the residual income neither in the operating expenses to revenues ratio. But there was a significant decrease in the employees to revenues ratio, as well as in the cost of goods sold to revenues ratio for the third year. Overall, they indicated that ERP adopting firms achieved some gains due to the increased efficiencies in some areas, however, these gains were offsetted by the increased costs in other areas.

Hunton et al. [9] used a matched pair design study to examine the longitudinal influence of adopting ERP systems on firms' performance. The study compared the financial performance of 63 ERP adopting firms with the financial performance of another 63 non- adopting firms. Overall, research results indicated that ERP adopters attained some significantly higher ratios of ROA, ATO, and ROI than non-adopters for the third year subsequent to the implementation of ERPs. Also, the 3-years average ROI and ROA were significantly higher for the adopting firms. Sub-analysis of the performance metrics from pre- to post-adoption showed that it did not significantly change for adopters, but it declined for the non-adopters over the same time window. ALSO, the study examined the interactive impact of the ERP adopter's financial health and size on its performance and found a significant interaction between health and size with the financial measures (ROS, ROA, and ROI). unhealthy / Large adopters had a better ROI than healthy / large adopters. However, healthy / Small adopters had a better performance in terms of the ROA, ROI, and ROS measures than unhealthy / small adopters. The study also used the ROS and ATO ratios to examine the profitability and the efficiency respectively. It indicated that the ATO ratio has significantly declined for non-ERP adopting firms, as well as the ROS, but the difference was not significant. Since ROS and ATO are the two components of the ROA; the study indicated that the improvement in the ERP adopters' performance is a result of the enhanced both profitability and efficiency. A comparison between ERP adopters and non-adopters suggested that financial gains attained from adopting ERPs result in lower prices to customers; therefore, the non-adopters' performance declines by comparison.

A study by Nicolaou and Bhattacharya [15] empirically examined the level of changes in ERP systems after implementation over a time-period affects firms' long-run financial performance. It also examined whether the ERP performance outcomes are affected by the system transformation timing and nature during the post-implementation period. Results indicated that, generally, subsequent modifications in the ERP system may solve or uncover the issues of implementation that affect the success and use of these systems. Specifically, firms which made early add-ons or upgrades to their ERP systems had significant changes in financial performance compared to other ERP-adopters. ERP adopting firms would suffer apparent differential performance deterioration in case of late improvements and both early and late abandonment.

Based on the previous discussion, a general hypothesis could be formulated as follows:

*The financial performance of cloud ERP adopters after the implementation is better than before implementation.*

## 3 Methodology

This section consists sample description, data collection, and measurement procedures.

### 3.1 *Population and Sampling*

Our initial sample contains all public shareholding companies in the services and industrial sectors listed on Amman Stock Exchange (ASE) and have implemented one of the known Cloud ERP systems including SAP, PeopleSoft, Oracle, Baan, and J.D. Edwards.

ERP adopting companies that went through exceptional changes during the time window of this study (i.e., made major acquisitions, unlisted, etc.) were excluded from the sample. The reason is that the firm's differences in performance ratios could potentially be driven by these changes rather than by the ERP system, or the impacts of the ERP system were affected by such changes.

The final sample consisted of 20 firms; 9 in the industrial sector, and 11 in the services sector.

### 3.2 *Data Collection*

Required financial data are collected from the financial statements for the three years study period that will be obtained from the web site of ASE. Ratios will be calculated for each firm in each year.

### 3.3 *Measurement*

This study examines the changes in firm performance from one year before to two years after the cloud ERP implementation. To test the main hypothesis, the following ratios will be calculated before and after the implementation of cloud ERP:

ROI, ROA, ROS, ATO, COGS/Revenues and SG&A/Revenues.

where;

ROI is the return on investment which is defined as the income before extraordinary items, divided by the sum of common equity, preferred stock, minority interest, and long-term debt.

ROA is the return on assets which is defined as the income before extraordinary items, divided by the average total assets.

ROS is the return on sales which is defined as the income before extraordinary items, divided by the net sales.

ATO is the asset turnover which is defined as the net sales, divided by the average of the total assets.

COGS is the cost of goods sold.

SG&A is the selling, general, and administrative expenses.

Performance was measured in two different periods of time; before and after the ERP adoption. The year that the cloud ERP was adopted, identified as year zero, served as the control year for comparing performance before and after Cloud ERP implementation. The pre-adoption period covered one year (t − 1), and the post-adoption period covered two years (t + 1 to t + 2).

To test for significant changes in the performance ratios, a t- test will be performed to compare the performance ratios before and after ERP implementation.

## 4   Results

The financial performance ratios for the study sample was indicated for one year before cloud ERP implementation and two years after it. The descriptive statistics of these ratios are as shown in Table 1.

Table 1 indicates that financial ratios for the year before the cloud ERP implementation have different amounts than the ratios after the implementation. The differences

**Table 1**  Descriptive statistics

| Financial ratio | Before cloud ERP | | | | After cloud ERP | | | |
|---|---|---|---|---|---|---|---|---|
| | Mean | Std. dev. | Max. | Min. | Mean | Std. dev. | Max. | Min. |
| ROI | 0.20 | 0.14 | 0.52 | 0.03 | 0.34 | 0.14 | 0.72 | 0.11 |
| ROA | 0.36 | 0.12 | 0.56 | 0.15 | 0.55 | 0.21 | 0.96 | 0.22 |
| ROS | 0.32 | 0.18 | 0.64 | 0.01 | 0.49 | 0.24 | 0.89 | 0.02 |
| ATO | 30.26 | 16.05 | 69.98 | 4.43 | 44.76 | 23.54 | 87.74 | 8.21 |
| COGS/Rev | 0.49 | 0.13 | 0.89 | 0.34 | 0.28 | 0.13 | 0.55 | 0.09 |
| SGA/Rev | 0.17 | 0.16 | 0.73 | 0.00 | 0.15 | 0.11 | 0.48 | 0.02 |

**Table 2** Independent samples t-test

| Financial ratio | Mean before cloud ERP | Mean after cloud ERP | Mean difference | $t$-value | Sig. |
|---|---|---|---|---|---|
| ROI | 0.20 | 0.34 | 0.14 | 3.718** | 0.001 |
| ROA | 0.36 | 0.55 | 0.19 | 4.432** | 0.000 |
| ROS | 0.32 | 0.49 | 0.17 | 3.012** | 0.004 |
| ATO | 30.26 | 44.76 | 14.50 | 2.805** | 0.007 |
| COGS/Rev | 0.49 | 0.28 | −0.21 | −5.841** | 0.000 |
| SGA/Rev | 0.17 | 0.15 | −0.02 | −0.617 | 0.543 |

** Significant at ($p \leq 0.01$)

where logical; the return ratios such as ROI, ATO, ROA, and ROS were increased after the ERP was implemented, while the cost ratios such as COGS/Rev and SGA/Rev were decreased.

To test the hypothesis of the study, which indicates that the financial performance of cloud ERP adopters after the implementation is better than before implementation, the independent samples t-test was performed for the study sample before and after the cloud ERP implementation. Table 2 showed that ROI, ROA, ROS, ATO and COGS/Rev are statistically significantly different (at $p \leq 0.01$) before and after implementing cloud ERP, while SGA/Rev is not significantly different (at $p \leq 0.01$). It is evident that the average return ratios are greater after the implementation than before. These results agreed with Elragal et al. [4], Nicolaou and Bhattacharya [15], Hunton et al. [9], Poston and Grabski [18].

Also, it is found that the average cost ratios are lower after the implementation than before, although the SGA/Rev is not significantly different. This result agreed with Gurbaxani and Wang [7], Galani et al. [5].

# 5 Conclusion

Although it is still new in Jordan, cloud ERP implementation was intensively researched around the world. This study builds on prior literature in this area and aimed at providing an evidence on whether cloud ERP adoptions improved the financial performance of companies that have adopted the cloud ERP systems.

A sample of 20 companies in the industrial and service sector in ASE, which adopted the cloud technology in implementing ERP, were tested in this study. The results found that the return-on-investment (ROI), the return-on-assets (ROA), the return-on-sales (ROS) and the assets-turnover (ATO) were higher for companies after than before the cloud ERP implementation. Also, it is evident that the ratio of cost-of-goods-sold-to-Revenues (COGS/Rev) was lower for companies after than before the cloud ERP implementation. That is, the return ratios increased, and the cost ratios decreased after implementing cloud ERP systems.

Overall, we can use this empirical evidence to recommend decision makers in companies to convert and highly utilize the cloud-based systems to maximize their benefits and minimize their costs. However, we recommend that future research and adopters may add to the results of this study by investigating the critical success factors of Cloud ERPs and their effects on the financial performance.

# References

1. Acar MF, Tarim M, Zaim H, Zaim S, Delen D (2017) Knowledge management and ERP: Complementary or contradictory? Int J Inf Manage 37(6):703–712
2. Chaudhari K (2020) The role of ERP in digital transformation. ERP NEWS. https://erpnews.com/the-role-of-erp-in-digital-transformation. Accessed on 17 Jan 2020
3. Davenport T (1998) Putting the enterprise into the enterprise system. Harvard Bus Rev (July–August)
4. Elragal A, Al-Serafi A (2011) The effect of ERP system implementation on business performance: an exploratory case-study. Commun IBIMA 2011(Article ID 670212)
5. Galani D, Gravas E, Stavropoulos A (2010) The impact of ERP systems on accounting process. World Acad Sci Eng Technol 66:418–423
6. Gupta S, Meissonier R, Drave V, Roubaud D (2019) Examining the impact of cloud ERP on sustainable performance: a dynamic capability view. Int J Inf Manage 51(c). https://doi.org/10.1016/j.ijinfomgt.2019.10.013
7. Gurbaxani V, Whang S (1991) The impact of information systems on organizations and markets. Commun ACM 34(1). https://doi.org/10.1145/99977.99990
8. Hult G, Ketchen D, Adams G, Mena J (2008) Supply chain orientation and balanced scorecard performance. J Manag Issues 20(4):526–544
9. Hunton J, Lippincottb B, Reck J (2003) Enterprise resource planning systems: comparing firm performance for adopters and nonadopters. Int J Account Inf Syst 4:165–184
10. Johansson B, Alajbegovic A, Alexopoulo V, Desalermos A (2015) Cloud ERP adoption opportunities and concerns: the role of organizational size. In: 48th Hawaii international conference on system sciences. IEEE, USA, https://doi.org/10.1109/HICSS.2015.504
11. Kale P, Banwait S, Laroiya S (2008) Enterprise resource planning implementation in Indian SMEs: issues and challenges. In: 12th annual international conference of society of operation management, pp 242–248
12. Migdadi M, Abu Zaid M (2016) An empirical investigation of knowledge management competence for enterprise resource planning systems success: insights from Jordan. Int J Prod Res 54(18):5480–5498. https://doi.org/10.1080/00207543.2016.1161254
13. Navaneethakrishnan C (2013) A comparative study of cloud based ERP systems with traditional ERP and analysis of cloud ERP implementation. Int J Eng Comput Sci 2(9):2866–2869
14. Nicolaou A (2004) Firm performance effects in relation to the implementation and use of enterprise resource planning systems. J Inf Syst 18(2):79–105. https://doi.org/10.2308/jis.2004.18.2.79
15. Nicolaou A, Bhattacharya S (2006) Organizational performance effects of ERP systems usage: the impact of post-implementation changes. Int J Account Inf Syst 7:18–35
16. O'Leary D (2000) Enterprise resource planning systems: systems, life cycle, electronic commerce, and risk. Cambridge University Press. https://doi.org/10.1017/CBO9780511805936
17. Peng G, Gala C (2014) Cloud ERP: a new dilemma to modern organisations? J Comput Inf Syst 54(4):22–30. https://doi.org/10.1080/08874417.2014.11645719
18. Poston R, Grabski S (2001) Financial impacts of enterprise resource planning implementations. Int J Account Inf Syst 2:271–294. https://doi.org/10.1016/S1467-0895(01)00024-0

19. Poston R, Grabski S (2000) The impact of enterprise resource planning systems on firm performance. In: Proceedings of the 21st international conference on information systems. Available from https://www.researchgate.net/publication/254579610_Impacts_of_Enterprise_Resource_Planning_Systems_Selection_and_Implementation
20. Salim S (2013) Cloud ERP adoption—A process view approach. PACIS 2013 proceedings. Paper 281. http://aisel.aisnet.org/pacis2013/281
21. Tsai W, Chen S, Hwang E, Hsu J (2010) A study of the impact of business process on the ERP system effectiveness. Int J Bus Manage 5(9)
22. Yassin M (2017) The determinants of internet financial reporting in Jordan: financial versus corporate governance. Int J Bus Inf Syst 25(4):526–556
23. Yassin M, Al-Khatib E (2019) Internet financial reporting and expected stock return. J Acc Finance Manage Strategy 14(1):1–28

# Analysis of Business Challenges and Opportunities Within the Era of COVID: Strategies for Sustainable Business

**Henry Karyamsetty and Hesham Magd**

**Abstract** The coronavirus pandemic has caught the world by surprise causing huge impact on the global economy leading to decline in GDP by over 1% approximately in 2020. There were huge implications in almost every business sector devastating mainly manufacturing, tourism, travel, service and trade of goods and commodities among nations. The pandemic has also caused a huge blow on global mobility restricting travel and movement. The study principally highlights the significant challenges and opportunities that can be exploited by business and proposes conceptual model to support establishments in realizing the benefits of sustainable business. Organizations has to potential utilize the role played by HR, supply chain in modulating the workflow and effectively engaging the workforce to efficiently deliver the outputs, by exploring the opportunities each organization can provide to business continuity. The success formula should always lay on the pillars of sustainable development emphasizing on economic, social and environment aspects in every stage of business operation.

**Keywords** Strategy · Business · Sustainable · Economy. Impact · Trade

## 1 Introduction

The entire world has sailed through a year and half with COVID-19 pandemic ever since the virus (SARS-CoV-2) was first detected and spilled out from China. Before every nation could anticipate the consequence, the disease has spread to every part of the world forcing the WHO to state COVID-19 as a global crisis and declare it as a pandemic in March 2020 [1]. Notwithstanding, COVID-19 has spread from epidemic level to pandemic situation in short period of time mainly due the airborne

H. Karyamsetty · H. Magd (✉)
Faculty of Business and Economics, Modern College of Business and Science, Muscat, Oman
e-mail: Hesham.Magd@mcbs.edu.om

H. Karyamsetty
e-mail: Henry.Karyamsetty@mcbs.edu.om

S. G. Yaseen (ed.), *Digital Economy, Business Analytics, and Big Data Analytics Applications*, Studies in Computational Intelligence 1010,
https://doi.org/10.1007/978-3-031-05258-3_16

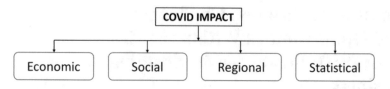

**Fig. 1** Categorizing the impacts, the global crisis has caused across all organizations

nature of the virus that gets transmitted easily and quickly through different media. As a result, the pandemic has caused huge loss effecting the $ 90 trillion global economy besides causing 4.0 million deaths globally reported by July 2021, and during the year 2020 the global trade declined by 5.3% from the pandemic situation [2]. Unknowingly, the employment rate was the lowest noted since the World War II where the global working hours have dropped by 10.5% in the 2nd quarter of 2020 equivalent to 305 million full time workers [3]. Further, studies indicate that more than 95 million people have suffered globally from the pandemic due to unemployment and poverty causing a huge global economic crisis ever known in the history. In addition, global economic forecasts predict a loss of $8.5 trillion in next few years due the prevailing pandemic condition [4]. Among them, tourism and travel are the most effected sectors globally due the ban on international travel and from closure of tourism locations which caused revenue losses and unemployment in associated sectors. Not to mention, the impact was also seriously felt on every industrial sector including educational institutions globally indicating 1.6 billion learners in 192 countries which represent 90% of global student population were affected from closure academic institutions, as well as on low-income generating countries and less developed nations are most effected [3, 5] (Fig. 1). The global economic impact shows global aviation industry has suffered a revenue loss of USD 371 bn in 2020 alone, from 60% reduction in international air passenger traffic causing − 66.3% reduction in airport revenues amount to USD 125 bn, bringing down the global economy (GDP) by −3.3% according to [6].

## 1.1  Background Research Problem

Currently almost everyone is desperate to come out of the present crisis due to COVID- 9 which has sunk the global economy to an unprecedented level. With all things considered the impact of the COVID-19 pandemic on business and economy were studied from the standpoint of many countries with reference to small and medium enterprises, larger firms across various sectors. Benjamin et al. [7] reported occurrence of high levels of anxiety, depression, and stress among health care workers due the pandemic outbreak as well as [8] confided prevalence of fear, causing disturbance in mental health on frontline health care workers. Also, the pandemic has affected the GCC construction industry by causing delays in project completion due to restriction of workforce and closure of borders and strict enforcement of laws [9] besides project suspensions, budget delays, labor shortage, job loss, time and cost overrun [10]. Even so the entire world is still captivated in such an infinite uncertainty, the pandemic occurrence has given scope to unfold the business opportunities to explore in every field amidst those insecurities and learn how organizations have moved forward in utilizing the opportunities strategically to retain sustainability in their operations especially during the post COVID-19 phase. Therefore, the objective of this research article is to highlight the challenges surrounding the business establishments and present the various measures taken by different business organizations globally and the strategic opportunities explored by different business establishments in overcoming the challenges in wake of the pandemic.

## 2 Research Methodology

### 2.1 Background Analysis

The present study draws mainly from reviewing the literature and published work on the impact the COVID-19 pandemic has caused on business establishments. In continuation, the authors intended to analyze the impact by studying the various implications and opportunities the business organizations can exploit during the crisis.

### 2.2 Sources

Details pertaining to the aspect of study are taken prominently from secondary sources of information published by various researchers and analyzing the findings interpreted in reference to specific context of industries. Research studies done by various researchers and available literature on the topic were extensively reviewed to critically evaluate the impact of COVID that is experienced by different industrial and business sectors globally. In addition, a thorough examination was also done by referring to the information and documents published by international organizations in websites and digital media. Further, observations disclosed by various business analysts on the impact of the pandemic are also taken into consideration concluding the study.

### 2.3 Analysis and Interpretation

The study predominantly follows content analysis approach which allows information to be analyzed systematically enabling the researcher to make conclusions with relation to the purpose of study [11]. By examining the challenges and opportunities different business organizations have reported the various challenges and opportunities were analyzed and specific challenges are identified to be considered as variables to evaluate the commonality and the impact of them in different organizations. Also, the most significant opportunities that can be exploited and adopted to suit every business establishment are examined. Following which the authors have come up in developing a conceptual model considering the challenges the pandemic has imposed and the opportunities arising. The proposed model acts as a pathway for any typical business establishment to adopt to comprehend the benefits of achieving sustainable business during the post COVID phase.

# 3 Literature Review

## 3.1 COVID-19 Pandemic Challenges on Global Business Establishments

The impact of the COVID-19 pandemic was felt in every sort of business such as tourism and hospitality, oil & gas industry, aviation, automotive, consumer products, manufacturing, and engineering etc. however the intensity of the affect varies, but the most severe effect was on travel and tourism industry largely and on small and medium enterprises (SMEs) [12]. SMEs are principally the largest employment providers to millions of people worldwide and employ over 50% of US workers, and the closure of many small businesses in the US has caused 32.7 million people jobless [13]. One of the leading pressing problems that brought shock to every nation was unemployment which resulting in pushing around 115 million people into poverty [14] (Fig. 2).

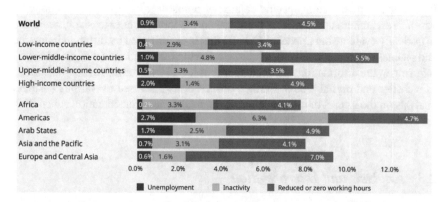

**Fig. 2** Working-hour losses into changes in unemployment, inactivity and reduced working hours, world and by income group and region, 2020. *Source* ILO [14]

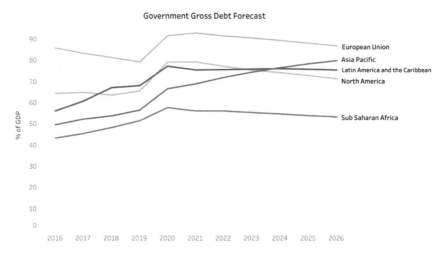

**Fig. 3** Gross debt suffered by different nations during the post COVID-19 phase. *Source* IMF, World Economic Outlook April

Subsequently due the increasing poverty across many countries has caused mounting debts amounting to USD 12 trillion representing 12% of global GDP, while the highest debt was experienced in EU countries with 90% of GDP in 2020 alone according to reports by IMF 2020 world economic outlook (Fig. 3). Further [15] also reported significant decline in active business owners which fell by 22% during the peak periods of COVID-19 in the US. Not only, but the effect on business establishments is also widely studied from the standpoint of different countries in each sector.

In the service and hospitality industry, the key problems lie on recruitment of staff and managing shortage of workforce and in the manufacturing sector, the challenges are mostly delays in raw material supply and decline in demand as the top key challenges faced by companies in the mainland China are decline in sales, commuting restrictions to work and insufficient liquidity. While reports by Brucal et al. [16] from the World Bank Group (WBG) shows the COVID-19 pandemic biggest blow was on South Asian firms as 34% of the firms are temporarily closed and sales declined up to 64% from 2019 records in comparison to other developing countries (Fig. 4).

Similarly, studies from [17] reveals that COVID-19 has shown potential impact on the operations and performance of SMEs in Chinese firms, in addition to these firms suffered from capital rotation, work delays, decline in market demand etc. [18]. Notwithstanding, the pandemic has also exposed entrepreneurs and startup business

**Fig. 4** Percentage of firms temporarily closed due to COVID-19

to all sorts of business risk as many startups business gradually closed during the crisis [19, 20] while the establishment of new startup firms declined by 25% during 2020 in France [21] (Fig. 5).

Moreover, as the intensity of global crisis from COVID-19 is escalating day by day, businesses across the world will be affected harder more than the level experienced during the early 2020 which obviously puts more risks on many lucrative businesses, driving them towards partially closure or reduced operation which caused the global trade to cease by 9.5% during 2020 in the first wave [2] (Fig. 6).

In summary, regardless of any business organization, the post COVID-19 has caused political, technological, and societal risks to the fore, where all of them

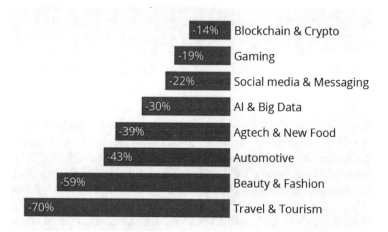

**Fig. 5** Effect of COVID on the revenues of some startups globally during December 2019 to June 2020 (Startup Genome: Statista 2020)

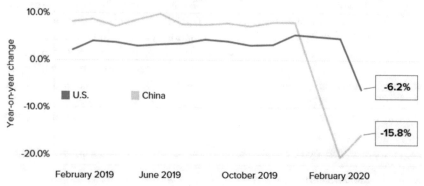

**Fig. 6** Sales decline seen in two largest world economies during COVID-19 (U. S. Census Bureau. National Bureau of Statistics of China)

are important to lead a resilient global economy [22]. Comparably, the crisis has also posed challenges to human resources (see Table 1) sector indicating every organization vulnerable to the effect should focus on taking care of the mental and physical well-being of their employees, mechanism to adopt to the changes in work pattern, communication among employees, creating a work environment at home as means to protect livelihoods [23]. Contemplating the promotion of remote working, the concept is not feasible to implement in certain sectors like manufacturing, construction, retail, transportation, and utilities where workers cannot work from home remotely, while according to survey by bureau of labor statistics only 29% of American workers could work from home [12].

Apparently, the pandemic has affected all the modes of transportation, for instance in the GCC countries there is a steep decline in the aviation sector and tourism industry including SMEs and educational institutions across the Saudi Arabia [24].

**Table 1** Business implications quoted by organizations globally during the post COVID-19 phase

| Business entity | Challenges for workforce | |
|---|---|---|
| | HR | Supply chain |
| PWC Malta | Establish crisis team | Effective communication |
| | Employee safety | Split team arrangement |
| | Contingency plans for critical roles | |
| | Flexibility work environment | |
| Acca global—China | Insufficient liquidity | Sales decline |
| | Employee commuting | Low production |
| Randstad—Canada | Employee safety | Competitive market |
| | Effective use of technology | |
| | Coping with change and uncertainty | |
| Anonymous | Money management | Client relations |
| | Work from home | Digitalization |
| | Following government regulations | |
| Rogers | Ensuring the mental and physical wellbeing of employees | |
| | Maintaining employee engagement, productivity and effectiveness | |
| | Creating an eco-system around work from home | |
| | Adapting to the new normal: facilitating change and shaping culture | |
| | Protecting employment and preserving livelihoods | |
| EFSE | Liquidity problems | Revenue decline |
| | Employment | |
| | Survival | |

In Dubai, the pandemic had caused 76% disruption in cash flow, 37% reduction supply chain demand and 41% of work premises closed indicating an overall 27% high risk in establishments going out of business [25]. Road transportation revenues from toll road tax have drastically fallen in many countries. Likewise, the pandemic has also caused obstacles to the economic growth of Oman which caused disruption mainly to the hydrocarbon industry impeding the GDP of 44% in 2012 to 30.1% in 2019 [26]. The major share of potential impact because of the crisis comes from oil and gas sector (36%) followed by real estate and construction (11%) manufacturing (10%) and retail sector of (7%) furthermore the total public expenditure reduced by 8.4% by June 2020 [27]. The downfall of Oman's GDP can also be attributed from the decline in revenue generation from tourism sector causing a total loss of $145.2 million during the crisis period over the year 2019 [28]. Not only that, but this also impacted the air travel industry and associated business such as hotels, restaurants, local private transportation etc. which recorded 78.4% decline that eventually led to unemployment, reduce income etc. in the region. Besides, global reports show other prominent sectors like manufacturing including the supply chain are also equally affected that led to slowing down of goods and commodities causing delays in meeting the demand [29] (for further illustration, see Table 1).

## 3.2 Strategic Opportunities Explored by Business Establishments for Sustainable Business

Amidst the unprecedented uncertainty of suffering from COVID-19 that has caused on global businesses more prominently on SMEs, the pandemic situation has opened gates for exploring new ways of sustaining businesses in organizations. Many firms have started to meticulously use the advancement in technology as a tool to promote their business operations through digitalization in marketing, sales etc. Also, business meets, and events have taken to digital media through virtual platforms and automation of systems are likely going to be augmented, employees are demanded to improve and acquire new skills etc. will be some of the long-term strategies industrial sectors are deemed to emphasize during post COVID-19. Studies report that digitalization has accelerated more than expected due to the pandemic, which can lend a helping hand to businesses in cost reduction and improving productivity. Despite the challenges, the global crisis has brought some significant opportunities in the cost savings on energy, low interest rates, cheaper public transport, and low rentals on housing etc. at the same time has increased business prospects by creating demand for personal and community hygiene products, medical equipment, enhanced logistic services, startups, telemedicine, telelaw services, e-commerce etc.

Alternatively, governments in different countries have stressed upon strategic planning approaches to all business sectors to improve performance and in turn overcome the uncertainties through laying emphasis on maintaining flexibility, technology, teamwork and communication and leadership competency [24]. However, in the corporate sector, COVID-19 has opened new perspectives to engage with partners and customers to attain business continuity, and every corporate industry have come up with their strategic plan, of which the most prominent ones every sector should improvise is on their ability to be remain prepared and being resilient, digitalization and prioritizing customer satisfaction. Nagem [30] highlighted the scope of attaining business success through implementing strategic cuts in organizations, sustaining innovation, customer intimation, streamlining processes and focus on lessons learnt. Likewise, many business analysts have proposed coping mechanisms for business establishments to maintain business continuity by practicing effective communication techniques with stakeholders and customers, optimization of resources and manpower and dealing competently with contracted parties.

Apart from the various strategies implemented by business establishments, there are some opportunities for human resources to explore by taking a key role in shaping the behaviors of employees for the new normal, promoting digitizing sales and marketing using technology, reshaping organization culture towards new changes, enhancing virtual learning and development, reworking on employee engagements etc. Another equally important thing to understand is until the world comes of the pandemic situation totally, businesses are going to encounter challenges, and hence HR should always remain perceptive to explore strategies and focus on opportunities that are feasible to maintain sustainable business. In such paradox, working remotely is going to be an integral part of work pattern for most of the organizations, which may require the human resources to prepare new working policies and check how such systems of working will maintain equal productivity and customer engagement. In contrast, organizations who have established policies for working remotely have fared well in increasing productivity, reducing space renting, travel costs in commuting and reducing the overall carbon footprint of employees and organization.

### 3.3 Conceptional Model

The coronavirus disease has inflicted very serious implications to all walks of life including both business and non-business sectors all over the world. As a result, the global economy has fallen paralyzing normal life, caused disruption in supply of commodities, led to delays in transportation of raw materials leading to disproportionate balance in supply and demand. In such an economic crisis looming largely on business activities, organizations must strategically plan measures to explore the opportunities for sustaining business. All the same time, organizations in advanced economies must evolve through the uncertainties by critically involving HR in preparing conceptual business models which are very important to sustain organization performance through integration of different approaches [31]. In principle, organizations must always strive to maintain consistency in business operations irrespective of the external pressures to traverse successfully through the pandemic crisis. For organizations to handle the post COVID situation successfully, the authors have

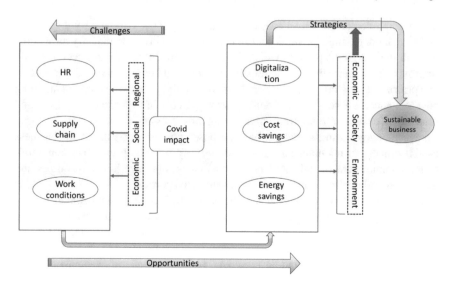

**Fig. 7** Conceptual model to achieve sustainable business through the post COVID-19 in organizations

proposed conceptual model (Fig. 7) that serves as framework to guide business establishments to sustain resilience. The intended model with help as pathway to determine the specific challenges and opportunities that can be explored in every business operation. The model will act a mechanism to drive any typical organization to focus on the key elements that are very essential to maintain sustainable business.

## 3.4 Interpretation and Discussion

The proposed model basically stands on the principle of sustainable development pillars integrated to organizational performance. Wide research studies are done in this perspective by various researchers with examples taken in large and small firms across the world. Many conceptual models were also proposed utilizing the concept of sustainable business performance in organizations and are successfully adopted by many leading firms globally [32]. Therefore, the conceptual model that is proposed in this study holds good significance in the context of challenges and opportunities that businesses are experiencing during post COVID, while it stands the same principle of sustainable economic development due the nature of success achieved through this concept. Obviously, the impact of the pandemic was widely felt in every organization causing economic, social, and regional (Fig. 1) implications to a wide scale and each of those impacts pose significant challenges to all business sectors principally holding human resources (HR), supply chain network and work conditions more vulnerable to the increasing crisis (Fig. 7). Arguably the pandemic has inevitably displaced the work conditions by shifting the working pattern of employees to remote or work from home nature, which brought significant challenges to economy, society both globally and regionally.

In the bottom line, the three entities are very vital to any organization to withstand fluctuations and for the business to remain afloat, opportunities that the three entities will offer have to be explored critically. However, the pandemic has deliberately opened opportunities to business giving way to digitalization of business operations and transactions which has led to reduction in cost bringing in some small savings, and changes in the work pattern has also led to energy savings in organizations that are intricately countering the challenges [33]. Indirectly the challenges that are currently surfacing all businesses have gradually led us to identify the nature of opportunities in the due course mainly when the workforce, digitalization, supply chain are given appropriate consideration [34, 35]. To lead business that are suffering with such challenges, the pathway recommends the organizations to exploit those opportunities that can bring a positive influence on the economy, society and environment concerned to the regional business and adopt strategies that will always integrate the sustainable pillars to achieve a successful sustainable business during the post COVID crisis.

## 4   Conclusion

The coronavirus disease has caused tremendous negative impact on the global economy, as the pandemic situation is still prevalent all over the globe, all walks of life are experiencing the agony of the crisis. The major cause of concern for every country is the decrease in international trade of goods and commodities, in addition to the impact felt on all categories of industries. Further, lower economic countries have suffered the most mainly from the closure of many SMEs and MSME leading to unemployment and loss of revenue. Due to the nature of the disease transmission, the possible ways to mitigate the impact has led many countries to impose restriction on movement, directing people to work from home, concept of working remotely became a new normal for most business establishments. Business analysts have exclusively studied broadly the nature of challenges business establishments are encountering post COVID, however the pandemic has also taught us to adapt and explore opportunities for sustaining business. From analyzing the various studies done by researchers and examining the literature, the conceptual model proposed by the authors will serve as framework to guide the business establishments to emerge as successful organisation during the post COVID-19 pandemic period.

## 4.1   Further Research

The conceptual study unquestionably opens demand to investigate further on the challenges and opportunities post COVID-19 in the context on nature, size, and geographical region of the organization. As the fact remains that not all organizations have felt similar impact, also the nature of challenges and opportunities arising vary depending on the influencing factors. Obviously, HR has a pivotal role to play in the entire game and studies should focus on how the management needs to tackle with the different partners involved in the business of an organization specially to identify the critical strategies that are required to run the business smoothly.

## 4.2 Limitations

The study in its present form has certain limitations that can be given due consideration in future research. Though the COVID-19 has gripped every nation in uncertainty, but the extent of impacts in terms of challenges that are faced by every business establishment are not well documented nor much research was carried out in this line. Further, there is also significant lack in availability of relevant literature on this topic that impedes business analysts, policy makers and planners in recommending strategic measures to establishments. Despite the implications the pandemic has caused, the opportunities the crisis can offer are not widely explored to suggest practical methods for sustaining businesses during the post COVID-19 situation.

## References

1. Chappell B (2020) COVID-19: COVID-19 is now officially a pandemic, WHO says. National Public Radio. https://www.npr.org/sections/goatsandsoda/2020/03/11/814474930/COVID-19-COVID-19-is-now-officiallya-pandemic-who-says. Accessed 15 July 2020.
2. Jackson JK, Weiss MA, Schwarzenberg AB et al (2021) Global economic effects of COVID-19. Congressional Research Service. https://crsreports.congress.gov. Accessed 1 June 2021
3. Committee for the Coordination of Statistical Analysis (2020) https://unstats.un.org/unsd/ccsa/. Accessed 25 May 2021
4. United Nations Department of Economic and Social affairs (2021) https://www.un.org/en/desa/covid-19-slash-global-economic-output-85-trillion-over-next-two-years. Accessed 20 May 2021
5. Kalogiannidis S (2020) Covid impact on small business. Int J Soc Sci Econ Invent 6:387–391
6. International Civil Aviation Organisation (2021). Economic impacts of COVID-19 on civil aviation. https://www.icao.int/sustainability/Pages/Economic-Impacts-of-COVID-19.aspx. Accessed 23 May 2021.
7. Benjamin YQ, Tan MD, Nicholas WS et al (2020) Psychological impact of covid-19 pandemic on health care workers in Singapore. Ann Intern Med 173:317–320
8. Ruiz MA, Gibson CAM (2020) Emotional impact of COVID-19 pandemic on U.S. Health Care workers: a gathering storm. Psychol Trauma Theory Res Pract Policy 12:S153–S155
9. Umar T (2022) The impact of COVID-19 on the GCC construction industry. Int J Service Sci Manage Eng Technol 13
10. Gamil Y, Alhagar A (2020) The impact of pandemic crisis on the survival of construction industry: a case of COVID-19. Mediterr J Soc Sci 11:122–128
11. Haggarty L (1996) What is content analysis? Med Teach 18:99–101

12. Surico P, Galeotti A (2020) The economics of pandemic: the case of COVID-19. Wheeler Institute of Business and Development, London Business School
13. Bartik AW, Bertrand M, Cullen Z et al (2020) The impact of COVID-19 on small business outcomes and expectations. Proc Natl Acad Sci USA 117:7656–17666
14. International Labour Organisation. ILO Monitor (2021) COVID-19 and the world of work. Seventh edition updated estimates and analysis. International Labour Organisation. Accessed 10 June 2021
15. Fairlie R (2020) The impact of COVID-19 on small business owners: evidence from the first 3 months after widespread social-distancing restrictions. J Econ Manag Strategy. 29:727–740
16. Brucal A, Grover A, Reyes S (2021) Damaged by disaster: the impact of covid-19 on firms in south Asia. Let's talk development. World Bank Blogs. https://blogs.worldbank.org/develo pmenttalk/damaged-disaster-impact-covid-19-firms-south-Asia. Accessed 5 May 2021
17. Sun T, Zhang WW, Dinca MS, Raza M (2021) Determining the impact of Covid-19 on the business norms and performance of SMEs in China. Econ Res Ekonomska Istraživanja. https:// doi.org/10.1080/1331677X.2021.1937261
18. Lu Y, Wu J, Peng J, Lu L (2020) The perceived impact of the Covid-19 epidemic: evidence from a sample of 4807 SMEs in Sichuan Province, China. Environ Hazards 19:323–340
19. Davidsson P, Gordon SR (2016) Much Ado about nothing? The surprising persistence of nascent entrepreneurs through macroeconomic crisis. Entrep Theory Pract 40:915–941
20. Salamzadeh A, Dana LP (2020) The coronavirus (COVID-19) pandemic: challenges among Iranian startups. J Small Bus Entrep. https://doi.org/10.1080/08276331.2020.1821158
21. Organisation for Economic Cooperation and Development (2020). Startups in the time of covid-19: facing the challenges, seizing the opportunities. OECD policy responses to coronavirus (COVID-19). https://www.oecd.org/coronavirus/policy-responses/start-ups-in-the-time-of-covid-19-facing-the-challenges-seizing-the-opportunities-87219267/ Accessed 3 May 2021
22. Klint C (2021) These are the top risks for business in the post covid world. The Davos Agenda 2021. World Economic Forum. https://www.weforum.org/agenda/2021/01/building-resilience-in-the-face-of-dynamicdisruption/ Accessed 23 Mar 2021
23. Rogers (2020) Leading human resources through the pandemic. Challenges, advice, evolution
24. Parveen M (2020) Challenges faced by pandemic covid 19 crisis: a case study in Saudi Arabia. Challenge 63:349–364
25. Dubai Chamber of Commerce and Industry (2020) Impact of COVID-19 on Dubai business community
26. Oxford Business Group (2021). Oman covid-19 recovery roadmap. https://oxfordbusinessg roup.com/sites/default/files/blog/specialreports/961045/OM21_MultisponsorCRRbooklet. pdf.
27. Al Amri T (2021) The economic impact of COVID-19 on construction industry: Oman's case. Eur J Bus Manage Res 6:146–152
28. Al Hasni ZS (2021) The economic impact of COVID-19 on the Omani tourism sector. Psychol Educ 58:824–830
29. Baldwin R, Mauro BW (2020) Economics in the time of COVID-19. Center for Economic Policy Research
30. Nagem S (2020) 5 business strategies for success during the pandemic. Association of international certified professional accountants. https://www.fm-magazine.com/news/2020/nov/bus iness-strategies-for-success-during-coronavirus-pandemic.html. Accessed 5 July 2021
31. Fontannaz S, Oosthuizen H (2007) The development of a conceptual framework to guide sustainable organizational performance. S Afr J Bus Manage 38:9–19
32. Gregurec I, Furjan MT, Pupek KT (2021) The impact of COVID-19 on sustainable business models in SMEs. Sustainability 13:1098
33. Alves JC, Lok TC, Luo Y et al (2020) Crisis challenges of small firms in Macao during the COVID-19 pandemic front. Bus Res China 14

34. Obrenovic B, Du J, Godinic D et al (2020) Sustaining enterprise operations and productivity during the COVID-19 pandemic: "Enterprise Effectiveness and Sustainability Model" Sustainability 12
35. Sinha D, Bagodi V, Dey D (2020) The supply chain disruption framework post COVID-19: a system dynamics model. Foreign Trade Rev 55:511–534

# Understanding User Acceptance of IoT Based Healthcare in Jordan: Integration of the TTF and TAM

Abeer F. Alkhwaldi and Amir A. Abdulmuhsin

**Abstract** The aim of this research is to suggest a unified model that integrates the task fit technology (TTF) model, technology acceptance model (TAM), IoT concerns, and social incentive to examine behavioural intentions to use the Internet of Things (IoT) applications in the healthcare context. A sample of 257 respondents in Jordan who are potential users of IoT participated in this research. Structural equation modelling (SEM) applied via AMOS 25.0 package is carried out to examine the study hypotheses. The findings illustrate that the theoretical model which integrated the TTF model for utility/advantage and TAM for the technology acceptance offers a better comprehensive insight of users' behaviour relevant to the study setting: (1) perceived usefulness (PU) and attitude (ATT) are significant to the behavioural intentions to use IoT; (2) PU has a critical mediation impact on the relationship between: perceived ease of use (PEoU), the TTF construct, security and privacy (SP), and social influences (SI) and behavioural intentions; (3) PEoU, TTF, and SI are revealed to play critical roles to predict behavioural intentions; (4) ITF, TTF, and SP affect the PEoU; (5) unexpectedly, PEoU and SI have no significant impact on ATT, and ITF and SP do not influence PU. The findings have theoretical and practical implications for IoT researchers, developers, and policymakers.

**Keywords** IoT acceptance · Healthcare · Jordan · TAM · TTF · IoT concerns · Social influence · Behavioural intention

A. F. Alkhwaldi (✉)
Department of Management Information Systems, College of Business, Mutah University, Karak, Jordan
e-mail: AbeerKh@mutah.edu.jo

A. A. Abdulmuhsin
Department of Management Information Systems, College of Administration and Economics, University of Mosul, Mosul, Iraq
e-mail: Dr.amir_alnasser@uomosul.edu.iq

© The Author(s), under exclusive license to Springer Nature Switzerland AG 2022
S. G. Yaseen (ed.), *Digital Economy, Business Analytics, and Big Data Analytics Applications*, Studies in Computational Intelligence 1010,
https://doi.org/10.1007/978-3-031-05258-3_17

# 1   Introduction

The Internet of Things (IoT) is a relatively state-of-the-art paradigm shift in internet services and applications and refers to a system of interconnected real-world objects with the ability of collecting, processing, and communicating data via the Internet network without any human interventions [1]. Peter T. Lewis introduced the IoT concept for the first time in September 1985 in his speech at U.S. Federal Communications Commission (FCC) [2]. According to [3], the IoT term was coined in 1999 by Kevin Ashton, a British technology pioneer. The aim of IoT technology is to extend the advantages of the ordinary internet; such technology currently uses the Internet to achieve smart identification, exchange information, allow remote control ability, and conduct different IoT functions [4]. A basic technology in the IoT application is radio frequency identification (RFID). Applying RFID, a unique identifying number such as an IP address could be assigned to each real object/device in the analogue domain [3]. According to Bsquare's Annual IoT Maturity Study [5], about 73% of all businesses plan to raise their investments in IoT, despite the respondents' acknowledgment regarding the complexity of IoT utilization [5]. It is estimated that more than tens of billions of objects will constitute an IoT by 2030 [6]. Getting lots of devices linked will affect different aspects of our daily life, society, and economy due to the considerable changes IoT heralds.

IoT technology has been offered wide attention in terms of both industry and research, also this technology has a large number of applications/uses in many fields [7]. The IoT technology is a worldwide internet-based architecture that is developing and growing dramatically not only in developed countries but also in developing countries; Jordan in particular [8, 9]. One of the high-impact policy interventions that Jordan implements to scale up data infrastructure in the country is to continually invest in modern IT infrastructure (e.g., IoT) with the aim of maintaining forward data momentum and ensure that the latest digital applications could be run seamlessly in the future [10–12]. Jordan has been looking for developments in vital digital technology such as IoT and that will provide opportunities to build a modern digital economy that improves economic activities over the following five years. The government in Jordan intends to develop IoT studies in its five-Year National Digital Transformation Strategy & Implementation Plan period, to create a comprehensive IoT systems and also has issued a set of internationally accepted standards and instructions for the IoT system in mid-2020 [13]. The country is working to support the IoT-associated communication, manufacturing, and e-services industry in addition to scale-up IoT applications to build an expanded value chain. Jordan seeks to master a range of self-developed main IoT technologies and to make initial applications in many areas such as education, transportations, social services, healthcare, smart cities, smart homes, and public safety [14].

E-healthcare presents a pivotal field for IoT technology [15]. IoT has many benefits to provide, for example, the Internet of Medical Things (IoMT) or the Internet of Healthy Things (IoHT). IoMT/ IoHT are applications of IoT technology for purposes related to medical and health status [16], where IoT devices are employed to enable remote processes of health monitoring (e.g., monitoring blood pressure, body temperature, and heart rate) [15]. Such applications can crucially enhance the quality of patients' life, particularly for chronic diseases, due to the ability to be monitored in non-clinical settings such as the patient's home. In addition, this innovative technology has been used during the COVID-19 pandemic for smart tracing of infected individuals and for replacing classic medical consultation with telehealth consultation and remote treatment [17].

Yet, IoT technology is still a fledgling field around the world, and Jordan is confronting different obstacles as IoT makes an attempt to obtain new ground. While Jordan continues to offer the support and fund to progressive research in the area of IoT, this IT is still inadequate in terms of application, as there are concerns regarding the use and information safety from the patients' perspective [7]. From the end-user perspective, this technology spark privacy and security issues as it is not clear who has access to the data collected, what is being done with it, and where such data goes [18]. Given the government investment in Jordan and low adoption rates of IoT applications/services, it has become crucial for researchers, decision-makers and, practitioners to understand the factors influencing the acceptance of IoT among Jordanian users since it is considered a vital step towards the development of a succeeded IoT-based healthcare system. Earlier studies tend to focus on the technical standpoint of implementing the IoT technology [3, 8, 19]. For example, Patel and Patel [20], discussed different key issues of the IoT technology, for example, the architecture, functional view, characteristics [20]. The majority of extant IoT research has examined the application of the technology business model from the organization perspective [21]. The literature on IoT acceptance from the user perspective, particularly in the healthcare context in developing countries, is yet in its early stages. Users' acceptance towards using IT is the key determining factor of actual use behaviour [22, 23]. In addition, investigating the users' acceptance of IT/IS has always been a significant issue in the literature of information management [24–26]. Yet, earlier studies have offered an inadequate understanding of the main predictors in user acceptance of state-of-the-art technologies (i.e., IoT). Taking into consideration the importance of the users' attraction and retaining of such technology, it is required to investigate the factors affecting users' acceptance of IoT applications in the context of healthcare. A comprehensive view of these variables provides the potential to derive valuable managerial implications concerning the implementation of marketing plans and strategies based on users' requirements, thus leading to better users' acceptance. In addition, the current research contributes to the body of knowledge and informing the future research directions on IoT adoption and use.

In view of the swift development and adoption of IoT technology for different services (e.g., healthcare), a study of the factors affecting users' adoption of IoT applications may reveal insights into its sustainability and viability. Though, limited studies have investigated the factors that influence IoT usage intention. Furthermore, IoT usage can be considered as the users' behaviour of obtaining, using, and diffusing the IoT applications and resources. Such behaviour implied two steps: (1) the user's perceptions of IoT by attitudes and adoption; (2) the level to which IoT can satisfy the requirements of the user, which indicates the usefulness and the advantages of IoT. However, to the best of the researchers' knowledge, the earlier literature that captures users' intentions to adopt IoT applications and services is too rough in terms of combining both perspectives. It is unconscionable to anticipate that a straightforward theoretical framework can be adapted to a continually changing IT environment without amendments and thus explaining user behaviour completely across a wide variety of technology adoption contexts.

Therefore, this research paper aims to determine to what extent and whether the aforementioned factors influence IoT usage intentions related to the two steps. Hence, a research framework was introduced to integrate the TAM for the technology adoption and the TTF model for the usefulness and the advantages of IoT.

## 2   Literature Review

### 2.1   IoT Technology

Real-world objects with the ability to communicate using the internet refer to IoT technology. From the classical standpoint, the IoT referred principally to RFID technology-tagged objects which are interconnected through the world wide web [27]. IoT has extensive applications in many fields at present, including tracking, payment, healthcare, smart homes, smart traffic management, workplace and home electronics, transport, and education [9, 15, 19]. IoT will come up with great efficiencies across different industries and its advantages to end-users are considerable [28]. IoT technology has many advantages, for example, waking the technology users to provide assistance with listening to the news based on selected headlines, controlling the air-conditioning or heating in order to save energy in workplaces and homes, and turning on the computers before the employees arrived at their workplaces [29]. From a healthcare viewpoint, IoT technology is expected to remind the users concerning any medicines they need to take, suggest exercise or sport they need to undertake with the aim of improving their wellbeing and offer daily recommendations of what they are allowed to eat, taking into consideration the users' health conditions and their daily schedules [29]. In this context, such innovative technology (i.e., IoT) will affect users' behaviour in various aspects of their daily life.

Current research has explored the technical issues of IoT implementation. For example, scholars [30] reviewed the security issues and the major challenges stunting the development of IoT-based networks. Also [31, 32] identified both the security and privacy concerns as the key obstacles for user-centric IoT services and applications. Many prior studies have focused on the usage of IoT technology from the industry or organization perspective, whilst limited efforts have been dedicated to understanding the acceptance of IoT from the users' point of view, particularly in developing countries. In addition, the previous literature has not yet examined the impact of factors influencing users' intentions to accept and use IoT technologies in the context of healthcare from the Jordanian users' perspective. Considering the lack of previous empirical research and the practical relevance, this study aims to build and examine an integrative research framework representing the determinants of users' acceptance and adoption of IoT technologies.

## 2.2 The Extension of TAM

Scholars have introduced a number of TAM extensions with different external variables to analyse the likelihood of IT/IS acceptance and adoption [33]. For instance, a study model on the basis of understanding the users' acceptance of libraries' self-issue and return systems was proposed [34]. Their model was based on the integration of the theory of planned behaviour (TPB) and TAM. To extend the basic TAM by analysing the influence of perceived behavioural control, trust, and subjective norm, a study examined the factors that affect employees' attitudes and behavioural intention (BI) to adopt IoT technology in a retail environment [35]. However, the Technology Acceptance Model only considers the short-term belief and attitudes after and prior to acceptance of new technology (e.g., IoT). A favourable outcome of IoT usage is anticipated when a well-fit between the IT and the tasks can be attained, which is the core of the TTF model. Accordingly, the theory of task-technology adaptation could help in compensating the TAM deficit in this regard. Combined theoretical models of the task technology fit and technology acceptance offer a better understanding of the variance in IT/IS usage than either the TTF model or TAM alone.

## 2.3 The TTF Model

This model is an extensively applied theoretical framework to evaluate how IT can lead to performing, judging the alignment between the characteristics of technology and task, and assessing utilization impact. Since the initial proposal of the TTF model, it has been enthusiastically used and investigated on a wide scope of ISs [36, 37]. Though the literature has examined the TTF model in several environments, little investigation has been carried out in the IoT setting. As yet, it is still uncertain whether a well fit between task and technology will affect users' intention to adopt IoT applications and services, particularly in the healthcare context. Regarding IoT technology, the TTF framework does not include a social factor, which could have a negative impact on the model's predictive power for social networking technologies. This limitation could be handled through the extension of TTF with a social incentive drawing perceptions from social influences.

## 3 Conceptual Framework and Hypotheses

The researchers propose a conceptual model based on the theoretical background of the TAM and TTF model integrated with the characteristics of IoT and social incentives. The proposed model identifies several predictors of IoT behavioral intention. A conceptual framework presenting the relationships between antecedents' factors and behavioral intentions to use IoT for healthcare purposes is shown in Fig. 1.

The simple assumption is that IoT behavioral intentions is cooperatively identified by perceived ease of use (PU) and attitude, which are functions of TTF, IoT characteristics, social incentives, and perceived usefulness (PU). In the framework: (1) both task-technology fit, and individual-technology fit are integrated; (2) security and privacy are the concerns of IoT; (3) social incentive represented by social influence (SI) construct is incorporated.

### 3.1   Constructs of the TAM

To study the dependent construct of behavioral intentions (BI) to use IoT applications, the TAM is employed to examine the relationship between PEoU, PU, attitude, and behavioral intention (BI).

**Perceived Usefulness** A failure in communicating clear benefits to potential IT users can have a negative effect on the diffusion of IoT applications and services. Rogers [38] stated that user is only willing to accept innovation if this innovation provides unique advantages compared to existing technological solutions [38]. In the context of the TAM, this perspective is described by the PU variable [39]. It refers to the individuals' subjective assessments of whether utilizing a certain technology or IS would improve their job performance [40]. The PU of IoT applications can be defined as the degree to which users believe that IoT applications can be steering

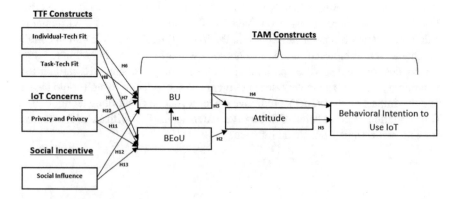

**Fig. 1** Proposed study model

forces toward attaining the technology goals in different contexts. IoT technology can supply healthcare with faster processes, lead to time reduction, and improved the quality of services perceived by technology users. Also, applied IoT technology, hospitals staff can receive data from the transponders installed in users' various devices through a mobile reader in their hand to decide whether they need any aid, thereby enhancing the efficiency of healthcare tasks. Accordingly, the PU of IoT technology is likely to be high. PU is a factor that has been repetitively found to affect the users' attitudes and also, is a direct determining factor of IT/IS behavioral intention [34]. In addition, PU mediates the impact of PEoU on BI, a relationship that has been confirmed by IT empirical research. In the IoT studies, for instance, it was supported that the behavioral intentions to utilizing IoT applications is significantly determined by PU. Hence, the following research hypotheses were proposed:

**H1**: Perceived usefulness (PU) has a positive impact on attitudes (ATT) towards using IoT applications and services in the healthcare context.

**H2**: Perceived usefulness (PU) has a positive impact on intentions to using IoT applications and services in the healthcare context.

**Perceived Ease of Use** From the perspective of IoT applications for healthcare, the PEoU can be described as the extent to which user believes that utilizing IoT applications will be free of efforts. The ease of acquiring skills of using IoT applications and services is an example of PEoU. Prior literature stated that PEoU is a significant determinant of attitude and the PU of using information technology [34]. According to Wu and Zhang [59], PEoU is significant for PU and attitudes toward using information systems. Likewise, PEoU could influence the behavioral intentions to accept IoT applications directly or indirectly over PU. Therefore, the following hypotheses are proposed:

**H3**: Perceived ease of use (PeoU) has a positive impact on the perceived usefulness (PU) of IoT applications and services in the healthcare context.

**H4**: Perceived ease of use (PeoU) has a positive impact on attitudes (ATT) towards using IoT applications and services in the healthcare context.

**Attitude** The TAM highlighted the relationship between users' attitude (ATT) and behavioral intention (BI), suggesting that ATT is defined as "an individual's positive or negative feeling (evaluative effect) about performing the target behavior" [41]. Thus, the ATT towards using IoT applications has been considered as the extent to which users perceive a negative/positive feeling associated with IoT applications. Previous studies have revealed that ATT is a critical predictor of behavioral intentions to utilize IT [34]. In the IoT context, it was reported that the ATT towards IoT technology was a significant determinant of behavioral intention to adopt it [15]. Thus, the following study hypothesis was proposed:

**H5**: ATT toward the use of IoT applications has a positive impact on intentions to use IoT applications and services in the healthcare context.

## 3.2  Constructs of the TTF Model

To explain the intentions to use IoT applications and services, not only users' inter-actions with the technology should be considered, but also task-focused on the action associated with the technology. The main issue to the users' evaluation of IoT technology lies in the fit of individuals-technology and tasks-technology.

**Individual-Technology Fit (ITF)**  Users' efficient usage of IoT applications depends on variables related to individual-technology fit, which implies the IoT based solutions that have been introduced to cater to the users' requirements of various information required for healthcare purposes [42], also the match between the using style with the content that the IoT applications provide. Thus, an individual's interaction with an IT is often associated with the user's individual-technology adaptation behaviour. Those IT functions are in alignment with individual abilities and task requirements. More experience with using information technology, which denotes individual-technology fit, is related to greater ease of use (EoU). Besides the significant impact of IT experiences on PEoU, IT experiences are also related to PU due to the fact that a more experienced user has a better ability to understand the usefulness/benefits of the technology. Hence, the following hypotheses are proposed:

**H6**: ITF has a positive impact on the PU of IoT applications and services in the healthcare context.

**H7**: ITF has a positive impact on the PEoU of IoT applications and services in the healthcare context.

**Task-Technology Fit (TTF)**  The "task-technology fit" variable is described as the extent to which the capabilities of the IT/IS are in alignment with the task that the users should perform, and it is a key construct to explain the levels of job performance [43]. Kim et al. [44] stated that the users' perceptions of whether a certain IT fits well with their current values, i.e., PU and PEoU, can be a foundation to form their perception of the IT actual use [44]. Furthermore, empirical findings have illustrated that PU and PEoU are influenced by task-technology fit; a user perceives the technology to be useful and easier to utilize for those tasks when the task-technology fit is higher. The prerequisites for the PU of IoT is that the user finds an alignment between tasks and technology. When a user enthusiastically decides to utilize IoT applications, the reason behind such a decision is very likely that the fit between the task and technology affects the perceived ease of use (PEoU) of IoT. Hence, the following hypotheses are proposed:

**H8**: TTF has a positive impact on the PU of IoT applications and services in the healthcare context.

**H9**: TTF has a positive impact on the PEoU of IoT applications and services in the healthcare context.

## 3.3 IoT Technology Concerns

Because privacy and security are prominent concerns of IoT, particularly in a health-care context, the current study addressed the causal impacts of such concerns on the two key factors TAM (i.e., PEoU and PU).

**Security and Privacy**
Concerns regarding security and privacy have increased with the growing utilization of the IoT, also researchers and practitioners anticipate this trend to continue. IoT users have fears regarding the adequate protection of their personal data, particularly due to the possible monitoring of their devices by unauthorized parties [29]. This reflects the pivotal importance of security and privacy when using IoT applications and services. IoT service providers can suffer the loss of their users' trust if they do not take into consideration appropriate procedures to maintain sufficient security and privacy levels with the aim of protecting their users' data. Alanazi and Soh [57] revealed that privacy concerns have a significant effect on the intentions to use IoT. In addition, perceived security protection and privacy concerns were found to be significant determinants of accepting IoT technology [45]. Also, security and privacy concerns inhibit the end-users from IT/IS adoption [46]. These two aspects will inevitably affect IoT practices such as PU and PEoU. Hence, the following hypotheses are proposed:

**H10**: Security and privacy concerns have a positive impact on the PU of IoT applications and services in the healthcare context.

**H11**: Security and privacy concerns have a positive impact on the PEoU of IoT applications and services in the healthcare context.

## 3.4 The Construct of Social Incentives

Since IoT represents an emerging technology, positive effect from other individuals exerts a crucial impact on the sustainable advancement of IoT. This research recognizes social incentives as a social influence (SI). As earlier research indicates, the behaviour of the individual user could be affected by others. Thus, it is required to investigate the effect of social incentives on IoT utilization.

**Social Influence (SI)**
IT/IS researchers have noticed that an individual may adopt a certain technology not due to personal persuasions but due to the other's views [22, 47]. The unified theory of acceptance and use of technology (UTAUT) suggests that SI is an important construct in understanding users' acceptance behaviour of IT [23]. SI can be defined as "the degree to which an individual perceives that important others believe he or she should use the new system" [23]. SI has also shown in various theoretical frameworks

of users' acceptance of ICT [48], and from an empirical viewpoint, SI has acquired strong support as a significant driver of users' behaviour.

This study considers SI as the degree to which users perceive those others in the group explicitly encourage and approve their participation in IoT practices and services for healthcare purposes. When a user notices that other individuals utilize IoT and perceive the advantages of its utilization, the user will become more prepared to utilize such technology, and this can lead to growth in the present and future use of technology. Similarly, the researchers expect that SI involves the user's perceptions of usefulness from other individuals in the group and plays a crucial role in driving ATT toward the use of IoT applications and services.). Hence, the following hypotheses are proposed:

**H12**: SI has a positive impact on the PU of IoT applications and services in the healthcare context.

**H13**: Social influence (SI) has a positive impact on ATT toward using IoT applications and services in the healthcare context.

# 4 Research Methodology

## 4.1 Data Collection

This research used the anonymous online survey approach developed (see Appendix 1) and operated via the web survey platform (https://www.surveymonkey.com) with the convenience sampling method. The survey questionnaire was posted through advertising the web link to the targeted respondents using various communication channels, such as social media groups and personal emails. Participation in the survey was completely voluntary. Since the development of IoT healthcare applications in Jordan is still in the initial phase, the participants were potential future users/adopters. In order to ensure that participants have the same understanding of the IoT applications and services in the healthcare context, a summarized definition is offered on the first page of the survey. Then, the participants are asked if they are familiar with the applications and services of IoT healthcare. If their answer is yes, they are asked to fill the questionnaire. If their answer is no, they are not allowed to complete the questionnaire. Data collection was carried out during June 2021; in the end, 257 valid surveys were received. Table 1 summarizes the demographic profile of the participants.

**Table 1** Demographic profiles of the study participants

| Demographic category | Description | Percentage | Frequency (%) |
|---|---|---|---|
| Gender | Male | 152 | 59.1 |
| | Female | 105 | 40.9 |
| Age | <20 | 37 | 14.4 |
| | 20–29 | 89 | 34.6 |
| | 30–39 | 52 | 20.2 |
| | 40–49 | 45 | 17.5 |
| | >50 | 34 | 13.3 |
| Education background | High school | 44 | 17.1 |
| | Diploma | 8 | 3.1 |
| | Undergraduate | 112 | 43.6 |
| | Masters | 52 | 20.2 |
| | Doctorate | 41 | 16 |
| Marital status | Single | 45 | 17.5 |
| | Engaged | 20 | 7.8 |
| | Married | 175 | 68.1 |
| | Widowed | 9 | 3.5 |
| | Divorced | 8 | 3.1 |
| Socio-professional status | Student | 113 | 43.9 |
| | Unemployed | 7 | 2.7 |
| | Employee | 80 | 31.1 |
| | Worker | 13 | 5.1 |
| | Freelance job | 25 | 9.7 |
| | Retired | 19 | 7.5 |

## 4.2 Questionnaire Development

The researchers employed a questionnaire instrument with three parts to examine the proposed study model. The first part provides an illustration of the IoT applications for healthcare uses. The second part provides demographic items about the respondents. While the third part provides items measuring the constructs related to respondents' intention to using IoT applications and services in the healthcare context. To consider the main aspects of IoT technology, the study model entailed eight variables. The variables were measured utilizing a 5-point Likert scale, anchored on "5 strongly agree" and "1 strongly disagree". Dawes [49] stated that there are no differences between 7-point and 5-point data sets [49].

**Table 2** Reliability and convergent validity

| Constructs | Cronbach's alpha | CR | AVE | Factor loading |
|---|---|---|---|---|
| Perceived usefulness | 0.9812 | 0.985 | 0.73 | 0.747–0.811 |
| Perceived ease to use | 0.9747 | 0.931 | 0.77 | 0.712–0.91 |
| Attitude | 0.8895 | 0.906 | 0.71 | 0.776–0.834 |
| Behavioural intention | 0.9313 | 0.937 | 0.77 | 0.782–0.865 |
| Individual-technology fit | 0.971 | 0.905 | 0.79 | 0.70–0.842 |
| Task-technical fit | 0.9676 | 0.834 | 0.69 | 0.779–0.857 |
| Security and privacy | 0.9815 | 0.888 | 0.81 | 0.788–0.935 |
| Social influence | 0.982 | 0.857 | 0.79 | 0.841–0.896 |

# 5  Data Analysis

## 5.1  Reliability and Convergent Validity

Reliability was evaluated based on Cronbach's alpha ($\alpha$). All multi-item variables have to satisfy the guideline for a Cronbach's alpha value (i.e., $\alpha > 0.70$) [50]. The ad-hoc tests applied in this research for convergent validity were: (1) items factor loadings should be significant and greater than 0.7; (2) composite reliability (CR) for each variable should be higher than 0.07; (3) Average variance extracted (AVE) where the acceptable threshold for each variable is greater than 0.5 [51]. Based on the above-illustrated criteria, all the indexes in the current research are acceptable (see Table 2).

## 5.2  Discriminant Validity

This index was evaluated based on the squared correlations between constructs and their AVE. An acceptable value of discriminant validity is thought to be attained if the square root of AVE for each variable (which is across the diagonal cells) was higher than the squared correlation between that variable and all other variables [52]. According to the correlation assessment listed in Table 3, the AVE values for the reflective constructs are higher than the off-diagonal squared correlations, implying acceptable discriminant validity for the study samples.

## 5.3  Measurement Model Analysis

According to Hair et al. [52] the measurement model "represents the theory showing how measured variables come together to represent constructs" [52]. The most

**Table 3** Discriminant validity

|       | PU    | PEoU  | ATT   | BI    | ITF   | TTF   | SP    | SI    |
|-------|-------|-------|-------|-------|-------|-------|-------|-------|
| PU    | **0.85** |       |       |       |       |       |       |       |
| PEoU  | 0.632 | **0.88** |       |       |       |       |       |       |
| ATT   | 0.617 | 0.270 | **0.84** |       |       |       |       |       |
| BI    | 0.065 | 0.53  | 0.183 | **0.88** |       |       |       |       |
| ITF   | 0.205 | 0.442 | 0.256 | 0.389 | **0.89** |       |       |       |
| TTF   | 0.379 | 0.221 | 0.115 | 0.322 | 0.448 | **0.83** |       |       |
| SP    | 0.245 | 0.347 | 0.397 | 0.444 | 0.671 | 0.225 | **0.90** |       |
| SI    | 0.025 | 0.379 | 0.304 | 0.132 | 0.064 | 0.502 | 0.441 | **0.89** |

Bold is used to show the square root of the AVE of each variable

**Table 4** Model fit indices

| Fit statistics          | Model value | Recommended value |
|-------------------------|-------------|-------------------|
| $X^2$/df (degree of freedom) | 0.138       | <3                |
| GFI                     | 0.934       | >0.900            |
| AGFI                    | 0.869       | >0.800            |
| CFI                     | 0.911       | >0.900            |
| NFI                     | 0.954       | >0.900            |
| NNFI                    | 0.959       | >0.900            |
| RMSEA                   | 0.074       | <0.080            |

popular indices applied to assess the fit of the measurement model include adjusted chi-square $X^2$/df, chi-square ($X^2$), Goodness of Fit Index (GFI), Adjusted Goodness of Fit Index (AGFI), Root Mean Square Error of Approximation (RMSEA), and standardized root mean square residual (SRMR); as is in line with the existing studies [53, 54]. Hair et al. [52] recommended reporting Chi-square statistics, also at least one incremental index, for example, Tucker-Lewis's coefficient (TLI), Normed Fit Index (NFI), and Comparative Fit Index (CFI); and one absolute index, for example, RMSEA and GFI. In consequence, in this research, the model fit was assessed based on the interpretation of different fit indices. The overall model fit was adequate, see Table 4.

## 5.4 Results of Hypotheses Testing

AMOS 25.0 package was employed to test the 13 hypotheses suggested above based on structural equation modelling (SEM). Table 5 illustrates the validation of the proposed study model, with variance ($R^2$), path coefficients, and significance of

**Table 5** Path coefficients and hypothesis testing for the study sample

| Hypothesis | Constructs' relationship | Path coefficient (β) | Results |
|---|---|---|---|
| H1 | PEoU → PU | 0.350*** | Validated |
| H2 | PEoU → ATT | 0.023* | Rejected |
| H3 | PU → ATT | 0.537*** | Validated |
| H4 | PU → BI | 0.50* | Validated |
| H5 | ATT → BI | 0.539** | Validated |
| H6 | ITF → PU | −0.584* | Rejected |
| H7 | ITF → PEoU | 0.191*** | Validated |
| H8 | TTF → PU | 0.193*** | Validated |
| H9 | TTF → PEoU | 0.363*** | Validated |
| H10 | SP → PU | −0.516* | Rejected |
| H11 | SP → PEoU | 0.229*** | Validated |
| H12 | SI → PU | 0.156*** | Validated |
| H13 | SI → ATT | −0.045* | Rejected |

**$R^2$(PU)** = 0.938, **$R^2$(PEoU)** = 0.458, **$R^2$(ATT)** = 0.88, **$R^2$(BI)** = 0.947
$^*p < 0.05$, $^{**}p < 0.01$, $^{***}p < 0.001$
Bold is used to show the variance

each relationship. PEoU is revealed to be significantly affected by the 3 external factors: ITF, TTF, and SP with a variance of 0.458. Likewise, PU is revealed to be significantly affected by the 2 external factors: TTF and SI, and as a result of the direct impact of PEoU, with a variance of 0.938. ATT is significantly defined by the PU, with a variance of 0.88. The dependent variable BI is significantly determined by PU and ATT, with a variance of 0.947. That is to say, the integrated impact of PU and ITT explains 94.7% related to the variance on the respondents' behavioural intentions to utilize IoT applications and services in the healthcare context. This result outperforms the variance resulted by any earlier theoretical models of IS/IT acceptance. Overall, 9 out of 13 hypotheses are validated by the data.

For the relationships in TAM (i.e., H1–H5) which are relevant to PU, PEoU, ATT, and BI, all the hypotheses are supported except H2. For the relationships between TTF and TAM (i.e., H6–H9) which explore the effect of exogenous variables of TTF on the key TAM variables (PU and PEoU), all the hypotheses are supported except H6. For the relationships between TAM and IoT concerns (i.e., H10 and H11) which explore the effect of exogenous variables of IoT concerns on the key TAM variables (PU and PEoU), H11 is supported. A significant relationship was validated between the security and privacy (SP) construct and PEoU.

# 6 Discussion

In terms of theory building, this research represents an endeavour to integrate the TTF model, the TAM, social incentive, and IoT concerns to investigate the causal determinants of users' behavioural intention to use IoT healthcare applications in Jordan. Data were drawn from current and potential users of IoT technology. The findings of the empirical data analysis strongly support 9 of the 13 hypotheses in the current research. Regarding each hypothesis, the researchers introduce the following interpretations into the TTF model, the TAM, social incentive, and IoT concerns, respectively.

The findings show that integrating the TTF and TAM variables offers a better illustration for the variance ($R^2$) in IoT usage than either the TTF model or TAM can offer by itself. In addition, this research suggests a better hybrid IT use theoretical framework explaining users' behavior concerning IoT. In contrast to what was suggested in hypothesis 6 regarding the impact of ITF on PU is not supported. Yet, a significant indirect relation is between ITF and PU, also PeoU has a mediation effect on this relation. A likely explanation for this impact is that better user experiences with IoT are prerequisite for ITF such that better-experienced individuals will have a better ability to realize the ease of use of IoT, and IoT might be realized to be useful only if users also perceive it to be easy to usage [55].

The current study examines the implications of the TTF model for TAM; the researchers highlighted the antecedents of the TAM core variables from a TTF standpoint. With respect to TTF as an exogenous variable, the direct impacts of ITF and TTF for IoT users were investigated. Also, the proposed research model presents modest support to what is subconsciously evident concerning the TTF model. As expected, harmonizing the performance of IoT to certain tasks (i.e., TTF), will allow users to perceive both the usefulness and the EoU of IoT applications. This finding agrees with the deductions of previous research [56], showing that TTF affects PU and PEoU in the IS environment.

Yet, ITF contributed to PU mediating by the PEoU in the current research. This different aspect could be because of the IoT setting under research. When the extent of ITF becomes higher, users realize IoT applications to be easier to use for specific tasks, and hence more useful. In sum, the higher the fit among the individuals (users), the tasks, and the information technologies utilized, the better likelihood that IoT applications would be realized in a positive manner.

In this regard, the current research offers additional evidence of the suitability and applicability of the TTF model and the TAM for the analysis of behavioral intentions to utilize IoT.

The results reveal that PEoU is a significant determinant of PU in that the more that users perceive IoT applications to be easy to use, the more possible they will be to consider the IoT applications as useful. This finding is in agreement with prior literature [57].

The missing link in hypothesis 2 between PEoU and ATT is not anticipated, which is in agreement with previous empirical findings; PEoU is a significant variable that influences attitudes toward using IT. A probable clarification for this finding could be that IoT applications are each accessible through internet-enabled devices and include similar features and capabilities to other applications the user is familiar with, which may have made IoT applications easy to use; therefore, the users' attitudes towards IoT acceptance and use depends entirely on the perceived usefulness of IoT applications.

Particularly, the indirect impact of PEoU on ATT through the PU of IoT applications was revealed to be evident. The impact of PEoU on ATT is more profound as user tends to concentrate on the functionality of the IoT applications by its own, rather than the EoU in terms of establishing attitudes toward IoT acceptance and use. Thus, PU acted as a significant mediating factor between PEoU and ATT toward IoT acceptance and use. That is if IoT applications offer critically required functionality, the user tends to accept some level of difficulty associated with usage.

PU and ATT were linked positively to behavioural intention (BI) to use IoT. This finding indicates that PU had an important positive impact on BI of IoT, which is consistent with the concept of IT acceptance as proposed by [40]. This research also illustrates that the impact of ATT on BI of IoT is both positive and significant, which is in line with the results of [40]. Particularly, ATT served as a critical mediator between PU on intentions to use IoT, because the indirect impact of PU on intentions to use through the ATT was revealed to be evident.

Consequently, the above-discussed findings correspond with the results of the TAM, demonstrating that the technology acceptance model is appropriate to study IoT applications.

This study considers the IoT concerns of security and privacy as a single independent construct linked with the TAM in the PU and PEoU of IoT applications. The results emphasized the significance of security and privacy (SP) as a contributor in explaining the EoU of IoT. This is a unique result in the field of IoT literature. The study finding holds vital implications for IoT practitioners and policymakers for strategic planning and also designing efficient IoT applications to enhance the performance of technology users.

The proposed research model tries to expand the role of social incentive in IoT applications by including the factor of social influences (SI). The SI exerted a significant positive influence on the PU of IoT. This result supports the findings of the previous literature, which has illustrated that users consider PU of IoT applications when they recognize that other individuals in their social networks have similar beliefs and values concerning the advantages of IoT applications. This finding is congruent with the prior findings of [58]. Therefore, management and practitioners' attention could be more effectively concentrated on the improvement of social incentives.

# 7   Conclusion

The IoT is a worldwide utilized IT that has spread to many various domains due to its several advantages. Increasing research has paid attention to healthcare to introduce the best solutions for individuals (i.e., patients). E-Healthcare is one such topic affecting by IoT that has vast potential. Therefore, the results of the current research are significant, with both theoretical research and practical implications, as discussed below.

Limited academic literature has investigated the factors that impact IoT acceptance and even fewer, in the context of healthcare, particularly in developing countries (e.g., Jordan). The conceptual framework proposed in the current study not only contributes in several ways to the existing knowledge in the field but also helps researchers and practitioners obtain a better understanding of users' behaviours in IoT. This study has value as it reveals multiple statistically significant relationships that explain why individuals select IoT and why they will accept to use it for healthcare purposes.

## 7.1 Theoretical and Practical Implications

Limited studies have investigated the factors influencing IoT acceptance and even fewer, in the context of developing countries (e.g., Jordan). The proposed model in this research not only provides contributions in many ways to the extant studies but also assists practitioners and academics obtain a better insight of users' behaviour towards IoT technology. This study has value as it acknowledged various statistically significant relationships that help to describe why users accept and use IoT applications.

From a theoretical perspective, the use of the TTF model and the TAM to the IoT technology delineated in this research not only offers more precise findings compared to the TTF model and the TAM proved separately but also improves the understanding of the TTF mechanisms due to its relation to nurturing IoT. In this regard, the researchers extended the TAM with the TTF framework, IoT concerns, and social incentives, this needs to be deemed a beneficial instrument for investigating users' behaviour in IoT settings. This research contributes to the extant literature in 3 important ways. (1) The researchers extended previous work on IoT by highlighting the significance of attaining task technology fit and individual technology fit and. The findings suggest that behavioural intentions to use IoT are indirectly influenced by PEoU, ITF, TTF, SP, and SI. The integrated research model offers richer insights and a better explanation compared to the independent perspective of the TTF model. The results indicated that in addition to TTF, ITF also has a significant influence on PU of IoT applications. The results of this research can be considered as a guide for future academic work on IoT technology with the TTF model. With respect to PU, the proposed framework also offers a tool to understand the influence of TTF on the PU of IoT. The results indicate that users are more probable to expend efforts to utilize IoT if they have the feeling that doing so is useful in terms of the fit between tasks and technology. (2) The researchers highlighted the significance of managing the perceived ease of use through the IoT concerns (i.e., security and privacy). These concerns affect users' PEoU, which could positively/negatively influence their behavioural intentions to use IoT applications. (3) The proposed research model introduces a social incentive for increasing the PU and ATT towards using IoT. The social incentive is represented by social influence. The findings indicate that SI has a significant influence on PU. These findings advance the knowledge on users' cognition, which has been principally investigated from a perspective of IT acceptance.

From a practical perspective, on the basis of the findings, a number of important guidelines and salient implications for IoT practitioners can be suggested. First, IoT experts should be aware that behavioral intentions depend not only on attitudes toward IoT applications but also on PU. Furthermore, the PU of IoT is a crucial mediator of the impact of PEoU, the TTF construct, security and privacy concerns, and social incentives on behavioral intentions. As PU is the most significant determinant of behavioral intentions, the behavioral intentions of users can be enhanced by promoting their belief about the efficiency of IoT. This result indicates that it is not adequate to build IoT applications with advanced user interfaces and a friendly screen to affect user's behavioral intentions. IoT technology practitioners must prioritize useful functions over the application's EoU.

Secondly, this research offers proof that the TTF of IoT identifies PU and PEoU, and ITF identifies PU mediated by PEoU. Hence, IoT applications need to be managed to explain the challenges and requirements of services provided, comprising the level of previous knowledge required and the devices' availability and other resources essential for users. IoT experts need to be especially aware of the significance of TTF and ITF, rather than the common IoT usability to better harmonize the individuals, tasks, and IT context. By providing opportunities relevant to users' particular tasks, IoT applications experts might have the ability to ensure fits between IoT applications and users' actual requirements.

Third, advanced security and privacy mechanisms are of the ways in which the providers of IoT applications and services can both differentiate themselves from other competitors and promote users' advantages of IoT to thus attract them for IoT acceptance and use. IoT practitioners need to focus on variables related to the PEoU by making complete utilization of obtainable security and privacy abilities for a better user experience. IoT developers can use the study findings to enhance the acceptance of such innovative and sophisticated applications by integrating the factors that contribute to trust development (e.g., security and privacy) in the design and development stages of IoT applications.

Forth, the results validated the significant influence of social incentive on perceived usefulness. IoT experts that leverage this insight are possible to attract, obtain and maintain users. At the same time, practitioners may distinguish their IoT applications from other service providers by making sure that they are useful for users. However, they need to consider the importance of the effects of social influences, for which they could depend on peers' influences to accelerate IT acceptance and use.

## 7.2 Future Work and Limitations of the Current Study

Though comprehensive and rigorous research was carried out, there are few limitations related to this research.

Firstly, this research was carried out in Jordan, where IoT technology is developing but is still in its initial stages. Thus, participants in the study sample took part of their own choice which could indicate an issue of "self-selection" bias. As the number of IoT users (study population) raises in Jordan, the possibility to use random probabilistic sampling will also be feasible in IoT future studies.

Secondly, cross-sectional research was conducted. However, users' behaviour is dynamic, thus a longitudinal research approach could offer better insights into the development of users' behaviour. Hence, longitudinal study will be a likely avenue for researchers in the future.

Thirdly, behaviour in IoT is a quite new area to academics. The results and implications introduced in this research need to be generalizable for external validity as they were gained from one research that investigated IoT and focuses on a particular sample in Jordan. Additional studies are required to support the generalizability of the results and discussion to involve various cultures and countries in which IoT is used.

## Appendix 1: Survey items

### a. Perceived Usefulness (PU): Adapted from [22, 59, 60]

**PU1**: I believe using IoT for healthcare would improve my performance.

**PU2**: I believe using IoT for healthcare would save my time.

**PU3**: I would use IoT for healthcare in any place.

**PU4**: I would find IoT for healthcare useful.

### b. Perceived Ease of Use (PEoU): Adapted from [22, 59, 60]

**PEoU1**: Learning to use IoT for healthcare purposes is easy for me.

**PEoU2**: Becoming skillful at using IoT for healthcare purposes is easy for me.

**PEoU3**: Interaction with IoT for healthcare purposes is easy for me.

**PEoU4**: I would find IoT for healthcare purposes easy to use.

### c. Attitude (ATT): Adapted from [44]

**ATT1**: I believe that using IoT for healthcare purposes is a good idea.

**ATT2**: I believe that using IoT for healthcare purposes is advisable.

**ATT3**: I am satisfied with using IoT for healthcare purposes.

### d. Behavioural Intention (BI): Adapted from [22, 23]

**BI1**: I intend to use IoT for healthcare purposes in the future.

**BI2**: I will always try to use IoT for healthcare purposes platforms.

**BI3**: I plan to use IoT for healthcare purposes frequently.

### e. Security and Privacy (SP): Adapted from [26, 61]

**SP1**: I fear to use IoT for healthcare purposes due to the loss of my personal data and privacy.

**SP2**: IoT for healthcare purposes offers a secure medium which sensitive personal information can be sent confidentially.

**SP3**: I find it Risky to disclose my personal details and health information to the IoT service providers.

**SP4**: I believe the information (personal and behavioural) being collected about me is not being used for purposes other.

**SP5**: I feel comfortable with the information being collected about me by the IoT service provider.

**SP6**: I do feel totally safe by providing personal privacy information through an IoT service provider.

### Social Influence: Adapted from [23]

**SI1**: People who are important to me think that I should use IoT for my healthcare.

**SI2**: People who are familiar with me think that I should use IoT for my healthcare.

**SI3**: People who influence my behaviour think that I should use IoT for my healthcare.

**SI4**: Most people surrounding me use IoT for their healthcare.

### f. Individual Technology Fit (ITF): Adapted from [56]

**ITF1**: I can independently and consciously use IoT for healthcare purposes.

**ITF2**: I actively engage in various types of IoT services and applications for healthcare purposes.

**ITF3**: I try to get outstanding performance when I use IoT for healthcare purposes.

### g. Task-Technology Fit (TTF): Adapted from [44, 56, 62]

**TTF1**: IoT services and applications are fit for the requirements of my healthcare.

**TTF2**: Using IoT services and applications fits with my healthcare practices.

**TTF3**: It is easy to understand which services, applications, or tools to use in IoT for healthcare.

**TTF4**: IoT applications are suitable for helping me get my healthcare services and transactions done.

**TTF5**: In general, the functions of HWDs fully meet my needs.

# References

1. Pattar S et al (2018) Searching for the IoT resources: fundamentals, requirements, comprehensive review, and future directions. IEEE Commun Surveys Tutor 20(3):2101–2132
2. Nnadi SN, Idachaba FE (2018) Design and implementation of a sustainable IOT enabled greenhouse prototype. IEEE
3. Gubbi J et al (2013) Internet of Things (IoT): a vision, architectural elements, and future directions. Futur Gener Comput Syst 29(7):1645–1660
4. Wang Y-H, Hsieh C-C (2018) Explore technology innovation and intelligence for IoT (Internet of Things) based eyewear technology. Technol Forecast Soc Change 127:281–290
5. Bsquare (2017) Annual IoT maturity survey. Available from: https://www.bsquare.com. Accessed 26 June 2021
6. Khan WZ et al (2016) Enabling consumer trust upon acceptance of IoT technologies through security and privacy model. Advanced multimedia and ubiquitous engineering. Springer, pp 111–117
7. AlHogail A, Al Shahrani M (2019) Building consumer trust to improve Internet of Things (IoT) technology adoption. In: Ayaz H, Mazur L (eds) Advances in neuroergonomics and cognitive engineering, Springer, Cham, pp 325–334
8. Al-Momani AM, Mahmoud MA, Ahmad MS (2018) Factors that influence the acceptance of internet of things services by customers of telecommunication companies in Jordan. J Organ End User Comput (JOEUC) 30(4):51–63
9. Solangi ZA, Solangi YA, Aziz MSA (2017) An empirical study of Internet of Things (IoT)—Based healthcare acceptance in Pakistan: PILOT study. In: IEEE 3rd international conference on engineering technologies and social sciences (ICETSS). IEEE, pp 1–7
10. World Bank (2021) Data practices in Mena, case study: opportunities and challenges in Jordan
11. Al-Mobaideen HO et al (2013) Electronic government services and benefits in Jordan. Int J Acad Res 5(6)
12. Alkhwaldi AF, Aldhmour FM (2022) Beyond the bitcoin: analysis of challenges to implement blockchain in the Jordanian public sector. Convergence of Internet of Things and blockchain technologies. Springer, pp 207–220
13. MoDEE ND (2021) Transform Strategy Implement Plan 2025:2021
14. TRC (2017) Green paper of "Internet of Things"
15. Sivathanu B (2018) Adoption of Internet of Things (IOT) based wearables for healthcare of older adults—A behavioural reasoning theory (BRT) approach. J Enabling Technol
16. Da Costa CA et al (2018) Internet of health things: toward intelligent vital signs monitoring in hospital wards. Artif Intell Med 89:61–69
17. Singh RP et al (2020) Internet of Things (IoT) applications to fight against COVID-19 pandemic. Diabetes Metab Syndr 14(4):521–524
18. Yildirim H, Ali-Eldin AMT (2019) A model for predicting user intention to use wearable IoT devices at the workplace. J King Saud Univ Compu Inf Sci 31(4):497–505
19. AlHogail A (2018) Improving IoT technology adoption through improving consumer trust. Technologies 6(3):64

20. Patel KK, Patel SM (2016) Internet of things-IOT: definition, characteristics, architecture, enabling technologies, application & future challenges. Int J Eng Sci Comput 6(5)
21. Lu Y, Papagiannidis S, Alamanos E (2018) Internet of Things: A systematic review of the business literature from the user and organisational perspectives. Technol Forecast Soc Change 136:285–297
22. Alkhwaldi AF, Absulmuhsin AA (2021) Crisis-centric distance learning model in Jordanian higher education sector: Factors influencing the continuous use of distance learning platforms during COVID-19 pandemic. J Int Educ Bus
23. Venkatesh V et al (2003) User acceptance of information technology: toward a unified view. MIS Q 425–478
24. Venkatesh V, Thong JYL, Xu X (2012) Consumer acceptance and use of information technology: extending the unified theory of acceptance and use of technology. MIS Q 36(1):157–178
25. Alkhwaldi A, Kamala M, Qahwaji R (2018) Analysis of cloud-based E-government services acceptance in Jordan: challenges and barriers. J Internet Technol Secur Trans (JITST) 7(2):556–568
26. Lian J-W (2015) Critical factors for cloud based e-invoice service adoption in Taiwan: an empirical study. Int J Inf Manage 35(1):98–109
27. Lin D, Lee CKM, Lin K (2016) Research on effect factors evaluation of internet of things (IOT) adoption in Chinese agricultural supply chain. In: IEEE international conference on industrial engineering and engineering management (IEEM). IEEE
28. Sharma V et al (2020) Security, privacy and trust for smart mobile-Internet of Things (M-IoT): a survey. IEEE Access 8:167123–167163
29. Alraja MN, Farooque MMJ, Khashab B (2019) The effect of security, privacy, familiarity, and trust on users' attitudes toward the use of the IoT-based healthcare: the mediation role of risk perception. IEEE Access 7:111341–111354
30. Kimani K, Oduol V, Langat K (2019) Cyber security challenges for IoT-based smart grid networks. Int J Crit Infrastruct Prot 25:36–49
31. Badii C et al (2020) Smart city IoT platform respecting GDPR privacy and security aspects. IEEE Access 8:23601–23623
32. Goyal P et al (2021) Internet of Things: applications, security and privacy: a survey. Mater Today: Proc 34:752–759
33. Dajani D, Yaseen SG (2016) The applicability of technology acceptance models in the Arab business setting. J Bus Retail Manage Res 10(3)
34. Chang K, Chang C-C (2009) Library self-service: predicting user intentions related to self-issue and return systems. Electron Libr 27(6):938–949
35. Patil K (2016) Retail adoption of Internet of Things: applying TAM model. In: IEEE international conference on computing, analytics and security trends (CAST). IEEE
36. Omotayo FO, Haliru A (2020) Perception of task-technology fit of digital library among undergraduates in selected universities in Nigeria. J Acad Librariansh 46(1):102097
37. Mohammed F et al (2017) Cloud computing adoption model for e-government implementation. Inf Dev 33(3):303–323
38. Rogers EM (1995) Diffusion of innovations. New York, p 12
39. Alkhwaldi A, Kamala M (2017) Why do users accept innovative technologies? A critical review of technology acceptance models and theories. J Multidisc Eng Sci Technol (JMEST) 4(8):7962–7971
40. Davis FD, Bagozzi RP, Warshaw PR (1989) User acceptance of computer technology: a comparison of two theoretical models. Manage Sci 35(8):982–1003
41. Ajzen I, Fishbein M (1975) Belief, attitude, intention and behavior: an introduction to theory and research. Addison-Wesley, Reading, MA
42. Sinha A et al (2019) Impact of Internet of Things (IoT) in disaster management: a task-technology fit perspective. Ann Oper Res 283(1):759–794
43. Goodhue DL, Thompson RL (1995) Task-technology fit and individual performance. MIS Q 213–236

44. Kim T et al (2010) Modelling roles of task-technology fit and self-efficacy in hotel employees' usage behaviours of hotel information systems. Int J Tour Res 12(6):709–725
45. Borhan SRFM et al (2019) Feasibility of IoT acceptance among Malaysian government agencies considering security factors. In: IEEE 6th international conference on research and innovation in information systems (ICRIIS)
46. Alkhwaldi A, Kamala M, Qahwaji R (2019) Security perceptions in cloud-based e-government services: integration between citizens' and IT-staff Perspectives. In: 12th International conference on global security, safety & sustainability (ICGS3-2019). IEEE, Northumbria University, London, England
47. Dajani D (2016) Using the unified theory of acceptance and use of technology to explain e-commerce acceptance by Jordanian travel agencies. J Comparat Int Manage 19(1):99–118
48. Alkhwaldi A (2019) Jordanian citizen-centric cloud services acceptance model in an e-government context: security antecedents for using cloud services. In: Faculty of engineering and informatics. University of Bradford, Bradford, UK
49. Dawes J (2008) Do data characteristics change according to the number of scale points used? An experiment using 5-point, 7-point and 10-point scales. Int J Mark Res 50(1):61–104
50. Sekaran U, Bougie R (2016) Research methods for business: a skill-building approach, 7th edn. Wiley, Chichester
51. Fornell C, Larcker DF (1981) Evaluating structural equation models with unobservable variables and measurement error. J Mark Res 18(1):39–50
52. Hair JF et al (2014) Multivariate data analysis: Pearson new international edition, always learning. Pearson Harlow, Essex
53. Shin D-H (2013) User centric cloud service model in public sectors: Policy implications of cloud services. Gov Inf Q 30(2):194–203
54. Dwivedi YK et al (2017) An empirical validation of a unified model of electronic government adoption (UMEGA). Government Information Quarterly
55. Dishaw MT, Strong DM (1999) Extending the technology acceptance model with task–technology fit constructs. Inf Manage 36(1):9–21
56. Yu TK, Yu TY (2010) Modelling the factors that affect individuals' utilisation of online learning systems: an empirical study combining the task technology fit model with the theory of planned behaviour. Br J Edu Technol 41(6):1003–1017
57. Alanazi MH, Soh B (2019) Behavioral intention to use IoT technology in healthcare settings. Eng Technol Appl Sci Res 9(5):4769–4774
58. Venkatesh V, Davis FD (2000) A theoretical extension of the technology acceptance model: four longitudinal field studies. Manage Sci 46(2):186–204
59. Wu B, Zhang C (2014) Empirical study on continuance intentions towards E-learning 2.0 systems. Behav Inf Technol 33(10):1027–1038
60. Venkatesh, V. and X. Zhang, Unified theory of acceptance and use of technology: US vs. China. Journal of global information technology management, 2010. 13(1): p. 5–27.
61. Alshurideh MT, Al Kurdi B, Salloum SA (2021) The moderation effect of gender on accepting electronic payment technology: a study on United Arab Emirates consumers. Rev Int Bus Strategy
62. Wang H et al (2020) Understanding consumer acceptance of healthcare wearable devices: an integrated model of UTAUT and TTF. Int J Med Inform 139:104156

# Digital Healthcare Provision Policies in United Arab Emirates (UAE) Amid COVID-19

Tahira Yasmin, Ghaleb A. El Refae, and Shorouq Eletter

**Abstract** The COVID-19 crisis has caused sudden pressure on health sector all around worldwide to meet the emergent safety and care needs. The pandemic has caused many governments to invest in technology to meet the increased healthcare demand. In the same lines, current study discussed the various digital solutions used in UAE healthcare sector amid pandemic. These solutions helped to facilitate people at wide scale while dropping the infection rate in the country. The adoption of technologies also displays the collaborated role of UAE government and health authorities. This unified approach has made UAE as a safest destination for nationals, residents and visitors. The innovative policies helped to meet the sudden health care challenges while paving the new road towards telemedicine. The study suggested that careful policies needed to have sustainable implementation of digital health approach.

**Keywords** COVID-19 · Digitalization · Health sector · Innovative policy · Telemedicine

## 1 Introduction

The sudden shock of COVID-19 pandemic has changed the world and lives of people as economically, socially and emotionally. The current crises has led to a digital move in all sectors from education to health and daily businesses. The nature of COVID-19 crisis posed a biggest challenge for health care sector services worldwide in last century. The economies were implementing various measures as strict lockdown, wearing masks and social distancing to control the daily infection rate. But at the same time, the hospitals were meeting every day challenges by adapting new strategies and solutions. The evolution of technology in health care was not new but it has tremendously accelerated the digital technologies usage with recent developments.

T. Yasmin (✉) · G. A. El Refae · S. Eletter
Al Ain Univeristy, Abu Dhabi, UAE
e-mail: tahira.yasmin@aau.ac.ae

215

The safety of health care workers is crucial at the time of COVID-19, so it has forced the healthcare providers and patients to move towards digital techniques. The use of information communication technology considered as digital healthcare adoption is becoming popular during quarantine and social distancing. Jnr [1] has discussed the role of telehealth, types and application of telehealth and digital care solutions. He presented the institutional, infrastructure and human determinants which influenced the digital healthcare and proposed some solutions. This point was justified by Wang et al. [2], they used the qualitative and quantitative data in China's Hubei province to explored the effectiveness of e-health during COVID-19. They justified that digital health has largely satisfied the patients and users in crisis period. By incorporating digital methods in healthcare like online medical consultation, registration and medical resources has enhanced the communication speed by reducing physical co-presence. They further concluded from their study that in some diseases need medical equipment for accurate diagnosis. They suggested that digital health became successful during high infection rate but this needs comprehensive analysis for long term usage in normal time. The e-health acted as functional tool amid pandemic and fostered digitization in medical system. Furthermore, Al-Shayea [3] highlighted that the use of artificial intelligence is very effective in medical field. They suggested a neural network for diagnostic purpose by using the data from UCI machine learning repository. They studies three diagnostic diseases in the simulation samples feed-forward back propagation network as for acute nephritis (100%), heart (88%) and disk hernia or spondylolisthesis (82%) respectively. In another study, Al-Shayea [4] evaluated the artificial intelligence in disease diagnosis and studies two cases as acute nephritis disease and heart disease respectively. He used feed-forward back propagation neural network to distinguish between infected and non-infected person. The study concluded the correct percent as in diagnosis of acute nephritis disease with 99% whereas in the heart disease it was 95% respectively.

Earnest and Young [5] has investigated the challenges which health and human services (HHS) faced in implementation of digital solutions. The sudden outbreak of COVID-19 was catastrophic in its effects and it put an extreme stress on HHS. Due to social distancing, lockdown measures and infection control compelled HHS organizations to shift from in person to digital services in overnight. Based on multi-country survey around 62% of HHS organizations increased the digital use during pandemic. Moreover, the use of digitalization has also enhanced the quality of service users and raised staff productivity. The findings also came up with the interesting facts that UAE is ahead in in terms of digital technologies adaptation while having 86% enhanced productivity. Figure 1 displays that prior to outbreak there was largely implementation of digital technologies with mainly three concerns as practitioner, ethics and IT adaptability. But amid pandemic the key factors in adaptation of digital technologies turned as alleviation of practitioner concerns, improved digital literacy and also mitigation of privacy concerns.

Moreover, Fig. 2 shows that there was significant use of digital technologies and data solutions in UAE during pandemic. The overwhelming percentage elaborates that digital tactics and infrastructure scale up in UAE with recent developments and

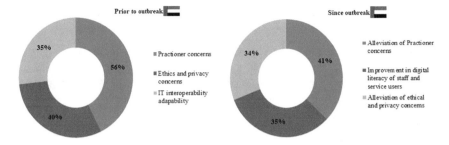

**Fig. 1** Pre and during COVID-19 concerns. *Source* [5]

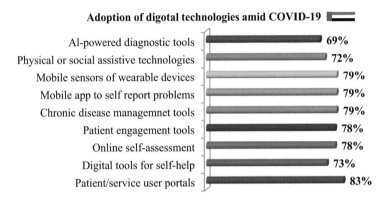

**Fig. 2** Adaption of digital technologies in UAE during COVID-19. *Source* [5]

investments in healthcare system. The collected data also shows the positive response by patients in using digital self-assessment tools and technologies.

Based on the above discussion this article highlights the digital methods which were employed in UAE to combat health concerns during Covid-19. The article then reviews the literature to shed light on digital approach in healthcare sector. After presenting the country's innovative ways of dealing with COVID-19, it has discussed that to follow social distancing, lockdown and huge number of patients the digital healthcare as telemedicine proved to be successful. The research objective of this study is to elaborate that how the adoption of recent digital health policies during COVID-19 harnessed the safety measures in UAE. Therefore, to address this objective this study adds the body of knowledge by reviewing effective policies to describe the role of telemedicine approaches. This study can offer some new insights on how ICT can help to manage the current pandemic crisis. Furthermore, it is urging the policy makers and health care sector decision makers to design a balanced approach which will be digitally designed with effective cost management and lead to high productivity.

## 2   Overview of Digital Healthcare Measures in UAE

UAE government has put various smart solutions to handle the drastic impact of COVID-19. These technologies help to protect frontline health workers, patients and overall community with minimization of physical interaction. Some of the these digital innovative techniques to handle COVID-19 are as below [6].

## 3   Virtual Doctor for COVID-19

To protect the community from this disease, Ministry of Health and Prevention (MoHaP) launched a chat box named as, "Virtual Doctor for COVID-19". Due to social distancing, this helped UAE people to detect their symptoms online that whether they are associated with novel coronavirus or not? At the same time, the system asked certain questions online from the person regarding travel history, close contact with someone who has COVID-19 and may suffer from certain symptoms. Upon answering the questions, the person will assist to a relevant doctor through the same service. Moreover, Department of Health Abu Dhabi also introduced DOH RemoteCare app which is another digital solution where people can get remote healthcare by booking appointments and getting tele-consultations with doctors via video or voice calls. Patients can also request the online prescription and medicines will be delivered to home.

This virtual doctor concept also known as, 'telemedicine', and this turned as an effective strategy in digital healthcare provision. Jnr [7] presented the extensive benefits of telemedicine, trends and recent policies in healthcare sector during COVID-19. He used extensive literature by considering the inclusion and exclusion criteria, quality assessment and data extraction to present the findings. He mentioned the integrated benefits of using telemedicine as it is suitable and effective during COVID-19. D'Souza et al. [8] also justified the adoption of telemedicine services at a tertiary care center in south India by using 456 patients satisfaction data. They reported maximum satisfaction among all age groups and this further lead various effective insights on best practices of telemedicine. Raparia and Husain [9] specifically discussed the use of telemedicine in retinal disease with significant advancements before and during COVID-19. They concluded that this telecare will continue expand with advanced diagnostic tools. Various researchers have supported the use of telemedicine is not a perfect substitute of health provision but it worked as effective strategy during COVID-19 pandemic in order to meet the increased healthcare demand [10–16].

## 4  The ALHOSN UAE App

The ALHOSN app is officially launched as an integrated part of digital healthcare for COVID-19 tests in UAE. Through this application the users can receive the result of COVID-19 test on smart phones and this application further detects if someone has close contact with the infected patient. All the users have unique QR code which includes information about user health. This application is encrypted so the data remained on the phone of user and can open and update anytime. This data further used by health authorities to identify the high risk patients and have also privacy protection through artificial intelligence. This app also combines the benefits of the, "Stay Home" and "TraceCovid" apps to enhance the safety of people. The Stay Home app helped to direct self-quarantine people in order to follow the requirements and also person can share the location to Department of Health. Whereas, the TraceCovid app helps in tracing COVID-19 cases by recording all the people who were near to COVID-19 patients. Furthermore, the Abu Dhabi Government is using the electronic bands as part of self-isolation and this is also linked with the ALHOSN app.

Jacob and Lawarée [17] highlighted the use of contact tracing application in Europe to monitor the contact with COVID patient. This digital contact tracing progressively been widely used as a solution to slow down the virus spread. They mentioned that despite the high acceptability rate there are certain technical issues relevant to digital tracing. Gupta et al. [18] analyzed the security features of these trace applications by considering Data Privacy, Security Vetting and design architecture. These aspects were further justified by different country studies by considering the public views and implications of such trace apps [19–21].

## 5  AI in Taxis to Curb the Spread of COVID-19

Artificial Intelligence (AI) technologies have employed by Dubai's Roads and Transport Authority (RTA) to monitor and report the COVID-19 violations. This technology used to monitor physical distancing and wearing masks in taxis by employing computer vision and machine learning algorithms. This initiative has helped to monitor the rules and regulations for the taxi drivers and for the riders.

## 6  OMAC COVID-19 Hackathon

The formation of One Million Arab Coders (OMAC) COVID-19 Hackathon aimed to build an innovative platform in education and health which emerged during COVID-19. This provides an opportunity for Arab coders to share their skills and knowledge employing in programming language to come up with new solutions. This initiative has also established a jury which will select 15 best teams to present their project and Dubai Future Foundation supported 5 selected projects with allocation of USD 50,000.

The COVID-19 pandemic urged the world to adapt new and immediate ways to combat various challenges. Sy et al. [22] mentioned that it is essential to promote various inter professional research works to come up with new policies and recommendations. This can result in valuable contribution as future research and information to combat future health crisis. You [23] argued that international collaboration needed to avoid future pandemic crisis. He used the South Korean government press briefings and public documents to highlight that how the country has shared the resources and expertise in combatting COVID-19. The author concluded that how much international cooperation is important in crisis management.

# 7 Smart Helmet to Monitor Coronavirus

UAE authorities used Smart helmets to monitor the people who may be infected with COVID-19. They introduced these helmets equipped with thermal cameras to detect the infected people from safe distance. Additionally, these helmets can read and create QR codes by recognizing vehicle license plat numbers. This proved to be an effective tool for patrols and police teams to handle the people in crowd and obtain vital information.

# 8 Concluding Remarks

Based on the research, it is concluded that [24] UAE aimed to provide the wide ranging digital health facilities around emirates. There are some crucial challenges in implementation of e-health so it is important to set the clear directions for privacy, safety and security in healthcare delivery. In this context, it is important to set the vision and priority areas for the health sector to leverage the digital technology successfully. There are four priority areas mentioned by DOH in line with international and national recommendations as person centricity, infrastructure, legal and ethical framework and flexible and adaptable regulation. These policies will help to develop a sustainable digital healthcare system to meet the needs while maintaining the high quality. As it has mentioned that since outbreak the UAE organizations have adopted digital technologies and one of the main factor was alleviation of practitioner concerns (refer Fig. 1).

During COVID-19, the continuous efforts and effective healthcare policies in time implementation has made UAE as a leading country in care and safety. Not only this, the continuous investment in this sector throughout the pandemic has turned as developed infrastructure and fundamental structural changes. UAE has proved to mitigate the risk and limit the mortality rate by effective healthcare policies with particular focus on telehealth. The use of digital methods helped all hospitals in UAE to provide the health facilities to patients while reducing the waiting and consultation times. There are extensive free healthcare services provisions in UAE such as free testing at airports, free digital applications and databases, quarantine digital watches and wide spread screening centers.

In the end, with ongoing efforts UAE has to think carefully that how to manage the digital healthcare for the long term as cost effectiveness, managing insurance and quality healthcare delivery. As the health care provision has gone through transition to a new advance phase by having widely accepted digital methods. There is continuous efforts need to monitor the health sector progress as COVID-19 crisis will gradually ends. But the question is that once the health sector conditions will normalize how the sector will follow long term sustainable digital plans with various strategies. It is vital that the traditional methods once replaced with digital ones also need customer satisfaction strategies as this may differs in some special care services. Furthermore, under the endorsement of Ministry of Health and Prevention has discussed digital healthcare roadmap in GCC. The main discussed aspects were cost efficiency, patient centric approach, hiring requirements and precision medicine in healthcare delivery system [25].

# References

1. Jnr BA (2021) Implications of telehealth and digital care solutions during COVID-19 pandemic: a qualitative literature review. Inform Health Soc Care 46(1):68–83. https://doi.org/10.1080/17538157.2020.1839467
2. Wang W, Sun L, Liu T, Lai T (2021) The use of E-health during the COVID-19 pandemic: a case study in China's Hubei province. Health Soc Rev. https://doi.org/10.1080/14461242.2021.1941184
3. Al-Shayea Q, El-Refae G, Yaseen S (2013) Artificial neural networks for medical diagnosis using biomedical dataset, Int J Behav Healthcare Res 4(1):45–63
4. Al-Shayea Q (2011) Artificial neural networks in medical diagnosis. Int J Comput Sci Issues 8(2):150–154
5. Earnest and Young (EY) (2021) Embracing digital: is COVID-19 the catalyst for lasting change?
6. UAE Government (2021). Smart solutions to fight COVID-19. https://u.ae/en/information-and-services/justice-safety-and-the-law/handling-the-covid-19-outbreak/smart-solutions-to-fight-covid-19
7. Jnr BA (2021) Integrating telemedicine to support digital health care for the management of COVID-19 pandemic. Int J Healthcare Manage 14(1):280–289. https://doi.org/10.1080/20479700.2020.1870354
8. D'Souza B, Rao SS, Hisham S, Shetty A, Sekaran VC, Pallagatte MC (2021) Healthcare delivery through telemedicine during the COVID-19 pandemic: case study from a tertiary care center in South India. Hosp Top. https://doi.org/10.1080/00185868.2021.1875277
9. Raparia E, Husain D (2021) COVID-19 launches retinal telemedicine into the next frontier. Seminars Ophthalmol 36(4):258–263. https://doi.org/10.1080/08820538.2021.1893352
10. Burroughs M, Urits I, Viswanath O, Simopoulos T, Hasoon J (2020) Benefits and shortcomings of utilizing telemedicine during the COVID-19 pandemic. Baylor Univers Med Center Proc 33(4):699–700. https://doi.org/10.1080/08998280.2020.1792728

11. Rauseo-Ricupero N, Henson P, Mica Agate-Mays M, Torous J (2021) Case studies from the digital clinic: integrating digital phenotyping and clinical practice into today's world. Int Rev Psychiatry 33(4):394–403. https://doi.org/10.1080/09540261.2020.1859465

12. Dash M, Shadangi PY, Muduli K, Luhach AK, Mohamed A (2021) Predicting the motivators of telemedicine acceptance in COVID-19 pandemic using multiple regression and ANN approach. J Stat Manag Syst 24(2):319–339. https://doi.org/10.1080/09720510.2021.1875570

13. Kaliya-Perumal A, Omar UF, Kharlukhi J (2020) Healthcare virtualization amid COVID-19 pandemic: an emerging new normal. Med Educ Online 25:1. https://doi.org/10.1080/10872981.2020.1780058

14. Leite H, Hodgkinson IR, Gruber T (2020) New development: 'Healing at a distance'— Telemedicine and COVID-19. Public Money Manage 40(6):483–485. https://doi.org/10.1080/09540962.2020.1748855

15. Palabindala V, Bharathidasan K (2021) Telemedicine in the COVID-19 era: a tricky transition. J Commun Hospital Internal Med Perspect 11(3):302–303. https://doi.org/10.1080/20009666.2021.1899581

16. Ranavaya M, Vieira DN (2021) COVID-19 pandemic and evolution of telemedicine to TeleIME. Forensic Sci Res. https://doi.org/10.1080/20961790.2021.1895411

17. Jacob S, Lawarée J (2021) The adoption of contact tracing applications of COVID-19 by European governments. Policy Design Practice 4(1):44–58. https://doi.org/10.1080/25741292.2020.1850404

18. Gupta R, Pandey G, Chaudhary P, Pal SK (2021) Technological and analytical review of contact tracing apps for COVID-19 management. J Location Based Services. https://doi.org/10.1080/17489725.2021.1899319

19. van Kolfschooten H, de Ruijter A (2020) COVID-19 and privacy in the European Union: a legal perspective on contact tracing. Contemp Secur Policy 41(3):478–491. https://doi.org/10.1080/13523260.2020.1771509

20. Samuel G, Roberts SL, Fiske A, Lucivero F, McLennan S, Phillips A, Hayes S, Johnson SB (2021) COVID-19 contact tracing apps: UK public perceptions. Crit Public Health. https://doi.org/10.1080/09581596.2021.1909707

21. Gasteiger N, Gasteiger C, Vedhara K, Broadbent E (2021) The more the merrier! Barriers and facilitators to the general public's use of a COVID-19 contact tracing app in New Zealand. Inform Health Soc Care. https://doi.org/10.1080/17538157.2021.1951274

22. Sy M, O'Leary N, Nagraj S, El-Awaisi A, O'Carroll V, Xyrichis A (2020) Doing interprofessional research in the COVID-19 era: a discussion paper. J Interprof Care, 34(5):600–606. https://doi.org/10.1080/13561820.2020.1791808

23. You J (2021) Advancing international cooperation as a strategy for managing pandemics. Asia Pac J Public Admin. https://doi.org/10.1080/23276665.2020.1866624

24. Department of Health (DOH) (2020) Digital policy on digital health. https://www.doh.gov.ae/en/resources/policies

25. Middle East Digital Health Forum (2021). https://dhf.khaleejtimesevents.com/

# Opportunities and Challenges of Applying Electronic Human Resources Management in Business Organizations an Applied Study in the Telecommunications Sector, Jordan

Abdul Azeez Alnadawi, Manal Abdulrahman, and Fatimah Omran

**Abstract** Electronic human resources management constitutes an important element in achieving success for business organizations today. The use of electronic human resources management techniques directly contributes to creating effective human resources, which is the decisive factor in achieving organizational goals. Business organizations are currently facing great challenges in various fields of their work, especially in the field of electronic human resources management and its role in providing appropriate and qualified talents to accomplish work and achieve success by taking advantage of the available opportunities and addressing the threats. The aim of this study is to identify the reality of electronic human resources management in business organizations and its role in contributing in the development of work and assisting organizations in taking advantage of the available opportunities and more effectively facing the challenges. The sample of the study amounted to (332) respondents who were selected from among the employees of the three Jordanian telecommunications companies. One of the most prominent findings of the study is the presence of a clear and statistically significant impact of electronic human resources management on the work of Jordanian telecommunications companies, taking advantage of the available opportunities and contributing to effectively facing the challenges.

**Keywords** Human resources management · Electronic human resources management · Information technology · Electronic human resources systems · E-management · E-business

## 1 Introduction

Electronic HRM is the new technological tool which is gaining widespread importance within business organizations around the world [1].

A. A. Alnadawi (✉) · M. Abdulrahman · F. Omran
Al Zaytoonah University of Jordan, Amman, Jordan
e-mail: a.alnadawi@zuj.edu.jo

© The Author(s), under exclusive license to Springer Nature Switzerland AG 2022
S. G. Yaseen (ed.), *Digital Economy, Business Analytics, and Big Data Analytics Applications*, Studies in Computational Intelligence 1010,
https://doi.org/10.1007/978-3-031-05258-3_19

Human resources management is the effective engine for the development of business organizations and it is the group of activities that seek to recruit, employ, develop and maintain the human component in business organizations, in addition to the training and evaluation processes. In light of the changes that the global economy is going through, human resources management is exposed to a fundamental transformation in its concepts, policies, strategies and practices. One of the most important factors that significantly affect human resources management is the impact factor of technology and information technology systems [17]. Organizations have witnessed huge technological developments in various fields with the invasion of information technology in all aspects of administrative work, this has led to the emergence of new terms in the environment of business organizations such as: E-Management, E-Business, E-Marketing, E-Learning, E-Government, E-Recruiting, and new names that relate to this development and the use of technology in the field of human resources management has resulted in what is called E-Human Resources Management (E -HRM) [4].

This study aims to analyze and discuss the term of electronic human resources management in order to understand the main opportunities arising from this term and the main challenges that are faced and which it can positively contribute in solving. Furthermore, this study attempts to provide organizations with recommendations to enhance the level of electronic human resources management application and the ability to take advantage of the benefits and opportunities it provides. This study is also considered a new reference to researchers and academics in Jordan and the Arab World, presented through a theoretical and practical framework, regarding electronic human resources management and its major and supportive activities while also presenting the opportunities that electronic human resources management can create and the main challenges that it can contribute to solving in organizations. Furthermore, this study forms the basis from which researchers can launch towards broader areas of investigation regarding the scientific variables identified in the study and the related challenges and their impact. Providing a modern and contemporary reference to administrative leaders and decision-makers in different organizations, through explaining the impact of electronic human resources management activities in business organizations, which include generating new ideas and solutions to treat the problems that organizations face, while reflecting positively on continuity and success. This study serves to highlight the most important factors that contribute to the achievement of organizational sustainability, as well as supporting competitiveness for the organization and helping to exploit available opportunities and avoid threats, reflecting positively on the growth, continuity and success of the organization [5].

Ultimately, this study was launched to answer the following main questions:

Is there a statistically significant impact of electronic human resources management to create opportunities and take advantage of them in business organizations? Is there a statistically significant impact of electronic human resources management on the contribution to facing the challenges of business organizations? What are the main opportunities generated by electronic human resources management facing business organizations?

## 2 Literature Review

### 2.1 Concept of Electronic Human Resources Management

Human resources management is considered the main factor for organizing the relationship between the organization and its employees, and between the organization and society. It aims to achieve a set of goals related to the organization and its employees through activities, programs and strategies related to the recruitment, selection, maintenance and development of distinguished human resources which in turn contributes to effectively and efficiently achieving the goals of the organization [5, 18].

In the recent modern era, human resources management practices and functions have been developing in accordance to the changes in the internal and external environments of the organization, resulting in the introduction of information technology in their practices, which has led to the creation of a set of new opportunities and challenges to human resources management and has contributed to the expansion of the role of human resources management to become more comprehensive in the organization. Human resources management activities and functions have become an integrated electronic system. This development has been reflected in the field of human resources management and practices, and a number of new names have emerged in this field: human resources [2, 10].

The concept of electronic human resources management is embodied in, and contributes to the following [11]:

- Using a set of techniques that assist managers and employees in the field of human resources management in carrying out their job in various fields such as recruitment, selection, training, etc.
- Creating a new methodology based on the conscious use of information and communication technologies in human resources management practices according to a new style of thinking and the use of effective mechanisms in reaching the specified objectives.
- Creating a new administrative style of work that is based on the use of information and communication technologies in the management and development of human resources within the organization in a manner that achieves complementarity between administrative work and information technology [7].

### 2.2 Elements of Electronic Human Resources Management

Electronic human resources management consists of a group of components that are electronically related to each other in achieving the desired goal. The following is a description of the most prominent components of electronic human resources management [23]:

- Human staff: It consists of a group of persons who are scientifically and technically qualified and who have the ability to innovate and adopt new ideas. They have the knowledge and ability to use modern technologies, as well as exploit capabilities to achieving the goals efficiently and at higher efficacy [16].
- Networks: It forms a group of communication systems that work to link technical devices to each other in order to achieve information sharing and exchange. These networks include:

  Local networks [20].
  Global Networks [2]

- Databases: A group of incubators dedicated to collecting information, which makes it easier for users to obtain information and deal with it in the process of storing and retrieving, in a way that leads to achieving ease and speed of work procedures.
- Internet networks: It is a vast group of communication networks that provide users with the required information in a rapid and less time-consuming manner, leading to excellent results in the areas of the organization's work.
- Information: A set of scientific material that is intended to be dealt with or shared among the members of the network and is in various forms such as: data, information, pictures, conversations, files and others.
- Computer devices: The group of devices through which communication and data exchanges between members of the network take place.

## 2.3   The Fields of Electronic Human Resources Management

The areas of electronic human resources management are grouped in to four and are defined as follows [23].

**Main Services**  It forms a group of major outputs for the work of electronic human resources management, which enables other non-specialists to participate and work to accomplish tasks such as:

1.  Employee self-service.
2.  Employee portal.
3.  Manager self-service.
4.  Strategic and collaborative.

The provision of basic electronic services achieves a set of benefits that can be identified as follows [15]:

1.  Keeping information more accurate and secure.
2.  Contributing to reducing administrative costs for human resources management.
3.  Reducing the use of paper and reducing its storage locations.
4.  Achieving integration in the employment process.
5.  Achieving performance efficiency in relation to administrative information.

6.   Achieving effective and efficient supervision in the various stages of work.

**Personnel Management** The process of using electronic tools in recruitment, testing and selection, where the internet and the intranet are used in the processes to fill vacancies in the organization. Organizations are also using electronic means to conduct interviews through the organization's electronic human resources management system, which also enhances the human resources process, thus, creating a set of opportunities for the organization to obtain various competencies, use extensive research, in addition to reducing the time and costs of the entire process [2].

**Electronic Training** The electronic and technological developments in the field of electronic human resources management constitute an opportunity to find mechanisms for employee electronic training programs that contribute to transferring the most prominent needs of workers in their field of work and developing their capabilities, while significantly reducing training expenses as well as shortening the time of the training process. Electronic training also provides vast opportunities for a large number of workers to participate in the e-training process, as well as overcoming many challenges in this field, including the ability to provide feedback that serves and develops the training process for workers [3].

**Electronic Evaluation** Electronic human resources management has created a possibility to use electronic evaluation and follow-up tools in various fields of business. This is done through the ability to automate many of the forms used in the evaluation process and through providing a system that contributes to obtaining integrated performance reports by electronically using various monitoring and follow-up systems. Electronic evaluation can be used more than once or continuously, and can also provide the ability to use a variety of criteria in the evaluation process. The use of electronic evaluation has provided many opportunities for success in obtaining a fair and effective evaluation process, while at the same time, it has worked to overcome many of the challenges and problems that managers faced in the classic evaluation process [1].

**Development of Job Benefits** The electronic human resources department has provided all workers in the organization with access to a set of advantages that can be obtained in a simple and easy manner and constitute a set of opportunities for workers to get acquainted with their entire job status through the use of a set of systems and technologies that are available in the field of electronic human resources management [12, 14].

# 3 Obstacles to the Implementation of Electronic Human Resources Management

The use of electronic human resources management faces many obstacles in its application, and we can define these obstacles within three groups [8]:

**First: Administrative Obstacles Related to the Administrative Issues of the Organization Include**

1. The lack of planning and coordination at the senior management level of electronic human resources management programs.
2. The lack of enthusiasm by upper management in evaluating and following up the implementation of electronic human resources management.
3. The complexity of administrative procedures and the lack of legislation and regulations governing electronic human resources management policies and programs.
4. The imprecise application of electronic human resources management may lead to paralysis in the administration's functions.

**Second: Human Obstacles Related to Workers Obstacles Include**

1. Aging technical know-how and the resistance to using modern technologies.
2. Lack of training programs in the field of modern technologies.
3. The awe of workers in dealing with electronic devices.

Fear of leakage of information related to people working in the organization, causing espionage that leads to great risks [6].

**Third: Financial Constraints Related to Financing Obstacles Include**

1. The high costs of purchasing and maintaining updated electronic and technical devices.
2. The lack of financial resources allocated to providing the electronic infrastructure necessary in the organization.
3. The limited budget allocated to the process of training workers in the field of information systems.
4. The high costs of developing electronic technologies and services supporting the electronic system.

# 4 Requirements for the Implementation of Electronic Human Resources Management

Electronic human resources management and its applications represent a comprehensive and significant development in management concepts and theories [8], as they exceed the methods, procedures and structures that underpin traditional management. Electronic human resources management is a complex process that requires the availability of an integrated system of technical, informational, financial, legislative, environmental and human components. In order to implement electronic human resources management, a set of integrated issues are required [9, 19, 21] and they are as follows:

- Support by top management of the organization.
- Adapting organizational structural elements.
- Achieving change in organizational culture.
- Developing the appropriate human personnel to implement electronic human resources management techniques.
- Providing the appropriate environment for the application of electronic human resources management.
- Ensuring information security and protection in electronic human resources management.

## 5 The Benefits of Applying Electronic Human Resources Management

The commitment by business organization of using and applying electronic human resources management that will result in many advantages and benefits to the organization and will contribute to excellence in the work of organizations [13, 16] are reflected as follows:

- Developing work mechanisms for electronic human resources management.
- Increasing the interconnection between electronic human resources management and the organization's senior management.
- Facilitating the participation process and reviewing the activities of electronic human resources management.
- Increasing the communication process between workers and between the electronic human resources department and other departments in the organization.
- Raising the efficiency of performance for workers in the electronic human resources department.
- Overcoming the daily work problems in the organization.
- Reducing the costs of administrative procedures.
- Reducing time to complete the specific tasks of human resources management.
- Achieving accuracy, objectivity and clarity in administrative processes related to human resources.
- Making correct decisions by providing continuously accurate information.

**Table 1** Distribution of population sample

| No | Company | No. of employees |
|----|---------|------------------|
| 1 | Orange | 527 |
| 2 | Zain | 416 |
| 3 | Umnia | 383 |
|  | Total | 1326 |

# 6 Field Study

## 6.1 Methodology

For the purpose of this study, the researchers followed a paradigm of a descriptive and analytical nature, using questionnaires to answer the research questions.

## 6.2 Population

The population of the study consists of the three (3) main telecommunications companies in the Hashemite Kingdom of Jordan, which are comprised of (Zain, Jordan Telecommunications Co. Orange and Umnia). The researchers chose the population from the companies consisting of top-level managers, middle level managers and human resources department employees in the above-mentioned telecommunications companies. The population consists of 1326 employees representing the human resource departments in the Jordan telecommunications companies. Table 1 clarifies the distribution of the study population according to the different telecommunications companies.

## 6.3 Sample

The study sample is made up of all top-level managers, middle level managers and human resources department employees in the Jordanian telecommunications companies. The simple random sampling method was used to choose the randomized sample for the research population. The researchers took an acceptable representative sample for the research community according to [22], which state that the acceptable sample size for such a population should be (300) samples or more. Accordingly, (400) questionnaires were distributed to the sample, of which the researchers received (356) responses, resulting in an (89%) response rate. After conducting different accuracy and reliability tests on the returned questionnaires, to ensure statistical usefulness, (24) incomplete questionnaires were excluded. The researchers adopted (332) as the final number of acceptable questionnaires.

## 6.4 Validity and Reliability of the Research Instrument

A survey questionnaire was used as the research instrument for collecting data from the sample population representing the human resources departments in the Jordanian Telecommunications Companies. In order to test the validity and reliability of the survey questions, they were presented to a number of specialized academic adjudicators to receive feedback on clarity and understanding of all survey paragraphs and questions. All observations and suggestions were taken into account and the appropriate corrections were made according to the adjudicators reflections and remarks, resulting in the increased strength and clarity of the design and formulation of the questions in the research instrument.

## 6.5 Testing the Reliability of the Research Instrument

Testing for the assurance of the reliability of the research instrument used was carried out by measuring of the variables included in the survey questionnaire through the use of Cornbrash Alpha factors measures of internal consistency. The result of the Cornbrash Alpha measures for the final sample was (0.868), which assures that the research instrument is characterized by consistency.

# 7 Statistical Analysis Results

## 7.1 Testing of Research Question Number 1

There is not a significant statistical impact of the use of electronic human resources management on creating and taking advantage of opportunities in the Jordanian telecommunications companies.

Table 2 clarifies the correlation coefficient (R) on the effect of using electronic human resources management on the creation of opportunities for the Jordanian telecommunications companies as (0.39) and the statistical value (t) as (8.31) as a function of (0.05) or less. Refuting the first null hypothesis (question-1) and accepting the alternate hypothesis (question 1) which states "there is an significant statistical impact of the use of electronic human resource management on creating and

**Table 2** Results of regression analysis to test the first main hypothesis

| Correlation coefficient R | Determination coefficient R2 | Beta coefficient B | (t) Value | (t) Significance |
|---|---|---|---|---|
| 0.39 | 0.15 | 0.39 | 8.31 | 0.00 |

**Table 3** Results of regression analysis to test the second main question

| Correlation coefficient R | Determination coefficient R2 | Beta coefficient B | (t) Value | (t) Significance |
|---|---|---|---|---|
| 0.52 | 0.27 | 0.52 | 11.90 | 0.00 |

taking advantage of opportunities in the Jordanian telecommunications companies." Furthermore, Table 2 clarifies that the beta coefficient (0.39) which expresses the positive impact, in other words every time the use of electronic human resources increased, it reflected positively on the ability to respond to the challenges faced by the Jordanian telecommunications companies. Additionally, the (R2) determination measure coefficient explains electronic human resources management (15%) variance is due to electronic human resources management and that (85%) variance is due to other unexplained factors.

## 7.2 Testing of Research Question 2

There is not a statistically significant impact of electronic human resources management on the contribution to facing the challenges of business organizations.

Table 3 clarifies that the correlation coefficient (R) on the effect of electronic human resources management to respond to the challenges faced by the Jordanian telecommunications companies was (0.52) and that the statistical value of (t) was (11.90) which is at the level of (0.05) or less. Refuting the second null hypothesis (question—2) and the acceptance of the alternate hypothesis (question 2) which states "there is an statistically significant impact of electronic human resources management on the contribution to facing the challenges of business organizations." Furthermore, Table 3 clarifies that the beta coefficient (0.52) which expresses the positive impact, in other words every time the use of electronic human resources increased, it reflected positively on the ability to respond to the challenges faced by the Jordanian telecommunications companies. Additionally, the (R2) determination measure coefficient explains electronic human resources management (27%) variance is due to electronic human resources management and that (73%) variance is due to other unexplained factors.

## 8 Conclusions

Upon analyzing the study results associated with the use of electronic human resources management, the following conclusions can be identified:

1. The use and application of electronic human resources management in business organizations constitutes an opportunity for the organization to achieve a set of benefits related to accomplishing their goals.
2. Electronic human resources management appears as the main and effective factor in the success of the business organization's work and its development.
3. Electronic human resources management combines distinct intellectual capital with advanced technological capital that contributes to achieving total quality management for the organization, while also achieving a set of opportunities for the organization.
4. Electronic human resources management can be expressed as a set of technology-based communications, information and networks that contribute to transforming manual work into integrated electronic administrative work.
5. The electronic human resources department focuses on the continuous development process through the use of electronic education.
6. Use of electronic human resources management has positively reflected on increasing the capacity and effectiveness of the organization, which contributes to developing human resources management practices that create a competitive advantage for the organization, which further leads to the organization outperforming other organizations.
7. The implementation of electronic human resources management contributes to reducing time and costs and an increase in accuracy in accomplishing the required tasks.
8. The existence of a positive effect of the use of electronic human resources management on creating opportunities for the Jordanian telecommunications companies.
9. The existence of a positive effect of the use of electronic human resources management on responding to challenges faced by the Jordanian telecommunications companies.

## 9 Recommendations

Based on the theoretical study of the concept of electronic human resources management and the conclusions reached by the field study, the following recommendations can be identified:

**First** The main recommendation is that organizations should use and apply electronic human resources management through:

1. Promoting of human and intellectual capital by adopting a supportive leadership style.
2. Re-engineering organizational structures by creating more flexible organizational structures capable of adapting to changing variables and meeting the requirements of electronic human resources management.

3. Creating an appropriate, compatible and stimulating organizational culture supporting the principle of using electronic human resources management and providing the requirements and procedures necessary for the implementation process.
4. Working to formulate a strategic vision capable of reflecting the concept of electronic human resources management in terms of planning, implementation and development, with a focus on the need to involve employees in formulating this vision.

**Second** Developing human capital capable of adopting and using electronic human resources management through:

1. Enhancing the capabilities of human capital, through training and development, to increase their ability to adapt to the requirements of the electronic human resources concepts and techniques.
2. Creating a spirit of cooperation in order to provide an interactive environment for transferring experiences between employees.
3. Providing a suitable work climate through holding a set of training programs and conferences to create competitive capabilities commensurate with the application of electronic human resources management.

**Third** Enhancing technology capital through:

1. Providing the required technological structure to use the electronic human resources management system by providing the necessary tools such as: Internet, Intranet, Networks, etc.
2. Providing technical and financial support that facilitates the use of technology capital such as: the software required for the uses of electronic human resources management, continuous maintenance of hardware, software and networks.
3. Formulating and implementing technological strategies that are based on: knowledge generation, organization, sustainability, transfer and technology.
4. Updating all types of technology and knowledge transfer formally and informally.
5. Applying technology management systems including: expert systems, neural network systems, logic systems, algorithm systems and thinking systems.

# References

1. Ahmad S (2015) Electronic human resource management: an overview. Int J Sci Technol Manage 4(1)
2. Al Shobaki MJ (2017) Impact of electronic human resources management on the development of electronic educational services in the universities. Hum Soc Sci
3. Alnidawy AB (2016) Learning organization impact on internal intellectual capital risks; an empirical study in the Jordanian pharmaceutical industry companies. Int Bus Res 9(10)

4. Alnidawy AB (2016) Human resources management activities adopted in the value chain model and their impact on the organizational. Int Bus Res 9(8)
5. Alnidaw AB (2021) An investigation into the impact of employee empowerment on organizational commitment: The case of Zajil International Telecom Company of Jordan. J Prod Qual Manage 12(3):327–344
6. Adhikari DR (2010) Human resource development (HRD) for performance management. Int J Prod Perform Manage
7. Armstrong M, Baron A (2018) Strategic HRM: the key to improved business performance. CIPD Publishing.
8. Bowen DE, Ostroff C (2014) Understanding HRM-firm performance linkages: the role of the "strength" of the HRM system. Acad Manage Rev 29(2)
9. Bondarouk TV, Ruël HJ (2009) Electronic human resource management: challenges in the digital era. Int J Human Resour Manage 20(3)
10. Çalişkan EN (2010) The impact of strategic human resource management on organizational performance. J Naval Sci Eng 6(2)
11. Dessler G (2018) Fundamentals of human resource management. Pearson.
12. Dianna L et al (2006) Factors affecting the acceptance and effectiveness of electronic human resource system. 16(2)
13. Dianna L et al (2013) Emerging issues in theory and research on electronic human resource management. Hum Resour Manage Rev 23(1)
14. Dianna L et al (2008) An expanded model of the factors affecting the acceptance and effectiveness of electronic human resource management systems. Hum Resour Manage Rev 19(2)
15. Hassan Al-Tamimi HA (2010) Factors influencing performance of the UAE Islamic and conventional national banks. Glob J Bus Res 4(2):1–9
16. Ma L, Ye M (2015) The role of electronic human resource management in contemporary human resource management. J Soc Sci 3(4)
17. Mohammad E et al (2018) The use of capital budgeting techniques as a tool for management decisions: Evidence from Jordan. Springer. ISSN 2195-4356
18. Noe RA et al. Human resource management, 10th edn. McGraw-Hill Education
19. Ruël H, Bondarouk T, Looise J (2004) E-HRM: innovation or irritation. An explorative empirical study in five large companies on web-based HRM. Manage Rev 15(3)
20. Rundquist J (2014) Knowledge integration in distributed product development. Int J Innov Sci 6(1):19–28
21. Strohmeier S (2007) Research in e-HRM: review and implications. Hum Resour Manage Rev 17
22. Sekaran U, Bougle R (2016) Research methods for business: a skill building approach. Wiley, New York
23. Yusliza M, Ramayah T (2018) Explaining intention to use E-HRM among HR professionals. Austr J Basic Appl Sci 5(8)

# An Investigation the Factors Affecting Towards Adoption of Digital Wallets in Iraq

Thaeir Ahmed Alsamman, Ali Abdulfattah Alshaher⬛,
and Amen Thaeir Ahmed Alsamman

**Abstract** Digital banking is a huge digital revolution in the Iraqi environment, as countries are now seen as cash-based societies. It is clear from the literature that the factors that could affect the intentions of individuals towards adopting digital wallets among Iraqis are unclear. This research aims to investigate the relationship between the variables represented by (perceived usefulness, perceived ease of use, social influence, compatibility, trialability, perceived trust, perceived security, intention to use digital wallet, adoption of digital wallets) with the intention to adopt digital wallets and by relying on the (TAM and IDT) model, perceived security factors and perceived trust. The hypothesized model is validated empirically via a questionnaire including 57-item based on 5-point Likert scales completed by 617 respondents (random sampled). Structural equation modeling was used to evaluate the proposed model by analyzing the confirmatory factor and path effects across the AMOS software. The results demonstrate that the indicators of model fit showed good fit. Also, the results obtained through this study also showed the importance and significance of the factors selected in this study in influencing individuals in the use and adoption of digital wallets. The study also revealed some practical effects that may be beneficial to managers, decision-makers and information technology specialists in a way that motivates them towards an interest in adopting digital wallets among Iraqis.

**Keywords** Digital wallets · Virtual wallets · Technology adoption · Digital transformation · TAM · IDT

T. A. Alsamman · A. A. Alshaher (✉)
Department of Management Information Systems, College of Administration and Economics, University of Mosul, Mosul, Iraq
e-mail: a.alshaher@uomosul.edu.iq

T. A. Alsamman
e-mail: thaieralsamman@uomosul.edu.iq

A. T. A. Alsamman
Nineveh Health Directorate, Mosul, Iraq

© The Author(s), under exclusive license to Springer Nature Switzerland AG 2022
S. G. Yaseen (ed.), *Digital Economy, Business Analytics, and Big Data Analytics Applications*, Studies in Computational Intelligence 1010,
https://doi.org/10.1007/978-3-031-05258-3_20

237

# 1 Introduction

The development in information technology on the one hand, and the considerations of the Covid 19 epidemic necessary for social distancing on the other hand, have become imperative considerations imposed on individuals towards the use of digital wallets to conduct payment transactions in the form of electronic money [44]. The widespread spread of digital currencies and the increasing electronic use of real currencies, means that there is an imperative need for a system that protects transactions in various types of these currencies, and this necessity requires the existence of safe and easy-to-access procedures for users, hence the idea of the electronic wallet as a guarantor for this purpose [48, 54]. In addition, the adoption of digital wallets was in line with the Iraqi government's plan and the goal of the Central Bank of Iraq to develop society into a cashless society. Despite the advantages and benefits achieved from the use of digital wallets [6], there are still some obstacles facing the adoption of digital wallets in the Iraqi environment [59], in addition to the previous studies conducted in this field showed that Iraqis do not have public awareness. For services provided by digital wallets [23]. Moreover, this trend has created competition between individuals who prefer cash wallets and those who prefer digital wallets. Therefore, this research is necessary to bridge the research gap and collect the necessary data to help the beneficiaries of digital wallets to make decisions that drive them towards adopting digital wallets within the framework of the two theories, TAM and IDT and some other factors, which are explained in the study model.

This study explains the process of adopting digital wallets through the integration between the TAM and IDT models and the security and trust factors to analyze the extent of individuals 'willingness to adopt digital wallets technology, and this study will be the first to use the two models in determining the factors affecting individuals' intentions towards adopting digital wallets in Iraq.

Hence, this study includes the following sections, Introduction of the study is in Sect. 1, While the theoretical background is in Sect. 2, the research model and hypotheses have been demonstrated in Sect. 3, the measurement method has been proposed in Sect. 4, while Sect. 5 presents the data analysis and discussion, lastly, Sect. 6 presents the study conclusions including implications and limitations.

# 2 Theoretical Background

The digital wallet is a system built on a digital basis to carry out digital exchanges and commercial transactions, and through the use of this wallet, purchases can easily be made through computers, smart phones or tablets, and in general, individuals' accounts in banks are linked with their digital wallet, in which they document and the protection of individual funds and business dealings from purchase and exchange. Thus, the digital wallet can be a means to authenticate and prove the identity of its owner and not only to make purchases over the Internet, and this comes from the wallet containing consumer money, the record of his commercial operations and his information, and it is worth noting that the digital wallet is suitable as a means of payment with many payment systems dedicated to smartphones thanks to their mutual support [51, 61].

Moreover, the digital wallet works to achieve a number of basic functions, including: secure application download, registration and access, saving purchase lists and trade discounts, storing multiple payment cards and other payment products on behalf of the user, completing financial transactions in two steps (the user and the financial firm in addition to being the provider for digital wallet service), but sometimes the service provider may be a technology company other than the financial company as a third party, storing information about different numbers of credit and debit cards, bank accounts, virtual currencies, managing multiple mobile payment services provided by different suppliers the funding from different sources, supporting of loyalty programs, person-to-person payments, and providing of bank statements [1, 3, 30]. On the other hand, there are many tasks for digital wallets [61]:

- Providing a secure storage place for credit card data and electronic cash.
- The primary mission of digital wallets is to make shopping more efficient.
- Digital wallets can serve their owners by tracking the purchases they want, obtaining receipts for these purchases, keeping books of the consumer's buying habits and suggesting what the consumer might find low in price for an item he buys regularly.
- Solve the problem of repeatedly entering shipping and payment information and filling out forms every time the consumer makes a purchase.

## 3    Research Model and Hypotheses

The study model and its hypotheses were developed based on the theories and models mentioned in previous studies with the addition of new factors, and accordingly we designed the conceptual model presented in Fig. 1 and its hypotheses as follows:

### 3.1    Innovation Diffusion Theory (IDT)

It is considered one of the pioneering theories invented by Rogers in the context of adopting information technology innovations and this study is still an important reference in the field of studying individuals' expectations and attitudes about the spread of innovation in information technology, and this theory is based on the idea that innovation is any technology that is unfamiliar or new to social systems [41, 42, 58, 63]. Rogers's [41] study focused on the innovation factor of relative advantage, compatibility, complexity, trialability, and visibility. For the purposes of this study, we relied on the factors of Compatibility and Trialability to construct hypotheses. Compatibility refers to "the extent to which an individual realizes that

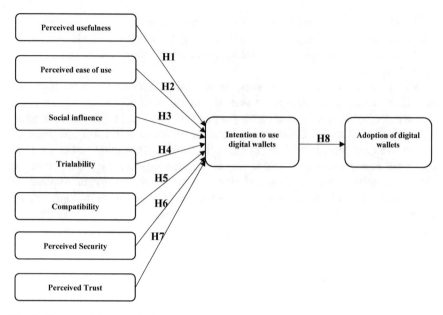

**Fig. 1** Conceptual framework of the study

digital wallets are compatible with current values" [41]. The basic idea here is based on creating a state of compatibility between the new idea and the existing system. As for "trialability it refers to the extent to which creativity can be experimented with according to certain foundations" [41]. The idea behind this factor is that new applications or ideas in the field of information technology may be in an experimental form and thus will be usable more quickly than those that have not been tried by reducing the uncertainty about these innovations. Previous studies [8, 24, 35, 37, 47, 51] regarding the two factors of compatibility and testability have indicated their positive and significant impact on the intention to use digital wallets. We derived the following hypotheses from this theory:

*H1: Compatibility positively influences intention to use digital wallets.*

*H2: Trialability positively influences intention to use digital wallets.*

## 3.2 Technology Acceptance Model (TAM)

The Technology Acceptance Model is one of the most popular models for studying the factors associated with the acceptance of any technology and the most used one. Fred Davis had built this model for the first time in 1989 and suggested that an individual's acceptance of a technology can be explained through factors: Perceived usefulness, Perceived ease of use [14, 44, 52, 60]. For the purposes of this study, we relied on the factors of perceived usefulness and perceived ease of use to construct hypotheses. The literature indicated that there is a strong and positive relationship between the factors of perceived usefulness and perceived ease of use in the intention to use digital wallets. In studies conducted by researchers such as [2, 26, 38, 44] they concluded that the relationship of these factors was positive and significant. From the above, we can formulate the first hypothesis:

*H3: Perceived usefulness positively influences intention to use digital wallets*

*H4: Perceived ease of use positively influences intention to use digital wallets.*

## 3.3 Social Influence

It refers to the degree to which the individual realizes that others believe the importance of using new technology [32]. In light of the development in electronic social media and the resulting gatherings or groups that have greatly affected the intentions of individuals towards adopting digital wallets [21]. Within this framework, previous studies indicated [8, 11, 21, 39, 49, 56] that the social influence positively and directly affected the intentions of individuals. Therefore, the hypothesis was set as follows:

*H5: Social influence positively influences intention to use digital wallets.*

## 3.4 Perceived Security

According to [39] security "is defined in digital wallets as the degree to which individuals believe that using digital wallets will remain safe". Individuals request digital wallets to be able to maintain the confidentiality of their information and to reduce the risks that they face in this direction [55]. Previous research has revealed [10, 33, 37, 49, 56] that the greater the security of digital wallets, the greater the intention of individuals to adopt them, and thus it was found that security positively and morally affects the intentions of individuals. Accordingly, the following hypothesis was formulated:

*H6: Perceived Security positively influences intention to use digital wallets.*

## 3.5 Perceived Trust

The technology of digital wallets requires that it be trusted by individuals, so the view of individuals towards digital wallets must have high trust in them, especially since these wallets require individuals to perform personal and financial information, meaning that individuals want to conduct their financial transactions with high trust. Previous research has shown [13, 15, 26, 54] that trust has a positive and significant effect on the intention to use digital wallets. Accordingly, the following hypothesis was formulated.

H7: *Perceived trust positively influences intention to use digital wallets.*

### 3.6 Intention to Use Digital Wallets

According to [16] opinion, the intention to use digital wallets represents a measure of the intensity of individuals' intention to use technology. The literature has indicated that there are many factors that explain an individual's intention and behavior towards using technology [27]. In studies conducted by [22, 34, 40, 62], it showed the positive and moral impact of intent on adopting digital wallets. Therefore, this study suggests the following hypothesis:

H8: *Intention to use digital wallets positively influences Adoption of digital wallets.*

## 4 Methods

### 4.1 Study Approach and Model

This study has adopted a casual approach to identify influences between constructs of study. Consequently, the literature review and the results of studies in this field have developed the study model that focuses on discovering the effects of the factors affecting the adoption of digital wallets in Iraq as an integrative model by adopting the TAM and IDT model, and the two factors of the perceived trust and perceived security as illustrated in Fig. 1 mentioned above.

### 4.2 Questionnaire Design and Source of Data

This study aims to explore the influences of factors affecting adoption of digital wallets in Iraq. The survey was also conducted from 14/7/2020 to 1/2/2020. 383 questionnaires received for analysis. As for the questionnaire, it was designed on the basis of the literature on this subject and according to a five-point Likert scale and the Table 1 clarifies the contents of the questionnaire.

**Table 1** Measurement items

| Variables | References |
|---|---|
| Compatibility | [53, 62] |
| Trialability | [53] |
| Perceived usefulness | [28, 31, 53, 62] |
| Perceived ease of use | [12, 53, 62] |
| Social influence | [28, 31, 53, 62] |
| Perceived security | [25, 31] |
| Perceived trust | [12, 25, 62] |
| Intention to use digital wallet | [12, 35, 53, 62] |
| Adoption of digital wallets | [35, 62] |

## 4.3 Pilot Test

In order to ensure the validity of the questionnaire and the metrics, the researchers conducted a pilot study on 35 randomly selected respondents, who were removed from the final survey. The suggestions resulting from the survey were taken into consideration in the final questionnaire of the study.

## 4.4 Statistical Methods

In order to perform the statistical analysis, structural equation modeling was used to test the causal relationship between the measurement variables in the research model. The main reason for using structural equation modeling is because it takes into consideration several equations simultaneously [57]. The software Amos was used for the analysis.

## 5 Data Analysis and Discussion

Before starting the hypothesis test, the validity of the scale should be checked, and for this reason a pilot study was conducted on a sample of 35 respondents, while they were excluded from the final survey. With regard to data analysis, SEM was used to verify the reliability and validity of data and then test the proposed hypotheses. Initial results showed that the model is reliable and valid for testing, except for some questions that were removed from the final data. Then data was obtained by building an electronic questionnaire for respondents, and Table 1 shows the demographic data for the study sample. It was found in Table 1 of the study sample that 78.93 was for males, while the rest of the percentage was for females. It is also clear that the largest

**Table 2** Demographic
Information

| Characteristics | Number of respondents | (N = 617) |
| --- | --- | --- |
| Gender | Male | 487 |
| | Female | 130 |
| Age | ≤40 | 224 |
| | 41–50 | 393 |

proportion of the sample was within the oldest age group, at a rate of 63.70 (Table 2).

Moreover, the Structural Equation Modeling (SEM) was used to analyze the empirical data and through Amos program.

## 5.1 Multicollinearity

For the purpose of hypothesis testing, the authors examined the issue of multi-collinearity so that the results are appropriate for regression analysis, and for this was used (VIF) and it was found that the values fall between 1.487 and 2.213 and this indicates there is no multicollinearity in this data.

## 5.2 Internal Consistency and Validity

The authors examined the quality of the model by assessing discriminant validity, content validity, and convergent validity as shown in Table 3.

According to what was mentioned in the above table, it should be verified whether it was according to the criteria specified in the literature, as the evidence indicated that the value of the loadings should be greater than 0.60 [36], while the values of CA and CR should be greater than 0.70, either the value of AVE must be greater than 0.50 [17, 19], so it is clear from this that the results specified in the table are greater than the criteria specified in the literature and therefore this is an indication of the validity of testing the hypotheses proposed in the study model.

## 5.3 Confirmatory Factor Analysis

Confirmatory factor analysis in (Amos) enables testing of the validity and accuracy of specific models that are built according to data and theoretical foundations. Confirmatory factor analysis is one of the applications of the Structural Equation Modeling (SEM) [5]. (SEM) is one of the important and contemporary research methodologies for analyzing data for behavioral studies, and this methodology enables the

**Table 3** Factor loadings, Cronbach's alpha, CR and AVE

| Constructs | | Loadings | Cronbach's alpha | CR | AVE |
|---|---|---|---|---|---|
| Perceived usefulness | PUS1 | 0.807 | 0.798 | 0.833 | 0.713 |
| | PUS2 | 0.801 | | | |
| | PUS3 | 0.804 | | | |
| | PUS4 | 0.825 | | | |
| Perceived ease of use | PUS5 | 0.744 | 0.796 | 0.807 | 0.703 |
| | PEU2 | 0.653 | | | |
| | PEU3 | 0.809 | | | |
| | PEU4 | 0.822 | | | |
| | PEU5 | 0.793 | | | |
| | PEU6 | 0.801 | | | |
| Social influence | SIN1 | 0.825 | 0.809 | 0.819 | 0.722 |
| | SIN2 | 0.756 | | | |
| | SIN3 | 0.789 | | | |
| | SIN4 | 0.850 | | | |
| | SIN5 | 0.843 | | | |
| Trialability | TRI1 | 0.781 | 0.801 | 00.812 | 0.702 |
| | TRI2 | 0.849 | | | |
| | TRI3 | 0.791 | | | |
| | TRI4 | 0.805 | | | |
| Compatibility | COM1 | 0.830 | 00.711 | 00.881 | 00.715 |
| | COM2 | 0.714 | | | |
| | COM3 | 0.720 | | | |
| | COM4 | 0.715 | | | |
| Perceived security | PSE1 | 0.826 | 0.833 | 0.867 | 0.754 |
| | PSE2 | 0.845 | | | |
| | PSE3 | 0.851 | | | |
| Perceived trust | PTR1 | 0.818 | 0.822 | 0.843 | 0.751 |
| | PTR2 | 0.768 | | | |
| | PTR3 | 0.834 | | | |
| | PTR4 | 0.856 | | | |
| | PTR5 | 0.831 | | | |
| | PTR6 | 842 | | | |
| Intention to use digital wallet | IND1 | 0.851 | 0.811 | 0.836 | 0.739 |
| | IND2 | 0.863 | | | |
| | IND3 | 0.819 | | | |
| | IND4 | 0.774 | | | |

(continued)

**Table 3** (continued)

| Constructs | | Loadings | Cronbach's alpha | CR | AVE |
|---|---|---|---|---|---|
| | IND5 | 0.735 | | | |
| | IND6 | 0.789 | | | |
| Adoption of digital wallets | ADW1 | 0.858 | 0.874 | 0.889 | 0.748 |
| | ADW2 | 0.855 | | | |
| | ADW3 | 0.757 | | | |
| | ADW4 | 0.861 | | | |
| | ADW5 | 0.864 | | | |

description of the interrelationships between the elements of the phenomenon that are designed and quantitatively studied, as well as its comprehensive interpretation without partition [43].

The goal of (SEM) is to determine the extent of the correspondence between the theoretical model of the study and the field data [9]. The following is a presentation of the most important indicators of conformity quality adopted in the confirmatory factor analysis through which to judge the quality of the model, and these indicators (fit indices) are presented in the Fig. 2 and Table 4 and as follows: CMIN/DF = 1.763 (the recommended value is smaller than 2), CFI = 0.925 (the recommended value is bigger than 0.9), GFI = 0.941 (the recommended value is bigger than 0.90), AGFI =

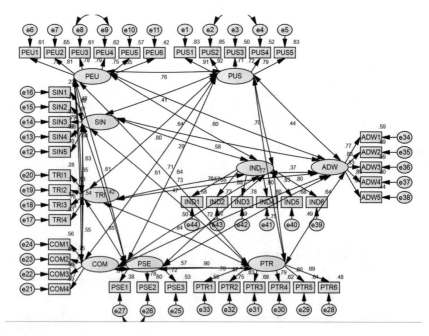

**Fig. 2** CFA for all constructs

**Table 4** The recommended and actual values of fit indices

| Fit indices | CMIN/DF | CFI | GFI | AGFI | IFI | NFI | RMSEA |
|---|---|---|---|---|---|---|---|
| Recommended value | <2 | >0.90 | >0.90 | >0.80 | >0.90 | >0.90 | <0.08 |
| Actual value | 1.763 | 0.925 | 0.941 | 0.922 | 0.952 | 0.971 | 0.03 |

0.922 (the recommended value is bigger than 0.90), IFI $= 0.952$ (the recommended value is bigger than 0.90), NFI $= 0.971$ (the recommended value is bigger than 0.90), and RMSEA $= 0.03$ (the recommended value is smaller than 0.08). These indices are within acceptable thresholds [4, 9, 18], indicating that the proposed model fits the data well.

Standardized Coefficients for the Proposed Measurement Model.

The abbreviations are: PUS: Perceived Usefulness; PEU: Perceived Ease of Use; SIN: Social Influence; TRI: Trialability; COM: Compatibility; PSE: Perceived Security; PTR: Perceived Trust; IND: Intention to Use Digital Wallet; ADW: Adoption of Digital Wallets.

## 5.4 Structural Model

After ensuring that the proposed study model conforms to the investigated sample data and arrives in the model to the required and specified fit indices standards through conducting the confirmatory factor analysis (CFA), it became possible to test the research hypotheses, as shown in Table 5 and Fig. 3, who present the results of hypothesis testing.

**Table 5** Overall Summary of the Hypotheses

| Hypothesis | | | | Estimate | S.E | C.R | P | Label | Study results |
|---|---|---|---|---|---|---|---|---|---|
| H1 | IND | <-- | PUS | 0.205 | 0.026 | 7.817 | *** | par_36 | Acceptance |
| H2 | IND | <-- | PEU | 0.144 | 0.025 | 5.851 | *** | par_39 | Acceptance |
| H3 | IND | <-- | SIN | 0.030 | 0.028 | 1.077 | 0.281 | par_38 | Reject |
| H4 | IND | <-- | TRI | 0.377 | 0.043 | 8.789 | *** | par_40 | Acceptance |
| H5 | IND | <-- | COM | 0.077 | 0.027 | 2.809 | 0.005 | par_41 | Acceptance |
| H6 | IND | <-- | PSE | 0.224 | 0.038 | 5.815 | *** | par_42 | Acceptance |
| H7 | IND | <-- | PTR | 0.304 | 0.035 | 8.592 | *** | par_43 | Acceptance |
| H8 | ADW | <-- | IND | 0.411 | 0.058 | 7.084 | *** | par_37 | Acceptance |

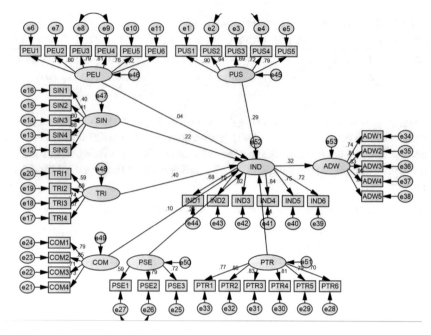

**Fig. 3** Results of SEM tests of study model

## 5.5 Results and Discussion

It is clear from the results of the above analysis that the study hypotheses are accepted with the exception of the hypothesis (H3). The hypotheses will be discussed as follows:

### H1: PUS → IND

It is clear from the path (PUS → IND) that there is a positive and significant effect of the perceived usefulness on the intention to use digital wallet, as evidenced by the value of the CR (7.817) which is greater than the critical value of (1.96) and the level of significance (0.000). The current result is in line with several studies as a study [12, 20, 31, 62]. Which indicated that the use of digital wallets increases the efficiency of completing individual transactions more quickly than the traditional method, and thus this led to an increase in the productivity of individuals.

### H2: PEU → IND

This path shows the significance of the relationship between the perceived ease of use and the intention to use the digital wallet, based on the value of the CR, whose value is (5.851), and the level of significance (0.000), and these results are in agreement with studies [12, 20, 31, 50, 53, 62]. This means that individuals do not face difficulty in using digital wallets and are able to interact with them, and therefore they make little effort towards these wallets.

### H3: SIN → IND

As for the results of testing this hypothesis, it indicated its rejection, that is, there is no significant impact on the social influence on the intention to use digital wallets, as the value of CR (1.077) is less than its standard value (1.96) and the level of significance (0.281). These results are consistent with studies of [29, 53]. This means that other individuals (such as friends, relatives, family, and those who are important to the individual) do not affect the individual's intention to use digital wallets.

### H4: TRI → IND

It is clear from the path (TRI → IND) that there is a positive and significant effect of the trialability on the intention to use digital wallet, as evidenced by the value of the CR (8.789) which is greater than the critical value of (1.96) and the level of significance (0.000). The current result is in line with several studies as a study [7, 11, 46]. Which indicated that the individuals have the knowledge and time to experiment with digital wallets.

### H5: COM → IND

This hypothesis shows the significance of the effect of the compatibility on the intention to use digital wallets, as evidenced by the amount of CR value (2.809) and the value of the level of significance of (0.005), and these results are in agreement with studies [45, 62]. This explains that digital wallets fit the lifestyle of individuals and their current situation, and therefore individuals prefer to deal with this technology.

### H6: PSE → IND

Also, this hypothesis shows the significance of the effect of the perceived security on the intention to use digital wallets, as evidenced by the amount of CR value (5.815) and the value of the level of significance of (0.000). The current result is in line with several studies as a study [31, 60, 62]. This explains that individuals do not have any concerns about this technology, and believe that it is safe and able to store their personal information on it.

### H7: PTR → IND

Similarly, the path (PTR → IND) indicates that there is a positive and significant effect of the perceived trust on the intention to use digital wallet, as evidenced by the value of the CR (8.592) which is greater than the critical value of (1.96) and the level of significance (0.000). The current result is in line with several studies as a study [31, 60, 62]. This explains that the transactions that can occur through this technology remain confidential and secure, as well as the support that the wallet service providers will provide in case of any problem, in addition to taking into account customers' considerations.

### H8: IND → ADW

As for this hypothesis represented by the path (IND → ADW), it shows the significant and positive effect of the intention to use digital wallets on adoption of digital wallets, as evidenced by the value of the CR of (7.084) at the level of significance (0.000).

The result of this hypothesis was consistent with previous studies represented by [27, 35, 62]. Consequently, this hypothesis explains that individuals intend to use digital wallets and, more importantly, continue to use them.

## 6 Conclusion

This study seeks to integrate the two models (TAM and IDT) with perceived trust and perceived security factors, in order to identify the intent of individuals to use and adopt digital wallets in the Iraqi environment. Emphasis was placed on individuals with experience in this field in order for the answers to be more realistic. Consequently, this study concluded that all the factors proposed in this study except for the social influence played an important and moral role in explaining the intentions of individuals to use and adopt digital wallets. Moreover, this study is considered a starting point for stakeholders to identify areas that enable them to focus on, in order to be a platform for influencing individuals' intentions towards increasing interest in this technology. Thus, the study enabled the government to take many measures that motivate individuals to use and adopt digital wallets, and perhaps the most important of which is educating and training individuals in a manner that increases their efficiency and effectiveness in this direction.

### 6.1   Theoretical Implications

This study relied on two models (TAM and IDT) to analyze the factors affecting individuals to use and adopt digital wallets. The theoretical model of the study depends on the integration of the factors mentioned in the theoretical framework. These factors were adopted due to the lack of interest by previous studies on these factors, which provided an opportunity for researchers to study in this direction.

### 6.2   Practical Implications

The results of this study contribute to assisting government agencies or digital wallet service providers in building policies or strategic decisions that will increase the use and adoption of digital wallets by individuals.

## 6.3   Limitations and Future Research

This study faces many constraints that should be paid attention to, including the size of the sample, as the study faced great difficulty in reaching this number, and this number must also be increased in order for the results to be generalized. Another challenge: This study did not include the moderate effects of age, gender, and educational level, which constitutes an obstacle to understanding the interaction between individuals according to their demographic characteristics. In addition, the study lacks perceived risk factor that plays a critical role in influencing individuals' intentions. Future studies should do such a study but with the addition of other variables. In addition to studying digital wallets from a blockchain perspective.

**Acknowledgements**   The authors are very grateful to the University of Mosul College of Administration & Economics for their provided facilities, which help to improve the quality of this work.

## Appendix 1: Measurement Items

Dear Sir/Madam:

We are writing an essay on "Investigate the Factors Affecting Towards Adoption of Digital Wallets in Iraq". We will be very grateful if you could spare a few minutes of your precious time to answer the following questions. I would like to assure you that your responses will be treated in strict confidentiality.

Regards

Demographic Characteristics:

Gender:

Age (Years):

| The scale | Strongly disagree | Disagree | Unsure | Agree | Strongly agree |
|---|---|---|---|---|---|
| Weight | 1 | 2 | 3 | 4 | 5 |

| R | Questions | 1 | 2 | 3 | 4 | 5 |
|---|---|---|---|---|---|---|
| *Perceived usefulness* | | | | | | |
| 1 | Using an e-wallet makes it easier for me to conduct my daily transactions | 1 | 2 | 3 | 4 | 5 |
| 2 | Using an e-wallet allows me to manage my transactions more efficiently | 1 | 2 | 3 | 4 | 5 |

(continued)

(continued)

| R | Questions | 1 | 2 | 3 | 4 | 5 |
|---|-----------|---|---|---|---|---|
| 3 | Using an e-wallet increases my productivity | 1 | 2 | 3 | 4 | 5 |
| 4 | Using an e-wallet enables me to accomplish tasks e.g., payments more quickly | 1 | 2 | 3 | 4 | 5 |
| 5 | Overall, I believe an e-wallet is more useful than traditional ways of conduct transactions | 1 | 2 | 3 | 4 | 5 |

*Perceived ease of use*

| | | | | | | |
|---|---|---|---|---|---|---|
| 1 | Learning how to use an e-Wallet is easy for me | 1 | 2 | 3 | 4 | 5 |
| 2 | My interaction with an e-Wallet is clear and understandable | 1 | 2 | 3 | 4 | 5 |
| 3 | I find an e-wallet easy to use | 1 | 2 | 3 | 4 | 5 |
| 4 | It is easy for me to become skillful at using an e-wallet | 1 | 2 | 3 | 4 | 5 |
| 5 | It is easy for me to remember how to perform task with an e-wallet | | | | | |
| 6 | I like the fact that payments done through an e-wallet require minimum effort | | | | | |

*Social influence*

| | | | | | | |
|---|---|---|---|---|---|---|
| 1 | People who influence my behavior think that I should use an e-wallet | 1 | 2 | 3 | 4 | 5 |
| 2 | People who are important to me think that I should use an e-wallet | 1 | 2 | 3 | 4 | 5 |
| 3 | e-wallets are widely used by people in my community | 1 | 2 | 3 | 4 | 5 |
| 4 | Almost all my friends use e-wallets | | | | | |
| 5 | My family members use e-wallets | | | | | |

*Compatibility*

| | | | | | | |
|---|---|---|---|---|---|---|
| 1 | Using e-wallet services is compatible with all aspects of my lifestyle | 1 | 2 | 3 | 4 | 5 |
| 2 | Using e-wallet services fits into my lifestyle | 1 | 2 | 3 | 4 | 5 |
| 3 | Using e-wallet services fits well with the way I like to purchase products and services | 1 | 2 | 3 | 4 | 5 |
| 4 | Using e-wallets is completely compatible with my current situation | 1 | 2 | 3 | 4 | 5 |

*Trialability*

| | | | | | | |
|---|---|---|---|---|---|---|
| 1 | I have many opportunities to try digital wallet technology | 1 | 2 | 3 | 4 | 5 |
| 2 | I have the knowledge where to try digital wallet technology | 1 | 2 | 3 | 4 | 5 |
| 3 | Before finally adopting any digital technology, we can experimentally use this technology | 1 | 2 | 3 | 4 | 5 |
| 4 | The period allowed for us to experiment with digital wallets before its final implementation is sufficient to satisfactorily identify it | | | | | |

*Perceived trust*

| | | | | | | |
|---|---|---|---|---|---|---|
| 1 | I trust that a transaction conducted through an e-wallet is secure and private | 1 | 2 | 3 | 4 | 5 |
| 2 | I trust payments made through e-wallet channels will be processed securely | 1 | 2 | 3 | 4 | 5 |
| 3 | I believe my personal information on an e-wallet will be kept confidential | | | | | |

(continued)

(continued)

| R | Questions | 1 | 2 | 3 | 4 | 5 |
|---|-----------|---|---|---|---|---|
| 4 | I believe e-wallet providers keeps customers' best interests in mind | | | | | |
| 5 | I believe that in case of any issue, the e-wallet service provider will provide me assistance | | | | | |
| 6 | I believe that the e-wallet service providers follow consumer laws | | | | | |

*Perceived security*

| 1 | I do not fear security issues with e-wallet | 1 | 2 | 3 | 4 | 5 |
|---|--------------------------------------------|---|---|---|---|---|
| 2 | I believe e-wallets are secure | 1 | 2 | 3 | 4 | 5 |
| 3 | I believe if I stored information on e-wallet it would be secure | 1 | 2 | 3 | 4 | 5 |

*Intention to use digital wallets*

| 1 | Assuming that I have access to e-wallet, I intend to use it | 1 | 2 | 3 | 4 | 5 |
|---|------------------------------------------------------------|---|---|---|---|---|
| 2 | I intend to use an e-wallet if the cost and times is reasonable for me | 1 | 2 | 3 | 4 | 5 |
| 3 | I intend to use an e-wallet in the future | 1 | 2 | 3 | 4 | 5 |
| 4 | I intend to increase my use of e-wallets in the future | | | | | |
| 5 | I intend to continue using an e-wallet more frequently in the future | | | | | |
| 6 | I intend to use an e-wallet in my daily life | | | | | |

*Adoption of digital wallets*

| 1 | I often use an e-wallet to manage my account | 1 | 2 | 3 | 4 | 5 |
|---|----------------------------------------------|---|---|---|---|---|
| 2 | I often use an e-wallet to transfer and remit money | 1 | 2 | 3 | 4 | 5 |
| 3 | I often use an e-wallet to make payments | 1 | 2 | 3 | 4 | 5 |
| 4 | Subscribe to financial products that are exclusive to mobile banking | | | | | |
| 5 | On average, how often have you used an e-wallet per month? (Never, 1–5 times; 6 to 10 times; 11 to 15 times; More than 15 times) | 1 | 2 | 3 | 4 | 5 |

# References

1. Aabye C, Weller K (2019) U.S. Patent No. 10,366,387. U.S. Patent and Trademark Office, Washington, DC
2. Abd Malik AN, Annuar SNS (2021) The effect of perceived usefulness, perceived ease of use, reward, and perceived risk toward e-wallet usage intention. In: Eurasian business and economics perspectives. Springer, Cham, pp 115–130
3. Aite (2016) The evolution of digital and mobile wallets. Mahindra Comviva, USA
4. Al-Arabi F (2018) The impact of justice and organizational trust on organizational commitment a proposed model: an empirical study on Algerian public hospital institutions, Ph.D. thesis in Management Sciences in Business Administration, Faculty of Economic and Commercial Sciences and Management Sciences, Abu Bakr Belkayed University
5. Al-Hawari A (2017) Verification of one-dimensional assumption using exploratory factor analysis versus. empirical factor analysis, comparative study. An-Najah Univ J Res 81(8):1423–1448
6. Alkhowaiter WA (2020) Digital payment and banking adoption research in gulf countries: a systematic literature review. Int J Inf Manage 53:102102

7. Aminu SA (2021) Technology acceptance model and motorists' Intention to adopt point of sale terminals for payment of petrol price in Lagos State, Nigeria. Govern Manage Review, 3(1).
8. Angelina C, Rahadi RA (2020) A conceptual study on the factors influencing usage intention of e-wallets in JAVA, Indonesia. Int J Acc 5(27):19–29
9. Azzouz AN, Al-Hashimi (2018) The use of modeling by structural equation in social sciences, Univ Sharjah J Hum Soc Sci 15(1):287–322
10. Brahmbhatt M (2018) A study on customers' perception towards e-wallets in Ahmedabad City. IUJ J Manage 6(1):11–15
11. Chakraborty S, Mitra D (2018) A study on consumers adoption intention for digital wallets in India. Int J Customer Relat 6(1):38
12. Chawla D, Joshi H (2020) Role of mediator in examining the influence of antecedents of mobile wallet adoption on attitude and intention. Glob Bus Rev. 0972150920924506
13. Chern YX, Kong SY, Lee VA, Lim SY, Ong CP (2018) Moving into cashless society: factors affecting adoption of e-wallet. B.A. Dissertation, University Tunku Abdul Rahman, Kampar
14. Davis FD (1989) Perceived usefulness, perceived ease of use, and user acceptance of information technology. MIS Q 319–340
15. Deaton T (2002). U.S. Patent application No. 09/874,745
16. Fishbein M, Ajzen I (2011) Predicting and changing behavior. Psychology Press Taylor & Francis Group, New York, NY, USA
17. Flynn BB, Schroeder RG, Sakakibara S (1994) A framework for quality management research and an associated measurement instrument. J Oper Manag 11(4):339–366
18. Gefen D, Straub DW, Rigdon EE (2011) An update and extension to SEM guidelines for administrative and social science research. Manage Inf Syst Quart 35(2):3–14
19. Hair JF, Anderson RE, Tatham RL, Black WC (1998) Multivariate data analysis, 5th edn. Prentice Hall
20. Halttunen V (2016) Consumer behavior in digital era: general aspects and findings of empirical studies on digital music with a retrospective discussion. Ph.D. thesis. Jyväskylä Studies in Computing (235)
21. Han JH (2020) The effects of personality traits on subjective well-being and behavioral intention associated with serious leisure experiences. J Asian Finance Econ Bus 7(5):167–176. https://doi.org/10.13106/jafeb.2020
22. Ing AY, Wong TK, Lim PY (2021) Intention to use e-wallet amongst the university students in Klang Valley. Int J Bus Econ 3(1):75–84
23. Jin CC, Seong LC, Khin AA (2019) Factors affecting the consumer acceptance towards Fintech products and services in Malaysia. Int J Asian Soc Sci 9(1):59–65
24. Junadi S (2015) A model of factors influencing consumer's intention to use e-payment system in Indonesia. Procedia Computer Science 59:214–220
25. Kim C, Tao W, Shin N, Kim K-S (2010) An empirical study of customers' perceptions of security and trust in e-payment systems. Electron Commer Res Appl 9(1):84–95. https://doi.org/10.1016/j.elerap.2009.04.014
26. Kurniawati HA, Arif A, Winarno WA (2017) Analisis minat penggunaan mobile banking dengan pendekatan technology acceptance model (TAM) yang telah dimodifikasi. E-Journal Ekonomi Bisnis Dan Akuntansi 4(1):24–29
27. Lim F-W, Ahmad F, Talib ANA (2019) Behavioural intention towards using electronic wallet: a conceptual framework in the light of the unified theory of acceptance and use of technology (UTAUT). Imp J Interdiscip Res 5:79–86
28. Lwoga ET, Lwoga NB (2017) User acceptance of mobile payment: the effects of user-centric security, system characteristics and gender. Electron J Inf Syst Dev Countries 81(1):1–24
29. Mannan B, Haleem A (2017) Understanding major dimensions and determinants that help in diffusion & adoption of product innovation: using AHP approach. J Glob Entrep Res 7(1):1–24
30. Marinova-Kostova K (2017) Mobile wallet-functions, components and architecture. In: 2nd conference on innovative teaching methods (ITM 2017), 28–29 June 2017, University of Economics Varna, Bulgaria

31. Mew J, Millan E (2021) Mobile wallets: key drivers and deterrents of consumers' intention to adopt. Int Rev Retail Distrib Consumer Res 31(2):182–210

32. Mugambe P (2017) UTAUT model in explaining the adoption of mobile money usage by MSMEs' customers in Uganda. Adv Econ Bus 5(3):129–136

33. Nedaa N (2020) A quantative approach to identifying factors that affect the use of e-wallets in Bahrain. Журнал Сибирского федерального университета. Гуманитарные науки 13(11)

34. Nikou SA, Economides AA (2017) Computers & education mobile-based assessment: investigating the factors that influence behavioral intention to use. Comput Educ 109:56–73

35. Nimansa AT, Kuruwitaarachchi N (2021) A study on finding the factors, hindering the use of digital wallets among youth in developing countries. Glob J Comput Sci Technol

36. Nunnally JC (1978) Psychometric theory, McGraw-Hill, New York

37. Padiya J, Bantwa A (2018) Adoption of E-wallets: a post demonetisation study in Ahmedabad City. Pac Bus Rev Int 10(10):84–95

38. Permana GPL, Rini HPS, Paramartha IGND (2021) Fintech Dari Perspektif Perilaku user Dalam Penggunaan E-wallet Dengan Menggunakan technology acceptance model (TAM). Widya Akuntansi dan Keuangan 3(1):50–70

39. Phan TN, Ho TV, Le-Hoang PV (2020) Factors affecting the behavioral intention and behavior of using E-wallets of youth in Vietnam. J Asian Finance Econ Bus (JAFEB) 7(10):295–302

40. Puspitasari I, Wiambodo ANR, Soeparman P (2021) The impact of expectation confirmation, technology compatibility, and customer's acceptance on e-wallet continuance intention. In: AIP conference proceedings, vol 2329, No 1. AIP Publishing LLC, p 050012

41. Rogers EM (1983) Diffusion of innovations, 3rd edn. The Free Press, New York

42. Rogers EM, Singhal A (2003) Diffusion of innovations. In: An integrated approach to communication theory and research, 5th edition. Salwen M, Stacks D (eds). LEA, Mahwah, NJ, pp 409–419

43. Sahrawi A, Bouslab A (2016) Structural modeling (SEM) and addressing the validity of standards in psychological and educational research: the global structure model of the relationships of administrative management competencies in the educational institution. J Psychol Educ Sci 3(2)

44. Saputra D, Gürbüz B (2021) Implementation of technology acceptance model (TAM) and importance performance analysis (IPA) in testing the ease and usability of e-wallet applications. arXiv:2103.09049

45. Shaw N, Sergueeva K (2019) The non-monetary benefits of mobile commerce: extending UTAUT2 with perceived value. Int J Inf Manage 45:44–55

46. Sikdar P, Kumar A, Alam MM (2019) Antecedents of electronic wallet adoption: a unified adoption based perspective on a demonetised economy. Int J Bus Emerg Mark 11(2):168–196

47. Singh G (2019) A review of factors affecting digital payments and adoption behaviour for mobile e-wallets. Int J Res Manage Bus Stud 6(4):89–96

48. Sohail A, Shobhit A, Saurabh S, Akhilesh V, Varma CP (2018) Development of advance digital mobile wallet. Int J Sci Res Dev 6:2758–2760

49. Soodan V, Rana A (2020) Modeling customers' intention to use e-wallet in a developing nation: extending UTAUT2 with security, privacy and savings. J Electron Commerce Organ (JECO) 18(1):89–114

50. Suhaimi AIH, Hassan MSBA (2018) Determinants of branchless digital banking acceptance among generation Y in Malaysia. Paper presented at the in 2018 IEEE conference on e-learning, e-management and e-services (IC3e). IEEE

51. Sukaris S, Renedi W, Rizqi MA, Pristyadi B (2021) Usage behavior on digital wallet: perspective of the theory of unification of acceptance and use of technology models. In: J Phys: Conf Ser 1764(1):012071 (IOP Publishing)

52. Tarhini A, Scott M, Sharma S, Abbasi MS (2015) Differences in intention to use educational RSS feeds between Lebanese and British students: a multi-group analysis based on the technology acceptance model. Electron J e-Learn 13(1):14–29

53. Tiong WN (2020) Factors influencing behavioural intention towards adoption of digital banking services in Malaysia. Int J Asian Soc Sci 10(8):450–457
54. To AT, Trinh THM (2021) Understanding behavioral intention to use mobile wallets in Vietnam: extending the TAM model with trust and enjoyment. Cogent Bus Manage 8(1):1891661
55. Tran VD (2020) The relationship among product risk, perceived satisfaction and purchase intentions for online shopping. J Asian Finance Econ Bus 7(6):221–231. https://doi.org/10.13106/jafeb.2020.vol7.no6.221
56. Trivedi J (2016) Factors determining the acceptance of e wallets. Int J Appl Mark Manage 1(2):42–53
57. Tsai JM, Cheng MJ, Tsai HH, Hung SW, Chen YL (2019) Acceptance and resistance of telehealth: the perspective of dual-factor concepts in technology adoption. Int J Inf Manage 49:34–44
58. Venkatesh V, Morris MG, Davis GB, Davis FD (2003) User acceptance of information technology: toward a unified view. MIS Q 425–478.
59. Widjaja EPO (2016) Non-cash payment options in Malaysia. J Southeast Asian Econ 398–412
60. Wong TKM, Man SS, Chan AHS (2021) Exploring the acceptance of PPE by construction workers: an extension of the technology acceptance model with safety management practices and safety consciousness. Saf Sci 139:105239
61. Wong WH, Mo WY (2019) A study of consumer intention of mobile payment in Hong Kong, based on perceived risk, perceived trust, perceived security and technological acceptance model. J Adv Manag Sci 7(2):33–38
62. Yang M, Mamun AA, Mohiuddin M, Nawi NC, Zainol NR (2021) Cashless transactions: a study on intention and adoption of e-wallets. Sustainability 13(2):831
63. Zhang X, Yu P, Yan J, Spil ITA (2015) Using diffusion of innovation theory to understand the factors impacting patient acceptance and use of consumer e-health innovations: a case study in a primary care clinic. BMC Health Serv Res 15(1):1–15

# Predictive Analysis of Energy Use Based on Some Forecasting Models

Ali AlArjani and Teg Alam

**Abstract** A prediction method is a way to estimate a sequence of values based on time-series. This study aims to anticipate Saudi Arabia's energy use using autoregressive integrated moving average (ARIMA), Holt-Winters (H-W), and artificial neural network (ANN) models. This study also examines the accuracy of forecasting methods. The study forecasts energy use time-series data from 1971 to 2014 using statistical software. According to the results, ARIMA (2, 1, 2) is suitable for predicting the Kingdom of Saudi Arabia's energy usage in 2025. The findings of the study will assist government agencies in forecasting energy use.

**Keywords** Saudi Arabia · Energy use · Autoregressive Integrated Moving Average (ARIMA) · Holt-Winters (H-W) · Artificial Neural Networks (ANN) · Forecasting methods

## 1 Introduction

This study aims to conduct a predictive analysis of Saudi Arabia's energy usage using forecasting models such as ARIMA, H-W, and ANN. Furthermore, the accuracy of predictive models is also investigated in this research.

Saudi Arabia's economy is based on the energy sector. Saudi Arabia's energy sector includes petroleum and natural gas production, consumption, exports, and electricity generation. The Kingdom is main producer of oil in the globe. Saudi Arabia has also tried to grow the energy sector and encourage more substantial investment, particularly from foreign firms. In addition, Saudi Arabia possesses one of the most significant natural gas reserves [1, 2].

According to Saudi government plans, multiple measures have been taken towards predicting the country's future. In this context, ARIMA (1, 0, 0), ARIMA (0, 1, 1), and ARIMA (1, 1, 2), and ANN suitable models were used for predicting the total

A. AlArjani · T. Alam (✉)
Department of Industrial Engineering, College of Engineering, Prince Sattam Bin Abdulaziz University, Al Kharj 16273, Kingdom of Saudi Arabia
e-mail: t.alam@psau.edu.sa

© The Author(s), under exclusive license to Springer Nature Switzerland AG 2022　　　257
S. G. Yaseen (ed.), *Digital Economy, Business Analytics, and Big Data Analytics Applications*, Studies in Computational Intelligence 1010,
https://doi.org/10.1007/978-3-031-05258-3_21

revenue and expenditure of Saudi Arabia [3]. Accurate electricity demand forecasting is necessary for policymakers to design power supply policies. On the other hand, limited data and variables rarely provide enough information to reach appropriate prediction accuracy. To overcome this problem, a new, more accurate grey forecasting model was developed [4]. Furthermore, A neural network model developed using statistical and analytical methods was compared to a series of neural network models developed using a multi-objective genetic algorithm [5]. In addition, algorithms for estimating electricity consumption for specific commodities and sectors were developed to address the shortcomings of existing modelling approaches [6].

The usefulness of a weather-free forecasting model based on a database with important information about previously produced power data and data mining techniques was investigated [7]. Besides, the problems were solved by creating an energy usage prediction model with a real-world application in Malaysia utilizing the Microsoft Azure cloud-based machine learning platform [8]. Based on the concept of model integration, a unique better integration model (stacking model) was proposed that can be used to predict building energy consumption [9].

Finally, the forecasting models are significantly important methods applied in numerous areas of scientific studies. For example, the researchers have used various prediction models in their studies to predict their goals [10–18]. In this study, the results of the forecasting models compare to find the best fit.

## 2   Methods and Materials

This article compares the forecasting models ARIMA, H-W, and ANN. As a result, the theoretical underpinnings of the ARIMA, H-W, and ANN forecasting models are presented below.

### 2.1   *Auto Regressive Integrated Moving Average (ARIMA)*

#### 2.1.1   Autoregressive (AR) Model

Consider regression models of the form

$$Y_t = \varphi_0 + \varphi_1 Y_{t-1} + \varphi_2 Y_{t-2} + \cdots + \varphi_3 Y_{t-3} + \varepsilon_t \tag{1}$$

where

$Y_t$ is the response-variable at time 't',

$Y_{t-p}$ is the independent variable at time lags, and these Y's in (1) serve as independent variables,

$\varphi_p$ is the assessed coefficients,

$\varepsilon_t$ is the residual term at time 't'.

Equation (1) uses time-lagged Y values as explanatory variables. As a result, the term Autoregression (AR) is commonly used to describe this type of model. Because of the following reasons, autoregression models should be treated differently than traditional regression models:

The explanatory variables in autoregression models are interdependent.
It is not always easy to decide how many $Y_t$ past values to include in the model.

### 2.1.2 Moving Average (MA) Model

The moving average (MA) model is a time series model that makes use of previous errors as an explanatory variable. Consider the following moving average.

$$Y_t = \mu_0 + \varepsilon_t - \omega_1\varepsilon_{t-1} - \omega_2\varepsilon_{t-2} \cdots - \omega_q\varepsilon_{t-q} \tag{2}$$

This model (2) is defined as the error series' moving average.

where

$\mu$ is the process constant mean,
$\omega_q$ is the assessed coefficients,
$\varepsilon_{t-q}$ is the residual at time 't'.

The AR and MA models are combined to generate the Autoregressive Moving Average (ARMA) model, a versatile and valuable class of time series models. When the data is stationary, these can be used.

By allowing the data series to difference, this class of models can be extended to non-stationary series. These are known as ARIMA, and ARIMA models are a type of linear model that can handle both stationary and non-stationary time series.

Keeping in mind that stationary processes vary around a fixed level and nonstationary processes lack a natural constant mean level, ARIMA (p, d, q) models provide an alternative approach to time series forecasting by attempting to describe the autocorrelations in the data. ARIMA consists of autoregressive (p), differentiable (d), and moving average (q) functions [19].

## 2.2 Holt-Winters (H-W)

When data contains a trend and a seasonal pattern, many organizations use the Holt-Winters model to generate short-term demand forecasts; additionally, this method is based on three smoothing equations: stationary component, trend, and seasonal. These equations may be additive or multiplicative.

If the time series data has a trend and seasonal effects, this method can be used. Besides alpha and beta smoothing-factors, a new parameter, gamma, is added to manage the impact on the seasonal element. The following are the steps in the Holt-Winters multiplicative technique [19].

Exponentially smoothed-series,

$$L_T = \alpha\left(Y_T \frac{1}{S_{T-M}}\right) + (1 - \alpha)(L_{T-1} - b_{t-1}) \tag{3}$$

Trend-estimate,

$$B_T = \beta(L_T - L_{T-1}) + (1 - \beta)B_{T-1} \tag{4}$$

Seasonality-estimates,

$$S_t = \gamma\left(\frac{Y_T}{L_{T-1} + B_{T-1}}\right) + (1 - \gamma)S_{T-M} \tag{5}$$

where

$L_T$ is the level-series,
$\alpha$ is the level-smoothing constant,
$Y_T$ is new observation,
$\beta$ is the trend estimate-smoothing constant,
$B_T$ is the trend-estimate,
$\gamma$ is the seasonality estimate -smoothing constant,
$S_T$ is the seasonal component-estimate,
$M$ is the total yearly seasons,
$S$ is the seasonality length (number of periods in the season),
$T$ is the time period,
and $0 \leq alpha \leq 1, 0 \leq beta \leq 1, 0 \leq gamma \leq 1$.

## 2.3 Artificial Neural Network (ANN)

The artificial neural network (ANN) in a computer system simulates how the human brain assesses and processes input.

It is the foundation of artificial intelligence (AI), which solves issues that would be impossible or difficult to solve using human or statistical standards. In addition, because ANNs can self-learn, they can produce better results as more data becomes available [20].

## 2.4 Measures of the Forecasting Methods' Accuracy

The errors that were investigated in this study are listed below.

**Table 1** Summary statistics

| Variable | Observations | Mean | Standard error | Standard deviation | Minimum | Maximum | Sum |
|----------|--------------|------|----------------|--------------------|---------|---------|-----|
| Energy use | 44 | 4136.64 | 253.44 | 1681.15 | 977.758 | 6905.7 | 182,012.2 |

$$\text{Mean Absolute Error (MAE)} = \frac{1}{n} \sum_{t=1}^{n} |e_t| \qquad (6)$$

$$\text{Mean Squared Error (MSE)} = \frac{1}{n} \sum_{t=1}^{n} (e_t)^2 \qquad (7)$$

$$\text{Root Mean Square Error (RMSE)} = \sqrt{MSE} \qquad (8)$$

$$\text{Mean Absolute Percentage Error (MAPE)} = \frac{1}{n} \sum_{t=1}^{n} \frac{|e_t|}{Y_t} \qquad (9)$$

$$Residual\ (e_t) = Observed\ value - Predicted\ value \qquad (10)$$

Following the error analysis, we compare the forecast accuracy measures. The best way to define the forecast is often to use less MAPE and less RMSE [19].

# 3 Result and Findings

## 3.1 Data

The data on the kingdom's energy usage was obtained from the US EIA. It is computed on a yearly basis and was measured in kilograms of oil equivalent per capita from 1971 to 2014. In addition, statistical software was used to generate summary statistics for the kingdom's energy use data, as shown in Table 1.

## 3.2 Predicted Value of Energy Use (kg of Oil Equivalent Per Capita) Using the Various Models

The purpose of this research is to evaluate Saudi Arabia's predicted energy use. The results of the implementation of various forecasting models for predicting Saudi Arabia's energy use are shown in Figs. 1, 2 and Table 2.

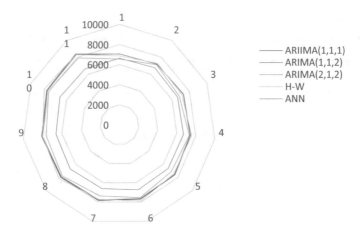

**Fig. 1** Predicted value of energy-use using the ARIMA, H-W and ANN models

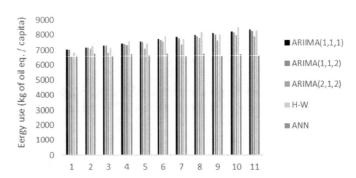

**Fig. 2** Predicted value of energy-use using the ARIMA, H-W and ANN models

Forecasts for future time-series data are created using ARIMA, H-W, and ANN models. Tables 2, 3 and Figs. 1, 2, 3 show that ARIMA (2, 1, 2) provides the best forecast based on accuracy measures. Thus, the ARIMA (1, 1, 1) model predicts energy use better than other models for this time-series data.

According to Fig. 3, the Kingdom's energy consumption has gradually increased since 1984. After 1984, it gradually increases and decreases until 1990, when it rises again and continues to rise until 2010.

The above statistics also show that energy use increased and decreased gradually between 2010 and 2014. After that, energy use slowly rises.

**Table 2**  Predicted value of energy-use using the following models

| Saudi Arabia | Predicted value of energy use (kg of oil equivalent per capita) using the following models | | | | |
|---|---|---|---|---|---|
| | ARIMA (1, 1, 1) | ARIMA (1, 1, 2) | ARIMA (2, 1, 2) | H-W | ANN |
| 2015 | 7047.97 | 7054.33 | 6552.6 | 6847.80 | 6622.55 |
| 2016 | 7183.67 | 7169.89 | 7114.7 | 7279.63 | 6756.26 |
| 2017 | 7317.14 | 7288.26 | 6829.9 | 7141.96 | 6623.91 |
| 2018 | 7449.85 | 7408.81 | 7332.7 | 7585.76 | 6714.21 |
| 2019 | 7582.29 | 7531.05 | 7103.7 | 7436.13 | 6647.12 |
| 2020 | 7714.64 | 7654.62 | 7554.8 | 7891.90 | 6742.37 |
| 2021 | 7846.96 | 7779.22 | 7373.8 | 7730.29 | 6596.89 |
| 2022 | 7979.28 | 7904.61 | 7780.2 | 8198.04 | 6732.65 |
| 2023 | 8111.58 | 8030.62 | 7640.8 | 8024.45 | 6594.34 |
| 2024 | 8243.89 | 8157.12 | 8008.6 | 8504.17 | 6719.46 |
| 2025 | 8376.2 | 8283.99 | 7905 | 8318.62 | 6626.01 |

**Table 3**  Measures of accuracy using the following methods

| Measures of accuracy | Measures of Accuracy using the following methods | | | | |
|---|---|---|---|---|---|
| | ARIMA (2, 1, 2) | ARIMA (1, 1, 2) | ARIMA (2, 1, 2) | H-W | ANN |
| MSE | 116,620.8 | 109,645.1 | 89,120.04 | 518,656.6 | 125,161.2 |
| RMSE | 341.4979 | 331.1269 | 298.298 | 720.1782 | 353.7814 |
| MAE | 266.9313 | 275.433 | 235.1379 | 482.397 | 284.3444 |
| MAPE | 0.073309 | 0.082255 | 0.068224 | 0.169608 | 0.061638 |

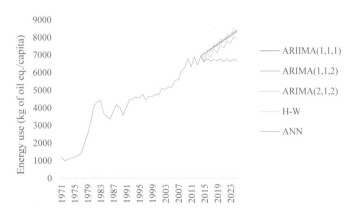

**Fig. 3**  Actual and forecasted graph using proposed forecasting models

## 3.3 Residual Analytic for Forecasting Models

This section of the study discusses the residual analytic for proposed forecasting models.

Figures 4, 5, 6, 7 and 8 show that the residuals of the models have analytical plots that assist in decision-making. Furthermore, the residual series is stationary, as shown by the time series plots in Figs. 4, 5, 6, 7 and 8. Therefore, all forecasting models mentioned in this study have a good fit for a linear model.

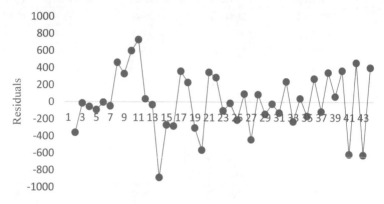

**Fig. 4** Residual plot of energy use data set using ARIMA (1, 1, 1)

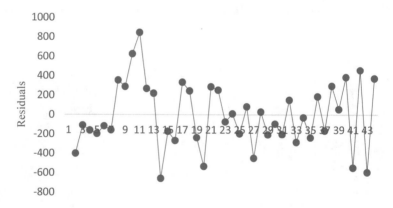

**Fig. 5** Residual plot of energy use data set using ARIMA (1, 1, 2)

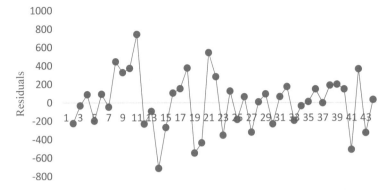

**Fig. 6** Residual plot of energy use data set using ARIMA (2, 1, 2)

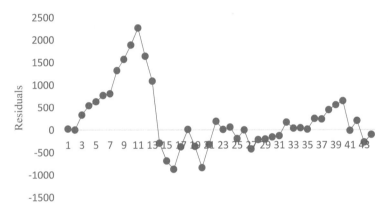

**Fig. 7** Residual plot of energy use data set using H-W

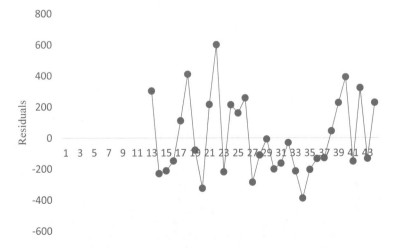

**Fig. 8** Residual plot of energy use data set using ANN

# 4 Conclusions

The main objective of the study is to predict energy use in the Kingdom of Saudi Arabia. ARIMA, H-W, and ANN models are the best models to predict the time series data. Energy use is essential because it operates several operations in different sectors such as schools, hospitals, airports, and towers. Therefore, every country must work hard to develop the energy sector to serve its citizens.

The Kingdom of Saudi Arabia has the largest sector of energy in the Middle East. Therefore, the Saudi government is changing and improving the energy sectors according to the 2030 vision. ARIMA, H-W, and ANN models are used to predict energy use comparatively.

In this study, MSE, RMSE, MAE, and MAPE were used to measure forecast accuracy, and therefore we compare the results to find the minimum error. In such a way, we can forecast with the lowest error, which is one of the main goals in this study. Therefore, the best forecast is the ARIMA (2, 1, 2) for this time-series data after comparing the accuracy measures. Thus, the Kingdom's energy use forecast value in 2025 will be 7905 (kg of oil equivalent per capita).

**Acknowledgements** We would like to express our sincerest appreciation to the College of Engineering at Prince Sattam bin Abdulaziz University in Al Kharj, Saudi Arabia, for their unwavering support and encouragement in our endeavor.

# References

1. The Embassy of the Kingdom of Saudi Arabia. https://www.saudiembassy.net/energy
2. Energy in Saudi Arabia. https://en.wikipedia.org/wiki/Energy_in_Saudi_Arabia#cite_ref-4
3. Alam T (2020) Predicting revenues and expenditures using artificial neural network and autoregressive integrated moving average. In: 2020 international conference on computing and information technology (ICCIT-1441). IEEE, pp 1–4
4. Li K, Zhang T (2018) Forecasting electricity consumption using an improved grey prediction model. Information 9(8):204
5. Khosravani HR, Castilla MDM, Berenguel M, Ruano AE, Ferreira PM (2016) A comparison of energy consumption prediction models based on neural networks of a bioclimatic building. Energies 9(1):57
6. Kalimoldayev M, Drozdenko A, Koplyk I, Marinich T, Abdildayeva A, Zhukabayeva T (2020) Analysis of modern approaches for the prediction of electric energy consumption. Open Eng 10(1):350–361
7. Sharma E (2018) Energy forecasting based on predictive data mining techniques in smart energy grids. Energy Inf 1(1):367–373
8. Shapi MKM, Ramli NA, Awalin LJ (2021) Energy consumption prediction by using machine learning for smart building: case study in Malaysia. Dev Built Environ 5:100037
9. Wang R, Lu S, Feng W (2020) A novel improved model for building energy consumption prediction based on model integration. Appl Energy 262:114561
10. Alam T (2019) Forecasting exports and imports through artificial neural network and autoregressive integrated moving average. Decis Sci Lett 8(3):249–260
11. Awel YM (2018) Forecasting GDP growth: application of autoregressive integrated moving average model. Empirical Econ Rev 1(2):1–16

12. Ersen N, Akyüz İ, Bayram BÇ (2019) The forecasting of the exports and imports of paper and paper products of Turkey using Box-Jenkins's method. Euras J Forest Sci 7(1):54–65
13. Jiang S, Yang C, Guo J, Ding Z (2018) ARIMA forecasting of China's coal consumption, price and investment by 2030. Energy Sources Part B 13(3):190–195
14. Jiang F, Yang X, Li S (2018) Comparison of forecasting India's energy demand using an MGM, ARIMA model, MGM-ARIMA model, and BP neural network model. Sustainability 10(7):2225
15. Kitworawut P, Rungreunganun V (2019) Corn price modeling and forecasting using Box-Jenkins model. Appl Sci Eng Progr 12(4):277–285
16. Nyoni T, Bonga WG (2019) Prediction of $CO_2$ emissions in India using ARIMA models. DRJ-J Econ Finance 4(2):01–10
17. Sen P, Roy M, Pal P (2016) Application of ARIMA for forecasting energy consumption and GHG emission: a case study of an Indian pig iron manufacturing organization. Energy 116:1031–1038
18. Urrutia JD, Abdul AM, Atienza JBE (2019) Forecasting Philippines imports and exports using Bayesian artificial neural network and autoregressive integrated moving average. In: AIP conference proceedings, vol 2192, no 1. AIP Publishing LLC, p 090015
19. Hanke, Wichern (2005) Business Forecasting, 8th edn. Pearson Prentice Hall (2005)
20. Artificial Neural Network (ANN), https://www.investopedia.com/terms/a/artificial-neural-networks-ann.asp

# Big Data Analytics and Audit Quality: Evidence from Canada

**Malik Abu Afifa, Yahya Marei, Isam Saleh, and Othman Hussein Othman**

**Abstract** This paper aims to understand better how Big Data and Big Data Analytics (BDA) affect professional judgement, audit performance and perceived audit quality in Canadian audit firms. Our findings are based on semi-structured interviews conducted with audit professionals firms. This research evidence suggests that auditors' skills and competence to perform engagement activities are assertively affected by BDA in audit methodology. Auditors benefit from being able to visualise audit evidence so they can use it to guide their professional judgement and decision making. We found evidence that the early stages of adopting data analytics and implementing are inefficient, but they save auditors' time as the tools get more familiarised. Finally, we documented that auditing professionals can use analytics to gain more insight into clients' business and offer them insights, which leads to confidence in clients.

**Keywords** Big Data · Big Data Analytics · Audit · Canadian audit

## 1 Introduction

Audit professionalism has been transformed in recent years due to most corporations' new sophisticated technology [15]. As a result of these technologies, auditors must understand the audit methods and update them on a regular basis to sustain public

M. Abu Afifa (✉) · I. Saleh
Accounting Department, Faculty of Business, Al-Zaytoonah University of Jordan, Amman, Jordan
e-mail: M.abuafifa@zuj.edu.jo

I. Saleh
e-mail: i.saleh@zuj.edu.jo

Y. Marei
Seneca College of Applied Arts and Technology, Toronto, Canada
e-mail: yahya.marei@senecacollege.ca

O. H. Othman
Accounting Department, Faculty of Business, Isra University, Amman, Jordan
e-mail: othman.othman@iu.edu.jo

© The Author(s), under exclusive license to Springer Nature Switzerland AG 2022
S. G. Yaseen (ed.), *Digital Economy, Business Analytics, and Big Data Analytics Applications*, Studies in Computational Intelligence 1010,
https://doi.org/10.1007/978-3-031-05258-3_22

trust in the audit process' efficiency [5, 7, 26]. Many corporations are being exposed to a massive amount of information, high velocity, and a fast number of data sets resulting from the continued use of the cloud, external data sources, and the internet [6]. Auditors have to sift through the large volumes of data presented by such events as emails, videos, records and texts to obtain the information they need [30].

It has been argued that current research in audit does not sufficiently address the skills required for examining and analysing data [11]. Furthermore, the audit team's structure, some researchers (such as [3]) suggest outsourcing a few audit processes in the Big Data's challenging environments, for example, outsourcing some data processing to Information Technology Corporations. However, this raises a series of issues regarding audit data's confidentiality and privacy [4, 27]. Several researchers suggest several factors are crucial to an audit's quality. These include the competence of an audit team and its experience [17] and the methodology [23]; prior researchers (such as [4, 27]) have, on the other hand, stated that there was a shortage of scientific research on the subject Big Data and auditors' skills from a conceptual perspective.

Besides, we examine how Big Data Analytics (BDA) can provide a more efficient internal control environment as well as how the minimum competencies and skill requirements for auditors can be affected by the use of BDA. As Big Data impose onerous on the external auditors, they also offer auditors opportunities to conduct prescriptive analytics [22]. External auditors are using data analytics in audit methodology as a result of the Big Data environment. Data analytics involves an investment in the Big Data environment, which includes hardware, software and skills that allows messy data coming from the client or third parties to be transformed thoroughly into valuable data. For example, an advanced program must interpret text and recognise voice and video in a hybrid workplace where personal devices, text, voice and video inter-connect with the corporate network. In addition to Deloitte's investment in text mining, firms like KPMG, EY and PWC have invested nearly $500 million to build analytic systems [21, 25, 27]. A survey conducted by Salijeni et al. [27] indicates that audit regulators welcome the use of data analytics in audit firms because it can reduce the risks associated with Big Data environments. Auditors welcome it, as well, since it reduces audit costs and enhances audit quality [1, 2]. In this research, it became apparent that audit skills have become one of the most important and relevant components of a successful business. These skills include knowledge of information technology and data science.

Then we examined the impact of Big Data and BDA on audit quality. Since Freddie Mac and WorldCom's substantial audit failures, investors, regulators and the general public have demanded higher audit quality. Audited quality is influenced by the quality of the evidence collected during the audit, as well as the auditor's professional judgement about the collected data. Simply gathering more data is not sufficient to enhance audit quality. In our research, we found auditing tools such as those used in this research were used to improve audit quality, not to improve efficiency, since this could be accomplished without sacrificing quality. Also, our results indicate that auditors strengthen their ability to use the tools efficiently and effectively once they understand how to use them. BDA tools are expected to improve the audit quality when included in the audit methodology in the future.

We also examined the auditors' professional judgement. Cao et al. [4] pointed out that Audit Data Analytics (ADA) help identify suspicious items in financial reports, which may be financial or non-financial. Auditors used to assess only a sample of financial report's data; however, with auditors using automated software to assess the entire report, higher audit quality is provided since ADA are able to reduce risks further. Additionally, auditors can devote more time to complex issues that require professional judgement. Therefore, this will reduce the overall allocation time as well as lower audit costs. However, many researchers have argued that BDA raises significant concerns, including the measurement of evidence [3], the structure of shreds of evidence [31], professional judgement [16] and auditors' knowledge [11]. When it comes to auditors' skills and knowledge, they start in business schools and academic syllabuses. Thus, in recent years, the American Accounting Association has increased its efforts to assist professors to incorporate Big Data into the classroom. A number of Big-4 accounting firms provide datasets and class resources for future auditors. The CPA Associations of Canada and USA developed a data analytics research initiative, partnering with business schools. In addition, audit courses are very full of information about information systems and technology in the audit [29]. We found evidence that business data analysis tools assist professionals in better understanding their clients' business environment and assessing risk.

Our research explores the issues of Big Data and audit. Auditors, clients, regulators and end-users of financial statements all have a vested interest in audit data analytics. In this era of digitalisation and fast-paced innovation, the use of data analytics is fast-growing. By contributing to the scant literature available on this topic, this research complements previous studies that have examined how BDA can affect the perceptions of auditor quality from practitioners' perspective. Being the first research on this topic in Canada and providing evidence on the impact of BDA on the audit method, this research contributes to an ongoing debate on the topic in the Canadian context. This research provides different stakeholders with valuable insights with respect to resource allocation for engagement teams and how data analytics influences auditing activities, and the skills and competence combinations needed to accomplish audit activities. Audit firms can rely on the results of this research to benchmark resource allocation to engagement teams.

## 2 Literature Review

Scholars have generally not reached a consensus on the definitions of Big Data concepts. The disagreement was rather conceptual rather than in terms of the procedures and processes for managing Big Data. The definition of Big Data is high volume, high velocity, and/or a wide range of information assets that necessitate novel processing methods to allow better decision-making, insight exploration, and process optimization [30, 32]. Volume measures the amount of information available in the Big Data, which differs based on the corporation's size and industry, where velocity indicates the speed of information. Variety indicates the speed of information, which differs based on the speed of information [3]. For instance, healthcare providers may have more volume than small car rental businesses. The velocity is the change or disappearance of information, which dimension has the highest potential risk to auditors [2]. The variety is related to data source and format—finally, the veracity is related to the reliability of new data added.

Audits are primarily used for receiving and then evaluating audit evidence to assess financial statements. The auditors' use of audit evidence differs from obtaining and applying other types of information to express their opinions about financial statements [32]. In the risk assessment process, auditors must use sufficient physical evidence according to audit standards. Further, the auditors should test physical evidence to ensure that the data are reliable. Because the nature of audit evidence is changing and it is no longer adequate due to technological advancement, traditional audit evidence becomes absolutely redundant. Big Data can help auditors in audit stages. For instance, the knowledge of the audit client and industry can provide additional insight into the engagement phase, facilitating the risk assessment phase in audit planning and conducting a fair value assessment based on non-financial data in the substantive test phase.

In the review phase, auditors can gain deeper knowledge of the client through taking a close look at the audit results. This knowledge will also help auditors to continually improve their client service.

A number of studies have postulated a convergence between Big Data implementation and auditors' skills. For instance, Salijeni et al. [27] documented in-depth interviews with individuals who understand the implementation, evolution and evaluation of business process audits. A review of published papers on BDA within the audit, it also explores how BDA impacts auditors and clients' relationships in analysing, storing and using financial as well as non-financial data. An examination of the challenges of adopting BDA on audit methodology, the findings of this analysis serve as the basis for future research, while Richins et al. [25] identify and highlight the importance of accountants. Professionals in the accounting profession have the skills and abilities to deal with unstructured data just as well as with structured data. Thus, BDA can supplement rather than replace the skills and knowledge of accountants. Yet, professional bodies and educational programmes must adapt to accommodate the challenges created by BDA. Earley [11] stated that auditors can conduct audit much more efficiently if they use data analytics to improve their audit skills, integrity and relevance of their available data and customers' expectations.

Auditors move between formal and informal audits according to various kinds of procedures. The formal approach is very structured, while the informal approach is based more on professional judgement. The difference is that auditing relies on "deeply rooted perspectives" (deeply embedded perspectives) rather than a set of technical steps [24]. In addition, it is responsible for a conflict between structure and judgement as they relate to a single organism and a single mechanism. The mechanism employs an integrative formal audit methodology that is reliant on algorithmic knowledge, while organisms emphasise understanding the whole rather than know everything. Even within a dominant structure audit approach, the structural judgement was found to be unable to represent a perfect approach [9].

Technology integration is used by some researchers to legitimise audit ideas [26]. Moreover, these technologies are associated with shared norms and cultural values in auditing, and are viewed as appropriate and desirable by main stakeholders including clients, professional bodies, educators, regulators and standard setters [27]. Curtis and Turley [7] found that audit firms are more likely to get pragmatic legitimacy if they can convince auditors of the benefits of using business risk audit in an audit. However, the auditor's concern is that BRA will impact their role in the audit process, generating the demand for moral legitimacy. The aforementioned implies that companies that fail to maintain pragmatic legitimacy can also achieve moral legitimacy through manipulation. In light of this, the adoption of new technology in audit requires at least moral legitimacy and pragmatic legitimacy as a cultural value. It is argued that integrating new ideas and technologies into an audit presents challenges and, consequently, requires careful consideration of the context and key audiences [27].

## 3 Methodology

This research examines the impact of Big Data and BDA on audit method, and the specific skills auditors need to perform their duties in an effective manner using BDA. Researchers contend that the qualitative method would be more appropriate than the quantitative method because BDA is new and exploited to their full extent. Quantitative approaches rely on an empirical data set which is insufficient to support the research objective. Obtaining qualitative data through interviews will allow the authors to produce a more in-depth, detailed report on the topic. So, the authors conducted an interview with open-ended questions as part of a semi-structured interview. Further questions may be asked if necessary, along with the semi-structured interview, which will give interviewees more opportunity to elaborate. In addition, an exploratory design was adopted in this research since it is more suitable for answering the research questions and fulfilling the research objectives based on the topic and the research question. The researchers have acquired several techniques, such as asking relevant questions, listening attentively, etc., that are fundamental to data collection.

To contact the appropriate respondents capable of providing relevant data, the researchers had to set criteria for choosing the respondents. In order to gather these formal data, we targeted respondents from the Big-four and mid-tier audit firms (see Appendix 1); because we believe that these audit firms utilize BDA more than others, and as a result, those who work in these firms have a greater understanding of BDA than others. Next, we prioritized responders with more than five years of auditing experience. Finally, our research interviewed 20 auditors in Canada who used BDA in audit work at different levels in the audit profession.

There are two kinds of data, primary and secondary data, described in many business research methodologies [12]. Therefore, the authors of this research are using the same categories of data as well. However, there is a difference between primary data gathered and secondary data that are used in the construction of the research outline and completing it. Primarily, users of primary data will rely on interviews to answer their research questions. The authors utilise interviews extensively as their primary data collection method. A semi-structured interview type is a preferred approach since it enables interviewees to express themselves more freely during the interview process than a structured interview method. They are also free to discuss any issues during the interview process. However, this research is based on the thematic analysis method using the interview transcriptions.

## 4 Results and Discussions

Based on the interviews used in this research, the following subsections present the results linked to the objectives and research questions.

### 4.1 *Skills and Competencies of Auditors*

It is accurate to say that competence is subject to more than one definition, and it may have different meanings as it refers to the ability, capability and effectiveness. Hartjen [18] draw our attention to the definition; in the review, competence is defined as a combination of skills, competencies and personal characteristics necessary to attain specific objectives. It also encompasses time management skills as they relate to performance. Hodge [19] documented competence as a system of attributes beyond skills, knowledge and tasks. It is a total and comprehensive picture of where an employee stands at the right time to perform key tasks. These attributes include cognitive attributes that must be present in order for an employee to perform key duties, such as knowledge, problem solving strategies, communication, pattern recognition and critical thinking. The generic approach builds on training and assessment processes that aim at training and assessing a candidate's attributes rather than achievement. These attributes are also assessed as separate from any achieved work. Consequently, future performance evaluation may fail to take into consideration those attributes [28].

During the evaluation process of auditors, individual behaviour, knowledge and abilities should be considered. The evaluation should also include an audit programme and its goal. Auditors should have sufficient competencies for meeting audit objectives; on the other hand, auditors in a team should not share the same competencies. Auditors should be selected based on the outcome of the evaluation audit process, receive training to ensure proficiency and be evaluated on a continuous basis.

As mentioned earlier, auditors need more understanding of analytical data tools in order to conduct high-quality auditing. The participants confirmed this belief by saying auditors should possess more advanced data analytical skills and competency.

Some participants have a different view; they believe it is more important to know about business and the industrial products related to Big Data than IT.

> Since the way in which we operate is that we understand the data more than the systems themselves, I would say that understanding the data is more important. At this time, though, it appears that the skills are related to both accounting and finance; it is more important for a person to possess a general business strategy and mindset than actual IT skill. [P 1]

> The implementation of a special person who acts as an analytics specialist for these large clients will also be more helpful to them as well, and this is the future, in my opinion. [P 2]

Despite this, all participants agreed that an analytical specialist on audit teams was supported in the firm even if he was not part of the audit team. Although it is a confirmation that companies in large firms do not outsource their analytical activities, it does not prove that smaller firms outsource analytical activities. Our findings also confirm that the Big Four audits are constantly preparing for analytic activities. Our participants stressed the importance of a specific program on programming languages, even with existing programs.

> In the future, it will be more analytically complex to audit using Big Data. In a traditional audit, all the work is generally handled using tick-and-tie methods, where samples are secured and then matched up with the general ledger data, and this is a straightforward approach. However, you would need some analytical skills to understand the distinction when you are using Big Data. [P 3]

Participants suggested the best approach to improving auditor skills and competency was through practice and hands-on. Participants also commented on new hiring in the audit department. They recommend that human resources recruit candidates with technical backgrounds rather than candidates from finance and business schools.

Besides, participants suggested that the school curriculums should contain a substantial amount of theoretical information about Big Data rather than practical information, because every firm has its tools. Several participants confirmed that some universities have already included courses and materials concerning Big Data within their curriculums.

> Fortunately, there are many examples out there and how-to's you can watch online to learn. Even if you don't believe me, maybe somebody shows me they have XYZ; you know there is a way to succeed in any project. [P 4]

> …however then, what if somebody can walk me through the steps of actually building this with this data, through this tool, and then actually demonstrate what result is. So yes, technical training may be required, but it also requires training in how we think about performing an audit. [P 5]

## 4.2 Professional Judgement

An auditor uses professional judgement as one of the most valuable skills during audit engagements, and they know that professional judgement varies from one auditor to another and is dependent on experience and competencies. The participants indicate that Big Data affect their professional judgement. Auditors' professional judgement is based mainly on the data's accuracy and reliability, so analytical tools improve the verified data. In fact, participants argued that the amount of evidence collected from Big Data is based on professional judgement, as it either increases their knowledge about business clients or gets them lost in the details. Lastly, participants dispelled the notion that Big Data will replace auditors' professional judgement by applying advanced analytical tools and robotics, which they indicated would only be useful for unstructured and non-financial data.

> The importance of skepticism as an auditor has definitely increased as a result of Big Data Analytics. I think data analytics has definitely increased the importance of professional judgement as an auditor. [P 5]

> I think it's a lot more accurate than a traditional audit since you have the results in front of you. It will also eliminate subjective judgement, which has never been easier as you see the results in front of you. [...] It greatly affects judgement right now and also into the future. [P 11]

> It is important to remember that using data analytics alone would not improve audit results. The integration of data analytics and professional judgement, aided by data analytics, has the potential to enhance audit efficiency. [P 13]

## 4.3   *Audit Efficiency*

Participants in our research have agreed on the costs of audit engagements and the length of audit engagements. While involvement with Big Data and analytic tools increases during the first year of implementation, the amount of time will decrease. Some participants believe that the time will not decrease since Big Data and audits are about the quality of reports rather than length of engagement. Different participants had a different percentage of time spent on Big Data during the engagement; however, they agree that the level of risk they faced drives the percentage of time spent on Big Data. Regarding audit fees and audit costs, participants agreed that the fees might not be reduced, and some argued that the cost might be higher due to the cost of implementing the analytical tools. Furthermore, participants agreed that implementing an audit for the entire population would be more expensive than sample-based audits.

> The cost of implementing the analytics software is an increase in time. In order to make the decision to move from sample-based to analytics, the auditor must consider quite a few things that usually take time. [P 7]

> When you start working with Big Data analysis for the first time, it may take more than two to three years before it becomes routine. [P 12]

> In the beginning, when implementing Big Data for the first time, performing the audit can be quite a time consuming, then it seems less effective as you need to do more complicated analysis. [P 17]

## 4.4   Audit Quality

The quality of audit reports has been a significant issue for decades. To examine the BDA on audit quality, we break down audit quality into four components: the inputs, outputs, the process and the interaction. Based on our research's findings, it appears possible to improve audit quality by leveraging analytical tools and having competent audit team members. The findings are supported by previous research, such as [8].

> ...the science and art of finding and evaluating trends, detecting inconsistencies, and extracting other valuable knowledge from data underlying or relating to an audit's subject matter by study, simulation, and visualisation for the purpose of preparing or conducting the audit play an important role in audit quality. [P 8]

> Audit firms use Big Data Analytics in audit methods to investigate possible advantages and better represent their customers. [P 9]

> It has a beneficial effect on audit results [....] Since, as I previously said, it provides a very clear picture. [P 6]

> Auditors may use Big Data Analytics to enhance risk assessment practices, substantive practices, and compliance monitoring testing. [....] Thus, audit quality will improve. [P 11]

> Currently, our analytical solutions are not intended to measure and analyse controls; rather, analytics can be seen as a substantive research strategy to improve audit quality. [P 1]

The input section covered the competencies of the audit members and the techniques they employed during the engagements. Our findings showed the audit team members have a wide range of expertise that helps them understand the clients' business environment. In line with this is the Financial Reporting Council [13], which indicates that understanding the client environment would enable the improvement of audit quality.

> The higher the audit team members' knowledge about the client environment, the higher the audit quality. [P 3]

> Using techniques in audit leads to improve the audit quality, but at the same time, this is dependent on the audit team members' proficiency in using it. [P 5]

Nevertheless, using planning phase tools will provide a better overall idea of the business and risks. Furthermore, the finding suggests auditors may not use these tools efficiently and effectively at the beginning, but, over time, they are becoming more efficient. It appears that tool use may be helpful in improving the quality of the audit, but the auditor's proficiency in performing the audit and extracting and interrupting the data can make it meaningful. This result is supported by International Federation of Accountants (IFAC) [20], which suggests that data analytical will not achieve audit quality independently.

> I'd say it's still mainly used in the risk assessment phase, followed by the audit preparation period. However, for areas such as revenue, it is becoming more normal to provide substantive analytical procedures as well as in the implementation part of the audit where we step away from sampling procedures to rely on analytics for the actual auditing. [P 4]

...It is used in two stages: preparation and implementation [...]. The real job, however, is in the implementation process. [P 6]

...I'd say it's more in the executions process because that's where you spend the most time, but it still assists with preparation. [P 8]

We used the analytics software all year, but I'd say it's more popular during the year-end audit. [P 11]

A process section includes elements linked to risk assessments and control tests and substantive tests [10]. It has been suggested that BDA may be used as an alternative to traditional sampling. The results of our research indicate that the internal control assessment is not relevant to the analysis. However, some participants say otherwise. Different sizes of clients may play a role in the reasons for the different findings. For small businesses, the internal control IT system is not complex; therefore, analytic tools may be preferable when auditing the internal controls. Another factor could be complex internal control systems resulting from the larger the business and, therefore, limited expertise with the analytical tools. They believed that the use of BDA impacted their sampling and risk assessments. In addition, they also agreed that BDA enables them to test the entire population, resulting in reduced exposure to risk.

Furthermore, auditors are capable of noticing unusual transactions and performing further investigations. Using BDA for content testing is the most significant part of the audit process. Auditors agreed that the phase that was affected the most by BDA was the planning.

The risk assessment process and the planning stage are still overwhelmingly used for this type of audit, but it is increasingly also being used in other areas.[....] the use of substantive analytical procedures has become increasingly common as well as the use of analytics during the execution of an audit, moving away from traditional sampling procedures. [P 3]

I think it might be deeper in the execution phase, since that is where you spend a lot of time, but as there is something to be said, the planning phase is also important... [P 15]

In terms of the first component, identifying risk, Big Data Analytics is now a big and critical part of our risk assessment procedures. Where we can divide the population into homogeneous subpopulations..... [P 11]

Our results show that the output of the audit elements positively impacted the audit methodology's analytical tools. Participants stated that the quality of audit reports improved when using data analytically, which implies clients' reporting was also improved. Additionally, audit firms provide improved audit opinions since they have more insight into audit shreds of evidence. Thus, our results are in line with previous research such as [10], which indicated that client quality and auditor reliability affected audit quality. In addition, participants who used the BDA mentioned that the tool makes it easier to detect fraud in non-financial data, but that it does not improve the quality of the data as more data are not necessarily more useful [14]. The use of non-financial data can help assess clients to determine the role they play in the company and support improvement initiatives.

Data analytics is about insights, and insights can lead to errors or the discovery of something that the customer is unaware of. [P 8]

> Looking through the whole population of data assists us in risk recognition and identifying outliers that, if accumulated, may contribute to material misstatements, outliers that we might have overlooked if we had used conventional auditing techniques. [P 13]

> Big Data analytics, in my opinion, boosts audit efficiency by allowing for a more rigorous and fact-based risk evaluation. This simply summarises what we're looking at, why we're looking at it, and how we're going through it. Second, I believe that using Big Data analytics in the audit process would result in a higher number of misstatements being found in the end. [P 2]

A key element of attaining audit quality is the relationship between shareholders. Our results have shown that the data analytics interacts positively with critical stakeholders, such as the audit committee and others on the engagement team. The participants stated that communication tools gave them the ability to communicate effectively and effectively with management about the foundation, primarily when visualising tools are used. As prior studies, such as [27] have demonstrated, the use of tools can ultimately enhance audit report reliability. Additionally, tools can facilitate communication with clients and audit teams since all team members have access to the same information at the same time.

> In some situations, the application of analytics may be a game-changer in terms of providing new information to management. For example, by examining reasonably basic GL data, we can elicit a wealth of knowledge that clients might not be aware of. [P 8]

> ….Another example is that when we do Big Data analysis, the results can include graphs or maps, or visualising may aid in understanding their market. As a result, our findings may serve as a useful reference for them. [P 4]

> Big Data, I believe, would have an effect on relationships with these various stakeholders. In other words, as you do the analysis, you submit fewer supporting materials, which means less work for our clients. [P 5]

> I don't believe it will boost interactions with all stakeholders; I believe it will benefit management in the sector more. [P 9]

## 5 Conclusion

This research examines the efficiency and quality of audits in Canada by examining how audits using business decision analysis such as BDA may affect audit efficiency and perceived audit quality. The research further intends to examine whether the use of BDA will affect the minimum competencies and skill requirements for auditors.

Researchers found that tools were used at both critical phases of the audit process—the planning and execution phases. Auditors performed a thorough risk assessment, identifying those areas of greater risk to focus on. Based on this research, it is clear that the importance and relevance of audit skills and competence have increased. This includes information technology and data science skills that are becoming increasingly important in the audit field, in addition to basic economic knowledge. According to the findings from the research, auditors should show they are proficient in Excel with a primary focus on IT skills. Auditor's skills should be upped accordingly with new demands in the job market. This suggests that academic institutions should redesign their curriculum to prepare graduates for the new job market demands.

A professional's judgement also relies extremely on the amount of experience and knowledge the auditor has, their ability to assess the client's organisation, and the quality of evidence provided. Researchers found that business data analysis tools aid in professional judgement in understanding the client's business environment and risk assessment. Further, the results demonstrate that the tools have not achieved full efficiency yet. Rather, implementing the tools for the first and second years will have an adverse effect on efficiency, increasing audit expenditures. The reason for this may be because introducing a new methodology takes more time for users to get accustomed to their new methods. Auditors will need further instruction when the tools are being first introduced.

However, efficiency has not been achieved because the main objective of using such tools in auditing is to improve audit quality. It has been argued that this cannot be accomplished at the expense of quality. The findings indicate an impact on the audit quality when BDA tools are incorporated into the audit methodology, which is expected to be high in the future. Auditors will strengthen their ability to use the tools efficiently and effectively as they understand how to use the tools. BDA tools give auditors a way to make a more significant impact on clients and company members. The visualisation tools improve communication among key stakeholders, including management, the audit committee and auditors; communication within the audit team is enhanced through BDA tools.

# 6 Implications

Our main contribution understands how data analytics affect the audit process and provides more insight into what skills and competence combinations are needed to accomplish the audit activities. This information will be of practical use for various stakeholders. Audit firms can use this research's findings as a benchmark for resource allocation to engagement teams. This research also identifies the key talents and skills that an auditor needs when performing an audit in a Big Data environment. Furthermore, this will help recruit new auditors by providing a pool of candidates with the necessary talents and skills. Finally, it was found that BDA can significantly increase audit quality and provide insight for clients. Through this, they can learn about BDA and how it can help them and their clients improve productivity. Clients will also benefit from the results by collaborating with auditors to find out more information about their business processes for better strategic planning.

The fact that researchers do not have enough time for conducting more interviews results in a lack of diversity in the empirical evidence. All of the interviewed participants are based on Canadian audit firms. Hence, the research is limited to information about the Canadian market. Due to this, the findings do not apply to firms not based within the Canadian market.

# Appendix 1. List of Interview Participants

| Code | Participant | Audit firm | Role(s) |
|------|-------------|------------|---------|
| P1 | Partner—audit methodology | Big four | BDA development and implementation |
| P2 | Partner—audit methodology | Big four | BDA development and implementation |
| P3 | Partner—audit methodology | Mid-tier audit firm | BDA development and implementation |
| P4 | Partner—audit assurance | Big four | BDA development and implementation |
| P5 | Partner—audit assurance | Big FOUR | BDA development and implementation |
| P6 | Partner—audit assurance | Mid-tier audit firm | BDA development and implementation |
| P7 | Partner—audit assurance | Mid-tier audit firm | BDA development and implementation |
| P8 | Partner—audit risk analytics | Big Four | BDA development and implementation |
| P9 | Partner—audit risk analytics | Big four | BDA development and implementation |
| P10 | Partner—audit risk analytics | Mid-tier audit firm | BDA development and implementation |
| P11 | Partner—data assurance | Big four | BDA development and implementation |
| P12 | Partner—data assurance | Big four | BDA development and implementation |
| P13 | Partner—data assurance | Big four | BDA development and implementation |
| P14 | Partner—data assurance | Mid-tier audit firm | BDA development and Implementation |
| P15 | Partner—data assurance | Mid-tier audit firm | BDA development and implementation |
| P16 | Partner—data assurance | Mid-tier audit firm | BDA development and implementation |
| P17 | Data analytics auditor | Big four | BDA development and implementation |
| P18 | Data analytics auditor | Big four | BDA development and implementation |
| P19 | Data analytics auditor | Mid-tier audit firm | BDA development and implementation |
| P20 | Data analytics auditor | Mid-tier audit firm | BDA development and implementation |

# References

1. Abu Afifa M, Alsufy F, Abdallah A (2020) Direct and mediated associations among audit quality, earnings quality, and share price: the case of Jordan. Int J Econ Bus Adm 8(3):500–516
2. Alles MG (2015) Drivers of the use and facilitators and obstacles of the evolution of big data by the audit profession. Account Horiz 29(2):439–449
3. Appelbaum D, Kogan A, Vasarhelyi MA (2017) Big data and analytics in the modern audit engagement: research needs. Audit J Pract Theory 36(4):1–27
4. Cao M, Chychyla R, Stewart T (2015) Big data analytics in financial statement audits. Account Horiz 29(2):423–429
5. Carpenter B, Dirsmith M (1993) Sampling and the abstraction of knowledge in the auditing profession: an extended institutional theory perspective. Acc Organ Soc 18(1):41–63
6. Cukier K, Mayer-Schoenberger V (2013) The rise of big data: how it's changing the way we think. Foreign Aff 92(3):1–13
7. Curtis E, Turley S (2007) The business risk audit—a longitudinal case study of an audit engagement. Acc Organ Soc 32(4–5):439–461
8. Dekeyser S, Gaeremynck A, Knechel WR et al (2021) The impact of partners' economic incentives on audit quality in big 4 partnerships. Account Rev (forthcoming):1–13
9. Dirsmith MW, Haskins ME (1991) Inherent risk assessment and audit firm technology: a contrast in world theories. Acc Organ Soc 16(1):61–90
10. Duh RR, Knechel WR, Lin CC (2020) The effects of audit firms' knowledge sharing on audit quality and efficiency. Auditing 39(2):51–79
11. Earley CE (2015) Data analytics in auditing: opportunities and challenges. Bus Horiz 58(5):493–500
12. Ferguson SL, Kerrigan MR, Hovey KA (2020) Leveraging the opportunities of mixed methods in research synthesis: key decisions in systematic mixed studies review methodology. Res Synthesis Methods 11(5):580–593
13. Financial Reporting Council (2017) Audit quality thematic review the use of data analytics in the financial statements. Report. The Financial Reporting Council (FRC), London
14. Fukukawa H, Mock TJ, Srivastava RP (2014) Assessing the risk of fraud at Olympus and identifying an effective audit plan. Jpn Account Rev 4(1):1–15
15. Gao Y, Han L (2021) Implications of artificial intelligence on the objectives of auditing financial statements and ways to achieve them. Microprocessors Microsyst (forthcoming):104036
16. Gordon J, Shortliffe EH (2008) A method for managing evidential reasoning in a hierarchical hypothesis space. Stud Fuzziness Soft Comput 219:311–344
17. Harris MK, Williams LT (2020) Audit quality indicators: Perspectives from Non-Big Four audit firms and small company audit committees. Adv Account 50:1–13
18. Hartjen RH (1974) Implications of Bandura's observational learning theory for a competency based teacher education model. In: The annual meeting of the American Educational Research Association, Chicago, IL, pp 1–23
19. Hodge S (2020) Book review: continuing professional education in Australia: a tale of missed opportunities, by B. Brennan. Adult Educ Q 70(2):196–198
20. International Federation of Accountants (IFAC) (2017) Handbook of international quality control, auditing, review, other assurance, and related services pronouncements. The International Federation of Accountants (IFAC), New York
21. Kokina J, Davenport TH (2017) The emergence of artificial intelligence: how automation is changing auditing. J Emerg Technol Account 14(1):115–122
22. Lee M, Cho M, Gim J et al (2014) Prescriptive analytics system for scholar research performance enhancement. Commun Comput Inf Sci 434(1):186–190
23. Marcolino MS, Alkmim MB, Pessoa CG et al (2020) Development and implementation of a methodology for quality assessment of asynchronous teleconsultations. Telemed E-Health 26(5):651–658
24. Power M (2013) The apparatus of fraud risk. Acc Organ Soc 38(6–7):525–543

25. Richins G, Stapleton A, Stratopoulos TC et al (2017) Big data analytics: opportunity or threat for the accounting profession? J Inf Syst 31(3):63–79
26. Robson K, Humphrey C, Khalifa R et al (2007) Transforming audit technologies: business risk audit methodologies and the audit field. Acc Organ Soc 32(4–5):409–438
27. Salijeni G, Samsonova-Taddei A, Turley S (2019) Big Data and changes in audit technology: contemplating a research agenda. Account Bus Res 49(1):95–119
28. Salman M, Ganie SA, Saleem I (2020) The concept of competence: a thematic review and discussion. Eur J Train Dev 44(6/7):717–742
29. Sledgianowski D, Gomaa M, Tan C (2017) Toward integration of Big Data, technology and information systems competencies into the accounting curriculum. J Account Educ 38:81–93
30. Warren JD, Moffitt KC, Byrnes P (2015) How big data will change accounting. Account Horiz 29(2):397–407
31. Zhang L, Pawlicki AR, McQuilken D et al (2012) The AICPA assurance services executive committee emerging assurance technologies task force: the audit data standards (ADS) initiative. J Inf Syst 26(1):199–205
32. Zhang J, Yang X, Appelbaum D (2015) Toward effective big data analysis in continuous auditing. Account Horiz 29(2):469–476

# Continuance Intention to Use YouTube Applying the Uses and Gratifications Theory

Saad G. Yassen, Dima Dajani, Ihab Ali El-Qirem, and Shorouq Fathi Eletter

**Abstract** The current research attempts to examine how gratifications predict the continuance intention to use YouTube among university students. Partial least squares structural equation modeling (PLS-SEM) was used to examine causal relations proposed in the research model. The sample comprised 446 students from selected universities in the United Arab Emirates. The research findings revealed that hedonic, escapism and social interaction gratifications have a significant impact on the continuance intention. However, it is unexpected to find that information seeking, mobility and information overload gratifications have an insignificant effect on the continuance intention.

**Keywords** Uses and gratification theory (UGT) · Continuance intention · Structural equation modeling · YouTube

## 1 Introduction

Social media sites have profoundly changed the way we live, communicate and work [1, 2]. Using social media and social computing, in general, involves rapidly disseminated user-generated content [3, 4]. Social media sites are connecting us and shaping us [5]. Although they attract billions of users around the world, they each have their unique features, functions, norms and culture [6–8]. In the realm of social

S. G. Yassen (✉) · D. Dajani · I. A. El-Qirem
Faculty of Business, Al-Zaytoonah University of Jordan, Amman, Jordan
e-mail: saad.yaseen@zuj.edu.jo

D. Dajani
e-mail: d.aldajani@zuj.edu.jo

I. A. El-Qirem
e-mail: i.elqirem@zuj.edu.jo

S. F. Eletter
College of Business, Al Ain University, Al Ain, UAE
e-mail: Shorouq.eletter@aau.ac.ae

media, YouTube is the second most popular social media platform with 1.9 billion users, and 500 h of video are uploaded every minute (brandwatch.com). It is dynamic, interactive and user-centric, and a go-to resource platform for seeking information and viewing videos [9]. Thus, YouTube users' generated content is diverse, global and easily uploaded and downloaded. YouTube is also the second-largest search engine and the most popular social media platform in the United States [5]. It is becoming an effective social and educational platform to enhance interaction and the learning process in a creative manner. One particularity of YouTube is characterized by the uploaded videos of content creators (YouTubers). It is a hub of participatory culture in which users generate video content [10]. An essential feature of YouTube is that it enables users to view uploaded videos and comment or share them on other social media sites [11]. Thus, YouTube content in various forms, such as videos, workshops, online courses, movies or social clips, provides students with effective social media tools that enable them to learn and share valuable content.

Students are likely to be familiar with YouTube technology and can be expected to be excited to use e-content. The YouTube platform tools allow students to make contact with their peers to form an online community [12]. However, although universities know that their students will be able to leverage social media content, intensive research is required to reveal how students gratify their needs through YouTube adoption and usage. The appeal of YouTube can be understood through the use and gratifications theory. Furthermore, despite recent literature about YouTube as a social media form, there is a vital need to elucidate how various types of gratification affect an individual's continuance intention to use YouTube.

The existing research is mainly devoted to exploring antecedents of YouTube advertising, YouTube learning, music listening and video sharing [9, 13–15]. To fill this research gap in the social media literature, empirical research is needed to understand how and why people use YouTube and persist with its use, and how gratifications subsequently enhance behavioral intention to use YouTube. To explore and predict continuance intention, the present research examines six categories of gratifications: hedonic gratification, social interaction gratification, escapism gratification, information seeking gratification, information overload gratification and mobility gratification [7, 16, 17]. Thus, based upon uses and gratifications theory (UGT), the current research aims to answer the following question: What and how do gratifications drive continuance intention to use YouTube among university students in the United Arab Emirates.

## 2   Literature Review

### 2.1   Uses and Gratification Theory

Uses and gratifications theory (UGT) is an influential paradigm that attempts to answer questions about how and why individuals actively select media to fulfill their needs and motives [18, 19]. It assumes that individuals thoughtfully select a specific media to satisfy their various needs and that the ultimate aim of actual use of social media is to obtain gratifications [20]. Thus, for UGT researchers, the basic argument remains the same—individuals' motivations are triggered by their own needs and expectations. According to UGT, people actively seek a specific type of media in terms of gratifications sought and gratifications obtained. Gratification sought refers to a user's expectations of using social media, while gratification obtained refers to a user's actual gratification when using a specific media [21, 22]. Gratifications are perceived to be the fulfillment of a need through social media [3].

Uses and gratifications theory researchers have identified several types of gratifications such as emotional gratification, cognitive gratification, social gratification, hedonic gratification, utilitarian gratification, technology gratification and habitual gratification [3, 21, 23, 24]. Thus, UGT has been successfully applied to study social networking sites [2, 7, 9, 18], smartphone usage [19], social media [3, 25], Facebook, Twitter, Instagram and SnapChat [22], Wechat [23], YouTube advertising [14], Twitter [2], consumption on YouTube [9], Facebook and kakao talk [16], mobile shopping [26] and music listening applications [27].

This theory provides a useful framework to analyze gratifications and needs for using social media platforms. It has the capacity to conceptualize people's intrinsic needs to choose specific social media platforms through various types of users' socio-psychological gratifications. Uses and gratifications theory assumes that users select a particular social media in order to satisfy one or more gratifications; namely, cognitive gratification (seeking information, understanding), personal gratification (self-concept) and tension release gratification (escapism). Wang et al. [3] provided four main needs and gratifications for using social media among college students: emotional, cognitive, social and habitual. Li et al. [28] used UGT for modeling online games. The study findings revealed that three categories of gratifications affect a user's continuance intention to use online games: hedonic (enjoyment, escapism), utilitarian (achievement) and social gratification (social interaction). According to Ifinedo [18], the gratification categories of entertainment, social enchantment, self-discovery and interpersonal connectivity through behavioral intention were found to have a significant impact on students' adoption of social networking sites. Regarding YouTube 'stickiness', Chiang [13] research findings indicated that continuance motivation and sharing behavior were antecedents of YouTube stickiness. However, as little is known about continuance intention regarding YouTube among university students, the current study aims to investigate the gratifications that students derive from using YouTube. Understanding what drives continuance intention to use YouTube is crucial in the context of students' relationships with a social media site such as YouTube.

In addition, although a few studies have participated in highlighting YouTube uses and gratifications, their contributions to the business literature are very limited [9, 13].

## 2.2 Continuance Intention

One of the main knowledge fields in IT literature is an intention-based model to predict IT tools or information systems' acceptance, adoption and usage. Social psychology postulates intention as the main construct for IT adoption research [29–35]. The premises of the theory of reasoned action and the theory of planned behavior postulate that a person's intention to adopt or perform or not perform a behavior is a function of their personal nature and social influence [36–38]. According to Ajzen and Fishbein [36, 37] behavioral intention predicts a user's actual behavior. It is the most significant predictor of usage behavior [11]. Technology acceptance models suggest there are belief–attitude–intention–behavior causal relationships [30–34]. Thus, research of the behavioral intention to adopt or use IT platforms or tools is undoubtedly mature [39]. However, scholars have suggested examining continuance intention to augment the expectation-confirmation theory [40]. One research stream employed the continuance intention construct to predict actual behavior and, in turn, focus on the antecedents of intention. Continuance intention to use YouTube is described as a measure of a student's behavioral intention to use YouTube. Continuance intention is the immediate antecedent of a student's actual use of YouTube [35, 41–44].

## 3 Research Model and Hypotheses

Drawing on uses and gratifications, we developed a conceptual framework of YouTube continuance intention among university students. The research model postulates that a student's continuance intention is predicted by hedonic gratification, social interactive gratification, information seeking gratification, mobility gratification, escapism and information overload gratifications. Continuance intention denotes students actively using and engaging in YouTube content. A hedonic gratification is an extent to which social media is enjoyable and entertaining [45]. It has been found to affect intention to adopt and use social media platforms in various settings [26–28, 46]. This gratification refers to the perceived entertainment, fun and enjoyment [47]. It is associated with the fulfillment of perceived enjoyment and passing time [48, 49]. Li et al. [28] found that three categories of gratifications influence a user's continuance intention to use social media games: hedonic, utilitarian and social gratifications. Wang et al. [3] revealed that social media use is significantly driven by four types of motivations: emotional, cognitive, social and habitual. Thus, the current research hypothesizes:

> H1: hedonic gratification has a positive effect on continuance intention to use YouTube.

> YouTube features allow students to receive information in different forms and sources [50]. Studies revealed that the use of social media sites is often driven by

the desire for sharing content or self-presentation [51]. Information seeking reflects a user's purposeful search for information to enhance awareness and knowledge and to self-educate [17, 47, 49]. This gratification is one of the primary motivations for using social media according to UGT literature. Hence, the following hypothesis is proposed:

H2: Information seeking has a positive effect on continuance intention to use YouTube.

Social interaction gratification refers to people's desire to build social relationships with others. Consistent with recent research findings, users interact and communicate on social media to gratify their social needs. Previous studies have shown that users are increasingly using social media platforms for interacting and socializing with others [21]. Users can interact through YouTube by sharing content, commenting on or reposting others' content [50]. In social media such as YouTube, users seek to receive attention from others to build their self-identity through social interaction gratification [8]. Thus, the following hypothesis is set forth:

H3: Social interaction has a positive effect on continuance intention to use YouTube.

Escapism refers to avoiding the real world, to leave the reality in which users live in a cognitive way [17, 47]. Social media research has indicated that escapism is characterized by greatly preventing users from thinking about their personal problems or feelings of stress. Chaouali [17] provided strong evidence that escapism determines satisfaction. Similarly, escaping is a crucial factor in mobile gaming for university students [47]. In addition, substantial research on social media has shown that escapism has a significant effect on an individual's intention to perform a specific behavior, especially when they wanted to get away from what they were doing [43, 52]. Hence, the following hypothesis is set forth:

H4: Escapism gratification has a positive effect on continuance intention to use YouTube.

Mobility reflects a remarkable feature of social media that may affect a user's continuance intention. Mobility refers to the extent to which users can access social media at anytime and anywhere [53]. It is a gratification opportunity that reflects the flexibility and accessibility of social media platforms [43]. For social media, mobility would seem to be a critical motivation because it allows the user access to social media services in real-time. Applying UGT to the YouTube context, the following hypothesis is set forth:

H5: Mobility gratification has a positive effect on continuance intention to use YouTube.

As the digital world provides massive amounts of data, information overload describes conditions in which users cannot process and perceive too much information in a real-time situation [54, 55]. Previous studies discuss different related constructs such as cognitive overload and knowledge overload [17, 54]. Information overload reflects a cognitive pressure that may influence an individual's continuance intention to use YouTube. Therefore, the next hypothesis is formulated as follows:

H6: Information overload has a positive effect on continuance intention to use YouTube.

Thus, building upon uses and gratifications theory (UGT), the current research model investigated students' gratifications as predictors of continuance intention to use YouTube.

# 4   Method

## 4.1   Sampling

The current research adopted a cross-sectional research design. The sample comprised private universities in Al Ain, Abu Dhabi and Dubai in the United Arab Emirates. The research unit of analysis included 446 students from selected universities. The research participants comprised 446 students, females 59.9% and 40.1% males. The students' age range was 14–24 years, 63.9%; 25–34 years, 2.89%; 35–44 years, 5.7%; 45 and above, 0.4%. Overall, 4.3% of students were enrolled in master's programs and 95.7% in B.Sc. programs.

## 4.2   Measures

Several measurement scales of the current model have long been used in previous research and have shown high validity. Consistent with social media research, all constructs were measured using well-validated multiple items. The measurements of uses and gratifications constructs were adapted from [16, 17] and [28]. Continuance intention was adopted from [30, 35, 40, 46]. Constructs of multiple items have been modified to fit the context of continuance intention to use YouTube. The empirical data was collected with a questionnaire survey. All instrument statements used a five-point Likert scale (1 = disagree with all, to 5 = strongly agree). The research's empirical field yielded a total of 446 completed valid responses from 520 distributed questionnaires. The response rate stood at 0.85%.

# 5   The Measurement Model

Partial least squares structural equation modeling (PLS-SEM) was used to examine the causal relationships proposed in the research model. Smart PLS3 was suitable for this research for the following reasons: (1) PLS-SEM has attracted much attention in business research; (2) it is a combination of factor analysis and path analysis, and (3) it is a variance-based SEM for small size samples [41, 56, 57].

The first step in evaluating PLS-SEM model results involves examining a reflective measurement model in terms of items' loadings, reliability of measures, composite reliability, Cronbach's alpha, convergent validity and discriminant validity. Loadings above 0.70 indicate that the construct explains more than 50% of the indicator's variance. The reliability of constructs was assessed by composite reliability with its minimum value of 0.70 and Cronbach's alpha with its minimum value of 0.60 [57]. Convergent validity is the extent to which a construct converges its individual indicators and can be measured by the average variance extracted (AVE) [58]. The AVE is measured by averaging the indicator's reliability of a construct. An acceptable threshold for the AVE is 0.50 or higher. Discriminant validity measures the distinctiveness of a latent construct. Table 1 shows the items' loadings range from 0.733

**Table 1** The measurement model results

| Construct | Items | Loadings | Cronbach's alpha | Composite reliability | Average value extracted |
|-----------|-------|----------|------------------|-----------------------|-------------------------|
| Continuance intention | C11 | 0.853 | 0.614 | 0.838 | 0.721 |
| | C12 | 0.846 | | | |
| Escapism | EG1 | 0.733 | 0.793 | 0.866 | 0.618 |
| | EG2 | 0.812 | | | |
| | EG3 | 0.825 | | | |
| | EG4 | 0.767 | | | |
| Hedonic | HG1 | 0.875 | 0.812 | 0.889 | 0.728 |
| | HG2 | 0.878 | | | |
| | HG3 | 0.804 | | | |
| Information overload | 101 | 0.865 | 0.862 | 0.915 | 0.783 |
| | 102 | 0.910 | | | |
| | 103 | 0.880 | | | |
| Information seeking | ISG1 | 0.826 | 0.798 | 0.881 | 0.712 |
| | ISG2 | 0.847 | | | |
| | ISG3 | 0.858 | | | |
| Mobility | MG1 | 0.738 | 0.729 | 0.848 | 0.650 |
| | MG2 | 0.845 | | | |
| | MG3 | 0.831 | | | |
| Social interaction | SIG1 | 0.737 | 0.739 | 0.836 | 0.561 |
| | SIG2 | 0.795 | | | |
| | SIG3 | 0.723 | | | |
| | SIG4 | 0.739 | | | |

to 0.910; composite reliabilities (CR) are greater than 0.7. The average variance extracted for each latent construct is greater than the threshold of 0.5.

Hair et al. [59] suggested that each construct's AVE should be compared to the squared inter-construct correlation of that same construct and all other relatively measured latent constructs [60, 61]. Table 2 illustrates the Fornell-Larcker criterion.

Sarstedt et al. [56] suggested the heterotrait-monotrait ratio (HTMT) of correlations. The HTMT is defined as the mean value of the indicators' correlations across constructs relative to the geometric mean of the average correlations of the constructs' indicators [56, 60]. Henseler et al. [62] suggested an HTMT value exceeding 0.90 refers to a lack of discriminant validity. Table 3 shows the heterotrait-monotrait ratio (HTMT).

**Table 2**  Fornell-Larcker criterion

|  | Continuance intention | Escapism | Hedonic | Information overload | Information seeking | Mobility | Social interaction |
|---|---|---|---|---|---|---|---|
| Continuance intention | 0.849 | | | | | | |
| Escapism | 0.469 | 0.786 | | | | | |
| Hedonic | 0.583 | 0.513 | 0.853 | | | | |
| Information overload | 0.396 | 0.454 | 0.376 | 0.885 | | | |
| Information seeking | 0.494 | 0.434 | 0.603 | 0.323 | 0.844 | | |
| Mobility | 0.508 | 0.472 | 0.610 | 0.388 | 0.741 | 0.806 | |
| Social interaction | 0.554 | 0.466 | 0.646 | 0.446 | 0.570 | 0.585 | 0.749 |

**Table 3**  Heterotrait-Monotrait ratio (HTMT)

|  | Hedonic | Information seeking | Mobility | Social influence | Continuance intention | Escapism | Information overload |
|---|---|---|---|---|---|---|---|
| Hedonic | | | | | | | |
| Information seeking | 0.743 | | | | | | |
| Mobility | 0.796 | 0.91 | | | | | |
| Social influence | 0.827 | 0.735 | 0.797 | | | | |
| Continuance intention | 0.824 | 0.702 | 0.759 | 0.817 | | | |
| Escapism | 0.631 | 0.538 | 0.617 | 0.609 | 0.669 | | |
| Information overload | 0.448 | 0.390 | 0.496 | 0.561 | 0.541 | 0.553 | |

An HTMT value exceeding 0.90 suggests a lack of discriminant validity [62, 63]. As a result, the measurement model assessment indicates its ability to model latent constructs while simultaneously taking into consideration all forms of measurement errors [64].

## 6   The Structural Model

After checking for potential collinearity, assessment of the construct cross-validated redundancy $Q^2$, coefficients of determination $R^2$, as well as the statistical significance and relevance of the path coefficients, was conducted. Table 4 illustrates the results of $Q^2$ and $R^2$ of the endogenous construct values.

The $R^2$ refers to the coefficient of determination. It is a measure of in-sample prediction of the continuance intention to use YouTube as an endogenous construct. Cross-validated redundancy $Q^2$ is referred to as blindfolding. Values larger than zero for a particular endogenous construct indicate that the path model's predictive accuracy is accepted, whereas values below zero indicate a lack of predictive relevance [58].

After analyzing the proposed research model's quality, the structural model was employed to test the hypotheses. Figure 1 displays the results of the structural equation modeling, showing the path coefficients along with their significance levels. Table 5 illustrates the path coefficients along with their significance levels.

The structural model findings support H1, H3 and H4 and fail to corroborate H6, H2 and H5. H4 predicts a positive and significant impact of escapism gratification on the continuance intention to use YouTube. Similar findings emerge for hedonic gratifications, which also have a positive and significant impact on the continuance intention (path coefficient = 0.261, p < 0.05). Hence, H1 also receives empirical support. The social interaction gratifications construct obtains empirical support from the data. Social interaction gratification proved to be a significant positive predictor (path coefficient 0.190; p < 0.05) for continuance intention to use YouTube. However, hedonic gratifications had the strongest impact on continuance intention (B = 0.261).

**Table 4**  Quality of structural equation

|  | SSO | SSE | $Q^2 (= 1 - SSE/SS0)$ | $R^2$ | Adjusted R |
|---|---|---|---|---|---|
| Continuance intention | 892.000 | 626.813 | 0.297 | 0.437 | 0.430 |
| Escapism | 1784.000 | 1784.000 | | | |
| Hedonic | 1338.000 | 1338.000 | | | |
| Information overload | 1338.000 | 1338.000 | | | |
| Information seeking | 1338.000 | 1338.000 | | | |
| Mobility | 1338.000 | 1338.000 | | | |
| Social interaction | 1784.000 | 1784.000 | | | |

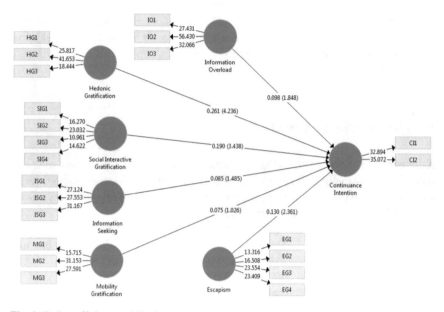

**Fig. 1** Path coefficients and T-values

**Table 5** The structural model results

| Path | Standardized coefficients | t-values | p-values | Hypothesis | Results |
|---|---|---|---|---|---|
| Escapism -> continuance | 0.130 | 2.361 | 0.019 | H4 | Support |
| Hedonic -> continuance | 0.261 | 4.236 | 0.000 | H1 | Support |
| Information overload -> continuance | 0.098 | 1.848 | 0.065 | H6 | Reject |
| Information seeking -> continuance | 0.085 | 1.485 | 0.138 | H2 | Reject |
| Mobility -> continuance | 0.075 | 1.026 | 0.305 | H5 | Reject |
| Social interaction-> continuance | 0.190 | 3.438 | 0.001 | H3 | Support |

The other path coefficients for escapism gratification and social interaction gratification equaled 0.130 for escapism and 0.190 for social interaction. Nevertheless, the current research findings failed to confirm information overload gratification, information seeking gratification and mobility gratification, as the effect of each construct was statistically insignificant. The results led to the rejection of H6, H2 and H5.

# 7  Conclusion, Discussion and Implications

The current research aims to examine the relationship between gratifications and continuance intention to use YouTube. The research findings confirm that uses and gratifications theory is antecedent to the continuance intention. The results reveal that hedonic gratification, escapism gratification and social interaction gratification have significant relations with continuance intention. This is in line with previous research findings [9, 17, 24, 48]. The relationship of hedonic gratifications with continuance intention is stronger in comparison to escapism and social interaction gratifications. In addition, the finding that information overload, information seeking and mobility gratifications have an insignificant relationship with continuance intention was unexpected. However, hedonic, escapism and social interaction gratifications accounted for 0.437% of continuance intention indicating that personal and social gratifications are the most remarkable needs of the university students.

The current research findings present some inconsistency with [17]. According to Chaouali [17], mobility gratification determines satisfaction and information overload affects emotional exhaustion. The possible reason may be that when students do not have any studying to undertake, they may prefer YouTube as a social media platform for their own entertainment and self-presentation needs. These empirical findings reveal that students are not concerned with information seeking or information overload problems. Students who have the continuance intention to use YouTube would do so only because they are gratifying their personal escapism, hedonic and social interaction needs. This result suggests some inconsistency with [65] findings which confirmed that entertainment, information seeking and academic learning are some of the main gratifications for using YouTube. In contrast, Lau [66] found that using social media for entertainment purposes such as video gaming negatively predicted academic performance among university students in Hong Kong. YouTube is meaningful and gratifies one or more of students' intrinsic needs. Most probably, YouTube becomes a habitual salient power that reinforces the relationship with students. YouTube offers potential capacity for gratifying a student's needs for hedonism, escapism and social interaction. Thus, in contrast to our expectations, information seeking, information overload and mobility gratifications were not found to have an effect on continuance intention.

However, the current research contributes to social media literature in several ways. The research has successfully verified the effect of hedonic, social interaction and escapism gratifications on the students' continuance intention. The present research provides new insights into a student's behavioral intention to use YouTube. That is, drawing on the theory of reasoned action and the theory of planned behavior; several models have been developed to predict behavioral intention to use technology. Most of these models suggest belief, attitude, perceived behavioral control and subjective norm as causal antecedents to an individual's intention. However, despite the fact that UGT is crucial to understanding how and why people use YouTube and persist with it, few studies have been conducted based on the user's gratifications and needs. Furthermore, an extensive review of social media research indicates that there has not been any attempt to empirically predict continuance intention to use YouTube from the uses and gratifications perspective in an emerging economy such as the United Arab Emirates. To fill this research gap in social media literature, this research posits that UGT is a useful theory for explaining students' continuance intention toward YouTube. In addition, comparing social networking sites or other social media platforms, YouTube has received limited attention and less research in the literature. Nevertheless, some limitations are inherent in this research. Because UGT is parsimonious, we excluded a number of gratifications identified in other studies. Future research should employ more diverse types of gratifications to represent a wider view of users' needs and motivations.

# References

1. Van den Eijnden RJJM, Lemmens JS, Valkenburg PM (2016) The social media disorder scale. Comput Hum Behav 61:478–487
2. Chen GM (2011) Tweet this: a uses and gratifications perspective on how active Twitter use gratifies a need to connect with others. Comput Hum Behav 27:755–762
3. Wang Z, Tchernev JM, Solloway T (2012) A dynamic longitudinal examination of social media use, needs and gratifications among college students. Comput Hum Behav 28:1829–1839
4. Susarla A, Jeong-Ha O, Tan Y (2012) Social networks and the diffusion of user-generated content: evidence from YouTube. Inf Syst Res 23(1):23–41
5. Kozinets RV (2019) Consuming techno-cultures: an extended JCR curation. J Consumer Res 46(3):620–627
6. de Bérail P, Guillon M, Bungener C (2019) The relations between YouTube addictions, social anxiety and parasocial relationships with YouTubers: a moderated-mediation model based on a cognitive-behavioral framework. Comput Hum Behav 99:190–204
7. Chang C-M (2018) Understanding social networking sites continuance: the perspectives of gratifications, interactivity and network externalities. Online Inf Rev 42(6):989–1006
8. Kim E, Lee JA, Sung Y, Choi SM (2016) Predicting selfie-posting behavior on social networking sites: an extension of theory of planned behavior. Comput Hum Behav 62:116–123
9. Khan ML (2017) Social media management: what motivates user participation and consumption on YouTube? Comput Hum Behav 66:236–247
10. Shifman L (2011) An anatomy of YouTube meme. New Media Soc 14(2):187–203
11. Lee DY, Lehto MR (2013) User acceptance of YouTube for procedural learning: an extension of the technology acceptance model. Comput Educ 61:198–208
12. Moor PJ, Heuvelman A, Verleur R (2010) Flaming on YouTube. Comput Hum Behav 26(6):1536–1546
13. Chiang H, Hsiao KL (2015) YouTube Stickiness: the needs, personal and environmental perspective. Internet Res 25(1):85–106
14. Dehghani M, Niaki MK, Ramezani I, Sal R (2016) Evaluating the influence of YouTube advertising for attraction of young customers. Comput Hum Behav 59:165–172
15. Jones T, Cuthrell K (2011) YouTube: educational potentials and pitfalls. Comput Sch 28(1):75–85
16. Ha YW, Kim J, Libaque-Saenz CF, Chang Y, Park MC (2015) Use and gratifications of mobile SNSs: Facebook and Kakao Talk in Korea. Telematics Inform 32(3):7425–7438
17. Chaouali W (2016) Once a user, always a user: enablers and inhibitors of continuance intention of mobile social networking sites. Telematics Inform 33:1022–1033
18. Ifinedo P (2016) Applying uses and gratifications theory and social influence processes to understand students' pervasive adoption of social networking sites: perspectives from the Americas. Int J Inf Manage 36:192–206

19. Joo J, Sang YM (2013) Exploring Koreans smartphone usage: an integrated model of the technology acceptance model and uses and gratifications theory. Comput Hum Behav 29:2512–2518

20. Rokito S, Choi YH, Taylor SH, Bazarova NN (2018) Over-gratified, under-gratified, or just-night? Applying the gratification discrepancy approach in investigates recurrent Facebook use. Comput Hum Behav. https://doi.org/10.1016/j.chb.2018.11.041

21. Al-Jabri I, Sadiq SM, Ndubisi NO (2015) Understanding the usage of global networking sites by Arabs through the lens of uses and gratifications theory. J Sci Manage 26(4):662–680

22. Phua J, Jin SV, Kim JJ (2017) Uses and gratifications of social Networking sites by bridging and bonding social capital: a comparison of Facebook Twitter, Instagram and Snapchat. Comput Hum Behav. https://doi.org/10.1016/j.chb.2017.02.041

23. Gan C, Hongxiu L (2018) Understanding the effects of gratifications on the continuance intention to use WeChat in China: a perspective on uses and gratifications. Comput Hum Behav 78:306–315

24. Alhabash S, Chiang YH, Huang K (2014) MAM & U&G in Taiwan: differences in the uses and gratifications of Facebook as a function of motivational reactivity. Comput Hum Behav 35:423–430

25. Gonzalez R, Gasco J, Liopis J (2019) University students and online social networks: effects and typology. J Bus Res 101:707–714

26. Huang J (2018) Timing of web personalization in mobile shopping: a perspective from users and gratifications theory. Comput Hum Behav. https://doi.org/10.1016/J.chb.2018.06.035

27. Krause AE, North AC, Heritage B (2014) The uses and gratifications of using Facebook music listening applications. Comput Hum Behav 39:71–77

28. Li H, Liu Y, Xu X, Heikkila J, Van Der Heijden H (2015) Modeling hedonic is continuance through the uses and gratifications: an empirical study in online games. Comput Hum Behav 48:261–272

29. Dajani D, Yaseen SG (2016) The applicability of technology acceptance models in the Arab business setting. J Retail Manage Res 10(3):46–56

30. Venkatesh V, Thong JYL, Xu X (2012) Consumer acceptance and use of information technology: extending the unified theory of acceptance and use of technology. MIS Q 36(1):157–178

31. Venkatesh V, Morris MG, Davis GB, Davis FD (2003) Users' acceptance of information technology toward a united view. MIS Q 27(3):425–478

32. Davis FD (1989) Perceived usefulness, perceived ease of use, and user acceptance of information technology. MIS Q 13(3):319–339

33. Davis FD (1993) User acceptance of information technology: systems characteristics, user perceptions and behavioral impacts. Int J Man Mach Stud 38(3):475–487

34. Bagozzi R (2007) The legacy of the technology acceptance model and a proposal for a paradigm shift. J Assoc Inf Syst 8(4):244–254

35. Yaseen SG, El Qirem I (2018) Intention to use-e-banking services in the Jordanian commercial banks. Int J Bank Market 36(3):557–571

36. Ajzen I, Fishbein M (1980) Understanding attitudes and predicting social behavior. Prentice-Hall, Englewood Cliffs, NJ

37. Fishbein M, Ajzen I (1975) Belief, attitude, intention, and behavior: an introduction to theory and research. Addison-Wesley, Reading, MA

38. Ajzen I (1991) The theory of planned behavior. Organ Behav Hum Decis Processes 50:179–211

39. Venkatesh V (2006) Where to go from here? Thought on future directions for research on individual-level technology adoption with a focus on decision making. Decis Sci 37(4):497–517

40. Venkatesh V, Thong JYL, Chan FKY, Hu PJ-H, Brown SA (2011) Extending the two-stage information systems continuance model: incorporating UTAUT predictors and the role of context. Inf Syst J. https://doi.org/10.1111/j.1365-2575-2011.00373

41. Al Omoush KS, Yaseen SG, Almaaitah MA (2012) The impact of Arab cultural values on online social networking. The case of Facebook. Comput Hum Behav 28:2387–2399

42. Yaseen SG, Al Omoush KS (2020) Mobile crowdsourcing technology acceptance and engagement in crisis management: the case of Syrian refugees. Int J Technol Human Interact 6(3):1–23
43. Grellhesl M, Punyanunt-Carter NM (2012) Using the uses and gratifications theory to understand gratifications sought through text messaging practices of male and female undergraduate students. Comput Hum Behav 28:2175–2181
44. Zhou Z, Fang Y, Vogel DR, Jin X-L, Zhang X (2012) Attracted to or locked in? Predicting continuance intention in social virtual world services. J Manag Inf Syst 29(1):273–306
45. Lin J (2014) The effects of gratifications on intention to read citizen journalism news: the mediating effect of attitude. Comput Hum Behav 36:129–137
46. Kuem J, Khansa L, Kim SS (2020) Prominence and engagement: different mechanisms regulating continuance and contribution in online communities. J Manag Inf Syst 37(1):162–190
47. Kaur P, Dhir A, Chen S, Malibari A, Almotairi M (2020) Why do people purchase virtual goods? A uses and gratifications (U&G) theory perspective. Telematics Inform. https://doi.org/10.1016/j.tele.2020.101376
48. Gan C (2017) Understanding WeChat users liking behavior: an empirical study in China. Comput Hum Behav 68:30–39
49. Whiting A, Williams D (2013) Why people use social media: a uses and gratifications approach. J Cetacean Res Manag 16(4):362–369
50. Gao Q, Feng C (2016) Branding with social media: user's gratifications usage patterns, and brand message content strategies. Comput Hum Behav 63:868–890
51. Brinkman CS, Gabriel S, Paravati E (2020) Social achievement goals and social media. Comput Hum Behav 111:106427
52. Hicks A, Comp S, Horovitz J, Hovarter M, Miki M, Bevan JL (2012) Why people use Yelp.Com: an exploration of uses and gratifications. Comput Hum Behav 28(6):2274–2279
53. Li Y, Yang S, Zhang S, Zhang W (2019) Mobile social media use intention in emergencies among Gen Y in China: an integrative framework of gratifications, task-technology fit, and media dependency. Telematics Inform 42:1012–1044
54. Beaudoin CE (2008) Explaining the relationship between internet use and interpersonal trust: taking into account motivation und information overload. J Comput-Mediat Commun 13:550–568
55. Schmitt JB, Debbelt CA, Schneider FM (2017) Too much information? Predictors of information overload in the context of online news exposure. Inf Commun Soc. https://doi.org/10.1080/1369118X.2017.1305427
56. Sarstedt M, Hair JF, Cheah JH, Becker J-M, Ringle CM (2019) How to specify, estimate and validate higher-order constructs in PLS-SEM. Australas Mark J 27:197–211
57. Huang YM (2019) Examining students continued use of desktop services: perspectives from expectation-confirmation and social influence. Comput Hum Behav 96:23–31
58. Hair JF, Howard MC, Nitzl C (2020) Assessing measurement model quality in PLS-SEM using confirmatory composite analysis. J Bus Res 109:101–110
59. Fornell C, Larcker DF (1981) Structural equation models with unobservable variables and measurement error: algebra and statistics, University of Michigan, working paper no. 266, 1–24
60. Hair JF, Risher JJ, Sarstedt M, Ringle CM (2018) When to use and how to report the results of PLS-SEM. Eur Bus Rev. https://doi.org/10.1108/EBR-11.2018-0203
61. Yaseen SG, Saib A, Nesrine A (2018) Leadership styles, absorptive capacity and firm's innovation. Int J Knowl Manage 14(3):82–100
62. Henseler J, Ringle CM, Sarstedt M (2015) A new criterion for assessing discriminant validity in variance-based structural equation modeling. J Acad Mark Sci 43(1):115–135
63. Sarstedt M, Ringle CM, Hair JF (2017) Partial least squares structural equation modeling. In: Handbook of market research, vol 26, pp 1–40
64. Hair JF, Hult TM, Ringle CM (2017) Mirror, mirror on the wall: a comparative evaluation of composite-based structural equation modeling methods. J Acad Mark Sci 45:616–632

65. Moghavvemi S, Suliman A, Jaafar NI, Kasem N (2018) Social media as complementary learning tool for teaching and learning: the case of YouTube. Int J Manage Educ 16:37–42
66. Lau WWF (2017) Effects of social media usage and social media multitasking on the academic performance of university students. Comput Hum Behav 68:286–291

# Assessing Service Quality and Customers Satisfaction Using Online Reviews

**Kholoud AlQeisi and Shorouq Eletter**

**Abstract** Prevailing communication technologies nowadays allow for consumers exchanging and sharing views over many products and services and the prevalence of the online media furtherly accelerated e-WOM influence. E-MOW is considered valuable for customers, both existing and potential, to extract in-depth information about marketing offers in an objectively reduced effort or cost. Online reviews project major influence on consumer decisions; correspondingly customers lacking previous experience with a particular food business operator is more likely to seek external information sources. From the business perspective, customers' reviews give insight around key performance flaws and reviews' analysis can enhance a business ability to serve customers and build a desired positioning based on the positive feedbacks. The current study is focused on take away food businesses in the UK through analyzing food delivery reviews posted by customers using clustering techniques and data mining in order to define key service performance aspects which lead to customer satisfaction/dissatisfaction. The finding indicate food quality and delivery services are among the key factors leading to customer satisfaction/dissatisfaction.

**Keywords** Online reviews · Delivery service key variables · Takeaway restaurants · Data mining · Consumer satisfaction

## 1 Introduction

Prevailing communication technologies nowadays allow for consumers exchanging and sharing views over various products and services online. Online reviews are considered valuable for customers, both existing and potential, to extract in-depth information about marketing offers in an objectively reduced effort or cost.

K. AlQeisi (✉) · S. Eletter
AlAin University, AlAin, UAE
e-mail: kholoud.alqeisi@aau.ac.ae

S. Eletter
e-mail: shorouq.eletter@aau.ac.ae

© The Author(s), under exclusive license to Springer Nature Switzerland AG 2022
S. G. Yaseen (ed.), *Digital Economy, Business Analytics, and Big Data Analytics Applications*, Studies in Computational Intelligence 1010,
https://doi.org/10.1007/978-3-031-05258-3_24

Service's inherent intangibility, requires searching for information before a purchase decision in order to lessen the risks. Moreover, services are primary experiential purchases, for instance, ordering a takeaway meal, where the evaluation of the takeaway cannot be completed until there is a trail consumption [1]. In this context, online reviews turn out to be major influential on consumers' decisions. Therefore, potential customers are more likely to seek external information sources or services not personally experienced yet. Hence, the written reviews as well as the start ratings left by actual consumers reflect perceptions of the various aspect of the service and provide details about experiences encountered, which offer clues for potential customers on what to expect. In addition, the critical reviews can serve as areas for improvement for management in pursue of customer satisfaction.

In the past few years, the food delivery services have grown very rapidly as takeaways business expanded. The breakout of the pandemic lead to the flourishing of the food delivery from takeaway food business through the staggering number of people using food delivery services on a daily basis. Food delivery is defined as a courier service to get the food to customers either from the food business or through a third party such as Just Eat, food Hub, Deliveroo and UberEATS [2].

In the UK, Just Eat is a major player in home deliver food services as for most Britain's food takeaway is a common form of dining and ordering fast food e.g. pizza is a leisure comparable to eating out in a restaurant. During 2019 the estimated foodservice delivery market was nearby eight billion pounds sterling mainly as a result of the dramatic rise of online ordering via apps and service deliveries. Furthermore, revenues generated from digital takeaway platforms are expected to continue growing with more customers turning to food ordering during the Covid-19 pandemic as witnessed at the early stages of the UK lockdown in March 2020 [3].

The review customers post on the third party portal regarding food delivery service attribute are rich with information that can be of great value to food business operators; the start rating in addition the textual reviews are reflective of customers' satisfaction or dissatisfaction with the service aspects. Therefore, the mining of the data posted present an important means for food business operators indicating areas for business improvement.

The current study is focused around sentiment analysis of textual reviews posted on Just Eat platform for three non-chain standalone food takeaway restaurants located in Telford & Wrekin—area within the UK. The insight derived from the text analysis can be of tremendous help in highlighting the service attributes that met customers' expectations and those that did not; which in turn reflect future intentions for current customers and likewise direct future customers' demand or buying behaviors. Most important, reviews offer opportunities for food takeaway managers to recover service displeasing encounters in means of decompensation for unsatisfied customers or extending valuable refunds. Therefore, examining online reviews has a significant value for food takeaway business operators.

## 2 Literature Review

### 2.1 Attributes of Food Delivery Service

In the context of takeaways and delivery business, many variables contribute to the overall assessment of the service from customers' perspective. Unlike eat in restaurants where aspects of physical environment such as ambience, décor, spatial layout and aesthetics present a key variable for overall service assessment the takeaway food business services assessment is more focused around factors related to food in terms of taste, value for money, consistency, quality and menu diversity; the service in terms of waiting time, server attentiveness/helpfulness and perceived service quality [4]. On the other hand, Teichert et al. [5] argue that speed is not the primary attribute in fast food delivery but rather four experiential factors are to be considered in the service mix decision, namely, the core and actual product, the brand satisfaction, payment process, and service handling. Chen et al. [6] report food quality and service quality being the most important factors of fast-food restaurant customer satisfaction, either in the normal situation or during the COVID-19 outbreak situation. In addition, service promptness is most important for customer satisfaction followed by staff politeness and safety protection.

### 2.2 Electronic Word of Mouth (e-WOM)

Electronic word-of-mouth, defined as any customers' positive or negative statement made available via the Internet to a mass of consumer segments [7], is a powerful means for explaining what persuade customers to consider adopting or using online offered products/services [8] and information gathered through e-MOW, or user generated content, is likely to carry more reliable inferences compared to claims obtained through other sources such as brand- generated-content [5] or restaurants' websites [9].

According to Chang et al. [10], reviewees, namely food business operators, are influenced by informational and numerical factors, that is the textual and star rating cues posted on the restaurant page. However, dissatisfied customers are behind the spread of negative MOW [11], which easily affect potential customers and may draw away from the service provider. On the other hand, Meek et al. [12] report moderate to positive reviews ratings are more helpful for potential customers than reviews that are negative or extremely positive.

Online reviews act as an important source of information for consumers, especially the customers who are unfamiliar with the business; the reviews impact the purchase decision-making process, the amount, frequency and propensity of purchasing depending on reviews power, volume and nature; hence businesses often take initiatives to get favorable ratings and reviews in order to leverage the benefits of positive reviews from customers whose decision relied on past online reviews.

## 2.3 Expectation and Satisfaction

Consumer behavior science refers to customers' value, expectation, and satisfaction as pillars to marketing success in terms of marketers' ability to offer value meeting customers' expectations and achieving satisfaction leading to loyalty. When customers make decisions of high involvement, tend to compare different alternative and build expectations around each option, hence value here from a customer perception is the difference between the gains and benefits of acquiring a product versus the costs incurred. Satisfaction or dissatisfaction is related to the outcomes of the purchase decision when the product performance meets, exceeds or not meet the anticipated performance, expectations [13]. The feeling of satisfaction and dissatisfaction are not distinctive from each other, it is argued that one can have mixed feelings towards a product or service where one can be satisfied about certain attributes and harper negative feelings around another [14].

Customer dissatisfaction or dissatisfaction is mainly generated by customer evaluation of the food business operator services and delivery. In cases of perceived failure on part of the food business operators, customers' evaluation is reflected in a negative e-MOW, issuing a complaint directly to the business and if not satisfied with the response or handling of the complaints, customers chose to switch to another provider [15]. As explained in previous section, many factors contribute to customers' overall evolution of the food operator business and mixed sentiments can be reflected in the textual review and ratings.

The notion that a customer's review is an indicator of a customer's satisfaction is examined in previous research [e.g. 16]. Xu [14] reports customer comment on different attributes namely the main service provider, the restaurant and the drivers along with the service platform. Moreover, customer tend to reveal positive or negative sentiment in the posted reviews when satisfied or dissatisfied with the restaurant and the overall consumption experience; in cases where customers were dissatisfied the textual review tends to be lengthy. Teichert et al. [5] reports aspects of speed delivery or unexpected early delivery results in exceeding customer expectations and brings on satisfaction while customers whose expectations were not met due to delays in delivery or not receiving items ordered leads to customers' dissatisfaction. Nguyen et al. [11] maintained that tangibles, responsiveness, and assurance flowed by reliability and empathy in dine in fast food restaurant are the most important drivers of customer satisfaction in the UK fast food industry.

The perceived service quality model [17] indicates that perceived service quality dimension: the *Technical Quality Outcome* (what the service process leads to for the customer in a "technical" sense) and the *Functional Quality Process* dimension (how the service process functions) has its impact on customer satisfaction in the service sector.

Satisfaction is a pursue for sustaining business development; moreover, in the case of takeaway food business, the pursue is critical since retention and patronage of food shops conditioned with services meeting expectation every time [6]. Therefore, this study aims at identifying which takeaway food and service delivery aspects contribute to customers' satisfaction/dissatisfaction.

# 3 Methodology

Online reviews are becoming a low-cost mean for customers' acquisition and retention [18]. This study is using an e-WOM dataset of delivery services on one the UK platform. The sampling approach used is convenience sampling targeting three takeaway restaurants who gave consent to data access. The dataset comprises text reviews related to customers' experience usually determined by customers' sentiments as well as a numeric rating that each customer assigned to the service to reflect customers' evaluation and feedback based on one's experience. The study aims to analyze e-WOM to find out which aspect of food delivery and other service quality dimensions makes customers satisfied or unsatisfied. The dataset comprised 1163 reviews.

The process started with data scraping from the Just Eat platform for the three takeaway restaurants with the restaurants permission. It is worth mentioning the menus for the restaurants are not identical. The reviews form Jan–Dec. 2020 was taken as an image, uploaded to excel, tabulated on sheets per month, date, textual review and star rating. Data preparation includes removing punctuation marks and converting letters to lower case, removing numbers, removing stop words and tokenization in order to obtain meaningful words from a sequence of characters. Clustering is an unsupervised data mining technique widely used for knowledge discovery [19, 20]. This study employs K-means clustering technique for text analysis for customer reviews. K-means partition the dataset into a number of clusters. The algorithm defines a centroid randomly. Then it measures the distance between each object and the centroid in order to map objects based on their distance from the nearest centroid. The process is repeated until the centroid reaches an optimum value [21]. The Rapid-Miner software is used for data analysis. The number of cluster is set to default k between 4 and 10 clusters (Fig. 1).

# 4 Results and Discussion

The initial start used (k = 4) revealed four different cluster for customer reviews (cluster_1 = 526; cluster_2 = 74; cluster_3 = 87; cluster_4 = 476). Table 1 showing the clusters.

The first cluster is focused around the food quality issues as can be seen in the table above. Customers reviews expressing food quality in terms of making or cooking being under cooked or overcooked or overdone, greasy, salty, spicy or even burned. In addition, the food status on arrival being cold or freezing, most logically due to late delivery. Other sentiments expressed about delivery issues in terms of ruined items

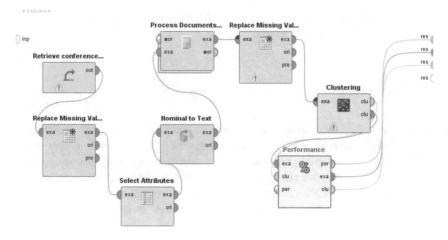

**Fig. 1** Clustering process using rapid miner

**Table 1** Clusters resulting from the first Rapid Miner run

| Cluster | Subject | |
|---|---|---|
| Cluster 1 | Product Quality issues (food, late, bad, cold, delay, late, delivery, missing, wrong, freezing, funny, awful, greasy, salty, chicken, overcooked, burned, disappointed, wrap ruined, spicy, terrible, shame, upset stomach, undercooked, overdone, poor service, rubbish, disgusting, rude, unhappy, wrong, price, incomplete, incorrect | 526 |
| Cluster 2 | Product dissatisfaction (bad, missing, incomplete, cold, late, burnt, delivery, arrive, service) | 74 |
| Cluster 3 | Ffood, delivery, minute, missing | 87 |
| Cluster 4 | Product enjoyment or brand satisfaction (lovely, good, excellent, nice, great, amazing, tasty, lovely, delicious) | 476 conveyed satisfaction |

indicating improper handing of order delivery, missing, wrong items or incomplete orders. Furthermore, delivery driver being rude in addition to poor service encounter.

The second and third cluster to some extent is focused around service issues such as late delivery leading to getting cold food or missing items while the forth cluster is expressing the aspect of the food/service that met expectations and lead to customer satisfaction as expressed in descriptive words such as lovely, good, excellent, nice, great, amazing, tasty, lovely, delicious.

The analysis is repeated and the number of clusters is set to 2 clusters. Cluster_1 = 687 and cluster_2 = 476. Most reviews in cluster_1 mirrored product issues that reflect customer dissatisfaction. Combing the clusters 1, 2, and 3 in Table 1. Cluster_2 reflects brand satisfaction issues identical to cluster 4 in Table 1.

Apart from the rapid mining technique, the preliminary coding of reviews reveals that customers apparently are vigilant and alert for tiny details around delivery service. Issues such as driver's ill manners, lack of courtesy, violation of social distancing, belated delivery, mixing or missing order items are among the key variables leading to customers' dissatisfaction with the services manifested in writing strong bad language feedback or using low stars' rate. On the other hand, customers who order on regular basis and apparently developed a liking for the menu, portions, and taste tend to leave positive feedback using sentiment words such as great, awesome or marvelous. Customer also report on unexpected delivery before promised time.

# 5  Implications

Fast-food takeaway business managers can benefit from paying attention to online customer reviews expressing thoughts and opinions about products/services. In order to survive and win in the market, takeaway food business customers are becoming increasingly dependent on online reviews making it critical for businesses to monitor and manage online reviews. The recommendations for takeaway food business operators is to be mindful of the key points raised by the customers in terms of delivery issues and drivers' manners towards customers from different age spectrum; food quality consistency; order accuracy and completeness. Future research can use numeric feedback analysis, star rating, to enrich the findings with more insights.

# 6  Conclusion

Online reviews present opportunities to examine customer satisfaction. The current study aims to identify factors that contribute to customer satisfaction through sentiment analysis of user generated online reviews. The online reviews aid prospect customers in making informative decisions around future food takeaway service choices. In addition, food business operators can derive more insight from customers' shared experience over platforms, which indicate areas for improvement to enhance business performance. The findings show that customers for fast food takeaway business are mainly focused around three aspects of the service: mainly food quality in terms of taste, temperature, and portions versus prices charged; delivery issue: mainly delivery time, manners of the delivery drivers and courtesy; and order issues: getting the right items and the receiving the entire order without missing on small details. These aspects when well delivered result in customer satisfaction and when miss represented or miss delivered cause customers dissatisfaction in the food takeaway services.

# References

1. Iacobucci D (2018) Marketing management, 5th edn. Cengage Learning, MA, USA
2. Taylor S (2020) How to convert your restaurant into a food delivery service? Available at https://www.highspeedtraining.co.uk/hub/convert-restaurant-into-food-delivery-service/. Accessed on 7 July 2021
3. Lock S (2020) Food delivery and takeaway market in the United Kingdom (UK)—Statistics & Facts. Available at https://www.statista.com/topics/4679/food-delivery-and-takeaway-market-in-the-united-kingdom-uk/. Accessed on 13 July 2021
4. Keller D, Kostromitina M (2020) Characterizing non-chain restaurants' Yelp star-ratings: generalizable findings from a representative sample of Yelp reviews. Int J Hosp Manag 86(1):1–12
5. Teichert T, Rezaei S, Correa JC (2020) Customers' experiences of fast food delivery services: uncovering the semantic core benefits, actual and augmented product by text mining. Br Food J 12(11):3513–3528
6. Chen W, Riantama D, Chen LS (2020) Using a text mining approach to hear voices of customers from social media toward the fast-food restaurant industry. Sustainability 13(1):1–17. Available online from https://doi.org/10.3390/su13010268
7. Cheung CMK, Thadani DR (2010) The effectiveness of electronic Word-of-Mouth communication: a literature analysis. [Online] Available: http://citeseerx.ist.psu.edu/viewdoc/download?doi=10.1.1.453.4915&rep=rep1&type=pdf. Accessed: 20 July 2021
8. Xu X (2019) Examining the relevance of online customer textual reviews on hotels' product and service attributes. J Hospitality Tourism Res 43(1):141–163
9. Huifeng P, Ha HY, Lee JW (2020) Perceived risks and restaurant visit intentions in China: do online customer reviews matter? J Hospitality Tourism Manage 43:179–189
10. Chang HH, Fang PW, Huang CH (2015) The impact of on-line customer reviews on value perception: the dual-process theory and uncertainty reduction. J Organ End User Comput 27(2):32–57
11. Nguyen Q, Nisar TM, Knox D, Prabhakar GP (2018) Understanding customer satisfaction in the UK quick service restaurant industry. Br Food J 120(6):1207–1222
12. Meek S, Wilk V, Lambert C (2021) A big data exploration of the informational and normative influences on the helpfulness of online restaurant reviews. J Bus Res 125:354–367
13. Kotler P, Armstrong G (2018) Principles of marketing: global edition, 17th edn. Pearson Education, Harlow, England
14. Xu X (2020) Examining an asymmetric effect between online customer reviews emphasis and overall satisfaction determinants. J Bus Res 106:196–210
15. Lee MJ, Sing N, Chan ESW (2011) Service failures and recovery actions in the hotel industry: a text-mining approach. J Vacat Mark 17(3):197–207
16. Chatterjee S (2019) Explaining customer ratings and recommendations by combining qualitative and quantitative user generated contents. Decis Support Syst 119:14–22
17. GroÈnroos C (2001) The perceived service quality concept—a mistake? J Serv Theory Pract 11(3):150–152
18. Korfiatis N, Stamolampros P, Kourouthanassis P, Sagiadinos V (2019) Measuring service quality from unstructured data: a topic modeling application on airline passengers' online reviews. Expert Syst Appl 116:472–486
19. Vishwakarma S, Nair PS, Rao DS (2017) A comparative study of K-means and K-medoid clustering for social media text mining. Int J Adv Sci Res Eng Trends 2(11):297–302
20. Aryuni M, Madyatmadja ED, Miranda E (2018) Customer segmentation in XYZ bank using K-means and K-medoids clustering. In: 2018 international conference on information management and technology (ICIMTech). IEEE, pp 412–416
21. Slamet C, Rahman A, Ramdhani MA, Darmalaksana W (2016) Clustering the verses of the Holy Qur'an using K-means algorithm. Asian J Inf Technol 15(24):5159–5162

# Factors Affecting Student Satisfaction Towards Online Teaching: A Machine Learning Approach

Ahmed Ben Said, Abdel-Salam G. Abdel-Salam, Emad Abu-Shanab, and Khalifa Alhazaa

**Abstract** During the outbreak of the Covid-19 pandemic, universities were forced to adopt technology and collaboration tools to reinforce online teaching and sustain their operations. This radical change pushes universities, researchers, educators, practitioners and decision makers to explore the perceptions of students and provide high quality online teaching operations. This study offers an understanding of the factors influencing students' satisfaction with online teaching. Using data from an institutional survey, a machine learning approach is developed along with feature importance analysis using Permutation Importance and SHAP. The two techniques yielded similar results, where quality, interaction, and comprehension were the most significant predictors of satisfaction while student class, gender and nationality were insignificant. Such results support previous research conducted on similar data but with different statistical techniques. Other factors might be significant in the online environment such as student support, academic experience, and assessment.

**Keywords** Online learning · Student satisfaction · Machine learning · Feature importance

A. Ben Said (✉) · A.-S. G. Abdel-Salam · K. Alhazaa
Student Experience Department, Student Affairs, Qatar University, Doha, Qatar
e-mail: abensaid@qu.edu.qa

A.-S. G. Abdel-Salam
e-mail: abdo@qu.edu.qa

K. Alhazaa
e-mail: khalifa.alhazaa@qu.edu.qa

E. Abu-Shanab
College of Business and Economics, Qatar University, Doha, Qatar
e-mail: eabushanab@qu.edu.qa

© The Author(s), under exclusive license to Springer Nature Switzerland AG 2022
S. G. Yaseen (ed.), *Digital Economy, Business Analytics, and Big Data Analytics Applications*, Studies in Computational Intelligence 1010,
https://doi.org/10.1007/978-3-031-05258-3_25

# 1 Introduction

The outbreak of the COVID-19 pandemic has pushed for transition from traditional to virtual education. This is not to say that online education started with this pandemic. On the contrary, it is a phenomenon that goes back to many years before 2019, where many studies investigated its limits and uncharted territories [11, 16]. Online education remains a channel that provides education to people who are busy, traveling, and cannot afford Face-to-Face education. In addition, higher education institutes adopted online technology for the same purpose, for reducing their cost, and for sustaining their market position during the pandemic. Satisfaction, being the cornerstone of any type of education, is a goal on its own. Several studies have embarked into the endeavor of evaluating it either by quantitative or qualitative measurements [15]. A number of factors have been found to affect student satisfaction with online teaching, including but not limited to: assessment, interaction, quality, academic experience, information technology (IT), comprehension, and student support.

The success of the educational process depends on student satisfaction. Many universities strive to attract more students and help in conducting the educational process, in a streamlined fashion. In addition, universities need to respond to the challenges and requirements of Covid-19. Based on that, this study attempts to explore the factors influencing student satisfaction with online environment. This study takes a machine learning approach to explore and test the research model.

# 2 Literature Review

Research related to online education attempted to explore various factors influencing students' satisfaction, where a list of predictors are reported. For instance, Jelena and Ana [16] define assessment as the feedback on the student learning process. They found, from a questionnaire they devised and distributed among 121 students in Serbia, that assessment is an essential dimension for online education. A good assessment plan must be flexible enough to accommodate all different dimensions of the learning process with regard to ethnicity, gender, location, and financial status [19]. Interaction, on the other hand, namely any form of engagement between teachers, students, and systems [17, 21], gains a sort of freedom when it is associated with online education. A student, who shies from participating in regular physical classes, is more prone to participate in online sessions due to an artificial sense of security. This increase in interaction in virtual media is bound to affect satisfaction positively. Systematic feedback collected on regular intervals is found to increase interaction in online teaching [19]. Higher educational institutes are in a continuous effort to reevaluate and reassess their online learning components [22] and build sustainable online learning strategies [18]. Indeed, it has been shown that exposing students to high standards is a key factor in online learning [23]. Thus, quality holds a vital role especially in students' perceptions [7], which might lead to higher enrollment

[8]. Another recommendation higher education institutes should keep in mind is to frequently review curricula and make sure it is on par with the most recent best practices of online education [5]. The importance of academic experience is highlighted in several studies in the literature [7, 13]. Online learning is inherently dependent upon the technical aspect. Hence, without doubt, IT is a big component of online teaching. It has been found that improvement in technology and online learning leads to direct improvement in the academic experience. In fact, online teaching can improve at a higher rate than physical class teaching [14]. To enhance the positive effect of technology on online education, it is recommended that higher education institutes keep abreast of ever changing technological needs and competencies in this field [22]. The aforementioned factors aim at enhancing and ensuring students' comprehension, which is to understand spoken or written text or visual material in general [24]. This process is complex and is a key step towards students' success in their university journey. Several studies found that hybrid classes with online components improve students' comprehension [4, 11]. Students studying online require special form of student support. Johnson et al. [18] suggest that students' support providers need better access to showcase their online material. In addition, an increase in intrusive student support is shown to increase student success as a whole [3]. It is worth mentioning that [1] have considered all the above dimensions to measure their effect on student perception and satisfaction. The authors used data collected for a validated student satisfaction survey developed in response to the COVID-19 pandemic and was conducted in Summer 2020. The survey included 45-items, and targeted a sample of 2354 students in a Gulf Cooperation Council (GCC) countries. Classical statistical approaches were used, including factor analysis, linear regression, and structure equation modeling. Their findings supported the role of quality, interaction, IT, academic experience, comprehension, and student support in predicting students' satisfaction. Still, assessment failed to be a significant predictor. Given the current worldwide health crisis and the rapid shift and adoption of online teaching, it is of paramount importance to dig into the aforementioned factors and explore how they contribute to the overall student satisfaction. This study leverages the power of machine learning techniques and their explanatory and interpretability powers to identify the influential factors among the given dimensions on student satisfaction.

## 3 Methodology and Analysis

### 3.1 Data and Techniques

The data were taken from a Student Satisfaction Survey in a national university comprising 45 items. Some items in the survey required demographic information, four items directly asked about the satisfaction about online learning, and other items measure student satisfaction towards the different services and facilities. The data were collected through a Two-Phase approach using the Post-stratified Sampling

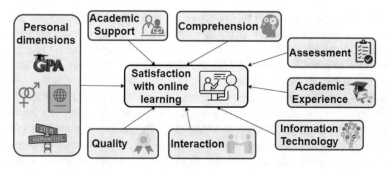

**Fig. 1** Factors associated to student satisfaction with online learning

design to reduce the non-response bias. The sampling frame was designed to be the enumeration of undergraduate students through Summer 2020, excluding students in the foundation program. In the first phase, data were collected from the population by sending the survey through email to 8794 students enrolled in the summer semester. Some students responded to the online survey after up to five reminders. The number of students who responded to the survey was 1687. In the second phase, a simple random sample of size 1000 was selected from the non-response subpopulation, and then the survey was sent by email followed by phone calls, including several incentives. The response rate for this phase was approximately 67%. The machine learning-based analysis aims at classifying students according to their satisfaction level using the factors depicted in Fig. 1. The trained classifier enables conducting a feature importance analysis. Feature importance refers to the techniques that assign a score to input features based on how useful they are at predicting a target variable. Feature importance scores play a critical role in a predictive modeling-based study, including providing insight into the data and the model, and is the basis for dimensionality reduction and feature selection which can improve the efficiency and effectiveness of a predictive model.

In this study, demographic characteristics that were considered are gender, nationality, major classification (STEM and Non-STEM), college, student class (Freshman, Sophomore, Junior, and Senior), and GPA.

## 3.2 Gradient Boosting Decision Tress (GBDT) for Student Satisfaction Classification

GBDT is a popular method for solving classification and regression problems. It improves the learning process by simplifying the objective and reducing the number of iterations to reach a sufficiently optimal solution. Gradient-boosted models have been widely used in various competitions grading on both accuracy and efficiency, making them a fundamental component in machine learning and data science-related projects. GBDT optimizes the predictive value of a model through successive steps in

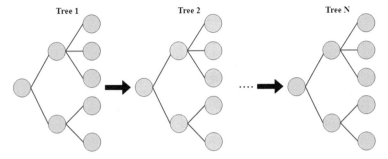

**Fig. 2** Illustration of GBDT: N trees are constructed. Each tree is trained on the residual of the previous tree

**Table 1** Classification results

| Metric | Value (%) |
|---|---|
| Accuracy | 85.6 |
| Recall | 85 |
| Specificity | 88 |
| Precision | 85 |
| F1 Score | 82 |

the learning process as depicted in Fig. 2. These factors are first preprocessed to unify the items related to each factor. The resulting feature is obtained by calculating the median value of all the items related to a given factor. The unified overall satisfaction is also calculated by computing the median of the items related to the satisfaction factors. This enables formulating a binary classification. As two different groups are categorized: students with the highest satisfaction. i.e., median value equal to 4 (class 1), and other students whose median satisfaction is less than 4 (class 0).

A loss measuring the difference between the actual and predicted values is minimized. The gradient is the incremental adjustment made in each step of the optimization process. Of the data, 80% is used for training and 20% for testing. Table 1 details the classification results using different evaluation metrics.

## 3.3 Feature Importance

Feature importance refers to the measure of the individual contribution of the corresponding feature for a particular classifier, regardless of the shape (e.g., linear or nonlinear relationships) or direction of the feature effect [9, 10]. This means that the feature importance of the input data depends on the corresponding classification model. Several techniques can be used to evaluate feature importance. In this

study, two main techniques are adopted: permutation importance [20] and SHapley
Additive exPlanations (SHAP) [12].

**Permutation Feature Importance** Permutation feature importance measures the
increase in the prediction error of the model after permuting the feature's values,
which breaks the relationship between the feature and the true outcome. A feature
is deemed important if shuffling its values increases the model error, as in this case
the model relied on the feature for prediction. On the other hand, a feature is deemed
unimportant if shuffling its values leaves the model error unchanged, as in this case,
the model ignored the feature for the prediction.

Figure 3 depicts the permutation importance results. The findings show that
Quality is the most significant feature, more than thrice important compared to the
second most important feature, namely Comprehension. They are followed by Interaction and Assessment. We notice that IT, Student Support and Academic Experience
are less significant. The College, Major Classification and GPA are insignificant for
GBDT as their corresponding weights are almost zeros. The rest of the features, i.e.
Student Class, Nationality and Gender have zero importance with respect to GBDT.

**SHAP Feature Importance** The goal of SHAP is to explain the prediction of a
data instance by computing the contribution of each feature to the prediction. The
SHAP explanation method calculates Shapley values from coalitional game theory.
The Shapley value is a solution concept in cooperative game theory. The feature
values of a data instance act as players in a coalition. Shapley values reflects how to
fairly distribute the payout (i.e., the prediction) among the features. One innovation
of SHAP is that the Shapley value explanation is represented as an additive feature
attribution method. Figure 4 depicts the results of SHAP feature importance analysis.
These findings are consistent with the permutation importance analysis. Indeed,
Quality is the most important feature followed by Interaction and Comprehension.
On the other hand, Assessment, Academic Experience, Student Support and IT have

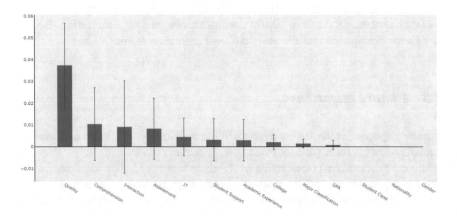

**Fig. 3** Permutation importance: feature weights

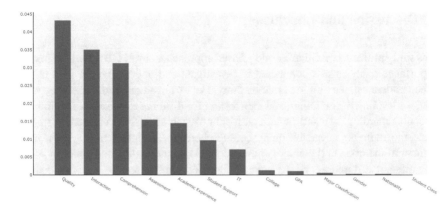

**Fig. 4** SHAP feature importance

a low contribution in the classification decision while the rest of features e.g. gender, major are almost insignificant.

Figure 5 depicts the summary plot of feature contributions. Features are sorted by the sum of SHAP value magnitudes over all samples, and SHAP values are used to show the distribution of the impacts each feature has on the model output. The color represents the feature value, red being high and blue being low. One can clearly notice two clusters for the top three features with Quality having the most separable clusters. High Quality values push the prediction towards class 1, i.e. high satisfaction while low quality pushes the model to predict the second class.

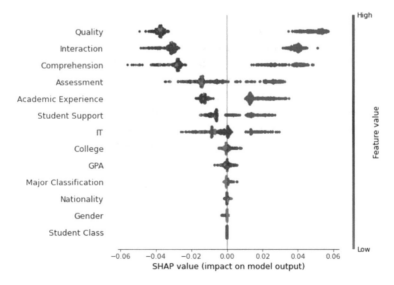

**Fig. 5** Summary of feature contributions

# 4   Discussion and Conclusion

This study utilized a machine learning-based approach, namely GBDT where feature importance analysis was conducted to investigate and understand the factors that affect student satisfaction in online learning. Data from an institutional survey were used for training the classifier, and then testing for differences among perceptions of students regarding their satisfaction with online teaching. Using Permutation Importance, the findings showed that quality, comprehension, and interaction were the most influential factors. On the other hand, student class, nationality, and gender yielded insignificant differences. SHAP analysis yielded similar results but with quality, interaction and comprehension as the most significant (sequence is different), while gender, nationality, and student class were the least significant. Results of this study were compared to another study where the same data were used to test a multiple regression test to predict satisfaction. The findings of the traditional method yielded the following significant predictors (ranked according to importance and beta value): quality, interaction, IT services, comprehension, student support, and GPA [1, 6]. Assessment, nationality, student class, and gender failed to report any significant prediction of satisfaction. Such conclusion supports the results obtained from our machine learning approach with minor differences. The accuracy of the results is attributed to many factors such as the adequacy of the training process, and the data size for machine learning techniques. This study yields important research implications, where machine learning approach resulted in minor differences in terms of the factors' importance. Nevertheless, quality remains the major predictor of satisfaction in all methods. In addition, regardless of the ranking of importance, and satisfying an efficient research model, the first few significant predictors were similar (quality, comprehension, and interaction being most significant). Finally, the other side of the coin (the lowest predictors) included gender, nationality and GPA (which might be a surprise based on previous research that supported such variables [2, 18, 19]. The second contribution of this study is its use of an institutional data that map students' perceptions regarding online teaching in the Covid-19 era. Universities needed to see how students' perceptions are influenced on one side, and the comprehensiveness of coverage of the set of factors related to the process on the other side. The study reported a set of important factors related to this topic, which serves as an excellent foundation for future research. Finally, exploring such topic from different perspectives, and utilizing different statistical techniques would greatly support the validity of research results.

# References

1. Abdel-Salam AG, Hazaa K, Abu-Shanab E (2021) Evaluating the online teaching experience at higher education institutes from students' perspective. In: Education, engineering education and instruction technology conference (EEEITC 21)
2. Abu-Shanab E, Al-Tarawneh H (2015) The influence of social networks on high school students' performance. Int J Web-Based Learn Teach Technol (IJWLTT) 10(2), April–June 2015:49–59
3. Abu-Shanab E, Anagreh L (2020) Contributions of flipped classroom method to students' learning. Int J Cyber Behav Psychol Learn (IJCBPL) 10(3):12–30
4. Abu-Shanab EA, Musleh S (2018) The adoption of massive open online courses: challenges and benefits. Int J Web-Based Learn Teach Technol (IJWLTT) 13(4):62–76
5. Adnan M, Anwar K (2020) Online learning amid the COVID-19 pandemic: students' perspectives. Online Submission 2(1):45–51
6. Al Hazaa K, Abdel-Salam GA, Ismail R, Johnson C, Al-Tameemi R, Romanowski MH, Ben Said A, Ben Haj Rhouma M, Elatawneh A, de AraÚjo G (2021) The effects of attendance and high school GPA on student performance in first-year undergraduate courses. Cogent Education 8(1)
7. Al-Fraihat D, Joy M, Sinclair J (2020) Evaluating E-learning systems success: an empirical study. Comput Hum Behav 102:67–86
8. Andrade MS, Miller RM, Kunz MB, Ratliff JM (2020) Online learning in schools of business: the impact of quality assurance measures. J Educ Bus 95(1):37–44
9. Breiman L (2001) Random forests. Mach Learn 45(1):5–32
10. Cassalicho G, Molnar C, Bischl B (2019) Visualizing the feature importance for black box models. Lect Notes Comput Sci 11051:655–670
11. Dyment J, Downing J, Hill A, Smith H (2018) I did think it was a bit strange taking outdoor education online: exploration of initial teacher education students' online learning experiences in a tertiary outdoor education unit. J Adventure Educ Outdoor Learn 18(1):70–85
12. Fisher A, Rudin C, Dominici F (2019) All models are wrong but many are useful: learning a variable's importance by studying an entire class of prediction models simultaneously. J Mach Learn Res 20(177):1–81
13. Gómez-Rey P, Fernández-Navarro F, Barbera E, Carbonero-Ruz M (2018) Understanding student evaluations of teaching in online learning. Assess Eval High Educ 43(8):1272–1285
14. Guest R, Rohde N, Selvanathan S, Soesmanto T (2018) Student satisfaction and online teaching. Assess Eval High Educ 43(7):1084–1093
15. Hammouri Q, Abu-Shanab E (2018) Exploring factors affecting users' satisfaction toward e-learning systems. Int J Inf Commun Technol Educ (IJICTE) 14(1):44–57
16. Jelena AL, Ana N (2019) Designing e-learning environment based on student preferences: conjoint analysis approach. Int J Cogn Res Sci Eng Educ 7(3):37–47
17. Jo I, Park Y, Lee H (2017) Three interaction patterns on asynchronous online discussion behaviours: a methodological comparison. J Comput Assist Learn 33(2):106–122
18. Johnson N, Veletsianos G, Seaman J (2020) US Faculty and administrators' experiences and approaches in the early weeks of the COVID-19 pandemic. Online Learn 24(2):6–21
19. Kung M (2017) Methods and strategies for working with international students learning online in the U.S. TechTrends: Link Res Pract Improve Learn 61(5):479–485
20. Lundberg SM, Su-In L (2017) A unified approach to interpreting model predictions. Advances in Neural Information Processing Systems. arXiv:1705.07874
21. Mehall S (2020) Purposeful interpersonal interaction in online learning: what is it and how is it measured? Online Learn 24(1):182–204

22. Rasheed RA, Kamsin A, Abdullah NA (2020) Challenges in the online component of blended learning: a systematic review. Comput Educ 144:103701
23. Tanis CJ (2020) The seven principles of online learning: feedback from faculty and alumni on its importance for teaching and learning. Res Learn Technol 28
24. VSG (2021) Comprehension. Department of Education and Training, Victoria State Government. Accessed from the Internet in 9 June 2021, from: https://www.education.vic.gov.au/Pages/default.aspx

# Effect of Implementing Building Information Modelling in Infrastructure Management of Smart Cities

Naser Musa AL Lozi

**Abstract** The present study focuses on the effect of implementing high end tool i.e. building information modelling tool for infrastructural development of Smart cities. BIM is a robust tool which efficiently helps in managing various aspects of an infrastructure management by simulating real life scenarios and providing a detailed insight about the activities beforehand so that any possible delay in work and any wastage of resources can be conserved. Smart city is a hub of infrastructure combined with information and communication technologies. This city enables the use of high-end internet services and modern-day devices, henceforth, the application of BIM in smart city management can have crucial influence of its performance. Present work considered several significant parameters involved in the infrastructural management of a smart city and highlights the role of BIM and concludes that building information modelling has a vital influence on numerous management activities involved in a smart city.

**Keywords** Smart city · Building information modelling · Infrastructure management

## 1 Introduction

Growth and development of a nation is majorly dependent on its infrastructure. The construction industry is amongst the supreme industries across the globe. Economic and social development is highly dependent on the performance of construction industry. One of the key factors about this industry is it complexity and intricacy. A lot of phases are there which needs to be addressed properly for completion of a project. Stating from planning and designing, construction and maintenance of buildings involves numerous complexities. A smooth completion of work is also highly dependent of several individuals like designer, engineer and contractor which

N. M. AL Lozi (✉)
Department of Engineering, University of Exeter, Exeter, UK
e-mail: na559@exeter.ac.uk

© The Author(s), under exclusive license to Springer Nature Switzerland AG 2022
S. G. Yaseen (ed.), *Digital Economy, Business Analytics, and Big Data Analytics Applications*, Studies in Computational Intelligence 1010,
https://doi.org/10.1007/978-3-031-05258-3_26

319

furthermore makes the task of building completion more complex. These complexities combine and affect the productivity of the construction industry. Therefore, an effective and robust model is required which has the potential to enhance the efficiency and value of any infrastructures, its qualities, reducing lifecycle costs, time and sustainability. An amount of these clarifications are obtainable for refining and determining accomplishment of projects which includes novel contracting blends, integrated project distribution, adopting innovative technologies in the "designing and construction processes such as 3-D coding and modelling" [3, 10, 29, 31, 32, 34, 35, 40, 41]. BIM referred as 'building information modelling' is one the widely used and accepted approach for addressing the aforementioned problems. This approach is very effective in simulating construction activities in a computer-generated setting. With the application of BIM, one can accomplish a comprehensive simulated model with the building requisite information. These models are further used for supporting designing, scheduling, planning, construction, and all the physical activities associated with building construction. Furthermore, BIM models are employed for facilities management and post completion maintenance. A simulation of real-life construction practices can also help in understanding and practicing sustainable construction. Sustainable construction is "a holistic process starting with the extraction of raw materials, integrated planning, design, and construction of buildings, along with their demolition and management of the resultant waste" [4, 14, 18, 22, 30]. It was encouraged to adopt civil engineering practices through sustainable construction based on six principles namely: "minimize resource consumption; maximize resource reuse; use renewable or recyclable resources; protect the natural environment; create a healthy, nontoxic environment; and pursue quality in creating the built environment" [8]. Although the terms "Sustainability" and "Green" are frequently used in the construction industry in an interchangeable manner, going green differentiates itself from sustainability in that it abstractly poises unsteadily on one leg of environmental health of the sustainability vitality, economic development, environmental health and social equity. Construction industry attempts to address the sustainability by developing guidelines for Smart City Building designs and construction that are not limited to resource efficiency confirming economic and environmental aspects, but is being expanded to disaster resilience which has social implications also [13].

Smart city is a modern perception that assimilates information and communication technology along with several modern-day devices connected to internet. Overall to summarize the concept of smart city one can conclude that it utilizes the internet facilities in a city to reduce human effort by controlling maximum function via remotely located devices. The modern devices and internet connectivity enhance the productivity of city services as well as processes and connect to citizens. In addition, this technology allows the officials and staffs to maintain a direct contact with the community, infrastructure of the city and helps in service and maintenance of the city up to a considerable extent. It also keeps a check on what is happening in the city and how the city infrastructures are growing significantly.

A unique area that customs diverse sorts of automated approaches and devices to gather data is referred as a "Smart City". Understandings multiplied after the data collection can be utilized to accomplish possessions, resources, and services

**Fig. 1** Insight of a typical
smart city [2]

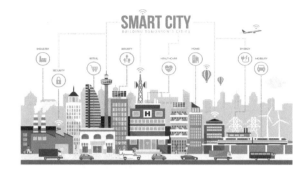

effectively. Contrary to that the data can be utilized to enhance several management procedures all along the smart city. Information and communication technologies adopted to augment superior performance and infrastructural services interactivity that leads to cost reduction, optimized resource consumption and increased interaction among citizens and management. Smart cities are mainly established with an aim to regulate "urban flows and allow real-time responses". Thus, smart city can be well equipped to react to encounters in relationship with its citizens. Irrespective of these facts, the term smart city remains indistinct to its essentials details and therefore it leads to numerous interpretations. These numerous facilities involved in a typical smart city are collection of data from citizens, structures, devices and possessions that is further administered and investigated to regulate and accomplish transportation systems, flow of traffic and power plants, along with several other utilities such as networks of water supply, waste management, crime exposure, information structures, schools, libraries, hospitals and other community services (Fig. 1).

Therefore, to manage the city based on such advanced technologies and information and communication system, a robust tool to manage these is an essential need of the hour. BIM is an innovative concept which processes the capability to predict simulated construction of a system preceding to its real construction. The simulation and modelling are crucial in any construction industry in order to reduce ambiguity, better safety and work out problems. In every single profession one can provide critical information into the model as an input before the construction phase and this obtains a detailed analysis of the whole construction services. Another key advantage of BIM is waste management. It tends to minimise wastage of resources and materials on-site and also assures the delivery of the product in-time. It also helps in avoiding stockpiles on-site. Number of materials can be determined easily, and this model is highly effective to prevent errors by enabling conflict and visual highlights to the team where parts of the building such as structural frame and building services pipes or ducts may intersect erroneously (Fig. 2).

**Fig. 2** Building information
modelling key features [36]

## 2   Literature Review

The literature studies applicable to this research consist of relevant works to sustainability in construction, building rating tools, the comparative studies, benchmarking, decision making, Fuzzy logic and governmental acts and policies. Preliminary research on building rating tools published so far show that more noteworthy accentuation is being set on the rating of new structures and those experiencing real renovation than existing structures. Although this is changing, as the majority of the rating schemes worldwide is presenting their versions for existing structures, yet the criteria considered still remain questionable. The essential point of this examination is to research the benchmarking procedure of existing structures as far as their basic structural adequacy, effectiveness, and sustainability challenge. The Stockholm Agreement in 1972 is considered as the landmark event for the awareness towards human environmental impacts due to population growth and the resulting socio-economic development. The contributions that technology and education can make to challenge environmental concerns were considered. "World Commission on Environment and Development (WCED)" by the United Nations and Brundtland Commission report published in the year 1987 on 'Our Common Future' endeavored the concerns of deteriorating human environment and the natural resources [19, 25, 38]. The published reports have opened avenue for the first Earth Summit in 1992, held in Rio, Brazil. A plan for administrations to implement actions towards wide range of ecological issues known as Agenda 21 [26]. The Kyoto protocol, 1997, set up a structure for nations to decrease their GHG discharges to an average of 5% below the levels they produced in 1990. The targets were to be achieved by 2012 by all the signing countries and also, low $CO_2$ producers could sell their allowances to high $CO_2$ producers called 'carbon trading'. United Nations Climate Change Conference (2007) in Bali, Indonesia gave an action plan called the 'Bali Road Map' under UNFCCC on strengthening international efforts to fight, mitigate and adapt to climate change. United Nations Climate Change Conference (2009) Copenhagen agreed on a climate deal that would come into force at the end of 2012 after the first commitment period of 1997 Kyoto Protocol [6, 7]. Construction methodologies of buildings, operation and maintenance and subsequent demolition are increasingly

identified as major source of environmental degradation as also, buildings are responsible of almost one third of energy consumption which includes both embodied and
operational energies. Consumption of water by buildings increased by almost 26%
between the years 1985 and 2005. Building demolition and construction accounted
for 60% of non-industrial waste [9]. Various characterized The Building information modelling (BIM) is covering different particular bases as it went for interfacing
natural things and their virtual portrayal with the target to utilize this association for
upgrading administrations and correspondence thoughts.

The BIM approach joins example and thoughts, instructed by Sensor Networks,
Ambient Intelligence, Grid Computing and Ubiquitous Computing. Their motivation exists to extract the administrations given by the Building information modelling
(BIM) and its application. In this way, it exhibited a meaning of BIM administrations and grouped them in view of the relationship to their life cycle and a physical
substance. They grouped BIM considering association with the elements. There are
4 substance relationship. They additionally ordered Building information modelling
in light of life cycles. BIM furnishes individuals centricity with innovation to convey
upgraded comes about. They displayed an overview on how Building information
modelling and Cloud assumes an indispensable part in social insurance. BIM is a
worldwide system of "savvy gadgets" that can detect and cooperate with their condition utilizing the web for their correspondence and cooperation with clients and
different frameworks. The fundamental idea behind each BIM innovation and usage
is "Gadgets are coordinated with the virtual universe of web and cooperate with it
by following, detecting and observing articles and their condition". The highlights
of a "savvy gadget" that can go about as an individual from BIM arrange are, gather
and transmit information, activate gadgets in light of triggers and get information
(from arrange or on the other hand web). The rise of infrastructural, operational and
natural issues, for example, environmental change, clamor contaminations, failing
has significantly expanded the requirement for strong, modest, operationally versatile, and shrewd checking frameworks. In this setting shrewd sensor systems are a
developing field of research which joins many difficulties of current software engineering, remote correspondence and versatile registering. In this paper an answer for
checking the clamor contamination levels in the infrastructural condition utilizing
remote inserted processing framework is proposed. The arrangement incorporates
the advances that have been rising in the field of portable registering and in addition to
Building information modelling (BIM) in light of their tremendous appropriateness.
Here, the detecting frameworks are associated with the established registering framework to screen the change of clamor contamination parameters, from their ordinary
conduct. This model is adaptable and distributive for any infrastructural condition
that requires persistent observing and behavioural examination. Execution of the
proposed display is assessed by utilizing model usage, comprising of Intel Galileo
and sensor sheets alongside inserted programming. The usage is tried for a few parameters and their behavioural examples as for client given details that gives a controlled
contamination checking to influence the earth to shrewd [12, 15, 16, 27, 33, 43].

There being many studies comparing LEED and BREEAM and other international
GBRS but no significant studies have been conducted in the Indian contexts which

appraise the green building programs followed in India. Since India is a developing country with highest density of population and growing urbanization, the sustainability of construction sector in India cannot be ignored. India has adopted LEED-India and GRIHA but they are in their nascent stages of implementation, it needs to be compared to BREEAM and US-LEED. It has turned out to be important to survey the rating frameworks since they are presently getting connected with governments' promotional strategies for green structures. Governments are arranging administrative commitments and authority motivation incentives to elevate rating of structures to give a push to the green building development. This makes performance-based evaluation of the GBRS to guarantee that they are conveying on expressed goals. GRIHA adopted by the Government of India seems to be adequate as far as rewarding sustainable actions in the development of good industry practice is concerned. The measure envisaged is simple, clear and impacts positively on the whole design. But the design and procurement team find it technically difficult to achieve any rating and the reward seems to be disproportional.

There is always a resistance as not enough information is available and the procedure cannot be outsourced. Also, it doesn't respond well to regional context. There are shortcomings observed in GRIHA EB as it applies only to apply to built-up area greater than 2500 $km^2$ which doesn't cover most residential buildings. The assessment of existing buildings should be integrated and holistic, not only green as buildings deteriorate structurally with time and change of usage. Sullivan [5], in the "General Services Administration's Office of Federal High-Performance Green Buildings commissioned a study of green building certification systems in accordance with the Energy Independence and Security Act of 2007 (EISA)". "Sustainable sites (construction related pollution prevention, site development impacts, transportation alternatives, storm water management, heat island effect, and light pollution) Water efficiency (landscaping water use reduction, indoor water use reduction, and waste water management strategies) Energy and atmosphere (commissioning, whole building energy performance optimization, refrigerant management, renewable energy use, and measurement and verification) Materials and resources (recycling collection locations, building reuse, construction waste management, and the purchase of regionally manufactured materials, materials with recycled content, rapidly renewable materials, salvaged materials, and FSC certified wood products) Indoor environmental quality (environmental tobacco smoke control, outdoor air delivery monitoring, increased ventilation, construction indoor air quality, use low emitting materials, source control, and controllability of thermal and lighting systems) Innovation and design process accredited professional and innovative strategies for sustainable design" [3]. The long-life Report is supposed to prepare essential know-how on "Guidelines of common standards and criteria" for work package, "Designing the Sustainable Prototype" it provides the necessary benchmarks for construction phase of long-life sustainable building pilot projects. Several researchers have identified the growing need and demand of smart cities and have addressed various challenges that needs to be overcome during the management of smart cities [1, 17, 37]. Yigitcantar et al. [42] highlighted that how a smart city technology and the concepts can be successfully perceived and utilized. Neves et al.

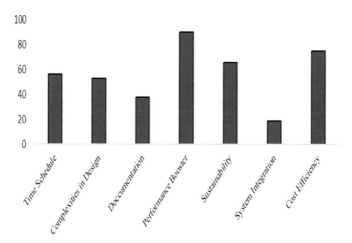

**Fig. 3** Numerous application of BIM as observed in literature

[28] studied the impacts of proper management as it provided a framework of open data initiatives on smart cities. Along with the role of several proposed framework for managing and handling smart cities there are still many uncertainties associated with it. To obtain an overall perspective for a smart city management one of the most efficient tools is BIM. The role and application have been well established in the past and literature highlights a significant contribution BIM plays in managing and designing infrastructures [11, 20, 23]. Marzouk and Othman [24] provides a detailed insight on planning efficacy of an infrastructure requirements for smart cities that are integrated in between with the GIS and BIM. The study highlights the efficiency of BIM in smart city infrastructural development.

Wang and Tian [39] in maintenance stage and commercial building operation pointed several obstacles in BIM application. Lin and Cheung [21] also provide a detailed study on role of BIM in smart cities. Its persistence of this research is to understand the credibility of smart cities and its dependency on BIM for effective and efficient management. Development of smart cities are known for its advancement movements and still has opportunity for enhancement. There is a necessity for the citizens to understand the management strategies of smart cities in a compelling way and this is the place BIM can offer help (Fig. 3).

## 3 Discussion

The present section discusses the major observation of the literature and elaborates the applicability of BIM in smart city. BIM have been extensively applied and used in various field of construction industry. The application of BIM involves certain limitations and challenges. In context of the above the present paper provides a detailed study on the BIM applications in smart cities. The influence of BIM in smart cities has always been a concern amongst the construction industry and it needs to be addressed. Construction industry attempts to address the sustainability by developing guidelines for Smart City Building designs and construction that is not limited to resource efficiency confirming economic and environmental aspects but is being expanded to disaster resilience which has social implications also [42, 43]. Smart city is a modern perception that assimilates information and communication technology along with several modern-day devices connected to internet. Overall to summarize the concept of smart city one can conclude that it utilizes the internet facilities in a city to reduce human effort by controlling maximum function via remotely located devices. The modern devices and internet connectivity enhance the productivity to city services, processes and connect to citizens. This knowledge also allows the official and staffs to maintain a direct contact with the community and infrastructure of the city. It also helps in service and maintenance of the city up to a considerable extent. It also keeps a check on the occurrence in the city and how the city infrastructures are developing. Contrary to that the data can be utilized to enhance several management procedures all along the smart city. Information and communication technologies adopted to improve the performance, superiority and interactivity of infrastructural services which leads to cost reduction, optimized resource consumption and increases interaction between management and citizens. The Smart city solicitations are established to accomplish efficient urban flow and permit real-time reactions. Irrespective of these facts the term smart city remains indistinct to its essentials details and therefore it leads to numerous interpretations. These numerous facilities involved in a typical smart city are collection of data from citizens, structures, devices and possessions that is further administered and investigated to regulate and accomplish transportation systems, flow of traffic and power plants, along with several other utilities such as networks of hospitals, water supply, crime exposure, waste management, schools, information structures, libraries, and numerous other community services. Therefore, to manage the city based on such advanced technologies and information and communication system a robust tool to manage these is an essential need of the hour. In actual construction the Role of BIM is an innovative concept which facilitates the capability to predict simulated construction of a facility prior. The simulation and modelling are crucial in any construction industry work out problems and improve safety in demand to reduce uncertainty. In every single profession one can provide critical information into the model as an input before the construction phase and obtained a detailed analysis of the whole construction services. It is one of the crucial advantages of BIM is waste management. It tends to minimize wastage of resource and materials on-site and assures the delivery of the product in-time. It also helps in avoiding stockpiles on-site. Therefore, in view of the above the present will emphasis on addressing the influence of BIM application in smart cities. The following subsections provide a detail insight on literature in relevant field and derive some conclusion on the basis which can be crucial for the fore coming research in this relevant field. Figure 4 depicts the probable application of BIM in construction of smart city.

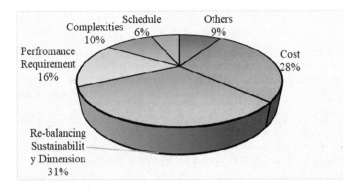

**Fig. 4** Pie-chart elaborating the probable application of BIM in smart city

## 4  Summary and Conclusion

This Study highlights the influence of BIM in the construction of smart cities. BIM is a robust tool which efficiently helps in managing various aspects of an infrastructure management by simulating real life scenarios and providing a detailed insight about the activities beforehand so that any possible delay in work and any wastage of resources can be conserved. Smart city is a hub of infrastructure combined with information and communication technologies. This city enables the use of high-end internet services and modern-day devices to steer numerous city management tasks with utmost ease. These tasks can be performed by remotely seated member and remotely located infrastructures. Henceforth, the application of BIM in smart city management can have crucial influence on its performance. It can help in cost cutting and will significantly help in optimized resource consumption. It will also efficiently deal with the design and performance complexities involved with several technologies implemented in the smart city. One of the crucial roles BIM can play in managing a smart city is by keeping a strict check on its time schedule of construction activities along with maintenance activities. For any type of infrastructure including smart cities maintenance, it is a significant part for better serviceability, and therefore by maintaining a proper schedule for maintenance activities BIM can significantly improve the performance of smart city. It will also aid in collision detection of several interfering activities and will identify the exact site of discrepancies. Considering the above-mentioned significant point, it will be completely fair to frame that building information modelling can have a vital influence on numerous management activities involved in a smart city.

# References

1. Ahad MA, Paiva S, Tripathi G, Feroz N (2020) Enabling technologies and sustainable smart cities. Sustain Cities Soc 61:102301
2. ArchDaily (2021) Smart cities paving the way to a smart future. [online] Available at: https://www.archdaily.com/936781/smart-cities-paving-the-way-to-a-smart-future. Accessed 1 Sept 2021
3. Azhar S (2011) Building information modeling (BIM): trends, benefits, risks, and challenges for the AEC industry. Leadersh Manag Eng 11(3):241–252
4. Aziz ND, Nawawi AH, Ariff NRM (2016) Building information modelling (BIM) in facilities management: opportunities to be considered by facility managers. Procedia Soc Behav Sci 234:353–362
5. Barlish K, Sullivan K (2012) How to measure the benefits of BIM—a case study approach. Autom Constr 24:149–159
6. Bazjanac V (2008) Impact of the US national building information model standard (NBIMS) on building energy performance simulation
7. Becerik-Gerber B, Rice S (2010) The perceived value of building information modeling in the US building industry. J Inf Technol Constr (ITcon) 15(15):185–201
8. Bryde D, Broquetas M, Volm JM (2013) The project benefits of building information modelling (BIM). Int J Project Manage 31(7):971–980
9. Dakhil A, Underwood J, Al Shawi M (2016) BIM benefits-maturity relationship awareness among UK construction clients. In: Proceedings of the first international conference of the BIM academic forum, Glasgow, UK, pp 13–15
10. Diaz PM (2016) Analysis of benefits, advantages and challenges of building information modelling in construction industry. J Adv Civil Eng 2(2):1–11
11. Follini C, Magnago V, Freitag K, Terzer M, Marcher C, Riedl M, Giusti A, Matt DT (2021) BIM-integrated collaborative robotics for application in building construction and maintenance. Robotics 10(1):2
12. Gupta S, Jha KN, Vyas G (2020) Proposing building information modeling-based theoretical framework for construction and demolition waste management: strategies and tools. Int J Constr Manage, 1–11
13. Henderson LK, Craig JC, Willis NS, Tovey D, Webster AC (2010) How to write a Cochrane systematic review. Nephrology 15(6):617–624
14. Hergunsel MF (2011) Benefits of building information modeling for construction managers and BIM based scheduling
15. Jung Y, Joo M (2011) Building information modelling (BIM) framework for practical implementation. Autom Constr 20(2):126–133
16. Kemp A (2020) The BIM implementation journey: lessons learned for developing and disseminating city information modelling (CIM). Built Environment 46(4):528–546
17. Lai CS, Jia Y, Dong Z, Wang D, Tao Y, Lai QH, Wong RTK, Zobaa AF, Wu R, Lai LL (2020) A review of technical standards for smart cities. Clean Technol 2(3):290–310
18. Lang TA, Lang T, Secic M (2006) How to report statistics in medicine: annotated guidelines for authors, editors, and reviewers. ACP Press
19. Latiffi AA, Mohd S, Kasim N, Fathi MS (2013) Building information modeling (BIM) application in Malaysian construction industry. Int J Constr Eng Manage 2(4A):1–6
20. Li S, Zhang Z, Mei G, Lin D, Yu J, Qiu R, Su X, Lin X, Lou C (2021) Utilization of BIM in the construction of a submarine tunnel: a case study in Xiamen city, China. J Civ Eng Manag 27(1):14–26
21. Lin YC, Cheung WF (2020) Developing WSN/BIM-based environmental monitoring management system for parking garages in smart cities. J Manag Eng 36(3):04020012
22. Lu W, Fung A, Peng Y, Liang C, Rowlinson S (2014) Cost-benefit analysis of building information modeling implementation in building projects through demystification of time-effort distribution curves. Build Environ 82:317–327

23. Luo W, Zhang J, Wang M, Wang K (2021) Research on transmission and transformation engineering cost system based on BIM 3D modelling technology. IOP Conf Ser: Earth Environ Sci 632(4):042029
24. Marzouk M, Othman A (2020) Planning utility infrastructure requirements for smart cities using the integration between BIM and GIS. Sustain Cities Soc 57:102120
25. Migilinskas D, Popov V, Juocevicius V, Ustinovichius L (2013) The benefits, obstacles and problems of practical BIM implementation. Procedia Eng 57:767–774
26. Mohandes SR, Omrany H (2015) Building information modeling in construction industry. J. Teknol 78:1
27. Muthusamy K, Chew L (2020) Critical success factors for the implementation of building information modeling (BIM) among Construction Industry Development Board (CIDB) G7 Contractors in the Klang Valley, Malaysia. In: 2020 IEEE European technology and engineering management summit (E-TEMS). IEEE, pp 1–6
28. Neves FT, de Castro Neto M, Aparicio M (2020) The impacts of open data initiatives on smart cities: a framework for evaluation and monitoring. Cities 106:102860
29. Pena G (2011) Evaluation of training needs for Building Information Modeling (BIM)
30. Pimentel LL (2016) BIM implementation—a bibliographic study of the benefits and costs involved. J Civil Eng Archit 10:755–761
31. Rokooei S (2015) Building information modeling in project management: necessities, challenges and outcomes. Procedia Soc Behav Sci 210:87–95
32. Sai Evuri G, Amiri-Arshad N (2015) A study on risks and benefits of building information modeling (BIM) in a construction organization
33. Sepasgozar SM, Hui FKP, Shirowzhan S, Foroozanfar M, Yang L, Aye L (2021) Lean practices using building information modeling (BIM) and digital twinning for sustainable construction. Sustainability 13(1):161
34. Smith P (2014) BIM implementation–global strategies. Procedia Eng 85:482–492
35. Terreno S, Anumba CJ, Gannon E, Dubler C (2015) The benefits of BIM integration with facilities management: a preliminary case study. In: Computing in civil engineering 2015, pp 675–683
36. The Associates (2021) 3 advantages of vertical integration In BIM implementation: BIM modelling services. [online] The AEC Associates Blog. Available at: https://theaecassociates.com/blog/3-advantages-vertical-integration-bim-implementation-bim-modelling-services/. Accessed 1 Sept 2021
37. Ullah Z, Al-Turjman F, Mostarda L, Gagliardi R (2020) Applications of artificial intelligence and machine learning in smart cities. Comput Commun 154:313–323
38. Wang X, Chong HY (2015) Setting new trends of integrated Building Information Modelling (BIM) for construction industry. Constr Innov 15(1):2–6
39. Wang Z, Tian H (2020) Research on obstacles of BIM application in commercial building operation and maintenance stage based on analytic hierarchy process. IOP Conf Ser: Earth Environ Sci 567(1):012037
40. Wong AK, Wong FK, Nadeem A (2010) Attributes of building information modelling implementations in various countries. Archit Eng Des Manage 6(4):288–302
41. Wong JK, Zhou JX, Chan AP (2018) Exploring the linkages between the adoption of BIM and design error reduction. In: Building information systems in the construction industry, p 113
42. Yigitcanlar T, Kankanamge N, Vella K (2020) How are smart city concepts and technologies perceived and utilized? A systematic geo-twitter analysis of smart cities in Australia. J Urban Technol, 1–20
43. Yu'e Q (2020) Research on the development of smart city and the innovation of internet of things industry in the era of "Internet+"

# Revisiting the Concept of Consumer Ethnocentrism After the Plague: Why Buying Local Matters

Saeb Farhan Al Ganideh

**Abstract** The pandemic took a heavy and instant toll on jobs all over the globe and represented the most severe crisis for employment since the 1930s Great Depression. How do ethnocentric tendencies towards local products and empathetic feelings towards local workers impact consumers' willingness to purchase local products during the pandemic is the main question that this paper grapples with at its most general level. Data were collected from 217 Jordanian subjects during the summer of 2020. Our findings show that COVID-19 pandemic outbreak contributed to build a new "buy local" line that might shift spending towards locally made products. Also, the results concluded that when Jordanian consumers express empathetic feelings towards local workers due the COVID-19, they are more willing to purchase local products. Finally, our study provides some interesting insights for managers on how to market their products during a severe crisis such as the pandemic.

**Keywords** Covid-19 · Local products · Empathetic feelings · Consumer ethnocentrism · Coronavirus

## 1 Introduction

The COVID-19 pandemic has reached almost every country in the world. Notably, the coronavirus pandemic has had profound impacts on societies and economies. The pandemic continues to impact all spheres of human lives all over the world, including economic, political, educational, scientific, and cultural contexts [1]. In general, the pandemic represents one of the most significant peacetime challenges facing businesses and governments in recent history [2].

Notably, the coronavirus pandemic and the lockdown measures have affected consumption behavior. While several studies have investigated the impact of COVID-19 pandemic on panic buying (i.e., [2–4]), impulse buying (i.e., [2]), green buying

S. F. Al Ganideh (✉)
Faculty of Business, Al Zaytoonah University of Jordan, Amman, Jordan
e-mail: saeb@zuj.edu.jo

© The Author(s), under exclusive license to Springer Nature Switzerland AG 2022
S. G. Yaseen (ed.), *Digital Economy, Business Analytics, and Big Data Analytics Applications*, Studies in Computational Intelligence 1010,
https://doi.org/10.1007/978-3-031-05258-3_27

(i.e., [5, 6]), fresh food buying (i.e., [1, 7]), healthy behavior and buying (i.e., [8]), it is still unclear how the epidemic impact local products' buying.

COVID-19 brought many negative consequences, including the closure of many businesses, economic vulnerability and job losses [9]. Empathic feelings toward local workers might impact positively individuals' ethnocentric tendencies toward purchasing local products [10].

Amidst a global pandemic, notably, many individuals attempt to support both locally owned businesses and domestic made products through purchases. The concept of consumer ethnocentrism is related to consumers' preferences of local products due to socio-psychological and socio-demographic variables. Highly ethnocentric consumers believe that buying home country made products is vital to support domestic economy and it is morally right [10]. Consumer ethnocentrism concept has been, also, "elaborated as the phenomenon of some consumers being ethnocentric, in that they tend to discriminate between products from the in-group and out-group" and purchase only local products [11]. Individuals with high level of consumer ethnocentric tendencies feel that domestic producers deserve help, and this justifies purchases of domestically produced products [11].

Overall, this paper aims to understand how consumers' willingness to purchase local products have shifted due to the COVID-19 pandemic. Closure of many local businesses, economic vulnerability and job losses, and reduced wages have "contributed to wide-ranging shifts in perceptions towards local products and buying behavior (i.e., [1, 9]). Specifically, how empathetic feelings towards local workers and ethnocentric tendencies towards local products impact individuals' willingness to purchase local products during the pandemic is the main question that this paper grapples with at its most general level.

## 2 Consumer Behavior in Crisis Times

To limit the spread of COVID-19, business and marketing activities that necessitate close physical proximity between customers and sellers have been restricted. The main dilemma that faced governments and businesses throughout the coronavirus pandemic is how to balance between protecting people's health and keeping an economy going. The COVID-19 pandemic outbreak "forced many businesses to close, leading to an unprecedented disruption of commerce in most industry sectors" ([13], p. 284). Notably, employees are losing their jobs at high rates, the highest level since the 1930's Great Depression [13].

Almost, "all countries that are trying to stimulate their economies to keep as much as possible of their necessary infrastructure intact and to keep citizens productive or ready to become productive once the pandemic has been overcome" ([13], p. 284).

Countries around the globe have adopted very different approaches to handle the current stress on the job markets and infrastructure. Some countries have chosen to support businesses in order to help them keep the workforce intact, but others with less financial strength cannot do the same [13]. Nonetheless, all countries, in order to keep their societies from deteriorating, encourage their citizens to buy local products.

The Theory of Reasoned Action (TRA) and the Theory of Planned Behavior (TPB) could represent a suitable theoretical background for the current study [14]. TRA and TPB have been used widely by researchers from various disciplines, including consumer behavior, health behavior and technology adaptation. Both TRA and TPB assume that "the best predictor of a behavior is behavioral intention, which in turn is determined by attitude toward the behavior and social normative perceptions regarding it" ([15], p. 231). Despite the TRA and the TPB only explain and predict relatively simple volitional behavior [14], strong evidence from literature show that changing TRA or TPB constructs leads to subsequent change in behaviors [15]. Overall, both theories have been perceived positively by wide number of researchers as it was successfully able to predict a variety of individuals' behaviors [16]. Nonetheless, a number of researchers argued that it is time for and TRA and TPB to retire as both theories are not only not able to explain sufficient variability in human behavior [17].

## 3 Research Model and Hypotheses

Theory and research have consistently documented links between ethnocentric tendencies for consumers and their willingness to purchase local products. Nonetheless, it is not clear how empathetic feelings towards local workers could mediate the relationship between consumer ethnocentric tendencies and (un)willingness to purchase products sourced from their home country and foreign made products. Despite previous literature has accredited the importance of understanding the intersection of consumer ethnocentrism and purchasing local products, very little is known how of such relationship can be mediated by empathetic feelings towards local workers during a human, economic and social crisis which is having a catastrophic effect on local workers.

Bearing in mind the inconsistency and scarcity in understanding consumers' willingness to purchase local products during the plague, it is vital to examine how empathetic feelings towards local workers can impact their willingness to purchase local products and their unwillingness to purchase foreign products.

The choice of domestic versus foreign products may be influenced by highly emotional factors such as empathetic feelings towards local workers and individuals' ethnocentric tendencies towards their home country. The concept of consumer ethnocentrism, originally developed by Shimp and Sharma [18], could explain why consumers express positive attitudes towards local products and they express negative attitudes towards foreign products [10]. Notably, consumers with high level of ethnocentrism believe that purchasing foreign products is wrong, as it has serious consequences on domestic workers, causes unemployment and it is plainly unpatriotic [18]. Purchasing foreign products is perceived by ethnocentric consumers as immoral since it has an adverse impact on the national economy and local workers [12].

Recently, many countries conducted national campaigns to encourage consumers to buy local products to minimize adverse effects on local workers and local businesses. Earlier, Ang et al. [19], argued that the 'buy local' campaigns attempt to increase consumers' awareness towards their local area and home country products. The COVID-19 pandemic outbreak contributed to build a new "buy local" line that might shift spending towards small businesses. Such cultural and spending shift can strengthen small businesses' role in generating more jobs in local communities, creating better community cohesion, and boosting local economies.

A conceptual model is proposed in Fig. 1 to show the anticipated relationships among our main four constructs and the hypotheses. Consumer ethnocentrism reflects a person's tendency to be ethnically centred and to blindly accept local culture ([10, 19]). Despite ethnocentrism was conceptualized originally as a sociological concept, then the concept was developed into a psychological construct with relevance to an individual level [20]. Shimp and Sharma [18] applied the ethnocentrism construct to the study of consumer behaviour and conceptualized the term "consumer ethnocentric tendencies" to reflect consumers' beliefs related to the appropriateness and morality of purchasing local made products [10, 19, 20].

Consumer ethnocentrism could clarify why some consumers have positive attitudes towards local products [20]. Consumers with high ethnocentric tendencies are most likely to perceive the positive aspects of domestic products and to discount the virtues of foreign products [18]. Consumers with high ethnocentric tendencies are willing to pay more to buy domestic products [10, 12, 19, 20]. Consumer ethnocentrism may change due to time, political status, economic crises or historical events [22]. Overall, consumers become more ethnocentric and more patriotic towards purchasing the national products during the crisis time [10, 12, 19]. Notably, the pandemic motivated many individuals to support local businesses and to purchase local products. Shoppers, during the pandemic, are more willing to support local workers and small businesses [5, 6, 8]. Thus, we hypothesize:

H1 Consumer ethnocentric tendencies towards local products are positively related to empathic feelings towards local workers.

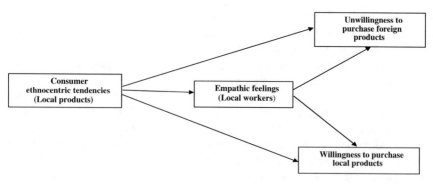

**Fig. 1** Research model

Ethnocentrism entails contempt for "other" culture of other people. The concept of ethnocentrism entails suggests that purchasing non-local products is inappropriate, since it has serious consequences for local economy, and may cause local workers to lose their jobs ([10, 12]). Ethnocentric consumers believe that purchasing foreign products is wrong, since it has serious consequences on their domestic economy, causes unemployment and it is plainly unpatriotic (i.e., [23]). The concept of consumer ethnocentrism could help to explain why consumers express positive attitudes towards local products [18]. Consequently, consumers with high ethnocentric tendencies believe that purchasing foreign made products is morally wrong; foreign made products may become objects of contempt. High ethnocentric consumers evaluate products based on the moral acceptability of purchasing foreign products [24]. They do not like to learn about foreign brands and products since their processing of foreign products is superficial [25]. It is obvious that ethnocentrism leads individuals to underestimate other cultures and, for other highly ethnocentric individuals, other ethnic groups become objects of dislike. In fact, the concept of consumer ethnocentrism purports emotional support and care for local workers and the domestic economy [10]. Based on the above arguments, we anticipate that:

H2 Consumer ethnocentric tendencies towards local products are positively related to their willingness to purchase local products.

Ethnocentric tendencies represent the emotional dimension of buying foreign products as well as the implication that the choice somehow threatens domestic industry and national security [26]. Different internal factors might encourage emerging ethnocentrism such as economic recession, high unemployment and rapid ethnological and organisational change [10, 12, 27]. Alternatively, the consequences of consumer ethnocentrism (whether ethnocentrism leads consumers to purchase their domestic products) have been well studied in previous research (i.e., [10, 12, 27]). Despite the concept of consumer ethnocentrism has been well researched in marketing literature, little known how do ethnocentric tendencies for consumers impact their empathetic feelings towards local workers [10, 29]. Purchasing foreign products is perceived by some consumers as immoral since it has an adverse impact on the national economy [30]. Alternatively, consumers with low ethnocentric tendencies will switch to foreign products instead of paying more [31]. Despite there is some evidence to show that some consumers may even prefer, foreign products over domestic products, previous recent studies have agreed that consumer ethnocentrism is negatively related to willingness to buy foreign products [12]. Thus, we anticipate that:

H3 Consumer ethnocentric tendencies towards local products are positively related to their unwillingness to purchase foreign products.

Empathy is significantly dynamic and individuals' empathetic feelings can change with new experiences and new circumstances [32]. It is not surprising that many local businesses are revamping their strategies and trying to survive the pandemic. To help small businesses, many countries conducted national campaigns to encourage consumers to buy domestic products from local businesses [8]. These campaigns have focused mainly on the importance of buying local products to keep local economy working and keep local jobs. Earlier, during the Asian crisis many countries conducted campaigns to encourage locals to buy their home country products and to boycott foreign brands in order to keep jobs [19]. Overall, there is still little research on the relationship between buying local and emphatic feelings towards local workers, however, a previous recent published work has strongly connected willingness to purchase local products with empathic feelings towards local workers [10]. Thus, we hypothesize:

H4 Empathic feelings towards local workers are positively related willingness to purchase local products.

Notably, empathy is a concept that refers to the emotional reactions of people to other individuals' experiences [32]. It is clear that showing empathy increases the likelihood of helping other people and understanding their bad experiences [32]. There is a positive connection between feeling empathic and being willing to support others [29]. Previous research concluded that "inducing empathy for a member of a stigmatized group can improve attitudes toward the group as a whole" ([29], p. 1656). Despite empathy improves attitudes toward the "other" and increases concern for people in need [34]. Individuals are more empathetic towards people who suffer from their context. Notably, a person who empathizes is more likely to support people in need financially [10, 34]. The main stream of research on willingness to purchase local products has focused more on prosocial and altruistic behaviour to explain this phenomenon rather on empathic feelings [29, 34]. Since the start of COVID-19 pandemic many countries have started to urge their citizens to buy local and to help local workers whose jobs are threatened due to the pandemic. Thus, we anticipate that:

H5 Empathic feelings towards local workers are positively related to unwillingness to purchase foreign products.

## 4 Method

### 4.1 Sample

Data were collected from 217 subjects who live in Amman city, the capital of Jordan during the summer of 2020. Jordan is a good representative country of other Arab countries and results obtained from the Jordanian context could be generalized on other Arab countries. Convenience sampling method were employed to approach individuals to participate in the current study. Electronic data collection is increasingly popular as a means to collect data due to social distancing and lockdown rules. A link for the survey (Google Docs) were sent to the subjects via Emails and WhatsApp messages. Of the 217 subjects, 90 were females, 127 were males. Moreover, the respondents under age 36 represented 47.9%, and approximately, 42% of the respondents above the age of 36. About 56% of our subjects were college educated, which was slightly greater than the percentage of subjects without a college education.

## 4.2 Measures

*Willingness to purchase local products*: We used two items (5-point Likert scale) originally developed by Grewal et al. [34] to measure consumers' intention to purchase local products (I would purchase local products when I am shopping next time; I would consider buying local products in any chance;).

*Unwillingness to purchase foreign products*: To measure the subjects' level of unwillingness to buy foreign products, two items (5-point Likert scale) developed originally by Darling and Arnold [34] used by Klein et al. [34] (I would feel guilty if I bought a foreign product, whenever possible, I avoid buying foreign products).

*Consumer ethnocentric tendencies (local products):* To measure subjects' ethnocentric tendencies towards purchasing local products, we used four items (5-point Likert scale) adopted from Shimp and Sharma [18] CETSCALE (it may cost me in the long run but I prefer to support local products; we should buy from foreign countries only those products that we cannot obtain within our own country; a real citizen should always buy local products; curbs should be put on all non-local products).

*Empathic feelings (local workers):* We measured our subjects' empathetic feelings towards local workers using four items (5-point Likert scale) originally developed by Granzin and Olsen [38] (I am really sorry about Jordanian workers' current situation; it is a shame that Jordanian workers are suffering because of COVID-19; it makes me feel bad to see the difficulties local workers are facing because of COVID-19; I feel really bad about local workers' losing jobs of COVID-19).

Correlation matrix for the study variables is presented in Table 1 and confirmed

**Table 1** Inter-correlations, Cronbach's alphas, CR, and AVE

| No | Scale | α | CR | AVE | 1 | 2 | 3 | 4 |
|----|-------|---|----|----|---|---|---|---|
| 1 | Consumer ethnocentric tendencies (local products) | 0.73 | 0.50 | 0.74 | 0.86[a] | | | |
| 2 | Empathic feelings (local workers) | 0.91 | 0.81 | 0.93 | 0.02 | 0.96 | | |
| 3 | Willingness to purchase local products | 0.73 | 0.59 | 0.74 | 0.67** | 0.15 | 0.86 | |
| 4 | Unwillingness to purchase foreign products | 0.81 | 0.68 | 0.81 | 0.30** | 0.12 | 0.21** | 0.90 |

*Notes* * $p < 0.05$ (two-tailed); ** $p < 0.01$ level (two-tailed); [a]Matrix diagonals represent the square roots of the AVE. The off-diagonal elements the represent the intercorrelations of measure scales. For constructs 1–4, this table supports the discriminant validity of measure scales as elements in the matrix diagonals (square roots of the AVEs) are greater in all cases than the off-diagonal elements (intercorrelations) in their corresponding row and column

discriminant validity for our measures [39].

## 5 Results

The conceptual proposed model of the current study was tested using structural equation modelling (SEM). Using SEM to test our proposed model was driven primarily as SEM allows the evaluation of the entire model [40]. To validate our proposed model, first, we have calculated Cronbach alpha, composite reliability (CR), and the average variance extracted (AVE) for our four used constructs. The usually desired levels are greater than 0.50 for the AVE, greater than 0.70 for the CR and greater than 0.6 for Cronbach alpha, though slightly lower levels are often also accepted ([39, 40]). The results, showed in Table 1, reveal that the values of AVE, CR and Cronbach's alpha are satisfactory and exceed the suggested minimum thresholds for these indicators ([40]). At the second stage, a confirmatory approach was chosen to test the hypothesized model using IBM-SPSS AMOS 22 [40].

The $\chi 2$/degrees of freedom ratio was 1.65, far below the common cut-off threshold of 5.0 suggested by Hair et al. [41]. The results for our study model were Comparative Fit Index (CFI) (0.98), Tucker-Lewis Index (TLI) (0.97), and Root Mean Square Error of Approximation (RMSEA) was 0.05 level. Our results confirmed the proposed model with adequate fit. The suggested model has far exceeded the threshold of 0.90 for CFI and the cutoff point of 0.07 or less for RMSEA [40]. The proposed relationships in our mode are depicted in Table 2.

## 6 Discussion and Conclusion

This study provides context and guidance to researchers seeking to understand the relationship between empathetic feelings towards local workers and purchasing local products during the COVID-19. Specifically, this study aims to understand links between ethnocentric tendencies for consumers, their empathetic feelings towards local workers and their willingness to purchase local products. While previous studies tend to focus more on exploring the construct of consumer ethnocentrism during normal economic and social situations, this study contributes to the existing base of knowledge by exploring the of concept consumer ethnocentrism during the Covid-19 pandemic. Overall, the concept of consumer ethnocentrism could clarify the level of loyalty consumers expresses towards locally made products.

Table 2 shows the test results of the proposed relationships depicted in the model. Results show consumer ethnocentric tendencies towards local products do not impact empathic feelings towards local workers (ß $= -0.01$, p $= 0.88$), rejecting our H1. As predicted by H2, cconsumer ethnocentric tendencies towards local products are positively related to their willingness of purchasing local products (ß $= 0.61$, p $<$ 0.01). Additionally, we expected that consumer ethnocentric tendencies towards local

**Table 2** Estimation results

| No | Path | Expected relationship | Estimate | P-value | Result |
|----|------|----------------------|----------|---------|--------|
| H1 | Consumer ethnocentric tendencies (local products) → Empathic feelings (local workers) | Significant (+) | −0.01 | 0.88 | Not supported |
| H2 | Consumer ethnocentric tendencies (local products) → Willingness to purchase local products | Significant (+) | 0.61 | 0.00 | Supported |
| H3 | Consumer ethnocentric tendencies (local products) → Unwillingness to purchase foreign products | Significant (+) | 0.27 | 0.00 | Supported |
| H4 | Empathic feelings (local workers) → Willingness to purchase local products | Significant (+) | 0.10 | 0.04 | Supported |
| H5 | Empathic feelings (local workers) → Unwillingness to purchase foreign products | Significant (+) | 0.14 | 0.14 | Not supported |

products are positively related to their unwillingness to purchase foreign products (H3). Our results ($\beta = 0.27$, $p < 0.01$ supported this hypothesis. H4 was supported as the results revealed that eempathic feelings towards local workers are positively related to their willingness to purchase local products ($\beta = 0.10$, $p < 0.05$). Finally, we expected empathic feelings towards local workers impact positively unwillingness to purchase foreign products. The results did not prove our expectations ($\beta = 0.14$, $p = 0.14$) rejecting H5.

Consistent with previous research findings, this study confirms a positive relationship between consumer ethnocentric tendencies towards local products are positively related to their willingness of purchasing local products and their unwillingness of purchasing foreign products (i.e., [10, 12, 24, 40]). Overall, international marketing and business literature, concluded that empathic feelings towards local workers are positively related towards purchasing local products (i.e., [10, 32, 42–48]). The findings from this study support this notion.

Our findings provide some interesting insights for local marketing managers on how to market their products during a severe crisis such as the COVID-19 pandemic. Specifically, our study reveals some useful and practical implications for local marketers and policymakers. We recommend that local marketers highlight local workers and job lost while communicating with their customers. Specifically, local marketers focus on the importance of buying local for the goodwill of all local workers.

It is clear that the COVID-19 pandemic outbreak gives local businesses a boost and contributes strongly to build a new "buy local" notion shifting spending towards local businesses. Such shift in spending, also, seems to support local workers. Local businesses should revamp their communication and marketing strategies to focus more on encouraging consumers to buy local products and to support local workers. Moreover, marketers are much recommended to develop marketing strategies and communications that suit consumer sentiments during the pandemic. Such marketing communications should activate empathic feelings for consumers.

# References

1. Hassen TB, El Bilali H, Allahyari MS, Berjan S, Fotina O (2021) Food purchase and eating behavior during the COVID-19 pandemic: a cross-sectional survey of Russian Adults. Appetite 165:105309
2. Hall MC, Prayag G, Fieger P, Dyason D (2020) Beyond panic buying: consumption displacement and COVID-19. J Serv Manage 32(1, 202):113–128
3. Hobbs JE (2020) Food supply chains during the COVID-19 pandemic. Can J Agric Econ/Revue Canadienne D'agroeconomie 68(2):171–176
4. He H, Harris L (2020) The impact of Covid-19 pandemic on corporate social responsibility and marketing philosophy. J Bus Res 116:176–182
5. Alexa L, Apetrei A, Sapena J (2021) The COVID-19 lockdown effect on the intention to purchase sustainable brands. Sustainability 13(6):3241
6. Naeem M (2020) Understanding the customer psychology of impulse buying during COVID-19 pandemic: implications for retailers. Int J Retail Distrib Manage 49(3):377–393
7. Chen J, Zhang Y, Zhu S, Liu L (2021) Does COVID-19 affect the behavior of buying fresh food? Evidence from Wuhan, China. Int J Environ Res Public Health 18(9):4469
8. Verma M, Naveen BR (2021) COVID-19 impact on buying behaviour. Vikalpa 46(1):27–40
9. Bradbury-Jones C, Isham L (2020) The pandemic paradox: the consequences of COVID-19 on domestic violence. J Clin Nurs 29(13–14):2047–2049
10. Al Ganideh SF (2017) Being Arab and American: understanding ethnocentric tendencies for Arab American consumers. J Glob Mark 30(2):72–86
11. Yang R, Ramsaran R, Wibowo S (2021) Do consumer ethnocentrism and animosity affect the importance of country-of-origin in dairy products evaluation? The moderating effect of purchase frequency. Br Food J (ahead-of-print). https://doi.org/10.1108/BFJ-12-2020-1126
12. Abdul-Latif S-A, Abdul-Talib A-N (2020) An examination of ethnic-based consumer ethnocentrism and consumer animosity. J Islamic Mark (ahead-of-print)
13. Donthu N, Gustafsson A (2020) Effects of COVID-19 on business and research. J Bus Res 117:284–289
14. Ajzen I, Fishbein M (1975) A Bayesian analysis of attribution processes. Psychol Bull 82(2):261
15. Montano DE, Kasprzyk D (2015) Theory of reasoned action, theory of planned behavior, and the integrated behavioral model. Health Behav Theory Res Pract 70(4):231
16. Conner M, Armitage CJ (1998) Extending the theory of planned behavior: a review and avenues for further research. J Appl Soc Psychol 28(15):1429–1464
17. Sniehotta FF, Presseau J, Araújo-Soares V (2014) Time to retire the theory of planned behaviour. Health Psychol Rev 8(1):1–7
18. Shimp TA, Sharma S (1987) Consumer ethnocentrism: construction and validation of the CETSCALE. J Mark Res 24(3):280–289
19. Ang S, Jung K, Leong S, Pornpitakpan C, Tan S (2004) Animosity towards economic giants: what the littlie guys think. J Consum Mark 21(3):190–207
20. Shimp TA (1984) Consumer ethnocentrism: the concept and a preliminary empirical test. Adv Consum Res 11:285–290

21. Vida I, Fairhurst A (1999) Factors underlying the phenomenon of consumer ethnocentricity: evidence from four Central European countries. Int Rev Retail Distrib Consum Res 9(4):321–337

22. Julie HY, Albaum G (2002) Sovereignty change influences on consumer ethnocentrism and product preferences: Hong Kong revisited one year later. J Bus Res 55(11):891–899

23. Roos J, Eastin I, Matsuguma H (2005) Market segmentation and analysis of Japan's residential post and beam construction market. For Prod J 55(4):24–30

24. Huddleston P, Good L, Stoel L (2001) Consumer ethnocentrism, product necessity and Polish consumers' perceptions of quality. Int J Retail Distrib Manage 29(5):236–246

25. Supphellen M, Gronhaug K (2003) Building foreign brand personalities in Russia: the moderating effect of consumer ethnocentrism. Int J Advert 22(2):203–226

26. Herche J (1992) A note on the predictive validity of the CETSCALE. J Acad Market Sci 20(3):261–264

27. Witkowski T, Beach L (1998) Consumer ethnocentrism in two emerging markets: determinants and predictive validity. Adv Consum Res 25:258–263

28. Sharma SH, Shimp TA, Shin J (1995) Consumer ethnocentrism: a test of antecedents and moderators. J Acad Market Sci 23(1):26–37

29. Batson CD, Chang J, Orr R, Rowland J (2002) Empathy, attitudes, and action: Can feeling for a member of a stigmatized group motivate one to help the group? Pers Soc Psychol Bull 28:1656–1666. https://doi.org/10.1177/014616702237647

30. Ahmed Z, Johnson J, Yang X, Fatt C, Teng H, Boon L (2004) Does country of origin matter for low-involvement products? Int Mark Rev 21(1):102–120

31. Lantz G, Loeb S (1996) Country of origin and ethnocentrism: an analysis of Canadian and American preferences using social identity theory. Adv Consum Res 23:374–378

32. Zaki J (2019) The war for kindness: building empathy in a fractured world. Crown

33. Myers MW, Hodges SD (2013) Empathy: perspective taking and prosocial behavior: Caring for others like we care for the self

34. Batson CD, Polycarpou MP, Harmon-Jones E, Imhoff HJ, Mitchener EC, Bednar LL, Klein TT, Highberger L (1997) Empathy and attitudes: can feeling for a member of a stigmatized group improve feelings toward the group? J Pers Soc Psychol 72(1):105

35. Grewal D, Monroe KB, Krishnan R (1998) The effects of price-comparison advertising on buyers' perceptions of acquisition value, transaction value, and behavioral intentions. J Mark 62(2):46–59

36. Darling JR, Arnold DR (1988) The competitive position abroad of products and marketing practices of the United States, Japan, and selected European countries. J Consum Market

37. Klein JG, Ettenson R, Morris MD (1998) The animosity model of foreign product purchase: an empirical test in the People's Republic of China. J Mark 62(1):89–100

38. Granzin KL, Olsen JE (1995) Support for buy American campaigns: an empirical investigation based on a prosocial framework. J Int Consum Mark 8(1):43–70

39. Frick V, Matthies E, Thøgersen J, Santarius T (2021) Do online environments promote sufficiency or overconsumption? Online advertisement and social media effects on clothing, digital devices, and air travel consumption. J Consum Behav 20(2):288–308

40. Kline RB (2015) Principles and practice of structural equation modeling. Guilford Publications

41. Hair JF, Celsi M, Ortinau DJ, Bush RP (2010) Essentials of marketing research, vol 2. McGraw-Hill/Irwin, New York, NY

42. Al Ganideh SF, Hamam MZ (2020) Is there a new "Facebook revolution" in the Arab World? Exploring young Jordanians' e-purchasing behavior. J Competitiveness Stud 28(1):56–70

43. Yaseen SG, Al-Janaydab S, Alc NA (2018) Leadership styles, absorptive capacity and firm's innovation. Int J Knowl Manage (IJKM) 14(3):82–100

44. Al Ganideh S, Yaseen S (2016) Arabia versus Persia: is this what the Arab Spring ended with? J Comparative Int Manage 19(1):1–18

45. Dajani D (2016) Using the unified theory of acceptance and use of technology to explain e-commerce acceptance by Jordanian travel agencies. J Comparative Int Manage 19(1):99–118

46. Al Ganideh SF, Jackson H, Marr NE (2007) An investigation into consumer ethnocentrism amongst young Jordanians. World J Retail Bus Manage 2007(2):40–51
47. Al Ganideh SF, Awudu I (2020) Arab-Muslim Americans' personality riddle and consumer ethnocentrism. J Glob Mark 34(2020):1–22
48. Ali N, Al Ganideh SF (2020) Syrian refugees in Jordan: burden or boon. Res World Econ 11(1):1–15

# The Determinants of Community Involvement Information Disclosure on Social Media by Malaysian PLCs

**Kar Seong Eng, Lian Kee Phua, and Char-Lee Lok**

**Abstract**  Today, social media has changed the mode of communication, and the way information is disseminated, bringing significant impact to individuals and organisations. Hence, the present study attempts to develop a model in line with Stakeholder Theory to examine stakeholder management as determinants of community involvement information disclosure on social media by Malaysian Public Listed Companies. The study's hypotheses are then tested using regression analyses on 114 Malaysian Public Listed Companies that owned official social media. The findings showed that companies' reputation and industry sensitivity have a positive relationship with the community involvement information disclosure on social media. Besides, companies under the Star stage of business life cycle also reported marginally significant result. The study attempts to contribute to voluntary disclosure literature by examining factors that motivate companies to engage with social media. This study also provides information to regulators interested in monitoring the corporate social responsibility disclosure on social media.

**Keywords**  Social media · Community involvement information · Voluntary disclosure · Malaysia

## 1 Introduction

Today, social media seems to have taken over the world. This statement is not hyperbole as one of the largest social media networking sites, Facebook, has more than 750 million active users [1]. However, many companies did not follow the crowd.

K. S. Eng (✉) · L. K. Phua · C.-L. Lok
School of Management, Universiti Sains Malaysia, 11800, Gelugor, Penang, Malaysia
e-mail: engks_0210@hotmail.com; kar.seong.eng@student.usm.my

L. K. Phua
e-mail: phualk@usm.my

C.-L. Lok
e-mail: lokcl@usm.my

© The Author(s), under exclusive license to Springer Nature Switzerland AG 2022    343
S. G. Yaseen (ed.), *Digital Economy, Business Analytics, and Big Data Analytics Applications*, Studies in Computational Intelligence 1010,
https://doi.org/10.1007/978-3-031-05258-3_28

Instead, they adopted a wait-and-see approach [2]. The first wave of companies started to join social media because of its significant benefits and impact [3]. Some of these companies have used social media as the medium to disseminate their corporate responsibility practices and activities. Corporate social responsibility (CSR) disclosure is the disclosure of information regarding the social responsibility activities that have been engaged by the companies [4]. Community Involvement is one of the most important categories of corporate social responsibility.

Through voluntary disclosure of CSR information on social media, the organisations can have two-way communications with the stakeholders. Unlike the traditional medium such as annual reports or websites, there is only one-way communication as the stakeholders cannot provide timely feedback to the organisations [5]. In addition, social media is a powerful and low-cost platform for organisations to globalise as social media is reachable to all people around the world. In contrast, the traditional medium is only reachable to a specific group of people [2].

Many Malaysian companies have started to use social media in disclosing their CSR activities. Companies like AMMB Holdings Berhad, Media Prima Berhad and Oriental Food Industries Holding Berhad disclosed their corporate social activities in helping the victims of the flood in Kelantan in 2014. These companies disclose their CSR information in social media because social media connects everyone in the world, and they can upload photos in social media, which is more attractive and appealing to the users and stakeholders [6]. During the Covid-19 pandemic, the use of social media has even increased significantly by all the parties, including organisations [7].

Although evidence suggests that many companies remain unconvinced as they are unclear of how the social media could be beneficial, our preliminary search on Malaysian listed companies revealed that more than 100 companies had established an official link to social media in their companies' websites. Among these early adopters, some have attempted to disseminate CSR information via their social media whereas the rest chose not to. Prior literature has identified various factors that motivated companies to disclose SCR in the traditional medium. Given the additional features provided by social media, such as fast and easy to reach out, interactive and low cost, it is interesting to explore what factors influence the early adopter companies to disclose CSR information on social media [5]. This study examines the relationships between stakeholder management and community involvement information disclosure on social media among the Malaysian public listed companies. The findings of this study can help investors, regulators and organisations to understand why most Malaysian companies are not using official social media to disclose corporate social responsibility. To date, corporate social responsibility disclosure is not mandatory in Malaysia. Thus, this study can provide evidence to the Malaysian government or Bursa Malaysia to make appropriate decision whether to make corporate social responsibility disclosure on social media mandatory. In addition, this study can provide insight for the policy makers to decide if there is a need to set the regulations on corporate social responsibility disclosure on social media.

## 2    Literature Review

Stakeholders are groups of individuals who are likely to be affected directly by the organisations' decisions [8]. Stakeholder theory differs from classical theory as it focuses not only on shareholders but also on different stakeholders [9]. Stakeholder theory considers all the impact of various stakeholders on corporate disclosure policies. It explains that the larger organisations have more effects and impacts on the community, which means they have a bigger group of stakeholders who can influence the organisations.

Over the last few decades, companies involved in corporate social responsibility have shared their corporate social responsibility endeavours through their corporate websites [1]. Anwar and Malik [1] illustrated that numerous companies had dedicated sections for corporate social responsibility on their website. However, it is argued that these sections are ineffective due to the lack of two-way communication.

In the Malaysia context, Thompson and Zakaria [10] found that about 81% of 257 companies selected for the year 2000 disclosed their CSR information in the annual reports. They found that most of the Malaysian organisations disclosed the CSR on the employees and human resources perspective (40%). Whereas for products and consumers is 24%, community involvement is 22% and the environment is only 16% [11].

Abdul Hamid and Atan [12] examined the level of CSR disclosure of the top three Malaysian Telecommunications Companies, namely Telekom Malaysia, Maxis Communication and Digi Communication, based on annual reports from the year of 2002 to 2005. A checklist is used to categorise CSR into four themes, i.e. environment and energy, human resource, product or services and community. They concluded that the CSR disclosure has significantly increased from 2002 to 2005. The result showed that Telekom Malaysia, the Government Link Company (GLC), focuses on the disclosure of community and human resources.

The empirical studies on CSR disclosure in Malaysia [10, 13–15] showed that even though the level of CSR activities is low among the Malaysian listed companies, the interest on CSR disclosure is increasing in Malaysia over time. The majority of the studies on CSR disclosure used content analysis.

## 3    Hypotheses Development

### 3.1    Profitability

Profitability is one of the factors that has been studied for corporate social responsibility disclosure by the researchers [16]. This is because companies' profitability indicates the efficiency and effectiveness of corporate management. Profitable companies have more incentives to disclose more information to their stakeholders to raise capital on the best available terms [17]. The investors are the group of stakeholders concerning the company's profitability because they have interest in investing in the company. Previous studies based on stakeholder theory suggest a positive relationship between the social responsibility disclosure and companies' profitability [18]. But there are mixed results for different studies by the researchers. Ameer and Othman [19] explained that there is a two-way relationship, positive and negative relationships, between the corporate social responsibility disclosure and companies' profitability. To determine the relationship between profitability from the investors' perspective and community involvement information disclosure on social media, the following hypothesis is tested.

H1: There is a positive relationship between profitability and the community involvement information disclosure on social media.

## 3.2   Ownership Concentration

When the organisations do not have a concentrated ownership structure, they have widely dispersed shareholders and investors. In other words, the organisations have to deal with more shareholders, and the absence of disclosure will increase the information asymmetry between the organisation and its many shareholders [20]. Whereas when the organisations have a concentrated ownership structure, they will be less motivated to disclose additional information about corporate social responsibility because shareholders can obtain the information directly from the company. Hence, when the organisations have less concentrated ownership, they will try to disclose the additional information, including corporate social activities, because they want to spread the information to more people [21]. Shareholders are the group of stakeholders concerned about ownership concentration because they are the board of the company. To determine the relationship between the ownership concentration from shareholders' perspective and community involvement information disclosure on social media, the following hypothesis is tested.

H2: There is a negative relationship between ownership concentration and the community involvement information disclosure on social media.

## 3.3   Companies' Reputation

The corporate social responsibility disclosure builds and maintains a positive company brand and image for the stakeholders [18]. This is because the perception of the public on the companies' reputation may be influenced by the corporate social responsibility disclosure. Social activities are important, but stakeholders need to be informed of what the organisations have done. The organisations are trying to disseminate the information about the social activities they have accomplished to more people and stakeholders [22]. Hence, social media is a good platform to disclose such information as it is reachable to the people around worldwide. Customers are the group of stakeholders that are concerning about the companies' reputation because they tend to buy products from reputable companies. Rufino and Machado [18] found that organisations with good reputations disclose more corporate social responsibility information than organisations that do not have a good reputation. As disclosing corporate social responsibility on social media can be reachable by users or stakeholders worldwide, it can improve and maintain the companies' reputation. To determine the relationship between the companies' reputation that are from customers' perspective and community involvement information disclosure on social media, the following hypothesis is tested.

H3: There is a positive relationship between the companies' reputation and the community involvement information disclosure on social media.

## 3.4 Leverage

Stakeholder theory explains the direct effect and power that stakeholders have on the disclosure of an organisation's strategy. The debtors are the group of stakeholders that are concerning the leverage ratio of a company because they would like to know if the company can pay back the debt. Brammer and Pavelin [20] argued that a company with a low level of leverage would have less pressure in corporate social responsibility because they will focus more on increasing the financial return. Hence, a negative association was found between the leverage ratio and the corporate social responsibility disclosure [23]. However, there is no significant evidence found to support that the leverage has a relationship with the corporate social responsibility disclosure [24, 25]. The hypothesis is:

H4: There is a negative relationship between the leverage ratio and the community involvement information disclosure on social media.

## 3.5 Company's Age

As an organisation matures, its reputation already been established and has a lot of involvement in corporate social responsibility activities. Hence, the stakeholders expect the organisations to continue disclosing and even disclosing more information on corporate social responsibility [26]. The stakeholders may react to any radical changes in strategies on disclosing the corporate social responsibility information. Suppliers are the group of stakeholders concerned about the company's age as they would be safer to provide their services and products to older companies due to the lower risk. Most of the previous studies concluded that the older companies would disclose more information [27–29]. Older companies tend to disclose more information to a wider group of stakeholders; hence, social media is a good platform for companies to disclose their corporate social responsibility information. Therefore, the following hypothesis is tested.

H5: There is a positive relationship between the age of a company and the community involvement information disclosure on social media.

## 3.6 Degree of Internationalisation

From the perspective of stakeholder theory, multinational firms have more diverse stakeholders and hence they need to fulfil more needs from the diverse stakeholders [30]. Firms that expand internationally need to consider the interest and expectations of wider stakeholders [31]. According to Feldman et al. [32], the firm's perceived risks are reduced significantly when the firm adopts the environmentally proactive posture. In addition, Brammer et al. [33] argued that the continuous involvement of stakeholders is being affected by socially responsible firms. Government and local authorities are concerned about the degree of internationalisation of the company as it may be part of government strategy or law enforcement. Previous studies found that internationalisation has a significant positive relationship with the corporate social responsibility rating [30, 33, 34]. To determine the relationship between the degree of internationalisation from government and local authorities' perspective and community involvement information disclosure on social media, the following hypothesis is tested.

H6: There is a positive relationship between the degree of internationalisation and the community involvement information disclosure on social media.

## 3.7 Business Life Cycle

The business life cycle has not been studied or has limited studies for corporate social responsibility disclosure. However, a similar determinant, the business model, has been studied in the academic field in the past decades [35]. For the relationship between business model and corporate social responsibility, some researchers suggest that some types of business models can foster stronger corporate social responsibility involvement and the types of business models can also increase corporate social responsibility disclosure [36]. Elkington [36] suggests that business models are the determinants of organisational behaviours, and they influence corporate social activities and disclosure. The idea of the business life cycle can be used to reconcile and conform to the divergent and different needs and demands of stakeholders in terms of the sustainable growth of the organisations. This is aligned with stakeholder theory, in which the organisations use a different business model in the different life cycles to address the needs and expectations of different stakeholders. Competitors are the group of stakeholders concerned on business life cycle as they want to position their company's business life cycle with their competitors to identify the strategy to be implemented. Whether or not to disclose the corporate social responsibility on social media depends on the business life cycle. To determine the relationship between the business life cycle from competitors' perspective and community involvement information disclosure on social media, the following hypothesis is tested.

H7: The community involvement information disclosure on social media is related to the business life cycle.

## 4 Methodology

### 4.1 Population and Sample

This study uses data of Malaysian Public Listed Companies for the financial year 2016 and 2017. The data were manually collected based on availability of the latest data during the data collection period. The population of the study is defined as all companies listed on the Main Board of Bursa Malaysia that have official social media. To determine the total population, the website link available at Bursa Malaysia for each Malaysian public listed company was accessed. The link to social media is then identified from the companies' website. Companies with no website link at Bursa Malaysia or do not have a link to the social media in the website are categorised as do not have official social media. As a result, only 114 companies have established official social media, and they are defined as the unit of analysis of this study.

The entire population is used for this study. According to Roscoe [37], in multi-variate research that includes multiple regression analyses, the sample size should be ten times or more than the number of variables in the study. To fulfill this requirement, the entire population is examined in this study.

### 4.2 Measurement of Corporate Social Responsibility Disclosure on Social Media

Content analysis is widely used to identify the corporate social responsibility disclosure in the annual reports of Malaysian companies as documented in several studies [38, 39]. According to Abbott and Monsen [40], content analysis translates qualitative information into quantitative scale.

After a meticulous review, a list of Malaysian community involvement information disclosure on social media measurement has been identified. An index measurement of 15 specific community involvement information disclosure items on social

**Table 1** Community involvement information disclosure on social media measurement

| Community involvement information (15 items) | 1. Donations to the charity, arts, sports, etc.<br>2. Relation with local population<br>3. Sponsoring educational seminars and conferences<br>4. Transportation for the employees' children<br>5. Establishment of educational institution(s)<br>6. Medical establishment<br>7. Corporate gifts<br>8. Public Hall and/or auditorium<br>9. Sponsoring education and scholarship for students<br>10. Providing job opportunities and helping in reducing the unemployment rate<br>11. Contribution toward community serving programs<br>12. Conducting projects in poor areas<br>13. Cash rewards<br>14. Financial assistance<br>15. Participating and financing community celebration |
| --- | --- |

media is adapted from Khasharmeh and Desoky [41]. The disclosure index is adapted because Khasharmeh and Desoky [41] investigated the online CSR disclosure, which is similar to the CSR disclosure on social media, as both are also internet-based. The checklist developed for the use of content analysis is shown in Table 1.

The dependent variable is community involvement information disclosure on social media by using the content analysis based on the number of sentences that contains CSR information. The community involvement information disclosure on social media was obtained until 31st December 2017. The use of 2017 data is because the data are manually collected based on the latest year during the data collection period. For companies that disclose any CSR information from any item on social media, a score of 1 will be given. Otherwise, 0 will be given to the company that does not disclose any information about the corporate social activities on social media.

A disclosure index is used to measure the extent of community involvement information disclosure on social media based on a total of 15 pre-determined items in the checklist. Each disclosure item is assigned with equal weight. This is consistent with the previous studies that the scoring method was based on an unweighted approach in measuring the extent of CSR information [15, 42]. This implies that all items are equal in importance. The unweighted approach has become the norm in disclosure studies because it reduces subjectivity [41, 43]. All the items included in the checklist have been used at least once in previous research and this increases the reliability of the index [41]. The reporting items were assessed based on the binary coding system, with "1" indicative of the presence of the reporting item, and "0" otherwise, so the maximum possible score is 15. Then, the extent of community involvement information disclosure on social media is scored by using the score of voluntary reporting items on social media divided by the number of items expected, which is 15.

## 4.3 Measurement of Independent Variables and Control Variables

The research model includes seven independent variables, i.e. profitability, ownership concentration, companies' reputation, leverage, company's age, degree of internationalisation and business life cycle, and two control variables, namely firm size and industry sensitivity. The measurements of these variables are presented in Table 2.

**Table 2** Measurement of independent and control variables

| Variables | Measurement | Sources |
|---|---|---|
| Profitability | Return on Assets (ROA) = Net Income/Total Assets | [18] |
| Ownership concentration | The percentage of ordinary shares held by a shareholding of 5% or more | [21] |
| Companies' reputation | A Dummy Variable, 1 for companies that are top 100 ranking in GradMalaysia, 0 otherwise | [18] |
| Leverage | Debt/Equity | [25] |
| Company's age | Number of years listed in Bursa Malaysia | [26] |
| Degree of internationalisation | Foreign Sales/Sales (FS/S) | [30] |
| Business life cycle | A Dummy Variable, 4 for companies that fall under Dogs 3 for companies that fall under Question Marks 2 for companies that fall under Cash Cows 1 for companies that fall under Stars | [35] |
| Firm size | Natural log of the book value of total assets Natural log of total sales | [18] |
| Industry sensitivity | A Dummy Variable, 1 for companies that are in sensitive industry which have involved in mining, chemicals, oil and gas, electricity generation, paper, pulp and metal, 0 for companies that are in insensitive industry which are involved in mining, chemicals, oil and gas, electricity generation, paper, electricity generation, paper, pulp and metal | [44] |

## 4.4   Multiple Regression

The regression equation is as follows:

$$
\begin{aligned}
\text{CSR\_ COM} = {} & \alpha + \beta_1 \text{PROFIT} + \beta_2 \text{CONCENTRATION} \\
& + \beta_3 \text{REPUTATION} + \beta_4 \text{LEVERAGE} + \beta_5 \text{AGE} \\
& + \beta_6 \text{INTERNATIONALISATION} \\
& + \beta_7 \text{BLC} + \text{Controls} + \varepsilon
\end{aligned} \tag{1}
$$

where

| | |
|---|---|
| CSR_COM | Extent of community involvement information disclosure on social media |
| PROFIT | Profitability |
| CONCENTRATION | Ownership Concentration |
| REPUTATION | Companies' Reputation |
| LEVERAGE | Leverage |
| AGE | Company's Age |
| INTERNATIONALISATION | Degree of Internationalisation |
| BLC | Business Life Cycle |
| Controls | Control Variables (i.e. Firm Size and Industry Sensitivity) |
| $\varepsilon$ | Error Term. |

## 5   Data Analysis and Results

### 5.1   Descriptive Statistics

The descriptive statistics reveal a few important observations. The average value was low and the standard deviation was not widely dispersed. This suggests that CSR disclosure on social media is still considered in its infancy stage. Community involvement information disclosure on social media has a mean score of 0.3175. There are companies that did not disclose any community involvement information disclosure on social media and resulted in the minimum value of 0. The maximum value could be seen on community involvement disclosure on social media, which is 0.7333. The standard deviation score is 0.1897. For the companies' reputation, it could be seen that it has a low mean score which was 0.22. It shows that most Malaysian public listed companies that have official social media did not fall under the top 100 reputable companies.

The findings indicate that most Malaysian companies have no official social media. Out of the 806 Malaysian public listed companies, only 114 Malaysian public listed companies had official social media. Most of the companies were using Facebook

**Table 3** Descriptive statistics analysis

| Variable | N | Minimum | Maximum | Mean | S.D. |
|---|---|---|---|---|---|
| CSR disclosure on social media—community involvement | 114 | 0.0000 | 0.7333 | 0.3175 | 0.1897 |
| Profitability | 114 | −0.3550 | 0.3113 | 0.0427 | 0.0851 |
| Ownership concentration | 114 | 0.0000 | 94.5200 | 50.5407 | 17.0572 |
| Companies reputation | 114 | 0 | 1 | 0.2200 | 0.4160 |
| Leverage | 114 | 0.0000 | 6.0000 | 0.6500 | 0.9920 |
| Companies age | 114 | 1 | 54 | 19.2600 | 11.8220 |
| Degree of internationalisation | 114 | 0.0000 | 100.0000 | 11.4455 | 23.2895 |
| Business life cycle_star | 114 | 0 | 1 | 0.0900 | 0.2840 |
| Business life cycle_cash cow | 114 | 0 | 1 | 0.0200 | 0.1320 |
| Business life cycle_question mark | 114 | 0 | 1 | 0.6000 | 0.4930 |
| Business life cycle_dog | 114 | 0 | 1 | 0.3000 | 0.4600 |
| Firm size | 114 | 28,272 | 485,379,581 | 14,318,263 | 55,696,690 |
| Industry sensitivity | 114 | 0 | 1 | 0.1400 | 0.3490 |

as their official social media. Out of the 114 Malaysian public listed companies, 105 companies had an official Facebook page, while 72 companies had an official Twitter account. Besides that, 45 companies had an official Instagram account, while 28 companies had an official LinkedIn account. Descriptive statistics analysis is presented in Table 3.

## 5.2 Empirical Results

The Variance Inflation Factor (VIF) has been used to test multicollinearity before multiple regression analysis is performed. In this study, the VIF value for all the variables was less than 2, which was in the range of 1.076–1.930. Hence, it indicated the absence of multicollinearity problems.

Multiple regression analysis is performed to examine the research model. The results indicate that Companies' Reputation showed a positive relationship with the extent of community involvement information disclosure on social media at the 5% significant level, which is 0.47. In addition, Star under Business Life Cycle Category is also found to have a marginal positive impact on the disclosure of community involvement information on social media at 10% significant level, which is 0.084. The results suggest that hypothesis 3 is supported, whereas hypotheses 1, 2, 4, 5, 6, 7 were rejected. The results support the view that the more reputable companies will disclose more community involvement information on social media. In addition,

**Table 4** Results of regression analysis

| Variable | DV: community involvement information (Coeff.) |
|---|---|
| Profitability | 0.322 |
| Ownership concentration | 0.200 |
| Companies' reputation | **0.470**\*\* |
| Leverage | 0.303 |
| Company's age | 0.460 |
| Degree of internationalisation | 0.211 |
| Business life cycle_star | **0.084**\* |
| Business life cycle_cash cow | 0.120 |
| Business life cycle_dog | 0.420 |
| Firm size | 0.131 |
| Industry sensitivity | **0.014**\*\* |
| Number of observations | 114 |
| Adjusted R-square | 0.156 |
| R-square | 0.238 |

\* Significant at the 10% level, \*\* significant at the 5% level, \*\*\* significant at the 1% level

industry sensitivity has a positive relationship with community involvement information disclosure on social media, which is 0.014. The results also provide further evidence to support the more sensitive the industry, the more disclosure on community involvement information on social media. The results of regression analysis are presented in Table 4.

The application of the stakeholder theory in explaining the factors influencing the disclosure of corporate social responsibility disclosure through social media by Malaysian public listed companies is not as strong as what has been anticipated. This could be due to social media is still not a popular platform for Malaysian companies as there are some costs incurred, being monetary or non-monetary, for owning the official social media.

Companies' reputation was found to be significantly and positively related to corporate social responsibility disclosure through social media. The findings of this study suggest that the more reputable the company is, the more corporate social responsibility disclosure. It indicates that the customers are the key stakeholder for the organisations to motivate a company to disclose the Community Involvement Information on social media. This is consistent with the previous study by Rufino and Machado [18]. Rufino and Machado [18] found that companies with the best reputation disclose more corporate social responsibility information than companies that do not have a good reputation.

The result of this study demonstrated that factors influencing the disclosure of corporate social responsibility through social media by Malaysian public listed companies were not strongly supported. This could be due to the low extent of corporate social responsibility disclosure on social media.

## 6   Conclusion

This study was conducted by identifying significant factors from previous studies and then examining their significance based on Stakeholder Theory. A multiple regression model was then composed, and the variables were tested.

The findings showed that companies' reputation and industry sensitivity have a significant positive relationship with the community involvement information disclosure on social media. Besides, companies under the Star stage of business life cycle also reported marginally significant result. The findings suggest that customers are the primary group of stakeholders that influencing the community involvement information disclosure on social media by Malaysian public listed companies.

Investors who are concerning on the profitability, shareholders who are concerning on ownership concentration, debtors who are concerning on leverage, suppliers who are concerning on company's age, government and local authorities who are concerning on degree of internationalisation and competitors who are concerning on business life cycle, are not having significant relationship with community involvement information disclosure.

There are several limitations to this study. First, this study only investigates Malaysian companies; hence, no comparison was made with other countries, especially with other Asian countries. Second, this study only studies Malaysian public listed companies; thus, there is no comparison made with the private companies, which Malaysian private companies could be actively using official social media.

Future research in the area of study can be extended in several directions. First, future research can use a time-series study to examine the corporate social responsibility disclosure trend on social media. In addition, an investigation for Asian countries can be made to compare with the Malaysian public listed companies. Furthermore, an investigation for Malaysian private companies can be compared with the Malaysian public listed companies. A case study can also be conducted to investigate in detail the extent of corporate social responsibility disclosure through social media. Lastly, future research can also investigate the official social media owned by the organisation's CEO as the descriptive analysis. For example, the CEO of AirAsia Group, Tony Fernandes, discloses many companies' information on his official personal social media.

## References

1. Anwar R, Malik JA (2020) When does corporate social responsibility disclosure affect investment efficiency? A new answer to an old question. SAGE Open 10(2):2158244020931121

2. Smart Insights (2021) Global social media research summary 2021. https://www.smartinsi ghts.com/social-media-marketing/social-media-strategy/new-global-social-media-research. Accessed 13 Aug 2021

3. Ellison NB, Steinfield C, Lampe C (2007) The benefits of Facebook "friends:" Social capital and college students' use of online social network sites. J Comput-Mediat Commun 12(4):1143–1168

4. Tan A, Benni D, Liani W (2016) Determinants of corporate social responsibility disclosure and investor reaction. Int J Econ Financ Issues 6:11–17

5. Garanina T, Aray Y (2021) Enhancing CSR disclosure through foreign ownership, foreign board members, and cross-listing: does it work in Russian context? Emerg Mark Rev 46:1–36

6. Said R, Zainuddin YH, Haron H (2009) The relationship between corporate social responsibility disclosure and corporate governance characteristics in Malaysian public listed companies. Soc Responsib J 5(2):212–226

7. Wong A, Ho S, Olusanya O et al (2021) The use of social media and online communications in times of pandemic COVID-19. J Intensive Care Soc 22(3):255–260

8. Sethi SP (1977) Dimensions of corporate social performance: an analytical framework. Calif Manage Rev 17(3):58–64

9. Langrafe TF, Barakat SR, Stocker F et al (2020) A stakeholder theory approach to creating value in higher education institutions. The Bottom Line 33(4):297–313

10. Thompson P, Zakaria Z (2004) Corporate social responsibility reporting in Malaysia: progress and prospects. J Corp Citizsh 13:125–136

11. Statista (2020) Number of social network users worldwide from 2010 to 2021 (in Billions). https://www.statista.com/statistics/278414/number-of-worldwide-social-network-users. Accessed 1 June 2021

12. Abdul Hamid FZ, Atan R (2011) Corporate social responsibility by the Malaysian telecommu-nication firms. Int J Bus Soc Sci 2(5):198–208

13. Abdullah SN, Mohamad NR, Mokhtar MZ (2011) Board independence, ownership and CSR of Malaysian large firms. Corp Ownersh Control 8(3):417–435

14. Amran A, Devi SS (2008) The impact of government and foreign affiliate influence on corporate social reporting: the case of Malaysia. Manag Audit J 23(4):386–404

15. Esa E, Mohd Ghazali NA (2012) Corporate social responsibility and corporate governance in Malaysian government-linked companies. Corp Gov 12(3):292–305

16. Naser K, Hassen Y (2013) Determinants of corporate social responsibility reporting: evidence from an emerging economy. J Contemporary Issues Bus Res 2(3):56–74

17. Juhmani O (2014) Determinants of corporate social and environmental disclosure on websites: the case of Bahrain. Universal J Account Finance 2(4):77–87

18. Rufino MA, Machado MR (2015) Determinants of voluntary social information disclosure: empirical evidence in Brazil. J Educ Res Account 9(4):367–383

19. Ameer R, Othman R (2012) Sustainability practices and corporate financial performance: a study based on the top global corporations. J Bus Ethics 108(1):61–79

20. Brammer SJ, Pavelin S (2008) Factors influencing the quality of corporate environmental disclosure. Bus Strateg Environ 17(2):120–136

21. Alqirem R, Abu Afifa M, Saleh I et al (2020) Ownership structure, earnings manipulation, and organisational performance: the case of Jordanian insurance organisations. J Asian Finance Econ Bus 7(12):293–308

22. Ahmad ZA (2016) Communicating CSR in the digital age: an exploratory study of a CSR award winning company in Malaysia. J Educ Soc Sci 4:252–257

23. Belkaoui A, Karpik PG (1989) Determinants of the corporate decision to disclose social information. Account Audit Accountability 2(1):36–51

24. Jennifer HL, Taylor ME (2007) An empirical analysis of triple bottom-line reporting and its determinants: evidence from United States and Japan. J Int Financ Manag Acc 18(2):123–150

25. Wang K, Sewon O, Claiborne MC (2008) Determinants and consequences of voluntary disclosure in an emerging market: evidence from China. J Int Account Audit Tax 17(1):14–30

26. Roberts RW (1992) Determinants of corporate social responsibility disclosure: an application of stakeholder theory. Account Organ Soc 17(6):595–612

27. Hossain M, Hammami H (2009) Voluntary disclosure in the annual reports of an emerging country: the case of Qatar. Adv Account Incorporating Adv Int Account 25:255–265

28. Prencipe A (2004) Proprietary costs and determinants of voluntary segment disclosures: evidence from Italian listed companies. Eur Account Rev 13(2):319–340

29. White G, Lee A, Tower G (2007) Drivers of voluntary intellectual capital disclosure in listed biotechnology companies. J Intellect Cap 8(3):517–537

30. Attig N, Boubakri N, El Ghoul S et al (2016) Firm internationalisation and corporate social responsibility. J Bus Ethics 134(2):171–197

31. Sanders GWM, Carpenter MA (1998) Internationalisation and firm governance: the roles of CEO compensation, top team composition, and board structure. Acad Manag J 41:158–178

32. Feldman SJ, Soyka PA, Ameer P (1997) Does improving a firm's environmental management system and environmental performance result in a higher stock price? J Invest 6:87–97

33. Brammer SJ, Pavelin S, Porter LA (2009) Corporate charitable giving, multinational companies and countries of concern. J Manage Stud 46:575–596

34. Kang J (2013) The relationship between corporate diversification and corporate social performance. Strateg Manag J 34:94–109

35. Witek-Hajduk MK, Zaborek P (2016) Does business model affect CSR involvement? A survey of polish manufacturing and service companies. Sustainability (Switzerland) 8(93):1–20

36. Elkington J (2004) Enter the triple bottom line. In the triple bottom line, does it all add up? Assess Sustain Bus CSR 1:1–16

37. Roscoe JT (1975) Fundamental research statistics for the behavioral sciences. Educ Psychol Measur 30(2):499–502

38. Adam NC (2011) Using preparation, proprietary and information costs in explaining non-reporting of social and environmental information by Malaysian companies. J Manage 36(2012):101–112

39. Alarussi AS, Jamefah MM, Selamat MH (2009) Internet financial and environmental disclosures by Malaysian companies. Issues Soc Environ Account 3(1):3–25

40. Abbott WF, Monsen RJ (1979) On the measurement of corporate social responsibility: self-reported disclosures as a method of measuring corporate social involvement. Acad Manag J 22(3):501–515

41. Khasharmeh HA, Desoky AM (2013) Online corporate social responsibility disclosures: the case of the Gulf Cooperation Council (GCC) countries. Global Rev Account Finance 4(2):39–64

42. Haji AA (2013) Corporate social responsibility disclosures over time: evidence from Malaysia. Manag Audit J 28(7):647–676

43. Ahmed K, Courtis J (1999) Associations between corporate characteristics and disclosure levels in annual reports: a meta-analysis. Br Account Rev 31(1):35–61

44. Kurniawan FXT, Pangesti C (2011) Parameter determinant of business who implement social responsibility. Econ J 16(1):107–118

# Cultural Value Orientations Among Managers of Travel Agencies

Dima Dajani, Saad G. Yaseen, and Samar Naseem Alqirem

**Abstract** This study aims to examine cultural value orientations on the work orientation of marketing managers of Jordanian travel agencies. Cultural Orientations are investigated depending on the theoretical framework proposed by Kluckhohn and Stodtbeck's (1961) model. Descriptive analysis and structural equation modeling approach are used to test the effect of cultural value orientations on work orientation. The study reveals that travel agency managers are characterized by some traditional values such as subjugation, individualism, future time orientation, being work orientation, and changeable human nature. Furthermore, managers are future-oriented, subjugated, and individualistic. They have a high tendency towards "being orientation" and changeable human nature. The study is one of the first to examine the influence of culture in the field of the hospitality industry in developing countries. It necessitates more studies to compare cultural values between Arab and western managers. However, the cross-sectional nature of the data is incapable of confirming the causal inference empirically.

**Keywords** Culture · Cultural value orientations · Jordanian travel agencies · Social values

## 1 Introduction

Culture has a great impact on how managers accomplish their duties and make decisions in organizations. Many researchers e.g., [1–5] have examined and confirmed the

---

**Fields of Specialization**: Culture and Organizational Behavior.

---

D. Dajani (✉) · S. G. Yaseen · S. N. Alqirem
Faculty of Business, Al-Zaytoonah University of Jordan, Amman, Jordan
e-mail: d.aldajani@zuj.edu.jo

S. G. Yaseen
e-mail: saad.yaseen@zuj.edu.jo

S. N. Alqirem
e-mail: sec_hospitality@zuj.edu.jo

vital role and effect of culture on management behavior. The globalized economy and IT innovation stress the importance of investigating culture and its effect on management behavior in different countries. People are no longer living in isolation and need to interact with each other to do business globally. Understanding people with different cultural values and orientations can be attained by analyzing the cultural values they have gained through their upbringing in their cultural context [6].

Hofstede [7, 8] argued that people possess various mental programs and belong to multiple subcultures in society at the same time. These subcultures may conflict with each other and control one's behavior. He also argued that the national culture influence workers' values and behavioral patterns. Hall [9] divided various cultures according to their ways of communication that are high context and low context. Trompenaars [10] identified several dimensions that differentiate cultures from each other which he defined as universalism versus particularism, collectivism versus individualism, affective versus neutral relationships, specificity versus diffuseness, achievement versus ascription, orientation towards time, and internal versus external control.

Furthermore, Schwartz [11] introduced a theory to explain the values by which cultures can be differentiated. He specified three cultural values: embeddedness versus autonomy, hierarchy versus egalitarianism, and mastery versus harmony. However, all of these theories have been developed for mapping and comparing national cultures. Consequently, it is vital to know that literature in cultural research is divided into two levels: (a) national culture and (b) organizational culture. The national culture includes the beliefs and values that humans acquire through their childhood and are very difficult to change. In contrast, the organizational culture includes practices and beliefs that can be easily altered [12]. Therefore, the present research attempts to explore how cultural value orientations can directly influence marketing work orientation.

Following [13] value orientations, this research examines cultural value orientations among marketing managers of the Jordanian travel agencies. The marketing department is primarily responsible for the business development of travel agencies. This includes product development, price setting, promotion and communication strategies, and distribution channels.

The past twenty years witnessed an improved amount of cross-cultural management research due to the importance of culture and its effect on management decisions globally. However, the majority of these studies were conducted in the developed nations, and some rising nations such as China, India, and parts of Eastern Europe [14, 15]. The impact of culture has been basic and national culture differences have been minimal, particularly in management and marketing literature [16]. Furthermore, only some studies have considered the effect of culture in the field of the tourism and hospitality industry [17–22]. Hence, there is concurrence that there is a necessity for a profound comprehension of the effect of cultural value orientations in developing countries and especially in the hospitality industry.

This research addresses the current gap in the literature by considering the effect of cultural value orientation on marketing work orientation in the tourism industry within the context of developing counties.

Tourism is a vital sector in the Jordanian economy and is considered to be the second-largest export industry. The generated revenues in this sector have exceeded the $5-billion mark in 2018, indicating an 8% improvement from the year 2017 [23]. Tourism accounts for approximately 13% of the country's GDP in 2019 and it is the largest producer of private sector employment in the country [24].

## 2 Literature Review

Cultural Value Orientations. It is implicit that culture is a multi-dimensional, complex, and fuzzy phenomenon to be researched [14, 25]. Culture has been referred to by many researchers as the way of doing things around us. It describes and shapes the behavior of human beings [26–28]. Schein [29] conceptualized culture as a way of solving issues and an adaptation to the environment. However, Hofstede [8] referred to culture as "the collective programming of the mind" and differentiated between human nature and personality. Human nature describes the features that human beings naturally share while personality gives a unique identity for each individual. He concluded that culture as a "collective programming of the mind" rests between human nature and personality. In addition, Solomon et al. [30] argued that culture provides the personality of a society. It includes both abstract ideas, such as values and ethics, and physical objects, such as clothing and food. It is a combination of shared ideas, traditions of an organization or society.

Culture can have a tremendous effect on organizational performance, efficiency, and effectiveness of employees [31]. It shapes the direction of work and how people think. Organizations are very much affected by the national culture where they exist [8, 32–34]. Ministry of Tourism and Antiquities [35] perceived culture as the essential influencer on the individuals' behavior, and each culture is composed of minor subcultures that offer more precise recognition and socialization for their constituents. Thus, researchers thrive to find a practical approach to study and examine specific characteristics of culture and analyze them. These characteristics are referred to as cultural dimensions or orientations [14, 36]. As a result, many conceptual frameworks tend to examine culture in different societies and organizations. These are based on the cultural dimensions of [8, 11, 13, 37, 38].

Hofstede [8, 12] introduced several culture dimensions that affected managerial values and human behavior in different cultures. His cultural dimensions were: (a) power distance that describes how people in different societies accept power relationships. (b) Individualism versus collectivism dimension describes the degree to which people are integrated into groups in different societies. (c) Uncertainty avoidance describes how people in different societies tolerate ambiguity. (d) Masculinity versus femininity dimension describes how certain societies and organizations focus on assertiveness and achievement by being more masculine than societies focusing on cooperation and caring. (e) Long term orientation versus short-term orientation describes how societies and organizations deal with time. Long-term societies tend to have a broader vision and sense of development whilst short-term-oriented societies tend to be less developed.

Trompenaars and Hampden-Turner [37, 39] presented a cultural model based on seven dimensions that differentiate societies from each other. These dimensions include universalism versus particularism; affective versus neutral relationships; specific versus diffuse; achievement versus ascription; attitudes towards time and attitudes towards the environment. The first five dimensions focus on the relationship among people and the last two focus on the relationship with time and environment.

Furthermore, Schwartz [11, 40] identified seven types of cultural values: (1) harmony that describes cultures that fits harmoniously into the environment. (2) Conservatism describes a traditional culture that respects social orders and traditions. (3) Hierarchy that describes a culture that accepts unequal distribution of power roles and resources. (4) Mastery that describes a daring and competitive culture. (5) Affective autonomy describes cultures that seek positive experiences and happiness. (6) Intellectual autonomy illustrates cultures that desire to pursue their ideas and being intellectual. (7) Egalitarianism describes cultures that seek equality, justice, and welfare for others.

In addition Hall [9, 41] proposed a model of culture that identified three dimensions; context (high–low), space (private–public), and time (monochromic–polychromic). The context dimension describes the degree to which the content of the message is vital as the message itself. While the space dimension describes the degree of comfort that people are willing to share their physical space with others. Finally, the time measurement indicates the degree to which people achieve multiple tasks at a time or one task at a time.

Kluckhohn and Strodtbeck Framework. A comprehensive cultural orientation framework from an epic perspective has been proposed by Kluckhohn and Strodtbeck [13]. They perceived culture as a unique human and social phenomenon and theorized three basic assumptions in developing their model. The first assumption stated that people encounter common problems for which they must-have solutions. The second assumption acknowledged that there are a variety of solutions to these problems. Finally, the third assumption affirmed that solutions were available in various societies but they were differently preferred. The diverse solutions that people choose to solve their problems reflected their own culture [13].

Based on these assumptions, Kluckhohn and Strodtbeck [13] developed an inclusive model for cultural analysis. They identified six dimensions of culture orientation: (1) the nature of people: which described what people think about others. The nature of human beings could be good, bad, or a combination of both. (2) The relationship with nature: this relationship can be expressed in terms of mastery, subjugation, or harmony. (3) Time orientation: it described the focus of people on past, present, or future. (4) Human activity: this dimension explained the mode of activity inside the society or organization. Some societies focus on living (being) others strive for goals (doing) and some are reflective (thinking). (5) Relationships among people: Individualism, Collectivism, and hierarchal structure. (6) Conception of physical space: this describes how space is treated in society. Societies could be private, public, or a mixture of both.

Kluckhohn and Strodtbeck [13] cultural orientation framework was selected because of its comprehensiveness and applicability in organizational settings [14, 42]. It has been recommended to assess the implication of culture and organizational practices [43]. The validity of these cultural orientations has been examined by several scholars [44, 45]. Moreover, Kluckhohn and Strodtbeck [13] work has formed the foundations of later researches in this field. Kluckhohn and Strodtbeck's framework provides Hofstede, Schwartz, and others present the etic approach to study cultural values from a psychological view. Hence, there are similarities between these orientations and models presented by Hofstede [8], Hall [9, 41], and Lo and Houkamau [46]. All of these cultural orientations are expected to be found in all societies.

# 3 Research Model and Hypotheses

Researchers have examined cultural value orientations in various settings [47–49]. Since culture forms beliefs about what is vital, helpful, and applicable in organizations, creates the rules of interactions that shape up the reactions of people among each other, and forms the distribution of power [50, 51], Jordanian culture should influence work orientation. Furthermore, the extensive literature stresses the effect of cultural value orientations in different contexts and settings [52–54]. Meanwhile, although researchers have addressed the importance of cultural value orientations, their contributions in general on the effect of various cultural value orientations on work activity orientation is very limited and directed to developed countries and other business contexts. Therefore, the adapted cultural value orientations framework of [13] consists of six major orientations: (1) Human nature: can be perceived as good, bad, or a combination of both. (2) Relationship to nature: this is expressed in terms of mastery, subjugation, or harmony. (3) Time orientation: the emphasis on the past, present, or future. (4) Human activity: it concentrates on being, doing, or thinking. (5) Relationships among people: having Individualistic, collective, or hierarchical structure. (6) Space: private, public, or mixture of both.

The nature of humans represents whether people are good or bad in their behavior in any setting. It also represents whether people change their behavior from good to evil or they do not change their behavior [13]. Whilst work activity orientation consists of three dimensions. The Doing dimension is based on the idea that people should be continuously doing tasks at work. The Thinking dimension stresses the fact that people should be rational in taking their decisions at work. Finally, the Being dimension stresses the importance of being spontaneous at work [44]. Hence, the current study hypothesizes:

H1: The human nature orientation is positively related to the work activity orientation of travel agent managers.

The above hypothesis tests whether the relationships among people affect work orientation. Kluckhohn and Strodtbeck [13] described three dimensions of relationships among people in various cultures; that is individualism, collectivism, and hierarchy. These dimensions were found to be influential at the workplace in different cultures [55]. Individualism is a used term to describe loose bonds among people in a specific culture [8]. In contrast, collectivism describes a strong bond among people in a society [56]. Finally, hierarchy is a term to describe the uneven distribution of power among people in a society. These cultural dimensions affect the achievement at work and management work orientation [2, 56, 57]. Thus, the study indicates that the relationship between people-orientation affects work orientation. Thus:

H2: The relational orientation is positively related to the work activity orientation of travel agent managers.

The environment orientation consists of mastery, subjugation, and harmony dimensions. In the mastery situation, people should manipulate and modify the environment around them. In contrast, in the subjugation culture, people should not alter their environment and they are influenced by it. The harmony situation is a balanced relationship between the environment and oneself [13]. Hence, the study suggests that environment orientation affects the work orientation of the managers of the travel agencies.

H3: The environment orientation is positively related to the work activity orientation of travel agent managers.

The construct of time consists of three dimensions. The Past orientation stresses the fact that our decisions at work should be influenced by our traditions. Whereas the Present orientation describes decisions at work should be based on present circumstances. Finally, the Future orientation stresses the idea of basing our decisions on future needs [13, 45]. Hence, the study proposes that time orientation affects the work orientation of managers of travel agencies.

H4: Time as a cultural value orientation is positively related to the work activity orientation of travel agent managers.

# 4 Methods and Data Collection

There are approximately 685 travel agencies in the area of Amman the capital of Jordan. The travel agency sector employs 4885 employees in the Kingdom out of which 3933 employees in Amman [24]. Furthermore, the number of package tours amounted to (2,131,614) in the year 2017. This is an indication that the sector is a large employer and revenue generator in the Jordanian economy. It is for the aforementioned reasons the study focused on the Jordanian travel agencies.

The sample of this study consists of managers of travel agencies. A simple random sample technique was used to collect data from three levels of managers (top management, middle management, and supervisors) in March 2020 before the outbreak of the COVID-19. The suggested sample size is 250 managers that is suitable for the size of the population according to Argyris and Schon [58]. Before the administration of the questionnaire, the researcher conducted a pilot study with 10 managers of travel agencies to test the clearness of the questionnaire and as a result, the construct of space orientation was removed because it was not understood by the managers.

Furthermore, the questionnaires were translated into Arabic language and back-translated to the English language to ensure accuracy. Subsequently, 270 questionnaires were distributed by the researchers, and 250 usable questionnaires were collected representing a 93% gross response rate which seems acceptable.

To test the effect of cultural value orientations on work activity orientation, a Partial Least Square (PLS) Structural Equation modeling approach was used. In addition, since the data were interval, it was possible to calculate the means and standard deviations of every cultural value at the aggregate level. Furthermore, internal reliability, composite reliability, discriminant, and convergent validity were tested.

# 5 Data Analysis and Results

The majority of the respondents were male (57.6%) versus (42.4) females. Most of the respondents were less than 30 years old representing 48.4% of the sample. 48% of the respondents hold a university degree and have 1–5 years of experience (43.2%). Finally, most of the interviewees (68%) had supervisory positions at the travel agencies.

Further descriptive analyses were made to allow the researchers to assess the respondents' cultural traits as was conducted in the study of Yeganeh and Su [14]. The researchers calculated the means and standard deviations of the latent variables in the structural model. The results are depicted in the Table 1. The results indicate that the managers of travel agencies believe that the nature of humans tends to change from good to evil and vise versa. Furthermore, they tend to be individualistic

**Table 1** Descriptive statistics for cultural orientations

| Constructs | Dimensions | Indicators | Mean | SD |
|---|---|---|---|---|
| Human nature | Evil | N1_3 | 3.60 | 0.175 |
| | Changeable | N2 | 4.00 | 0.036 |
| Relationship among people | Individualism | R1<br>R1-2 | 3.59 | 0.042 |
| Relation with environment | Subjugation | E2<br>E2-2 | 3.00 | 0.256 |
| Time | Present | T2 | 3.56 | 0.104 |
| | Future | T3 | 3.60 | 0.161 |
| Work orientation | Being | W3<br>W3-2 | 3.04 | 0.344 |

in their relationship with people at the organization. This means that their primary responsibility is for themselves first then everything else comes second. Further, managers believe that they should not change the environment surrounding them and they are not in control of nature. They believe that their decisions should be led by future needs and circumstances yet they are spontaneous and accomplish tasks on time.

The Measurement Model. The measurement of the research model incorporated all of the exogenous and endogenous cultural orientation constructs and their indicators were tested by smart PLS (version 2.0). The research constructs were measured as reflective constructs. The analysis of individual item reliability confirms that the bulk of the measures have factor loadings greater than 0.7 [59] as appeared in Table 4. All factor loadings are significant at level ($\alpha = 0.05$). A filtration procedure was performed for items that did not reach this value to sustain cautiousness (Hair et al., 2013). Items with the lowest loadings were eliminated to improve the loadings for the other measures as well as the $R^2$ value.

Furthermore, the reliability of the scale, convergent validity, and discriminant validity were evaluated. Table 2 shows that all variables are consistent with the acceptable levels of composite reliability index above (0.7), which offers clear confirmation of the reliability of the model's measurements. Further, the analyses show that the average variance extracted (AVE) ranges between (0.6460) and (0.8057), which exceeds (0.5) in all constructs [60]. Consequently, convergent validity is fulfilled.

Finally, Discriminant validity for the research construct is measured. The analysis in Table 3 shows that the associations between the constructs are lower than the square root of the AVE [60], thus verifying discriminant validity and the model fitness for this study.

The Structural Model. The structural model and its associated hypotheses are tested based on the calculated values of path coefficients ($\beta$) and significance level (t-values), and the model goodness of fit is determined by ($R^2$) for the dependent

**Table 2** The measurement model

| Activity | Constructs | | | | |
|---|---|---|---|---|---|
| | Indicators | Loadings | AVE | Composite reliability | R square |
| Work orientation | W3 | 0.853 | 0.738411 | 0.849518 | 0.290901 |
| | W3-2 | 0.865 | | | |
| Nature of humans | N1-3 | 0.905 | 0.660628 | 0.793231 | |
| | N2 | 0.709 | | | |
| Relationships among people | R1 | 0.797 | 0.646073 | 0.784976 | |
| | R1-2 | 0.810 | | | |
| Relation to broad environment | E2 | 0.876 | 0.772384 | 0.871575 | |
| | E2-2 | 0.882 | | | |
| Time orientation | T2 | 0.950 | 0.805747 | 0.892064 | |
| | T3 | 0.841 | | | |

**Table 3** Discriminant validity for research constructs

| Activity | Construct name | | | | |
|---|---|---|---|---|---|
| | Work orientation | Relation to broad environment | Nature of humans | Time | Relationships among people |
| Work orientation | **0.859** | | | | |
| Relation to broad environment | 0.356489 | **0.878** | | | |
| Nature of humans | 0.194569 | 0.268085 | **0.812** | | |
| Time | 0.1333828 | 0.206072 | 0.200933 | **0.897** | |
| Relationships among people | 0.399228 | 0.034924 | 0.027009 | 0.077920 | **0.803** |
| AVE | 0.738411 | 0.772384 | 0.660628 | 0.805747 | 0.646073 |

**Table 4** Results of proposed hypotheses

| | Path coefficient (B) | T-value (bootstrapping) | Results |
|---|---|---|---|
| Activity -> Nature of humans | 0.119 | 2.127462 | Supported |
| Activity -> Relationships among people | 0.3904 | 8.025969 | Supported |
| Activity -> Relation to broad environment | 0.3075 | 5.646499 | Supported |
| Activity -> Time | −0.0160 | 0.243722 | Not supported |

variable. In general, the study model represents 0.291 of the variance of work orientation. The fit measurements suggest that the entire proposed structural model is acceptable, the prognostic relevance of the constructs is positive, and has a significant explanatory capacity in terms of the variance of the impact of cultural value orientations on the work orientation in Jordanian Travel agencies. Furthermore, Table 4 illustrates that the nature of humans, the relationship among people, and relation to the broad environment have a statistically significant effect on the work orientation with path coefficients at 0.119, 0.3904, and 0.3075 respectively. All the relationships were significant at the 0.05 level ($p < 0.05$), except the relationship between time and work orientation.

# 6 Discussion

It is apparent that for the past two decades, the significance of culture in cross management research has led to an increase of studies in this area [14]. However, few research papers have been carried out to examine the effect of cultural orientations on management behavior in the tourism industry. There is a paucity of research examining the effect of cultural orientation in developing countries [61]. Therefore, this research draws on [13] cultural orientation model to examine the value orientations among Arab managers in the Jordanian travel agencies. The current research also examines the relationship between these cultural orientation values. Arab cultural values that are interwoven with Islamic religion are assumed to affect Arab managers in various disciplines and have been validated by many researchers [62–64]. The results of the current research indicated that Jordanian managers' orientations are future in time, subjugation to nature, individualistic to people, being inactivity, and changeable in human nature.

Furthermore, the results indicated a positive significant relationship between the nature of humans, relationships among people, relationship to the broad environment, and work activity orientation.

For the time orientation, Jordanian managers tend to be future-oriented. This is not consistent with the study of Nydell [65] and Yeganeh and Su [14] that indicated that Islamic Iranian culture is past Oriented. Although Arab culture shares basic Islamic value orientations, Sabri [66] argued that value orientations vary among cultures and within the same culture. This implies that Arab managers could be culturally grounded or deviates from their culture according to their circumstances [67]. This is congruent to the theory of action introduced by Uddin [68] that provided two types of action: espoused theory and theory in use. Espoused theory reflects what people believe in and theory in use is what people act at individual and organizational levels. In the Islamic Arab culture, espoused Islam is future-oriented but Islam in use is past-oriented. This explains the futuristic orientation of Jordanian managers [61].

The result of the study implies that Jordanian managers seemed to be subjugated. Subjugation entails that life is regulated by external strength and people cannot truly amend their fate. Since Jordanian culture has been encompassed by Islamic religion and belief in God, Jordanian managers have a propensity for subjugation and fatalism. While the demonstration of fatalism in the Islamic religion is plentiful, the Koran declares that humans can choose and interfere in their destiny. Likewise, Oyserman [69] found evidence that both fatalism and personal choices were evident in the work-related orientations of Islamic managers.

Hofstede [8] viewed societies as individualistic in which behaviors and beliefs are determined individually; whereas collective society's behavior is determined by loyalty for groups. He completed his argument by stating that most Arab societies have collective cultural values. In contrast, the result of the study showed that Jordanian managers are characterized by a high degree of individualistic orientation. This result is different from those of [8] that ranked Jordan as a collective country. However, according to Chin [61], they argued that Arab managers have a variety of value orientations between espoused Islam and Islam in use. This means that there is a discrepancy between the ideal and real Islam. This might be the reason for considering Jordan as a collective society but when it comes to work setting managers are individualistic in their orientation.

In addition, the results of the current research suggested that the Jordanian managers have a high tendency towards being orientation at work and changeable human nature orientation. The Being culture stresses peace of mind, immediate enjoyment, and spontaneous behavior [13]. This is possible because the Jordanian managers are affected by Islamic culture that considers life as a transient step to eternal life which does not worth hard work and stress. Also, the changeable orientation allows managers to have a mixture of good and evil orientations. This changeable orientation is explained in the Holy Quran. "And by the soul and He who proportioned it, and inspired it with discernment of its wickedness and its righteousness, He has succeeded who purifies it, and he has failed who instills it" (The Qur'an 91:7–10). In addition, the Quran differentiates between good and evil in this verse "And have shown him the two ways" (The Qur'an 90:10).

This means that human beings can perceive good and evil and differentiate between them. Good and evil powers are inside all human beings. Those who use the good power over the evil power will purify their souls and vise versa. Furthermore, the changing economic, social conditions and the uncertainty in the Middle East area could have a psychological effect on managers of travel agencies. This reflects their changing human nature.

Moreover, the current research conducted further analysis to examine the effect of several cultural value orientations on work activity orientation. The result indicated that relational orientation, relation to nature and nature of humans has a positive effect on work orientation. In contrast, the time orientation did not have any significant effect on work orientation. The results suggested that relational orientation had the strongest effect on work activity orientation representing (39.0%) of the variance followed by relation to nature presenting (30.7%) of the variance and human nature presenting (11.9%). The results convey that managers of travel agencies agree that the strength in the relationship among people, relation to nature, and nature of humans are imperative determinants of work activity orientation. Perhaps the individualism, subjugation, and the changeable human nature of the travel agents affect their way of work and allow them to be comfortable and peaceful at work. Individualism allows them to work independently with referring to a group that usually takes more time and effort to communicate with. The subjugation orientation makes them less rebellious at work therefore they feel more comfortable and less stressed. Finally, the changing human behavior gives them a chance to react to different situations at work without any fuss.

# 7  Conclusion, Implication and Future Research

The current research examines the cultural value orientations among Jordanian travel agencies using cultural orientations proposed by Kluckhohn and Strodtbeck [13]. The results indicate that the nature of humans, the relationship among people, and the relation to the broad environment have a positive effect on work orientation. In contrast, the time construct does not affect work orientation. This implies that the marketing manager should organize and priorities the time spent on marketing activities inside the travel agencies. Further, marketing managers should create and apply more training programs to their employees stressing the importance of group work and working with different cultures. This training usually increases the experience of the employees and changes the routine work resulting in a happier and efficient work atmosphere. Further research should examine other value orientations that could be valid in the context of Arab culture. Moreover, future research should examine the cultural value orientations in different organizations that might result in different conclusions. A comparison between cultural orientations in the public and private sectors may produce interesting results that could be generalized. Finally, this research has put the first step in examining cultural values in tourism organizations in Jordan. Further studies are needed to examine these values in various Arab countries for a proper generalization in this sector.

# References

1. Hofstede G (1980) Culture and organizations. Int Stud Manag Organ 10(4):15–41
2. House R, Javidan M, Hanges P, Dorfman P (2002) Understanding cultures and implicit leadership theories across the globe: an introduction to project GLOBE. J World Bus 37(1):3–10
3. Trompenaars F (1985) The organization of meaning and the meaning of organization. Unpublished Ph.D. Dissertation, The Wharton School of the University of Pennsylvania
4. Batjargal B, Webb JW, Tsui A, Arregle JL, Hitt MA, Miller T (2019) The moderating influence of national culture on female and male entrepreneurs' social network size and new venture growth. Cross Cult Strateg Manage
5. Nguyen PT, Yandi A, Mahaputra MR (2020) Factors that influence employee performance: motivation, leadership, environment, culture organization, work achievement, competence and compensation (A study of human resource management literature studies). Dinasti Int J Digital Bus Manage 1(4):645–662
6. Kurnaedi E, Agustina K, Karyono O (2020) Strategy for improving service performance through organizational culture and climate. Budapest Int Res Critics Inst-J (BIRCI-J), 1360–1368.
7. Hofstede G (2001) Culture's consequences: comparing values, behaviors, institutions and or organizations across nations. Sage
8. Hofstede G (1980) Values and culture. Culture's consequences: international differences in work-related values
9. Hall ET (1976) Beyond culture. Dobleday & Company, NY
10. Trompenaars F (1996) Resolving international conflict: culture and business strategy. Bus Strateg Rev 7(3):51–68
11. Schwartz SH (1999) A theory of cultural values and some implications for work. Appl Psychol 48(1):23–47
12. Hofstede G (1984) Cultural dimensions in management and planning. Asia Pac J Manage 1(2):81–99
13. Kluckhohn FR, Strodtbeck FL (1961) Variations in value orientations
14. Yeganeh H, Su Z (2007) Comprehending core cultural orientations of Iranian managers. Cross Cult Manage Int J 14(4):336–353
15. Jamal T, Robinson M (2009) The SAGE handbook of tourism studies. Sage
16. Adler NJ, Jelinek M (1986) Is "organization culture" culture bound? Hum Resour Manage 25(1):73–90

17. Muskat B, Muskat M, Blackman D (2013) Understanding the cultural antecedents of quality management in tourism. Manag Serv Qual Int J 23(2):131–148
18. Dimanche F (1994) Cross-cultural tourism marketing research: an assessment and recommendations for future studies. J Int Consum Mark 6(3–4):123–160
19. Kim C, Lee S (2000) Understanding the cultural differences in tourist motivation between Anglo-American and Japanese tourists. J Travel Tour Mark 9(1–2):153–170
20. Reisinger Y (2005) Guest editorial: leisure travel: national trends, cultural differences
21. Dajani D (2016) Using the unified theory of acceptance and use of technology to explain e-commerce acceptance by Jordanian travel agencies. J Comparative Int Manage 19(1):121–144
22. Ma X, Wang R, Dai M, Ou Y (2021) The influence of culture on the sustainable livelihoods of households in rural tourism destinations. J Sustain Tour 29(8):1235–1252
23. The Jordan Times (2019) Tourism revenues surpass $5-billion mark in 2018. Retrieved 1 October 2019 from www.jordantimes.com/news/local/tourism-revenues-surpass-5-billion-mark-2018
24. Ministry of Tourism and Antiquities (2019) Tourism statistical [Online]. Available at: https://www.mota.gov.jo/Contents/stat2019.aspx. Accessed: 15 June 2020
25. Wales W, Gupta VK, Marino L, Shirokova G (2019) Entrepreneurial orientation: international, global and cross-cultural research. Int Small Bus J 37(2):95–104
26. Maher C (2020) Career needs and career values: the mediating role of organizational culture. In: Recent advances in the roles of cultural and personal values in organizational behavior. IGI Global, pp 240–260
27. Kilmann RH, Saxton MJ, Serpa R (1985) Gaining control of the corporate culture. Jossey-Bass
28. Deal TE, Kennedy AA (1982) Corporate cultures the rites and rituals of corporate life
29. Schein EH (1992) Organizational culture and leadership. Jossey-Bass, San Francisco, CA, 418 p
30. Solomon M, Russell-Bennett R, Previte J (2012) Consumer behaviour. Pearson Higher Education AU
31. Halisah A, Jayasingam S, Ramayah T, Popa S (2021) Social dilemmas in knowledge sharing: an examination of the interplay between knowledge sharing culture and performance climate. J Knowl Manag
32. Denison DR (1990) Corporate culture and organizational effectiveness. Wiley
33. Zhong L, Wayne SJ, Liden RC (2016) Job engagement, perceived organizational support, high-performance human resource practices, and cultural value orientations: a cross-level investigation. J Organ Behav 37(6):823–844
34. Popli M, Akbar M, Kumar V, Gaur A (2016) Reconceptualizing cultural distance: the role of cultural experience reserve in cross-border acquisitions. J World Bus 51(3):404–412
35. Keller KL, Kotler P (2012) Branding in B2B firms. In: Handbook of business-to-business marketing. Edward Elgar Publishing
36. Hofstede G (2009) Geert Hofstede cultural dimensions
37. Trompenaars F, Hampden-Turner C (1993) The seven cultures of capitalism. Piatkus, London, p 25
38. Hall ET, Hall MR (1990) Understanding cultural differences: [Germans, French and Americans], vol 9. Intercultural Press, Yarmouth, ME
39. Trompenaars F, Hampden-Turner C (1998) Riding the waves of culture: understanding diversity in global business, 2nd edn
40. Schwartz S (2006) A theory of cultural value orientations: explication and applications. Comp Sociol 5(2–3):137–182
41. Hall ET (1959) The silent language. Doubleday, New York, NY
42. Darley WK, Blankson C (2008) African culture and business markets: implications for marketing practices. J Bus Ind Mark 23(6):374–383
43. Nakata C, Sivakumar K (2001) Instituting the marketing concept in a multinational setting: the role of national culture. J Acad Mark Sci 29(3):255–276
44. Maznevski ML, Gomez CB, DiStefano JJ, Noorderhaven NG, Wu PC (2002) Cultural dimensions at the individual level of analysis: the cultural orientations framework. Int J Cross-cult Manag 2(3):275–295

45. Lo KD, Houkamau C (2012) Exploring the cultural origins of differences in time orientation between European New Zealanders and Māori
46. Hampden-Turner C, Trompenaars A (1993) The seven cultures of capitalism: value systems for creating wealth in the United States, Japan, Germany, France, Britain, Sweden, and the Netherlands. Doubleday
47. Ali AJ, Al-Kazemi AA (2007) Islamic work ethic in Kuwait. Cross Cult Manage Int J 14(2):93–104
48. Ali AJ (2008) Business and management environment in Saudi Arabia: challenges and opportunities for multinational corporations. Routledge, New York, NY
49. Dedoussis E (2004) A cross-cultural comparison of organizational culture: evidence from universities in the Arab world and Japan. Cross Cult Manage Int J 11(1):15–34
50. De Long DW, Fahey L (2000) Diagnosing cultural barriers to knowledge management. Acad Manag Perspect 14(4):113–127
51. Mufune P (2003) African culture and managerial behaviour: clarifying the connections. S Afr J Bus Manage 34(3):17–28
52. Barakat H (1993) The Arab world: society, culture, and state. University of California Press
53. Cleveland M, Laroche M, Hallab R (2013) Globalization, culture, religion, and values: comparing consumption patterns of Lebanese Muslims and Christians. J Bus Res 66(8):958–967
54. Feghali E (1997) Arab cultural communication patterns. Int J Intercult Relat 21(3):345–378
55. Okpara JO (2014) The effects of national culture on managers' attitudes toward business ethics: implications for organizational change. J Account Organ Chang 10(2):174–189
56. Hofstede G, Bond MH (1984) Hofstede's culture dimensions: an independent validation using Rokeach's value survey. J Cross Cult Psychol 15(4):417–433
57. Hofstede G (1994) Business cultures. UNESCO Cour 30(4):17–14
58. Sekaran U (2003) Research method for business: a skill building approach, 4th edn
59. Chin WW (1998) The partial least squares approach to structural equation modeling. Mod Methods Bus Res 295(2):295–336
60. Fornell C, Larcker DF (1981) Structural equation models with unobservable variables and measurement error: algebra and statistics
61. Yaseen SG, Eirefae GA (2019) Islamic work ethics for Arab managers: the missing paradigm between espoused Islam and Islam-in-use. Int J Econ Bus Res
62. Kumar N, Che Rose R (2010) Examining the link between Islamic work ethic and innovation capability. J Manage Dev 29(1):79–93
63. Nydell M (1996) Understanding Arabs: a guide for westerners. Intercultural Press Yarmouth, ME
64. Sabri HA (2012) Re-examination of Hofstede's work value orientation perceived leadership styles in Jordan. Int J Commer Manag 22(3):202–218
65. Dastmalchian A, Javidan M, Alam K (2001) Effective leadership and culture in Iran: an empirical study. Appl Psychol 50(4):532–558
66. Uddin M (2009) Cross-cultural comparison of marriage relationship between Muslim and Santal communities in rural Bangladesh. World Cult ejournal 17(1)
67. Oyserman D (2016) What does a priming perspective reveal about culture: culture-as-situated cognition. Curr Opin Psychol 12:94–99
68. Argyris C, Schon DA (1974) Theory in practice: increasing professional effectiveness. Jossey-Bass
69. Latifi F, Kiani GR (1997) An expressive model to interpret multi-faceted cultures: an application for strategic planning in Iran. Working Paper Series-Henley Management College HWP

# An Investigation on Mobile Service Quality of Food Delivery Provider from Customers' Experiences

Siti Nurhayati Khairatun

**Abstract** The hit of the COVID-19 pandemic has impacted numerous business sectors, particularly food service establishments, globally. In an effort to curb the infection rate, many countries have imposed movement control orders which, among others, restrict or ban dining at restaurants and mass gatherings. Many studies reported that the food service sector has recorded billions of income losses during this pandemic. As part of survival strategy, most food service businesses switched to delivery services to make ends meet. This paper aimed to investigate the mobile service quality of one major food delivery provider in Malaysia. A quantitative approach was employed in this study with a convenient sampling plan to gather data. An online survey was administered to customers in Klang Valley who have been using mobile delivery apps for ordering their foods. The results indicated that high service quality increases the levels of customers' satisfaction while using the food delivery service. This study is significant to the food businesses and the food delivery providers in understanding the customers' expectations and how to improve their delivery services.

**Keywords** Mobile service quality · Customers' experiences · Food delivery provider · Online survey · Malaysia

## 1 Introduction

The online food delivery market has an increase of 40% in the U.S. [1]. This is because smartphones had become soul mates for youngsters. About 97.7% of youngsters with age 15 years and above have been using smartphones in carrying out their daily activities [2]. This trend resulted in changing purchase habits of youngsters from offline to online buying either for clothes, gadgets, as well as food and beverage, etc. There is a rapid growth of online purchases by consumers in China as modern

S. N. Khairatun (✉)
Department of Food Service and Management, Faculty of Food Science and Technology, Universiti Putra Malaysia, Seri Kembangan, Malaysia
e-mail: snkhairatun@upm.edu.my

© The Author(s), under exclusive license to Springer Nature Switzerland AG 2022
S. G. Yaseen (ed.), *Digital Economy, Business Analytics, and Big Data Analytics Applications*, Studies in Computational Intelligence 1010,
https://doi.org/10.1007/978-3-031-05258-3_30

gadgets provide a lot of free-of-charge applications for services [3]. Therefore, this trend has agitated the development of online food delivery (OFD) or food delivery apps (FDA) in the food & beverage industry.

Online food delivery (OFD) is a system where customers make orders through websites or apps from various F&B premises. In Malaysia, internet providers' expansion and excessive smartphone penetration have driven the development and prosperity of various local food delivery apps (FDA) such as Foodpanda, GrabFood, Dahmakan, DeliverEat, PickUpp, Mammam Deliveries, Meal2U, FoodTime, and so on. There is a slight difference between OFD and FDA whereby OFD service providers usually have their restaurant and customers can do the ordering on their website. In other words, some of the restaurants will provide their OFD services such as fast- food restaurants like McDonald's, KFC, Pizza Hut, and Domino's. However, FDA, also known as multi-restaurant mobile applications, is an intermediate between various restaurants and customers [4]. The orders for FDA can be made only through mobile apps. FDAs such as Foodpanda, GrabFood, and Dahmakan cooperate with various restaurants and they provide orders, monitors, deliveries, payment, and tracking, but they are not responsible for the actual preparation of food [5]. The expansion of FDAs had become more and more significant nowadays. As proof, FDAs created revenues of US$ 95.4 billion in 2018 alone, and most revenues in China amounted to US$ 38.4 billion in 2018 [6].

*Online Food Delivery Application During Pandemic*

Besides the growth of internet and smartphone usage, consumer behavior also contributes to the accelerating expansion of FDAs. People prefer high convenience services due to the hectic and hustle in life. A previous study stated that the habit of eating out becomes more common due to women's presence in the workplace, growing urbanization, and increased family incomes [5]. Hence, OFD and FDA are the best inventions for every busy urban consumer to get their meal anytime anywhere. According to [7], the main concern that motivates customers to use technology-based self-service is time savings. People can access websites or apps easily, choose whatever they want, place an order, key in the delivery details, get the food and do the payment without getting out. FDAs which can meet the customers' expectations and needs can attract loyalty from customers and compete with others. The satisfaction of the customers can determine the success of every business organization [8]. A poor service quality brings a loss of an average of 12% of the consumers [9]. That being said, customers' satisfaction is the key component for FDAs to maintain long-term affiliation.

COVID-19 pandemic has changed the business operation model for most players to survive during a hard time. In Malaysia, the government has imposed Movement Controlled Order (MCO) in many phases to flatten the infection rate. During this lockdown, only essential services were allowed to continue operating with strict observation of standard operating procedures (SOP). The 17 essential services were including (1) food and beverage, including for animals; (2) health care and medicine including dietary supplements, animal care and veterinary services; (3) water; (4)

energy; (5) security and safety, defense, emergency, welfare and humanitarian assistance; (6) solid waste management, public sanitization and irrigation; (7) land, water and air transportation; (8) port, shipyard, and airport services including unloading, lightering (cargo transfer between vessels), cargo handling, pilotage and commodity storage or bulking; (9) communications including media, telecommunications and internet, postage, courier as well as broadcasting (for purposes of information dissemination, news and related matters; (10) banking, insurance, takaful and stock market; (11) community credit (pawnshop); (12) e-commerce and information technology; (13) extraction, distilling, storage, supply and distribution of fuel and lubricants; (14) hotels (only for purposes of quarantine, isolation, work for essential services and not for tourists), (15) critical construction, maintenance and repairs; (16) forestry services (limited to enforcement) and wildlife; and (17) logistics limited to delivery of essential services.

For restaurant businesses, switching the dine-in practices to take-away or delivery modes are crucial to continue operating during the pandemic. That being said, food delivery services are the most preferred method of buying food to minimize human contact. Numerous food delivery companies exist to cope with the high demand for food delivery services. Due to the high competition among Food apps such as Foodpanda, Grab Food, Dahmakan, etc., each organization knows the importance of providing the best customer service in gaining their popularity and profitability. Customers' satisfaction can directly influence the revenue of a business [10]. The key component for the success of the business is customers' satisfaction and it can help to expand its market value [8]. A high level of customers' satisfaction is associated with a decrease in customers complaints and further increases customers' loyalty [11]. Even though food apps are simple and easy, there are some problems of dissatisfaction from consumers when their expectations differ from the product or service actual performance. Based on a study conducted by [12], most of the online customers complained about unsatisfactory customer service (33.8%), product failures such as product performance or product quality (26.2%), dissatisfied price factor (15%), and others. All these factors might influence the level of customers' satisfaction, especially customer service. There are many studies have been conducted in the area of service quality with customers' satisfaction in different setting and countries such as the healthcare industry [13], automobile repair services [14], hotel industry [15], retails [16, 17], telecoms and cellular [18–20], tourism industry [21], public transport [22], gaming industry [23], banking [24], and food industry [25–29]. The literature found a significant relationship between service quality and customers' satisfaction. In a study on factors of the intention for the continued acceptance and use of mobile crowdsourcing to participate in refugee crisis management, the findings showed that multiple variables including the individual and crowd performance expectancy, the social influence, and perceived risks on the individual and crowd levels have a significant influence on the intention for the continued acceptance and engagement in mobile crowdsourcing to participate in refugee crisis management [37].

This study is conducted to evaluate customers' satisfaction among Foodpanda users based on service quality. Foodpanda is the leading global online food delivery marketplace which is founded globally in 2012 with the support of Rocket Internet.

It is headquartered in Berlin, Germany [30, 31]. At present, the organization is active in over 246 cities around the globe, operating in 11 countries with a partnership with over 115,000 global restaurants [32]. Foodpanda was then acquired by the German competitor Delivery Hero in December 2016. In February 2016, Foodpanda was merged with Foodora which provides technical support for Foodpanda [30]. In Malaysia, Foodpanda was operating successfully in more than 85 cities in East and West Malaysia [33].

Mobile Service Quality (M-S-QUAL) introduced by Huang et al. [34] is adopted in this study to analyze the relationships between customers' perceived service quality of Foodpanda and its influence on customers' satisfaction. Due to different settings, different sectors, different geographical locations, and different samples, results from previous studies may not truly reflect the customers' satisfaction with the service quality of Foodpanda. Literature that implements M-S-QUAL to study the relationship between perceived service quality and customers' satisfaction is also limited and never applied on Foodpanda. Therefore, this study attempts to examine the correlation between service quality and customers' satisfaction in Foodpanda and to determine the most significant variable in M-S-QUAL which affects the most on customers' satisfaction towards Foodpanda.

## 2 Methodology

This study employed a quantitative approach to examine the relationship between service quality and customers' satisfaction in using Foodpanda service. Figure 1 shows the data collection process completed in this study.

The first step in data collection was designing a research framework to provide accurate and valid answers to the research questions. Next, the population and

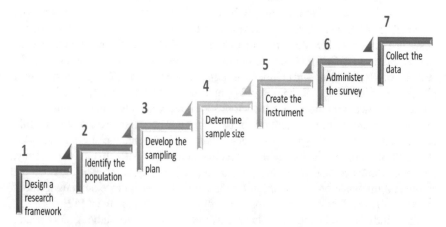

**Fig. 1** Data collection process

sampling plan should be determined. After a sample size was identified, the survey was administered and all data gathered were analyzed.

## 2.1 Sampling Plan

The target population of this study was Foodpanda users in Klang Valley. This was because Klang Valley has the highest population in Malaysia. Klang Valley is a combination of Selangor, Kuala Lumpur, and Putrajaya. According to the Department of Statistics Malaysia, the population of Klang Valley was 8.41milllion (Selangor: 6.53million, Kuala Lumpur: 1.78 million, Putrajaya: 0.10 million), which made around 25.8% of Malaysia's population (32.6 million) [35]. Besides, previous media reports had mentioned that Foodpanda has 1500 partner restaurants across the Klang Valley, Johor Bahru, and Penang with around 1800 freelance riders [36]. These two points showed that Klang Valley has a high proportion of Foodpanda users. Foodpanda users in Klang Valley had been chosen as a target population as it is representative of the population. Convenience sampling was used to collect the data. The survey questionnaire was distributed to the public residing in Klang Valley. However, before the respondents answer the questionnaire, there was an inclusion statement to ensure that they have been using the Foodpanda service.

## 2.2 Survey Instrument

A structured questionnaire consisted of four main sections: frequency usage of Foodpanda apps, a rating scale for Foodpanda, factors that influence service quality, and demographics, described as following.

Section A: Foodpanda application usage was designed to ensure the respondents had used Foodpanda before and how frequently they used the delivery service. This section consisted of three questions of multiple choices and the respondents have to choose the most relevant answer.

Section B: The rating scale for Foodpanda was developed to determine customers' satisfaction towards Foodpanda based on their experiences. In this section, respondents were required to mark their best answer by using the linear scale from 1 = Strongly Disagree to 5 = Strongly Agree.

Section C to F were created to measure factors that influence service quality such as efficiency, fulfillment, responsiveness, and contact, adopted from a scale of Mobile Service Quality (M-S-QUAL). Respondents marked the best answer based on the linear scale from 1 = Strongly Disagree to 5 = Strongly Agree.

Lastly, Section G was about demographic characteristics of respondents which included gender, age group, ethnicity, religion, nationality, educational level, and occupation.

The data collected was analyzed by using IBM Statistical Package for Social Sciences (SPSS) version 25.0. Pearson's correlation analysis was used to determine the relationship between service quality and customers' satisfaction.

## 3   Results and Discussion

### 3.1   Reliability Analysis

In this study, reliability analysis was conducted to determine the internal consistency for all of the items. Table 1 showed the reliability analysis for two parts which consisted of Part A: Customers' Satisfaction and Part B: M-S-Qual. To test whether the questions to measure the variables are reliable, the coefficient of Cronbach's Alpha must at least larger than 0.5. Based on the outcomes shown in Table 1, all attributes were reliable as their coefficient of Cronbach's Alpha, $\alpha > 0.5$.

In Part A, the Cronbach's Alpha coefficient of customer rating is 0.700. According to the Rule of Thumb, if the score can reach 0.7 or above will be better. Even though the alpha value had been tried to increase to 0.7 and above by deleting one of the items, none of the values increase to 0.7 and above. Through the analysis, the alpha value became 0.420 if the item 'I enjoy when using the Foodpanda apps' was deleted. These values indicate that the particular item cannot be deleted as it will cause the results to become unreliable.

For Part B, there are four attributes of M-S-Qual in this study. Table 1 shows that the attribute which has the highest Cronbach's Alpha value was the efficiency (0.895), followed by fulfillment and responsiveness with $\alpha = 0.853$ respectively and lastly $\alpha = 0.828$ by contact. Four of the attributes showed a high degree of consistency. According to the literature, it showed a very high degree of reliability too for all of these four factors [34]. There was no difference in the value if any one item was deleted for all the attributes in Part B.

### 3.2   Demographic Data

A total of 267 responses were received but only 178 were usable (66.7%). There were 37% of male and 63% of female respondents. They were made up of different age groups where the majority was the younger age group as 57.9% of the respondents were 18–25 years old. It was followed by a group of 26–35 years old (40.4%), 36–45 years old (1.1%), and 46–55 years old (0.6%). Among the respondents, there were 97.2% of Malaysians and 2.8% of non-Malaysians. The major group was Chinese respondents (82.0%) as compared to Malay (12.9%), India (3.9%), and others (1.2%). For the educational level, the majority of the respondents were Bachelor degree graduates (73.1%), followed by Diploma holders (14.1%), SPM (O-Level) holders

**Table 1** Reliability analysis

| Items | Cronbach's Alpha | Cronbach's Alpha if item deleted |
|---|---|---|
| *Part A: Customers' Satisfaction Customer Rating (3 items)* | | |
| I am very satisfied with Foodpanda | | 0.676 |
| I enjoy when using the Foodpanda apps | 0.700 | 0.420 |
| I feel I am getting good food product at a reasonable price through Foodpanda | | 0.692 |
| *Part B: M-S-Qual Efficiency (7 items)* | | |
| This mobile site makes it easy to find what I need | | 0.874 |
| It is easy to get anywhere on the mobile site | | 0.890 |
| The site enables me to complete the transaction quickly | | 0.905 |
| Information at this mobile site is well organized | 0.895 | 0.871 |
| The site loads its pages quickly | | 0.866 |
| The mobile site is simple to use | | 0.875 |
| The mobile site is well organized | | 0.872 |
| *Fulfillment (9 items)* | | |
| The site delivers orders when promised | | 0.860 |
| This mobile site makes items available for delivery within a suitable time frame | 0.853 | 0.826 |
| The site quickly delivers what I order | | 0.873 |
| The site correctly sends out the items ordered | | 0.836 |
| The site has the items that the company claims to have in stock | | 0.837 |
| The site is truthful about its offerings | | 0.822 |
| The site makes accurate promises about the delivery of products | | 0.811 |
| When the order is completed, the order information is sent on time | | 0.829 |
| When the order is completed, the service provider can provide customized information | | 0.838 |
| *Responsiveness (7 items)* | | |

(continued)

**Table 1**   (continued)

| Items | Cronbach's Alpha | Cronbach's Alpha if item deleted |
|---|---|---|
| The mobile site provides me with convenient options for returning items | | 0.823 |
| This mobile site handles product returns well | | 0.820 |
| This mobile site offers a meaningful guarantee | | 0.873 |
| The site tells me what to do if my transaction is not processed | 0.853 | 0.832 |
| This mobile site provides a telephone number to reach the company | | 0.823 |
| This mobile site has a customer service representative available online | | 0.816 |
| The site offers the option to speak with a live person if there is a problem | | 0.829 |
| *Contact (4 items)* | | |
| The service agents are friendly when receiving complaints | | 0.803 |
| The service agents provide consistent advice | 0.828 | 0.763 |
| The customer service representatives are polite | | 0.789 |
| The call center personnel can help with problems | | 0.780 |

(5.6%), Master's degree (2.8%), and Ph.D. holders (2.2%). For occupation data of respondents, 28% were self-employed, private sector (24.7%), the government sector (3.4%), and retirees (0.6%). A summary of the data is tabulated in Table 2.

## 3.3   Frequency Usage of Foodpanda Apps

Among the 178 respondents, only 5.6% of respondents used the Foodpanda service via apps almost every day, 41.6% of respondents used the apps not even once a month, 33.7% used it at least once a month, and 19.1% used it at least once a week. Besides that, the majority of the respondents spent RM10-RM20 (61.2%), 36% of respondents always spent more than RM20, and only 2.8% of respondents spent less than RM10 per transaction (see Table 3).

**Table 2** Respondents' demographic data

| Variables | N = 178 | |
|---|---|---|
| | Frequency (n) | Percentage (%) |
| *Gender* | | |
| Male | 66 | 37.1 |
| Female | 112 | 62.9 |
| *Age group* | | |
| 18–25 | 103 | 57.9 |
| 26–35 | 72 | 40.4 |
| 36–45 | 2 | 1.1 |
| 46–55 | 1 | 0.6 |
| 56–65 | 0 | 0 |
| >65 | 0 | 0 |
| *Nationality* | | |
| Malaysian | 173 | 97.2 |
| Non-Malaysian | 5 | 2.8 |
| *Ethnicity* | | |
| Malay | 23 | 12.9 |
| Chinese | 146 | 82.0 |
| Indian | 7 | 3.9 |
| Others | 2 | 1.2 |
| *Religion* | | |
| Islam | 24 | 13.5 |
| Buddhism | 124 | 69.7 |
| Hinduism | 6 | 3.3 |
| Christianity | 24 | 13.5 |
| Others | 0 | 0 |
| *Educational level* | | |
| Primary school | 4 | 2.2 |
| SPM | 10 | 5.6 |
| STPM | 0 | 0 |
| Diploma | 25 | 14.1 |
| Bachelor degree | 130 | 73.1 |
| Master of degree | 5 | 2.8 |
| Ph.D. | 4 | 2.2 |
| Others | 0 | 0 |
| *Occupation* | | |
| Student | 77 | 43.3 |
| Self-employed | 50 | 28.0 |

(continued)

**Table 2** (continued)

| Variables | $N = 178$ | |
|---|---|---|
| | Frequency (n) | Percentage (%) |
| Government sector | 6 | 3.4 |
| Private sector | 44 | 24.7 |
| Housewife | 0 | 0 |
| Retired | 1 | 0.6 |
| Others | 0 | 0 |

**Table 3** Respondents' usage of Foodpanda

| Variables | $N = 178$ | |
|---|---|---|
| | Frequency (n) | Percentage (%) |
| *Have you ever used Foodpanda apps before?* | | |
| Yes | 178 | 100 |
| No | 0 | 0 |
| *How frequently do you use Foodpanda apps?* | | |
| Not even once a month | 74 | 41.6 |
| At least once a month | 60 | 33.7 |
| At least once a week | 34 | 19.1 |
| Almost everyday | 10 | 5.6 |
| *How much do you usually spend on Foodpanda apps?* | | |
| Less than RM10 | 5 | 2.8 |
| RM10-RM20 | 109 | 61.2 |
| More than RM20 | 64 | 36.0 |

## 3.4   Levels of Customers' Satisfaction

From the data collected, the statement 'I am very satisfied with Foodpanda' had a mean of 3.92 with 0.6% disagree, neutral (23.6%), agree (59.5%), and strongly agree (16.3%). For the second statement 'I enjoy when using the Foodpanda apps' 19.1% of respondents strongly agree, agree (57.3%), neutral (19.7%), and disagree (3.9%) with a mean score of 3.92. Lastly, the statement 'I feel I am getting good food product with a reasonable price through Foodpanda' gets the lowest mean score 3.71 with 9.6% of respondents disagree, neutral (26.4%), agree (47.1%), and strongly agree (16.9%). Overall, the level of customers' satisfaction towards Foodpanda is moderate as all means were not close to M = 5 (most satisfied) (see Table 4).

**Table 4** Levels of customers' satisfaction

| Items | Mean | Percentage (%) | | | | |
|---|---|---|---|---|---|---|
| | | Strongly disagree | Disagree | Neutral | Agree | Strongly agree |
| I am very satisfied with Foodpanda | 3.92 | 0 | 0.6 | 23.6 | 59.5 | 16.3 |
| I enjoy when using the Foodpanda apps | 3.92 | 0 | 3.9 | 19.7 | 57.3 | 19.1 |
| I feel I am getting good food product at a reasonable price through Foodpanda | 3.71 | 0 | 9.6 | 26.4 | 47.1 | 16.9 |

## 3.5 Factors Influencing M-S-Qual of Foodpanda Service

Four factors that are being tested including efficiency, fulfillment, responsiveness, and contact. In terms of efficiency, 37.1% of respondents strongly agree and 46.6% of respondents agree with the statement 'The site enables me to complete the transaction quickly with the highest mean of 4.19. This shows that the payment system of Foodpanda apps was functioning. Besides, the statement 'The mobile site is simple to use' obtained a second higher mean (4.08) with 30.9% of strongly agree and 51.7% of agree. However, the statement 'The site loads its pages quickly' had the highest disagreement where 6.2% of respondents feel that the speed to load the site is slow (M = 3.84) among other statements. The second highest disagreement in this section was the statement 'The mobile site is well organized' which contributed a mean of 4.01 where 1.1% of respondents were strongly unsatisfied with it and 3.9% of respondents were unhappy with it. If the site loads slowly, it is also difficult to browse the site freely.

There were 9 items in the fulfillment section. Both statements 'The site delivers orders when promised' and 'The site correctly sends out the items ordered' obtained the highest mean (M = 4.11). More than 80% of respondents agreed that they received their orders correctly. The result showed the statement 'The site quickly delivers what I order' had only a mean of 3.83. Based on this data, it impliedly indicated that the delivery speed was not enough to fulfill respondents' needs.

The overall mean for efficiency and fulfillment was in the range of 3.8–4.2. However, in the section of responsiveness, the range of mean was lower (3.4–3.7). There was a higher percentage of respondents choosing "Neutral". This might be due to some of the respondents never had experienced getting the wrong food, facing problems in the ordering process, never contacted with customer service of Foodpanda, hence they did not have much idea about it. Regarding responsiveness, the highest percentage of agreement was the statement 'This mobile site offers a meaningful guarantee' (M = 3.71). Next was the 'This mobile site has customer service representatives available online' (M = 3.67). The lowest mean went to the statement

**Table 5** Pearson's correlation on customers' satisfaction and M-S-QUAL

| Hypothesis | Remarks | Pearson correlation | Strength |
|---|---|---|---|
| H1: There is a positive relationship between efficiency and customers' satisfaction | Supported | 0.538 | Moderate |
| H2: There is a positive relationship between fulfillment and customers' satisfaction | Supported | 0.621 | Strong |
| H3: There is a positive relationship between responsiveness and customers' satisfaction | Supported | 0.542 | Moderate |
| H4: There is a positive relationship between contact and customers' satisfaction | Supported | 0.512 | Moderate |

'The site offers the option to speak with a real person if there is a problem' (M = 3.44).

In the contact section, there were also a lot of respondents who answered "Neutral". Among four items, the highest mean and agreement was by the statement 'The customer service representatives are polite' (M = 3.63). Apart from that, 'The service agents provide consistent advice' scored the lowest mean (M = 3.46) with only 9.6% of respondents strongly agreed to it.

Overall, it can be concluded that the efficiency and fulfillment in Foodpanda apps were quite satisfying. However, the after-sales service from Foodpanda required improvement as responsiveness and contact of Foodpanda received a low degree of agreement (M = 9.67).

Results of Pearson's Correlation analysis showed that the relationship between the mobile service quality and customers' satisfaction, was supported and there was a positive relationship between each attribute included efficiency, fulfillment, responsiveness, and contact, with customers' satisfaction. It also indicated that fulfillment was the most significant variable in M-S-Qual which dominated the effect on customers' satisfaction towards Foodpanda. A summary of the analysis is illustrated in Table 5.

Based on a guide, the absolute value of r can represent the strength of the correlation [38]. H1: There is a positive relationship between efficiency and customers' satisfaction was accepted with a moderate strength (r = 0.538). The result is similar to the study conducted previously which stated that customers' satisfaction exhibited a high correlation with the efficiency of online food delivery services [2]. A highly efficient mobile site such as high transaction speed, easy to use, and well organized can provide an enjoyable experience to the users as they can easily get what they want [39]. This can increase their intention to use. Hence, efficiency is one of the main factors to assess the satisfaction level of customers. Besides, the results also showed a positive relationship between fulfillment and customers' satisfaction with strong strength, the highest r-value, 0.621. H2 was accepted. Some similar studies also obtained the results that fulfillment plays a crucial role in customers' satisfaction [2, 4]. Customers expect fast and easy when they do online ordering. This is probably because they only had a short break from their work and they do not wish any delay which might mess up their time and performance [3]. As a result, order fulfillment variables especially on-time delivery was always the main concern among users. H3 was accepted too where responsiveness showed a positive relationship with customer service but with moderate strength (r = 0.542). Based on the previous study, there was a positive and significant correlation between responsiveness and customers' satisfaction [2]. It is very important for FDA to provide its service responsively to gain loyal customers. It does not matter if the service fails to deliver the correct food, but the company must assist the users with explanation, replacement, or compensation to show that they care for every customer. This can make the user satisfied with the service and might make an order again with the apps.

Lastly, the contact variable also had a positive relationship with customers' satisfaction with moderate strength at r = 0.512 which is the lowest. Based on the past study, the contact variable did not show a positive correlation with customers' satisfaction but it exhibited a high correlation with customer loyalty [2]. This might explain the lowest strength of contact with customers' satisfaction as people would not focus on this factor when first time use, or make it as a main concern in choosing the FDA platform. However, it still had the relationship as if the customer service representative for the particular FDA is polite, willing to help, and portray a good image and service, it can build a good impression for users and increase the satisfaction level and intention to use next time.

## 4 Conclusion

It can be concluded that high service quality increases the levels of customers' satisfaction while using the food delivery service. In order to sustain itself in the competitive market, Foodpanda should do some improvements. First, it can improve the delivery time as most of the respondents were not satisfied with the current delivery time. Besides that, Foodpanda should invest more in customer service including chatting systems, real-time assistance, and refund procedures. These can help Foodpanda in gaining customers' attention and retaining them.

The limitations of this study are the recruitment of a small sample size and time constraints. For representative data, larger sample size is needed. This study used respondents residing in Klang Valley, Malaysia only. It is recommended that future studies could recruit samples from other big cities such as Penang, Johor Bahru, Kuching, or Kota Kinabalu. Due to time constraints, the survey was carried out via an online platform. This method was only accessible to those who can afford internet connection and compatible devices. Future research may consider the usage of the face-to-face survey to obtain a bigger dataset.

The implication of this current study is threefold: the field of research, practice, and society. The results from this study would add valuable information to the literature in research relating to online food delivery service application and customers' satisfaction, particularly during the pandemic. The practical implication in this study was to outline some recommendations to the online food delivery service providers on how to improve their mobile service quality. Next, society as one of the stakeholders in this business, will benefit from this study because this study offers a baseline reference for a better understanding of their needs.

**Acknowledgements** The author received no financial support for the research, authorship, and/or publication of this article.

# References

1. Sherly K, Philipp L (2011) Online, mobile and text food ordering in the U.S restaurant industry. Cornell University: Cornell Hospitality Report (2011)
2. Yusra H, Agus A (2019) The influence of online food delivery service quality on customer satisfaction and customer loyalty: the role of personal innovativeness. J Environ Treatment Tech, 6–12
3. Liu W, Florkowski W (2017) Online meal delivery services: perception of service quality and delivery speed among Chinese consumers, pp 1–23
4. Sjahroeddin F (2018) The role of E-S-Qual and food quality on customer satisfaction in online food delivery service. In: 9th industrial research workshop and national seminar
5. Pigatto G, Machado JG, Santos Negreti A, Machado LM (2017) Have you chosen your request? Analysis of online food delivery companies in Brazil. Br Food J 119(3):639–657
6. Statista Report (2018) eServices Report 2019—online food delivery
7. Ding TU, Verma R, Iqbal Z (2007) Self-service technology and online financial service choice. Int J Serv Ind Manag 18(3):246–268
8. Khadka K, Maharjan S (2017) Customer satisfaction and customer loyalty. Centria University of Applied Sciences Pietarsaari
9. Riscinto-Kozub KA (2008) The effect of service recovery satisfaction on customer loyalty and future behaviour intention: an exploratory study in the luxury hotel industry. Doctor of Philosophy thesis, Auburn University
10. Best R (2005) Strategies for growing customer value and profitability. In: Best R (ed) Market-based management. Pearson Prentice Hall, pp 7–20
11. Johnson MD (2001) The evolution and future of national customer satisfaction models. J Econ Psychol, 217–245
12. Cho Y, Im I, Hiltz R, Fjermestad J (2002) An analysis of online customer complaints: Implications for web complaint management. In: Proceedings of the 35th Hawaii international conference on system science
13. Yesilada F, Direktor E (2010) Health care service quality: a comparison of public and private hospitals. Afr J Bus Manage 4(6):962–971
14. Izogo EE, Ogba IE (2015) Service quality, customer satisfaction and loyalty in automobile repair services sector. Int J Qual Reliab Manage 32(3):250–269
15. Dedeoglu BB, Demirer H (2015) Differences in service quality perceptions of stakeholders in the hotel industry. Int J Contemp Hosp Manag 27(1):130–146
16. Anselmsson J, Johansson U (2014) A comparison of customer perceived service quality in discount versus traditional grocery stores: an examination of service quality measurement scales in a Swedish context. Int J Qual Serv Sci 6(4):369–386
17. Omar MW, Shaharudin MR, Jusoff K, Ali MN (2011) Understanding the mediating effect of cognitive and emotional satisfaction on customer loyalty. Afr J Bus Manage 5(17):7683–7690
18. Ahmad Z, Ahmed I, Nawaz M, Shaukat M, Ahmad N (2010) Impact of service quality of short messaging service on customers retention: an empirical study of cellular companies of Pakistan. Int J Bus Manage 5(6):154–160
19. Ali I, Rehman K, Yilmaz A, Nazir S, Ali J (2010) Effects of corporate social responsibility on consumer retention in cellular industry of Pakistan. Afr J Bus Manage 4(4):475–485

20. Omotayo O, Joachim AA (2008) Customer service in the retention of mobile phone users in Nigeria. Afr J Bus Manage 2(2):26–31
21. Debata BR, Patnaik B, Mahapatra SS, Sree K (2015) Interrelations of service quality and service loyalty dimensions in medical tourism: a structural equation modeling approach. Benchmark Int J 22(1):18–55
22. Kumar KS (2012) Expectations and perception of passengers on service quality with reference to public transport undertakings. IUP J Oper Manage XI 3:67–81
23. Wu HC (2014) The effects of customer satisfaction, perceived value, corporate image and service quality on behavioral intentions in gaming establishments. Asia Pac J Mark Logist 26(4):540–565
24. Malik S (2012) Customer satisfaction, perceived service quality and mediating role of perceived value. Int J Mark Stud 4(1):68–76
25. Marinelli N, Simeone M, Scarpato D (2015) Does quality really matter? Variables that drive postmodern consumer choices. Nutri Food Sci 45(2):255–269
26. Kafetzopoulos DP, Gotzamani KD, Psomas EL (2014) The impact of employees' attributes on the quality of food products. Int J Qual Reliab Manage 31(5):500–521
27. Bujisic M, Hutchinson J, Parsa HG (2014) The effects of restaurant quality attributes on customer behavioral intentions. Int J Contemp Hosp Manag 26(8):1270–1291
28. Jang S, Ha J (2014) Do loyal customers perceive the quality of restaurant attributes differently? A study of Korean restaurant customers. J Foodserv Bus Res 17(3):257–266
29. Wettstein N, Hanf JH, Burggraf C (2011) Unshakable loyalty in the food sector: sustainable customer retention. Empirical study of organic food consumers in Germany. J Consum Prot Food Saf 6(3):359–365
30. Hassan M (2018) Effect of rebranding on the customer satisfaction of Foodpanda Bangladesh Limited
31. Jashim UA, Asma A (2018) Foodpanda: changing the way Bangladeshis eat meal. Sage
32. Foodpanda (2020). https://www.foodpanda.com/about/
33. Foodpanda Malaysia (2020). https://www.foodpanda.my/
34. Huang ET, Lin SW, Fan YC (2015) M-S-QUAL: mobile service quality measurement. In: Electronic commerce research and applications. Elsevier, pp 126–142
35. The Department of Statistics Malaysia (2019). https://www.dosm.gov.my/
36. Rosli L (2018) Foodpanda records 100 pct growth in 2017. New Straits Times
37. Saad GY, Al Omoush KS (2020) Mobile crowdsourcing technology acceptance and engagement in crisis management: the case of Syrian refugees. Int J Technol Human Interact (IJTHI) 16(3):1–23
38. Evans JD (1996) In straightforward statistics for the behavioral sciences. Thomson Brooks/Cole Publishing Co.
39. Kedah Z, Ismail Y, Haque AA, Ahmed S (2015) Key success factors of online food ordering services: an empirical study. Malays Manage Rev, 19–36

# Online Hotel Booking Continuous Intention in the Digital Transformation Age: Case of Vietnam

Bui Thanh Khoa, Nguyen Hoang Nam, Dam Thi Nguyet, and Bui Thi Bich Phung

**Abstract** With the remarkable development of information technology, especially in tourism, many hotels have increased the need to integrate technology into their overall business. The study aims to determine the relationship between perceived value, belief in third-party booking sites, trust in hotel, and continuous online hotel booking intention. This study carried out qualitative research, including face-to-face interviews; and quantitative research with 543 respondents who had experienced booking on the website. The results pointed out that trust in hotels and belief in third-party booking sites positively influenced the online hotel booking continuous intention. Moreover, the perceived value negatively impacted the relationship between belief in third-party booking sites and online hotel booking continuous intention. Some managerial implications were proposed for accommodation businesses to develop effective business strategies.

**Keywords** Perceived value · Belief in third-party booking sites · Hotel trust · Online hotel booking continuous intention

## 1 Introduction

The emergence and rapid growth of hotels use information technology to improve their overall business. An example of this is the emergence and development of the website online hotel reservations. Because these sites allow consumers to book hotels online, they become more favorable [1]. Another example is the mobile platform, which has helped consumers book hotels using their smartphones [2, 3]. In the previous study, perceived value is considered the main factor in predicting buying behavior [4]. Perceived value is related to the buying behavior of individuals when booking hotels. For example, Lien et al. [5] showed that perceived value can influence consumer decisions on hotel reservations and how much they will pay for it. Moreover, trust has a critical role in the development of online transactions more

B. T. Khoa (✉) · N. H. Nam · D. T. Nguyet · B. T. B. Phung
Industrial University of Ho Chi Minh City, Ho Chi Minh City, Vietnam
e-mail: buithanhkhoa@iuh.edu.vn

© The Author(s), under exclusive license to Springer Nature Switzerland AG 2022
S. G. Yaseen (ed.), *Digital Economy, Business Analytics, and Big Data Analytics Applications*, Studies in Computational Intelligence 1010,
https://doi.org/10.1007/978-3-031-05258-3_31

than traditional transactions. It is also beneficial for consumers as they can easily avoid purchasing without being subjected to the uncertainties and risks involved in online transactions. Many shreds of evidence from prior studies have emphasized the positive effect of trust or beliefs on consumer behavior in the online market [6]. Furthermore, while consumers should be based on information provided by the site online booking of third party suppliers to reservations, the information related to the accommodation, i.e., price, service, feedback, is helpful for customers [7–9]. This research notes more about faith in the booking website of a third party; this will significantly affect the intent of reservation. In general, consumers tend to trust a hotel provider and sound policy, predicted intention future reservations of individuals [10].

While previous studies have assessed the relationship between perceived value and trust on tourists' behavior in booking [5, 11], customers who have used the service at the hotel will change their beliefs on how the hotel and with third parties related to book accommodation service for online booking further in the future? This study examined the impact of customer trust sources, i.e., hotel and third-party booking websites, on online hotel booking continuous intention. Furthermore, this study examined the moderator of perceived value in online booking behavior, i.e., online hotel booking continuous intention. The research result was the basis for proposing the hotel businesses' managerial implications for increasing the business performance through customers' online room booking. This study began with a literature review of the research model and the research hypothesis. In addition, research methodology pointed out the research method, the data collection, and the data processing. The data analysis was presented as the basis for discussion and conclusion.

## 2   Literature Review and Hypotheses

Customer trust is the readiness of a customer to be susceptible in the sellers' operation activities that the seller will do a significant transaction to the trustor [12]. Trust is the means of creating long-term relationships between seller and buyer [4] and is a critical factor in acquiring a certificate [13]. It is one of the essential components of buyer and seller interactions, and it has been a deciding factor in whether or not to buy a certificate for a long time. Many researchers in marketing said trust is the psychological state that intends to accept errors or gaps between the expectation and performance of the consumers after using the service [14]. According to Bijlsma-Frankema and Woolthuis [15], consumers prefer to buy from a trusted salesperson over a distrusted salesperson. The previous study examined the difficulties consumers face when booking a hotel in traditional ways, such as travel. Hence, the hypothesis H1 was proposed:

Hypothesis 1.    *Hotel trust positively influences the online hotel booking continuous intention.*

When customers believe in an online booking website, they can save time due to reducing the information search period to make a decision [16, 17]. In contrast, a lack of belief will limit the ability of customers to transact with businesses [18]. Many consumers, who plan to travel, concern about the reliability of third-party booking sites [1]. Belief in a recommendation website will create the transactions from the customer with the recommended business [5, 19]. The young tourist often relies on third-party booking websites or mobile apps to decide where they will rent in their tours [2]. Hotel information and ratings are available on online hotel booking websites, which are required for any booking. Gharib et al. [20] emphasized the belief that booking sites will influence consumers' online intentions, as websites provide consumers with information to help them decide on the continuous booking. Hence, the hypothesis H2 was proposed as:

Hypothesis 2.    *The belief in third-party booking sites has a positive influence on the online hotel booking continuous intention.*

Perceived value is commonly considered as a tradeoff between the costs and benefits [21, 22]. It is considered a prominent factor that affects customers' decision-making process and plays an essential role in determining customer satisfaction, decision-making, and behavior [23]. For example, value acts as a mediator between quality and purchase decisions [24]. Most studies have found a relationship between value and online booking intention and customer loyalty [25]. The group also that the relationship between individually perceived value can be predicted as the positive antecedent of online hotel booking continuous intention. Moreover, perceived value has a positive relationship with customer trust as the higher the perceived value is, the higher trust is [26, 27]. Customers' positive experiences increase their belief about the rating or recommendations from the booking site [10]. Therefore, this study proposed that:

Hypothesis 3.    *Perceived value positively influences the relationship between trust in hotels and online hotel booking continuous intention.*

Hypothesis 4.    *Perceived value positively influences the relationship between belief in third-party booking sites and online hotel booking continuous intention.*

As noted in Fig. 1, belief in third-party booking sites and trust in the hotel impacts the perceived value that directly affects the intention to book a hotel online continuously. The perceived value has the moderator effect on the relationship between trust in hotels and online hotel booking continuous intention and the relationship between belief in third-party booking sites and online hotel booking continuous intention.

# 3    Research Methodology

This study carried out qualitative research, including face-to-face interviews; and quantitative research with 550 individuals who had experienced booking on the

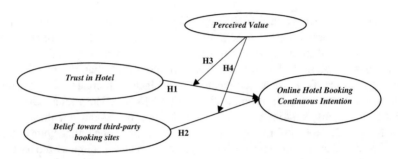

**Fig. 1** Research model

website. The research sample was selected by convenience sampling method via direct questionnaire and an online survey. The advantage of this method is that it is easy to approach the survey object and is often used when limited in space, time, and cost [28]. All questions are fixed in a 5-point Likert, and the answer is assessed from 1 (strongly disagree) to 5 (strongly agree).

The measurements were adopted from previous studies and presented in Table 1. The perceived value was adapted from Chiang and Lee [23], including 4 items. In this study, belief in third-party booking sites used three items to compare compatibility, developed from research by Bilgihan et al. [29]. Trust in the hotel was measured by 5 items, revised from Sparks and Browning [30]. Finally, online hotel booking continuous intention was measured with 4 items, developed by Kim et al. [31].

After removing inappropriate responses, this study ultimately resulted in 543 valid respondents, as Table 2.

Table 2 showed the demographic profile of respondents. Collected data included 48.1% male and 51.9% female; respondents age of 20–40 was 408 (75.1%). According to the data collected from the survey, the graduated level accounted for the largest percentage of survey respondents with 393 copies, accounting for 72.3%. There are four main occupational groups: office workers with 48.6%, followed by business and trade groups with 23.2%. The third occupational group is students, with 108 people accounting for 19.9%; the last occupational group is housewives, with 45 respondents (8.3%). Among 550 respondents, the main income group is 500–650 USD, accounting for 31.5% (171 respondents), followed by a group of 300–500 USD with 135 respondents, accounting for 24.9%.

## 4   Result

Table 3 showed the three independent constructs, including PV, TR, BEL, as well as a dependent construct (BIN), was reliable, all scales have Cronbach's alpha coefficient greater than 0.7, and corrected item-total correlation was more outstanding than 0.3 [32], so it is concluded that the scales are standard and statistical significance.

**Table 1** Measurement scales

| Code | Item |
|------|------|
| *Perceived value (PV)* | |
| PV1 | Compare to the service quality; I feel this hotel offers a fair price |
| PV2 | Compare to the cost; it is worth continuous hotel booking from the website |
| PV3 | There are many benefits compared to the cost of continuous hotel booking from the website |
| PV4 | Continuous hotel booking from the website brings much positive performance more than my expectation |
| *Belief in third-party booking sites (BEL)* | |
| BEL1 | This third-party booking site is honest with me |
| BEL2 | I believe the information on this third-party booking site |
| BEL3 | This third-party booking site always cares about its audiences |
| *Trust in hotel (TR)* | |
| TR1 | This hotel is reliable |
| TR2 | This hotel will take full responsibility |
| TR3 | I trust this hotel |
| TR4 | I know this hotel always serves me good quality service |
| TR5 | I will say positive things if I discuss the hotel with others |
| *Online hotel booking continuous intention (BIN)* | |
| BIN1 | If I plan to book this hotel, I will book with the price shown |
| BIN2 | There is a high chance that I will book this hotel |
| BIN3 | Willingness to let me book this hotel is very high |
| BIN4 | I will book the hotel from this website |

The study carried out Exploratory Factor Analysis (EFA) to evaluate the convergent validity of the independent scale. According to Table 4, the KMO value was 0.773 (>0.5). Significant in Bartlett's Test of Sphericity was 0.00 (<0.05); hence, observed items were correlated with each other in the population. The eigenvalue coefficient was 2.025 (>1), and the extracted factor has a good summary of information. The total variance extracted was 78.829% (>50%), which shows the three extracted factors. The results of the EFA were shown in Table 4.

The result in the EFA analysis for the online hotel booking continuous intention was satisfying. The KMO value was 0.748, higher than 0.5; Bartlett's sig was 0.00, less than 0.05. The eigenvalue coefficient was 2.588, more significant than 1.0; the total variance explained was 64.703 (>50%), which shows that one extraction factor explains 64.703% of the variation of the observed data.

Linear regression result in Table 5 confirmed the relationship between BEL, TR, and BIN, gave the following results:

The multicollinearity test through VIF pointed out there was no multicollinearity between the independent variables in this model with the VIF of TR and BEL were

**Table 2** Demographic statistic

|                 | Value           | n   | %    |
|-----------------|-----------------|-----|------|
| Gender          | Male            | 261 | 48.1 |
|                 | Female          | 282 | 51.9 |
| Age             | <20             | 87  | 16.0 |
|                 | 20–30           | 195 | 35.9 |
|                 | 31–40           | 213 | 39.2 |
|                 | >40             | 48  | 8.8  |
| Education level | High school     | 42  | 7.7  |
|                 | Graduated       | 393 | 72.3 |
|                 | Post-graduated  | 108 | 19.9 |
| Occupation      | Student         | 108 | 19.9 |
|                 | Businessman     | 126 | 23.2 |
|                 | Office worker   | 264 | 48.6 |
|                 | Housewife       | 45  | 8.3  |
| Income          | <300 USD        | 84  | 15.5 |
|                 | 300–500 USD     | 135 | 24.9 |
|                 | 500–650 USD     | 171 | 31.5 |
|                 | 500–850 USD     | 84  | 15.5 |
|                 | >850 USD        | 69  | 12.7 |

**Table 3** Reliability of the scales in the study

|      | Corrected item-total correlation | Cronbach's Alpha if item deleted |      | Corrected item-total correlation | Cronbach's Alpha if item deleted |
|------|------|------|------|------|------|
| *Perceived value (Cronbach's Alpha = 0.927)* | | | *Trust in hotel (Cronbach's Alpha = 0.913)* | | |
| PV1  | 0.785 | 0.92  | TR1  | 0.754 | 0.898 |
| PV2  | 0.877 | 0.889 | TR2  | 0.792 | 0.89  |
| PV3  | 0.876 | 0.89  | TR3  | 0.917 | 0.865 |
| PV4  | 0.786 | 0.92  | TR4  | 0.727 | 0.907 |
|      |       |       | TR5  | 0.723 | 0.904 |
| *Belief in third-party booking sites (Cronbach's Alpha = 0.851)* | | | | | |
| BEL1 | 0.695 | 0.817 | *Online hotel booking continuous intention (Cronbach's Alpha = 0.817)* | | |
| BEL2 | 0.725 | 0.79  | BIN1 | 0.598 | 0.79  |
| BEL3 | 0.756 | 0.763 | BIN2 | 0.604 | 0.786 |
|      |       |       | BIN3 | 0.704 | 0.739 |
|      |       |       | BIN4 | 0.658 | 0.761 |

**Table 4** The EFA result of independent constructs

|      | 1     | 2     | 4     | Initial eigenvalues | Extraction sums | Factor |
|------|-------|-------|-------|---------------------|-----------------|--------|
| TR3  | 0.952 |       |       | 4.123               | 34.357          | Trust in hotel |
| TR2  | 0.873 |       |       |                     |                 |        |
| TR1  | 0.847 |       |       |                     |                 |        |
| TR4  | 0.822 |       |       |                     |                 |        |
| TR5  | 0.792 |       |       |                     |                 |        |
| PV2  |       | 0.939 |       | 3.312               | 61.956          | Perceived value |
| PV3  |       | 0.936 |       |                     |                 |        |
| PV4  |       | 0.876 |       |                     |                 |        |
| PV1  |       | 0.870 |       |                     |                 |        |
| BEL3 |       |       | 0.906 | 2.025               | 78.829          | Belief in third-party booking sites |
| BEL2 |       |       | 0.880 |                     |                 |        |
| BEL1 |       |       | 0.819 |                     |                 |        |

**Table 5** Results of regression analysis

|            | Unstandardized coefficients (B) | Standardized coefficients (Beta) | Sig. | VIF |
|------------|--------------------------------|----------------------------------|-------|-------|
| (Constant) | 0.552                          |                                  | 0.000 |       |
| TR         | 0.586                          | 0.715                            | 0.000 | 1.069 |
| BEL        | 0.189                          | 0.231                            | 0.000 | 1.069 |
| Adjusted R Square = 0.647; Durbin-Watson = 1.861 |

1.069, which was less than 3. Through the Adjusted $R^2$, the dependent variable's degree changes under the impact of independent variables. In Table 5, the adjusted $R^2$ was 64.7%; therefore, 64.7% of the change in BIN was explained by TR and BEL. Durbin-Watson in this study was 1.861, belonged to the range of (1,3), so the regression model does not violate the autocorrelation. The ANOVA analysis table showed that the F value = 318.501 and sig. = 0.000, which is less than 0.05. The included variables were statistically at the 5% level of significance.

In assessnment of the influence of PV as a moderator on the relationship between TR and BIN with 95% confidence level as Fig. 2, TR had a significant effect on BIN (B = 0.7403, sig. = 0.00), and PV significantly impacted on BIN (B = 0.2286, sig. = 0.004); however, TR * PV had no effect on BIN (B = −0.031, sig. = 0.145). Therefore, PV did not moderate the relationship between TR and BIN.

An assessment of the influence of PV as a moderator on the relationship between BEL and BIN with 95% confidence level as Fig. 3, BEL had a significant effect on BIN (B = 1.7063, sig. = 0.00), and PV significantly impacted on BIN (B = 1.4464, sig. = 0.00); moreover, BEL * PV had negative effect on BIN (B = −0.3449, sig. = 0.00). Therefore, PV has a negative regulation of the correlation between BEL and BIN.

```
Model Summary
           R         R-sq        MSE          F          df1          df2              p
        .7987       .6380       .1302     316.6373     3.0000     539.0000          .0000


Model
                coeff          se           t          p          LLCI        ULCI
constant        .2318        .3233        .7171      .4736       -.4032       .8669
TR              .7403        .0845       8.7585      .0000        .5743       .9064
PV              .2286        .0810       2.8232      .0049        .0696       .3877
TR*PV          -.0310        .0210      -1.4725      .1415       -.0723       .0103
```

Fig. 2 The moderating result of PV on the relationship between TR and BIN

```
Model Summary
           R         R-sq        MSE          F          df1          df2              p
        .5997       .3596       .2303     100.9033     3.0000     539.0000          .0000


Model
                coeff          se           t          p          LLCI        ULCI
constant      -3.4734       .5228      -6.6445      .0000      -4.5003     -2.4465
BEL            1.7063       .1371      12.4448      .0000       1.4369      1.9756
PV             1.4464       .1288      11.2328      .0000       1.1935      1.6993
BEL*PV         -.3449       .0337     -10.2266      .0000       -.4112      -.2787
```

Fig. 3 The moderating result of PV on the relationship between BEL and BIN

# 5 Discussions

The research model proposed two kinds of trust affecting the online hotel booking continuous intention: (1) Belief in third-party booking sites; (2) Trust in hotel. Furthermore, the perceived value was considered the moderator on the relationship between trust and online hotel booking continuous intention. The results pointed out that trust in hotel and belief in third-party booking sites positively influenced the online hotel booking continuous intention. Furthermore, perceived value negatively influences the correlation between belief in third-party booking sites and online hotel booking continuous intention.

*Firstly*, hypothesis 1 (H1) was accepted with a 99% confidence level. Trust in hotel had the strongest and most positive impact on the online hotel booking continuous intention (B = 0.586 > 0; Beta = 0.821). This result shows that trust in a hotel is crucial, affecting visitors' continuous intention to book through the website [1]. Trust is always considered the antecedent of loyalty [33], which is necessary for e-commerce because customers cannot perceive the products sold online by senses, except sight. Therefore, when customers have confidence in the hotel brand they seek, they will overcome their fears about safety or quality.

*Secondly*, belief in third-party booking sites is the next influencing factor on continuous online hotel booking intention (Beta = 0.198, sig. = 0.00); hence, hypothesis 2 (H2) was supported with a 99% confidence level. The results are also consistent with Ozturk et al. [34]. This result makes perfect sense as electronic trust and loyalty are prerequisites for customers to book online through a booking website [35]. Because often booking online, customers often have to pay online, but when they have not yet arrived at the hotel they are staying at, they have accepted to spend money to book a room, showing their trust in the website that customers can pre-order.

*Lastly*, research results indicated that perceived value has a moderating role in the relationship between belief ins third-party booking sites and online hotel booking continuous intention; hence, hypothesis H4 was accepted. There is a moderating effect on the relationship above (B $= -0.3449$, sig. $= 0.00$) as shown in Fig. 4. This result means that when PV increases by 1 unit, the relationship between BEL and BIN will decrease by 0.3449 units, and vice versa. Meanwhile, the research confirmed that PV has no moderating role in the relationship between TR and BIN; therefore, H3 is rejected. According to the customer's perception, when the price and quality are suitable, they will spend money because tourists often compare the price and quality to go together or make a booking decision. Therefore, perceived value strongly influences the purchasing decision of repeat customers compared to potential customers. However, customers who have used hotel services in the past and think that what they get compared to the cost is positive will not be interested in the information on the website for online booking or reduce trust in outside information to make a decision [36, 37].

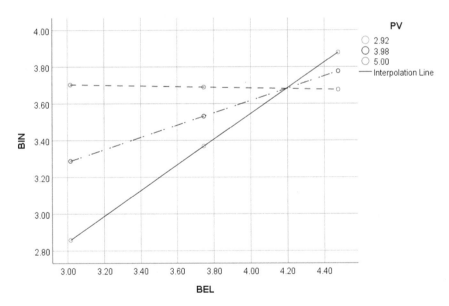

**Fig. 4** The relationship between BEL and BIN under the moderating role of PV

# 6 Conclusion

This study has made theoretical and practical contributions for hotel service businesses that allow online booking. Firstly, this research has shown that customers will continue to book hotels online if they trust the hotel and third-party apps or websites. In addition, a significant finding was that perceived value is a negative moderating of the relationship between belief in third parties and intention to continue to book online. Besides these research implications, this research also has some inferences for practicality. First, the research advises hotel owners in the hotel industry to establish a firm trust in customers because this is the factor that has the most substantial impact on booking intention. Next, the study also advises that hoteliers focus on enhancing perceived value for guests by providing quality products, as value has been shown to influence booking intention. Managers build better services on operational strategies to understand consumers' perceived value and enhance online hotel bookings. Therefore, homeowners and managers must remember the perceived value promoted by adding price and quality consumers' benefits. The result also stressed that booking consumers also impact trust for third-party booking sites, so both third-party hotels and websites should be strange. Try to keep your brand image to encourage customers to gain more trust through online hotel booking. Managers can build their beliefs by strengthening and promoting information on websites to efficiently meet customer requirements, handle customer feedback, and build a promotional reason combined. Therefore, administrators should build an interactive evaluation system that will receive visitors' comments to improve customer trust and create a healthy online booking environment.

As with any research paper, this study also has many limitations. Firstly, the sample in this research was selected according to convenience sampling, so the collected data may not have high reliability. This method is easy to implement and less expensive, but it has low confidence in representativeness. Therefore, further studies should be conducted with one of the probability sampling methods. Secondly, the research was done during the pandemic, so the customer's booking continuous intention has mainly changed, leading to the sample because the customer recalls his experience, so it is not authentic. Finally, this study shows two factors that affect customers' continuous intention to book online, so there may be many more online reviews, hotel communication, etc.

# References

1. Mahat F, Abdullah D, Bahari KA, Che Azmi N, Mohd Kamal SB, Zainol N (2020) A conceptual model of online hotel booking: the role of online review and online trust towards online booking intention. ESTEEM J Soc Sci Humanit 4:83–92
2. Khoa BT, Ha NM, Ngoc BH (2022) The accommodation services booking intention through the mobile applications of generation Y: an empirical evidence based on TAM2 model. In: Ngoc Thach N, Ha DT, Trung ND, Kreinovich V (eds) Prediction and causality in econometrics and related topics. Springer International Publishing, Cham, pp 559–574. https://doi.org/10.1007/978-3-030-77094-5_43

3. Ashrianto PD, Yustitia S (2020) The use of social media in searching for information about Papua. Jurnal The Messenger 12. https://doi.org/10.26623/themessenger.v12i2.1939
4. Dong Y, Ling L (2015) Hotel overbooking and cooperation with third-party websites. Sustainability 7:11696–11712
5. Lien C-H, Wen M-J, Huang L-C, Wu K-L (2015) Online hotel booking: the effects of brand image, price, trust and value on purchase intentions. Asia Pac Manag Rev 20:210–218
6. Abdullah D, Jayaraman K, Kamal SBM (2016) A conceptual model of interactive hotel website: the role of perceived website interactivity and customer perceived value toward website revisit intention. Procedia Econ Finance 37:170–175
7. Kimery KM, McCord M (2002) Third-party assurances: mapping the road to trust in e-retailing. JITTA: J Inf Technol Theory Appl 4:63
8. Fattah RA, Sujono FK (2020) Social presence of Ruangguru in social media during Covid-19 pandemic. J Messenger 12. https://doi.org/10.26623/themessenger.v12i2.2276
9. Zainal M, Al-Eideh BM (2021) Modeling ethical decision-making behaviors through using information index. J Syst Manag Sci 11:15–28
10. Kim SY, Kim JU, Park SC (2017) The effects of perceived value, website trust and hotel trust on online hotel booking intention. Sustainability 9:2262
11. Wang L, Law R, Guillet BD, Hung K, Fong DKC (2015) Impact of hotel website quality on online booking intentions: eTrust as a mediator. Int J Hospit Manag 47:108–115
12. Rouibah K, Al-Qirim N, Hwang Y, Pouri SG (2021) The determinants of eWoM in social commerce: the role of perceived value, perceived enjoyment, trust, risks, and satisfaction. J Glob Inf Manage (JGIM) 29:75–102
13. Lee SK, Min SR (2021) Effects of information quality of online travel agencies on trust and continuous usage intention: an application of the SOR model. J Asian Finance, Econ Bus 8:971–982
14. Propheto A, Kartini D, Sucherly S, Oesman Y (2020) Marketing performance as implication of brand image mediated by trust. Manag Sci Lett 10:741–746
15. Bijlsma-Frankema K, Woolthuis RK (2005) Trust under pressure: empirical investigations of trust and trust building in uncertain circumstances. Edward Elgar, Cheltenham
16. Kim H-W, Xu Y, Gupta S (2012) Which is more important in Internet shopping, perceived price or trust? Electron Commer Res Appl 11:241–252
17. Jin Z, Lim C-K (2021) Structural relationships among service quality, systemic characteristics, customer trust, perceived risk, customer satisfaction and intention of continuous use in mobile payment service. J Syst Manag Sci 11:48–64. https://doi.org/10.33168/JSMS.2021.0204
18. Valdez LE (2021) Socially responsible buyers' online trust on the website and their level of satisfaction. Handbook of research on reinventing economies and organizations following a global health crisis. IGI Global, pp 80–97
19. Everard A, Galletta DF (2005) How presentation flaws affect perceived site quality, trust, and intention to purchase from an online store. J Manag Inf Syst 22:56–95
20. Gharib RK, Garcia-Perez A, Dibb S, Iskoujina Z (2019) Trust and reciprocity effect on electronic word-of-mouth in online review communities. J Enterp Inf Manag 33:120–138. https://doi.org/10.1108/jeim-03-2019-0079
21. Sweeney JC, Soutar GN (2001) Consumer perceived value: the development of a multiple item scale. J Retail 77:203–220. https://doi.org/10.1016/s0022-4359(01)00041-0
22. Khoa BT, Huynh LT, Nguyen MH (2020) The relationship between perceived value and peer engagement in sharing economy: a case study of ridesharing services. J Syst Manag Sci 10:149–172. https://doi.org/10.33168/JSMS.2020.0210
23. Chiang C-C, Lee L-Y (2013) An examination of perceived value dimensions of hotel visitors: using exploratory and confirmatory factor analyses. J Int Manag Stud 8:167–174
24. Khoa BT (2020) The impact of the personal data disclosure's tradeoff on the trust and attitude loyalty in mobile banking services. J Promot Manag 27:585–608. https://doi.org/10.1080/10496491.2020.1838028
25. El-Adly MI (2019) Modelling the relationship between hotel perceived value, customer satisfaction, and customer loyalty. J Retail Consum Serv 50:322–332

26. Chen YS, Chang CH (2012) Enhance green purchase intentions: the roles of green perceived value, green perceived risk, and green trust. Manag Decis 50:502–520
27. Sharma VM, Klein A (2020) Consumer perceived value, involvement, trust, susceptibility to interpersonal influence, and intention to participate in online group buying. J Retail Consum Serv 52:101946
28. Khoa BT, Hung BP, Mohsen H (2022) Qualitative research in social sciences: data collection, data analysis, and report writing. Int J Public Sector Perform Manag 9. https://doi.org/10.1504/ijpspm.2022.10038439
29. Bilgihan A, Nusair K, Okumus F, Cobanoglu C (2015) Applying flow theory to booking experiences: an integrated model in an online service context. Inform Manag 52:668–678
30. Sparks BA, Browning V (2011) The impact of online reviews on hotel booking intentions and perception of trust. Tour Manage 32:1310–1323
31. Kim SH, Bae JH, Jeon HM (2019) Continuous intention on accommodation apps: integrated value-based adoption and expectation–confirmation model analysis. Sustainability 11:1578
32. Nunnally JC, Bernstein I (1994) The assessment of reliability. Psy Theo 3:248–292
33. Khoa BT (2020) Electronic loyalty in the relationship between consumer habits, groupon website reputation, and online trust: a case of the groupon transaction. J Theor Appl Inf Technol 98:3947–3960
34. Ozturk AB, Bilgihan A, Nusair K, Okumus F (2016) What keeps the mobile hotel booking users loyal? Investigating the roles of self-efficacy, compatibility, perceived ease of use, and perceived convenience. Int J Inform Manag 36:1350–1359
35. Schlosser AE, White TB, Lloyd SM (2018) Converting web site visitors into buyers: how web site investment increases consumer trusting beliefs and online purchase intentions. J Market 70:133–148. https://doi.org/10.1509/jmkg.70.2.133
36. Paulose D, Shakeel A (2021) Perceived experience, perceived value and customer satisfaction as antecedents to loyalty among hotel guests. J Qual Assur Hosp Tourism 1–35. https://doi.org/10.1080/1528008x.2021.1884930
37. Mai Chi VT, Paramita W, Ha Minh Quan T (2021) Does customer experience always benefit company? Examining customers' epistemic motivation and interaction with service contexts. Australas Market J. https://doi.org/10.1177/1839334921998867

# Simulated Annealing Algorithm as Heuristic Search Method in the Weibull Distribution for Investment Return Modelling

**Hamza Abubakar and Shamsul Rijal Muhammad Sabri**

**Abstract** In this paper, a modified internal rate of return (MIRR) has been presented on the assumption of Weibull distribution to investigate the investment's attractiveness in the Malaysian property development sector (MPDS). The research intends is to produce parameters estimates of the Weibull distribution for investment analysis for a long-time investment period. The MIRR data were obtained from the company financial report for 5 years investment period from 2010–2014. The Maximum likelihood estimation method has been incorporated with the Simulated annealing algorithm (SA) in estimating the parameters of Weibull distribution. The shape parameter of the Weibull distribution reflects the effectiveness in maximizing the investment return based on MIRR with lower returns and is represented as the slope of the fitted line on a Weibull probability plot. The Weibull parameter estimated using Simulated annealing (SAA) has been compared with the existing Weibull parameter estimation methods. The finding reveals that the proposed methods have good agreement with other methods used for Weibull parameter estimates based on MIRR data. The research is expected to provide an overview of the investment behaviour for the long term investment period. Therefore, the new approach based on SA in estimating the parameters Weibull function can serve as a good alternative approach for the assessment of the rate of return on investment potential.

**Keywords** Modified internal rate of return · Weibull distribution · Maximum likelihood · Simulated annealing algorithm · Goodness-of-fit tests

H. Abubakar (✉) · S. R. M. Sabri
School of Mathematical Sciences, Universiti Sains Malaysia, 11800 Pulau Penang, Malaysia
e-mail: zeeham4u2c@ikcoe.edu.ng

H. Abubakar
Department of Mathematics, Isa Kaita College of Education, Dutsin-Ma, Katsina State, Nigeria

© The Author(s), under exclusive license to Springer Nature Switzerland AG 2022
S. G. Yaseen (ed.), *Digital Economy, Business Analytics, and Big Data Analytics Applications*, Studies in Computational Intelligence 1010,
https://doi.org/10.1007/978-3-031-05258-3_32

# 1  Introduction

This article is intended to assess the financial performance of the Malaysian property development sector (MPDS) based on their internal rate of return (IRR), which will be used to further describe their investment behaviour for a short or long time investment period. Modelling the behaviour of stock investments is a critical and daunting financial objective. Investors looking to optimize their profits and regulators need predictive models. The investment valuation is an approach use in assessing and describing the company's performance in terms of the rate of return [1, 2]. Some studies have been published by various authors in evaluating businesses' performance include the work of [3–7]. Choosing the most valid criterion for evaluating a business' success is one of the most attractive and fascinating areas of research in the field of investment and finance, which enables investors to identify investments with high returns [8]. Such investments may not have been identified until any steps to reduce the cost of return on investment have been taken too late. Besides, forecasting an investment return enables the organization to estimate the projected cost of managing returns. When an investor faces a variety of options and choices while considering investment decisions, he needs to decide on the number of asset selections, as well as the amount of investment. As a result, making an informed decision about how much money to invest in each asset is critical for investors [9]. Investments can also be made by generating value from savings, but because of market volatility and uncertain financial conditions, several difficulties might arise when making decisions. Financial analysts research on pricing and investment efficiency adopt various measures to investigate the profitability of their financial investment this include the Return on Investment (ROI), Internal rate of return (IRR), Net Present Value (NPV), Return on Equity (ROE), and return on asset (ROA).

   This study employed MIRR data generated from the MPDS financial report based on the assumption of Weibull distribution. The Weibull distribution (WD) parameters have been estimated based on the MLE incorporated with Simulated annealing (SA) to investigate the behaviour of IRR to evaluate the potentiality of investment return in the MPDS. The contributions of our work include; (1) Presents modified internal rate of return (MIRR) modelling techniques; (2) Propose an investment return modelling on the assumption of Weibull distribution for Malaysian property development companies; (3) Estimate Weibull distribution parameters based on the maximum likelihood method (MLE) incorporated with the simulated annealing algorithm (SAA); (4) Compare the simulated annealing algorithm (SAA) results with the existing methods. The effectiveness of the SA in Weibull distribution (WD) parameters estimate based on a real data set will be explored based on goodness-of-Fit. The present study will be beneficial to investment decision-makers in an investment making appropriate investment decisions with minimum risk and higher investment returns.

## 2  Related Studies

This study lacks literature related to statistical modelling of internal rate of return data. However, a recent work by Sabri et al. in [2] proposed a new technique for developing an adjusted internal rate of return to model portfolio investment valuations. Following the work of Sabri et al. [1, 2] was the Markov chain simulation technique developed by [1] to forecast the investment performance of construction companies. Another similar study was conducted purposely to determines the ROA's transfer likelihood matrix in forecasting the long-term trend in the construction sector [10]. The approach used the Markov chains technique as stochastic analysis, considering that the value of Markov's property changes with their state vector. A simulation study on Modified internal rate of return modelling was conducted in [11]. The purpose was to examine the investment behaviour for the long term investment period.

The Weibull Distribution has been widely in the modelling of various types of data since its introduction in 1951 Wallodi [12]. These studies in diverse applications areas include reliability engineering [13] and lifetime analysis in [14]. Various studies have been conducted on the popularity of the Weibull distribution in fitting different kinds of data set, especially in emerging areas of stock prices, wind speed and finance and actuarial data application [15–21]. The general challenge faced by researchers in applying Weibull distribution in their data analysis is the difficulty in estimating the parameters. These problems have been addressed with the advent of powerful metaheuristics algorithms such as the Genetics algorithm, Simulated annealing etc.

Various metaheuristics algorithms have been applied as a standalone model in investing/portfolio selection problems, this includes a hybrid artificial bee colony that was proposed to select the best portfolio [22]. The proposed method guides the investors to overcome the constraint of the functions used in the portfolio optimization include configuration setting which is a time-consuming, non-linear problem-solving method, complex problems and intangible calculation parameters and inflexibility of the variables. A fuzzy theory for portfolio selection model was proposed based artificial bee colony algorithm [23]. A genetic algorithm (GA) was used in the portfolio optimization problem revealed that investment with fewer assets has a better performance outcome [24]. Investment decision techniques based on evolutionary computing known as hybrid metaheuristic algorithms have been proposed in [25]. The method was found to be good based on the analysis of variance of the obtained. Another investment selection method was proposed in financial engineering studies based on particle swarm optimization (PSO) for data for weekly prices in [26]. Erana-Diaz et al. in [27] combined the SA with another two machine learning (ML) methods in minimizing a business risk response. The method is constructed based on the computational mathematics model with avoiding, mitigate, transfer, and accept as the risk factor responses. Recently Abubakar, estimated the parameter of extended Weibull distribution using a simulated annealing algorithm in [11]. The results were reasonable with less time and computational resources than other estimation methods. Although Simulated annealing (SAA) has been applied in various optimization areas, it recorded tremendous achievement in parameters estimation of a linear and non-linear function [28–32]. In our work, Simulated annealing (SA) will be utilized in maximizing the likelihood of the Weibull distribution for optimal investment return in the MPDS. However, MIRR data has never been analyzed based on the assumption of Weibull distribution (WD), therefore, this study is brand-new focusing on the stock investment modelling using the MIRR data of the MPDS on the assumption of the WD.

## 3  Research Methods

### 3.1  Modified Internal Rate of Return Modelling Framework [2]

The investment strategy for holding the stock is by allocating a level amount of contribution at the beginning of the years for $K$ years. If we wish to hold the stock for the company chosen in the long term period, the stock valuation can be also seen by computing the MIRR. At the same time, if the company declares dividends yearly, the cash dividends are reinvested and together deposited with the level contribution to enlarging the share units. At the end of $K$ years, we let all our share units earn the share capital which indicates the profit of our investment for $K$ years. If our share capital is less than our total contribution, we may expect our MIRR to be in a negative form. The detailed procedure of the investment return was documented in [33, 34] and [2]. The Net Present Value (NPV) of stock investment is computed at time zero as follows,

$$NPV = \left[ S_K^{(2)} P_{u_{K+1,2}} + B_K + \delta_K \right] (1+r) - \frac{u_{K+1,1} - u_{1,1}}{365}$$
$$- C \sum_{k=1}^{K+1} \mu_k (1+r) - \frac{u_{k+1,1} - u_{1,1}}{365} \tag{1}$$

$$\equiv F(K)(1+r) - \frac{u_{K+1,1} - u_{1,1}}{365} - \sum_{k=1}^{K+1} C_k^* (1+r) - \frac{u_{k+1,1} - u_{1,1}}{365} \tag{2}$$

where $k = (1, 2, 3, \ldots, K)$ and $S_k^{(2)}$ is accumulated share unit after share issuance at the end of the year $k$ which can be computed as follow;

$$S_k^{(2)} = \psi_k \times S_k^{(1)} \tag{3}$$

where $\psi_t$ is the function of share issuance, $S_k^{(1)}$ is the share units at the beginning of the year $k$, and $F(K)$ is the terminal value investment fund to be let at the end of the year $K$ which can be computed as follows;

$$F(K) = S_K^{(2)} P_{u_{k+1,2}} + B_K + \delta_K \tag{4}$$

where $u_{K,1}$ represent the date of share purchased and sold, $u_{k,2}$ is the date of dividend and share issued based on the stock reported on year $k$, $P_{u_{k,2}}$ defined the stock price at the date $u_{k,2}$, $B_k$ represents the cash balance at the year $k$, $\delta_K$ defined as a cash dividend at year $k$, $r$ represents the modified IRR of the project development companies, $C$ is the yearly fixed contribution which can be computed as follows:

$$C = \mu_k \frac{1}{C_k^*} \tag{5}$$

It is very important to choose the best potential stocks to hold in a holding term. Furthermore, holding a stock for a $K$-years period of the investment may vary in terms of MIRR. Some might choose the best point of time to start investing, but it is very difficult to identify it as the MIRR measure can only be observed yearly. Therefore, by assuming the MIRR for all starting times to invest are common, we may define the MIRR, denoted as $R_{tiK}$, as a random variable (RV) with the mean $E(R_K)$ and variance and $Var(R_K)$. In investment, we may obtain a positive value of profit (or even be greater than our capital investment) as well as earning nothing. This indicates our capital of investment, $C$, could be infinite or even zero value. For some time $K$, our terminal investment $C(1 + R_K)^K$ is in between 0 to infinity. Hence,

$$0 < C(1 + R_K)^K <\to -1 + R_{tiK} \tag{6}$$

Since, $R_{tiK} > -1$, we define a non-negative transformed rate of return, $X_{tiK}$ such that

$$X_{tiK} = 1 + R_{tiK} \tag{7}$$

## 3.2 Transformed Modified Internal Rate of Return on Weibull Distribution

This study considered investment returns for 62 companies from the Malaysian property development sector. The investment behaviour based on MIRR distribution from one (1) to five (5) year investment periods have been considered (i.e. $K = (1, 2, 3, 4, 5)$. The periods under study starts from 2010 and lasts until 2014. We set $t_1 = 2010$ and $t_T = 2014$ and hence, $T = 5$ Furthermore, all companies' MIRR is counted in our study. For example, for a one-year investment period (i.e.($K = 1$), a maximum of $T = 5$ has been obtained by multiplied with 62 companies understudy to obtain the sample size of 310 of MIRR data. The transform MIRR data is assumed to follow the WD. The WD is flexible and easy to apply in the modelling of many different forms of data [35].

In this study, the transformed MIRR data $X_{itK}$ is assumed to come from the Weibull distribution (WD). According to the observed MIRR data, the WD can be described based on PDF as follows.

$$
f_{X_{tiK}}(\alpha_K, \beta_K, \eta_K) = \begin{cases} \frac{\beta_K}{\eta_K}\left(\frac{X_{tiK}-\alpha_K}{\eta_K}\right)^{\beta-1} e^{-\left(\frac{X_{tiK}-\alpha}{\eta_K}\right)^{\beta}}, & X_{tiK} > \alpha \\ 0, & X_{tiK} \leq \alpha \end{cases}
$$ (8)

where $\alpha_K, \beta_K$ and $\eta_K$ are parameters to be determined. From Eq. (9), the three parameters WD mean is defined as,

$$
E(X_{tiK}) = \eta_K \Gamma\left(1 + \frac{1}{\beta_K}\right) + \alpha_K
$$ (9)

where $\Gamma$ signifies a gamma function defined in Eq. (10)–(13). The variance of the three parameters Weibull variance is a function of the shape and scale parameters deduced as follows.

$$
Var(X_{tiK}) = \eta^2\left[\Gamma\left(1 + \frac{2}{\beta_K}\right) - \Gamma^2\left(1 + \frac{1}{\beta_K}\right)\right]
$$ (10)

a higher variance would generally provide a lower reliability performance of MIRR at the same point in time. The MIRR distribution is as follows

$$
f_{R_{tiK}}(R_{tiK}; \alpha_K, \beta_K, \eta_K) = \frac{\beta_K}{\eta_K}\left(\frac{(R_{tiK}+1) - \alpha_K}{\eta_K}\right)^{\beta-1} e^{-\left(\frac{(R_{tiK}+1)-\alpha_K}{\eta_K}\right)^{\beta}}, R_{\tau iT} > -1
$$ (11)

This indicates that the mean and variance of Eq. (11) can be re-written as in Eq. (12) and (13) respectively as follows

$$
E(R_{tiK}) = \eta_K \Gamma\left(1 + \frac{1}{\beta_K}\right) + \alpha_K
$$ (12)

$$
var(R_{tiK}) = \eta_K^2\left[\Gamma\left(1 + \frac{2}{\beta_K}\right) - \Gamma^2\left(1 + \frac{1}{\beta_K}\right)\right]
$$ (13)

for a period of investment, $K$, the MLE, a random sample of $x_1, x_2, x_3, \ldots, x_{K*}$ size $K*$, where $K* \in K$ has been considered. The likelihood function is presented as,

$$L = \prod_{K=1}^{K*} \frac{\beta_K}{\eta_K} \left( \frac{(R_{tiK} + 1) - \alpha_K}{\eta_K} \right)^{\beta_K - 1} e^{-\left( \frac{(R_{tiK}+1)-\alpha_K}{\eta_K} \right)^{\beta_K}}, R_{tiK} > -1 \quad (14)$$

The log-likelihood function of Eq. (14) is given as follows,

$$InL = \sum_{K=1}^{K*} \left[ \begin{array}{c} In(\beta_K) + (\beta_K - 1)In((r_{tiK} + 1) - \alpha_K) \\ - \beta_K In(\eta_K) - \left( \frac{(r_{tiK} + 1) - \alpha_K}{\eta_K} \right)^{\beta_K} \end{array} \right] \quad (15)$$

It is quite difficult and exhaustive to derive the gradient of the Weibull model in Eq. (15) to attain the complicated objectives function as observed by many scholars. To address the complication involved in estimating the parameters of WD, various metaheuristics algorithms have been incorporated including [16, 36–38] and [40]. A simulated annealing procedure has been adopted according to [16] in estimating the MIRR distribution on the assumption of Weibull distribution.

## 3.3 The Goodness of Fit Measure

In this section, statistics are employed to determine how closely a particular distribution matches the associated distribution for a particular dataset [40]. The distribution with the best fit will be the one whose goodness-of-fit statistic is the lowest [42]. To examine whether a theoretical probability density function is suitable to describe the transformed MIRR distributions $(X_{tiK})$ data during the investment period or not, several tests are used for validating the accuracy of the predicted MIRR distribution obtained from the Weibull probability density function; we employed goodness-of-fit (GOF) presented in Eq. (16)–(19) as follows

$$KS = \max_{1 \leq i \leq n} \left| \widehat{F}(X_{tiK}) - F(X_{tiK}) \right| = \max_{1 \leq i \leq n} \left| \widehat{F}(X_{tiK}) - P_i \right| \quad (16)$$

$$AIC = 2k - 2InL\left( \widehat{\xi} \right) \quad (17)$$

$$RMSE = \sqrt{\frac{1}{n} \sum_{i=1}^{n} \left[ \left( \widehat{F}(X_{tiK}) - \overline{F} \right)^2 \right]} \quad (18)$$

$$R^2 = \frac{\sum_{i=1}^{n} \left[ \left( \widehat{F}(X_{tiK}) - \overline{F} \right)^2 \right]}{\sum_{i=1}^{n} \left[ \left( \widehat{F}(X_{tiK}) - \overline{F} \right)^2 \right] + \left[ \sum_{i=1}^{n} \left( F_n(X_{tiK}) - \widehat{F}(X_{tiK}) \right)^2 \right]} \quad (19)$$

## 4 Results and Discussion

The unknown parameters of the WD have been estimated via a simulated annealing algorithm. The results obtained has been compared with numerical methods (NM), moments methods (MM) and regression methods (RM) in term of goodness statistical measure using the MIRR data for five years investment period. Based on the estimated parameters of the WD, the histogram of transformed MIRR data and the fitted Weibull Probability density function (WPDF) from 1 to 5 years investment period have been presented in Figs. 1, 2, 3, 4 until 5 respectively, whereby Figs. 6, 7, 8 and 9 displayed the goodness of fitness statistical measure of Weibull distribution, based on different estimation methods in terms of the AIC, K-S, RMSE and $R^2$ for the stock size from 62, up to 310.

### 4.1  Analysis of Findings

The result of the GOF statistics revealed that the methods of estimates employed in this study performed well with the coefficient of determination much closer to 1. On the other hand, the RMSE and K-S test of all methods utilized in this study are closer to zero which indicates the efficiency of the estimation methods performance within

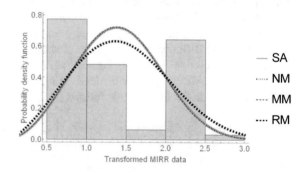

**Fig. 1** Fitted PDFs of MIRR data for 1 year investment period

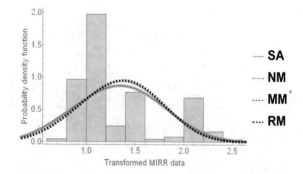

**Fig. 2** Fitted PDFs of MIRR data for 2 year investment period

**Fig. 3** Fitted PDFs of MIRR data for 3 year investment period

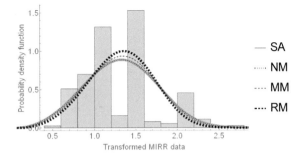

**Fig. 4** Fitted PDFs of MIRR data for 4 year investment period

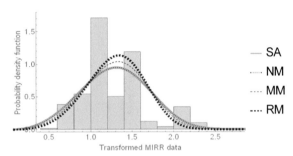

**Fig. 5** Fitted PDFs of MIRR data for 5 year investment period

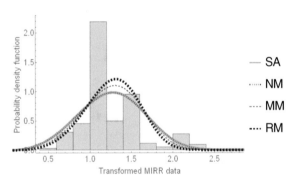

**Fig. 6** AIC of various Weibull parameter estimation methods

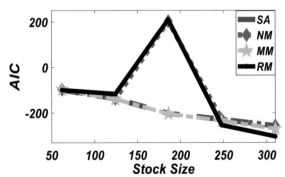

**Fig. 7** KS test of various
Weibull parameter
estimation methods

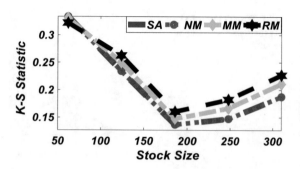

**Fig. 8** RMSE of various
Weibull parameter
estimation methods

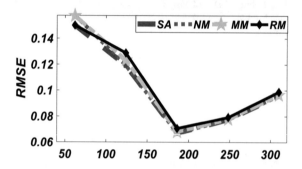

**Fig. 9** R-squared of various
Weibull parameter
estimation methods

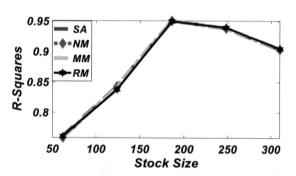

the same limit. However, the SAA outperforms the numerical methods (NM) in terms
of accuracy and efficiency when optimizing unknown parameters of the distribution.
The results displayed in Figs. 1 2, 3, 4, 5, 6, 7 and 8 were supported by the GOF
test which indicated that there are virtually no differences between the estimates
obtained via SAA and other estimation methods. This is not surprising because the
objective function in both cases is the log-likelihood function. The difference in
these methods being only in the way the objective function is maximized; one is
through differentiation and the other by optimization. It further observed that the
estimates obtained by maximum likelihood (ML) and method of moments (MM) are
quite close. The regression method (RM) revealed better performance when stock

size (SS) is small. That is, the RM estimates are possible values for those of the maximum likelihood method. Estimated parameters obtained are consistently better as the complexity increases, that is sample size increases. The main estimation theory states unequivocally that the larger the sample size, the better the estimate. However, as the SS grows, more parameters are involved, the fitness function (likelihood function) to be optimized becomes more difficult. As a result, deciding on sample size is a matter of compromise [16, 36, 42]. However, it is quite clear that the estimation methods under study displayed similar behaviour for all performance metrics used in this study. This again indicates that the WD is a better fit for the MIRR data of the MPDS.

Reporting on the MIRR distribution analysis using the Weibull distribution produced objective performance indicators. The Weibull analysis formulates the best applicable investment behaviour to reduce risk in the MPDS. Also, by using a probabilistic Weibull plot, it's possible to predict or forecast the behaviour of risk involved in the company and to plan a particular action to minimize the risk. If the MIRR data does describe a straight line it generally means, there is a steady rate of return in the investment. Now, if the slope of the fitted line is greater than one, it indicates an increase in a monotonic decrease in the investment return over time. The analysis considers the slope of the line (if straight) and gleans a few trends about investment behaviour. Detailed risk analysis reveals the impacts of a short time or long time investment period, yet the longer the investment period, the less risk on the investment. This revealed that purchasing shares of this company may have a lower return for a short term period. This situation characterizes a risky investment. One of the objectives is to maximize the likelihood of the transformed MIRR. The longer the investment period, the high the return of the investment for this sector. This is not to say that the yield hike in the Malaysian property production market would take any specific form. Instead, it acts as a model for the investors to follow to generate a suitable probability distribution. To build a better distributions model, such an analysis can be generalized to other datasets. This was supported by the magnitudes of the RMSE and the correlation coefficient $R^2$, which were very close. As a result, the MLE-SAA is recommended for estimating MIRR distributions to eliminate the complexity and uncertainties associated with the Weibull model. The Weibull distribution model gives an adequate representation of the MIRR data for the MPDS.

## 5 Conclusion

In this study, the Weibull distribution parameters have been estimated based on the maximum likelihood method (MLE) incorporated with Simulated annealing (SA) in investigating the behaviour of investment return in the MPDS. The investment modelling techniques were successfully presented on the assumption of Weibull distribution in fitting the modified internal rate of return (MIRR) data of the MPDS. The data have allowed determining the distribution of MIRR from short time to long term investment period (1–5 years). Based on SAA in maximum likelihood estimation is a viable method of obtaining parameter estimates of Weibull distribution, especially considering the small variance attained. We believe that using SAA becomes more meaningful when the likelihood surface is more rugged (and has more dimensions) than the usual distributions. The Weibull parameters were evaluated by a set of indicators defined from the MIRR distribution.

The study will be extended to three parameters Weibull distribution and extended Weibull distribution using a Simulated annealing algorithm to investigate the investment behaviour of MPDS based on the transformed MIRR data for a longer investment period. Extension from our study, the different classes of statistical distribution can be adopted such as Gamma distribution, lognormal distribution, Tukey-Lambda Distribution, Rayleigh distribution etc. to investigate the fit to MIRR data. This study is currently in progress and will be reported in a future article.

# References

1. Mustafa W, Sabri SRM (2020) A simulation study: obtaining a sufficient sample size of discrete-time Markov chains of investment in a short frequency of time 10:906–919. https://doi.org/10.18488/journal.aefr.2020.108.906.919
2. Sabri SRM, Mustafa Sarsour W (2019) Modelling on stock investment valuation for long-term strategy. J Invest Manag 8:60. https://doi.org/10.11648/j.jim.20190803.11
3. Doganaksoy N (2004) Weibull models. Technometrics. https://doi.org/10.1198/tech.2004.s226
4. Erik Karl'en CW (2017) Eturn ate rediction
5. Genschel U, Meeker WQ (2010) A comparison of maximum likelihood and median-rank regression for Weibull estimation. Qual Eng. https://doi.org/10.1080/08982112.2010.503447
6. Sgarbossa F, Zennaro I, Florian E, Persona A (2018) Impacts of weibull parameters estimation on preventive maintenance cost. IFAC-PapersOnLine. https://doi.org/10.1016/j.ifacol.2018.08.369
7. Teimouri M, Gupta AK (2013) On the three-parameter Weibull distribution shape parameter estimation. J Data Sci
8. Jamei R (2020) Investigating the mathematical models (TOPSIS, SAW) to prioritize the investments in the accepted pharmaceutical companies in Tehran Stock Exchange 5:215–227. https://doi.org/10.22034/amfa.2020.1880616.1312
9. Raei R, Bahrani Jahromi M (2012) Portfolio optimization using a hybrid of fuzzy ANP, VIKOR and TOPSIS. Manag Sci Lett https://doi.org/10.5267/j.msl.2012.07.019
10. Sarsour WM, Sabri SRM (2020) Forecasting the long-run behavior of the stock price of some selected companies in the Malaysian construction sector: a Markov chain approach. Int J Math Eng Manag Sci 5:296–308. https://doi.org/10.33889/IJMEMS.2020.5.2.024
11. Abubakar H, Sabri SRM (2021) Simulation study on modified Weibull distribution for modelling of Investment return. Partanika J Sci Technol 29
12. Weibull W (1951) A statistical distribution function of wide applicability. J Appl Mech 18(18):293–297
13. Peng X, Yan Z (2014) Estimation and application for a new extended Weibull distribution. Reliab Eng Syst Saf. https://doi.org/10.1016/j.ress.2013.07.007
14. Lawless JF (2003) Statistical models and methods for lifetime data, 2nd edn
15. Elmahdy EE, Aboutahoun AW (2013) A new approach for parameter estimation of finite Weibull mixture distributions for reliability modeling. Appl Math Model. https://doi.org/10.1016/j.apm.2012.04.023

16. Abbasi B, Eshragh Jahromi AH, Arkat J, Hosseinkouchack M (2006) Estimating the parameters of Weibull distribution using simulated annealing algorithm. Appl Math Comput. https://doi.org/10.1016/j.amc.2006.05.063

17. Nadarajah S, Kotz S (2006) The modified Weibull distribution for asset returns. Quant Financ 6:449

18. Lai CD, Xie M, Murthy DNP (2003) A modified Weibull distribution. IEEE Trans Reliab. https://doi.org/10.1109/TR.2002.805788

19. Malevergne Y, Pisarenko V, Sornette D (2006) The modified Weibull distribution for asset returns: Reply

20. Almetwally EM, Almongy HM (2019) Estimation methods for the New Weibull-Pareto distribution: simulation and application 17, 613–632. https://doi.org/10.6339/JDS.201907

21. Akdağ SA, Dinler A (2009) A new method to estimate Weibull parameters for wind energy applications. Energy Convers Manag. https://doi.org/10.1016/j.enconman.2009.03.020

22. Rahmani M, Eraqi MK, Nikoomaram H (2019) Portfolio optimization by means of meta heuristic algorithms 4:83–97. https://doi.org/10.22034/amfa.2019.579510.1144

23. Gao W, Sheng H, Wang J, Wang S (2019) Artificial bee colony algorithm based on novel mechanism for fuzzy portfolio selection. IEEE Trans Fuzzy Syst. https://doi.org/10.1109/TFUZZ.2018.2856120

24. Chang TJ, Yang SC, Chang KJ (2009) Portfolio optimization problems in different risk measures using genetic algorithm. Expert Syst Appl. https://doi.org/10.1016/j.eswa.2009.02.062

25. Bavarsad Salehpoor I, Molla-Alizadeh-Zavardehi S (2019) A constrained portfolio selection model at considering risk-adjusted measure by using hybrid meta-heuristic algorithms. Appl Soft Comput J. https://doi.org/10.1016/j.asoc.2018.11.011

26. Ni Q, Yin X, Tian K, Zhai Y (2017) Particle swarm optimization with dynamic random population topology strategies for a generalized portfolio selection problem. Nat Comput. https://doi.org/10.1007/s11047-016-9541-x

27. Erana-Diaz ML, Cruz-Chavez MA, Rivera-Lopez R, Martinez-Bahena B, Avila-Melgar EY, Heriberto Cruz-Rosales M (2020) Optimization for risk decision-making through simulated annealing. IEEE Access. 8:117063–117079. https://doi.org/10.1109/ACCESS.2020.3005084

28. Javidrad F, Nazari M (2017) A new hybrid particle swarm and simulated annealing stochastic optimization method. Appl Soft Comput J. https://doi.org/10.1016/j.asoc.2017.07.023

29. Zhang W, Maleki A, Rosen MA, Liu J (2018) Optimization with a simulated annealing algorithm of a hybrid system for renewable energy including battery and hydrogen storage. Energy. https://doi.org/10.1016/j.energy.2018.08.112

30. Abubakar H, Rijal S, Sabri M, Masanawa SA, Yusuf S (2020) Modified election algorithm in hopfield neural network for optimal random k satisfiability representation. Int J Simul Multidisci Des Optim 16:1–13

31. Abubakar H, Danrimi ML (2021) Hopfield type of artificial neural network via election algorithm as heuristic search method for random Boolean kSatisfiability. Int J Comput Digit Syst 10:659–673. https://doi.org/10.12785/ijcds/100163

32. Ghadiri Nejad M, Güden H, Vizvári B, Vatankhah Barenji R (2018) A mathematical model and simulated annealing algorithm for solving the cyclic scheduling problem of a flexible robotic cell. Adv Mech Eng. https://doi.org/10.1177/1687814017753912

33. Kellison S (2009) stephen-kellison-theory-of-interest-3e.pdf

34. Protter P, Capinski M, Zastawniak T (2004) Mathematics for finance: an introduction to financial engineering

35. Thomas GM (1995) Weibull parameter estimation using genetic algorithms and a heuristic approach to cut-set analysis

36. Abbasi B, Niaki STA, Khalife MA, Faize Y (2011) A hybrid variable neighborhood search and simulated annealing algorithm to estimate the three parameters of the Weibull distribution. Expert Syst Appl. https://doi.org/10.1016/j.eswa.2010.07.022

37. Yonar A, Yapici Pehlivan N (2020) Artificial bee colony with levy flights for parameter estimation of 3-p Weibull distribution. Iran J Sci Technol Trans A Sci. https://doi.org/10.1007/s40995-020-00886-4

38. Yang F, Ren H, Hu Z (2019) Maximum likelihood estimation for three-parameter Weibull distribution using evolutionary strategy. Math Probl Eng 2019. https://doi.org/10.1155/2019/6281781
39. Jiang H, Wang J, Wu J, Geng W (2017) Comparison of numerical methods and metaheuristic optimization algorithms for estimating parameters for wind energy potential assessment in low wind regions
40. Sultana T, Muhammad F, Aslam M (2019) Estimation of parameters for the lifetime distributions 12:77–92
41. Lei J (2016) A goodness-of-fit test for stochastic block models. Ann Stat 44:401–424
42. Tashkova K, Šilc J, Atanasova N, Džeroski S (2012) Parameter estimation in a nonlinear dynamic model of an aquatic ecosystem with meta-heuristic optimization. Ecol Modell. https://doi.org/10.1016/j.ecolmodel.2011.11.029

# The Relationship Between the Innovative Marketing Mix Elements and the Firms' Performance

Abdul Razzak Alshehadeh, Ghaleb Awad Elrefae, Farid Kourtel, Abdelhafid Belarbi, and Ihab Ali El-Qirem

**Abstract** This study explored the innovative practices related to marketing mix elements and investigates their impact on firms' performance. Validated questionnaires (n = 60) were distributed to four of the biggest malls in Amman. The main finding of this study is that the growing complexity of customers' needs and the changeable businesses environments promoted investors to use innovative marketing tools, which improves the organization's performance in terms of efficiency and effectiveness. Also, the innovative marketing tools is not limited to a single element of the marketing mix but include all the marketing strategies related to the product, pricing, distribution, and promotion. Implementing more innovative marking mix elements in the organization will add value and be beneficial to the customers. It is recommended to give more attention to identifying innovative marketing tools that were ignored by other competitors, then build integrated marketing mix strategies, considering these overlooked aspects, which is expected to enable them to maximize their performance internally and externally.

**Keywords** Innovative · Innovative marketing mix · Performance · Jordan

## 1 Introduction

Innovation in the marketing mix is considered one of the modern trends that reveals the development of the marketing vision. This field is no longer limited to the study of market needs. However, it also refers to the "value creation" level by altering and convincing the potential customer to buy the innovative product when presented

A. R. Alshehadeh (✉) · I. A. El-Qirem
Faculty of Business, Al-Zaytoonah University of Jordan, Amman 11733, Jordan
e-mail: abdulrazzaqsh@zuj.edu.jo

G. A. Elrefae · A. Belarbi
College of Business, Al Ain University, 64141 Al Ain, UAE

F. Kourtel
Department of Marketing, Setif University, Setif, Algeria

© The Author(s), under exclusive license to Springer Nature Switzerland AG 2022
S. G. Yaseen (ed.), *Digital Economy, Business Analytics, and Big Data Analytics Applications*, Studies in Computational Intelligence 1010,
https://doi.org/10.1007/978-3-031-05258-3_33

in the market, which challenges the local and international modern firms. These challenges affect their efficiency in using the available resources and the effectiveness in planning and goals' achievement [1]. It was noticed that the most critical variable affecting the customer's buying decision is the prices value, as they keep looking for the product with the lowest price and reasonable quality. It was shown that innovation is not limited to the product but includes prices, innovative payment plans, irresistible promotion tools, and distribution methods. Therefore, the firms are needed to adopt innovative marketing tools to avoid the threat of being out of the market, as happened with Nokia, which did not believe in innovation. Thus, other companies like Apple Inc. took their share, using the innovative marketing mix widely [2]. This study revealed the relationship of using innovative marketing mix elements (product, price, distribution, and promotion) on the Jordanian firms' performance.

## Innovative Marketing Mix

The innovative marketing mix represents a set of interactive elements that aim at improving the firms' performance that depend basically on how far the changes and the keep updating of such elements, considering risk element and the possibility of facing internal or external conflict [3].

## Innovative Product

It is essential to provide new and innovative products to the market. These products could have many shapes with different purposes; they could be explicitly designed to enhance the customers' satisfaction or reflect various forms to invade new and available markets sectors. Further, the firms interested in adopting innovative marketing mix elements should be aware of the relation between the risk and the diversity in markets target [4]. Additionally, the innovative product could be defined as providing a significant item, new service, or improvement in the product used to the technology [5]. In addition, the innovative product is also defined as presenting new items to the market to improve the item's use and quality [6]. Thus, the innovative product is characterized by added value in its usages, technical descriptions, or the technology employed in the design and presenting it in the market [7]. However, providing innovative service is more complex than innovative product, as the services must provide the motivation, which faces a risk of the ability to provide reasons that can achieve the success of the new service. In contrast, in the field of product, there are research centres developing designs and models and new benefits of the item [3].

The success and growth of any firm are linked with the customers' acceptance of its products, its ability to face challenges and other firms' competition [8]. The process of creating a new innovative product requires good planning and scientific study of the internal and external environment, further evaluation of its points of strength and weakness with the seeking of opportunities in the market, besides revise and the analysis of every stage of the innovation process starting by the generating idea, passing by the analysis and development business process, and ending by testing and presenting the products to the market [9].

## Innovative Pricing

Innovation pricing is considered an effective way to achieve a good and distinguished place in the market. The well-studied and innovative price is the winning card as it is directly affecting the customers' behaviour. However, as part of innovation elements, pricing did not attract great attention compared to the product, distribution, and promotion [10]. The innovative price highlights the entrepreneurial process of defining a product and a price level to make the product chargeable. Other shapes for real innovation are instalment payment or vehicle insurance instalment [11].

In pricing, the innovation patterns vary based on the firms' goal, activity, product nature, targeted market, or the competitive environment. Innovation pricing could be in many ways, including Self-pricing [4], pricing based on the product cost, bundle product pricing [12], at-and off-peak time pricing [13], psychological pricing, activity-based pricing [14].

## Innovative Distribution

Innovation distribution is the strategy based on grabbing the opportunities and getting over competitors' sales by adopting appropriate distribution channels [15]. Further, it is one of the firms' essential activities characterized by high speed, accuracy performance and the design of distributive channels that suit the nature of the product. Recently, many firms have adopted new and developed distribution activities [16]. The innovative distribution activity includes many stages, i.e., designing, defining channels, and transporting the product. The firms must adopt distinguished, innovative distribution methods compatible with their capabilities, goals, competitors' distribution patterns, or target market [17]. There are many innovative distribution methods such as postal, Electronic, placing products on shelves and supply shopping carts [18].

## Innovative Promotion

Promotion is the art of communication, attraction, and convinces of the products displayed or presented in the market [5]. Also, innovative promotion activities should be designed professionally, and the font should be new and attractive, and it needs to suit the nature of the product and target customer [19]. Thus, the innovative promotion activity is supported by designing the promotion message, which clarifies the goal to be achieved. Therefore, promotion should be attractive, distinguished, and unique to the customer desire to try the product. The innovative promotion tries to make the customer believe that this product is designed to meet his needs. The practical promotion can justify the high price of the product [19].

Furthermore, innovative advertising is one of the marketing mix elements that aims to attract the customer's interest and attention to watch the advertisement. The advertising words show that the product is unique, and it is difficult to have an alternative. Thus it will last longer in the customer's mind and motivate him to have that product [20]. Any innovative advertisement must have the innovation in the main title, subtitle, the content of advertising, closure of advertising, cartoon, images & music, person advertising, innovative show, the following elements [20].

## Performance and Its Measurement Indicators

Performance is the success of achieving the firms' defined goals, mainly linked with efficiency and effectiveness as indicators [1]. Performance can also be defined as a system that processes input into outcomes representing the institution's vision and goals [20]. Some researchers consider that firms with good performance can establish and maintain value for their customers and receptively to their shareholders and employees. Undoubtedly, the excellent performance of marketing is the one that adopted the innovative marketing tools [3].

## Impact of Innovative Marketing Mix on the Firms' Performance

Innovative marketing indicates the firms' ability to create multiple opportunities for growth and continuity through increased profitability and sales, besides reducing the cost. The impact of innovative marketing mixes on the firms' performance could be illustrated by addressing the four leading indicators of performance: profitability, cost, sales, and return on investment [17].

## Impact of Innovative Marketing Mix on Profitability

Profitability is the profit resulting from firms' innovation in different fields, including production, finance, marketing, and management affairs. In other words, all the institution's activities can help in achieving profitability. The innovative products help achieve two related advantages: the highest and the lowest cost [14, 21].

- The highest price advantage:

When the firms come up with new innovative products or make some improvements to some products that are compatible with the customers' needs, the firms may temporarily monopolize the market or part of it because of the modernity element that resulted from innovation and compatibility [22].

- The lowest price advantage:

The innovation products can be the first to be launched or introduced improvement in its products, which is the lowest cost. At the same time, the competitors were trying to catch up with the firms through imitation. Further, the firms utilized the time to achieve the lowest cost advantage through learning or the experience acquisition in this type of innovation that reduces the production cost through the continuous improvement of production methods or by benefiting the new technology [14].

Impact of innovative marketing mix on the reduction of cost and the sales increase innovatively providing the same product that other competitors present in the market with lower prices through reducing the cost, relying on less raw materials, or adopting the more productive technical process. Thus, the institution's sales will be increased. Thus, the profits will get higher. An innovative marketing mix in presenting and developing innovative products will increase return on investment by increasing the sales volume [23].

## 2   Method

The study population included the four largest malls in Amman. Sixty validated questionnaires were distributed from March to May of 2021 to the purchasing managers, their deputies, advertising managers, executive directors, department managers, and their deputies. The analysis was done by the SSPS program as done previously [24].

**Study Hypotheses**

HA: There is a relationship between using the innovative marketing mix elements (product, price, distribution promotion) and firms' performance.

Ha1: There is a relationship between the innovative product and firms' performance.

Ha 2: There is a relationship between innovative price and firms' performance.

Ha 3: There is a relationship between innovative distribution and firms' performance.

Ha 4: There is a relationship between innovative promotion and firms' performance.

## 3   Results and Discussion

Variables were assessed by measures of central tendency (mean) and measures of dispersion or variation (standard deviation). Further, the Likert scale was ordered from 1 to 5, as 1 indicated of strongly disagrees and 5 indicated of strongly agree). They were then graded according to the mean as done previously [25].

As it is shown in Table 1, the mean of all the items of the innovative marketing mix ranged between (3.54–4.53), and item No. (10) ranked first in terms of the degree of approval, as it stated, "Innovative distribution is based on a strategy of capturing opportunities and gaining sales from competitors by adopting a distribution mechanism that is more appropriate than other options available", with a high grade and a standard deviation (0.56), and this indicates the presence of convergence and agreement in the answers of the study sample to the items in the field of the study tool, while paragraph No. (15) ranked the lowest among In terms of the degree of approval, it states, "Innovative promotion is modern, renewed and has never been introduced before, in line with what is on the market and features the right promotional message", with a high grade and standard deviation (0.97), this also indicates the presence of convergence and agreement in the answers of the study sample to the items of the study tool field.

As shown in Table 2, the value of (R) was 0.78, and the value of ($R^2$) was 0.86, which indicate a strong relationship between the innovative marketing mix and the performance of the firms. There is a statistically significant relationship between the independent and dependent variables, where the value of (f) was 647.3, and the $P$-value was 0.00. Thus, the study's main hypothesis (Ha) is accepted, which states that

**Table 1** The mean and standard deviations of the innovative marketing mix elements (product, price, distribution promotion) and firms' performance

| No. | Items | Mean | SD | Grade |
|-----|-------|------|-----|-------|
| *Innovative product* | | | | |
| 1 | The innovative product can enhance customer satisfaction and exploiting opportunities in new markets | 4.22 | 1.02 | High |
| 2 | An innovative product can provide a new commodity or service or significantly improve product characteristics of used technology and product-making materials | 3.94 | 0.96 | High |
| 3 | The innovative product is a product that can deliver everything new to the market in a way that reflects a significant improvement in the level of use or quality of the product on offer | 4.17 | 0.92 | High |
| 4 | The innovative product reflects every material or moral benefit, characterized by innovation and excellence from what competitors offer | 3.95 | 1.03 | High |
| *Innovative price* | | | | |
| 5 | Innovative pricing is an effective way to achieve excellence and create a distinct market position; the innovative price is a trump card if carefully developed and carefully studied | 4.27 | 0.78 | High |
| 6 | An innovative price is a promotional tool that significantly impacts consumer behaviour and is an essential source of cost control | 4.12 | 0.88 | High |
| 7 | innovative price should include the use of an instalment strategy or easy allocation | 4.10 | 0.93 | High |
| 8 | Innovative pricing encourages purchasing patterns that depend on rational decision-making rather than emotional | 3.96 | 0.96 | High |
| *Innovative distribution* | | | | |
| 9 | Innovative distribution is a vital activity based on speed, performance accuracy, and immediate availability of the product provided | 4.23 | 0.92 | High |
| 10 | Innovative distribution is based on capturing opportunities and gaining sales from competitors by adopting a distribution mechanism that is more appropriate than other options available | 4.53 | 0.56 | High |
| 11 | Innovative distribution is based on the design of distribution channels that fit the nature of the product | 3.91 | 0.82 | High |
| 12 | Innovative distribution is based on the design and use of distribution methods and transportation based on the idea of renewal, excellence, and individual performance, which positively reflects the degree of consumer satisfaction | 4.13 | 0.96 | High |
| *Innovative promotion* | | | | |
| 13 | Innovative promotion is based on the art of communication, attraction, and persuasion of what is on the market or what will be displayed | 3.66 | 1.05 | High |
| 14 | A successful innovative promotion must be characterized by an attractive rhythm and formulation that aligns with the nature of the product and the target consumer | 3.81 | 0.11 | High |

(continued)

**Table 1** (continued)

| No. | Items | Mean | SD | Grade |
|-----|-------|------|-----|-------|
| 15 | Innovative promotion is modern, renewed and has never been introduced before, in line with what is on the market and features the right promotional message | 3.54 | 0.97 | High |
| 16 | Innovative promotion must be an attractive, unique, and unique influence, creating consumer curiosity and arousing the product experience, and making the target of innovative promotion see that the organization has understood its need and desire | 3.86 | 0.70 | High |
| *Performance of the firms* | | | | |
| 17 | The good performance of the organization is judged by the innovative profit resulting from the marketing innovation of the products | 3.85 | 0.52 | High |
| 18 | The organization is evaluated by the higher price and lower cost features | 4.10 | 0.14 | High |
| 19 | Performance is assessed by reducing the cost of adopting a more productive technological process, | 3.57 | 0.85 | High |
| 20 | The good performance of the organization is judged by achieving an appropriate return on investments and optimizing the resources available to the organization | 3.87 | 0.95 | High |

**Table 2** Multiple regression analysis, revealing the relationship between the innovative marketing mix elements and the firms' performance

| Innovative marketing mix | performance of the firms | | | | | | | | | |
|---|---|---|---|---|---|---|---|---|---|---|
| | β | t | Sig | Tolerance | VIF | f | Sig. | R | $R^2$ | Durbin-Watson |
| Product | 0.33 | 19.62 | 0.000 | 0.96 | 1.03 | 657 | 0.0 | 0.78 | 0.86 | 0.68 |
| Price | 0.43 | 22.26 | 0.000 | 0.91 | 1.12 | | | | | |
| Distribution | 0.41 | 21.83 | 0.000 | 0.92 | 1.08 | | | | | |
| Promote | 0.45 | 23.87 | 0.000 | 0.95 | 1.04 | | | | | |

there is a relationship between the use of the elements of the innovative marketing mix (product, price, distribution, promotion) and the performance of the firms, which is justified because the innovative marketing mix includes several interactive and integrated activities to raise and improve the firms' overall performance by adopting an innovative approach based on continuous change and renewal, considering the risk factor.

Table 2 shows differences between the innovative product and the performance of the firms, where the value of (t) was 19.6 with a $p$-value $\leq$ of 0.05. Thus, there is a relationship between the innovative product and the performance of the firms. Similarly, there was a significant relationship between innovative price, promotion, and distribution with the performance of the firms. Their fur the sub-hypothesis (Ha1, Ha2, Ha3, and Ha4) were also accepted.

# 4 Conclusion

The innovative marketing mix is a renewed activity as it is based on providing the benefit to the customer even before he realizes his need for such a product. Also, it refers to the use of new or nontraditional ideas in marketing. Therefore, it is not limited to the product but includes the other marketing mix elements, i.e., products, price, promotion, and distribution. This study is limited to a specific type of firm which does not reflect the specialized marketing companies. However, the significance of this study lies in addressing the subject of innovation in the marketing mix elements by highlighting the effects of innovation in developing and improving the firms' performance and expanding its competitive scope. The results of an innovative marketing mix could be explored concerning different performance indicators, which are selected based on the firms' goals, such as market share, sales growth in a specific region, or customer satisfaction. The firm has the right to choose the best method to identify and assist its performance.

# 5 Recommendations

- Modern firms need to consider adopting an innovative marketing mix to empower the firms to achieve targets in the market besides improving the firms' performance.
- Getting the benefit of the firms' experience, which are pioneers in marketing mix innovation to face any deviation in the performance to keep innovative marketing a beneficial tool for the customer and improve the firms' performance at the same time.
- There should be a specific and specialized unit in the firm that keeps looking for what others have missed in their products in terms of planning and designing an integrated innovative marketing mix that includes four main innovative marketing mix elements. As a result, the firms' performance will be improved.

# References

1. Ungerman O, Dedkova J, Gurinova K (2018) The impact of marketing innovation on the competitiveness of enterprises in the context of industry 4.0. J Competitiveness 10(2):132
2. Jia J, Yin Y (2015) Analysis of Nokia's decline from marketing perspective. Open J Bus Manag 3(4):446
3. Zakaria R (2016) The effect of marketing innovation on the quality of banking services—case study of Gulf Bank Algeria. Master thesis, University of Mohammed Khiedr—Biskra, Algeria
4. Kalogiannidis S, Melfou K, Papaevangelou O (2020) Global marketing strategic approaches on multi-national companies product development. Int J Sci Res Manag 8(12)
5. Nadjme A (2015) Leadership and innovation management, 2nd edn. Dar-Alzahra'a Publishing, Amman-Jordan

6. Jamous A (2013) Knowledge management in business organizations and their relationship to modern management approaches—analytical portal, 1st edn, Wael Publishing, Amman, Jordan
7. La'aqaqnah K (2016) The role of bank marketing mix in improving the quality of banking service. Master's thesis, Department of Economic Sciences, Mohammed Bo-Khaydar University, Biskra, Algeria
8. Hussain I, Mu S, Mohiuddin M, Danish RQ, Sair SA (2020) Effects of sustainable brand equity and marketing innovation on market performance in hospitality industry: mediating effects of sustainable competitive advantage. Sustainability 12(7)
9. Pride W (2008) Marketing. Cengage Learning, ISBN 0-547-16747-4
10. Obeidat MA (2008) The basics of pricing in contemporary marketing. Dar Al-Masiera for Distribution and Printing, Amman
11. Hamid T (2008) Product development and pricing. Dar-Elyazori for Scientific Publishing, Amman, Jordan
12. Alfareis S, Machus D (2006) The effect of prices on consumer sensitivity. J Econ Legal Sci 28(3), Tishreen University, Syria
13. Chen S, Zhou F, Su J, Li L, Yang B, He Y (2020) Pricing policies of a dynamic green supply chain with strategies of retail service. Asia Pac J Market Logistics
14. Brueckner JK, Flores-Fillol R (2020) Market structure and quality determination for complimentary products: alliances and service quality in the airline industry. Int J Ind Organ 68:102557
15. Miller D (2020) Advertising, production and consumption as the cultural economy. Routledge, pp 75–89
16. Janabi T, Abboud H (2000) The impact of the purchaser and competition in determining market share in business companies. Master thesis Submitted To The Faculty of Management And Economics, University of Kufa
17. Muangkhot S, Ussahawanitchakit P (2015) Strategic marketing innovation and marketing performance: an empirical investigation of furniture exporting businesses in Thailand. Bus Manag Rev 7(1):189
18. Lambin J, Moerloose C (2008) Marketing Stratégique Et Opérationnel, 7th edn, Dunod, Paris, ISBN 978-2-10-053858-4
19. Feldman Barr T, Mcneilly KM (2003) Marketing: is it still just advertising? The experiences of accounting firms as a guide for other professional service firms. J Serv Mark 17(7):713–729
20. Cake DA, Agrawal V, Gresham G, Johansen D, Di Benedetto A (2020) Strategic orientations, marketing capabilities and radical innovation launch success. J Bus Ind Market
21. Al-Chahadah A, Qasim A, El Refae G (2020) Financial inclusion indicators and their effect on corporate profitability. AAU J Bus Law 4(2)
22. Al-Chahadah AR, El Refae GA, Qasim A (2020) The impact of financial inclusion on bank performance: the case of Jordan. Int J Econ Bus Res 20(4)
23. Ndesaulwa AP, Kikula J (2016) The impact of innovation on performance of small and medium enterprises in Tanzania: a review of empirical evidence. J Bus Manag Sci 4(1):1–6
24. Musa K, AlShehadeh A, Alqerem R (2019) The role of data mining techniques in the decision-making process in Jordanian Commercial Banks. In: 2019 IEEE Jordan international joint conference on electrical engineering and information technology (JEEIT). https://doi.org/10.1109/JEEIT.2019.8717461
25. Al Chahadah A, El Refae G, Qasim A (2018) The use of data mining techniques in accounting and finance as a corporate strategic tool: an empirical investigation on banks operating in emerging economies. Int J Econ Bus Res 15(4)

# How the Cryptocurrencies React to Covid-19 Pandemic? An Empirical Study Using DCC GARCH Model (2019–2021)

**Naima Bentouir, Ali Bendob, Mohammed El Amine Abdelli, Samir. B. Maliki, Mourad Kertous, and Afef Khalil**

**Abstract** The paper aims to examine the dynamic conditional correlation between three major cryptocurrencies: Bitcoin, Litecoin, and Ethereum, during the Covid-19 pandemic. Our sample includes two panels: before/during Covid-19 covering the periods of 01/01/2019 to 12/31/2019 and from 01/01/2020 to 01/03/2021 using hourly data and the DCC GARCH model. The empirical results found that the Covid-19 pandemic positively impacts the cryptocurrencies prices starting from the second semester of 2020. Furthermore, these digital currencies became more vulnerable and showed a high level of volatility, especially Ethereum and Litecoin. Besides, we concluded that there was no significance on the short-run volatility effect during the Covid-19 period, which means the continuity of persistence in the long term.

N. Bentouir (✉)
Finance and Accounting Department, LMELSPM Laboratory, Ain Temouchent University, Ain Temouchent, Algeria
e-mail: naima.bentouir@univ-temouchent.edu.dz

A. Bendob
Economic Faculty, LMELSPM Laboratory, Ain Temouchent Unviversity, Ain Temouchent, Algeria
e-mail: bendobali4@gmail.com

M. El Amine Abdelli
IAE of Bretagne Occidentale, Laboratory of LEGO, University of Western Brittany-UBO, Brest, France
e-mail: abdelli.univ@gmail.com; amineabdelli@uasl.es

Samir. B. Maliki
Economics and Management Faculty, Mecas Laboratory, University of Tlemcen, Tlemcen, Algeria
e-mail: samir.maliki@univ-tlemcen.dz

M. Kertous
Law, Economic and Management AES Faculty, AMURE Laboratory, UBO-Brest University, Brest, France
e-mail: mourad.kertous@univ-brest.fr

A. Khalil
University of Carthage, Tunis, Tunisia
e-mail: Akhalil@utctunisie.com

© The Author(s), under exclusive license to Springer Nature Switzerland AG 2022
S. G. Yaseen (ed.), *Digital Economy, Business Analytics, and Big Data Analytics Applications*, Studies in Computational Intelligence 1010,
https://doi.org/10.1007/978-3-031-05258-3_34

**Keywords** Covid-19 · Financial markets · Cryptocurrencies · Co-movement · DCC GARCH model · Pandemic

**JEL Classification** C22 · G01 · G14 · G1

# 1 Introduction

Today, the world suffers from a deadly Covid-19 pandemic that is rapidly expanding geographically with high rates of contaminations. Reference [1] stated that Bitcoin prices had a volatile relationship with the stock markets during the analysis period. This situation disrupted the global activity of the economy. Reference [2] showed that the volatile cryptocurrency is closely related to global economic activity. Reference [3] stated that the appearance of Bitcoin was associated with high volatility and high risk in trading. The last week of February was the worst week since the 2008 financial crisis when investors and financial markets lost trillions of dollars. The first week of March increased losses that put investors in instability.

Cryptocurrency research has advanced significantly since this product was developed over the past decade. Reference [4] examined the causality between liquidity and volatility in equity markets. The outbreak starts in December 2019 due to the lack of information on how long it could last if China can control 40–4; the rapid expansion around the world of Covid-19 quickly affected financial markets [5]. The performance of financial markets began to react and showed the worst performance at the end of February. The following figure shows the coronavirus outbreak and total confirmed cases worldwide from January 22nd to May 3rd. We can see that the virus most infects China, the USA, Spain, and Italy.

The entire market capitalization has entered the exponential growth area [6]. The volatility and uncertainty continue to affect financial markets, which have been under severe pressure. Reference [7] concluded that herding led to increased insecurity. The coronavirus is spreading faster compared to previous epidemics such as (SARS, swine flu). The coronavirus epidemic was not the first virus outbreak that threatened to affect the financial markets [8]. The impact of this pandemic became global when imports from China immediately affected the export economy around the world. Commodity markets had also seen a drop like oil prices due to the lower level of demand.

The twenty-first century is known for the digital century, and it is dictating a new set of rules, as dealerships were introduced as futures contracts in the regulated derivatives market [9]. The Covid-19 outbreak did not affect only the real ones. Financial markets but also impacts cryptocurrency. The value of these digital currencies continues to rise rapidly in recent years compared to traditional currencies. Voluntary participants of the parties govern digital currencies. They are not controlled by any life cycle of cryptocurrency is in an early stage, and its price is unstable in the future. We can say it has been the most exciting and promising development of the currency.

The Covid-19 is the second significant impact since the global financial crisis of 2008 on Bitcoin prices. When the effects of his pandemic appeared earlier last year in digital currencies, it has been enriched by a significant recession due to the continued spread of the virus. Bitcoin prices had fallen by 10.08% last February, with a decrease of $2000 in just two weeks. Our paper will discuss two other cryptocurrencies, Ethereum and Litecoin, which launched in 2013 and 2011. They have mining procedures nearly the same as Bitcoin. There is also a blockchain for creating a decentralized application with an open-source. Although the difference between these two digital coins is that Litecoin is very complex than Ethereum, it has much more in common with Bitcoin with an open-source. The rest of this paper is organized as follows: Section two presents the data and methods, the findings of the third section, the main discussion, and finally, the conclusion.

## 2  Literature Review

Covid-19 spread opens the door for new research investigations on many topics and fields. Reference [10] focused on international haven properties of digital currencies with the equity indices. The sample includes the five most active indexes; S&P, FTSE100, MSCI, IBEX, FTSEMIB, and CSI300. They empirically investigate the effect of covid-19 on the trading safety of the indexes using the modified value at risk approach. The main results highlighted that Bitcoin and Ethereum could not present a haven in this period for the selected sample. Reference [11] set out to examine the existence of volatility spillover, using developed data of the Chinese financial markets during the Covid-19 outbreak. The results indicate there is a significant impact of Covid-19 on directional spillover in markets of Bitcoin. Reference [12] concluded that the erratic behavior of 51 cryptocurrencies is affected by the Covid-19 pandemic. Reference [13] studied the correlation between Bitcoin and two benchmarks' indexes, S&P500 and VIX, to measure market uncertainty covering September 14, 2014, and April 14, 2020. They concluded that Bitcoin is good for diversification and the S&P500 but in normal and calm times. The main finding shows that Bitcoin has a weak possibility to be a haven. Reference [14] Investigated the performance of the cryptocurrencies before and after the Covid-19 pandemic using daily returns and the self-similarity of digital currency. The main finding highlights that covid-19 has a positive effect on the efficiency of cryptocurrency markets. Reference [15] in this research paper, the authors applied the wavelet analysis to examine the impact of The Covid-19 on the volatility of both currencies and digital currencies.

The study covers January/May 2020. The main finding of this research paper indicates that the cross-currency hedging strategy can be eligible which enables the currencies to perform better on the time of crises. Reference [16] aimed to test the effects of Covid-19 on Bitcoin and gold prices doing June 24, 2019, to 22 May 2020, Using OLS and Granger causality test. The results revealed a significant relationship between gold and Bitcoin during the study before and after the Covid-19 pandemic. Reference [14] this paper studied the herding biases of Bitcoin returns using self-similarity intensity. The authors empirically examine the efficiency of digital currency under multifractal analysis; as a result, they proved that Covid-19 showed a positive impact on cryptocurrency markets' efficiency. Reference [17] tried to examine the asymmetric efficiency of digital currencies using hourly data and four cryptocurrencies prices. The authors concluded that Bitcoin and Ethereum are the strong ones and more responsive during the Covid-19. Reference [18] focused on the digital complementary currencies during the Covid-19 pandemic and its effect on the public policies. Reference [19] aimed to test the efficiency of 45 digital currencies and 16 benchmarks stock indexes under two time series periods before and after the Covid-19. They concluded that digital prices become more volatile than the stock markets.

## 3   Data and Methodology

To achieve our investigation, we collected data which contain three crypto-currencies (Bitcoin, Litecoin, and Ethereum) historical prices, divided into two panels; before the Covid-19 pandemic and during the Covid-19 covering the period of January 01st, 2019 to December 31st, 2019 and from January 01st, 2020 to January 03rd, 2021. The data was collected from Binance exchange data.

### 3.1   The Econometric Model

The conditional variance equation in the simplest case is given as follows:

$$Y_t = X_t'\Theta + \varepsilon_t$$
$$\sigma_t^2 = \omega + \alpha\varepsilon_{t-1}^2 + \beta\sigma_{t-1}^2 \tag{1}$$

The Dynamic Conditional Correlation (DCC) GARCH model is defined as:

$$r_t = \mu_t + a_t$$
$$a_t = H_t^{1/2}z_t$$
$$H_t = D_t R_t D_t \tag{2}$$

The DCC GARCH model is based on the hypothesis that the conditional [20] returns are normally distributed with zero mean and conditional covariance matrix $H_t = E[r_t r_t' | I_t - 1]$ the covariance matrix is expressed as follows:

$$H_t = D_t R_t D_t \tag{3}$$

where $D_t = diag\left[\sqrt{h_{1t}}, \sqrt{h_{2t}}\right]$ is a diagonal matrix of time-varying standard deviations issued from the estimation of univariate GARCH processes j:

$$h_{it} = \omega_i + \alpha_i \varepsilon_{t-1}^2 + \beta_i h_{t-1}$$

$$R_t = \begin{bmatrix} 1 & q_{21t} \\ q_{21t} & 1 \end{bmatrix} \tag{4}$$

This matrix is decomposed into:

$$R_t = Q_t^{n-1} Q_t Q_t^{n-1}$$

With $Q_t$ is the positive definite matrix containing the conditional variances.

## 4 Results and Discussion

We divided our data into two panels; the first panel covers the period before the Covid-19 pandemic, and the second panel considers the pandemic spread.

The descriptive statistics.

To detect the characteristics of the data, we put the descriptive statistics on two panels before and during Covid-19 for three cryptocurrencies: Bitcoin, Ethereum, and Litecoin. According to Table 1, Litecoin showed high volatility with 1.35%, followed by Bitcoin and Ethereum 0.98%, 0.95%, respectively, before the Covid-19 pandemic.

The second panel represents the descriptive statistics of the digital currencies during the Covid-19; from the standard deviation, we can say that Bitcoin returns moved from its mean by 1.45%, followed by the Ethereum 1.35% and Litecoin with

**Table 1** Descriptive statistics of the cryptocurrencies returns

| | Panel 01 | | | Panel 02 | | |
| | B-Covid19 | | | Covid-19 | | |
| | RLTC | RETH | RBTC | RLTC | RETH | RBTC |
|---|---|---|---|---|---|---|
| Mean | −3.29E−05 | −0.000209 | −4.53E−06 | 0.000173 | 0.000169 | 0.000136 |
| Median | −0.000145 | −0.000187 | −4.27E−05 | 0.000000 | 0.000000 | 0.000000 |
| Std. Dev | 0.013576 | 0.009571 | 0.009820 | 0.009226 | 0.013518 | 0.014558 |
| Skewness | 23.12704 | −0.222376 | 2.457528 | 2.411235 | 1.177295 | 1.380684 |
| Kurtosis | 1254.556 | 28.38752 | 87.92927 | 67.84678 | 43.34606 | 60.74150 |
| Jarque–Bera | 4.77E+08 | 195,969.2 | 2,199,789 | 1,556,037 | 601,072.4 | 1,229,748 |
| Probability | 0.000000 | 0.000000 | 0.000000 | 0.000000 | 0.000000 | 0.000000 |

*Source* Author's calculations

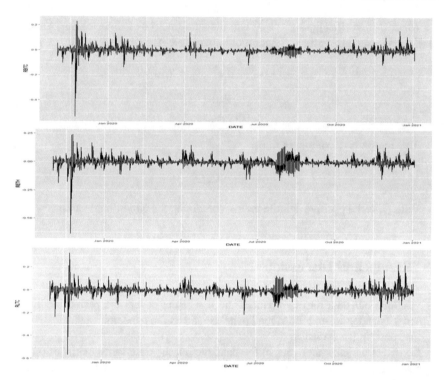

**Fig. 1** The cryptocurrency returns during the whole period study. *Source* Authors' calculations using R software

0.92%, that means the Bitcoin prices are more affected by the Covid-19 outbreak from January 01st, 2020, to 03rd January 2021.

Figure 1 shows the Bitcoin, Ethereum, and Litecoin return from January 01st, 2019, to January 03rd, 2021. We can note that the figure represents our descriptive statistics when the Bitcoin, Ethereum, and Litecoin return represent high volatility during covid-19 at the beginning of the year (January 2020) and in the period from August to the end of the year (Table 2).

For testing the existence of the Heteroskedasticity, we used the ARCH and Breusch-Pagan-Godfrey tests; the result highlights that there is an ARCH effect according to the *P*-value, which is less than 5% of significance on both models (0.00), so could run to the GARCH (1,1) model, which indicated all the parameters are significant at 5% of significance; that means the previous errors and the conditional variance of prices can forecast the continuity of fluctuations, in another word the existence of the positive relationship between Covid-19 and cryptocurrencies prices spatially in the second semester of 2020.

According to the results of the ARCH and GARCH (1.1) model we present in Fig. 2, the historical prices of the cryptocurrencies using intraday data (hourly data), we can observe that there was a negative shock at the beginning of the Covid-19

**Table 2** Heteroskedasticity Breusch-Pagan-Godfrey and ARCH Tests

| | B-Covid-19 | Covid-19 |
|---|---|---|
| *Heteroskedasticity test: ARCH* | | |
| F-statistic | 7.1126 | 27.463 |
| Obs*R-squared | 7.1077 | 27.384 |
| Prob. F(1,7292) | 0.0077 | 0.0000 |
| Prob. Chi-Square(1) | 0.0077 | 0.0000 |
| *Breusch-Pagan-Godfrey* | | |
| F-statistic | 3044.369 | 81.588 |
| Obs*R-squared | 3319.499 | 160.271 |
| Scaled explained SS | 374,664.2 | 5204.69 |
| Prob. F(2,8832) | 0.0000 | 0.0000 |
| Prob. Chi-Square(2) | 0.0000 | 0.0000 |
| Prob. Chi-Square(2) | 0.0000 | 0.0000 |
| *GARCH TEST B-Covid-19* | | |
| | Coefficient | *P*-value |
| RESID $(-1)^2$ | 0.355698 | 0.0000 |
| GARCH $(-1)$ | 0.552537 | 0.0000 |
| *GARCH TEST Covid-19* | | |
| | Coefficient | *P*-value |
| RESID $(-1)^2$ | 0.279545 | 0.0000 |
| GARCH $(-1)$ | 0.296127 | 0.0000 |

*Source* Author's calculations

appearance (early of 2020). Still, starting from the middle of the year the prices rise till the end of the year (Table 3).

Table 2 shows the strong correlation between Bitcoin's returns and Ethereum and Litecoin's returns, wherein the correlation arranges between 73 and 62% in the period before the Covid-19 outbreak. Furthermore, the table also represents the augmentation of the correlation level between the cryptocurrencies, which is arranged between 88% for the pair of RBTC/RLTC and 68% for the couple of RETH/RBTC, while the correlation between the pair RETH/RLTC increased from 46% before the Covid-19 to 77% in the covid-19 pandemic period.

Table 4 also shows the dynamic correlation between the cryptocurrencies pairs for two periods before and during the Covid19 outbreak, using the Angle methodology Dynamic conditional correlation GARCH model and the multivariate normal distribution, our results highlight the existence of the dynamic correlation according to the *P*-value, which is less than 5% of significance. We can explain that by the existence of the sensibility of cryptocurrency prices to the Covid-19 shock.

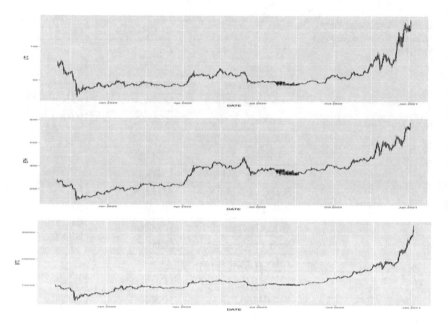

**Fig. 2** The cryptocurrency prices for the whole study period. *Source* Author's calculation using R software

**Table 3** The covariance correlation

|  | RBTC | RLTC | RETH |
|---|---|---|---|
| *B-covid-19* | | | |
| RBTC | 1.0000000 | **0.7260044** | **0.6154441** |
| RLTC | **0.7260044** | 1.0000000 | **0.4558791** |
| RETH | **0.6154441** | **0.4558791** | 1.0000000 |
|  | RBTC | RLTC | ETH |
| *Covid-19* | | | |
| RBTC | 1.0000000 | **0.8761557** | **0.6774934** |
| RLTC | **0.8761557** | 1.0000000 | **0.7702203** |
| RETH | **0.6774934** | **0.7702203** | 1.0000000 |

*Source* Author's calculations the correlation value is significant at 1% level of significance

## 5 Discussion

The estimation results from the Heteroskedasticity tests based on two models; ARCH LM and Breusch-Pagan-Godfrey, highlight that we can run the GARCH model according to the $P$-value, which is less than 5% that means it enables to estimate GARCH (1.1) model that has showed the significance of the parameters at 5%,

**Table 4** Dynamic conditional correlation GARCH estimation

| Panel A covid-19 | | | Panel 2 B-Covid-19 | | |
|---|---|---|---|---|---|
| | Coefficient | *P*-value | | Coefficient | *P*-value |
| Mu1[a] | 0.000172 | 0.009521 | Mu1 | 0.000004 | 0.886434 |
| Omega | 0.000000 | 0.309436 | Omega | 0.000000 | 0,970,655 |
| Alpha1 | 0.014750 | 0.000000 | Alpha1 | 0.064312 | 0.000000 |
| Beta1 | 0.984250 | 0.000000 | Beta1 | 0.910336 | 0.000000 |
| Mu2 | 0.000212 | 0.135005 | Mu2 | 0.000137 | 0.147171 |
| Omega | 0.000001 | 0.158957 | Omega | 0.000006 | 0.000000 |
| Alpha1 | 0.014840 | 0.000000 | Alpha1 | 0.125533 | 0.000000 |
| Beta1 | 0.982150 | 0.000000 | Beta1 | 0.811936 | 0.000000 |
| Mu3 | 0.000168 | 0.197351 | Mu3 | 0.000006 | 0.960061 |
| Omega | 0.000001 | 0.196279 | Omega | 0.000001 | 0.724568 |
| Alpha1 | 0.020021 | 0.000000 | Alpha1 | 0.058779 | 0. 011,724 |
| Beta1 | 0.977448 | 0.000000 | Beta1 | 0.936294 | 0.000154 |
| Dcca1 | 0.044137 | **0.361567** | Dcca1 | 0.030122 | **0.000154** |
| Dccb2 | 0.927477 | **0.000000** | Dccb2 | 0.957102 | **0.000000** |

*Source* Author's calculation. [a] Our variables results are sorted respectively, Bitcoin, Litecoin, Ethereum

which means the previous errors and the conditional variance of prices can forecast to the continuity of fluctuations; we can explain this results by the existence of the positive relationship between the Bitcoin, Ethereum and Litecoin prices and the Covid-19. The correlation matrix confirmed that by the augmentation of the correlation between the cryptocurrency's pairs, which is arranged between 88% for of RBTC/RLTC pairs and 68% for of RETH/RBTC pairs, wherein the correlation between the pair RETH/RLTC increased from 46% before the Covid-19 to 77% in the covid-19 pandemic period. These results explain the positive response of the cryptocurrencies to the Covid-19 period.

GARCH model coefficients higher than ARCH coefficients mean the present information has more effect than the historical information's. In other words, the investors take into account the actual pieces of information for these decisions. The total of GARCH+ARCH is less than one (1), which means the continuity of volatility shocks' persistence equal to 0.35/0.27 means the existence of the responsive of cryptocurrencies prices to the shocks in the short term. In contrast, the values equal to 0.55/0.29 indicated that the variance of the high level of volatility will be followed by other very high variance in the next period for both f periods before and during the Covid-19, respectively. Wherein the total of the and in both tables are less than one, which means the shocks in the Bitcoin Litecoin and Ethereum returns are characterized by continuity and take a short time to disappear. The dynamic correlation between cryptocurrencies during and before covid-19 has been tested by Angle methodology Dynamic conditional correlation GARCH model and the multivariate normal distribution, and it's indicated that the existence of the dynamic correlation according to the P-value which is less than 5% of significance. According to dcca1/dccb1 for the period before Covid-19, we can say that there is a short and long-run volatility impact on cryptocurrencies in that period. In contrast, the dcca1/dccb1 for the period of Covid-19 highlighted the absence of short-run volatility effect and the existence of the long-run volatility effect for the cryptocurrencies against the Covid-19.

# 6   Conclusion

Based on the high-frequency data (intraday data), this research paper employs the DCC GARCH model to testing the effective transmission of Covid-19 to the significant cryptocurrencies Bitcoin, Ethereum and Litecoin by dividing our study period onto two panels; before the Covid-19 ad during the Covid-19 outbreak using hourly data. Running also ARCH and GARCH models; the outputs reveal that the volatility of the return of cryptocurrencies is bidirectional (BTC/ETH) before the Covid-19 pandemic; other findings showed that there is both long and short-run volatility that means short/large volatility. According to the results of the DCC GARCH for the second period (Covid-19), the cryptocurrencies are not significant in the long-run term for the pairs BTC/ETH and BTC/LTC. These results highly suggest investigating how to manage cryptocurrencies portfolio, especially in this higher volatility period. The obtained results support the previous studies such as [12].

# References

1. Shaen C (2020) The contagion effects of the COVID-19 pandemic: evidence from gold and cryptocurrencies. Available at SSRN: https://ssrncom/abstract=3564443 or http://102139/ssrn3564443:1-12
2. Christian C (2018) Long- and short-term cryptocurrency volatility components: a GARCH-MIDAS analysis. J Risk Finan Manag. https://doi.org/10.3390/jrfm11020023
3. Dehua S (2019) Forecasting the volatility of Bitcoin: the importance of jumps and structural breaks. Eur Finan Manag, Wiley. https://doi.org/10.1111/eufm.12254
4. Bedowska B (2019) Causality between volatility and liquidity in cryptocurrency market. University of Zurich, 1–14. https://doi.org/10.13140/RG.2.2.32637.61927
5. Akther U (2020) Bitcoin a hype or digital gold? Global evidence. Wiley, 1–17. https://doi.org/10.1111/1467-8454.12178
6. ElBahrawy E (2017) Evolutionary dynamics of the cryptocurrency market. R Soc Open Sci 4:170623. https://doi.org/10.1098/rsos.170623
7. Bouri E (2018) Herding behaviour in the cryptocurrency market. Department of Economics Serirs Papers, University of Pretoria, Working paper N 2018-34
8. Bouoiyour J, Selmi R (2020) Coronavirus spreads and Bitcoin's 2020 Rally: is there a link? 2020. ffhal-02493309
9. Shynkevich A (2020) Bitcoin futures, technical analysis and return predictability in Bitcoin prices. J Forecast 1–40. https://doi.org/10.1002/for.2656

10. Conlon T, Corbet S, Mcgree R (2020) Are cryptocurrencies a safe haven for equity markets? An international perspective from the COVID-19 pandemic. Res Int Bus Finan 54. https://doi.org/10.1016/j.ribaf.2020.101248

11. Sa C (2021) Pandemic-related financial market volatility spillovers: evidence from the Chinese COVID-19 epicentre. Int Rev Econ Financ 71:55–81. https://doi.org/10.1016/j.iref.2020.06.022

12. Na J (2021) Changes to the extreme and erratic behaviour of cryptocurrencies during COVID-19. Phys A: Stat Mech Its Appl 565. https://doi.org/10.1016/j.physa.2020.125581

13. Kristoufek L (2020) Grandpa, Grandpa, tell me the one about Bitcoin being a safe Haven: new evidence from the COVID-19 pandemic. Front Phys 8. https://doi.org/10.3389/fphy.2020.00296

14. Mnif E, Jarboui A, Mouakhar K (2020) How the cryptocurrency market has performed during COVID 19? A multifractal analysis. Finan Res Lett 36. https://doi.org/10.1016/j.frl.2020.101647

15. Umar Z, Gubareva M (2020) A time–frequency analysis of the impact of the Covid-19 induced panic on the volatility of currency and cryptocurrency markets. J Behav Exp Finan 28. https://doi.org/10.1016/j.jbef.2020.100404

16. Al-Naif KL (2020) Coronavirus pandemic impact on the Nexus between gold and Bitcoin prices. Int J Finan Res 11(5):442–449. https://doi.org/10.5430/ijfr.vlln5p442

17. Naeem MA, Bouri E, Peng Z, Shahzad SJH, Vo XV (2021) Asymmetric efficiency of cryptocurrencies during COVID19. Phys A: Stat Mech Its Appl 565. https://doi.org/10.1016/j.physa.2020.125562

18. Gonzalez L, Cernev AK, de Araujo MH, Diniz EH (2020) Digital complementary currencies and public policies during the covid-19 pandemic. Revista de Administracao Publica 54(4):1146–1160. https://doi.org/10.1590/0034-761220200234x

19. Lahmiri S, Bekiros S (2020) The impact of COVID-19 pandemic upon stability and sequential irregularity of equity and cryptocurrency markets. Chaos, Solitons and Fractals 138. https://doi.org/10.1016/j.chaos.2020.109936

20. Orskaug E (2009) Multivariate DCC-GARCH model with various error distributions. Master of Science in Physics and Mathematics, Norwegian University of Science and Technology

# The Impact of COVID-19 on Amman Stock Market (ASE) Performance: An ARDL Approach

Hamad kasasbeh, Marwan Alzoubi, Ayman Abdalmajeed Alsmadi, and Ala'a Fouad Al-dweik

**Abstract** The aim of this paper is to investigate the impact of Covid-19 on Amman Stock Exchange performance (ASE) in Jordan using the Auto Regressive Distributed Lag Model (ARDL) and daily observations over the period of January 3, 2021 to April 25, 2021. The variables included in this study are the ASE index, the Covid-19 positive cases, the Covid-19 death cases, the Covid-19 recovered cases, the world Oil prices and the Euro/JD exchange rate. The long-run results reveal that positive cases have negative effect on ASE performance, recovered cases and exchange rate have positive effects. The positive reaction of the performance of ASE to movements in exchange rate may be attributed to the fact that imports largely exceeds exports. Therefore, an appreciation of the JD makes imports less costly which goes in line with economic theory as gross consumption in Jordan is mainly attributed to foreign products. Furthermore, positive cases are also highly significant while death cases are not which could be attributed to the fact that positive cases may have spill-over effect while death cases may not. Also, the magnitude of positive cases is much larger than that of death cases. In the short-run, all variables except recovered cases cause the performance of ASE. The system adjusts in the current day to any short-run departure in the previous day from long-run equilibrium in the previous day, which confirms the evidence of the long-run cointegration relationship.

**Keywords** Stock market · Jordan · Performance · Covid-19 · ARDL

H. kasasbeh · M. Alzoubi · A. A. Alsmadi (✉) · A. F. Al-dweik
Department of Banking and Financial Sciences, Alzaytoonah University of Jordan, Amman, Jordan
e-mail: ayman.smadi@zuj.edu.jo

H. kasasbeh
e-mail: hamad.k@zuj.edu.jo

M. Alzoubi
e-mail: m.alzoubi@zuj.edu.jo

A. F. Al-dweik
e-mail: alaa.dweik@zuj.edu.jo

© The Author(s), under exclusive license to Springer Nature Switzerland AG 2022
S. G. Yaseen (ed.), *Digital Economy, Business Analytics, and Big Data Analytics Applications*, Studies in Computational Intelligence 1010,
https://doi.org/10.1007/978-3-031-05258-3_35

# 1   Introduction

Coronavirus (COVID-19) pandemic has caused unprecedented problems for human health on one hand, and the economic problems that have arisen because of it on the other hand [1]. The latest global financial stability reports showed that the global system has actually been affected by COVID-19 significantly and that the intensification of the COVID-19 may affect the stability of the global financial system.

Covid-19 appeared for the first time at the end of 2019 in China and later spread to most countries. Jordan registered the first case of the virus on March 2, 2020 and the Jordanian government took several proactive measures to fight the disease. On March 17, the number of confirmed cases reached 40. This led the Jordanian government to take many additional precautionary measures, including activating the Defense Law issued in 1992. This law required the following: (i) suspension of all institutions, official departments and the private sector; (ii) citizens were prevented from leaving homes except in cases of extreme necessity; (iii) mass transit and transportation was suspended, malls, commercial centers and airports were shut down. It is anticipated that in the short term, emerging markets and developing economies such as Jordan, are likely to be the most economically affected. These countries suffer from weak health systems, and rely heavily on tourism, remittances and exports of primary commodities, in addition to the financial weaknesses. Jordan is now more vulnerable to high levels of debt than it was before the crisis, and as such is under more financial pressures.

Logically, the Corona pandemic is likely to affect the performance of financial markets in general, which leads to a negative impact on economic growth and economic development, especially in developing countries. For example, World Bank reports forecasted a decline in economic growth from about 2.4% in 2019 to $-2.1$ and $-5.9\%$ in 2020 in sub-Saharan Africa. In addition, World Bank reports predicted that there will be losses in economic output of up to 79 billion US dollars [2]. Moreover, the United Nations Economic Commission for Africa (UNECA) expects that growth in Africa will slow in the year 2020 to about 1.8% in the best cases, and it may witness a contraction of about 2.6%, which may lead to the arrival of nearly 29 million citizens to the state of poverty Extreme.

One of the main concerns of both financiers and economists is the extent of the impact of the Corona pandemic on the stock market, and this is what was also paid attention to in the aftermath and during the global financial crisis of 2007–2009. This could create fears for both local and international investors which could lead to selling what they have sharply [3]. As the World Health Organization and other public health officials communicate, the number of confirmed cases as well as the risk of a COVID-19 outbreak for the general public and investors will shape their feelings about the disease and significantly affect the stock market globally.

For example, when stock prices are rising, investors' behavior is more effective and they have high optimism and vice versa so that when stock prices tend to go down, the investors' degree of effectiveness is low and therefore their feelings are surrounded by a high degree of pessimism. Thus, they tend to wait for the market to go back to a safe environment [4–6]. These situations, which can occur in stock markets, can often lead to abnormal and exaggerated reactions in the short term. Therefore, investors will resort to keeping assets that are safe for them [3]. These imbalances can lead to a price change in the stock exchange leading to a decline in the performance of the market in general.

In 2020, the emergence of the Corona pandemic led to the collapse of stock markets globally, as this pandemic was classified as the most influential after what happened in 1918 in Spain. The emergence of the Corona pandemic led to an increase in fear and global closures that affected the economy, which helped in the collapse of global stock markets.

In mid-February, the working day in the stock markets in Asia ended with an increase in prices while the European markets ended up and down together. The Dow Jones index ended its trading with an increase in prices, while revenues on US Treasury bonds for 10 years and 30 years fell to 1.54 and 2.02%. On the eleventh of February, the stock markets in Asia and the Pacific and European stock markets closed their dealings with an increase in prices while The Dow Jones Industrial Average closed flat in prices while there was an unprecedented increase in prices in the Nasdaq and the S&P 500. In addition, oil prices rose by 2% while revenues on US 10-year and 30-year Treasuries rose to 1.59% and 2.05% respectively.

In addition, several studies have analyzed the impact of the COVID-19 pandemic on the performance of financial markets, such as [7–12]. All of those studies argue that the occurrence of the pandemic led to negative performance and higher price fluctuations. Because Jordan is an open and small economy, Covid-19 affected real and financial sectors including ASE. Figure 1 indicates a negative relationship between Covid-19 daily positive cases and ASE performance during the period January 3, 2021-March 25, 2021.

The current paper will contribute to the existing literature in many ways; First, most studies have focused on China, USA and Europe. This study is the first of its kind on Jordan. Second, most studies examined the impact of Covid-19 on stock market using standard multiple regression method and only two studies used ARDL without using control variables. This study uses daily oil prices and exchange rate as control variables. Thus it is the aim of this study is to investigate whether the COVID-19 affected ASE performance. We look forward to the results of this study to be helpful for decision-makers.

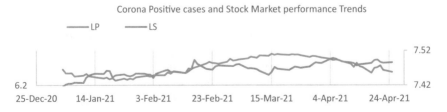

**Fig. 1** Trend of COVID 19 positive daily cases and stock market performance (in Ln Form)

The rest of this paper is organized as follows: Sect. 2 discusses the related literature, Sect. 3 describes the data and economic model, Sect. 4 shows the empirical results and Sect. 5 presents the conclusion.

## 2   Literature Review

Recent empirical studies confirm that the Covide-19 has had huge effects on the stock markets around the world. [13] Observes that the reports of new Corona cases in China have mixed impacts on financial volatility. Reference [14] Report a negative impact of Covide-19 on stock prices of China. Reference [15] Note that all the stock market indexes in India have witnessed downturns during the Covid-19 outbreak.

Reference [5] In their study, which aimed to know the impact of the Corona pandemic on stock prices in China and by using the method of studying the event, they found that there is a negative impact due to the Covid-19 outbreak on stock price movements in the Chinese tourism industry [16]. Studied the relationship between the instability of stock return expectation and price volatility in the United States with Covid-19 and during the period from January 1, 2019 to June 30, 2020. The results of the study showed that the expectation of return and price volatility for both the S&P 500 and the DJIA underwent a break single structure.

Reference [16] Revealed that the Dow Jones index showed a significant decline in the average stock value during the COVID-19 period, in addition, the China Stock Exchange Composite Index witnessed a significant increase in average stock values during the COVID-19.

Moreover, some countries in Asia witnessed unexpected and abnormal negative returns compared to several other countries, and that those negative returns of the Corona pandemic on stock markets gradually decreased in the middle and end of April [17]. The spread of communicable diseases is one of the news that can have a negative impact on stock prices in different markets locally and globally. Many studies have scientifically examined the effects of the spread of the disease on stock prices in the market [18]. Discussed that SARS has affected the stock markets in many countries such as Canada, China, the Philippines, and others. They found that the spread of SARS has a significant impact on both China and Vietnam.

In a study by [11] on the impact of Covid-19 on stock markets for the period from March 10, 2020 to April 30, 2020. The results of this study showed that the negative impact of the Corona pandemic on emerging stock markets has gradually decreased and started to decline by mid-April [8]. In their study shows that there is an unprecedented impact in the stock market of Covid-19. They noted that there had been no previous outbreak of an infectious disease, including the Spanish flu, which had affected the stock market as strongly as the Covid-19 pandemic. They also found that government restrictions on economic activities as well as social distancing are the main reasons why the US stock market has reacted more strongly to Covid-19 than previous outbreaks.

Reference [19] Investigate the impact of Covid-19 on stock markets of MENA region. Corona virus spread is measured in this study by cumulative total cases, cumulative total death and new deaths, while stock market return is measured by the change in the stock market index. Results show that stock market returns in the MENA countries were negatively affected by corona virus. According to [13] where they studied the impact of COVID-19 on the stock market globally, their study included data from several countries, namely, China, South Korea, Italy, Germany and Spain. Based on co-integration testing and data from January 23, 2020 to March 13, 2020, the study results show a long-term relationship between the number of deaths from COVID-19 and stock markets in all countries. In addition, the study found a long-term relationship between some cases of COVID-19 and most stock markets.

Reference [5] aimed in their to know the extent of the short-term impact of the Corona pandemic on stock markets in 21 countries in the world, including Japan, Singapore, the United States and the United Kingdom. This is based on data for the period between February 21, 2019 to March 18, 2020. Based on the results of the research, it was found that the Corona pandemic had a negative impact on the return of stock markets belonging to the countries under study. In addition, the results of the study showed that the stock market in Asia was responding more and faster to COVID-19 and that some of them are recovering quickly as well.

Reference [9] Examine the impact of the Covid-19 outbreak on the Malaysian stock market over the period December 31, 2019 to April 18, 2020. Their results reveal that the higher number of Covid-19 cases in Malaysia tended to adversely affect the performance of the KLCL index [20] study the effect of the corona virus pandemic on the performance and effectiveness of the Nigerian money market and foreign exchange market. The results show that there is a low positive correlation between Covid-19 and capital market of Nigeria [21]. Examine the impact of Covid-19 on the performance of Pakistani stock market. He finds that only Covid-19 recovery cases are influencing the performance of Pakistani stock market.

Reference [17] Used ARDL model to examine the effect of Covid-19 pandemic on stock market of the Kingdom of Saudi Arabia over the period March 2, 2020-May 20, 2020. Their findings indicate that there is a negative impact of Covid-19 on stock market only on the long run. Indeed, we can conclude that Covid-19 has negatively affected stock market prices and performance in all developed and developing countries.

This study examines both the long-run and short-run relationships between ASE performance and Covid-19 in Jordan. Based in the above literature review the following hypotheses are tested:

H01: the performance of ASE is positively influenced by the Covid-19 positive cases.

H02: the performance of ASE is positively influenced by the Covid-19 death cases.

H03: the performance of ASE is negatively influenced by the Covid-19 recovered cases.

# 3   Data and Methodology

## 3.1   Data

This section analyses the empirical impact of Covid-19 on Amman Stock Exchange (ASE) in Jordan, using daily observation over the period January 3, 2021 to April 25, 2021. The variables included are the ASE index LS, the Covid-19 positive cases LP, the Covid-19 death cases LD, the Covid-19 recovered cases LR, the world Oil prices LO and the Euro/JD exchange rate LE. All variables are in natural logarithm form. The data was collected from various source. The dependent variable is the daily closing price of ASE as a proxy for market performance. The independent variables are the daily number of positive Covid-19 cases, the daily number of Covid-19 death cases, the daily number of Covid-19 recovered cases. The daily world oil prices and the daily Euro/Jordan dinar exchange rate represents the control variables.

The data for Covid-19 cases are collected from www.worldometer.com. Daily world oil prices are collected from the USA Energy Information Administration (EIA), the daily Euro/Jordan dinar exchange rate data was collected from www.exchangerate.com. The data for daily closing price index is obtained from daily trading data of ASE.

## 3.2   Descriptive Statistics

Table 1 represents the descriptive statistics of the variables. The number of observations is 81 covering the period from January 3, 2021 to April 25, 2021. The mean value of Amman stock exchange performance (S) daily Covid-19 positive cases (P), daily Covid-19 recovered cases (R), daily world oil prices (O), and the Euro/Jordan Dinar exchange rate (E), are 1747, 3929, 42, 3623, 55 and 0.8566 with standard deviation of 32.1, 2865.8, 32.2, 2781.1, 6.4 and 0.007, respectively.

**Table 1**  Descriptive statistics

| | Amman stock market performance (S) | Daily COVID 19 data in Jordan | | | Daily oil prices (O) | Euro-JD exchange rate (E) |
|---|---|---|---|---|---|---|
| | | Positive cases (P) | Deaths (D) | Recovered (R) | | |
| Mean | 1746.927 | 3929.309 | 41.96296 | 3622.617 | 55.11235 | 0.856617 |
| Median | 1749.360 | 2790.000 | 26.00000 | 2442.000 | 53.20000 | 0.857640 |
| Maximum | 1799.570 | 9717.000 | 109.0000 | 9535.000 | 66.20000 | 0.874920 |
| Minimum | 1662.110 | 776.0000 | 6.000000 | 641.0000 | 45.50000 | 0.840520 |
| Std. Dev. | 32.14085 | 2865.785 | 32.21934 | 2781.141 | 6.420580 | 0.007257 |
| Observations | 81 | 81 | 81 | 81 | 81 | 81 |

## *3.3    Graphical Representation of Data*

Figure 2 shows the graphical representation of data at their level. All series S, P, D, R, O and E are none stationary at their level. Thus, if such data used to estimate the relationship between the variables, the results will be spurious. Therefore, in order to make them stationary, it is essential to confirm the order of the data at first difference. Figure 3 displays the insight view of stationary series in their difference. Figure 2 shows a negative association between Corona positive cases. Death cases do not seem to go in line with performance, both variables move together for the most part; this may be due to the simultaneity of their movements. The reaction of the market for death cases may take longer time. As for exchange rate, when the Euro goes down (JD appreciates), the market seems to react positively, which goes in line with economic theory as aggregate consumption is mainly attributed to foreign products.

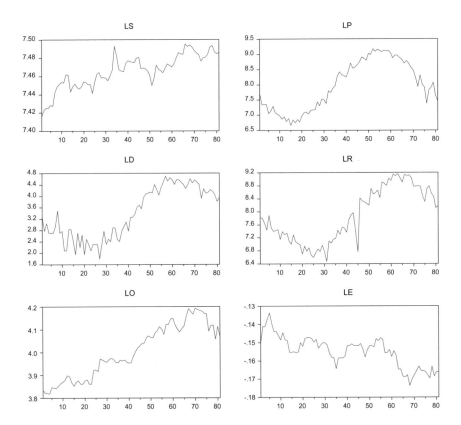

**Fig. 2**  Graphical representation of data on their level form

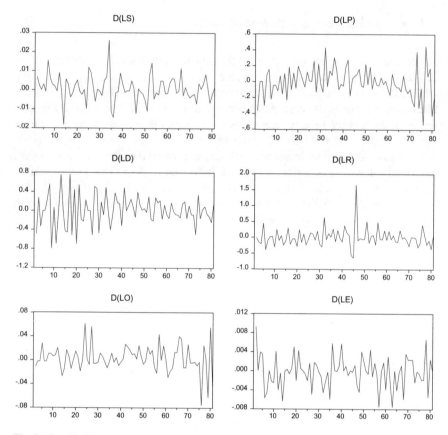

**Fig. 3** Graphical representation of data on first difference

## 3.4 The Model

To examine the impact of Covid-19 on ASE performance, we specify the functional form of the model as follows:

$$St = F(P_t, Dt_t, R_t, O_t, E_t) \qquad (1)$$

Model (1) can be specified in a linear form as follows:

$$S_t = \lambda_0 + \lambda_1 P + \lambda_2 D_t + \lambda_3 R_t + \lambda_4 O_t + \lambda_5 E_t + e_t \qquad (2)$$

where
$S_t$ is the daily closing price of (ASE) index, as a proxy for market performance.
$P$ is the daily Covid-19 positive cases.
$D$ is the daily Covid-19 death case.

$R$ is the daily Covid-19 recovered cases.
$O$ is the daily world Oil prices.
$E$ is the daily Euro/Jordan dinar exchange rates.
$e_t$ is the error term.
Equation (2) in ARDL model form can be presented as:

$$\Delta LS_t = \beta_0 + \beta_1 (LS)_{t-1} + \beta_2 (LP)_{t-1} + \beta_3 (LD)_{t-1}$$
$$+ \beta_4 (LR)_{t-1} + \beta_5 (LO)_{t-1} + \beta_6 (LE)_{t-1} + \sum_{i=1}^{j} \beta_7 \Delta (Ls)_{t-i}$$
$$+ \sum_{i=1}^{q} \beta_8 \Delta (LP)_{t-i} + \sum_{i=1}^{q} \beta_9 \Delta (LD)_{t-i} + \sum_{i=1}^{q} \beta_{10} \Delta (LR)_{t-1}$$
$$+ \sum_{i=1}^{q} \beta_{11} \Delta (LO)_{t-i} + \sum_{i=1}^{q} \beta_{12} \Delta (LE)_{t-i} + e1_t \tag{3}$$

where $\Delta$ is Difference operator, L is natural logarithm, $\beta_0$ is the constant. $\beta_1$, $\beta_2$, $\beta_3$, $\beta_4$, $\beta_5$ $\beta_6$ are the long run coefficient, $\beta_7$, $\beta_8$, $\beta_9$, $\beta_{10}$, $\beta_{11}$, and $\beta$ 12 are the short run coefficients. $J$ is the optimum lag of dependent variable, $q$ is the optimum lag of independent variables. To establish relationship between variables Eq. (3) is used to bound test to cointegration.

In case of no cointegration the ARDL model is specified as:

$$\Delta LS_t = \beta_0 + \sum_{i=1}^{j} \beta_7 \Delta (Ls)_{t-i} + \sum_{i=1}^{q} \beta_8 \Delta (LP)_{t-i} + \sum_{i=1}^{q} \beta_9 \Delta (LD)_{t-i}$$
$$+ \sum_{i=1}^{q} \beta_{10} \Delta (LR)_{t-i} + \sum_{i=1}^{q} \beta_{11} \Delta (LO)_{t-i} + \sum_{i=1}^{q} \beta_{12} \Delta (LE)_{t-i} + e1_t \tag{4}$$

If there is exit cointegration, the error correction model (ECM) is specified as:

$$\Delta LS_t = \beta_0 + \sum_{i=1}^{j} \beta_7 \Delta (Ls)_{t-i} + \sum_{i=1}^{q} \beta_8 \Delta (LP)_{t-i} + \sum_{i=1}^{q} \beta_9 \Delta (LD)_{t-i}$$
$$+ \sum_{i=1}^{q} \beta_{10} \Delta (LR)_{t-i} + \sum_{i=1}^{q} \beta_{11} \Delta (LO)_{t-i}$$
$$+ \sum_{i=1}^{q} \beta_{12} \Delta (LE)_{t-i} + \varphi ECT_{t-i} + e1_t \tag{5}$$

**Table 2** Unit root test

| Variable | | Augmented Dickey-Fuller test equation Schwars info criteria-intercept | | | Integration order |
| | | T static value | Prob | Test critical values at 5% level | |
|---|---|---|---|---|---|
| S | Level | −126.4820 | 0.0001 | −2.898623 | I(0) |
| P | Level | −1.361036 | 0.5971 | −2.899115 | I(1) |
| | First difference | −15.47610 | 0.0001 | −2.899115 | |
| D | Level | −1.939502 | 0.3128 | −2.900137 | I(1) |
| | First difference | −7.471340 | 0.0000 | −2.900137 | |
| R | Level | −1.315887 | 0.6185 | −2.899115 | I(1) |
| | First difference | −13.73279 | 0.0001 | −2.899115 | |
| O | Level | −1.529613 | 0.5136 | −2.898623 | I(1) |
| | First difference | −10.03080 | 0.0000 | −3.516676 | |
| E | Level | −1.541004 | 0.5078 | −2.898623 | I(1) |
| | First difference | −9.748915 | 0.0000 | −2.899115 | |

where $\varphi$ is the speed of adjustment of parameter with a negative sign, $ECT$ is the error correction term $\beta 7, \beta 8, \beta 9, \beta 10, \beta 11, \beta 12$ are the coefficients of the short run dynamic models.

# 4   Empirical Findings

## 4.1   Unit Root Test

The study aims to examine the impact of Covid-19 on Amman stock exchange performance by applying the ARDL model. The ARDL model requires that variables are integrated of order $I(1)$ or $I(0)$ but not integrated of order more than 1. The results of the Augmented Dickey Fuller (ADF) reveal that all the variables are of order $I(0)$ and $I(1)$. Table 2 shows that all independent variables are non-stationary at level and become stationary at first difference $I(1)$, while the data for dependent variable is stationary at level $I(0)$.

## 4.2   Selection of Lags

Table 3 reports the VAR Lag Order Selection Criteria and shows a lag of order 1 according to (FPE, AIC, SC and HQ) while LR requires 2 lags.

**Table 3** VAR, lag order selection criteria

| Lag | LogL | LR | FPE | AIC | SC | HQ |
|---|---|---|---|---|---|---|
| 0 | 457.7737 | NA | 1.70e−13 | −12.37736 | −12.18911 | −12.30234 |
| 1 | 724.7180 | 482.6937 | 3.04e−16[a] | −18.70460[a] | −17.38680[a] | −18.17944[a] |
| 2 | 760.2683 | 58.43881[a] | 3.14e−16 | −18.69228 | −16.24494 | −17.71697 |

[a]Indicates lag order selected by the criterion
LR—sequential modified LR test statistic (each test at 5% level)
FPE—final prediction error
AIC—akaike information criterion
SC—Schwarz information criterion
HQ—Hannan-Quinn information criterion

## 4.3 Auto Regressive Distributed Lag Model (ARDL) Estimation

The results from ARDL model, based on the Akaike information criteria (AIC), are given in Table 4 which shows a maximum of 7 lags and the selected model is ARDL (6, 5, 3, 3, 7, 7).

To find out whether the explanatory variables with their lags in the estimated ARDL model presented in Table 4 affect the dependent variable in the short-run, we apply Wald-test. Our results reveal that all the explanatory variables jointly affect ASE performance (Table 5).

Having estimated ARDL model as shown above, we now turn to check there if is an evidence of the long-run cointegration association among the variables in levels. This can be done by using Bounds test proposed by [22]. Table 6 shows that the value of F-statistic (3.74) is greater than the upper bound of Pesaran test statistics at 10 (almost significant at 5%). This implies that there is long run cointegration in the study variables.

## 4.4 ARDL Cointegration and Long Run Form

Table 7 presents the estimation of the long-run levels model and the error correction model. Panel 1 shows the results for the short-run error correction representations of estimated ARDL model and Panel 2 reports the coefficients of variables for long-run.

The error correction term (CointEq (−1) or ECT) is negative (−0.253) and significant at 1% level. The negative sign implies that if the system in the previous day is moving out of equilibrium in one direction, it is going to pull it back to long-run equilibrium at a speed of 25.3%. Meaning that 25.3% of short-run departures from long-run equilibrium are corrected each period, which confirms the evidence of the long-run cointegration relationship.

**Table 4** Estimated ARDL model

Selected model: ARDL(6, 5, 3, 3, 7, 7), dependent variable LS

| Variable | Coefficient | Std. error | t-statistic | Prob.* |
|----------|-------------|------------|-------------|--------|
| LS(−1) | 0.746896 | 0.085907 | 8.694246 | 0.0000 |
| LS(−2) | −0.231699 | 0.104468 | −2.217887 | 0.0328 |
| LS(−3) | 0.031939 | 0.106274 | 0.300533 | 0.7655 |
| LS(−4) | −0.113467 | 0.115329 | −0.983855 | 0.3316 |
| LS(−5) | 0.178700 | 0.118861 | 1.503442 | 0.1412 |
| LS(−6) | −0.528518 | 0.099389 | −5.317683 | 0.0000 |
| LP | −0.009879 | 0.003982 | −2.480625 | 0.0178 |
| LP(−1) | −0.010453 | 0.004471 | −2.337905 | 0.0249 |
| LP(−2) | 0.006946 | 0.004553 | 1.525495 | 0.1356 |
| LP(−3) | 0.002964 | 0.004691 | 0.631791 | 0.5314 |
| LP(−4) | 0.003999 | 0.004719 | 0.847312 | 0.4023 |
| LP(−5) | 0.021365 | 0.004520 | 4.727008 | 0.0000 |
| LD | 0.005718 | 0.002421 | 2.361829 | 0.0236 |
| LD(−1) | 0.002034 | 0.002287 | 0.889562 | 0.3794 |
| LD(−2) | 0.002500 | 0.002287 | 1.093401 | 0.2813 |
| LD(−3) | 0.005641 | 0.002297 | 2.455121 | 0.0189 |
| LR | −0.004609 | 0.002503 | −1.841386 | 0.0736 |
| LR(−1) | −0.006018 | 0.002753 | −2.185590 | 0.0353 |
| LR(−2) | −0.005678 | 0.002458 | −2.309836 | 0.0266 |
| LR(−3) | −0.013928 | 0.002389 | −5.830060 | 0.0000 |
| LO | −0.026734 | 0.029144 | −0.917289 | 0.3649 |
| LO(−1) | 0.001633 | 0.036446 | 0.044794 | 0.9645 |
| LO(−2) | 0.033488 | 0.038551 | 0.868669 | 0.3906 |
| LO(−3) | −0.038908 | 0.040893 | −0.951460 | 0.3475 |
| LO(−4) | −0.076748 | 0.044137 | −1.738876 | 0.0904 |
| LO(−5) | 0.007399 | 0.044184 | 0.167451 | 0.8679 |
| LO(−6) | 0.024468 | 0.044338 | 0.551845 | 0.5844 |
| LO(−7) | 0.089833 | 0.032049 | 2.803020 | 0.0080 |
| LE | 0.106285 | 0.204453 | 0.519851 | 0.6063 |
| LE(−1) | −1.087852 | 0.237110 | −4.587963 | 0.0000 |
| LE(−2) | −0.186269 | 0.248726 | −0.748893 | 0.4587 |
| LE(−3) | 0.355619 | 0.241068 | 1.475183 | 0.1486 |
| LE(−4) | 0.338273 | 0.235489 | 1.436470 | 0.1593 |
| LE(−5) | −0.764096 | 0.239222 | −3.194091 | 0.0029 |

(continued)

**Table 4** (continued)

Selected model: ARDL(6, 5, 3, 3, 7, 7), dependent variable LS

| Variable | Coefficient | Std. error | t-statistic | Prob.* |
|----------|-------------|------------|-------------|--------|
| LE(−6) | −0.715712 | 0.236903 | −3.021121 | 0.0045 |
| LE(−7) | 0.660712 | 0.195149 | 3.385673 | 0.0017 |
| C | 6.648663 | 0.947925 | 7.013912 | 0.0000 |
| | R-squared | 0.934189 | Mean dependent var | 7.469119 |
| | Adjusted R-squared | 0.870157 | Prob(F-statistic) | 0.000000 |
| | F-statistic | 14.58932 | Akaike info criterion | −7.360564 |

**Table 5** Short-run causality, based on Wald test

| Regressor | Hypothesis | Chi-square prob | Result |
|-----------|------------|-----------------|--------|
| LP | C(7)= ⋯ = C(12) = 0 | 0.000 | Reject |
| LD | C(13)= ⋯ = C(16) = 0 | 0.014 | Reject |
| LR | C(17)= ⋯ = C(20) = 0 | 0.000 | Reject |
| LO | C(21)= ⋯ = C(28) = 0 | 0.000 | Reject |
| LE | C(29)= ⋯ = C(36) = 0 | 0.000 | Reject |

**Table 6** ARDL bounds test (the critical values are obtained from [22]

| Test statistic | Value | k |
|----------------|-------|---|
| F-statistic | 3.741571 | 5 |
| *Critical value bounds* | | |
| Significance (%) | I0 bound | I1 Bound |
| 10 | 2.26 | 3.35 |
| 5 | 2.62 | 3.79 |
| 2.5 | 2.96 | 4.18 |
| 1 | 3.41 | 4.68 |

The long-run cointegration results from panel 2 are to a large extant consistent with our expectations. Positive covid-19 daily cases have a long-run significant negative impact on the performance (an increase of 1% in positive cases lead to a decline of 2.65% in performance); we reject the null hypothesis H01. Recovered cases have a positive and significant impact on performance (an increase of 1% in recovered cases lead to an increase of 0.12% in performance); we reject the null hypothesis H03. The Euro/JD exchange rate has the biggest impact and significant at the 1% level. When the Euro goes down (JD appreciates), the market seems to react positively, which

**Table 7** ARDL (6, 5, 3, 3, 7, 7)—based on AIC

| Dependent Variable: LS | | | | |
|---|---|---|---|---|

*Panel 1: cointegrating form*

| Variable | Coefficient | Std. Error | t-Statistic | Prob |
|---|---|---|---|---|
| D(LP) | −0.231699 | 0.104468 | −2.217887 | 0.0328 |
| D(LP(−1)) | 0.113467 | 0.115329 | 0.983855 | 0.3316 |
| D(LP(−2)) | −0.178700 | 0.118861 | −1.503442 | 0.1412 |
| D(LP(−3)) | 0.528518 | 0.099389 | 5.317683 | 0.0000 |
| D(LP(−4)) | 0.009879 | 0.003982 | 2.480625 | 0.0178 |
| D(LD) | −0.010453 | 0.004471 | −2.337905 | 0.0249 |
| D(LD(−1)) | −0.002964 | 0.004691 | −0.631791 | 0.5314 |
| D(LD(−2)) | −0.003999 | 0.004719 | −0.847312 | 0.4023 |
| D(LR) | 0.021365 | 0.004520 | 4.727008 | 0.0000 |
| D(LR(−1)) | −0.002034 | 0.002287 | −0.889562 | 0.3794 |
| D(LR(−2)) | −0.002500 | 0.002287 | −1.093401 | 0.2813 |
| D(LO) | 0.005641 | 0.002297 | 2.455121 | 0.0189 |
| D(LO(−1)) | 0.006018 | 0.002753 | 2.185590 | 0.0353 |
| D(LO(−2)) | 0.005678 | 0.002458 | 2.309836 | 0.0266 |
| D(LO(−3)) | 0.013928 | 0.002389 | 5.830060 | 0.0000 |
| D(LO(−4)) | 0.026734 | 0.029144 | 0.917289 | 0.3649 |
| D(LO(−5)) | −0.001633 | 0.036446 | −0.044794 | 0.9645 |
| D(LO(−6)) | −0.033488 | 0.038551 | −0.868669 | 0.3906 |
| D(LE) | −0.038908 | 0.040893 | −0.951460 | 0.3475 |
| D(LE(−1)) | −0.007399 | 0.044184 | −0.167451 | 0.8679 |
| D(LE(−2)) | −0.024468 | 0.044338 | −0.551845 | 0.5844 |
| D(LE(−3)) | −0.089833 | 0.032049 | −2.803020 | 0.0080 |
| D(LE(−4)) | −0.106285 | 0.204453 | −0.519851 | 0.6063 |
| D(LE(−5)) | 1.087852 | 0.237110 | 4.587963 | 0.0000 |
| D(LE(−6)) | 0.186269 | 0.248726 | 0.748893 | 0.4587 |
| CointEq(−1) | −0.253104 | 0.085907 | −2.946256 | 0.0055 |

Cointeq = LS − (−2.6587*LP + 0.0137*LD + 0.1249*LR −0.0640 * LO −4.5902 * LE+1.4050)

*Panel 2: long run coefficients*

| Variable | Coefficient | Std. error | t-statistic | Prob. |
|---|---|---|---|---|
| LP | −2.658687 | 1.152150 | −2.307587 | 0.0267 |
| LD | 0.013653 | 0.022580 | 0.604654 | 0.5491 |
| LR | 0.124922 | 0.050488 | 2.474283 | 0.0181 |
| LO | −0.064021 | 0.122695 | −0.521785 | 0.6049 |
| LE | −4.590184 | 1.508637 | −3.042604 | 0.0043 |

(continued)

**Table 7** (continued)

| Dependent Variable: LS | | | | |
|---|---|---|---|---|
| C | 1.405033 | 0.995508 | 1.411373 | 0.1665 |

goes in line with economic theory as gross consumption in Jordan is mainly attributed to foreign products. Death cases and oil prices are not significant.

## 4.5 Residual Diagnostics Tests

The diagnostic tests show that the ARDL (6, 5, 3, 3, 7, 7) model has the proper econometric properties. The functional form is correct, the residuals are not serially correlated, normally distributed, and the Heteroscdasticity problem is not present.

1. Based on the Jarque–Bera test, Fig. 4 shows that the probability value is 0.867 implying that we cannot reject the null hypothesis at 5% (Ho: the residuals are normally distributed) and conclude that the residuals are normally distributed (Fig. 4).
2. Based on Breusch-Godfrey Serial correlation LM test; the residuals are free from serial correlation. Table 8 shows that the R-squared probability value is

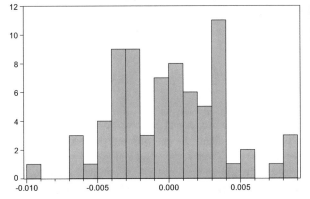

**Fig. 4** Normality test

**Table 8** Breusch-Godfrey serial correlation LM test

| F-statistic | 0.692790 | Prob. F(2, 35) | 0.5069 |
|---|---|---|---|
| Obs*R-squared | 2.817955 | Prob. Chi-Square(2) | 0.2444 |
| Heteroskedasticity test: Breusch-Pagan-Godfrey | | | |

**Table 9** Heteroskedasticity test: Breusch-Pagan-Godfrey

| F-statistic | 1.096877 | Prob. F(36,37) | 0.3902 |
|---|---|---|---|
| Obs*R-squared | 38.20333 | Prob. Chi-Square(36) | 0.3696 |
| Scaled explained SS | 8.454497 | Prob. Chi-Square(36) | 1.0000 |

0.244, not significant at 5% level. Therefore, we cannot reject the null hypothesis (Ho: no serial correlation).

3.  Based on Breusch–Pagan-Godfrey heteroscedasticity test; the residuals are free from Heteroscedasticity. Table 9 shows that the R-squared probability value is 0.367 not significant at 5% level. Therefore we cannot reject the null hypothesis (Ho: no heteroscedasticity).

## 4.6   Stability Test

In order to investigate the stability of long-run and short-run parameters of the selected error correction ARDL model, the cumulative sum (CUSUM) and cumulative sum of squares (CUSUMsq) are employed as suggested by [22]. The results of both (CUSUM) and CUSMsq are presented in Figs. 5 and 6, respectively. The plots of the two figures are between critical bounds of 5% level of significance. The null hypothesis cannot be rejected (Ho: all parameters are stable) which means that the model is stable.

## 5   Conclusion

Results of the long-run reveal that the coefficients of positive cases, recovered cases and exchange rate are statistically significant in determining the performance of ASE. An increase of 1% in positive cases, recovered cases and exchange rate lead to a decline of 2.65%, an increase of 0.12% and a decline 4.59% in performance. It is worth noting that the greatest impact on performance is coming from exchange rate followed by positive cases. The positive reaction of the performance of ASE may be attributed to the fact that imports largely exceeds exports. Therefore, an appreciation of the JD makes imports less costly which goes in line with economic theory as gross consumption in Jordan is mainly attributed to foreign products. Furthermore, positive cases are also highly significant while death cases are not which could be attributed to the fact that positive cases may have spillover effect while death cases may not. Also, the magnitude of positive cases is much larger than that of death cases.

The error correction term (CointEq $(-1)$ or ECT) is negative $(-0.253)$ and significant at 1% level. The negative sign implies that if the system in the previous day

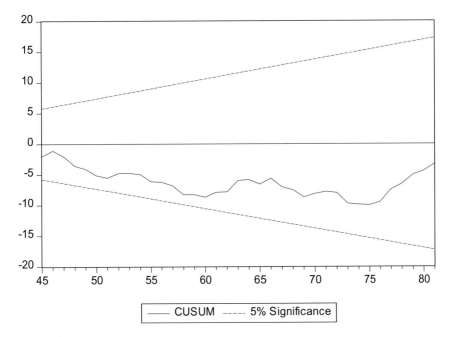

**Fig. 5**  CUSUM

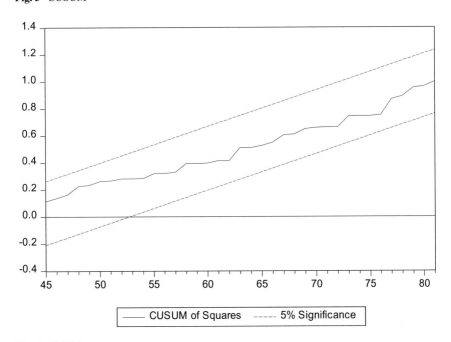

**Fig. 6**  CUSMsq

is moving out of equilibrium in one direction, it is going to pull it back to long-run equilibrium at a speed of 25.3% meaning that 25.3% of short-run departures from long-run equilibrium in the previous day are corrected in the current day, which confirms the evidence of the long-run cointegration relationship.

In addition, we would like to say that the results of our paper are confirmed with previous research such [13, 19, 20].

# References

1. Adrian T, Natalucci F (2020) COVID-19 crisis poses threat to financial stability. The International Monetary Fund blogs (IMF)
2. Burns W (2012) Risk perception and the economic crisis: a longitudinal study of the trajectory of perceived risk. Risk Anal: Int J
3. Hong et al (2021) COVID-19 and instability of stock market performance: evidence from the U.S. Finan Innov
4. He Q et al (2020) The impact of COVID-19 on stock markets. Econ Polit Stud
5. Lu X, Lai K (2020) Relationship between stock indices and investors' sentiment index in Chinese financial market. Syst Eng Theory Pract
6. Mazur M (2020) COVID-19 and the March 2020 stock market crash: evidence from S&P500. Finan Res Lett
7. Ashraf BN (2020) Stock market's reaction to COVID-19: cases or fatalities? Res Int Bus Finan
8. Baker S et al (2020) The unprecedented stock market reaction to COVID-19. Rev Asset Pricing Study
9. Liu H (2020) The COVID-19 outbreak and affected countries stock markets response. Int J Environ Res Public Health
10. Topcu M, Gulal O (2020) The impact of COVID-19 on emerging stock markets. Financ Res Lett. https://doi.org/10.1016/j.frl.2020.101691
11. Narayan PK (2020) Has COVID-19 changed exchange rate resistance to shocks? Asian Econ Lett
12. Scott R et al (2020) The unprecedented stock market reaction to COVID-19. National Bureau Of Economic Research 1050 Massachusetts Avenue Cambridge
13. Albulescu C (2020) Coronavirus and financial volatility: 40 days of fasting and fear. Available at SSRN: https://ssrn.com/abstract=3550630 or https://doi.org/10.2139/ssrn.3550630
14. Amr A, Nader A (2020) The impact of coronavirus pandemic on stock market return: the case of the MENA region. Int J Econ Finan
15. Bora D, Basistha D (2021, Feb) The outbreak of COVID-19 pandemic and its impact on stock market volatility: evidence from a worst-affected economy. J Public Aff 11:e2623. https://doi.org/10.1002/pa.2623. Epub ahead of print. PMID: 33786019; PMCID: PMC7995228
16. Kelvin Y et al (2020) Impact of Covid-19: evidence from Malaysian Stock Market. Int J Bus Socety
17. Wu W et al (2020) The impact of the COVID-19 outbreak on Chinese-listed tourism stocks. Financ Innov
18. Lee C, Chen M (2020) The impact of COVID-19 on the travel & leisure industry returns: some international evidence. Tour Econ
19. Alber N, Saleh A (2020) The impact of Covid-19 spread on stock markets: the case of the GCC countries. Int Bus Res
20. Pesaran H et al (2001) Bounds testing approach to the analysis of level relationships. J Appl Econ
21. Duro et al (2020) Covid-19 and tourism vulnerability. Tour Manag Perspect

22. Marcoa et al (2020) COVID-19 pandemic and lockdown measures impact on mental health among the general population in Italy. Front Psychiatry
23. Alok K et al (2020) Does the Indian financial market nosedive because of the COVID-19 outbreak, in comparison to after demonetisation and the GST? Emerg Markets Finan Trade
24. Balde R et al (2020) Labor market effects of COVID-19 in sub-Saharan Africa: an informality lens from Burkina Faso, Mali and Senegal. United Nations University-Maastricht Economic and Social Research Institute on Innovation and Technology (MERIT)
25. Ahmed S (2020) Impact of COVID-19 on performance of Pakistan stock exchange. Retrieved from: https://doi.org/10.20944/preprints202007.0083.v1
26. Wu et al (2021) The impact of the COVID-19 outbreak on Chinese-listed tourism stocks. Finan Innov
27. Xiaolin H, Zhigang Q (2020) How does China's stock market react to the announcement of the COVID-19 pandemic lockdown? Econ Polit Stud

# Firm's Sharia-Compliant Status, Cross-Sectional Returns, and Malaysian Cross-Border Acquisition

**Md. Mahadi Hasan, Jimoh Olajide Raji, A. T. M. Adnan, Ezaz Ahmed, and Mohammad Irfan**

**Abstract** The results suggest that acquirers' agency costs influence CBA performance. Those with reduced agency expenses, as defined by sharia compliance, greater ethnic diversity on the board, and greater gender diversity on the board, demonstrate superior performance. Additionally, the performance is influenced positively by the acquirer's prior year profitability and diversification of acquisition strategy. Because the sample was restricted to the CBA in Malaysia, the findings could not be generalized to other countries. The findings are plausible to lead policymakers in developing acceptable policies on sharia compliance, female representation, and representation of diverse races on corporate boards of directors. Additionally, the findings could be used to encourage suitable investment policy decisions that improve enterprises' long-term success. The unique novelty of this work is to illustrate the application of corporate finance theories that attempt to explain CBA effectiveness in emerging economies where CBA is vigorously sought and significant rate of failure are observed using long-term datasets and robust performance indicators and analytics.

Md. Mahadi Hasan (✉)
Anwer Khan Modern University, Dhaka, Bangladesh
e-mail: mahadihasan@sefb.uum.edu.my; mahadi@akmu.edu.bd

J. O. Raji
Universiti Utara Malaysia, Kedah, Malaysia
e-mail: raji@uum.edu.my

A. T. M. Adnan
BGMEA University of Fashion and Technology, Dhaka, Bangladesh
e-mail: atmadnan@buft.edu.bd

E. Ahmed
Columbia College, Columbia, USA
e-mail: eahmed@columbiasc.edu

M. Irfan
CMR Institute of Technology, Bangalore, Karnataka, India
e-mail: irfan.m@cmrit.ac.ind

**Keywords** Cross-border acquisition · Shareholders' wealth effect · Buy-and-hold abnormal returns · Agency cost · Bursa Malaysia

# 1  Introduction

Despite the prevalence of cross-border acquisition-based development strategies (CBA), there are substantial execution problems, with failure rates ranging from 40 to 80% stated by multiple sources [31, 43]. Furthermore, it was reported that global businesses spend more than $2 trillion on acquisitions of all sorts each year, with failure rates ranging from 70 to 90%. [15, 20]. More precisely, PwC's analyses of the Malaysian market indicate that 70% of mergers and acquisitions fail in general.

The high failure rate of the CBA has prompted numerous research on the wealth effect of the CBA. However, the findings are contradictory. For example, Francis et al. [28] demonstrated positive returns in the United States; Banerjee et al. [8] demonstrated positive returns in India; and Khin et al. [40] demonstrated good returns in Malaysia. In comparison, Wang et al. [48] demonstrated negative returns or values in Asia, while Bertrand and Betschinger [12] demonstrated negative returns or values in Russia. Basuil and Datta [9] in the United States, Chakrabarti [17] in India, and Conn and Michie [21] in the United Kingdom all demonstrate that acquiring enterprises generates a negligible or neutral return.

In addition to cultural similarities and distance [18], predecessors' acquisition activity [28], bidders' prior experiences [9], as well as the degree of R&D engaged [29], another possible cause might be agency costs at the time CBA decisions were made. According to Jensen and Meckling [38], agency problems arise from management's competing interests with shareholders, and managers tend to over-invest extra capital within their control. This disagreement leads to agency expenses, which lower business worth. In the framework of CBA, managers may prioritize international expansion over shareholder value development. As a result, acquirers with higher agency expenses may have poorer share returns.

To the author's knowledge, no in-depth research has been done on long-term SWE and its determinant connected to agency cost variables after a CBA in Bursa Malaysia (BM). Only a few studies have been done on the long-term shareholder wealth effect (SWE) after a CBA, and most of them are in industrialized countries. However, heterogeneity prevents universal application of the study's conclusions. As a result, long-term SWE studies are inconclusive. To the author's knowledge, the wealth effect was not previously studied in the BM setting using Sharia-compliant status, ethnic diversity, gender diversity, payment form, and acquisition technique. Thus, incorporating some or all of the above criteria into CBA studies assessing long-term SWE will help fill crucial research gaps.

Using Buy and Hold Abnormal Return (BHAR) and other proxy indices to estimate agency costs, this study seeks to determine if agency costs or corporate governance characteristics explain Malaysian acquirers' performance post-CBA. From 2004 to 2015, 178 CBA deals involving publicly traded acquirers were studied.

## 2   Literature Review

Many scholars, including [3, 5, 13, 33, 39], and [30], claim that Sharia-compliant enterprises outperform conventional firms in times of crisis. Moreover, other studies claim that Sharia-compliant enterprises provided crisis protection due to their low leverage [13] and that they are suited for risk-averse investors (see [3]). Dewandaru et al. [23] even suggest that Sharia-compliant enterprises are a distinct asset class due to their low leverage. This is surprising because Sharia-compliant corporations have low leverage, and low leverage equities should be safer, according to normal financial theory (see [14]). Firms' low financial leverage may be mitigated by their strong operating leverage.

Thus, Sharia-compliant enterprises represent more informed investors [26]. Maybe it is because an Islamic label and low leverage are positive traits (for example, higher investor awareness). Some research shows that a Sharia index is a strong signal of increased profitability (e.g. [16, 24]). This may be especially true for Sharia-compliant enterprises, though it can also be the case. The study anticipates Sharia-compliant enterprises to be less profitable than non-compliant firms because leverage increases return on equity. However, Sharia-compliant means that the company's actions are legal. This would have a tremendous impact on long-term SWE post-CBA. Thus, this analysis suggests that acquiring enterprises' Sharia-compliant status will have a positive impact on their performance following CBA.

(a)   *Assuming all other factors remain constant, the Sharia-compliant status of acquiring firms is positively related to shareholders' wealth effect following CBA in the short-term (i.e., post-12 month acquisition event).*

(b)   *Assuming all other factors remain constant, the Sharia-compliant status of acquiring firms is positively related to shareholders' wealth effect following CBA in the long-term (i.e., post-24 month acquisition event).*

(c)   *Assuming all other factors remain constant, the Sharia-compliant status of acquiring firms is positively related to shareholders' wealth effect following CBA in the long-term (i.e., post-36 months post acquisition).*

## 3   Methodology

### 3.1   *Sample and Data Sources*

Data from the Bursa Malaysia website, prospectus, and annual report of each business were used to assess the acquiring firm's sharia-compliant status, racial diversity, and gender distinctiveness. Thomson Reuter's database was used to collect CBA features, such as payment mode and control level of the target firm. In addition, monthly stock return indexes, market capitalization, market-to-book value, and yearly cash flow statements were obtained from the Datastream database.

The study's sample comprises CBA deals involving acquiring corporations from 2004 to 2015. Pending and withdrawn deals, deals with an uncertain status, intended rumours, dismissed rumours, and deals with an intent to withdraw were deemed uncompleted by financial companies and hence excluded from the sample. Numerous transactions from the post-period were also removed. The overall number of participants in 12-, 24-, and 36-month performance studies is 178, including their respective companies.

## 3.2 Abnormal Return Calculation

The results of a long-run event study are sensitive to both the technique and the standard used [4, 22, 35, 41, 44]. This is why the two most essential parts of evaluating the long-run wealth effect are adopting an acceptable technique to compute anomalous returns and matching them to a suitable standard. We employed the BHAR method to post-event performance with the characteristics-based benchmark in this study.

**Buy-and-hold Abnormal Return (BHAR).** The very first step in computing BHAR is to compute company i's holding period returns for the period studied in months (T),

$$BHR_{i,T} = \prod_{t=1}^{T} (1 + r_{it}) - 1 \tag{1}$$

where $r_{it}$ is the company i's monthly raw return in month $t$. The holding period return for the benchmark $b$ is calculated using the same formula.

$$BHR_{b,T} = \prod_{t=1}^{T} (1 + r_{bt}) - 1 \tag{2}$$

After benchmark correction, the buy-and-hold abnormal returns for each firm $I$ in month $t$ are the difference between the firm's and the benchmark's buy-and-hold returns:

$$BHAR_{it} = BHR_{it} - BHR_{bt} \tag{3}$$

The value weighted ($w_i$) mean of the buy-and-hold abnormal return (BHAR) for month t was calculated as follows using Eq. (3):

$$\overline{BHAR_t} = \sum_{i=1}^{n_t} BHAR_{it} \tag{4}$$

**Benchmark Firms Constructed** Identifying benchmark businesses based on size and the book-to-market ratio is common in long-run event study research [31, 35]. Before recognizing any business as a benchmark candidate firm, certain requirements must be met. The research builds on a number of empirical investigations [11, 45], and [47] that employ Euclidean distance as a matching estimate. Following these investigations, the researcher computes the Euclidean distances between each of the sample acquirers and the benchmark candidates depending on the firm's size as assessed by market capitalization and market-to-book ratio.

# 4 Results and Discussion

## 4.1 Regression Model

Following the research Ibrahim et al. [35], the baseline equation was used to examine the influence of agency cost proxies on acquirers' post cross-border acquisition performance.

$$BHAR_{it} = \alpha + \beta_1 SCS + \beta_i CV + \varepsilon \tag{5}$$

where $BHAR_{it}$ = acquirers' long-run post-cross-border acquisition performance, which represents a 12 month return, a 24 month return, and a 36 month return. $\alpha$ = intercept term, SCS = Sharia-complaint status, CV = Control Variables such as Board ethnic diversity, BGD = Board gender diversity, AS = Acquirer size, AP = Acquirer previous year profit, AI = Acquirer Industry, MP = Mode of payment, and DA = diversifying acquisition.

## 4.2 Measurement of Explanatory Variables

Guided by prior related empirical literature, the following explanatory variables are measured in Table 1.

## 4.3 Descriptive Statistics

The descriptive statistics of the variables utilized in the investigation are shown in Table 2. When it comes to value-weighted buy-and-hold abnormal, BHAR3Y has the highest maximum value when compared to BHAR1Y and BHAR2Y, while BHAR2Y has the lowest maximum value.

**Table 1** Measurement of the agency cost variables

| Variables | Measures | Supported studies |
|---|---|---|
| Sharia-complaint status (SCS) | Whether the acquiring firm has Sharia-complaint status | Abdul-Rahim and Che-Embi [1], Rahim et al. [43] |
| Board ethnic diversity (BED) | The Her-findahl Hirshman Index is used to assess ethnic diversity. (HHI)[a] | Cheong and Sinnakkannu [19] |
| Board gender diversity (BGD) | If the firm's board of directors includes at least one female member, the value is 1; otherwise, the value is 0 | Terjesen et al. [46] |
| Acquirer profit (AP) | Return on assets (ROA) of the previous of CBA event of the acquiring firm | Du and Boateng [25] |
| Firm Size (AS) | The natural log of the firm's entire asset | Ibrahim and Minai [36], Ibrahim and Hwei [37] |
| Mode of payment (MP) | If the manner of payment is cash, then 1 otherwise 0 | Khin et al. [40], Barbopoulos et al. [10] |
| Diversifying acquisitions (DA) | "Diversifying" when the first two-digits of the main industry code of the bidder and the target are not the same | Aybar and Thanakijsombat [7] |

[a]The Herfindahl Index is calculated by subtracting one from the sum of the squared proportions of each ethnic group

The average Sharia-compliant acquiring firm (SCS) is 73%, whereas non-compliant acquiring is 27%. The table demonstrates that acquiring corporations' boards are ethnically and gender diverse. Malay, Chinese, and Indians are the primary ethnic groupings in Malaysia. Several members of the board are Chinese or Malay. On average, 15.4% of these ethnicities are recognized on the board, with women accounting for 46.6% of them. The increased female board presence aligns with the Malaysian government's goal of increasing female board representation to 30% by 2020.

In terms of the amount of control in target businesses and manner of payment, Malaysian acquirers have a greater level of control in target enterprises abroad, with around 70%. Furthermore, cash is the most common form of payment, accounting for around 68% of all transactions.

## 4.4 Correlation Statistics

Correlation analysis was performed on all variables utilized to see whether substantial high correlations among variables may result to a multicollinearity problem. The findings of the correlations are shown in Table 3. It can be observed that none of the

**Table 2** Descriptive analysis

| Variable | Obs. | Mean | Std. Dev. | Min | Max |
|---|---|---|---|---|---|
| BHAR1Y | 176 | 0.01 | 0.72 | −1.72 | 5.52 |
| BHAR2Y | 176 | −0.03 | 1.04 | −2.29 | 8.16 |
| BHAR3Y | 176 | −0.12 | 1.27 | −5.53 | 7.93 |
| BED | 176 | 0.38 | 0.18 | 0.00 | 0.66 |
| AS | 176 | 12.73 | 1.66 | 7.86 | 17.50 |
| AP | 176 | 0.08 | 0.25 | −2.32 | 0.64 |
| SCS | | | | | |
| 0 | 176 | 0.27 | 0.45 | 0 | 1 |
| 1 | 176 | 0.73 | 0.45 | 0 | 1 |
| BGD | | | | | |
| 0 | 176 | 0.65 | 0.48 | 0 | 1 |
| 1 | 176 | 0.35 | 0.48 | 0 | 1 |
| DA | | | | | |
| 0 | 176 | 0.17 | 0.38 | 0 | 1 |
| 1 | 176 | 0.83 | 0.38 | 0 | 1 |
| MP | | | | | |
| 0 | 176 | 0.32 | 0.47 | 0 | 1 |
| 1 | 176 | 0.68 | 0.47 | 0 | 1 |

**Table 3** Correlation statistics

| | BHAR1Y | BHAR2Y | BHAR3Y | BED | AS | AP | SCS | BGD | DA | MP |
|---|---|---|---|---|---|---|---|---|---|---|
| BHAR1Y | 1.00 | | | | | | | | | |
| BHAR2Y | 0.79 | 1.00 | | | | | | | | |
| BHAR3Y | 0.62 | 0.81 | 1.00 | | | | | | | |
| BED | 0.04 | 0.00 | −0.04 | 1.00 | | | | | | |
| AS | −0.08 | −0.07 | −0.01 | 0.06 | 1.00 | | | | | |
| AP | 0.08 | 0.05 | 0.11 | 0.00 | 0.23 | 1.00 | | | | |
| SCS | 0.03 | 0.00 | 0.06 | −0.26 | −0.06 | −0.08 | 1.00 | | | |
| BGD | 0.03 | 0.06 | 0.09 | −0.20 | −0.07 | 0.07 | 0.07 | 1.00 | | |
| DA | 0.02 | 0.10 | 0.08 | −0.05 | 0.17 | 0.15 | 0.03 | −0.05 | 1.00 | |
| MP | −0.12 | −0.17 | −0.18 | 0.06 | −0.08 | −0.02 | −0.1 | 0.07 | −0.02 | 1.0 |

correlations are more than 0.7, which does not suggest a multicollinearity concern, according to Hair et al. [32].

Due to heteroskedasticity, the study uses WLS regression. It is acknowledged that selecting an appropriate weight for data analysis is not an easy undertaking. Agung [6] recommended choosing alternative weight variables based on adjusted R-squared, SSRs, Values for the Akaike information criterion (AIC) and the Schwarz

**Table 4** WLS for BHAR1Y, BHAR2Y, and BHAR3Y model

| Model | BHAR1Y | BHAR2Y | BHAR3Y |
|---|---|---|---|
| SCS | 0.35*** | 0.35*** | 0.50*** |
| | (0.06) | (0.08) | (0.10) |
| BED | 1.47*** | 1.55*** | 1.70*** |
| | (0.30) | (0.41) | (0.46) |
| AS | −0.03 | −0.07** | −0.02 |
| | (0.02) | (0.03) | (0.04) |
| AP | 1.79* | 5.17*** | 4.50*** |
| | (0.95) | (1.30) | (1.46) |
| BGD | 0.15* | 0.15 | 0.31** |
| | (0.09) | (0.12) | (0.14) |
| DA | 0.13 | 0.45** | 0.60** |
| | (0.14) | (0.20) | (0.22) |
| MP | 0.13* | 0.02 | 0.13 |
| | (0.07) | (0.10) | (0.11) |
| Constant | −0.82** | −0.88* | −1.96*** |
| | (0.34) | (0.47) | (0.53) |
| Obs | 176 | 176 | 176 |
| $R^2$ | 30.13% | 34.56% | 31.65% |
| Adj. $R^2$ | 27.22% | 31.84% | 28.80% |

*Notes* ***, ** and * represent 1%, 5% and 10% level of significance, respectively. The figure in parenthesis represents standard error

information criterion (SIC). To illustrate this point, each independent variable (except dummy variables) is assigned a weight in the model. By comparing weight variables, we chose the model with the best adj. R2, SSRs, AIC, and SIC (see [6]). In this study's three models, only the AP variable meets these conditions for a good fit model using WLS regression. Therefore, Table 4 shows the outcomes of utilizing WLS regression using AP as a weighted variable.

## 4.5 Regression Statistics

Table 4 shows that both sharia-complaint status (SCS) and board gender diversity (BGD) influence Malaysian acquirers' long-term performance. SCS had a considerable positive impact on 12-, 24-, and 36-month long-run performance at the 1% level, validating the earlier claim. In other words, more firm participation leads to more active monitoring and control of investee enterprises. As a result, sharia-compliant enterprises have better corporate governance, decrease agency expenses, and better long-term performance of CBA decisions. Moreover, BED has a 1% favorable influence on Malaysian enterprises' three-year returns. Firms with a diverse board of directors have reduced agency costs, which leads to greater alignment of CBA decisions with shareholder wealth [38]. It demonstrates this point. According to our findings, board ethnic diversity (BED) and board gender diversity (BGD) are key predictors of shareholder wealth. The results reveal that BED has a considerable positive impact on three Malaysian acquirers' long-term earnings. It follows that the greater the number of ethnicity featured on board, the better the post-CBA performance. Our findings coincide with Erhardt et al. [27] and [2] who claim that ethnic diversity is linked to company performance. Similarly, BGD has a considerable impact on Malaysian acquirers' one-year and two-year performance. The conclusion corroborates [34] and [2] findings that female board members increase board deliberations on complicated problems and company performance. Previous acquirer profitability (AP) has a considerable favorable impact on Malaysian acquirers' long-term returns. A 1% influence of BGD and DA on post-CBA long run performance in models with Long-term performance measures for two and three years. The longer the purchaser returns, the better for Malaysian CBA. This result agrees with [25]

# 5 Concluding Remarks

In this study, we look at the influence of sharia-compliant status, female board members, and board ethnic diversity on Malaysian acquirers' long-term profits in order to validate the agency cost associated with CBA. CBA's long-term success is measured by buy and hold returns (BHAR). Our findings from the Weighted Least Square regression indicate that the acquirers' agency cost has an effect on the CBA performance. Those having lower agency costs, as shown by a higher SCS, a higher BGD, and a higher BED, perform better. Additionally, AP and DA have a favorable effect on performance. However, the study discovers a detrimental effect of AS on a Malaysian firm's post-cross-border mergers and acquisitions performance.

The study's findings are important for the corporate governance literature because they show how various governance characteristics impact the wealth of Malaysian acquirers through cross-border mergers and acquisitions. Good corporate governance, according to agency theory-based research, can increase shareholders values by minimizing conflicts between shareholders and agents (managers). We believe, based on our study, that efficient corporate governance can result in low internal costs associated with aligning stakeholders' risk preferences and interests and managing the organization's activities across borders.

The goal of this research is to identify the factors that influence acquirer returns in Malaysian post-cross-border mergers and acquisitions. This has ramifications for affected businesses and authorities. For instance, our findings indicate that adherence to sharia is a critical component in sourcing value development for cross-border mergers and acquisitions in Malaysia, an expanding market. Due to the fact that a firm's sharia-compliant presence strengthens its corporate governance structure and decreases agency costs, regulators must ensure that sharia governance is attracted in order to ensure effective monitoring of firms engaged in cross-border mergers and acquisitions.

Furthermore, the findings highlight the critical importance of ethnic and gender diversity on corporate boards of directors in determining acquirers' long-term performance following cross-border mergers and acquisitions. Given Malaysians' advocacy for the removal of racial barriers and the promotion of an equal share of economic well-being, companies' commitments to ethnic diversity are likely to match social norms that increase enterprises' reputation. As a result, policymakers and regulators in Malaysia should place a higher focus on the racial makeup of Malaysian boards of directors.

Finally, our results on the role of female directors in business performance show that having female directors on boards enhances acquirers' profits on Malaysian post-cross-border mergers and acquisitions. The substantial positive effect of board diversity on Malaysian companies' post-cross-border mergers and acquisitions performance suggests that female director participation on boards of directors is a crucial problem of corporate governance that requires extra attention.

# References

1. Abdul-Rahim R, Che-Embi NA (2013) Initial returns of Shariah versus non-Shariah IPOs: are there any differences? Jurnal Pengurusan (J Manag) 39:37–50
2. Adams RB, Ferreira D (2009) Women in the boardroom and their impact on governance and performance. J Financ Econ 94(2):291–309
3. Adamsson H, Bouslah K, Hoepner A (2014) An Islamic equity premium puzzle, ICMA Working Paper, Henley Business School, University of Reading
4. Agrawal A, Jaffe JF, Mandelker GN (1992) The post-merger performance of acquiring firms: a re-examination of an anomaly. J Financ 47(4):1605–1621
5. Al-Khazali O, Lean HH, Samet A (2014) Do Islamic stock indexes outperform conventional stock indexes? A stochastic dominance approach. Pac Basin Financ J 28:29–46
6. Agung IGN (2011) Cross section and experimental data analysis using EViews. Wiley, Singapore
7. Aybar B, Thanakijsombat T (2015) Financing decisions and gains from cross-border acquisitions by emerging-market acquirers. Emerg Mark Rev 24:69–80
8. Banerjee P, Banerjee P, De S, Jindra J, Mukhopadhyay J (2014) Acquisition pricing in India during 1995–2011: have Indian acquirers really beaten the odds? J Bank Finance 38:14–30
9. Basuil DA, Datta DK (2015) Effects of industry-and region-specific acquisition experience on value creation in cross-border acquisitions: the moderating role of cultural similarity. J Manage Stud 52(6):766–795
10. Barbopoulos L, Paudyal K, Pescetto G (2012) Legal systems and gains from cross-border acquisitions. J Bus Res 65(9):1301–1312
11. Berry H, Guillén MF, Zhou N (2010) An institutional approach to cross-national distance. J Int Bus Stud 41(9):1460–1480
12. Bertrand O, Betschinger M-A (2012) Performance of domestic and cross-border acquisitions: empirical evidence from Russian acquirers. J Comp Econ 40(3):413–437
13. Bhatt V, Sultan J (2012) Leverage risk, financial crisis, and stock returns: a comparison among Islamic, conventional, and socially responsible stocks. Islam Econ Stud 20:87–143

14. Battisti E, Bollani L, Miglietta N, Salvi A (2020) The impact of leverage on the cost of capital and market value: evidence from Sharīʿah-compliant firms. Manag Res Rev

15. Bunce J (2013) M&A next generation integrators: critical attributes for success. In 13 (ed). ModalMinds Inc., San Francisco

16. Cai J (2007) What's in the news? Information content of S&P 500 additions. Financ Manage 36(3):113–124

17. Chakrabarti R (2007) Do Indian acquisitions add value? Available at SSRN 1080285

18. Chakrabarti R, Gupta-Mukherjee S, Jayaraman N (2009) Mars–Venus marriages: culture and cross-border M&A. J Int Bus Stud 40(2):216–236

19. Cheong CW, Sinnakkannu J (2014) Ethnic diversity and firm financial performance: evidence from Malaysia. J Asia-Pac Bus 15(1):73–100

20. Christensen CM, Alton R, Rising C, Waldeck A (2011) The big idea: the new M&A playbook. Harv Bus Rev 89(3):48–57

21. Conn C, Michie J (2001) Long-run share performance of UK firms engaging in cross-border acquisitions. Citeseer

22. Cui H, Leung SCM (2020) The long-run performance of acquiring firms in mergers and acquisitions: does managerial ability matter? J Contemp Account Econ 16(1):100185

23. Dewandaru G, Masih R, Bacha OI, Masih AMM (2017) The role of Islamic asset classes in the diversified portfolios: mean variance spanning test. Emerg Mark Rev 30:66–95

24. Denis DK, McConnell JJ, Ovtchinnikov AV, Yu Y (2003) S&P 500 index additions and earnings expectations. J Financ 58(5):1821–1840

25. Du M, Boateng A (2015) State ownership, institutional effects and value creation in cross-border mergers & acquisitions by Chinese firms. Int Bus Rev 24(3):430–442

26. Elliott WB, Van Ness BF, Walker MD, Warr RS (2006) What drives the S&P 500 inclusion effect? An analytical survey. Financ Manage 35(4):31–48

27. Erhardt NL, Werbel JD, Shrader CB (2003) Board of director diversity and firm financial performance. Corp Gov: An Int Rev 11(2):102–111

28. Francis BB, Hasan I, Sun X, Waisman M (2014) Can firms learn by observing? Evidence from cross-border M&As. J Corp Finan 25:202–215

29. Francoeur C (2006) The long-run performance of cross-border mergers and acquisitions: evidence to support the internalization theory. Corp Ownersh Control 4(2):312–323

30. Gati V, Nasih M, Agustia D, Harymawan I (2020) Islamic index, independent commissioner and firm performance. Cogent Bus Manag 7(1):1824440

31. Hasan MM, Ibrahim Y, Olajide RJ, Minai MS, Uddin MM (2017) Malaysian acquiring firms' shareholders' wealth effect following cross-border acquisition. J Account Financ Emerg Econ 3(2):147–158

32. Hair JF, Anderson RE, Babin BJ, Black WC (2010) Multivariate data analysis: a global perspective 7

33. Hussein K, Omran M (2005) Ethical investment revisited: evidence from Dow Jones Islamic indexes. J Invest 14(3):105–126

34. Huse M, Solberg AG (2006) Gender-related boardroom dynamics: how Scandinavian women make and can make contributions on corporate boards. Women Manag Rev 21(2):113–130

35. Ibrahim Y, Minai MS, Hasan MM, Raji JO (2019) Agency costs and post cross border acquisition performance of Malaysian acquires. Int J Bus Soc 20(3):870–887

36. Ibrahim Y, Minai MS (2009) Islamic bonds and the wealth effects: evidence from Malaysia. Investment Manag Financ Innov 6(1):184–191

37. Ibrahim Y, Hwei KL (2010) Firm characteristics and the choice between straight debt and convertible debt among Malaysian listed companies. Int J Bus Manag 5(11):74

38. Jensen MC, Meckling WH (1976) Theory of the firm: managerial behavior, agency costs and ownership structure. J Financ Econ 3(4):305–360

39. Jouaber-Snoussi K, Ben Salah M, Rigobert MJ (2012) The performance of Islamic investment: evidence from the Dow Jones Islamic indexes. Markets, Bankers Investors 121:4–16

40. Khin EWS, Lee LY, Yee CM (2012) Cross-border mergers and acquisitions: Malaysian perspective. Actual Prob Econ 134(8)

41. Pontiff J, Woodgate A (2008) Share issuance and cross-sectional returns. J Financ 63(2):921–945
42. Rahim KF, Ahmad N, Ahmad I, Rahim FA (2013) Determinants of cross border merger and acquisition in advanced emerging market acquiring firms. Procedia Econ Finan 7:96–102
43. Rahim RA, Yong O (2010) Initial returns of Malaysian IPOs and Sharia-compliant status. J Islamic Account Bus Res 1(1):60–74
44. Rau PR, Vermaelen T (1998) Glamour, value and the post-acquisition performance of acquiring firms. J Financ Econ 49(2):223–253
45. Swaminathan V, Murshed F, Hulland J (2008) Value creation following merger and acquisition announcements: the role of strategic emphasis alignment. J Mark Res 45(1):33–47
46. Terjesen S, Couto EB, Francisco PM (2016) Does the presence of independent and female directors impact firm performance?" A multi-country study of board diversity. J Manag Gov 20(3):447–483
47. Van HJ, Gijsbrechts E, Pauwels K (2008) Winners and losers in a major price war. J Mark Res 45(5):499–518
48. Wang S-F, Shih Y-C, Lin P-L (2014) The long-run performance of Asian commercial bank mergers and acquisition. Mod Econ 5(4):341

# Challenges and Opportunities of Blockchain Integration in the Egyptian Banks: A Qualitative Analysis

**Shahinaz Gamal and Mayada M. Aref**

**Abstract** The purpose of the research is scanning the opportunities and threats of implementing blockchain technology in one of the major applications in the banking sector in Egypt: cross border payment and settlement. To reach this aim a qualitative analysis was conducted in which a semi-structured interviews approach was used. The main findings manifested the ability of this technology to prevent fraud, enhance transparency, and ensure financial stability and accountability. It enables a secure environment to store and transmit data between organizations and can reduce uncertainty. The process is faster, and this can promote customer experience. The results disclosed that integrating blockchain technology within the current systems in the Egyptian banks should undergo a transitional phase in which the technology can be outsourced to ensure security and efficiency. The originality of the research lies in the attempt to reconnoiter the readiness of the Egyptian banking sector to embrace innovative technologies.

**Keywords** Egyptian banking sector · Blockchain · Cross borders payments settlements · Digital IT adoption · Digital economy

## 1 Introduction

Throughout time, innovative technologies threaten the traditional ways of conducting business inside any organization that adopts those novel technologies. Zahra and Covin [1] defined innovation as *the lifeblood of corporate survival and growth*. After several years O'Sullivan and Dooley [2] gave a more detailed definition; in their words, innovation is *the process of making changes, large and small, radical*

S. Gamal (✉) · M. M. Aref
Sociocomputing Department, Faculty of Economics and Political Science, Cairo University, Giza, Egypt
e-mail: shahinazg@cu.edu.eg

M. M. Aref
e-mail: mayadaaref@feps.edu.eg

© The Author(s), under exclusive license to Springer Nature Switzerland AG 2022
S. G. Yaseen (ed.), *Digital Economy, Business Analytics, and Big Data Analytics Applications*, Studies in Computational Intelligence 1010,
https://doi.org/10.1007/978-3-031-05258-3_37

*and incremental, to products, processes, and services, that results in the intro-duction of something new for the organization that adds value to customers and contributes to the knowledge store of the organization.* Accordingly, the role of inno-vation is enduring a competitive advantage and generating values as well. Therefore, it is crucial to find adequate solutions to the innovator's dilemma [3]. The inno-vator's dilemma states that any organization possesses relatively limited capabilities compared to the ones needed by novel technologies. One way out of this dilemma is adopting a strategy that ensures a smooth transition from a traditional centralized organization to a more distributed and decentralized ecosystem.

When scanning the chronological order of the emergence of digital innovative technologies, from telebanking to fintech, it can be interpreted that the obvious advantages of those technologies were; convenience, availability, speed, efficiency, effectiveness and transparency of the banking processes as well as facilitating trans-actions across various sectors [4]. The recognition of the importance of innovations caused a swift and a cursory dependence from the banking sector all over the world on the rising information and communication technology (ICT). This dependence offered an alternate way to conduct some functions such as the banks business struc-ture and the customer relationship management [5]. One of the most important rising ICT is the Blockchain Technology (BCT). BCT can be comprehended as a radical digital paradigm that can cause disruption [6–8]. Blockchain, an emerging infras-tructural technology, can fundamentally transform the ways people transact, trust, collaborate, and identify themselves [9].

BCT has received exceptional attention in the business, regulators as well as academic circles. Since the inception of BCT, it was affined to the financial domain and can be a concrete approach to unify the process of collection and verification of data in a manner that decreases costs to the minimal. In addition, BCT ascertain that the information is neither altered nor forged. This fact can ameliorate the reliability of the collected data. It mutated the financial sector and took over the currently used traditional business models and technologies [10]. Another major advantage is the undoubtful effect of BCT on the shape, size, and the conduction of business in the banking and finance industry [11].

While the focus of several studies, covering BCT, was about implementation and the technical challenges [12, 13], those challenges should not be the only concern. Technological design and implementation might constitute a less threatening chal-lenge if compared to business processes related challenges [14, 15]. Although BCT can offer myriad benefits to the banking system, some constraints can slow down or even hinder the thorough adoption of this technology in both macro aspects (such as requiring large investments in infrastructure, a need of a solid legal foundation, etc.) and micro aspects (such as the acceptance, knowledge and ability to use new technology, etc.).

This unjustified bias towards technological adaptation when studying the BCT resulted into an empirical and theoretical gap. This research attempts to contribute to that field by addressing adoption challenges that organizations might face when they merge a novel technology with their information systems. A special focus is given to how the banking industry employees evaluate the challenges regarding the infusion of BCT in banks.

In this research, in an attempt to better understand the underlying potentials and challenges of the use of BCT in the banking sector; interviews with practitioners and experts were conducted. The interviews main aim was to analyze the perceived importance and potentials of BCT functions in the banking industry. It is widely believed that one of the strength points of qualitative research is supporting the researcher in understanding the nature and complexity of the studied phenomenon. Advocates of the usage of qualitative research believe that when the texts are quantified, the phenomenon under study can be better analyzed [16]. This research contributes to the emerging research on how BCT can enhance the existing banking process; and discuss the readiness of the Egyptian banking sector to embrace technology, especially in the process of Cross Border Payment Settlement (CBPS).

Accordingly, the research is organized as follows: after this introductory section, section two presents the conceptual background of BCT. Section three discusses the potential benefits and challenges of the applications of BCT in the banking sector with a special focus on CBPS. The methodology of data collection and interview analysis are presented in section four. Section five is a discussion of the research outcomes, while section six concluded the research.

## 2 Blockchain Technology Concepts

The inception of blockchain implementation began in 2008 when Nakamoto [17] proposed the eminent cryptocurrency "Bitcoin" and discussed that such technology can aid in solving the double-spending issue. A blockchain is given that name because of its features and structure. It is mainly made up of blocks of information that are chained together through a code, in each block this code relates to the previous one, similar to a timestamp, provides the link to the previous block making it impossible to modify or corrupt previously recorded transactions and thus leading to an unalterable ledger. Blockchains are constantly growing over time as more blocks are added, so they actually form a chain-like structure tracking back to the root block [18].

BCT came with a new scheme where data validation and responsibility of data insertion are distributed among all nodes of the network [6]. The validity of the history can only be achieved if, and only if, a consensus is reached all over the system that the transaction history is completely correct and comply with the rest of the data stored all over the nodes of the network. Therefore, BCT enhance transparency through ensuring the validity of an entire history of the transaction [19, 20]. BCT can be seen as an analogue to a distributed ledger that is maintained, updated, and verified in a harmonized matter by each of the nodes involved in the transactions within a given network. No single node can maintain the database; rather each one has a copy of it.

The concept of storing data in a centralized system cause the data to be prone to hacking and fraud. A problem that blockchain distributed decentralized mechanism can overcome since all transaction must be verified by different nodes making fraudulent transactions implausible [21]. In addition, each node of the network stores a replica of the data, so data is hardly lost [22]. The decentralization feature of BCT gives rise to the replication of data all over the nodes of the distributed network. Authorized users can retrieve data whenever needed and hence data availability is guaranteed even if some setbacks affect a number of nodes. Transactions that occur in the network are recorded in automatically and in real time, which make it hectic to any organization to forge date as this, will entail the simultaneous modification of the entire copies of the ledger [7].

If BCT is utterly adopted, it will undoubtedly maintain a reasonable level of security and transparency of data. Unauthorized or mischievous users are not able to modify or add blocks to the blockchain without being perceived because simply the rest of the nodes will promptly notice the deceitful behavior. Thus, it is impossible to threaten the integrity of the blocks of the ledger [23]. Even in the case of hacking the ledger, the actual verified network copies will overwrite the hacked versions returning the records to its original state [24]. It can be seen that the monitoring process is switched to networked computation rather than human entities; this feature adds trustworthiness to the process and is a genuine feature of BCT.

## 2.1  Classification of Blockchain Systems

Although all blockchains are analogous in structure, they differ in many other aspects [25, 26]. The differentiation between blockchain systems can depend on factors such as participants who can access data, immutability, control, and the degree of openness of the blockchain. Blockchains can be broadly divided into public, private or consortium networks. In a public network, access is granted to all nodes; any node can join, act, and leave the network without distressing the mechanism of the network or affecting the generation of new blocks. The distinguishing feature of decentralization of the public BCT causes any procedure to be completely uncontrollable by one single organization. Such type of networks is characterized by the presence of a considerable number of participants, so any attempt to alter or counterfeit data is almost impossible. Public blockchain guarantees transparency to all transactions taking place in the network, it might not seem applicable for all industries [27].

In private networks, access to the network is granted only to authorized nodes. The operator of the private blockchain holds the responsibility of the entire network and determines the privileged nodes who are permitted to join the network and specify their roles. Based on their assigned roles, entitled nodes will verify the transaction and accordingly the new transaction will be visible to the rest of the nodes in the network. Communication between nodes possessing a copy of the ledger is necessary for reviewing, writing and hence approving new transactions. The ledger is updated each time a newly submitted or newly verified transactions occurs. Moreover, the operator can abolish inactive nodes, so nodes are required to possess a pre-determined minimum number of connections to be considered active and to continue as a part of the network. Private blockchains are typically designed and configured to have better performance and scalability than the public ones [18].

In the consortium blockchain, a sub-category of private blockchains, the accessibility of data and some transaction processes are only assigned to a group of participants instead of only one that is why it is considered to be partially centralized. Similar to private ones, consortium blockchains are designed for higher performance, scalability, and confidentiality if compared to public ones [28].

# 3 Blockchain Technology Applications in the Banking Sector

BCT played a crucial role in the paradigm shift caused by the amalgamation between financial technology and the financial sector [29]. In financial services, data security is indispensable. Hence, the usage of a private or consortium blockchains is a congruent option [30] because it gives the moderators an adequate control over the nodes participating in submitting and verifying transactions. The applications of BCT in the banking industry can be categorized into Fraud Reduction [31], Trade Finance [32], Know your Customer [33] and Cross Border Payment and Settlement [34, 35].

Tapscott and Tapscott [8] believe that intrinsic properties of BCT can easily disrupt the core functions of any financial system through altering the way they work. Holotiuk et al. [36] added that the BCT would result into a more efficient payment system. It will not only ease the cross-border transactions, but it will minimize intermediary costs as well. Santander being the first UK bank to introduce BCT for international payments predicted that this technology could reduce banks' infrastructure costs related to cross-border payments, securities trading, and regulatory compliance [37]. The introduction of blockchain at Barclays has cut down the time necessary to execute a capital exchange from ten days to less than one day [38].

In the traditional centralized banking procedure, the transfer of money across borders is safe and secure but it is also slow and expensive. An important application of BCT in the banking sector is improving the clearing and settlement procedures, an inter-banks payment settlement for both domestic and cross borders transactions. This step allows banks to align the transaction participants without the need for an intermediary institution to control the process [36]. Applying this procedure converts the settlement process into an instant one. This research focuses on a prevalent application of BCT: Cross Border Payment and Settlement (CBPS).

## 3.1 Cross Border Payment and Settlement (CBPS)

The globalization process is deepening; international trade is increasing rapidly and the flow of factors of production between countries is increasing as well according to the OECD statistics, those facts necessitates a revision to the CBPS procedures. Interbank payments are needed to perform such procedures and they are conventionally conducted through intermediary clearing firms. The intermediary clearing firm undergo a series of steps starting from bookkeeping followed by the transaction and balance reconciliation, then payment initiation and the process goes on from a step to another. This process is money and time consuming. In the CBPS the clearing procedures for each country is different, a remittance for example requires several days to arrive. In order to overcome those drawbacks and inefficiencies BCT can be used in a point-to-point payment scheme. If such proposal can be implemented, transaction costs of banks will be reduced. Banks will be able to offer a swift and a satisfying payment clearing services for cross-border commercial activities instead of the tedious traditional risk management functions [28].

J. P. Morgan applied BCT in various applications in Asian countries; *Confirm* (that globally validate accounts information) and *PayDirect* (the global clearing solution) to transfer payments through the most efficient route. J. P. Morgan added a step to improve international funds transfers between banks through the BCT, and those transfers cover a myriad of payment transactions. The cost reduction accompanied by this procedure is inevitable since the number of rejected transactions caused by mismatched payment are almost negligible and the transactions are done near-real time [39]. Furthermore, those applications will boost both transparency and customers' payment experience.

The experimental usage of BCT platforms in CBPS is dispersing worldwide and an increasing number of financial institutions are joining the novel field. Al Rajhi Bank has also used the Ripple BCT for the first time in Saudi Arabia. The bank used the platform to undergo a secure, cross-border money transfer. Furthermore, The Saudi Arabian Monetary Authority is encouraging domestic banks to indulge in the RippleNet allied banking blockchain network, it expected to make monetary transfers with regional banks faster, cheaper, and more secure. It will also ensure for domestic banks' customers faster, cheaper, and more transparent cross border transactions [40].

Standard Chartered bank used the Ripple blockchain platform to manage its first cross-border transaction. The transaction completed in less than 10s including the foreign exchange process. This is likely unbeaten. The existing banking system and traditional network would have taken up to two days to perform the same operation. The National Australia Bank has also used Ripple's ledger technology to transfer funds between two banks [28]. Enterprise software provider R3 is teaming up with Mastercard, this partnership is expected to provide faster payments and to establish a blockchain solution for CBPS through augmenting worldwide connectivity. Finally, according to the results of Evan's econometric model, BCT adoption has positive and significant relationship with the financial market development [41].

In Egypt, cross borders payments are seen in remittances. Egypt is in the top five nations globally in terms of remittance flows from expat communities. Given the important role that remittances can play in the Egyptian economy, the National Bank of Egypt (NBE) is continuously aiming to develop and enhance the infrastructure that pertains to this line of business. Hesham Elsafty, group head for Financial Institutions and International Financial Services at NBE: confirmed that joining RippleNet will provide the bank with *cheaper, quicker and more reliable payments* [42].

NBE took major steps in adopting BCT. It was the first Egyptian bank to join the R3 blockchain Consortium. R3, the enterprise blockchain technology company, is leading an ecosystem of more than 300 firms working together to build distributed applications for usage across various industries such as financial services, insurance, healthcare, trade finance, and digital assets. NBE main target group is Egyptian expats who constantly send their remittances to Egypt. By improving its blockchain platform through this step, NBE can improve liquidity management and have access to real-time financial records. The higher productivity that will result from joining the R3 blockchain consortium will enable NBE to expand its network and activities not only in Egypt but also throughout North Africa, and the Middle East. The bank seeks to expand its client base by offering more efficient and faster CBPS. Hisham Okasha, NBE chairperson, clearly stated that *by joining this initiative together with world banks and companies, we will be able to closely monitor and engage directly in global blockchain developments* [43].

## 3.2 The Potential Benefits of BCT

### 3.2.1 Enhance Data Integrity and Availability

Data integrity refers to the accuracy and consistency of data, integrity is closely related to availability [22]. When a customer faces a service failure or denial, this may cause or be caused by an integrity violation. In the banking sector, in particular, the importance of data integrity is a paramount issue. Unauthorized alteration of data is unacceptable. Data must stay accurate and consistent all over the process through setting predefined rules that ensures the correctness and validity of the database. Data integrity will ensure that crucial financial data can be recovered if lost, searchable and traceable [25, 44]. To safeguard the immutability feature of BCT and data remains unaltered; an encryption process is undergone using both public and private key procedures. The decryption mechanism can be done by authorized entities. When BCT verifies a transaction, and this transaction is added in a chronological order to the database, a time stamp is appended to this transaction making the tracing of the transaction an easy task [45].

### 3.2.2 Enhance Data Security and System Resilience

The details of transactions lead to a tampering free data because the machine-based algorithm ensures the correctness of data checking and security and replaces the human-based checking. The less the human intervention the more secure the transactions. The BCT can trace any invalid transaction or any information that deviate from the network consensus and declare it as an invalid transaction [11]. BCT enhances information transparency and security: two themes that lowers the risks in the banking sector. Data security has always been a dilemma for the banking industry. Encrypting information using the BCT can aid in this context where sharing information between several nodes of the network is secured and protected. Once data is added to the block it becomes inherently unaltered. The BCT reduces the risk associated with centralized ledger approaches; there is no central point of failure. Banks using blockchain can eschew cyber-related crimes including hacking and fraud [46].

### 3.2.3  Shared Infrastructure and Cost Reduction

All the nodes of the network who represent the peers of the blockchain initiate a shared infrastructure. Using a shared infrastructure reduces cost and time and is an important advantage [47]. BCT systems can minimize the intermediaries in the payment processing system. The less the intermediaries the higher the security. Another benefit of reducing intermediaries in the payments process can be seen in the cost reduction as well [36].

### 3.2.4  Increase the Efficiency of the Processes

When the permitted nodes update the records, the entire blockchain is updated making the monitoring and analysis procedures more effective for the banking sector on the aggregate level. BCT is used in CBPS to overcome the disadvantages of traditional ways and reduce the transaction risks [28]. In addition, the automation of financial applications leads to a ledger that is less prone to errors.

### 3.2.5  Improved Customer Experience

BCT can also enable consolidated, accurate repositories of customer information that can be accessed by all parties in the network. Using BCT, banks may be able to serve customers far more quickly than with traditional systems. Smart contracts can be substantial nodes in BCT based networks in the banking sector. BCT enables deploying code with predefined rules that automatically execute when conditions are met through smart contracts mechanism. Smart contracts automatically execute across all ledgers. The enforcement of such contracts can restrain defaults [48].

However, while recognizing the potentials of BCT its crucial to address the potential risks and challenges associated with it. It is worth mentioning that minimizing risks and avoiding technical, social, and political aspects of failure can increase the pace of adopting the innovative technology [49].

## 3.3 The Challenges Facing BCT

To reap the weal of the BCT in the financial sector several technical as well as social aspects must be addressed. The mainstream of research focuses mainly on the technical challenges and strive for solutions. While in this research, the emphasis is given to the social side. The social domain includes concerns regarding data governance and how flexible is the regulatory framework.

### 3.3.1 Data Governance

Data governance is the process of managing the availability, usability, integrity and security of the data in an organization in order to support the decision making process. BCT is a new technology and is not integrated in many areas of governance yet. It still brings up a number of governance issues as it both removes the need for a centralized authority and second as it is a permanent data storage mechanism [28]. To reiterate, the problem lies in the very nature of BCT, it is a distributed and an immutable ledger. Those intrinsic features of BCT adds to the complexity of the governance process. Reaching a consensus about a certain aspect within a massive number of participants with different regulations and understanding makes the decision-making process a complex one.

### 3.3.2 Regulation

Governments seek to regulate any innovative procedure specially if it is widely spreading and touches upon financial sector [12]. The absence of regulations may hinder the diffusion of BCT within organizations [50]. Countries are getting ready with their regulatory settings to encounter the potential illegal activities that may accompany the usage of BCT [33]. Jurisdictional readiness and legal structure for BCT applications is not very mature yet. BCT needs to be thoroughly understood by the regulators in order to offer a feasible as well as an efficient regulation framework [29].

### 3.3.3 Behavioral Change

Change is constant, but it faces resistance. In the presence of trusted third parties (BCT), customers need to comprehend that their electronic transactions are safe and secured [51].

As an emerging technology, BCT is expected to validate itself and gradually take over traditional banking processes. Table 1 compiles the main points of comparison between the latter and the former.

**Table 1** Comparison between traditional and BCT-based banking processes

|  | Traditional banking processes | BCT- based banking processes |
|---|---|---|
| Mode of control | Centralized | Decentralized |
| Efficiency | Redundancy and duplication of tasks | Improve payments efficiency and flexibility of transactions |
| Speed | Slower | High speed |
| Cost | Less in energy consumption but higher costs in terms of bank fees, exchange rate, book-keeping and paper work | Lower operational and administrative costs |
| Legal framework | Well established framework that abides to national rules | Evolving framework that might conflict with domestic and international institutions |
| Governance | Central governance | Decentralized governance that leads to higher complexity in the governance process |
| Security | Vulnerable, data might be corrupted, forged. Malicious transactions can increase | BCT is immutable and irreversible ensuring higher degrees of security |

# 4  Data Collection and Analysis

## 4.1  Data Collection Methodology

BCT is another milestone in the Egyptian banking sector development, a new unprecedented business model that needs comprehensive investigation and analysis in order to assess its potential advantages and threats. In this research, data was collected using semi-structured interviews approach. This approach enables the researcher to grasp hidden and implicit information. The role of the researcher is to ensure that the questions and responses are both adequately understood [52]. The interviews were flexible and adaptable. Participants possess a high degree of freedom in expressing their views and sharing their experiences with the researchers. Four broad concepts were discussed with each participant. The first concept covers the organization strategy in adopting latest technologies and adaptability. The second concept addresses the challenges imposed when a new technology is adopted; especially the readiness of bank's employees, challenges regarding the existing regulatory framework, and data governance. The third concept addresses the benefit of adopting new information technology and the expected potentials of BCT. While in the fourth concept, interviewees were asked to mention three factors that they believe essential for the banks' success.

To analyze the outcome of the interviews the research used purposeful sampling, which is a qualitative research technique mainly used to identify and select information-rich cases. The steps of this technique are; first identify and select knowledgeable and/or experienced participants who understand the phenomenon under study. Second step involves the availability and willingness to participate. Finally, the participant should be able to deliver knowledge and experience, beside his own opinion, in a comprehensible and accurate method [53]. The sample included respondents from both public and private banks in Egypt. Solid information was obtained from experienced and knowledgeable participants. The participants were from several departments and varied from decision makers to IT specialists. Seven interviews were conducted with an average time of 40 min, three of which were conducted face to face while the rest were performed virtually due to the COVID-19 situation. Interviews were performed during May and June 2020. To foster the competency of qualitative research, more than one data collection method should be used in the study and [54]. The researchers added two focus group; each one consisted of nine bankers. The concepts were discussed with them for one hour.

A qualitative analysis was conducted with the interview synthesis methodology involving several steps. First, the interviews were broken down into quotes. Next, the quotes were coded; similar quotes were consolidated into a single code. Finally, a conclusion of the ideas that encompassed all the relevant points for a certain category was developed. MAXQDA analytics was used to code the script and to visualize the output.

## 4.2 Interview Analysis

### 4.2.1 Organization Strategy and Adaptability Towards Latest Technologies

Respondents affirmed that in order to be able to compete in the global market, latest technology should be adopted. They believe that technology is one of the main factors that enables a bank to remain viable and competitive in the market. Moreover, leading banks with highest market shares should be the pioneers in using the latest technology and popularize it in Egypt. The main obstacle in using the BCT is the expected resistance to change. The respondents stated that:

> The bank made a huge leap in terms of technology … introduced new services that are totally under the umbrella of Internet banking.
>
> Technology is a part of the competitive edge of international banks.
>
> The banking sector is heading to technological based initiatives; digital signature is one example so no foreseen barriers can stop adoption and implementation of blockchain.

### 4.2.2 Organization Strategy Towards Training and Development of Employees

Respondents acknowledged that their banks offer the training needed to increase the efficiency of their staff. The training is provided all around the year and it is a priority within the banking sector to provide efficient training to employees in different topics. Training topics include banking related topics such as finance or monetary policy or general topics that are offered to improve the client/employee relationship. Respondents made it clear that:

> The policy of banks is to train the employees in order to get the best of them through properly designed training workshops.
>
> It is a global strategy to invest in employees.

### 4.2.3    Organization Strategy Towards Collaboration

Tapscott and Tapscott [55] emphasized the importance of collaboration even with competitors. This paradoxical, co-opetition is a necessity in the age of globalization. To cope with higher uncertainties, many companies have turned to co-opetition where both co-operation and competition can take place. The concept discussed in this part is how respondents perceive the challenges of collaborating with other competing banks. Respondent clarified that the idea of collaboration is not new, and that it is already implemented. Respondents were aware of the benefits of collaboration such as reducing cost, enabling better services to the customers, spillovers and sharing experience. Two main concerns were mentioned; the first issue was data privacy and governance and the second issue is the uneven distribution of collaboration rewards.

> No challenges are involved cooperation is always needed especially if this will entail efficiency and lower costs.

### 4.2.4    Blockchain Technology Awareness and Banks' Success Factors

Figure 1 displays graphically the codes to BCT related from all the Interviews. The frequency of the code is proportional to its font size. According to the respondents,

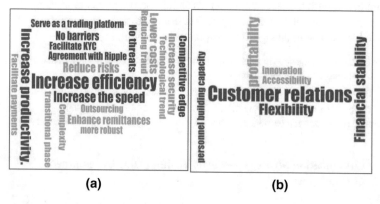

**(a)**                                    **(b)**

**Fig. 1** Textual Analysis of the interviews' answers concerning: **a** blockchain technology, **b** banks success factors

in pane A it is stated that the most crucial benefits were efficiency and productivity followed by speeding the processes and cost reduction. Other important factors included risk and fraud. The fact that BCT can enhance the ability to compete with rival banks is among the benefits that were stated. Moreover, the application of BCT might act as a trading platform and facilitate remittances and cross borders payments. Integrating BCT within the prevailing systems should undergo a transitional phase in which the technology will be outsourced to ensure security. According to the analysis of the interviews, the regulatory framework is well prepared and ready to BCT usage and the banks are not expected to face hindering barriers in that context. While in Pane B the major three influential factors that supplements to the success of the bank institute according to the analysis of textual responses of the participants were customer relationship management, financial stability, and profitability.

## 5   Discussion

BCT is rather a new phenomenon and its implementation in the Egyptian banking sector is in its outset. The analysis of the interviews concluded that banks recognize the potential benefits of the usage of BCT, but it is obvious that while some banks are already on track, others are still in the planning phase. A sound implementation of BCT necessitates not only the engagement of employees; but also, the understanding of the main sources of threats. Lack of knowledge, lack of trust, and complexity of the system are all threatening factors that entail a negative effect on adoption [6]. It is essential to increase the society's knowledge and understanding of new technologies and reassure users about security threats in order to succeed in the new system.

One of the dimensions that must be addressed in that context is the appropriate training strategies that can foster employees' abilities. The importance of training was clear in the respondents' views: *If employees are properly trained, it is like any other technology; it will be easy.* The research analysis also emphasized that the main three factors that can contribute to the bank success and sustainability can be enhanced through the adoption of BCT. The use of BCT will eliminate the redundancy of data collection and hence will improve customer convenience, which, is the most important factor impacting customers' experience in the banking sector. Maklan et al. [56] added that the speed of the process is another factor that adds to the customers' experience. From another angle, improving data reliability will enhance customers' experience too. This improvement can be achieved because of the decentralization manner of BCT and the encrypted data, which permits only, authorized entities to access stored data.

An increasing strand of empirical studies demonstrates that adoption and diffusion of ICT may improve profitability and banking performance could be enhanced through technological progress [5]. Several areas of expected cost reduction and increased profitability from using BCT can be excepted and CBPS is not an exception. This cost saving will result from reducing the cost of identification and onboarding processes [57]. Likewise deploying BCT in the banking sector in particular can offer a spectrum of opportunities from the ability to compete in the market to the efficient performance of the bank. Those factors are expected to consolidate financial stability. On the other hand, the respondents added that the main threats were the inadequate infra-structure in Egypt and the governance process of the blockchain. Finally, theoretically speaking a concrete legal framework is essential in the implementation phase of a newly introduced technology, the respondents affirmed that Egypt possess an appropriate framework that complies with the needs of BCT but neoteric regulations might be needed in the future. As a step towards improving the legal framework and in order to acclimate with the BCT requirements and other rapid developments in financial technology, The Central Bank of Egypt (CBE) launched its Fintech regulatory "Sandbox" to ensure faster and easier access to new financial solutions without threatening financial stability and consumer protection [58].

The research findings are harmonious with several studies. Chang et al. [59] concluded that adaptive financial institutions that are able to embrace new technologies, such as BCT, are more likely to withstand competitive environment. The study added that continuous employees training is a corner stone in building a successful blockchain based system. Similarly, Almahirah [60], who examined the effect of blockchain smart contracts on the financial and banking services in Jordan, confirmed the benefits of BCT in terms of enhanced efficiency, boosting customer experience and reducing bank's operational costs. The research findings bolstered the importance of raising the awareness of both employees and clients in order to succeed in accelerating the technology adoption rate. On the other hand, a line of research [28] emphasized the role of top managers in hasting the acceptance and adoption of BCT, Almahirah [60] added that managers' role is undeniable in encouraging employees to comprehend BCT potentials.

## 6   Conclusion and Future Work

One of the keys to success in a competitive and a volatile market is innovation. A viable organization can comply with abrupt opportunities and banks are not an exception. For a bank to compete and succeed it is obliged to adopt new technologies. BCT would ensure competitive edge in a rapidly progressing technological environment and is already gaining momentum and causing paradigm-disruption effects in the banking sector around the world. The impact of BCT on the traditional business model of banks is clear. BCT allows banks to transform their business model significantly. It is able to eliminate redundant procedures, allowing for swift transactions with lower costs. It also improves security levels. However, to apply this technology in the economy, banks need to explore and probe its potentials in both economic and social applications. The benefits and potentials of blockchain are undeniable; the research suggests that BCT is one of the latest approaches that have the ability to enhance decentralization, transparency, and accountability. It enables a secure and transparent environment to store and transmit data between organizations without a central point of control with the potency to reduce uncertainty and insecurity in transactions.

The research examined the opportunities and challenges facing BCT adoption in the Egyptian banks. Future research can expand the methodology and use empirical method to analyze the influencing factors. The research can act as an onset for building a theoretical framework for evaluating and disbanding the discussed challenges.

# References

1. Zahra S, Covin J (1994) The financial implications of fit between competitive strategy and innovation types and sources. J High Technol Manag Res 5:183–211
2. O'Sullivan D, Dooley L (2009) Applying innovation. SAGE Publication Inc., Thousand Oaks
3. De Rossi L, Abbatemarco N, Salviotti G, Gaur A (2020) Beyond a blockchain paradox: how intermediaries can leverage a disintermediation technology. In Proceedings of the 53rd Hawaii international conference on system sciences
4. Peters G, Panayi E (2016) Understanding modern banking ledgers through blockchain technologies: future of transaction processing and smart contracts on the internet of money. In: Banking beyond banks and money. Switzerland. Springer International Publishing, Cham, pp 239–278
5. Campanella F, Peruta M, Giudice M (2017) The effects of technological innovation on the banking sector. J Knowl Econ 8:356–368
6. Attaran M, Gunasekaran A (2019) Blockchain-enabled technology: the emerging technology set to reshape and decentralize many industries. Int J Appl Decis Sci 12:424–444
7. Beck R, Müller-Bloch C (2017) Blockchain as radical innovation: A framework for engaging with distributed ledgers. In: HICSS 2017 proceedings, pp 5390–5399
8. Tapscott D, Tapscott A (2017) How blockchain will change organizations. MITSloan Manag Rev 58
9. Angelis J, Ribeiro da Silva E (2019) Blockchain adoption: a value driver perspective. Bus Horiz 62:307–314
10. Glaser F (2017) Pervasive decentralization of digital infrastructures: a framework for blockchain enabled system and use case analysis. Waikoloa, Hawaii, U. S.A. Paper presented at the 50th Hawaii international conference on system sciences
11. Osmani M, El-Haddadeh R, Hindi N et al (2021) Blockchain for next generation services in banking and finance: cost, benefit, risk and opportunity analysis. J Enterp Inf Manag 34:884–899
12. O'Leary DE (2017) Configuring blockchain architectures for transaction information in blockchain consortiums: the case of accounting and supply chain systems. Intell Sys Acc Fin Mgmt 24:138–147
13. Mendling J, Weber I, Aalst W et al (2018) Blockchains for business process management: challenges and opportunities. ACM TMIS, 9
14. Chong A, Lim E, Hua X et al (2019) Business on chain: a comparative case study of five blockchain-inspired business models. JAIS 20:1310–1339
15. Zhang L, Xie Y, Zheng Y et al (2020) The challenges and counter measures of blockchain in finance and economics. Syst Res Behav Sci 37:691–698
16. Miles M, Huberman A, Saldana J (2014) Qualitative Data analysis: a methods sourcebook. SAGE Publications Inc.
17. Nakamoto S (2008) Bitcoin: a peer-to-peer electronic cash system
18. Werner F, Basalla M, Schneider J et al (2020) Blockchain adoption from an inter-organizational systems perspective: a mixed-methods approach. Inf Syst Manag 1–16

19. Lin I, Liao T (2017) A survey of blockchain security issues and challenges. Int J Netw Secur 19:653–659
20. Tasca P, Tessone CJ (2019) A taxonomy of blockchain technologies: principles of identification and classification. Ledger 4:1–39
21. Puthal D, Malik N, Mohanty S et al (2018) The Blockchain as a decentralized security framework. IEEE Consum Electron Magasine 7:18–21
22. De Rossi LM, Abbatemarco N, Salviotti G (2019) Towards a comprehensive blockchain architecture continuum. In Proceedings of the 52nd Hawaii international conference on system sciences
23. Aste T, Tasca P, Di Matteo T (2017) Blockchain technologies: the foreseeable impact on society and industry. Computer 50:18–28
24. Xu J (2016) Are blockchains immune to all malicious attacks? Finan Innov 2:25
25. Zheng Z, Xie S, Dai H et al (2018) Blockchain challenges and opportunities: a survey. Int J Web Grid Serv 14:352–375
26. Casino F, Dasaklis TK, Patsakis C (2019) A systematic literature review of blockchain-based applications: current status, classification and open issues. Telemat Informa 36:55–81
27. Beck R, Müller-Bloch C, King (2018) Governance in the blockchain economy: A framework and research agenda, J Assoc Inf Syst 19:1020–1034
28. Guo Y, Liang C (2016) Blockchain application and outlook in the banking industry. Finan Innov 2:1–12
29. Ali O, Ally M, Clutterbuck et al (2020) The state of play of blockchain technology in the financial services sector: a systematic literature review. Int J Inf Manag 54:102199
30. Yermack D (2017) Corporate governance and blockchains. Rev Finan 21:7–31
31. Phan L, Li S, Mentzer K (2019) Blockchain technology and the current discussion on fraud. Comput Inf Syst J Articles. Paper 28
32. Chang S, Luo H, Chen Y (2020) Blockchain-enabled trade finance innovation: a potential paradigm shift on using letter of credit. Sustainability 12:188
33. Kumar M, Nikhil P (2020) A blockchain based approach for an efficient secure KYC process with data sovereignty. Int J Sci Technol Res 9
34. Chen Y, Bellavitis C (2020) Blockchain disruption and decentralized finance: the rise of decentralized business models. J Bus Ventur Insights 13
35. Yao X, Zhu T (2017) Blockchain is to create a new ecology of cross-border payment. Finan Expo (Wealth) 5:46–48
36. Holotiuk F, Pisani F, Moormann J (2017) The impact of blockchain technology on business models in the payments industry. In: Wirtschaftsinformatik 2017 proceedings
37. Belinky M, Rennick E, Veitch A (2015) The Fintech 2.0 Paper: rebooting financial services: Santander InnoVentures, Oliver Wyman and Anthemis Grou
38. Hooper A, Holtbrügge D (2020) Blockchain technology in international business: changing the agenda for global governance. Rev Int Bus Strategy 30:183–200
39. Morgan J (2021) J. P. Morgan uses blockchain technology to help improve money transfers https://www.jpmorgan.com/news/jpmorgan-uses-blockchain-technology-to-help-improve-money-transfers. Accessed 10 July 2020
40. Ripple (2018) Ripple and Saudi Arabian Monetary Authority (SAMA) Offer Pilot Program for Saudi Banks. https://ripple.com/insights/ripple-and-saudi-arabian-monetary-authority-offer-pilot-program-for-saudi-banks/ Accessed 1 May 2021
41. Evans O (2019) Blockchain technology and the financial market: an empirical analysis. Actual Prob Econ 211:82–101
42. Palmer D (2021) Egypt's Largest Bank Joins Ripple network for cross-border payments: https://finance.yahoo.com/news/egypt-largest-bank-joins-ripple-095855366.html. Accessed 8 July 2020
43. DailyNewsEgypt (2018) NBE joins one of largest global blockchain initiatives https://dailynewsegypt.com/2018/04/22/nbe-joins-one-largest-global-blockchain-initiatives/ Accessed 12 Aug 2020
44. Iansiti M, Lakhani K (2017) The truth about blockchain. Harv Bus Rev 95:118–127

45. Ismail LM (2019) A review of blockchain architecture and consensus protocols: use cases, challenges, and solutions. Symmetry 11
46. Carminati B, Ferrari E, Rondanini C (2018) Blockchain as a platform for secure inter- organizational business processes. In: IEEE 4th international conference on collaboration and internet computing
47. Cocco L, Pinna A, Marchesi M (2017) Banking on blockchain: costs savings thanks to the blockchain technology. Future Internet 9:25
48. Osba A, Miller A, Shi E, et al (2016) The blockchain model of cryptography and privacy-preserving smart contracts. In: IEEE Symposium on security and privacy, pp 839–858
49. Yeoh P (2017) Regulatory issues in blockchain technology. J. Finan Regul Compliance 25(2):196–208
50. Gupta A, Gupta S (2018) blockchain technology application in Indian banking sector. Delhi Bus Rev 19(2):75–84
51. Crosby M, Pattanayak P, Verma S et al (2016) Blockchain technology: beyond bitcoin. Appl Innov 2:71
52. Kulic K (2015) The crisis intervention semi-structured interview. Brief Treat Crisis Interv 5:143–157
53. Cresswell V, Plano Clark L (2011) Designing and conducting mixed method research, 2nd edn. Sage, Thousand Oaks, CA
54. Kruger R (2015) Casey M (2015) Focus groups: a practical guide for applied research. Sage, Thousand Oaks (CA)
55. Tapscott D, Tapscott A (2016) Blockchain revolution: how the technology behind Bitcoin and other cryptocurrencies is changing the world. Penguin Book, London
56. Maklan S, Antonetti P, Whitty S (2017) A better way to manage customer experience: lessons from the royal bank of Scotland. Calif Manage Rev 59:92–115
57. Marito P (2021) Blockchain and banking how technological innovations are shaping the banking industry. Palgrave, Macmillan, Pisa, Italy
58. The Central Bank of Egypt (CBE). https://fintech.cbe.org.eg/home/sandbox?en (2020) Accessed 20 Aug 2021
59. Chang V, Baudier P, Zhang H et al (2020) How Blockchain can impact financial services—the overview, challenges and recommendations from expert interviewees. Technol Forecast Soc Change 158
60. Almahirah M (2021) The effect of smart blockchain contracts on the financial services industry in the banking sector in Jordan Ilkogretim 20:1845–1853

# Impacts of Financial Technology on Profitability: Empirical Evidence from Jordanian Commercial Banks

**Abdul Razzak Alshehadeh, Ghaleb Awad Elrefae, Mohammed Khudari, and Ehab Injadat**

**Abstract** This study aimed to illustrate the effects of the financial technology tools as a base for reinforcing financial inclusion indicators on profitability in the Jordanian commercial bank listed in the Amman stock exchange. A panel of data from 16 banks listed in the Amman stock exchange was used between 2010 and 2019. The financial technology changed the structure of the overall financial services, therefore, increase the availability for a wider social group. Further, the financial technology tools significantly reinforce the financial inclusion indicators over the studied profitability indicators, including return on assets and equity. It is recommended to adopt effective and modern financial, technological strategies that provide the marginalized social groups access to the financial services and products that meet their needs. Thus, getting the added value, which would increase the financial inclusion indicators.

**Keywords** Financial technology · Profitability · Financial inclusion · Commercial banks

## 1 Introduction

Financial technology (FT) refers to any business that uses technology to enhance or automate financial services and processes. The term is a broad and rapidly

A. R. Alshehadeh
Faculty of Business, Al-Zaytoonah University of Jordan, Amman 11733, Jordan
e-mail: abdulrazzaqsh@zuj.edu.jo

G. A. Elrefae
College of Business, Al Ain University, 64141 Al Ain, UAE
e-mail: president@aau.ac.ae

M. Khudari (✉)
College of Graduate Studies, Universiti Tenaga Nasional, 43000 Kajang, Selangor, Malaysia
e-mail: khudari@uniten.edu.my

E. Injadat
Business Administration, Sur University College, 411 Sur Sultanate, Sur, Oman
e-mail: dr.ehab@suc.edu.om

© The Author(s), under exclusive license to Springer Nature Switzerland AG 2022
S. G. Yaseen (ed.), *Digital Economy, Business Analytics, and Big Data Analytics Applications*, Studies in Computational Intelligence 1010,
https://doi.org/10.1007/978-3-031-05258-3_38

growing industry serving both consumers and businesses. FT represents digitizing and modelling the data to become the main income source and an economic value [1].

FT includes all the technological innovations in the financial sector. As the one that updates the financial services of clients and their investments. In this light and based on the well-developed financial industry over the last few decades, thus it forms an excellent place for innovations in the FT field [2]. For example, the successful adaption of the "e-Fawateercom" platform by The Central Bank of Jordan indicates the importance of effectively using innovative and technological tools in the financial sector [3].

Using FT tools was raised after the world financial crisis in 2008 to upgrade the traditional financial sector [4]. Many studies used different variables to investigate the dynamic relationships among banks' profitability and their determinants using Pooled mean group (PMG) and dynamic fixed effect (DFE) models [5]. This study aimed to explore the effect of financial technology in enforcing the financial inclusion indicators. Further, analyze the impact of the digitalized financial inclusion on the profitability indicators among Jordanian commercial banks listed in the Amman Stock Exchange (ASE).

## 2   Study Problem

The Arab countries still recording low levels of financial inclusion over the wild world. It was reported that only about a third of the adults (37%, n = 160) in the Arab countries have a bank account, which means about two-thirds (63%) have been excluded from the funding and governmental financial services. On the other hand, in 2017, the percentage of having a bank account was relatively high in Arab united Emarat, Bahrain, and Kuwait (82%, 83%, and 80% respectively), this percentage was limited to 25% among Yemen, Sudan, Mauritania, Iraq, and Syria [6]. Financial inclusion means that individuals and businesses can access valuable and affordable financial services delivered responsibly and sustainably. And adapting it has become a vital program locally and internationally. Previous studies that focused on financial inclusion showed that more than half of the adults (59%) in the wild world could not have tools to control credit, insurance, and savings. In Jordan, only ten per cent of the Jordanians, who has a bank account, have access to the tools to control the lending sources [6].

Further, recent studies considered FT one of the essential tools to merge individuals and corporations under the financial inclusion umbrella [7]. Thus, offering the official financial services for all the social classes meets their needs [2]. Offering accessible and affordable FT tools to all community individuals will achieve social welfare and financial inclusion [8]. From the above, the study tries to answer the following question: Is there an impact of improving financial inclusion indicators by using FT tools on profitability indicators in commercial banks listed on the ASE?

# 3   Financial Technology and Financial Inclusion

## 3.1   The Importance of Financial Technology and Financial Inclusion

Financial inclusion is an effective, high-priority tool that aids countries' social, economic, and strategic development. The importance of financial inclusion is manifested in three main aspects:

(a)   Social aspect: Reinforcing financial inclusion helps reduce poverty and financial exclusion and raise awareness among people about effectively using their savings [9].

(b)   Economic aspect: Several studies established a relationship between financial inclusion, financial stability, and economic growth. Increase financial services, and financing of small and medium-sized enterprises contribute to increasing bank deposits, which supports economic growth and helps achieve financial stability [10, 11].

Several studies have supported the importance of FT, especially in the banking sector, through the following [12]:

(a)   FT covers a wide range of financial services, which are unable to provide by traditional banking.

(b)   FT changes the overall financial services structure and provides banking services to the clients, making them faster, cheaper, safer, and more accessible.

(c)   FT plays a critical role in increasing the banking operations area and expanding access to financial inclusion services for a wide range of marginalized groups that are traditionally inaccessible.

## 3.2   The Relationship Between FT and Financial Inclusion

Financial inclusion is defined as the availability and equality of opportunities to access financial services. It is increasing interest in contributing to economic and financial development and promoting growth and income. The World Bank considers financial inclusion a key enabler to reduce extreme poverty and boost shared prosperity [7]. Also, financial inclusion is regarded as one of the most important lessons learned after the global financial crisis due to its essential role in maintaining financial stability [13], which allowed it to prioritize international policy strategies [14]. Many studies expected an increased adaption of financial inclusion by many countries [15].

Nowadays, FT is considered one of the essential tools for community transformation to integrate individuals and institutions within the umbrella of financial inclusion [16]. FT offers financial services to all social individuals and institutions through official channels, including bank accounts, online payment and transfer services, insurance, finance and credit, and other innovative financial services at competitive prices. Further, FT allows avoiding using informal channels that are not subject to supervision [8]. Recently, the importance of FT awareness has been increased, and it was found that 94% of consumers used at least one FT tool, which was online transfer money, and 75% of them used FT services for online payment. However, the global average percentage adoption of FT services is 46%. Online money transfers and payments have the most common used FT services, as the percentage of consumers (18%) in 2015 increased to (50%) in 2017 and to (75%) in 2019 [15].

A previous study that included 12 countries in the Middle East, North Africa, Afghanistan, and Pakistan region found that the number of startups in FT has increased sevenfold since 2009, with investments focused on Egypt, Jordan, Lebanon, and the United Arab Emirates. These startups have emerged side by side and compete with banks, which also use digital technology to move to more customer-focused business models [17].

Also, FT plays a vital role in promoting and facing the critical challenges of financial inclusion, growth, and diversification of economic activity by providing innovational financial services to many clients, especially those who do not deal with the banking system. Further, facilitating the availability of alternative sources of financing to meet the needs of small and medium-sized enterprises [16].

Emphasizing the importance of using FT by the financial firms in general and banks in specific Abbasi & Weigand studied the impact of FT on the diversity of financial services, operational efficiency, financial stability of the bank, the client's protection ways, the results of which can be summarized by the following points [18]:

1. FT contributed to the diversity of financial services provided to customers and institutions by supporting online and third-party payments, accepting electronic deposits, providing digital credit services, developing online lending platforms, and supporting investment processes through e-commerce operations.
2. FT is enhancing the efficiency of operational processes, mainly by reducing transaction costs, cash management and credit operations.
3. FT is enhancing financial stability by reducing the negative impacts of many risks, including credit, liquidity, and operating risks, as well as lowering operational challenges associated with the banking system's infrastructure, in addition to achieving more excellent profitability rates.
4. FT affects client's protection by strengthening security systems, fraud risks, confidentiality protection and clints data privacy, and avoiding the risk of unfair exclusion or distinction between clients.

### 3.3 Availability and Use of Financial Services

Table 1 shows banking services access methods, including ATMs or branches of Jordanian banks listed in the ASE (2010–2019). It was found that there is an increase in the ATMs and branches numbers, which indicates an increase in the volume of operations and clients, besides the financial stability and operational efficiency of

**Table 1** Total access points for banking services

| # | Year | ATMs number | Branches number | Yearly ATMs increasing number (%) | Yearly branches increasing number (%) |
|---|------|-------------|-----------------|-----------------------------------|---------------------------------------|
| 1 | 2010 | 420 | 630 | 4.21 | 3.65 |
| 2 | 2011 | 466 | 706 | 4.65 | 4.66 |
| 3 | 2012 | 542 | 785 | 8.18 | 5.85 |
| 4 | 2013 | 612 | 876 | 7.54 | 5.58 |
| 5 | 2014 | 647 | 958 | 3.76 | 5.03 |
| 6 | 2015 | 705 | 1132 | 6.25 | 10.68 |
| 7 | 2016 | 765 | 1241 | 6.45 | 6.68 |
| 8 | 2017 | 810 | 1389 | 4.85 | 8.99 |
| 9 | 2018 | 846 | 1487 | 3.92 | 6.10 |
| 10 | 2019 | 929 | 1630 | 9.11 | 8.78 |

these banks. Further, these data demonstrate the commercial banks' willingness to expand their business to the regions and places of marginalized groups.

Further, the number of bank service access points is constantly increasing, providing services to the largest segment of clients. There was an increase of 10 branches per year during the period from 2010 to 2019, raising the number of the branches and ATMs, indicating an increase in the interest of clients in the use of electronic banking services, especially online banking, as besides showing the necessity for developing of the financial inclusion indicators and increasing the number of digital access points to meet the client's needs.

Table 2 shows that the loans-to-deposits ratios were close to each other during the studied period. The highest loan-to-deposit ratio (75.82%) was in 2017, while the lowest (64.58%) was in 2014. Furthermore, there is a downward deviation in the deposits-to-assets ratios, the highest ratio (68.72%) was in 2015, while the lowest (63.70%) was in 2010. Besides, banks' loan-to-asset ratios indicate efficient operating in credit facilities, reaching the highest ratios (51.21%) in 2017, while the lowest ratio (43.10%) was in 2013. Overall, there is a stable proportion of the volume of loans granted by banks to the volume of deposits. Also, there is an acceptable size for banks' deposits to the total assets available in banks.

Table 3 demonstrate that the profitability indicators of Jordanian commercial banks were close during the studied years. The return on assets ratio in 2013 was the highest (1.41%), and the lowest rate (0.97%) was in 2015 and 2016. Regarding the Return on Equity ratio, the highest rate (10.19%) was in 2018, whereas the lowest ratio (6.63) was in 2010, indicating financial and economic stability in the banks' operations and acceptable return on equity ratio for shareholders and investors.

**Table 2** Indicators of the use of financial services for banks listed on the ASE

| # | Year | Deposits | Credit loans | Total assets | credits-to-deposit ratio | credits-to-asset ratio | Deposit-to-asset ratio |
|---|------|----------|--------------|--------------|--------------------------|------------------------|------------------------|
| 1 | 2010 | 30,882,139,663 | 21,408,367,027 | 48,477,966,019 | 69.32 | 44.16 | 63.70 |
| 2 | 2011 | 32,564,590,886 | 22,020,352,390 | 50,516,950,642 | 67.62 | 43.59 | 64.46 |
| 3 | 2012 | 32,985,132,373 | 22,617,592,951 | 50,850,261,215 | 68.57 | 44.48 | 64.87 |
| 4 | 2013 | 32,094,064,857 | 21,444,088,508 | 49,754,654,527 | 66.82 | 43.10 | 64.50 |
| 5 | 2014 | 35,505,052,281 | 22,929,321,927 | 52,938,354,453 | 64.58 | 43.31 | 67.07 |
| 6 | 2015 | 37,519,378,093 | 24,592,364,772 | 54,600,231,585 | 65.55 | 45.04 | 68.72 |
| 7 | 2016 | 36,199,739,061 | 26,106,818,552 | 53,468,221,280 | 72.19 | 48.83 | 67.70 |
| 8 | 2017 | 37,672,169,227 | 28,562,616,466 | 55,773,699,747 | 75.82 | 51.21 | 67.54 |
| 9 | 2018 | 39,072,490,429 | 29,365,501,431 | 58,125,593,366 | 75.16 | 50.52 | 67.22 |
| 10 | 2019 | 40,529,736,041 | 29,285,041,375 | 60,446,481,868 | 72.26 | 48.45 | 67.05 |

**Table 3** Profitability indicators for Jordanian commercial banks

| # | Year | Profitability indicators | |
|---|------|------------------|------------------|
|   |      | Return on assets | Return on equity |
| 1 | 2010 | 0.98 | 6.63 |
| 2 | 2011 | 1.17 | 8.00 |
| 3 | 2012 | 1.24 | 8.27 |
| 4 | 2013 | 1.41 | 9.49 |
| 5 | 2014 | 1.22 | 8.89 |
| 6 | 2015 | 0.97 | 7.28 |
| 7 | 2016 | 0.97 | 7.28 |
| 8 | 2017 | 1.07 | 7.59 |
| 9 | 2018 | 1.36 | 10.19 |
| 10 | 2019 | 1.23 | 9.26 |

# 4 Method

The study is based on the descriptive analytical approach using panel data of Jordanian commercial banks from 2010 until 2019, using the statistical program (SPSS) as described previously [19, 20]. The data were extracted from the annual report of the ASE (https://www.exchange.jo/en). This study included all the Banks listed in ASE (n = 16). The current study explored the impact of financial technology indicators on enhancing financial inclusion in ASE-listed commercial banks by testing two independent variables.

1. **Availability of financial services:** This variable was assessed by the number of branches and ATMs of Jordanian commercial banks.
2. **Use of financial services:** This variable was assessed by the credits-to-deposits ratio, the deposits-to-assets ratio, and the credits-to-assets ratio of Jordanian commercial banks.

The dependent variable was assessed as follows:

1. **Return on assets:** This variable was measured by equating net profit after tax to the average total assets of Jordanian commercial banks.
2. **Return on Equity:** This variable was measured by equating net profit after tax to the average total Equity of Jordanian commercial banks.

The impact of FT tools as a strengthener of financial inclusion indicators to improve the return on assets and return on equity ratios were measured using the following Eqs. (1 and 2, respectively), which consisted of the five independent variables for each Bank (i) and during a defined period (t) in addition to the margin of error (E):

$$\ln ROA_{it} = ROA_{it-1} + No.ATMs_{it} + NO.Branches_{it}$$
$$+ Deposits\,to\,assets_{it} + Credits\,to\,deposit_{it}$$

$$+ \, Credits \, to \, Assets_{it} + E_{it} \tag{1}$$

$$\ln ROE_{it} = ROE_{it-1} + No.ATMs_{it} + NO.\,Branches_{it}$$
$$+ \, Deposits \, to \, assets_{it} + Credits \, to \, deposit_{it}$$
$$+ \, Credits \, to \, Assets_{it} + E_{it} \tag{2}$$

## Study Hypotheses

The study's hypotheses are as follows:

$HA_1$: Financial technology tools have a significant impact as a strengthener of financial inclusion indicators to improve the return on assets ratio in ASE-listed commercial banks.

$HA_2$: Financial technology tools have a significant impact as a strengthener of financial inclusion indicators to improve the return on equity ratio in ASE-listed commercial banks.

## Testing Hypotheses

The multiple regression analysis (Table 4) showed a significant relationship ($P$-value $\leq 0.05$) between the FT variable and the financial inclusion indicators to improve the return on assets ratio as one of the profitability indicators in the commercial banks. Further, the $R^2$ indicating that the independent variables (financial inclusion indicators) accounted for 53.7% of the subordinated variable, i.e., return on assets ratio, in ASE-listed commercial banks. Thus, the first hypothesis can be accepted where there is a significant impact of financial technology tools as a strengthener of financial inclusion indicators to improve the return on assets ratio.

**Table 4** Multi-regression analysis of the impact of FT tools as a strengthener of the financial inclusion indicators to improve the return on assets ratio

| Independent variable | Return on assets | | | |
|---|---|---|---|---|
| | B | Beta | T | Sing |
| Number of ATMs | −1.041 | −2.005 | −2.474 | 0.001** |
| Number of branches | −0.004 | −0.348 | −0.621 | 0.000** |
| Deposits to assets | 1.035 | 32.745 | 0.284 | 0.009** |
| Credits to deposits | −0.037 | −6.246 | −3.251 | 0.012* |
| Credits to assets | 0.171 | 14.251 | 4.145 | 0.003** |
| Multiple correlation coefficient | 0.561 | | | |
| $R^2$ | 0.537 | | | |
| F | 7.354 | | | |
| Sing | 0.013* | | | |

*$P$-value $\leq 0.05$; **$P$-value $\leq 0.01$

**Table 5** Multi-regression analysis of the impact of FT tools as a strengthener of the financial inclusion indicators to improve the return on equity ratio

| Independent variable | Return on equity ratio | | | |
|---|---|---|---|---|
| | B | Beta | T | Sing |
| Number of ATMs | −0.031 | −0.356 | −0.558 | 0.046* |
| Number of branches | 0.023 | 2.564 | 3.157 | 0.040* |
| Deposits to assets | 0.432 | 13.681 | 1.142 | 0.042* |
| Credits to deposits | 0.154 | 4.520 | 1.587 | 0.026* |
| Credits to assets | −0.951 | −10.621 | −2.741 | 0.045* |
| Multiple correlation coefficient | 0.637 | | | |
| $R^2$ | 0.589 | | | |
| F | 10.168 | | | |
| Sing | 0.011* | | | |

*$P$-value $\leq 0.05$

We have found a significant relationship (Table 5) between the FT tools and financial inclusion indicators, thus improving the return on equity ratio and the ability of a firm to generate profits from its shareholders' investments in the bank. The independent variables, i.e., FT indicators, interpreted 58.9% of the dependent variable, i.e., return on equity ratio. Thus, the second hypothesis of research hypotheses can be accepted as a significant impact of the FT tool as a strengthener of financial inclusion indicators to improve the return on equity ratio.

# 5 Conclusion

This study aimed to explore the impact of FT tools on strengthening financial inclusion indicators and banks' profitability in light of Jordan's experience. The most important findings of the study are that FT, by its various tools, led to a change in the structure of financial services and the way and method of providing and diversifying banking services, leading to promote and increase the availability of financial services for community groups that had no access to the traditional financial services.

The study results also showed a significant impact of FT tools on strengthening financial inclusion and profitability indicators, including the Jordanian commercial banks' return on assets and equity. It is recommended to adopt and employ more advanced and digitalized FT tools to improve financial inclusion indicators further and provide financial services to all marginalized social groups, like online payment, transfer, insurance, finance and credit services, and other innovative financial services at affordable prices. Using practical FT tools will play a role in obtaining an added value of the collected financial transaction data, thus increasing the financial inclusion indicators and improving the banks' profitability and income.

# References

1. Leong K, Sung A (2018). FinTech (financial technology): what is it and how to use technologies to create business value in a fintech way? Int J Innov, Manag Technol 9(2)
2. Kim Y, Park YJ, Choi J, Yeon J (2015) An empirical study on the adoption of "Fintech" service: focused on mobile payment services. Adv Sci Technol Lett 114(26)
3. Al-Dmour H, Nweiran M, Al-Dmour R (2017) The influence of organizational culture on E-commerce adoption. Int J Bus Manag 12(9)
4. Baber H (2020) Impact of FinTech on customer retention in Islamic banks of Malaysia. Int J Bus Syst Res 14(2)
5. Bekhet HA, Alsamadi AM, Khudari M (2021) Dynamic linkages between profitability and its determinants: empirical evidence from Jordanian Commercial Banks. J Asian Finan, Econ Bus 8(6):687–700. https://doi.org/10.13106/JAFEB.2021.VOL8.NO6.0687
6. Demirguc-Kunt A, Klapper L, Singer D, Ansar S (2018). The Global Findex Database 2017: measuring financial inclusion and the fintech revolution. World Bank Publications
7. Sadłakowski D, Sobieraj A (2017) The development of the FinTech industry in the Visegrad group countries. World Sci News (85)
8. Abdel Reda M, Jawad M, al-Karim H (2020) The role of financial technology in enhancing the financial inclusion strategy exploratory research for a sample of Baghdad Commercial Bank and Business Bay employees. Warith J Sci Res 2(1)
9. Pushkar B, Gupta A (2019). Impact of E-banking its growth and future in India. In: Proceedings of 10th international conference on digital strategies for organizational success
10. Kabiraj S, Siddik M (2018) Does financial inclusion induce financial stability? Evidence from cross-country analysis. Australas Account, Bus Finan J 12(1)
11. Chahadah AA, Refae GAE, Qasim A (2018) The use of data mining techniques in accounting and finance as a corporate strategic tool: an empirical investigation on banks operating in emerging economies. Int J Econ Bus Res 15(4):442–452
12. Yao M, Di H, Zheng X, Xu X (2018) Impact of payment technology innovations on the traditional financial industry: a focus on China. Technol Forecast Soc Change 135
13. Al Chahadah A, El Refae G, Qasim A (2018) The use of data mining techniques in accounting and finance as a corporate strategic tool: an empirical investigation on banks operating in emerging economies. Int J Econ Bus Res 15(4)
14. Cihak M, Mare DS, Melecky M (2016) The Nexus of financial inclusion and financial stability: a study of trade-offs and synergies. Policy Research Working Paper; No. 7722
15. Mwongeli Musau S (2018) Financial inclusion and stability of commercial banks in Kenya. Doctoral dissertation in philosophy, Business Administration (Finance) in Kenyatta University
16. Shehata MM (2019) A proposed accounting model for measurement and disclosure of information financial technology innovations as a basis for enhancing financial inclusion and its impact on banking performance rates: with an applied study. J Account Res, Tanta University (1)
17. Belouafi A (2021) Fintech in the MENA region: current state and prospects. Islamic FinTech, pp 335–366
18. Abbasi T, Weigand H (2017) The impact of digital financial services on firm's performance: a literature review. Arxiv Preprint Arxiv:1705.10294
19. Musa K, AlShehadeh A, Alqerem R (2019) The role of data mining techniques in the decision-making process in Jordanian Commercial Banks. In: 2019 IEEE Jordan international joint conference on electrical engineering and information technology (JEEIT), pp 726–730. https://doi.org/10.1109/JEEIT.2019.8717461
20. Al-Chahadah A, Qasim A, El Refae G (2020) Financial inclusion indicators and their effect on corporate profitability. AAU J Bus Law 4(2)

# Bank Capital and Reputational Risk

Mais Sha'ban, Claudia Girardone, and Anna Sarkisyan

**Abstract** Environmental, social, and governance risks are becoming increasingly relevant for bank regulators and stakeholders due to market developments, regulation, and social attitudes. This paper investigates the relationship between equity capital and reputational risk related to Environmental Social Governance (ESG) issues for the European Economic Area's listed banks. We find that banks with higher reputational risk face higher costs of financial distress and therefore tend to have more capital. Additionally, in line with the corporate finance literature, we find that equity capital is negatively associated with size and positively with profits and market return volatility risk.

**Keywords** Bank capital · Capital structure · Reputational risk · ESG risk

## 1 Introduction

Banks are the most regulated financial institutions, and the financial sector is among the most regulated in the economy [1]. This is motivated, among other factors, by the need to ensure stability and protect consumers and results in the provision of government safety nets. Post-crisis international regulators have focused on banks' leverage, liquidity, and, in particular, quantity and quality of their capital due to its important function as a buffer to absorb losses in case of crisis. Additionally, the "specialness" of banking firms and their remarkable growth in asset size in recent years have highlighted the importance of understanding their capital structure and risk-taking. Hence, it is crucial for banks to continuously assess the risks they are exposed to and the suitability of their capital to absorb potential losses.

M. Sha'ban (✉)
Al-Zaytoonah University of Jordan, Amman, Jordan
e-mail: m.shaban@zuj.edu.jo

C. Girardone · A. Sarkisyan
Essex Business School, University of Essex, Colchester, UK

© The Author(s), under exclusive license to Springer Nature Switzerland AG 2022
S. G. Yaseen (ed.), *Digital Economy, Business Analytics, and Big Data Analytics Applications*, Studies in Computational Intelligence 1010,
https://doi.org/10.1007/978-3-031-05258-3_39

Environmental, social, and governance (ESG) risks are becoming increasingly relevant for bank regulators and stakeholders due to market developments, regulation, and social attitudes [2]. ESG factors influence the reputation and performance of financial institutions and are currently included in their business models. The literature is mainly focused on the relationship between ESG factors and corporate profitability and performance. In this paper, we assess the impact of reputational risk, measured by the RepRisk index, on the capital structure of the European Economic Area's listed banks. The aim is to examine whether ESG incidents impact bank decision to hold more capital. We find a positive relationship between equity capital and banks' reputational risk related to ESG issues. The result provides evidence for the trade-off theory, which states that firms with higher risk face higher costs of financial distress therefore tend to have more capital. It also provides evidence to the regulatory view of capital, which predicts that equity capital held by banks depends on the probability of their capital falling below the regulatory minimum requirements; hence, riskier banks hold higher equity capital. We also investigate other determinants including profitability, size, and market return volatility risk. Overall, the findings suggest that size and market-related factors (i.e., market return volatility risk and reputational risk) play a crucial role in the capital structure decision.

The paper is structured as follows. Section 2 reviews the relevant theoretical and empirical literature and discusses the hypotheses. Section 3 explains the data and main methodology used in the paper. Section 4 presents and discusses the results. Finally, Sect. 5 concludes.

## 2 Literature Review and Hypothesis Development

In this section, we review relevant theories and empirical studies that relate bank capital structure and risk including reputational risk.

Literature on the determinants of banks' capital structure considers two alternative views: on one hand, the corporate finance view that extends the conventional determinants of capital structure for non-financial firms to banks. An alternative view is the buffer/regulatory view according to which banks hold capital buffers above the regulatory minimum requirements in order to avoid the high costs associated with issuing equity capital at short notice.

In relation to risk, the regulatory (or buffer) view of capital predicts that equity capital held by banks depends on the probability of their capital falling below the regulatory minimum requirements; hence, riskier banks hold higher equity capital. Similarly, the trade-off theory of corporate finance assumes that firms with higher risk face higher costs of financial distress therefore tend to have more capital. Accordingly, both the buffer view and the corporate finance view predict a positive impact of risk on banks' equity capital.

Empirically, a number of studies find a significant relationship between equity capital and different types of risk including market volatility risk, credit risk, and portfolio risk [3–7]. To the best of our knowledge, there are no previous studies investigating the impact of reputational risk related to ESG issues on banks' capital structure. However, it is well documented in the literature that firms with better corporate social responsibility (CSR) have lower cost of equity financing and lower risk [8–10]. Reference [8] examine the impact of CSR on the cost of equity for a sample of U.S. firms. The authors find that better CSR performance, mainly related to investment in enhancing responsible employee relations, environmental policies, and product strategies, is associated with lower cost of equity capital. Reference [11] investigate the determinants of reputational risk for a sample of banks in Europe and the U.S. for the period 2003–2008. Their results show that in case of operational loss the reputational damage is lower for well-capitalized banks. Moreover, the authors find that reputational damage is positively associated with the bank's size and profitability.

Hence, it is expected that banks with lower reputational risk related to ESG issues can obtain cheaper equity financing [8]. Therefore, we expect banks that are less involved in ESG issues to have lower cost of capital and consequently to hold lower capital as they can obtain better price when issuing equity at short notice. On the contrary, banks that are more involved in ESG issues are expected to have higher equity-issuing costs and, consequently, to hold higher equity capital.

*H1. The relationship between equity capital and reputational risk is positive and significant.*

## 3 Methodology

### 3.1 Empirical Model

Following [5], we use the baseline model borrowed from the corporate finance literature for non-financial firms to explain the determinants of banks' capital structure and analyze the relationship between capital and reputational risk.

Our standard capital structure regression can be presented as follows:

$$
\begin{aligned}
Capital_{ict} = {} & \beta_0 + \beta_1 RepRisk_{ict-1} + \beta_2 MarketRisk_{ict-1} \\
& + \beta_3 Size_{ict-1} + \beta_4 Profitability_{ict-1} + \beta_5 MarkettoBook_{ict-1} \\
& + \beta_6 Collateral_{ict-1} + \beta_7 Dividend_{ict} + \beta_8 GDPgrowth_{ct} \\
& + \beta_9 Inflation_{ct} + c_t + u_{ict}
\end{aligned}
\tag{1}
$$

The dependent variable is the book equity capital ratio (*Capital*) of bank $i$ in country $c$ at time $t$; the explanatory variables include the reputational risk index (*RepRisk Index*) which quantifies reputational risk exposure related to ESG issues, market return volatility risk (*Market Risk),* natural logarithm of total assets (*Size*), profitability (*Profitability*), market-to-book ratio (*Market to Book*), tangibility (*Collateral*), and dividend dummy (*Dividend*). All bank-level explanatory variables are lagged by one year to control for potential endogeneity issues. We include a set of macroeconomic variables to control for heterogeneity across countries including GDP growth and inflation. The model also includes time fixed effects ($c_t$) to account for heterogeneity across time. Standard errors are clustered at the bank level to control for serial correlation of errors and heteroscedasticity [12]. The model is estimated using ordinary least squares (OLS).

We now provide a description of our variables and the expected relations between the independent variables and the equity capital in line with the predictions of the corporate finance view and the regulatory view of capital. Table 1 displays the definition of bank-specific and macroeconomic variables as well as the data sources used in the study.

Our dependent variable is the banks' book equity capital ratio measured as a ratio of the book value of equity to the book value of assets.

Our key independent variable is banks' reputational risk measured by reputational risk index (RepRisk Index). The RepRisk Index is an algorithm that captures and quantifies reputational risk exposure related to environmental, social and governance

**Table 1** Definition of variables and data sources

| Variables | Definition | Source |
|---|---|---|
| *Dependent variables* | | |
| Capital | Book value of equity/book value of assets | BankScope |
| *Independent variable* | | |
| RepRisk | Reputational risk index | RepRisk |
| *Bank-level control variables* | | |
| Market Risk | Log of annualized standard deviation of daily stock returns * (market value of equity/market value of bank) | DataStream |
| Size | Log (total book value of assets) | BankScope |
| Profitability | Return on average assets (ROA) | BankScope |
| Market-to-Book | Market value of assets/book value of assets | BankScope/DataStream |
| Collateral | (Total securities + cash and due from banks + fixed assets)/book value of assets | BankScope |
| Dividend | A dummy variable equal to 1 if the bank pays dividends in a given year, 0 otherwise | DataStream |
| *Macroeconomic control variables* | | |
| GDP Growth | Annual percentage change in gross domestic product | World Bank Development Indicators and Eurostat database |
| Inflation | Annual percentage change in average consumer price index | World Bank Development Indicators |

issues. RepRisk identifies ESG incidents through screening media and stakeholder information sources executed in 15 languages including print and online media every day. The incidents considered in the index are: (i) environmental including climate change, global pollution, ecosystems, waste issues, and animal mistreatment; (ii) social including community relations (human rights abuses, social discrimination) and employee relations (child labor, discrimination in employment, and poor employment conditions); and (iii) corporate governance including corruption, bribery, money laundering, executive compensation issues, misleading communication, fraud, tax evasion, and anti-competitive practices. These incidents are used to calculate an ESG risk exposure score, the RepRisk Index, for each bank in the sample. Major incidents are distinguished from minor incidents, based on the severity, reach, and novelty of an incident. RepRisk Index decays to zero over a maximum period of two years if no risk incident has appeared for a company. The current RepRisk Index used in our study denotes the level of media and stakeholder exposure of a bank related to these issues; we use its end-of-year value. It ranges from zero (lowest) to 100 (highest), with the higher value indicating higher risk exposure to ESG issues. It is worth mentioning that the RepRisk Index data are available from the beginning of year 2007 and for 74 banks in our sample. Data availability for the RepRisk index determines our chosen sample and time span (2007–2014).

Our measure of reputational risk is expected to have a positive effect on equity capital according to both the buffer view and corporate finance view. Under the trade-off theory assumption, firms with higher risk and higher volatility in cash flows face higher costs of financial distress and therefore tend to have more capital. As for the buffer view, riskier banks are required to have more equity capital as they have higher probability of their capital falling below the minimum regulatory capital.

Turning to other independent variables, our second measure of risk is the market return volatility risk used in [5]'s model and measured as the annualized standard deviation of daily stock returns multiplied by the market value of equity over the market value of bank. Market return volatility risk is expected to have a positive impact on banks' use of equity capital. Riskier banks with higher market return volatility are expected to hold higher capital ratios to decrease the probability of insolvency and the costs of bankruptcy. Reference [6] show that there is a positive relation between risk-weighted assets as a measure of capital requirements and volatility.

Size calculated as the logarithm of total assets is expected to have a negative impact on equity capital according to the trade-off theory that predicts that larger firms tend to have more leverage capacity. However, according to the buffer view the relation between size and equity capital is ex-ante ambiguous [5]. Larger banks may hold larger buffers to compensate for higher complexity and asymmetric information; alternatively, it is possible that larger banks hold smaller buffers because they are better known to investors and able to issue equity with less cost at a short notice.

Next, we measure profitability as return on the book value of average assets. The pecking order theory of [13] predicts a positive relationship between profitability and equity capital as profitable firms prefer to use internal financing rather than debt; whereas the agency theory expects firms with higher profitability to rely more on debt financing to discipline managers and decrease free cash-flow [14]. Reference [15] report that most empirical corporate finance studies find a positive relation between profitability and equity capital, which validates the pecking order theory. On the other hand, the buffer view of capital predicts a negative relationship between profitability and equity capital. Profitable banks are better known to investors and can obtain a better price when issuing equity at short notice so they do not need to hold higher equity capital.

We use a market-to-book ratio as a measure of growth opportunities. It has been found to have a positive relation with the equity capital of financial and non-financial firms [5, 15, 16], which is in line with the sign predicted by the trade-off theory where higher growth opportunities increase the costs of financial distress and consequently less debt is used (growth is an intangible asset that cannot be used as collateral). On the other hand, the buffer view predicts that banks with higher growth opportunities tend to hold less equity capital. Based on the same argument given for profitability, banks with higher growth opportunities are better known to investors and can obtain better price when issuing equity at short notice [5].

Further, we include collateral as a measure of tangibility, which is expected to have a negative relation with equity capital as according to the trade-off theory tangibility enhances the lenders' willingness to provide debt financing to borrowers. [16, 17], and [15] argue that tangibility reduces the costs of financial distress and hence motivates higher debt financing.

As for the dividend variable, we use a dummy that takes the value of one if the bank pays dividends in a given year and zero otherwise. Corporate finance studies support the pecking order theory that dividend-paying firms with higher profits prefer internal financing rather than debt financing; hence, a positive relation is expected. Alternatively, the buffer view expects dividend-paying banks to have lower equity-issuing costs and, consequently, to hold lower equity capital.

Finally, we include macroeconomic variables to control for the anticipated exposure of banks' activities to the economic conditions. These variables are the GDP growth which is a measure of the annual percentage change in gross domestic product and inflation which is a measure of the annual percentage change in the average consumer price index.

## 3.2   Data Sources

The data for the analysis are drawn from the following sources: banks' financial statements data from the BankScope database of the Bureau Van Dijk; market data (stock prices, dividends, and number of shares) from Thompson Financial's DataStream database; reputational risk index from RepRisk database; country-level economic data from the World Bank Development Indicators and Eurostat database.

The sample period spans from 2007 to 2014 due to data availability. To select the sample, we start with listed commercial banks and bank holding companies in the European Economic Area excluding Iceland (to avoid the Icelandic financial crisis) and Liechtenstein (as there are no listed banking institutions). The focus is on the 28 EU countries, Norway, and Switzerland. To avoid the possibility of outliers driving the results, we follow [18] and winsorize all bank-level variables at the 1% level.

# 4 Results

## 4.1 Descriptive Statistics

Table 2 provides descriptive statistics for the main variables used in the study. The data in the table show that the mean of book capital ratio for our sample is approximately 9%, suggesting a relatively high leverage of the sample banks. The results also show high variation in the banks' book capital ratios, which contradicts the traditional view that the amount of capital held by banks is determined by regulatory requirements and suggests low capital dispersion among banks falling under the same regulatory regimes. The mean of total book assets is 171 billion euro; the smallest bank in the sample has a total asset value of 45 million euro and the largest bank has a total asset value of 1970 billion euro which exhibits significant heterogeneity in the sample. The sample banks, on average, appear to earn low returns during the sample period as suggested by the mean return on assets of about 0.5%.

Figure 1 illustrates the relationship between the sample banks' average assets and average RepRisk Index for the period 2007–2014. Banks that have the highest reputational risk exposure over the period seem to be the largest banks in terms of assets (e.g., Barclays Plc, Royal Bank of Scotland Group Plc, BNP Paribas SA,

**Table 2** Descriptive statistics

| Variable | Mean | Std. Dev. | Min | Median | Max | Observations |
|---|---|---|---|---|---|---|
| Capital | 8.898 | 5.167 | 1.080 | 7.620 | 29.670 | 1324 |
| RepRisk | 15.171 | 17.821 | 0.000 | 12.000 | 71.000 | 592 |
| Market risk | 3.793 | 4.526 | 0.100 | 2.140 | 25.630 | 1251 |
| Size (Euro billions) | 171 | 390 | 0.045 | 9.804 | 1970 | 1324 |
| Profitability | 0.461 | 1.461 | −6.560 | 0.550 | 4.150 | 1324 |
| Market-to-Book Ratio | 102.096 | 11.230 | 85.060 | 99.340 | 161.750 | 1324 |
| Collateral | 28.741 | 14.022 | 3.280 | 27.010 | 68.640 | 1313 |
| Dividend | 0.559 | 0.497 | 0.000 | 1.000 | 1.000 | 1485 |
| GDP growth | 1.147 | 3.065 | −14.810 | 1.600 | 11.090 | 1477 |
| Inflation | 2.139 | 1.606 | −4.480 | 2.120 | 12.350 | 1490 |

*Note* The table presents summary statistics for the full sample of publically traded commercial banks and bank holding companies in EEA (excluding Iceland), covering the period from 2007–2014. Bank-level variables are winsorized at 1% level

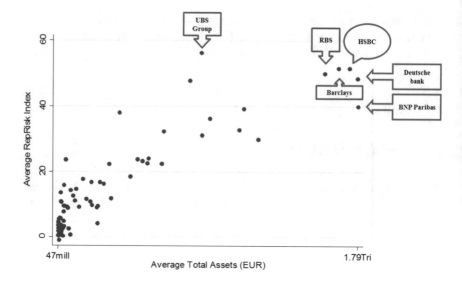

**Fig. 1** The figure shows the relationship between sample banks' average assets (*Average Total assets (EUR)*) and average RepRisk Index (*Average RepRisk Index*) for the period 2007–2014

Credit Suisse Group AG, HSBC Holdings Plc, UBS Group AG, Deutsche Bank AG, and Societe Generale SA).

## 4.2  Empirical Results

In this section we analyze the results derived from estimating Eq. (1) testing our hypothesis on the full sample. The results are reported in Table 3. In order to determine whether to apply a random or a fixed effects estimator, we use the Hausman (1978) test. The test suggests that the random effect assumption cannot be accepted; hence, we use the fixed effects estimator. The model is estimated with macroeconomic variables and time fixed effects.

The results show that the estimated coefficient of reputational risk index is positive and statistically significant at 5% level suggesting that banks with higher risk related to ESG issues tend to hold more capital. This is in line with the buffer view that suggests that these banks have a higher cost of capital [8, 9] and consequently hold higher capital as it can be costly for them to issue equity at short notice. The result is also consistent with the trade-off theory assumption that banks with higher risk face higher costs of financial distress and therefore tend to have more capital. The regression in Table 3 enables us to accept the hypothesis that the relationship between capital and reputational risk is positive and significant.

**Table 3** Equity capital ratios and reputational risk

| Dependent variable | Book capital ratio |
|---|---|
| RepRisk | 0.0213** |
| | (0.009) |
| Market Risk | 1.595*** |
| | (0.220) |
| Size | −0.666*** |
| | (0.100) |
| Profitability | 0.354* |
| | (0.204) |
| Market-to-Book | −0.0569** |
| | (0.024) |
| Collateral | −0.011 |
| | (0.013) |
| Dividend | 0.280 |
| | (0.255) |
| GDP growth | 0.034 |
| | (0.079) |
| Inflation | 0.159* |
| | (0.084) |
| Constant | 21.47*** |
| | (3.392) |
| Time fixed effects | Yes |
| Clustering (bank) | Yes |
| Adjusted R-squared | 0.568 |
| Number of observations | 470 |

*Note* The table reports the regression results of estimating the relation between banks' capital ratio and reputational risk controlling for a set of bank-specific and country-level factors. The dependent variable is book capital ratio. The independent bank-specific variables include: (i) RepRisk Index; (ii) market risk; (iii) size; (iv) profitability; (v) market-to-book ratio; (vi) collateral; and (vi) dividend; all lagged by one year. The independent country-level variables include: (i) GDP growth; and (ii) inflation. The regressions are run on a sample of 74 publicly traded commercial banks and bank holding companies in EEA (excluding Iceland) covering the period of 2007–2014. Standard errors clustered at bank level are reported in parentheses. *, **,***Indicate significance at 10%, 5%, and 1% levels, respectively

The coefficients of market return volatility risk and size variables also support both the buffer view and the corporate finance view of capital. The buffer view suggests that riskier banks hold higher equity to avoid their capital falling below the regulatory minimum, while larger banks take advantage of being known to the public and hence able to issue equity at a lower cost and at short notice. The corporate finance view suggests that riskier banks tend to have more equity capital to lower the costs of financial distress and larger banks tend to rely more on debt as they are less likely to face default risk.

In terms of profitability, it seems that profitable banks prefer to use internal financing rather than debt which validates the pecking order theory. The results also show that banks with higher growth opportunities (higher market to book ratio) hold less equity capital since they can issue equity at a lower cost and at short notice.

# 5 Conclusions

In this paper, we examine the relationship between bank equity capital and reputational risk exposure related to ESG issues for the European Economic Area's listed commercial banks and bank holding companies. We contribute to the literature as to the best of our knowledge there are no previous studies investigating the impact of reputational risk on banks' capital structure.

We find that banks with higher ESG reputational risk exposure tend to hold more equity capital. Our results suggest that banks with higher reputational risk face higher costs of financial distress and therefore tend to have more capital. Additionally, from the buffer view perspective, riskier banks tend to have more equity capital as they have higher probability of their capital falling below the minimum regulatory capital.

The study offers potentially important policy implications as the debate on the optimal capital structure of banks is still ongoing. The results suggest that reputational risk is an important factor in bank capital management.

# References

1. Chortareas GE, Girardone C, Ventouri A (2011) Regulation and bank performance in Europe. In: Bank performance, risk and firm financing. Palgrave Macmillan, London, pp 154–173
2. La Torre M, Leo S, Panetta IC (2021) Banks and environmental, social and governance drivers: follow the market or the authorities? Corp Soc Responsib Environ Manag (2021)
3. Brewer E, Kaufman GG, Wall LD (2008) Bank capital ratios across countries: why do they vary? J Finan Serv Res 34:177–201
4. Jokipii T, Milne A (2008) The cyclical behaviour of European Bank Capital Buffers. J Bank Finance 32:1440–1451
5. Gropp R, Heider F (2010) The determinants of bank capital structure. Rev Finan 14:587–622
6. Vallascas F, Hagendorff J (2013) The risk sensitivity of capital requirements: evidence from an international sample of large banks. Rev Finan 17:1947–1988

7. Distinguin I, Roulet C, Tarazi A (2013) Bank regulatory capital and liquidity: evidence from Us And European publicly traded banks. J Bank Finan 37:3295–3317
8. El Ghoul S, Guedhami O, Kwok CC, Mishra DR (2011) Does corporate social responsibility affect the cost of capital? J Bank Finance 35:2388–2406
9. Dhaliwal D, Li OZ, Tsang A, Yang YG (2014) Corporate social responsibility disclosure and the cost of equity capital: the roles of stakeholder orientation and financial transparency. J Account Public Policy 33:328–355
10. Valter D, Alain F (2017) ESG sustainability and financial capital structure: where they stand nowadays. Int J Bus Soc Sci 8:116–126
11. Fiordelisi F, Soana M-G, Schwizer P (2013) The determinants of reputational risk in the banking sector. J Bank Finance 37:1359–1371
12. Petersen MA (2009) Estimating standard errors in finance panel data sets: comparing approaches. Rev Finan Stud 22:435–480
13. Myers SC, Majluf NS (1984) Corporate financing and investment decisions when firms have information that investors do not have. J Financ Econ 13:187–221
14. Jensen MC (1986) Agency costs of free cash flow, corporate finance, and takeovers. Am Econ Rev 76:323–329
15. Frank MZ, Goyal VK (2009) Capital structure decisions: which factors are reliably important? Financ Manage 38:1–37
16. Rajan RG, Zingales L (1995) What do we know about capital structure? Some evidence from international data. J Financ 50:1421–1460
17. Titman S, Wessels R (1988) The determinants of capital structure choice. J Financ 43:1–19
18. Beltratti A, Stulz RM (2012) The credit crisis around the globe: why did some banks perform better? J Financ Econ 105:1–17

# Bank Interest Rate Spread: Operating Expenses and Noninterest Income Beyond Risk

**Marwan Alzoubi and Nawaf Salem**

**Abstract** This study investigates the impact of Operating Efficiency, Noninterest Income and risk on net interest margin (NIM) of Jordanian banks during the period 2007–2017 by applying the random effect model in a panel setting. Our results confirm that bankers are risk-averse; they require risk premiums by widening the spread whenever their risks are higher. Operating efficiency (Operating expenses to total assets) is highly significant with the expected positive sign. This may be due to lack of competition and requires additional measures by regulators such as licensing new banks. To examine this further, we find that concentration is significant and positively associated with bank spread. Likewise, since noninterest income helps reduce the margin, banks could improve their performance with better diversification. Such as providing consultancy services, cash management, leasing, venture capital financing, insurance (through subsidiaries), investment banking (underwriting) and trust services in addition to the off-balance sheet activities.

**Keywords** Operating efficiency · Net interest margin · Noninterest income · Panel analysis

## 1 Introduction

Net interest income is the main source of income for banks that operate according to the traditional model of banking which depends heavily on credit and deposits services. Jordanian banks are not an exception, they still operate according to the traditional model even though the banking law is open to more services but many of them require prior approval from the Central Bank of Jordan. High interest margins may result from improved activities or from higher loan rates and lower deposit rates,

M. Alzoubi (✉) · N. Salem
Department of Banking and Financial Sciences, Alzaytoonah University of Jordan, Amman, Jordan
e-mail: m.alzoubi@zuj.edu.jo

N. Salem
e-mail: N.Salem@zuj.edu.jo

© The Author(s), under exclusive license to Springer Nature Switzerland AG 2022
S. G. Yaseen (ed.), *Digital Economy, Business Analytics, and Big Data Analytics Applications*, Studies in Computational Intelligence 1010,
https://doi.org/10.1007/978-3-031-05258-3_40

which have implications on operational and managerial efficiencies and impede the intermediation process. It is long argued that high interest rate spreads indicate a more concentrated and less efficient banking sector. It is unclear however, why are these spreads so high. Several factors may help explain this phenomenon such as competition, operating expenses, risk-aversion, and regulatory cost including reserve requirements and capital adequacy, noninterest income, and the macroeconomic environment.

Clearly, the lower the spread, the higher is the bank activity and income and the better is the intermediation. This is especially needed at times of economic slow-downs, which have been prevailing in Jordan for the last ten years. Yet this is not the case, spreads are high at all times, which indicates that the level of economic activity may have a minimal effect. The contribution of this paper is manifested in explaining the persistent nature of the net interest margin (NIM) which seems to be unique. This may be due to the high level of operating expenses and noninterest income (NII).

Banks are special firms, they trade capital in the form of deposits and loans as dealers of capital and their spreads are the rewards they receive for providing immediacy. The deposit function makes them different and very sensitive. Deposits usually comprise over 70% of their sources of funds, this is specifically why central banks regulate them very heavily, but many times this comes at a high cost due to the requirements of licensing, statutory reserves, capital adequacy, allowances, concentration, on-site and off-site supervision (the regulatory cost). Many of these requirements add to the spread. The regulatory cost is usually reflected in the form of higher interest rate spreads.

Moreover, banks face a number of specific risks. 'Borrow short and lend long' risk, or simply interest rate risk, is inherently on top of the list in addition to the credit, liquidity, market, operational, exchange rate risks, etc. They work under many constraints, some of them are internal while the others are external imposed by their supervisors and the macroeconomic conditions. The aim here is to focus on operating efficiency and noninterest income in addition to risk. Jordanian banks have been able to sustain their profitability even during times of internal and external Instability and the slowdown of economic activity. Therefore, the question of this paper is about the impact of operating efficiency and noninterest income on bank spread. The control variable are liquidity, risk-aversion, the regulatory cost including reserve requirements and capital adequacy and the macroeconomic environment including interest rate and market structure (concentration).

In less competitive environments, inefficiency becomes a serious issue and mani-fest itself in higher spread levels due to higher levels of operating expenses. In such markets, it is imperative that regulators should interfere to limit those expenses espe-cially executive salaries and benefits. An alternative way to enhance performance is through diversification. Noninterest products include advisory services, cash management, leasing, venture capital financing, insurance (through subsidiaries), and investment banking (underwriting) and trust services in addition to off-balance sheet activities. Banks can compensate for any loss in spread income by noninterest income.

This paper is organized as follows; Sects. 2 and 3 represent a summary of the literature survey, and the theoretical model, respectively. In Sect. 4, the data sources and empirical results are analyzed and Sect. 5 concludes.

## 2 Literature Review

The net interest margins (NIM) is one of the key metrics for Banks, and bank performance. It is the main source of bank income and represents gross profit for the banking institution. Operating expenses and noninterest income (NII) are two of the most important factors that drive NIM. As operating expenses go up, banks tend to increase the NIM and as NII goes up, the spread declines.

The original model of intermediation, called the dealership model to analyze the determinants of bank interest margin was introduced by [12]. They propose that banks act as intermediaries between suppliers and demanders of funds. Therefore, they require positive interest margins or noninterest income as "the price of providing immediacy for both depositors and borrowers in the face of the uncertainty generated by random deposit supplies and loan demands". The model indicates that the spread has positive and statistically significant association with four factors: the level of managers' risk aversion; the market structure of the banking system; the average size of bank transactions; and the variance of interest rates. They confirm that even in a world of highly competitive banking markets, positive margins will tend to exist. Interest margins cannot disappear, under quite reasonable assumptions, as long as transactions uncertainty is present. Thus, they call the margin due to transactions uncertainty the "pure spread".

Their pioneering model has been extended by several empirical studies such as [18]. They conclude that interest margin is positively related to the degree of risk aversion. Reference [1] examines different types of credits and deposits, which lead to the diversification benefits of financial intermediaries. The author demonstrates that diversification plays an important role in reducing pure interest spreads. Reference [3] extends the model by taking into account credit risk and interest rate risk and tests whether the existence of short-term assets have unique effect on NIM. He finds that the default risk has a greater impact on the interest margin of large banks than the interest rate risk.

Reference [2] employ the panel data technique to identify the main determinants of the Brazilian bank interest spreads based on the [12] model. They suggests that operating costs and the ratio of non-interest bearing deposits to total operational assets act to increase the bank interest margin. Reference [17] suggest that operating costs positively affect NIM. They also extend the original model of [12] by analyzing the fundamental elements affecting the interest margin in the principal European countries (Germany, France, the United Kingdom, Italy and Spain) for the period 1993–2000 using a panel of 15,888 observations. They conclude that interest rate risk and credit risk have positive impact on the interest margin. They also find that the increase in the degree of concentration causes upward pressure on interest margins.

Reference [9] also investigates the determinants of net interest margins of banks in four Southeast Asian countries (Indonesia, Malaysia, Philippines and Thailand) based on the dealership model. Their results reveal that NIM is partially explained by bank-specific factors namely operating expenses, capital, loan quality, collateral liquid assets, short-term interest rates. In addition, low competition has a positive impact on NIM. Bank capital and operating cost have a positive impact on NIM in Southeast Asia with the exception of Thailand. Reference [23] suggest that competition lowers interest margin and fee income compensates banks.

Reference [23] identify the determinants of bank margins in European banks using the basic [12] model. The results show that net interest margin and profitability tend to be positively associated with overhead expenses, the level of capital and increased inflation rate. Reference [4] examine the impact of bank characteristics, the indicators of the financial structure and the macroeconomic environment on net interest margins and profitability in Tunis for the 1980–2000 period. They conclude that high net interest margin and profitability tend to be associated with banks that hold a relatively high amount of capital, large overheads and high inflation rates. Surprisingly, they conclude that concentration ratio has a negative and significant impact on net interest margins. This implies that concentration is less beneficial to the commercial banks in Tunis than competition.

Reference [6] conduct a systematic comparative analysis of the determinants of the interest margins of Central and Eastern European banks versus banks operating in developed Western European economies. They conclude that concentration, capital adequacy and risk behavior have positive and statistically significant effect on bank interest margin in both West and East while operational efficiency has a negative and statistically significant effect in the West and in the Accession countries only. Likewise, [22] claim that the most important driver of interest margins is the credit risk in Central and Eastern Europe. Their results reveal that lower operating costs (increased efficiency), positive economic developments and financial deepening lead to lower net interest margins.

Reference [13] test the impact of commercial bank-specific variables on NIM in the old European Union member countries and candidate countries between 1995 and 2006. All the variables are statistically significant in the first sub-period (1995–2000). More specifically, bank size, managerial efficiency, economic growth (GDP), and capitalization are negatively associated to NIM whereas the market power, operating costs, capital adequacy, default risk, credit risk, implicit interest payments, and inflation are all positively and significantly related to NIM.

Reference [11] investigate the drivers of interest rate spreads of commercial banks in Pakistan and find that the share of non-remunerative deposits in total deposits and administration expense in total expense are positively associated with interest rate spreads, while the share of non-interest income NII in total income negatively affects banking spreads. In addition, market concentration and macroeconomic variables (such as real GDP and interest rates) also have a positive effect. Reference [21] examine the determinants of net interest margins of Lebanese commercial Banks and demonstrate seven major determinants; opportunity cost, credit risk, bank capitalization, market structure, off-balance, size and economic growth. All the variables positively drive net interest margin, except credit risk, which has a negative and statistically significant impact.

On the other hand, [8] explore the determinants of 372 Swiss commercial bank profitability before and during the crisis for the period 1999 to 2009. Apart from the net interest margin, they use return on average assets (ROAA) and return on average equity (ROAE) as alternative proxies for bank profitability and cost of financial intermediation. They conclude that profitability is primarily determined by five factors: operational efficiency, growth of total loans, funding costs, business model and the effective tax rate. Reference [19] reports a negative impact of NII on NIM for the period 1997–2004. This result is obtained from banks operating in 28 industrial countries. Reference [16] studies the determinants of bank interest spread in Estonia for the period 1998–2011. The results show that interest spread is positively and significantly correlated with interest rate risk.

Reference [10] investigate the main determinants of the net interest margin of banks operating in Central and Eastern European countries during 1999–2010. Prior to 2008, the net interest margins essentially declined due to strong capital inflows and stable macroeconomic environment. After 2008, significant growth in government debt accompanied by the increased macroeconomic risks and abating capital inflows increased net interest margins. Low credit demand, higher capitalization and significantly increased share of non-performing loans are associated with lower margins. The results also show the important contribution of higher efficiency to lowering banks' interest margins. Reference [14] provide evidence of negative association in Chinese banks for the period 1998–2012.

Reference [20] show that both short-term funding and foreign bank presence have negative effects on the net interest margins of Chinese local banks. Reference [15] shows that NII has a negative influence on NIM of banks operating in Vietnam. Reference [7] argue that interest rates in many advanced economies have been low for almost a decade and are often expected to remain so. They examine the effects of (low) interest rates on bank net interest margins and profitability based on a sample of 3385 banks from 47 countries between 2005 and 2013. They conclude that low interest rates have a significantly greater impact on bank net interest margins than high interest rates. A recent study by [] analyze a large panel data set on all United State commercial and savings bank insured by the Federal Deposit Insurance Corporation for the period 2001–2015. They conclude that low interest rates negatively affect NIM.

# 3  The Empirical Model

The following multiple regression model is estimated to test the reliability of the theoretical relationship between the dependent and independent variables of Jordanian banks:

$$NIM_{it} = \beta_0 + \beta_1\, TRisk_{it} + \beta_2\, CAR_{it} + \beta_3\, OETA_{it} + \beta_4\, NNIIGI_{it}$$
$$+ \beta_5\, CRTA_{it} + \beta_6\, DepA_{it} + \beta_7\, HHI_{it} + \beta_8\, rGDPgr_{it} + \mu_{it} \qquad (1)$$

where $\beta0$ is the intercept, $\beta i$ represents the coefficients, NIM is Net Interest Margin, TRisk is Total Risk defined as risk-weighted assets to total assets, CAR is Equity to asset, OETA is Operating expenses to total assets, NNIIGI is Net noninterest income to gross income, CRTA is Liquid Reserves (Cash Reserves and Dues from Other Banks/Total Assets), DepA is Statutory reserves held in the Central Bank to total assets HHI is Herfindahl–Hirschman index of bank market power, rGDPgr = Real GDP growth rate, $\mu$ is error term, i is t individual factor and t is the time factor.

## 3.1 Variable Definitions

Table 1 reports the dependent and independent variables' definitions and measurements.

### 3.1.1 The Dependent Variable

There are several ways to measure net interest margin, which include the following:

(1)  NIM = (Interest Revenues − Interest Expenses)/Total Assets
(2)  NIM = (Interest Revenues/Loans) − (Interest Expenses/Deposits)
(3)  NIM = (Interest Revenues/*Interest Sensitive Assets*) − (Interest Expenses/*Interest Sensitive* liabilities)

This study employs the first definition for two reasons, the first is that it captures net revenue generated by the core operations of Jordanian banks represented by loans and deposits and the second is that it reflects the level of managerial efficiency as it represents the amount of net revenue generated on bank total assets. The literature also provides other definitions that include net noninterest income in addition to net interest income.

### 3.1.2 The Independent Variables

As indicated above, the influence variables are divided into two groups, internal and external. The internal variables are taken from financial statements and the external are macroeconomic variables represented by real GDP growth rate as a proxy for the demand for banking services and market interest rate. Market rate is dropped from the model, as it did not show any statistical significance.

The Target Variables

**Total Risk versus Credit risk**. Credit risk impact is straightforward; when banks are faced with higher credit risk, they raise their margin. Banks are rational risk-averse organizations, which implies that any additional risk-taking will not be welcomed without a significant increase in return. Although credit risk is the most important banking risk, this study applies the total risk concept introduced by the Basel accord which is defined as the risk-weighted assets to total assets, it captures the market risk, the operational risk in addition to credit risk. The sign is expected to be positive. **Operating expenses**. We expect a positive association between operating expenses and spread. **Net noninterest income**. Net noninterest income is the commission and fees income of a bank. In a competitive environment, banks lean on noninterest income to compensate for any decline in interest income, this behavior is well documented in the literature, especially in advanced countries. The Jordanian banking system, however, is criticized by being fragmented and therefore banks operating on a less competitive environment. A negative relationship is expected between net interest income and net noninterest income.

The Control Variables

**Capital adequacy**. Capital adequacy is a measure of bank risk aversion. However, the higher the ratio, the higher is the opportunity cost of equity capital. A positive sign is expected based on the increased cost. Banks compensate the increased cost by raising their interest rate spread. This especially true in Jordanian banks due to the fact that the majority of them are overcapitalized. **Cash ratio to Total Assets**. This variable is defined as cash reserves and dues from other banks to total assets. As this ratio increases, banks will be facing higher opportunity cost of holding liquid assets and forced to raise their spreads further to compensate for holding idle funds. A positive sign is anticipated. **Statutory Reserves held at the Central Bank of Jordan**. Reserves held at the central bank of Jordan are not remunerated. Therefore, any increase in the reserve requirement implies higher opportunity cost. The effect of this variable is similar to that of the cash ratio as both represent idle funds which is costly. Therefore the sign is also positive.

**Market Concentration**. Market concentration is a proxy for market power. It expected to positively impact bank spread. Market concentration is measured by the Herfindhal-Hirschman Index which is equal to the sum of squared individual banks market share. If market share is 100% (the market consist of one bank), the squared value will be 10,000, the highest possible value. If we have ten banks with equal market share, i.e., 10% each, then the square value for each is 100 and the total value is therefore 1000, the market is not concentrated. If the number is 100 banks with equal market share, for example, then the total value is 100 which is very small but indicate a highly competitive market. Therefore, the higher the value, the higher the concentration level and the lower is the competition level. **Gross domestic product**. As far as economic activity, GDP is intuitively a driving factor of performance. As real growth rate goes up, banks stand ready to finance and accept the increased deposits. The net effect is an increase in net interest income. A positive sign is expected. On the other hand, improved economic environment implies less uncertainty and less spreads, a negative sign.

## 3.2 Descriptive Statistics

Table 2 presents the descriptive statistics. Total risk, defined as risk-weighted assets to total assets, averaged 68.5% across banks with a maximum value of 96.3% and a minimum value of 37.4%. Total risk accounts for credit, market and operation risks. The mean is on the upper side, which helps explain the high margins that prevail in Jordan. Similarly, banks are solvent with a clear risk aversion behavior.

This is apparent from the high levels of capital to assets ratio, which averaged 14.3%. High solvency positions drive the net interest margin up. The market structure suffers from a significant degree of concentration as evident from the Herfindhal Hirschman Index.

The Herfindhal-Hirschman Index is calculated for the period 2007–2017. Its average is 28.8% which indicates also that there is some degree of concentration. Table 3 presents the correlation coefficients among the independent variables. It shows that the coefficients are generally acceptable and within normal levels except for the correlation between HHI and real growth rate of GDP (rGDPgr) of 69%. Pooled Regression, Fixed Effect and Random Effect.

## 3.3 Pooled Regression, Fixed Effect and Random Effect

### 3.3.1 Wald-Test and Hausman-Test

Table 6 reports the regression results of the pooled, fixed and random effects. The variables TRisk, OETA, NNIIGI and DepA in the pooled regression model are significant at the 5% level. In the fixed and random models, all variables are significant except DepA and rGDPgr. To determine which model is most reliable, Wald-test is used to select between the pooled and the fixed models. The model includes 13 banks, therefore the fixed effect includes 12 dummy variables.

$$\begin{aligned} NIM = \ &\beta_0 + \beta_1 \text{ TRisk} + \beta_2 \text{ CAR} + \beta_3 \text{ OETA} + \beta_4 \text{ NNIIGI} + \beta_5 \text{ CRTA} \\ &+ \beta_6 \text{ DepA } \beta_7 \text{ HHI} + \beta_8 \text{ rGDPgr} + \beta_9 d_2 + \beta_{10} d_3 + \beta_{11} d_4 + \beta_{12} d_5 \\ &+ \beta_{13} d_6 + \beta_{14} d_7 + \beta_{15} d_8 \beta_{16} d_9 + \beta_{17} d_{10} + \beta_{18} d_{11} + \beta_{19} d_{12} + \beta_{20} d_{13} \quad (2) \end{aligned}$$

Results of Wald-test are shown in Table 4. The Null and alternative hypotheses are:

H0: All dummy variables are equal to zero, the pooled effect is most appropriate

$$B9 = \beta 9 = \cdots = \beta 20 = 0$$

HA: Fixed effect is appropriate.

Given that the probability values of F-statistic and Chi-square are very small, less than 5 percent, the null hypothesis is rejected and the fixed effect model is the most appropriate. Hausman-test is used to select between the fixed effect and the random effect models. The results of Hausman-test are reported in Table 5.

H0: The random effect is most appropriate.

HA: Fixed effect is appropriate.

Based on the $P$-Value of Chi-square statistic (0.9784), the null hypothesis cannot be rejected and therefore, the random effect model is the most appropriate. The Wald-test is recommending the fixed effect model but after running the Hausman-test between random and fixed, the test is recommending the random effect. The final decision based on both tests is to use the random effect.

## 3.4 Residual Diagnostics

To examine whether the variances of the residuals are homoscedastic, that is having equal variances, the Jarque–Bera test is used. The results show that its value is 4.65 and the $P$-value is 0.0976, which leads to failing to reject the null hypothesis (H0: the residuals are normally distributed) and conclude that the residuals are normally distributed. Breusch-Pagan LM is employed as the residuals cross-section dependence test to examine whether there is a serial correlation in the residuals (to avoid the bias in the standard error and the less efficient coefficients). The null hypothesis states that there is no serial correlation. The results show that the value of the Breusch-Pagan LM is 98.25 with a $P$-value of 0.0604 (greater than 5%). The null hypothesis is therefore not rejected and conclude that the residuals are not serially correlated. From those two tests, we can conclude that the estimation of the parameters is efficient which leads to confirming the reliability of the random effect model.

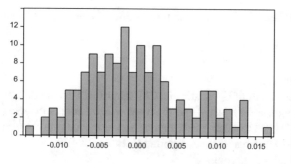

| | | |
|---|---|---|
| Series: Standardized Residuals | | |
| Sample 2007 2017 Observations 143 | | |
| Mean | 5.34e-19 |
| Median | -0.000875 |
| Maximum | 0.016271 |
| Minimum | -0.013530 Std. |
| Dev. 0.006305 | |
| Skewness | 0.391641 |
| Kurtosis | 2.590810 |
| Jarque-Bera | 4.653259 |
| Probability | 0.097624 |

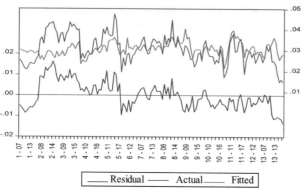

—— Residual —— Actual —— Fitted

# 4   Data Sources and Empirical Results

## 4.1   Data Sources

This study estimates a balanced data simple linear regression using a random effect model of 13 Jordanian commercial banks for the period 2007–2017 comprising 143 observations. The annual data is collected from financial statements submitted to the Amman Stock Exchange (ASE) as they are all public shareholding companies and listed on the ASE. There are 25 banks operating in Jordan, 8 are branches of foreign banks, 4 Islamic and 13 national commercial banks. These banks are selected based on their structures and functions. All of them are conventional commercial banks, and each has a license of a Jordanian national bank. The other banks are either foreign branches or Islamic.

## 4.2 Empirical Results

The results of the regression estimates show that the explanatory variables are generally significant with the predicted signs. The results are presented in Table 6 below using NIM as a dependent variable. Table 6 shows that total risk ratio (TRisk), operating expenses ratio (OETA) and concentration ratio (HHI) are statistically significant at the 5% level while capital ratio (CAR), net noninterest income ratio (NNIIGI) and cash reserves ratio (CRTA) are significant at the 1% level. Reserves held in the central bank and real GDP growth rate are not significant.

Operating expenses, our first target variable, trigger an increase in NIM when they go up, a sign of inefficiency. When they go down, NIM goes down, that is, when banks are being more efficient, lower operating expenses ratio prevails this result is obtained [2, 4, 6, 8, 9, 11, 13, 17, 22, 23] and [10].

The second target variable, which is net noninterest income, has a negative influence on NIM as anticipated. Similar result reported by [11, 15, 19] and [14]. The impact of total risk, measured as the risk-weighted assets to total assets as defined according to the Basel Accord is in line with the theory. More risk requires more return, i.e., higher NIM. This is similar to the results obtained by [4, 6, 9, 12, 17, 18, 22, 23], and [21]. Cash reserves levels when they go up, NIM goes down. Our results show that when those reserves increase, less capital will be available for banks to lend therefore NIM declines this is also shown by [9]. Finally, the concentration rate (HHI) is positively associated with NIM as expected. This result is reported by [6, 17] and [11] Market interest rate is not a determinant of the margin and dropped from the estimation.

## 5 Summary and Conclusions

A number of bank characteristics such as total risk, managerial efficiency, noninterest income, risk aversion, concentration, and opportunity cost of holding liquid reserves affects the performance of Jordanian bank. Our focus in this paper is operational efficiency and noninterest income, beyond the risk factor. Banks are risk-averse agents; they require additional premiums to increase their risk (operational, market, credit). To do that, they raise their interest rate margins. Operational efficiency is highly significant with the expected positive sign telling that a significant proportion of the spread is explained by the efficiency factor (defined as Operating expenses to total assets). This implies that competition is not optimal and the system is fragmented, which can be corrected through licensing.

Banks can improve their performance with better diversification not by increasing the levels of fees and commissions. Noninterest income is the most significant and carry the expected sign. Banks need to think outside the box and innovate new instrument and services in addition to credit facilities. Such as providing consultancy services, cash management, leasing, venture capital financing, insurance (through subsidiaries), investment banking (underwriting) and trust services.

Risk aversion (capital adequacy ratio) and concentration drive bank margin positively while Cash Reserves and Dues from Other Banks (to total Assets) drives it negatively. Raising capital levels does not reduce risk rather it is a protection cushion against the existing increased levels of risks and anticipated risks. However, this is not a call to minimize capital, but rather to keep it as high as possible in order protect deposit holders. The results confirm that more concentration implies more power and higher margins. Increasing the degree of competition is therefore a necessity for efficiency purposes and better performance.

# 6 Appendices

See Tables 1, 2, 3, 4, 5 and 6.

**Table 1** Variable definitions and measurements

| Variables | Measures |
|---|---|
| A. Dependent | |
| Net interest margin (**NIM**) | Interest income-interest expenses/total asset |
| B. Independent | |
| Total risk (**TRisk**) Solvency and risk aversion (**CAR**) management Efficiency (**OETA**) Net noninterest Income (**NNIIGI**) Liquidity level (**CRTA**) Regulatory cost (**DepA**) Market concentration (**HHI**) Economic activity | Risk-weighted assets/total assets Total capital/total assets Operating expenses/total assets Net noninterest income/gross income Cash reserves and dues from other banks/total assets Statutory Reserves at the Central Bank of Jordan/total assets Herfindhal-Hirschman index Real GDP growth rate at market prices |

**Table 2** Descriptive statistics

| | NIM | TRisk | CAR | OETA | NNIIGI | CRTA | DepA | HHI | rGDPgr |
|---|---|---|---|---|---|---|---|---|---|
| Max | 0.048 | 0.963 | 0.219 | 0.211 | 0.395 | 0.373 | 0.105 | 0.341 | 0.082 |
| Min | 0.015 | 0.374 | 0.091 | 0.010 | 0.049 | 0.073 | 0.022 | 0.192 | 0.020 |
| Mean | 0.031 | 0.685 | 0.143 | 0.027 | 0.159 | 0.201 | 0.043 | 0.288 | 0.037 |
| STD | 0.007 | 0.108 | 0.027 | 0.017 | 0.061 | 0.065 | 0.011 | 0.041 | 0.021 |

**Table 3** Correlation matrix

|        | Trisk | CAR   | OETA  | NNIIGI | CRTA  | DepA  | HHI  | rGDPgr |
|--------|-------|-------|-------|--------|-------|-------|------|--------|
| TRisk  | 1.00  |       |       |        |       |       |      |        |
| CAR    | 0.28  | 1.00  |       |        |       |       |      |        |
| OETA   | 0.21  | −0.03 | 1.00  |        |       |       |      |        |
| NNIIGI | 0.30  | 0.07  | 0.14  | 1.00   |       |       |      |        |
| CRTA   | −0.43 | 0.00  | 0.00  | 0.21   | 1.00  |       |      |        |
| DepA   | 0.07  | 0.00  | 0.04  | −0.01  | −0.02 | 1.00  |      |        |
| HHI    | −0.15 | 0.10  | 0.03  | 0.14   | 0.44  | 0.03  | 1.00 |        |
| RGDPGR | −0.17 | 0.04  | −0.10 | 0.19   | 0.51  | 0.14  | 0.69 | 1.00   |

**Table 4** Wald-test

| Test statistic           | Value | P-value |
|--------------------------|-------|---------|
| F-statistic Chi-square   | 29.4  | 0.0000  |
|                          | 352.6 | 0.0000  |

**Table 5** Hausman-test

| Test statistic | Value | P-value |
|----------------|-------|---------|
| Chi-square     | 7.62  | 0.472   |

**Table 6** Panel regression results

|                      | Pooled regression | | Fixed effect OLS | | Random effect | |
|----------------------|-------------|---------|-------------|---------|-------------|---------|
|                      | Coefficient | P-value | Coefficient | P-value | Coefficient | P-value |
| *A-Internal variables* |           |         |             |         |             |         |
| TRisk                | 0.010       | 0.049   | 0.007       | 0.027   | 0.007       | 0.024   |
| CAR                  | 0.004       | 0.845   | 0.072       | 0.000   | 0.069       | 0.000   |
| OETA                 | 0.076       | 0.016   | 0.037       | 0.031   | 0.039       | 0.025   |
| NNIIGI               | −0.043      | 0.000   | −0.027      | 0.000   | −0.028      | 0.000   |
| CRTA                 | −0.006      | 0.552   | −0.021      | 0.001   | −0.021      | 0.001   |
| DepA                 | 0.141       | 0.003   | 0.012       | 0.686   | 0.018       | 0.558   |
| HHI                  | 0.025       | 0.145   | 0.020       | 0.027   | 0.020       | 0.025   |
| *B-External variables* |           |         |             |         |             |         |
| RGDOGR               | −0.030      | 0.401   | −0.007      | 0.718   | −0.008      | 0.677   |
| Adjusted $R^2$       | 0.26        | –       | 0.81        | –       | 0.44        | –       |
| F-statistic          | 5.89        | 0.0000  | 25.98       | 0.0000  | 13.09       | 0.0000  |

# References

1. Allen L (1988) The determinants of bank interest margins: a note. J Finan Quant Anal 231–235
2. Afanasieff TS, Lhacer PM, Nakane MI (2002) The determinants of bank interest spread in Brazil. Money Affairs 15(2):183–207
3. Angbazo L (1997) Commercial bank net interest margins, default risk, interest rate risk and off balance sheet activities. J Bank Finance 55–87
4. Ben Naceur S, Goaied M (2008) The determinants of commercial bank interest margin and profitability: evidence from Tunisia. Front Finan Econ 5(1):106–130
5. Carbo Valverde S, Rodríguez Fernández F (2007) The determinants of bank margins in European banking. J Bank Finance 31(7):2043–2063
6. Claeys S, Vander Vennet R (2008) Determinants of bank interest margins in central and eastern Europe: a comparison with the west. Econ Syst 32(2):197–216
7. Claessens S, Coleman N, Donnelly M (2017) "Low-for-long" interest rates and banks' interest margins and profitability: cross-country evidence. J Financ Intermediation 35:1–16
8. Dietrich A, Wanzenried G (2011) Determinants of bank profitability before and during the crisis: evidence from Switzerland. J Int Finan Markets Inst Money 21(3):307–327
9. Doliente JS (2005) Determinants of bank net Interest margins in southeast Asia. Appl Financ Econ Lett 1(1):53–57
10. Dumičić M, Rizdak T (2013) Determinants of banks' net interest margins in central and eastern Europe. Financ Theory Pract 37(1):1–30
11. Hasan Khan M, Khan B (2010) What drives interest rate spreads of commercial banks in Pakistan? Empirical evidence based on panel data. SBP Res Bull 6(2):15–36
12. Ho TJS, Saunders A (1981) The determinants of bank interest margins. J Finan Quant Anal 16(4):581–600
13. Kasman A, Tunc G, Vardar G, Okan B (2010) Consolidation and commercial bank net interest margins: evidence from the old and new European Union members and candidate countries. Econ Model 27(3):648–655
14. Liya L, Minghui L, Sha S, Jinqiang Y (2014) The Relationship between net interest margin and non-interest income for Chinese banks. Econ Res J 07
15. Le TD (2017) The interrelationship between net interest margin and non-interest income: evidence from Vietnam. Int J Manag Finan 1–16.
16. Männasoo K (2012) Determinants of bank interest spread in Estonia. Eesti Pank. Working Paper Series 1
17. Maudos J, de Guevara JF (2014) Factors explaining the interest margin in the banking sectors of the European Union. J Bank Finance 28(9):2259–2281
18. McShane RW, Sharpe IG (1985) A time series/cross section analysis of the determinants of Australian trading bank loan/deposit interest margins: 1962–1981. J Bank Finance 9(1):115–136
19. Nguyen J (2012) The relationship between net interest margin and noninterest income using a system estimation approach. J Bank Finance 36(9):2429–2437
20. Qi M, Yang Y (2016) The determinants of bank interest margins: a short-term funding perspective. Appl Econ Finance 4(1):127–137
21. Saad W, El-Moussawi C (2010) The determinants of net interest margins of commercial banks in Lebanon. J Money, Investment Bank 23:118–132
22. Schwaiger MS, Liebig D (2009) Determinants of the interest rate margins in central and Eastern Europe. Oestereichische National bank, Financial Stability Report, No. 14
23. Valverde SC, Fernández FR (2005) The determinants of bank margins revisited: a note on the effects of diversification. Department of Economic Theory and Economic History of the University of Granada. 05/11

# The Case of Analyst Ratings in US Equity Mutual Funds

Luis Otero-Gonzalez, Renato Correia-Domingues, Pablo Duran-Santomil, and Manuel Rocha-Armada

**Abstract** This paper evaluates the usefulness of analyst assessments to select mutual funds, using the Morningstar Analyst ratings. These notes are forward-looking qualitative and quantitative analyses of mutual funds about five pillars: Process, Performance, People, Parent and Price, that includes factors like the cost, past performance, quality of management or interest alignment. Our sample contains US equity funds covering the period August 2012 to August 2016. Our conclusions support the ability of Gold ratings to select funds that will behave better in terms of future performance. We have found little evidence that, on average, funds with a better Analyst Rating (Gold) have a better performance in terms of risk adjusted measures (alpha and sharpe). The predictability is observed in several analyses done in one year ahead but not for three-years. This evidence is more relevant in the case of the analysis made by investment style´s category.

**Keywords** Mutual funds · Performance · Persistence · Ratings · Morningstar

L. Otero-Gonzalez (✉) · P. Duran-Santomil
Department of Finance and Accounting, University of Santiago of Compostela, Av. do Burgo das Nacións, s/n, Santiago de Compostela, Spain
e-mail: oterogonzalezluis@gmail.com

P. Duran-Santomil
e-mail: pablo.duran@usc.es

R. Correia-Domingues
Unit Department of Business Science of Polytechnic Institute of Tomar, Estrada da Serra, Campus da Quinta do Contador, Tomar, Portugal
e-mail: renato.domingues@ipt.pt

M. Rocha-Armada
Campus de Gualtar - Edifício 8 - 2.22, Braga, Portugal
e-mail: mjrarmada@gmail.com

© The Author(s), under exclusive license to Springer Nature Switzerland AG 2022          523
S. G. Yaseen (ed.), *Digital Economy, Business Analytics, and Big Data Analytics Applications*, Studies in Computational Intelligence 1010,
https://doi.org/10.1007/978-3-031-05258-3_41

# 1 Introduction

Mutual funds are one of the most important assets used by investors to build their portfolios. Investors are very interested in the criteria to select and identify mutual funds with good expectations about future performance. In this sense, it is quite common to use ratings in the asset allocation process.

Most popular ratings are Morningstar Star ratings and some empirical studies such as [1, 2], among others, have shown that the downgrade or upgrade of quantitative ratings influence the flows of mutual funds. Furthermore, Stars Rating (backwards looking) are able to explain short-term out-of-sample performance and preserve the investors wealth. Morey and Gottesman [3], Müller and Weber [4], Meinhardt [5], Otero et al. [6] have also supported the performance persistence of quantitative ratings but only in the short term. However, the selection of funds based exclusively on historical performance or quantitative ratings, excludes a set of qualitative factors that can explain future performance. This explains the reason why it was important add to quantitative ratings, ratings based on the opinions of analysts who evaluate mutual fund features, as: Governance, Process, People, Parent, Board Quality, Corporate Culture, Fees, Manager Incentives or Regulatory Issues. Despite the several studies that have been done on quantitative ratings, qualitative ratings are not that popular and very little research has been done about the ability to select good funds based on qualitative ratings. In the specific case of Morningstar, there are two alternatives: Analyst Ratings and Stewardship Grade. As far as we know, very few researches have been conducted on this subject.

References [7–12] studied the effects on performance and flows of Stewardship Grades. On the other hand, [12–14] and [15] are the unique authors that focus their research on Analyst ratings. Analyst rating was launched in September 2011 by Morningstar and is a forward-looking measure based on analyst's expectation about of the mutual fund's future performance relative to peers and over the long term. The rating reflects the valuation of analyst in five dimensions which includes factors like the cost, past performance, quality of management, interest alignment, etc., Otero-González and Durán-Santomil [15] find "higher abnormal flows to funds receiving higher ratings suggesting that the average retail investor values the analyst's subjective views when allocating their wealth".

Thus, investors take into account both stars (backward-looking) and analyst (forward-looking) to take their decisions of investment. In this paper we assess to what extent selecting mutual funds based on ratings criteria has an impact on the financial and risk performance of investors. In particular, we endeavor to answer this essential question: Do good analyst (forward-looking) ratings outperform non-recommended ones in the short and long term? Previous research on analyst ratings is scarce and generally includes the entire universe of mutual funds with rating, despite the heterogeneity in terms of investment area, exchange rate risk, period of analysis, etc. The fact that previous studies have been conducted with very limited samples or assuming questionable hypothesis about rating persistence, explains why our research contains analyst ratings for rated funds from August 2012 to August 2016. Further, the results of the literature about the ability of "good analyst rated" to outperform "bad analyst rated" have shown mixed evidence. The empirical research concerning these questions takes particular interest to asset managers, financial advisors and investors using ratings to take their portfolios decisions. Our results are in line with [13, 14] and [15]. In this way, we have found a small evidence that, on average, funds with a better Analyst Rating (Gold) have a better performance in terms of risk adjusted measures alpha and Sharpe). Predictability is observed in several analyses performed after one year, but not in those carried out after three-years. This evidence is more relevant in the case of the analysis made by investment style's category. In the analysis of the pillars in which the analyst ratings are broken down we do not find evidence that future performance is related to any of these specific dimensions. The paper is organized as follows. First, we review and summarize the main existing research about qualitative ratings and mutual fund's performance. In the fourth section, we present the empirical analysis, the statistical models and the main results. Finally, we summarize the main conclusions.

## 2 Previous Research

There is a debate in literature about the power of ratings to predict future performance. Most of the studies employ quantitative ratings to examine the ratings' capacity by choosing the best to pick the best funds.

The quantitative aspects of ratings have some limitations. Quantitative ratings cannot quantify aspects of qualitative ratings. It is true that when ratings are made based on historical performance, the ability of managers has implications, as well as other qualitative aspects. However, there is no evidence that past performance has this capacity to capture the ability of managers to obtain results above the market. On the other hand, there are aspects that have to do with ethics, correct procedures, appropriate strategies, as well as whether fund managers are hedging the fund's wealth in the medium and long term. Certain qualitative aspects, such as the quality of management and procedures, cannot be captured by past performance and it is in this sense that qualitative ratings can provide very useful information to choosing the best funds. Tufano and Sevick [16] conducted the first empirical study which analyzed the quality of the board of directors of mutual funds. They examined the relationship between the composition and compensation of boards of directors of U.S. mutual funds and the fees charged to their investors. In the last years, authors as [17–32] investigated the quality of governance in mutual funds and they found that, in general, funds with better governance perform better in the future. There are other papers that focus on the influence of Stewardship Grade or Analyst Rating on fund flows. Wellman and Zhou [7] found that investors were concerned about these ratings and used them to invest their money. The better classified ratings can attract more money than the worsts ratings. Lai et al. [9] suggest that investors react more strongly to poor fund performance by withdrawing funds when the board quality component of

Morningstar's Stewardship Grade is perceived to be bad. Otero-González and Durán-Santomil [15] concluded that higher Analyst ratings (Gold and Silver) receive higher abnormal flows[1], particularly in retail funds. These studies report evidence on the importance for qualitative ratings to investors. As a result of the relative newness of Morningstar's qualitative ratings, compared to the development of Star Ratings' investigation, there was not much research on this subject and few authors had studied it. In the case of Stewardship, we highlight [7–10, 23] and [11]. For Analyst, only [13] and [14] had studied their effect on future performance.

QL rating was an assessment of fund administration, investment management, product and company capabilities and strengths reported as Business and Management Strength rating and Sector Strength rating. The author used data from two of the largest fund subcategories Australian Equity Trusts– General (AET) and Superannuation–Australian Equity Trusts (SAET). To our knowledge only [14, 15] and [16] have analyzed the effect on Morningstar Analyst on future performance. Armstrong et al. [14] studied Morningstar Analysis Ratings at July 2013 (1159 individual mutual funds: equity fund, fixed-income funds, etc.) and concluded that there is a significant positive relationship between these ratings and the future performance as measured by the 3-year Alpha applying OLS and quantile regression. Results for quantile regression show that, for better performing funds, higher Analyst Rating does not necessarily predict better performance in the future. She also found that the People pillar of these ratings has a significant predictive power for funds' future performance. However, the author cautions in her work that "whether the Analyst Ratings can predict future fund performance, we need more data, which is not available as of yet, because these are relatively newer ratings, with a long-term focus". Finally, she found that Sharpe Ratio and Analyst Ratings are significantly positively related to contemporaneous fund performance. The People and Process Ratings are also individually significantly related to the Sharpe Ratio. Armstrong et al. [14] tried to understand the extent to which Morningstar Analysis Ratings have the power to influence investors in terms of flows, as well as the ability of these ratings to provide above-average performance. The authors employed a sample of 412 equity funds from September 2011 to June 2014 and examined if the analyst ratings contained information about the rated funds' future performance. They computed out-of- sample performance using cumulative risk-adjusted returns measured over the 6, 12, 18, and 24 months following each fund's initial rating and identified out-perform peer funds horizons of up of 18 months or more comparing Gold, Silver and Bronze rated funds with Not Recommended funds. Finally, they showed that a portfolio of Gold rated funds has significantly higher alphas than a portfolio that contains all funds that are not rated Gold. Otero-González and Durán-Santomil [15] found that mutual funds classified as Gold behave better than "Not recommended" funds with 5-factor alpha in the short term for the United States investment area. and data from the period 2012–2018. Morningstar Direct reports from 2011 the analyst rating but the sample is very limited till half 2012 when there are around 200 mutual funds rated for the United States investment area. Previous work was carried out with several limitations, such as: all the universe of mutual funds with rating is included, the heterogeneity in terms of investment area is not considered, mutual funds have different level of exchange rate risk, limited period of analysis, etc.

# 3 Data and Sample

That's why in our research we only include the analyst ratings for rated funds from August 2012 to August 2016, as previous research was done with very limited samples or assuming questionable hypothesis. We restrict our selection to the funds focused on the investment area of United States, with stars ratings and included in the nine common categories resulting from combining size and value[2]. In line with [14] we also eliminate from the selection multiple share class, including only one equivalent for each mutual fund and we eliminate the funds less than two years old and less than five million of net assets, to avoid incubation bias. The sample contains 10.772 monthly observations and an average of 220 mutual funds rated, with a good representation in each level except for the negative case, where surprisingly, any mutual fund has achieved this grade. Similarly, to [14] we distinguish between "Not recommended" and "Recommended" mutual funds, where recommend is composed by the categories from Gold to Bronze, and the others are classified into "Not recommended". More than 80% are classified as "Recommended" because they have a gold, bronze or silver rating and this is explained by Morningstar by the fact that they prioritize high quality funds and in general they are bigger and with lower expenses and turnover ratios (see Table 1).

Table 2, summarizes the fund characteristics across the rating categories, sorting the rated funds into the five analyst rating categories from the highest rating of Gold to the lowest rating. We observe that, on average, Gold, Silver and Bronze rated funds are larger than Neutral and Under-review rated funds. Gold mutual funds have the lower levels of expense ratios, low rotation and higher levels of risk adjusted performance than the other mutual funds. At the same time, Neutral mutual funds are the worst in performance and downside risk, measured through Value at Risk, despite lower expenses ratios and bigger size than Silver and Bronze.

To measure out-of-sample performance we use two risk-adjusted metrics: Alpha and Sharpe ratio. Table 3 contains the out-of-sample performance and downside risk

**Table 1** Analyst rating distribution of the period (August 2021–August 2016)

| Rating analyst | Freq | Percent | Cum |
|---|---|---|---|
| Gold | 1923 | 17.85 | 54.71 |
| Bronze | 397 | 36.85 | 36.85 |
| Silver | 2811 | 26.1 | 99.61 |
| Neutral | 2026 | 18.81 | 73.51 |
| Under Review | 42 | 0.39 | 100 |
| Total | 10,772 | 100 | |

**Table 2** Fund characteristics by analyst rating

| Variable | Gold | Silver | Bronze | Neutral | Under review |
|---|---|---|---|---|---|
| ManagerTen_ | 93.78 | 97.22 | 84.95 | 60.54 | 36.40 |
| Netexp_ | 0.70 | 0.93 | 0.99 | 0.91 | 0.92 |
| Turnover_ | 20.83 | 33.08 | 45.28 | 62.83 | 53.84 |
| VaR99_ | 5.19 | 5.10 | 4.96 | 5.30 | 4.00 |
| Sharpe12_ | 1.13 | 1.10 | 1.07 | 0.95 | 1.65 |
| Sharpe36_ | 1.09 | 1.07 | 1.06 | 1.00 | 1.13 |
| Alpha12_ | 0.13 | -0.40 | -1.16 | -1.77 | -0.40 |
| Alpha36_ | 0.68 | 0.33 | -0.78 | -1.94 | -0.59 |
| Return12_ | 18.27 | 18.47 | 18.50 | 17.10 | 18.23 |
| Return36_ | 18.48 | 18.15 | 18.04 | 17.56 | 18.22 |
| LogSize | 10,400.00 | 2,450.00 | 2,510.00 | 3,020.00 | 1,370.00 |

*Note* ManagerTen is the number of months that a manager(s) has been at the helm of the fund. Netexp is the net expense ratio declared. Turnover is the turnover ratio, i.e., the percentage of the mutual fund that have been replaced with other holdings in a given year. VaR99 is the value at risk at a confidence level of 99%. Sharpe12 (sharpe36) is the ratio of Sharpe is the yearly risk-adjusted return calculated over a period of 12 (36) months. Alpha12 (Alpha36) is the excess return of a fund relative to the return of the benchmark index calculated over a period of 12 (36) months. Return12 (Return36) is the net return of a fund calculated over a period of 12 (36) months. LogSize is the logarithm of the assets of the fund

**Table 3** Conditional three years performance and value at risk by Rating

| Performance (t+36) | Analyst rating (t) | | | | |
|---|---|---|---|---|---|
| | Negative | Neutral | Bronze | Silver | Gold |
| Sharpe_ | - | 0.823 | 0.7288 | 0.740 | 0.787 |
| Return_ | - | 16.801 | 15.730 | 15.005 | 15.794 |
| Alpha_ | - | -1.687 | -1.954 | -1.684 | -0.825 |
| VaR99_ | - | 7.17. | 6.369 | 6.080 | 6.385 |

*Note* Sharpe is the Sharpe Ratio calculated in a three-annual basis, Alpha is the beta-adjusted return over a three-year period, Return is the net 3 years return. VaR99 is the value at risk at a confidence level of 99%

measures of the mutual funds three years after the initiation. Thus, we try to analyze if those good rated mutual funds have in general a better ex-post good out of sample performance. We also differentiate between Analyst and Stars ratings to assess the importance of backward looking or forward looking explaining future performance. In the particular case of star ratings, we observe a monotonic decrease across the categories for all the performance metrics, showing that in general, better rated funds obtained better performance and less downside risk 36 months later. Surprisingly, for the Analyst rated funds we observe that Neutral mutual funds obtained in general greater performance than "Recommended", except in terms of alpha, but with a higher downside risk. Comparing Gold, Silver and Bronze; we observe greater performance for the best ratings and the contrary in terms of downside risk, but with minor differences.

## 3.1 Models

In the section, we look at whether the analyst rating can help identify products that will outperform their peers in the next period after the initial rating. At the same time, we also include separately stars ratings to compare both alternatives, taking into account that investors take their decisions based on backward looking ratings. Finally, we use both ratings to assess if decision making combining good stars and analyst mutual funds can help in selecting outperformers. Firstly, we estimate some models to evaluate out of sample performance based exclusively on the analyst rating and their main pillars. If analyst rating is a forward-looking measure that reflects the expectations of analyst about future performance in the long run, we expect that higher ratings will obtain higher future performance. We calculate out of sample risk adjusted returns for 12 and 36 months after the initiation rating and then we regress the different metrics using Gold, Silver and Bronze indicators variables. For 12 months models, we estimate a panel data regression model (random effects) and for 36 months we use OLS regression because we only have one period of three years after the initial grade. The methodology based on panel data can control for individual effects with advantages like the reduction of collinearity and efficiency, among others [14]. The following equations are estimated:

$$Y_{i,t+k} = \alpha_i + \beta_1 \text{Gold}_{it} + \beta_2 \text{Silver}_{it} + \beta_3 \text{Bronze}_{it}$$
$$+ \sum \text{Month}_t + \sum \text{Category}_j + \varepsilon_{it} \tag{1}$$

$$Y_{i,t} + k = \alpha_i + \beta_1 \text{Gold}_{it} + \beta_2 \text{Silver}_{it}$$
$$+ \beta_3 \text{Bronze}_{it} + \beta_4 \text{Netexp}_{it} + \beta_5 \log \text{Size}_{it}$$
$$+ \beta_6 \text{ManagerTen ant}_{it} + \sum \text{Month}_t$$
$$+ \sum \text{Category}_j + \varepsilon_{it} \tag{2}$$

where $Y_{i,t} + k$ is the performance obtained by the fund i for 12 of 36 months after the initial rating. As performance measures we use Alpha, Sharpe and Total Return. Gold, Silver and Bronze are indicator variables considered as recommended funds, that take the value of 1 when the mutual fund is rated as one of these categories and 0, otherwise. Category are dummy variables for the nine categories considered in the study. Finally, $\alpha i$ and $\beta 1$, $\beta 2$, $\beta 3$, $\beta 4$, $\beta 5$, and $\beta 6$ are parameters of the regression and $\varepsilon i$ the term error. We estimate the models first controlling by month and categories, and then including some additional controls like expenses, size and age. Following [11] and [15], we also include as control variables the manager tenure (ManagerTen) of the fund, the costs measured by the net expenses ratio (Netexp) and the size of the mutual fund (LogSize).

## 3.2 Results for Analyst Ratings (Forward Looking)

Table 4 shows the results of regressing the risk adjusted return with the analyst indicators variables after twelve months since the rating is available, through a robust random effects panel data model. As in previous works, the models were initially estimated without control variables to evaluate the effect of selecting funds based exclusively on the rating analyst. Subsequently, other explanatory variables such as costs, size and management experience have been included. In all cases, category and time control variables have been included. As you can see, the results depend on the metric used, but in general, only funds classified as gold show better 12-month performance than the "Not recommended" funds in terms of Sharpe's ratio. Over 36 months the results are more disappointing and the only significant sign is negative. Therefore, our analysis reveals the inability of analyst ratings to identify funds that outperform their peers except in the case of gold funds, where the results show better performance than those classified as not recommended. The differences with previous studies may be due to the fact of considering in our study a different sample focused exclusively on US, the panel data methodology and the different period considered.

## 4   Robustness

We conducted some additional robustness tests to check the consistency of our results and to provide other complementary analysis. In addition to panel data we also use the quantile regression to extend the regression model to conditional quantiles of the different performance metrics because it is more appropriate for a heterogeneous mutual fund universe where strategies and objectives can vary [11]. This model let us capture information about the coefficients at different quantiles of the dependent variable given the set of endogenous variables (star rating). In addition, the conditional quantile regression developed by [33] deals well with skewed distributions of fund performance. In particular, we adopt the bootstrapping method proposed by [34] and implemented in the software Stata 12.

Given Yi as the different performance metrics used in this paper, and Xi as a vector of exogenous variables representing the rating of the fund, the quantile model can be written as:

$$y_i = X'\beta_\phi + u_{\phi i}$$

**Table 4** Analyst ratings and out-of-sample performance after 12 and 36 months

| Variable | Alpha-12 | Sharpe-12 | AlphaC-12 | SharpeC-12 | Alpha-36 | Sharpe-36 | AlphaC-36 | SharpeC-36 |
|---|---|---|---|---|---|---|---|---|
| Gold | 0.7204 | 0.2023*** | 0.7288 | 0.3572*** | 0.311 | 0.0368 | 0.024 | −0.012 |
| Silver | −0.1788 | 0.0219 | −0.2108 | 0.1581 | −0.3122 | −0.0945* | −0.1202 | −0.0711 |
| Bronze | −0.1094 | −0.0334 | 0.4072 | 0.1216 | −0.1893 | −0.047 | 0.1439 | −0.0055 |
| Netexp_ | – | – | −0.7281 | −0.0824 | – | – | −0.8025 | −0.1673*** |
| logSize | – | – | 0.2006 | −0.0123 | – | – | 0.4997*** | 0.0520*** |
| ManagerTen_ | – | – | −0.0118*** | −0.0039*** | – | – | −0.0095** | −0.0009** |
| USlargeblend | −1.8096*** | 0.2660*** | −2.7008*** | 0.2364*** | −2.2694*** | 0.1259* | −2.8618*** | 0.0785 |
| USlargegrowth | −2.6385*** | 0.1472** | −2.0243*** | 0.2421*** | −0.9392 | 0.2351*** | −1.3944** | 0.1900*** |
| USlargevalue | −1.2823*** | 0.2819*** | −1.5891*** | 0.1575 | −1.2142** | 0.1176* | −2.0041*** | 0.037 |
| USmidcap | −2.1359*** | 0.1582** | −1.9125*** | 0.2656*** | −2.0007*** | 0.1896** | −2.3070*** | 0.1563** |
| _cons | 0.6169 | 0.9160*** | −1.7682 | 1.6383*** | 0.8999 | 1.1514*** | −8.1729** | 0.2718 |
| N | 790 | 852 | 529 | 562 | 172 | 182 | 161 | 170 |
| r2 | 0.0003 | 0.007 | 0.0756 | 0.4327 | 0.1203 | 0.1195 | 0.2466 | 0.2983 |

This table reports the coefficients for Panel Data models for the Alpha and Sharpe performance measures after 12 and 36 months. Gold, Silver and Bronze are dummies to control the medal of a fund obtained from Morningstar Analyst Rating. ManagerTen, the manager tenure, Netexp, the net expense ratio and LogSize, the logarithm of the assets of the fund, are control variables. Finally, USlargeblend, USlargegrowth, USlargevalue and USmidcap are dummies to control for categories. N is the number of observations, r2 is a measure of the goodness of fit. *Significant at 10%; ** significant at 5% and *** significant at 1%

**Table 5** Pillars and out-of-sample performance after 12 and 36 months

| Variable | Alpha 12 | Sharpe 12 | Alpha C-12 | Sharpe C-12 | Alpha 36 | Sharpe 36 | Alpha C-36 | Sharpe C-36 |
|---|---|---|---|---|---|---|---|---|
| Parentpos | 0.5001 | 0.0688* | −0.0211 | −0.0089 | 0.21 | 0.0294 | −0.1735 | −0.0307 |
| Peoplepos | −0.543 | −0.0881 | −0.1532 | −0.0304 | 0.1531 | 0.0431 | −0.0274 | 0.007 |
| Processpos s | −0.284 | −0.0275 | 0.3184 | 0.0293 | −0.4298 | −0.1147* | 0.4842 | 0.0356 |
| USlargebl end | −2.1777*** | 0.2538*** | −2.873*** | 0.1849*** | −2.5693*** | 0.0906 | −3.2014*** | 0.0239 |
| USlargegr owth | −1.8044*** | 0.2255*** | −1.94*** | 0.2335*** | −1.1939 | 0.1937** | −1.4232** | 0.1829** |
| USlargeva lue | −1.1168* | 0.2754*** | −1.7632*** | 0.1844** | −1.2380* | 0.1343 | −2.2574*** | 0.0195 |
| USmidcap | −1.5612** | 0.2857*** | −1.6983** | 0.2659*** | −1.7527** | 0.1782* | −1.9586** | 0.1650* |
| Netexp_ | – | – | −0.9361* | −0.1591*** | – | – | −0.8113 | −0.1491** |
| logSize | – | – | 0.4038** | 0.0470** | – | – | 0.6220*** | 0.0736*** |
| ManagerT en_ | – | – | −0.0088** | −0.0009** | – | – | −0.0095* | −0.0010* |
| cons | 3.3915*** | 0.3681*** | 0.01 | 0.01 | 1.1183 | 1.1884*** | −11.0519** | −0.2456 |
| N | 527 | 565 | 478 | 493 | 130 | 130 | 123 | 123 |
| r2 | 0.1586 | 0.8135 | 0.1647 | 0.8243 | 0.1219 | 0.0974 | 0.2539 | 0.3118 |

This table reports the coefficients for Panel Data models for the Alpha and Sharpe performance measures after 12 and 36 months. Parentpos, Peoplepos, Processpos are dummies which take the value 1 in case the Pillars Parent, People and Process of the Morningstar Analyst Rating is positive, 0 otherwise. ManagerTen, the manager tenure, Netexp, the net expense ratio and LogSize, the logarithm of the assets of the fund, are control variables. Finally, USlargeblend, USlargegrowth, USlargevalue and USmidcap are dummies to control for categories. N is the number of observations, r2 is a measure of the goodness of fit of the model. *Significant at 10%; ** significant at 5% and *** significant at 1%

**Table 6** Quantile regression (Out of sample performance after 12 months)

| Variable | Alpha-12 | Sharpe-12 | AlphaC-12 | SharpeC-12 |
|---|---|---|---|---|
| q25 | | | | |
| Gold | 0.7348 | 0.1542** | −0.7879 | 0.0494 |
| Silver | −1.3122** | −0.0175 | −1.4756** | −0.0429 |
| Bronze | −0.3791 | 0.0083 | −0.0466 | 0.0386 |
| Netexp_ | − | − | −2.6220*** | −0.1997*** |
| logSize | − | − | 0.2811* | 0.0165 |
| ManagerTen | − | − | −0.0056 | −0.0008 |
| cons | −3.9877*** | 0.0974 | −4.4653 | 1.2791** |
| q50 | | | | |
| Gold | 0.1641 | 0.0577 | −0.4749 | 0.0068 |
| Silver | −0.3934 | 0.0066 | 0.066 | 0.0096 |
| Bronze | −0.1826 | −0.0113 | 0.3432 | 0.0671 |
| Netexp_ | − | − | −0.4713 | −0.1107 |
| logSize | − | − | 0.4343*** | 0.0399 |
| ManagerTen | − | − | −0.0104** | −0.001 |
| _cons | −0.3282 | 0.4845*** | −7.6554** | 1.2672* |
| q75 | | | | |
| Gold | 1.1570** | 0.0809 | 0.8035 | −0.0494 |
| Silver | 1.1729** | 0.0853 | 1.0237 | 0.0723 |
| Bronze | 0.4998 | 0.0378 | 0.6718 | 0.0846* |
| Netexp_ | − | − | −0.2892 | −0.1315 |
| logSize | − | − | 0.1479 | 0.0543** |
| ManagerTen | − | − | −0.0025 | −0.0014*** |
| _cons | 2.4698*** | 0.6826*** | −0.6839 | 1.5432** |
| N | 790 | 852 | 529 | 562 |

This table reports the coefficients for Quantile regression. Alpha12 is the beta-adjusted return over a one- year period; Sharpe12 is the yearly risk-adjusted return. Gold, Silver and Bronze are dummies to control the medal of a fund obtained from Morningstar Analyst Rating. ManagerTen, the manager tenure, Netexp, the net expense ratio and LogSize, the logarithm of the assets of the fund, are control variables. N is the number of observations. *Significant at 10%; ** significant at 5% and *** significant at 1%

Assuming that:

$$Quant_\phi(y_i|X_i)$$
$$X'\beta_\phi$$
$$Quant_\phi(u_{\phi i}|X_i) = 0$$

As can be seen in Table 6, the results of the quantile regression show that in general the signs are not significant for most of the ratings, with gold being the only ones that outperform those not recommended in quartile 25 and 75. When included the management costs, the gold ratings that are significant, cease to be, and it is therefore reasonable to think that both the size of the funds and the costs can explain the differences in performance more than analyst ratings.

One reason why analyst ratings were not very significant in previous analysis might be because "Morningstar analyst rating is a qualitative, forward-looking measure that reflects the analyst's expectation of the future performance relative to its peers over a business cycle" and thus the analysis is only coherent when we compare peers, or the same categories. Thus, in this section we try to make the analysis only for large-scale categories because they have enough number of mutual funds. As can be seen in the table, the analysis by categories shows that gold funds outperform neutral or negative funds in the 12-month term and partially in the 36-month term in the case of the large-blend category. The same happens in the Growth category, where gold funds return again outperform to those not recommended in most of the metrics, terms and including control variables. This capacity is not observed in the case of Value, where most of the signs are positive but not significant. Therefore, gold ratings allow us to identify funds that exceed their peers, mainly for a term of one year, but not in all categories. Comparing the results with the main model we observe that, in general, for some categories we obtain that the gold ratings are more significant, but in general, very few differences were obtained.

## 5 Conclusion

Many investors select their investments in mutual funds based on quantitative rating. However, the selection of funds based exclusively on quantitative ratings, excludes a set of qualitative factors that can explain future performance. Morningstar has two systems to classify mutual funds based on qualitative aspects: Morningstar Analyst Rating and Morningstar Stewardship Grade. Morningstar Analyst Ratings are forward-looking qualitative and quantitative analyses of mutual fund about five pillars: Process, Performance, People, Parent and Price, that includes factors such as cost, past performance, quality of management, interest alignment, etc. Morningstar Analyst Ratings are based on convictions that funds will outperform their benchmarks over the long term. Although there are several studies on quantitative ratings, Analyst Ratings are not that popular and little research has been done on them. In this paper, we assess in what extent selecting mutual funds is based on Morningstar Analyst Ratings. In particular, we attempt to answer whether good analyst ratings outperform non-recommended ones in the short (12 month) and long term (36 month) and whether it is useful to combine both ratings in the screening process to identify good future performers. The data was collected from Morningstar Direct database covering the period August 2012 to August 2016. We selected US equity funds with the previously mentioned ratings. Our findings support the ability of Gold ratings to select funds that will perform better in terms of future performance. Our results are in line with previous empirical evidence found in [13, 14] and [15] but we have found little evidence that, on average, funds with a better Analyst Rating (Gold) have a better performance in terms of risk adjusted measures (alpha and Sharpe). Predictability is observed in several analyses done after one year, but not at the end of three-years. This evidence is more relevant in the case of the analysis done by investment style category. When analyzing whether the pillars on which analysts' ratings are broken down, we find no evidence that future performance is related to any of these specific dimensions. The inclusion of other variables such as costs, size and manager tenure reflect the importance of considering other variables for fund´s selection. Nevertheless, in many estimations, Gold ratings are still significant in explaining performance, indicating that costs are not the only factor determining the predictive power of qualitative ratings. Our results support the use of qualitative ratings in the investment fund selection process, accompanied by other variables.

**Funding Information** This paper is financed by National Funds of the FCT – Portuguese Foundation for Science and Technology within the project UIDB/03182/2020.

# References

1. Blake C, Morey M (2000) Morningstar ratings and mutual fund performance. J Finan Quant Anal 35(3):451–483
2. Del Guercio D, Tack PA (2008) Star power: the effect of morningstar ratings on mutual fund flow
3. Morey M, Gottesman A (2006) Morningstar mutual fund ratings redux. J Investment Consult 8(1):25–37
4. Müller S, Weber M (2014) Evaluating the rating of stiftung warentest: how good are the ratings and can they be improved? Eur Financ Manag 20(2):207–235
5. Meinhardt C (2014) Ratings and performance of german mutual funds—a comparison of feri trust, finanztest and fondsnote. Available at SSRN: http://ssrn.com/abstract=1943577
6. Otero L, Duran-Santomil P, Correia-Domingues R (2020) Do investors obtain better results selecting mutual funds through quantitative ratings? Spanish J Finan Account 49(3):265–291. https://doi.org/10.1080/02102412.2019.1622066
7. Wellman JW, Zhou J (2008) Corporate governance and mutual fund performance: a first look at the morningstar stewardship grades. Available at SSRN: https://ssrn.com/abstract=1107841
8. Ng WS (2009) Does morningstar shine in the universe of mutual funds? A study on morningstar mutual fund ratings, Singapore Management University Master Thesis, Available at: http://ink.library.smu.edu.sg/etd_coll/10
9. Lai S, Tiwari A, Zhang Z (2010) Mutual fund flows, performance persistence, and board quality. Available at SSRN: https://ssrn.com/abstract=1570525
10. Chen CR, Huang Y (2011) Mutual fund governance and performance: a quantile regression analysis of morningstar's stewardship grade. Corp Governance: Int Rev 19(4):311–333
11. Gottesmam A, Morey M (2012) Mutual Fund corporate culture and performance. Rev Finan Econ 21(2):69–81
12. Cao XJ, Ghosh A, Goh CYJ, Ng WS (2014) Governance matter: morningstar stewardship grades and mutual fund performance. Research Collection School of Economics. Available at: http://ink.library.smu.edu.sg/cgi/viewcontent.cgi?article=2718&context=soe_research

13. Kamal R (2013) Can morningstar analyst ratings predict fund performance? J Appl Bus Res 29(6):1665–1672
14. Armstrong W, Genc E, Verbeek M (2019) Going for gold: an analysis of morningstar analyst ratings. Manage Sci 66(5):1949–2443
15. Otero-González L, Durán-Santomil P (2021) Is quantitative and qualitative information relevant for choosing mutual funds? J Bus Res 123:476–488
16. Tufano P, Sevick M (1997) Board structure and fee-setting in the US mutual fund industry. J Financ Econ 46:312–335
17. Del Guercio D, Dann LY, Partch MM (2003) Governance and boards of directors in closed-end investment companies. J Financ Econ 69:111–152
18. Qian M (2006) Whom can you trust? A Study on Mutual Fund Governance (October 2006). Available at SSRN: https://ssrn.com/abstract=685543
19. Meschke F (2007) An empirical examination of mutual fund boards. Available at SSRN: http://ssrn.com/abstract=676901
20. Khorana A, Servaes H, Meschke F (2007) An empirical examination of mutual fund boards. Available at SSRN: http://ssrn.com/abstract=676901
21. Ferris SP, Xuemin YS (2007) Do independent directors and chairmen really matter? the role of boards of directors in mutual fund governance. J Corp Finan 13:392–420
22. Boyd DP, Yilmaz M (2007) Stewardship as a factor in the financial performance of mutual funds. J Bus Econ Res 5(3):11–17
23. Gerrans P (2006) Morningstar ratings and future performance. Account Finan 46(4):605–628
24. Trahan EA (2008) Mutual fund governance and fund performance. Corp Ownersh Control 5(4):384–392
25. Evans AL (2008) Portfolio manager ownership and mutual fund performance. Financ Manage 37:513–534
26. Kong SX, Tang DY (2008) Unitary boards and mutual fund governance. J Finan Res 31:193–224
27. Cremers M, Driessen J, Maenhout P, Weinbaum D (2009) Does skin in the game matter? Director incentives and governance in the mutual fund industry. J Finan Quant Anal 44:1345–1373
28. Adams JC, Mansi SA, Nishikawa T (2010) Internal governance mechanisms and operational performance: evidence from index mutual funds. Rev Finan Stud 23(3):1261–1286
29. Chou W-H, Ng L, Wang Q (2011) Do governance mechanisms matter for mutual funds? J Corp Finan 17:1254–1271
30. Ding B, Wermers R (2012) Mutual fund performance and governance structure: the role of portfolio managers and boards of directors. Available at SSRN: https://ssrn.com/abstract=2207229
31. Hazenberg ZZ (2012) Investment fund governance: an empirical investigation of Luxembourg UCITS. Thesis at University of Luxembourg. Available at http://orbilu.uni.lu/handle/10993/15562
32. Kryzanowski L, Mohebshahedin M (2016) Board governance, monetary interest, and closed—endfund performance. J Corp Finan 38:196–217
33. Mamatzakis E, Xu B (2017) Does corporate governance matter in fund management company: the case of china, Available at https://mpra.ub.unimuenchen.de/76138/1/MPRA_paper_76138.pdf
34. Koenker R, Basset G (1978) Regression quantiles. Econometrica 46(1):33–50
35. Efron B (1979) Bootstrap methods: another look at the jackknife. Ann Stat 7(1):1–26

# ICT, Financial Development and Carbon Emissions in Sub-Saharan African Countries

Damilola Felix Eluyela, Uwalomwa Uwuigbe, and Francis O. Iyoha

**Abstract** This research study analyzes the moderating effect of information and communication technology and the role of financial development on carbon emissions intensity for sub-Saharan African nations between 1996 to 2019. The econometric approach employed is the generalized method of moments (GMM). ICT and financial development play a substantial role on carbon emissions intensity, according to empirical data. This means that a rise in ICT and financial growth will raise carbon emissions by 0.013 and 0.002%, respectively, for internet access and mobile phone subscriptions. The study proposes that the government and financial service providers in the SSA area encourage financial incentives for green technology adoption, which will reduce carbon emissions intensity.

**Keywords** Carbon emissions · Financial development · Internet access · Mobile phone subscription · Generalised method of moments (GMM)

## 1 Introduction

Information and communication technology and financial growth have boomed due to globalization, which is one of its distinctive characteristics [11]. The influence of ICT expansion on many factors has been studied from a variety of viewpoints in the developing literature. Unquestionably, information and communication technology (ICT) has a two-fold perspective. First, ICT is suggested as a catalyst for industrial development, which by extension stimulates economic development. On the flip hand,

D. F. Eluyela (✉)
Landmark University, Omu-Aran, Kwara State, Nigeria
e-mail: eluyela.damilola@lmu.edu.ng

U. Uwuigbe · F. O. Iyoha
Covenant University, Ota, Nigeria
e-mail: uwalomwa.uwuigbe@covenantuniversity.edu.ng

F. O. Iyoha
e-mail: francis.iyoha@covenantuniversity.edu.ng

© The Author(s), under exclusive license to Springer Nature Switzerland AG 2022
S. G. Yaseen (ed.), *Digital Economy, Business Analytics, and Big Data Analytics Applications*, Studies in Computational Intelligence 1010,
https://doi.org/10.1007/978-3-031-05258-3_42

ICT could also have a negative effect on the environment due to increased industrialization. Industrialization is explicitly dependent on the increased consumption of energy, which results in the degradation of the environment and a negative effect on humanity. The connection between $CO_2$ emissions and levels of financial development has become the subject of a broad range of studies in recent years. Established, are two conflicting perspectives on the linkage that exists between $CO_2$ emissions and financial expansion. They claim that economic growth may considerably lower $CO_2$ emissions. Via the systems of foreign direct investment and growth, financial development, according to previous research, may likely contribute to a considerable increase in $CO_2$ emissions. According to this viewpoint, financial development can draw foreign direct investments and higher research and development investments, reduced environmental degradation and resulting in increased economic growth. Among the methods through which ICT influences carbon emissions are financial development and economic expansion, according to [30]. Financial development is viewed as a crucial component of economic development's implementation. A nexus between the deficit and surplus sectors of a country's economy is the financial sector. Growth in finances increases fund mobilization, use of funds and tracking [22]. Also, ICT could help the financial development sector grow even faster. For example, internet use promotes investment practices, lowers bank loan costs, and can improve trading, resource allocation, and resource monitoring. This is similar to how the financial sector promotes businesses and industries to invest in new, environmentally sustainable technologies [5, 17].

The second viewpoint contends that a mature, financially developed system could increase $CO_2$ emissions. Financial development, according to studies, can amplify levels of environmental degradation by industrialization and energy consumption channels [1, 2]. These findings claim that as a result of financial growth, households are more readily able to obtain credit, purchasing high energy-consuming home gadgets and automobiles, culminating in increased usage of energy and $CO_2$ emissions [23, 31]. Although there are substantial connections between $CO_2$ emissions and financial development, there have been very few observational studies on the influence of financial development on the levels of carbon emissions particularly in sub-Saharan African nations. The majority of recent study in industrialized nations has focused on Latin America and Asia [9, 10, 14, 19, 20].

Most recent research on this topic has focused primarily either on the causal link between $CO_2$ emissions and financial development or on the connection between energy use and $CO_2$ emissions. Therefore, our study seeks to fill this lacuna by evaluating the moderating effect of ICT and financial growth on $CO_2$ emissions in forty-five SSA countries from 1996 to 2019. So, we looked at the net effect of ICT and financial development interactions on $CO_2$ emissions. This research adds to the corpus of knowledge in a variety of ways: This study is the pioneer work that has examined the influence of ICT and financial development on carbon emissions, according to our understanding. Interaction calculated coefficients result in marginal or conditional effects, as well as an unconditional impact. The net impacts of financial development and ICT on levels of carbon emissions are calculated based on earlier studies [7].

## 2 Literature Review

There have been several studies conducted especially in recent years that have looked at the link between $CO_2$ emissions and financial development. Several research works have discovered a negative correlation between levels of financial development and extent of $CO_2$ emissions, whereas others have discovered a clear, substantial positive correlation. In the middle of both extremes, studies have been unable to uncover any significant connection between extent of $CO_2$ emissions and levels of financial development. When it comes to BRIC nations, [29] utilized panel data analysis to examine if financial development and greater economics lead to environmental degradation. According to their findings, financial development brought about decreases in carbon emissions levels in the nations examined. [16] utilized the Autoregressive Distributed Lag (ARDL) bounds testing approach to understanding the effect of growth, electricity, and financial development on the environment in China. There was less pollution as a result of financial development. While studying the effect of financial development, coal use, trade openness and economic growth on the levels of $CO_2$ emissions in South Africa, [27] revealed that financial development has the potential to minimize carbon emissions.

In twelve Middle Eastern and North African nations, [20] investigated the relationship that exists between financial development levels and extent of carbon emissions (MENA). Deploying the simultaneous-equation panel data model, they discovered that greater financial development levels might cut carbon emissions through stimulating technological breakthroughs and increasing investment in conservation of energy research and development. Some research work pertaining to the BRICS context has also found that internet usage, which is the proxy for ICT, has led to increases in carbon emissions. As a result, the consistency of the atmosphere suffers [8, 15]. Studies focusing on the Asian region came up with similar findings [18, 19]. ICT might be utilized to offset the possibly negative impacts of globalisation on environmental deterioration, according to [8]. Analyzing panel data from 1990 to 2015 for twenty emerging nations Per [25], an increase in ICT use of 1% translates with a 0.16% rise in carbon dioxide emissions. In an analysis of industrial expansion and financial development's effect on the quality of the environment, [33] discovered that financial development might help decrease carbon emissions in the nations they investigated. Others [3, 4, 10, 19, 27, 28] have shown a positive link between extent of $CO_2$ emissions and the levels of financial development in addition to the research mentioned above.

## 3 Research Methods

In this study, forty-five (45) SSA nations' panel data from 1996–2019 are used [13, 21]. Using the panel data model, carbon emissions intensity is utilized as the dependent variable [6, 7]. ICT and levels of financial development are the main

**Table 1** Variables Measurement

| Variables | Signs | Net inflows (% of GDP) |
|---|---|---|
| Carbon emissions intensity | $CO_2IT$ | $CO_2$ emissions per GDP |
| Internet access | INT | Internet penetration (per 100 people) |
| Mobile phones | MOB | Mobile penetration subscription (per 100 people) |
| Financial development | FIN | Domestic credit to the private sector (% of GDP) |
| Gross domestic product | GDP | GDP growth rate (annual %) |
| Population | POP | Population growth rate (annual %) |
| Foreign direct investment | FDI | Net inflows (% of GDP) |

explanatory factors. Internet access and mobile phone subscriptions serve as proxies for information and communication technology. Financing development is measured by domestic credit (% of GDP). The data was sourced from WDI (World Development Indicators report). For stability of the result, all data obtained were converted to their natural logarithm forms. The measurement for all variables is presented in Table 1. A two-step system GMM model was used as the primary econometric estimation. This model was adopted from the study of [7]. Unlike the one-step procedure of GMM, which only controls for the existence of homoscedasticity, the two-step GMM model approach handles the issues pertaining to heteroscedasticity.

## 3.1 Model Specification

To comprehensively evaluate and assess the moderating effect and role of ICT and financial development is on carbon emissions in selected sub-Saharan African countries, we used the model of [7]. The implicit and functional model is stated in Eq. (1).

$$y = f(ICT, FIN, ICT * FIN, W) \tag{1}$$

where y is the dependent variable representing $CO_2$ emission intensity, ICT is proxied by internet access and mobile phone subscription. FIN represents financial development, ICT*ENG denotes the interactions between the ICT penetration and financial development; $W$ represents the vector of the control variables (Foreign direct investment, GDP growth, and population growth). The implicit model in Eq. (1) is converted to the following equations in level and first difference, as shown in Eqs. (2) and (3), respectively. This summarizes the standard system GMM estimation technique.

$$CO_{2_{i,t}} = \sigma_0 + \sigma_1 CO_{i,t-\tau} + \sigma_2 ICT_{i,t} + \sigma_3 FN_{i,t}$$

$$+ \sigma_4 ICT^* FIN_{i,t} + \sum_{i=1}^{2} \delta_h W_{hi,t-\tau} + \eta_t + \xi_t + \varepsilon_{it} \tag{2}$$

$$CO_{2_{i,t}} - CO_{2_{i,t-\tau}} = \begin{bmatrix} \sigma_1 \left( CO_{i,t-\tau} - CO_{i,t-2\tau} \right) + \sigma_2 \left( ICT_{i,t} - ICT_{i,t-\tau} \right) \\ + \sigma_3 \left( FIN_{i,t} - FIN_{i,t-\tau} \right) \\ + \sigma_4 \left( ICT^* FIN_{i,t} - ICT^* FIN_{i,t-\tau} \right) \\ + \sum_{h=1}^{2} \delta_{hi,t-\tau} \left( W_{hi,t-\tau} - W_{hi,t-2\tau} \right) + \left( \xi_t - \xi_{i,t-\tau} \right) + \left( \varepsilon_{i,t} - \varepsilon_{i,t-\tau} \right) \end{bmatrix} \tag{3}$$

where $CO_{2i,t}$ is $CO_2$ emission intensity of the country i in period t, $\sigma_0$ is constant, ICT represents the technology penetration indicators (i.e., internet penetration subscription (per 100 people); mobile cellular subscription (per 100 people), FIN represent financial development (measured by domestic credit to the private sector- a percentage of GDP); ICT*FIN denotes the interaction between the ICT and financial development; W represents the vector of the control variables (population growth, GDP growth, and foreign direct investment), $\tau$ stands for the coefficient of autoregression which is one as explained within this study's framework and that is because a year lag possesses the ability to completely take into consideration, past information, $\xi_t$ represents the time-specific constant, $\eta_t$ denotes the country-specific effect, and finally $\varepsilon_{i,t}$ denotes the stochastic term.

## 4 Results and Discussion

Carbon emissions intensity ($CO_2IT$) is used as a dependent variable in the system GMM results shown in Table 2. Columns 1 and 2 are dedicated to internet access and cell subscription regressions, respectively. The specification diagnostics indicate no indication of second-order autocorrelation due to the non-significance of the AR (2) coefficients. As a result of the Hansen statistics, the instruments set is not over-identified. And the number of instruments (27) is significantly fewer than the total number of groupings (45). A viable GMM model is confirmed. The moderating effect of ICT and financial development (ICT x FIN) on carbon emissions intensity ($CO_2IT$) shows a significant positive relationship.

A shift in the moderating role of ICT and that of financial development will result in 0.002% and 0.013% rise in carbon emissions intensity ($CO_2IT$) for selected sub-Saharan African nations. Ceteris paribus, this is significant and positive at a 1% probability level (holding all other regressors in the model constant). As a result, the moderating influence of ICT, and financial development on carbon emissions intensity ($CO_2IT$) displays a positive connection. This suggests that ICT and financial development (FIN) would raise the levels of carbon emissions intensity ($CO_2IT$) for selected sub-Saharan African nations. Net impacts are estimated to analyze the total

**Table 2** GMM Result on ICT, Financial Development and $CO_2$ Emissions Intensity

| | Dependent Variable: $CO_2$ Intensity | | | | | |
| --- | --- | --- | --- | --- | --- | --- |
| | Internet Access | | | Mobile Phone Penetration | | |
| | INT | MOB | FIN | INT | MOB | FIN |
| , .$CO_2$IT | 0.925*** | 1.014*** | 0.920*** | 0.928*** | 1.016*** | 0.933*** |
| | (0.00165) | (0.000754) | (0.00249) | (0.00116) | (0.000483) | (0.00131) |
| FIN | | | 0.0120*** | | | 0.00747*** |
| | | | (0.00196) | | | (0.00181) |
| INT | − 0.00324** | 0.00740*** | 0.0184*** | | | |
| | (0.00130) | (0.00241) | (0.00316) | | | |
| MOB | | | | − 0.00164*** | 0.00132*** | 0.00189*** |
| | | | | (0.000551) | (0.000371) | (0.000540) |
| FINxINT | | | 0.00013*** | | | |
| | | | (0.0000225) | | | |
| FINxMOB | | | | | | 0.00002*** |
| | | | | | | (0.00000498) |
| GDP | − 0.0421 | − 0.358*** | − 1.831*** | 0.642*** | − 0.187*** | − 0.822*** |
| | (0.158) | (0.0968) | (0.343) | (0.117) | (0.0447) | (0.234) |
| POP | − 0.789*** | 0.202*** | 0.638*** | − 1.660*** | 0.131*** | − 0.0161 |
| | (0.283) | (0.0510) | (0.213) | (0.143) | (0.0380) | (0.183) |
| FDI | − 0.0110** | − 0.00255 | − 0.0181** | − 0.00648*** | 0.000607 | − 0.00325 |
| | (0.00449) | (0.00172) | (0.00700) | (0.00140) | (0.000907) | (0.00388) |
| Constant | 6.490*** | 1.122* | 8.420*** | 7.836*** | 0.459 | 5.926*** |
| | (1.763) | (0.586) | (2.099) | (0.825) | (0.346) | (1.381) |
| Net Effects | – | – | 0.0159 | – | – | – |
| AR(1)_P-value | 0.303 | 0.00557 | 0.303 | 0.303 | 0.0115 | 0.303 |
| AR(2)_P-value | 0.381 | 0.743 | 0.364 | 0.375 | 0.727 | 0.367 |
| Hansen P-value | 0.452 | 0.217 | 0.682 | 0.205 | 0.133 | 0.732 |
| Fisher | 1,524,000 | 6,418,000 | 151,728 | 1,740,000 | 8,630,000 | 369,281 |
| Number of Groups | 45 | 45 | 45 | 45 | 45 | 45 |
| No. of Instruments | 27 | 27 | 27 | 27 | 27 | 27 |
| Observations | 531 | 525 | 508 | 591 | 578 | 560 |

Standard errors in parentheses. *** $p < 0.01$, ** $p < 0.05$, * $p < 0.1$

effect of the relationship between factors of ICT, and financial development levels in $CO_2$ emissions intensity. The net impact of the role of ICT in mitigating the impact of financial development levels on the extent of carbon emissions intensity ($CO_2IT$) is 0.0159 [(19.591*0.00013) + (0.0184)]. In the computation, 644.355 is the mean value of financial development (see appendix 1); the unconditional effect of internet access is 0.0184, while the conditional impact of the interaction between financial development and internet access is 0.00013. This suggests that financial growth and internet availability have a beneficial influence on carbon emissions intensity ($CO_2IT$). The interplay between internet access and financial development will raise carbon emission intensity ($CO_2IT$).

# 5 Conclusion, Recommendations, and Policy Implications

The effect of ICT and financial sector growth in reducing carbon emissions in selected SSA nations was examined in this study. It looked at 45 SSA nations between 1996 and 2019. ICT factors such as internet and mobile phone subscriptions were examined. As a proxy for financial development, domestic credit to the private sector is used as a percentage of GDP. In conclusion, the GMM estimate technique findings suggest that an increase in the interaction role of internet access and financial development would result in a 0.013% rise in carbon emissions intensity for chosen SSA nations.

Moreover, the positive effects of mobile phone subscriptions are lesser than those of internet access. A percentage increase in moderating the effect of mobile phone subscription and financial development will increase carbon emissions intensity by 0.002%. The overall net effects also show a positive and significant output. The net effects of ICT in moderating the effect of financial development levels on the measure of carbon emissions intensity is 0.0159. This implies that a percentage increase in the net impact of ICT role and financial development levels will bring about a 1.59% rise in the carbon emissions intensity. This result is consistent with prior literature, where [25] found that a rise of 1% in ICT consumption culminates in a 0.16% increase in $CO_2$ emissions. The empirical findings of this study recommend the following policy implications. Government and financial service providers in the SSA region should promote financial incentives for green technology adoption, reducing carbon emissions intensity. To mitigate and reduce the high rate of carbon emissions in the SSA area, policymakers need to concentrate more on tax policies and the development of alternative business models, as well as enabling new ICT equipment services and operations. This clever program will offer revolutionary improvements to smartphone users such as online banking, shopping online, and online purchasing. These intelligent applications will minimize people's travel and the transportation of products, which will reduce carbon emissions. There are several limitations to this research. First, another proxy for ICT, such as fixed broadband, may be used to assess the validity of the findings. In addition, globalization, commerce, and levels of institutional quality may be incorporated into the model in order to provide policymakers greater information.

# References

1. Abbasi F, Riaz K (2016) CO2 emissions and financial development in an emerging economy: an augmented VAR approach. Energy Policy 90:102–114
2. Acheampong AO (2019) Modeling for insight: does financial development improve environmental quality? Energy Econ 83:156–179
3. Ali HS, Law SH, Lin WL, Yusop Z, Chin L, Bare UAA (2019) Financial development and carbon dioxide emissions in Nigeria: evidence from the ARDL bounds approach. Geo J 84(3):641–655
4. Al-Mulali U, Ozturk I, Lean HH (2015) The influence of economic growth, urbanization, trade openness, financial development, and renewable energy on pollution in Europe. Nat Hazards 79(1):621–644
5. Andrianaivo M, Kpodar K (2012) Mobile phones, financial inclusion, and growth. Rev Econ Inst 3(2):1–30
6. Apergis N, Payne JE, Rayos-Velazquez M (2020) Carbon dioxide emissions intensity convergence: evidence from central American countries. Frontiers Energy Res 7(158):1–7. https://doi.org/10.3389/fenrg.2019.00158
7. Asongu S (2018) ICT, openness and $CO_2$ emissions in Africa. Environ Sci Pollut Res 25(10):9351–9359. https://doi.org/10.1007/s11356-018-1239-4
8. Balsalobre-Lorente D, Driha OM, Bekun FV, Osundina OA (2019) Do agricultural activities induce carbon emissions? the BRICS experience. Environ Sci Pollut Res 26(24):25218–25234
9. Bekhet HA, Matar A, Yasmin T (2017) CO2 emissions, energy consumption, economic growth, and financial development in GCC countries: dynamic simultaneous equation models. Renew Sustain Energy Rev 70:117–132
10. Cetin M, Ecevit E, Yucel AG (2018) The impact of economic growth, energy consumption, trade openness, and financial development on carbon emissions: empirical evidence from Turkey. Environ Sci Pollut Res 25(36):36589–36603
11. Chavanne X, Schinella S, Marquet D, Frangi JP, Le Masson S (2015) Electricity consumption of telecommunication equipment to achieve a tele-meeting. Appl Energy 137:273–281
12. Dogan E, Turkekul B (2016) $CO_2$ emissions, real output, energy consumption, trade, urbanization, and financial development: testing the EKC hypothesis for the USA. Environ Sci Pollut Res 23(2):1203–1213
13. Eluyela DF, Asaleye AJ, Popoola O, Lawal AI, Inegbedion H (2020) Grey directors, corporate governance and firms' performance nexus: evidence from Nigeria. Cogent Econ Financ 8(1):1–16. https://doi.org/10.1080/23322039.2020.1815962
14. Hao Y, Zhang ZY, Liao H, Wei YM, Wang S (2016) Is $CO_2$ emission a side effect of financial development? an empirical analysis for China. Environ Sci Pollut Res 23(20):21041–21057
15. Haseeb A, Xia E, Saud S, Ahmad A, Khurshid H (2019) Does information and communication technologies improve environmental quality in the era of globalization? an Empirical analysis. Environ Sci Pollut Res 26(9):8594–8608
16. Jalil A, Feridun M (2011) The impact of growth, energy and financial development on the environment in China: a cointegration analysis. Energy Econ 33(2):284–291
17. Latif Z, Yang M, Pathan ZH, Jan N (2017) Challenges and prospects of ICT and trade development in Asia. Hum Syst Manag 36(3):211–219
18. Lee JW, Brahmasrene T (2014) ICT, CO2 emissions and economic growth: evidence from a panel of ASEAN. Glob Econ Rev 43(2):93–109
19. Lu WC (2018) The impacts of information and communication technology, energy consumption, financial development, and economic growth on carbon dioxide emissions in 12 Asian countries. Mitig Adapt Strat Glob Change 23(8):1351–1365

20. Omri A, Daly S, Rault C, Chaibi A (2015) Financial development, environmental quality, trade, and economic growth: what causes what in MENA countries. Energy Econ 48:242–252

21. Ozordi E, Eluyela DF, Uwuigbe U, Uwuigbe OR, Nwaze CE (2020) Gender diversity and sustainability responsiveness: evidence from Nigerian fixed money deposit banks. Problems Perspect Manag 18(1):119–129. https://doi.org/10.21511/ppm.18(1).2020.11

22. Raheem ID, Oyinlola MA (2015) Financial development, inflation, and growth in selected West African countries. Int J Sustain Econ 7(2):91–99

23. Sadorsky P (2010) The impact of financial development on energy consumption in emerging economies. Energy Policy 38(5):2528–2535

24. Saidi K, Mbarek MB (2017) The impact of income, trade, urbanization, and financial development on $CO_2$ emissions in 19 emerging economies. Environ Sci Pollut Res 24(14):12748–12757

25. Salahuddin M, Alam K, Ozturk I (2016) The effects of Internet usage and economic growth on CO2 emissions in OECD countries: a panel investigation. Renew Sustain Energy Rev 62:1226–1235

26. Shahbaz M, Hye QMA, Tiwari AK, Leitão NC (2013) Economic growth, energy consumption, financial development, international trade, and $CO_2$ emissions in Indonesia. Renew Sustain Energy Rev 25:109–121

27. Shahbaz M, Shahzad SJH, Ahmad N, Alam S (2016) Financial development and environmental quality: the way forward. Energy Policy 98:353–364

28. Tamazian A, Rao BB (2010) Do economic, financial, and institutional developments matter for environmental degradation? Evidence from transitional economies. Energy Econ 32(1):137–145

29. Tamazian A, Chousa JP, Vadlamannati KC (2009) Does higher economic and financial development lead to environmental degradation: evidence from BRIC countries. Energy Policy 37(1):246–253

30. Tsaurai K, Chimbo B (2019) The impact of information and communication technology on carbon emissions in emerging markets. Int J Energy Econ Policy 9(4):320

31. Xing T, Jiang Q, Ma X (2017) To facilitate or curb? the role of financial development in China's carbon emissions reduction process: a novel approach. Int J Environ Res Public Health 14(10):1222

32. Xiong L, Tu Z, Ju L (2017) Reconciling regional differences in financial development and carbon emissions: a dynamic panel data approach. Energy Procedia 105:2989–2995

33. Zafar MW, Saud S, Hou F (2019) The impact of globalization and financial development on environmental quality: evidence from selected countries in the organization for economic co-operation and development (OECD). Environ Sci Pollut Res 26(13):13246–13262

# Role of Accounting Information Systems (AIS) Applications on Increasing SMES Corporate Social Responsibility (CSR) During COVID 19

**Adel M. Qatawneh and Hamad Kasasbeh**

**Abstract** Main aim was to highlight role of AIS applications in increasing CSR during COVID 19. (132) individuals responded to an online questionnaire. Results indicated that AIS supported CSR among SMEs in Jordan attributed to credibility and reliability of AIS outcomes. In other words, main hypothesis was accepted and there was a positive impact of AIS applications on CSR in the fields of legal, economic, ethical and philanthropy and indicated that organizations were to support the community during the pandemic through organizing their own finances, avoid corruption and support the society through funding, donations and charity works as they were aware of their profits and abilities.

**Keywords** Accounting information systems · Corporate social responsibility · Philanthropy · Donations · Reliability · Covid 19

## 1 Introduction

With the development of organizational practices and the increased amount of data, there appeared a need to regulate, organize and properly store such data in way that guarantees full and accurate access to it when needed [16]. AIS appeared as an approach to organize and regulate financial data in way that enables accountants and specialist to access financial data, store it, restore it and rely on it in the best way possible [19]. In light of these social and economic changes, in addition to the appearance of AIS in various fields and at all levels, there existed no limits to technology except the limits of the capabilities of the human being who uses it [21].

The emergence of the new Corona virus at the end of 2019 and its continuation until the moment led to an increased dependence on technology in all its forms,

A. M. Qatawneh (✉)
Accounting, Faculty of Business, Al-Zaytoonah University of Jordan, Amman, Jordan
e-mail: a.qatawneh@zuj.edu.jo

H. Kasasbeh
Department of Banking and Financial Scenics, Al-Zaytoonah University of Jordan, Amman, Jordan

© The Author(s), under exclusive license to Springer Nature Switzerland AG 2022
S. G. Yaseen (ed.), *Digital Economy, Business Analytics, and Big Data Analytics Applications*, Studies in Computational Intelligence 1010,
https://doi.org/10.1007/978-3-031-05258-3_43

as countries—as a means to control the virus—imposed quarantine and multiple closures on many industrial, commercial, and educational sectors [18].

This has led to the transformation of commercial work to remote work, and education to distance education, and many of the activities that the individual was previously doing have become things that must be done remotely based on technology. Perhaps it is logical to look at accounting practices, which in turn have become remote, and the need for organizations to stand by the community in order to mitigate the severe damage caused by the virus, which was represented in adopting positions of social responsibility for organizations [20].

From that point, current research sought to examine role of accounting information systems (AIS) applications on increasing SMEs corporate social responsibility (CSR) during COVID 19. Researcher adopted variables of CSR that including (economic, legal, ethical and philanthropic) and the influence that might motivate these dimensions when the organization is basing its financial activities on AIS applications.

## 1.1 Literature Review

### 1.1.1 Accounting Information Systems AIS

Accounting is a worldwide language that is used as a language among organizations in the whole world; this language is understandable within organizational environment whether it was governmental, private, industrial, or any different business area [3]. The main concept of AIS applications is to examine, review and plan accounting reports in order to highlight financial performance on the organization and support decision making and taking processes.

AIS are basically a tool that screen, process and analyze accounting and monetary information and data that are utilized by the inner clients of the organization (employees, leaders and managers), and supports them with dependable accounting information that enables them to perform better from a financial perspective [17]. The main idea behind AIS application is based on turning the conventional accounting practices to a technological environment that facilitates the adequacy of financial reports in terms of quality, time and speed [9].

Ali et al. [4] indicated that the move towards adopting AIS applications within different sizes of organizations is attributed to the growth in size, practices, activities and innovation of these organizations; this managed to increase the need for a tool that is higher in speed and more accurate on processing financial data.

The fundamentality of AIS applications before COVID 19 pandemic were used as sub-systems that supports the conventional financial practices of an organizations, but after the pandemic, things changed, and the reliance on AIS applications grew higher due to the lockdown and the need to work remotely [5, 11]. On the other hand, With the hit of COVID 19 pandemic, and the transition to total lockdown, managers depended a lot on AIS in terms of registering expenditures, costs, financial practices and financial lists stored on the cloud [14]. AIS applications in all its forms are designed to be a sort of risk management strategies that appears vividly in hard situations like pandemics, wars, and natural disasters, it is meant to give access to entitled personnel to sensitive data like accounting details, predict financial risks and profit reports [12].

### 1.1.2 Corporate Social Responsibility

Corporate social responsibility is a term that has arisen in a clear and vivid way especially during the previous decade. The main concept of CSR is based on the notion of the fact that organizations are ought to give to the local society through building a social care management approaches that justifies their being among the community [2]. In an academic approach, CSR is nothing but a type of corporate self-regulation infused within the business model. CSR is basically voluntary practices that are adopted by the organization in order to operate in an environment that is economic, social and environmental as the same time [1, 13].

## 1.2 Hypotheses Development

### 1.2.1 Corporate Social Responsibility in Accounting

The notion of CSR is based on the fact that organizations are ought to framework their effect on society by using its assets to support local community, achieve sustainable development, address social issues and improve its image among society members. Adopting CSR means for the organization to be part of the community, present a part of its profits and return to upgrade social status, give back to the community one way or another and support itself through supporting individuals outside its premises [17].

The gathering between CSR and accounting practices came from the increase in the activities and sizes of organizations, this increase managed to throw its implication on society in terms of the financial, economic and social impact [17]. From that point, there appeared the concept of social accounting (disclosure) which required the organization to contribute to achieving welfare of society through improving environmental conditions and reducing negative effects of its activities like pollution, in addition to reaching economic development.

### 1.2.2 Relationship Between AIS and CSR

It is believed that what gathers between AIS applications and CSR is connected to the so-called social accounting, this concept is a branch of accounting science that focuses on controlling organizational activities, study its impact on society and build a contractual relationship derived from the rules of the social contract that combines the interests of both (organization and society) [6]. From another perspective, what gathers between AIS and CSR is the fact that AIS facilitates decision making based on reliable data, so, it defines that interest and requirements of an organization within the society, control its interaction with the community, and framework its relation to the environment it works in [15]. At the same point, authors argued that when we talk about AIS and CSR, we give an indication that the organization commitment won't be limited to protecting interest of stakeholders, but at the same time, protect other members of society, in other words, the whole idea isn't about matching profits with costs, but it goes longer into realizing social costs and benefits, supporting local community and be a part of the society as a whole.

### 1.2.3  AIS Supports Philanthropic Orientations of Organizations

Historically, corporate philanthropy encompassed organizations seeking and identifying charitable groups and donating services, products or money. Today, organizations are strategic giving by closely linking corporate social responsibility and philanthropy effort to their goals and mission. This is achieved by these organizations donating to communities within their business coverage areas. Companies and in particular SMEs are using AIS as a technology to increase this philanthropic work during the Covid-19 period [8]. The economy, through financial and commercial organization entered the sphere of COVID 19 crisis in order to mitigate its severity, such organizations played a human role dictated by the rules and standards of CSR and managed to achieve for itself gains that are not necessarily financial or immediate, but rather gains that can be granted later on [7].

The reliance on accounting information systems in achieving social responsibility appeared through the aforementioned idea (social accounting), which was created by the applications of accounting information systems through the ability of specialists to access information and financial statements regardless of the lockdown and wide closures, and thus gave accountants opportunity to determine the extent to which organizations can contribute to alleviating the burdens and severity of the pandemic on society.

There are many areas in which accounting information systems were able to represent social responsibility, such as determining and controlling costs, accounting disclosure to show the extent of the organization's ability to contribute to community support based on its financial statements for the previous year, charitable and voluntary works, reducing working hours for employees based on quarantine and closures, in addition to supporting the various strata of society, whether financially or in any other way [22].

Based on what was mentioned before, researcher was able to reach following set of hypotheses:

**Main Hypothesis**

H: AIS application increased CSR among SMEs during COVID 19 pandemic.

**Sub-Hypotheses**

H1: AIS applications increased economic support from SMEs during COVID 19.

H2: AIS applications increased legal support from SMEs during COVID 19.

H3: AIS applications increased ethical support from SMEs during COVID 19.

H4: AIS applications increased philanthropic support from SMEs during COVID 19.

## 2 Methods

Realizing aim of current research was done depending on quantitative approach which was based on gathering numerical data and process them mathematically in order to reach numerical results that can be generalized. As a continuation of methods; a 5 Likert scale questionnaire was developed in two sections, the first contained demographics of individuals while the later contained variables of research (ethical, economic, legal and philanthropic). The questionnaire was uploaded online—due to COVID 19 health precautions—through Google Forms.

Population of study consisted of financial and accounting managers and officers within SMEs in Jordan. A sample of (150) employee was chosen to respond and represent population, after application process (132) questionnaire were suitable for statistical analysis which gave a ratio of 88% as statistically accepted. SPSS V 27th was used in order to screen and process gathered primary data. The Cronbach alpha test was also used to test tool reliability, as it was found that the alpha = 0.937 is higher than the acceptable ratio of 0.60, which indicates the reliability of the scale. Also, multiple regression, simple regression and descriptive statistics were used through analysis.

### 2.1 Demographic Analysis

Frequency and percentage were used to analyze respondents' answers to demographic variables (gender, age, and experience). It was found out through analysis that males formed 95.5% of total sample who were within age range of 40–50 (43.2%) with an experience of more than 15 years in accounting (72.7%) with bachelor degree (63.6%).

As for the questionnaire analysis, mean and standard deviation were calculated. It was seen that all respondent held a positive attitude towards statements of questionnaire as all means were above 3.00. Also, the most positively answered statement was articulated "*Ethics are always implemented within finances of the organization*" with a mean of 4.77/5.00 compared least positive statement "*Well organized finances gives better image for the organization*" with a mean of 3.77/5.00. Results also indicated

**Table 1** Testing main hypothesis

| Model summary | | | | | |
|---|---|---|---|---|---|
| Model | R | $R^2$ | F | t | Sig |
| 1 | 0.841[a] | 0.707 | 313.906 | 17.717 | 0.000 |

that "legal" implications was the most influenced by AIS practices and application scoring a mean of 4.60 compared to "economic" scoring a mean of 4.41.

**Hypotheses Testing**

**Main Hypothesis**

**H: AIS application increased CSR among SMEs during COVID 19 pandemic**

Table 1 showed the results of linear regression for main hypothesis, Pearson correlation coefficient was 0.841 reflected a **high correlation relationship**, the value of the determination coefficient of 0.707 and independent variable explained 70.7% of the variance in the dependent variable. F value of 313.906 was significant at 0.05 level, which reflected the significance of the regression. That meant AIS application increased CSR among SMEs during COVID 19 pandemic.

**Sub-Hypotheses**

**H1: AIS applications increased economic support from SMEs during COVID 19**

Table 2 indicated results of linear regression for 1st hypothesis, Pearson correlation coefficient of 0.772 reflected **a high correlation relationship**, and value of the determination coefficient of 0.596 showed independent variable explained **59.6%** of variance in dependent variable. F value of 192.171 was significant at 0.05 level, reflected the significance of regression. That meant AIS applications increased economic support from SMEs during COVID 19.

**H2: AIS applications increased legal support from SMEs during COVID 19**

Table 3 indicated results of linear regression test for 2nd hypothesis, Pearson correlation coefficient of 0.815 reflected a **high correlation relationship**, and value of determination coefficient of 0.664 that independent variable explained **66.4%** of variance in dependent variable. F value of 257.404was significant at 0.05 level, which

**Table 2** Testing 1st hypothesis

| Model summary | | | | | |
|---|---|---|---|---|---|
| Model | R | $R^2$ | F | t | Sig |
| 1 | 0.772[a] | 0.596 | 192.171 | 13.863 | 0.000 |

**Table 3** Testing 2nd hypothesis

| Model summary | | | | | |
|---|---|---|---|---|---|
| Model | R | $R^2$ | F | t | Sig |
| 1 | 0.815 | 0.664 | 257.404 | 16.044 | 0.000 |

**Table 4** Testing 3rd hypothesis

| Model summary | | | | | |
|---|---|---|---|---|---|
| Model | R | $R^2$ | F | t | Sig |
| 1 | 0.641 | 0.410 | 90.456 | 0.41096 | 0.000 |

**Table 5** Testing 4th hypothesis

| Model summary | | | | | |
|---|---|---|---|---|---|
| Model | R | $R^2$ | F | t | Sig |
| 1 | 0.722 | 0.521 | 141.463 | 11.894 | 0.000 |

reflected the significance of the regression. That meant AIS applications increased legal support from SMEs during COVID 19.

**H3: AIS applications increased ethical support from SMEs during COVID 19.**
Table 4 indicated results of linear regression test for 3rd hypothesis, Pearson correlation coefficient of 0.641 reflected a **high correlation relationship**, value of determination coefficient of 0.41 that independent variable explained **41%** of variance in the dependent variable. F value of 90.456 was significant at 0.05 level, which reflected the significance of the regression. That meant AIS applications increased ethical support from SMEs during COVID 19.

**H4: AIS applications increased philanthropic support from SMEs during COVID 19**

Table 5 indicated results of the linear regression for main hypothesis, Pearson correlation coefficient of 0.722 reflected a **high correlation relationship**, value of the determination coefficient of 0.521 that independent variable explained **52.1%** of variance in dependent variable. F value of 141.463 was significant at 0.05 level, which reflected significance of regression. That meant AIS applications increased philanthropic support from SMEs during COVID 19.

# 3 Discussion

Current research was meant to highlight the role of AIS application in increasing CSR of Jordanian SMEs during COVID 19 pandemic. Results indicated that AIS application increased CSR among SMEs during COVID 19 pandemic scoring a variance of 70.7%. Also, it appeared that AIS applications increased legal support for SMEs during COVID 19 with a variance of 66.4% as the most influenced variable by AIS applications. In the 2nk rank appeared the variable of economics which scored a variance of 59.6%. Followed by economic, it appeared that philanthropic was also influenced in the 3rd rank scoring a variance of 52.1%. In the last rank, ethical dimension appeared with the variance of 41% but revealing a high correlation with AIS applications.

Results of the study demonstrated that the adoption of accounting information systems applications contributed to increasing the social responsibility of organizations—[7, 15, 17, 18]—through the support provided to them at multiple levels, including (economic, legal, ethical and philanthropic). Based on the analysis, the legal aspect of social responsibility was most affected by the adoption of accounting information systems, where accounting disclosure contributed to supporting organizations' attitudes towards making some decisions that could lead to the imposition of social responsibilities on the organization, and based on these legal positions, and depending on the outputs of the accounting information systems, the organization was able to determine the scope of its assistance to the community, the limits of its social responsibilities, and the scope of services that it can provide [10].

## 4 Conclusion and Recommendation

Generally speaking, social responsibility of many SMEs emerged directly through providing financial aid to small entrepreneurial organizations and businesses, financial support provided for research efforts aimed at finding a vaccine for the Corona virus, supporting health sectors, and standing by the media sector and non-profit organizations. It also appeared indirectly through reducing the working hours of employees, providing health services, and maintaining the relationship with suppliers and customers. Study recommended the need to increase employees' awareness to pay attention to social responsibility as it is a basic requirement to improve the social and economic life of individuals and society.

## References

1. Abbas J (2020) Impact of total quality management on corporate green performance through the mediating role of corporate social responsibility. J Clean Prod 242:118458
2. Agudelo MAL, Jóhannsdóttir L, Davídsdóttir B (2019) A literature review of the history and evolution of corporate social responsibility. Int J Corp Soc Responsib 4(1):1–23
3. Al-Wattar YMA, Almagtome AH, AL-Shafeay KM (2019) The role of integrating hotel sustainability reporting practices into an accounting information system to enhance hotel financial performance: evidence from Iraq. African J Hospitality Tourism Leisure 8(5):1–16

4. Ali M, Hameedi K, Almagtome A (2019) Does sustainability reporting via accounting information system influence investment decisions in Iraq?. Int J Innov Creativity Change 9(9):294–312
5. Arif D, Yucha N, Setiawan S, Oktarina D, Martah V (2020) Applications of goods mutation control form in accounting information system: a case study in sumber indah perkasa manufacturing, Indonesia. J Asian Finan Econ Bus 7(8):419–424
6. Dewi IGAAO, Dewi IGAAP, Kustina KT, Prena GD (2018) Culture of tri hita karana on ease of use perception and use of accounting information system. Int J Soc Sci Humanit 2(2):77–86
7. Diao X, Aung N, Lwin WY, Zone PP, Nyunt KM, Thurlow J (2020) Assessing the impacts of COVID-19 on Myanmar's economy: a social accounting matrix (SAM) multiplier approach, vol 1. Intl Food Policy Res Inst
8. Dobbbins R, Witt S (2020) Social accounting. In Handbook of financial planning and control. Routledge, pp 277–285
9. Ganyam AI, Ivungu JA (2019) Effect of accounting information system on financial performance of firms: a review of literature. J Bus Manag 21(5):39–49
10. García-Sánchez IM, García-Sánchez A (2020) Corporate social responsibility during COVID-19 pandemic. J Open Innovation: Technol Market Complexity 6(4):126
11. Ha VD (2020) Impact of organizational culture on the accounting information system and operational performance of small and medium sized enterprises in Ho Chi Minh City. J Asian Finan Econ Bus 7(2):301–308
12. Haleem AH, Kevin LLT (2018) Impact of user competency on accounting information system success: banking sectors in Sri Lanka. Int J Econ Financ Issues 8(6):167
13. He H, Harris L (2020) The impact of Covid-19 pandemic on corporate social responsibility and marketing philosophy. J Bus Res 116:176–182
14. Hutahayan B (2020) The mediating role of human capital and management accounting information system in the relationship between innovation strategy and internal process performance and the impact on corporate financial performance. Benchmarking: An Int J
15. Jawabreh O, Saleh M, Alsarayreh M, Gharaibeh A (2020) The application of social accounting in Jordanian hotels, the role in accounting disclosure and its impact on the quality of the financial statements. Transylvanian Rev 27(47)
16. Jovanović T, Dražić-Lutilsky I, Vašiček D (2019) Implementation of cost accounting as the economic pillar of management accounting systems in public hospitals–the case of Slovenia and Croatia. Econ Res-Ekonomska Istraživanja 32(1):3754–3772
17. Khaghaany M, Kbelah S, Almagtome A (2019) Value relevance of sustainability reporting under an accounting information system: Evidence from the tourism industry. Afr J Hospitality Tourism Leisure 8:1–12
18. Kowalski PA, Szwagrzyk M, Kielpinska J, Konior A, Kusy M (2021) Numerical analysis of factors, pace and intensity of the corona virus (COVID-19) epidemic in Poland. Eco Inform 63:101284
19. Nagirikandalage P, Binsardi B (2017) Inquiry into the cultural impact on cost accounting systems (CAS) in Sri Lanka. Manag Auditing J
20. Primahendra R, Sumbogo TA, Lensun RA (2020) Handling corona virus pandemic in the Indonesian political context: a grounded theory study. Euro J Mol Clin Med 7(8):113–129
21. Qatawneh A (2020) The role of computerized accounting information systems (cais) in providing a credit risk management environment: moderating role of it. Acad Account Finan Stud J 24(6):1–17
22. Rusconi G (2021) Health, economics, education and stakeholders: some ethical insights for public and private management and social accounting. Almatourism-J Tourism Cult Territorial Dev 12(23):125–137

# Application of Corporate Governance Mechanisms to Protect the Value of Shareholders: Evidence of the Banking Sector in Kosovo

Esat Durguti and Emine Gashi

**Abstract** The persistence of this research is empirically to investigate the influence of corporate governance instruments on the wealth of financial intermediaries in wide-ranging. The study intertwined a combination of corporate governance variables and some of the important financial indicators, utilizing panel data for 9 banks, including the period 2013–2020. The study to achieve the set objectives applied the 2SLS dynamic approach to test the impact of control variables on the natural logarithm of assets. The results provided by the econometric analysis show that board size, sovereign committees, NIM, NPL's, and equity to liabilities have an important impact on the protection of the assets of financial institutions. While surprising results have been generated in the composition of the board structure in terms of gender diversity, they have turned out to be insignificant.

**Keywords** Banks' indicators · Corporate governance · 2SLS

## 1 Introduction

A large number of businesses around the world at the beginning of the twenty-first century, have encountered difficulties and on the threshold of collapse, as a result of rapid change and the impact of the covid-19 pandemic. Finally, it has been observed that inadequate application of corporate governance is considered a crucial factor of failure. However, in many studies, it has been reported that the non-implementation of corporate governance mechanisms has affected not only non-financial businesses but also financial ones, respectively banks. Given the fact that financial institutions, respectively banks have a significant impact on intermediation between different stakeholders, which directly affects economic growth. Also, [12] have found that the

E. Durguti (✉)
University "Isa Boletini", Street Ukshin Kovaçica 40000, Mitrovica, Republic of Kosovo
e-mail: esat.durguti@umib.net

E. Gashi
College for International Management, "GLOBUS", Street Bedri Pejani no 1, 10000 Prishtina, Republic of Kosovo

© The Author(s), under exclusive license to Springer Nature Switzerland AG 2022
S. G. Yaseen (ed.), *Digital Economy, Business Analytics, and Big Data Analytics Applications*, Studies in Computational Intelligence 1010,
https://doi.org/10.1007/978-3-031-05258-3_44

economic growth of some countries is supported by the banking industry because it helps finance businesses. However, when we are discussing between two stakeholders, businesses on the one hand and the financial industry on the other, there are arguments and counter-arguments regarding asymmetry information. To eliminate this concern, many studies have been conducted on information asymmetry, which directly affects the components of corporate governance. Therefore, [14] documented that the importance of corporate governance initially attracted the attention of the US authorities due to accounting scandals in firms Enron and WorldCom, etc., further arguing that the weak corporate governance system creates gaps for conflict of interest concerning the evaluation process. Therefore, to conduct this research we have used the data panel for 9 banks that operate in the Kosovo market. Explicitly, in this research we have posed two research questions: (a) can adequate corporate governance mechanisms influence the increase of the value of the banking institution, and (b) are these mechanisms applied by banks in Kosovo.

The contribution of the study is expected to be in several aspects. First, in advancing and raising the debate among scholars about corporate governance mechanisms. Second, the application of the dynamic approach through the 2SLS method, to eliminate the dilemmas between the studies conducted, and third, in terms of policy-making implications. The paper is organized with sections such as introduction, review of the empirical literature, analysis, and econometric findings, and the final part is presented as part of the conclusions.

## 2 Theoretical Background

Several academic and empirical pieces of research have been devoted to the impact of CG components on the capital structure of a company. Conferring to the literature, the main components of CG recognized to affect funding decisions are the following: board size, board independence, executive compensation, and executive entrenchment. Nevertheless, the results are varied and questionable. Corporate governance reconciliation studies offer a wide range of qualitative and quantitative analyzes that reveal the degree, scope, and levels of compliance [25, 26], and [20], in addition to its importance to the bank's performance and value [23, 27], and [24]. Besides this, the most recent study in the field of CG, by [6], analyzing the influence of CG in Romania and Italy, using financial industry also CG components in productivity and shareholder value protection, turns out to be significant in both cases. This study's sample included 34 Romanian banks as well as over 350 Italian banks that used the dynamic VAR techniques, which were integrated at various levels.

Theoretically, revisions are based on the hypothesis that banks with poor CG should have inferior ratings compared to banks with effective CG, as stakeholders do not tolerate a higher risk of expropriation without receiving a premium for such investments [17], and [18]. A confident association among quality of governance and performance has been observed in European studies [16–18, 21], and [10], Japanese [4], and [9] companies. Garcia et al., [11] assessed the effects of board structure on default risk for European banks, such as the composition of the board, sovereign commissions, representation of women on boards, and Earnings quality. The GMM dynamic approach was used to conduct an empirical investigation. The study's findings indicate that the components chosen for the study have a significant impact.

As distinguished earlier, various opinions regarding board size dominate. [7] argued that firms with large board numbers have low debt. This fact, they argue that the size of the board can raise pressure on management to decrease the level of debt, and to increase the productivity of the firm. However, [25] have contested the findings, arguing that there is a consistent connection between CG components and capital structure. Moreover, [8] argue that the size of the board appears to have a strong confident impact on the bank's performance and assets, utilizing a sample of 34 banks from 2009 to 2018 employing the OLS approach. The findings regarding the size of the board are contradictory among different revisions and so far, we do not have any common consensus. Therefore, in this study we have hypothesized:

*H1: There is a significant relationship between board size and bank assets.*

Additionally, about the configuration of the board construction in terms of gender, studies are smaller in number, but the prevailing opinion is that a board that has a mixed structure of females and males, has a predisposition to be more effective. [2]. Consequently, some recent studies on governing boards in terms of gender were conducted, and all conclusions show that mixed boards in terms of gender inspire additional efficiency to the team. These conclusions are supported by a current study from [28], [3] and [19]. Therefore, our hypotheses regarding diversity in terms of gender are:

*$H_2$: There is no significant relationship between female gender and bank assets.*
*$H_3$: There is no significant relationship between male gender and bank assets.*

And the last component but not of importance is the variable specific commissions created by investors, as a result of the request of supervisory bodies, which in the scientific works are known as sovereign boards. Numerous revisions have confirmed the effectiveness of the independence of these boards in controlling management and protecting the assets of investors, [25] documented a negative association between sovereign boards and the asset structure of companies. Almoneef and Samontaray., [1] reached the same conclusions after analyzing the components of CG and the productivity in the Saudi banking system, arguing empirically that board independence harms the bank's productivity, correspondingly in ROE. Therefore, our hypothesis presented is:

*$H_4$: There is a significant relationship between sovereign boards (logSBC) and banks' assets.*

The study also applied some of the core financial factors to measuring the strength of financial sustainability. These indicators are presented as control variables to investigate their effect on the assets of the firm, respectively the banks. For these control variables, no hypothesis will be presented, as the emphasis is oriented on the components of corporate governance and their impact on increasing the value of the firm, respectively the bank.

## 3 Econometric Analysis

The sample contained within the analysis consists of 9 commercial banks licensed by the Central Bank of Kosovo, out of a total of 10 banks operating in Kosovo. Kosovo's banking sector is a relatively new sector, dominated by over 90 percent of banks with foreign capital. In this study, panel data were used, including the period 2013–2020, and this data was provided by audit reports for each bank in particular, and then processed to suit the research. Based on previous studies conducted by different authors, different techniques have been used to come to the most accurate conclusions.

Therefore, our study includes an adequate combination of factors including corporate governance parameters and key financial indicators. Various authors have applied different models to test the impact of CG parameters on productivity and shareholder protection. The models that are most suitable for this study are the GMM estimator, respectively 2SLS as this method calculates the endogeneity of the data and the robustness of the instruments to achieve the assessment of interdependence among the parameters of CG and protection of shareholder capital. Therefore, in the following, we will present the general equation of 2SLS.

$$Y_{it} = \sum_{j=1}^{p} \alpha_j Y_i, t - j + X_{it}\beta_1 + \vartheta_i + \varepsilon_{it} \tag{1}$$

In addition, because of the equations presented beyond, we will present the concrete equation for our research.

$$\begin{aligned} \text{LogAssets}_{i,t} = {} & \alpha + \mu(\text{NIM})_{i,t} + \beta_1(\text{NPL}'s_{i,t}) + \beta_2(\text{ELR}_{i,t}) + \beta_3(\text{LogBS}_{i,t}) \\ & + \beta_4(\text{LogCM}_{i,t}) + \beta_5(\text{LogCF}_{i,t}) + \beta_6(\text{LogSBC}_{i,t}) + \vartheta_i \end{aligned} \tag{2}$$

Empirical studies on the application of the dynamic approach, respectively 2SLS estimation have determined the $N > T$ condition that is a fundamental requirement. This condition in our case is met in accurateness and that the panel data are well-adjusted.

# 4 Econometric Findings

The early finding of the study is that the financial intermediaries in Kosovo are implementing with precision and high efficiency, the component of CG that derive as an obligation set by the Central Bank of Kosovo, as well as the regulatory set of the Basel Committee. It is worth noting that the governing structures of the regulatory authority have released regulations on the establishment of committees, which are an integral part of the CG component, according to the requirements of the Basel Committee. Table 1 gives us a detailed description of the factors applied in this study, starting with the number of observations, the lowest, largest, mean, and standard deviation.

The data used are panel data, and this data as such before applying the econometric model is preferable to do some preliminary diagnostic tests on the suitability of the 2SLS estimation. One such preliminary test is multicollinearity, known as the correlation matrix. We have applied this test to evaluate the degree of interrelationship between variables.

Furthermore, the outcomes from the correlation analysis exposed in Table 2 show that the factors correlate them, and it is noted that the problem with multicollinearity

**Table 1** Descriptive statistics

|       | L_Assets | NIM   | NPL's | ELR   | LogBS | LogCM | LogCF | LogSBC |
|-------|----------|-------|-------|-------|-------|-------|-------|--------|
| Obs   | 72       | 72    | 72    | 72    | 72    | 72    | 72    | 72     |
| Mean  | 6.515    | 0.041 | 0.057 | 0.158 | 0.803 | 0.687 | 0.225 | 0.642  |
| Std.D | 1.385    | 0.014 | 0.042 | 0.171 | 0.135 | 0.142 | 0.203 | 0.178  |
| Min   | 5.029    | 0.001 | 0.015 | 0.076 | 0.698 | 0.477 | 0.000 | 0.000  |
| Max   | 9.479    | 0.077 | 0.244 | 1.556 | 1.041 | 0.903 | 0.477 | 1.041  |

*Source* Author's calculations

**Table 2** Correlation matrix

|          | L_Assets | NIM    | NPL's  | ELR    | LogBS | LogCM | LogCF | LogSBC |
|----------|----------|--------|--------|--------|-------|-------|-------|--------|
| L_Assets | 1.000    |        |        |        |       |       |       |        |
| NIM      | −0.064   | 1.000  |        |        |       |       |       |        |
| NPL's    | 0.280    | -0.039 | 1.000  |        |       |       |       |        |
| ELR      | 0.110    | -0.334 | -0.065 | 1.000  |       |       |       |        |
| LogBS    | 0.059    | -0.239 | 0.099  | 0.296  | 1.000 |       |       |        |
| LogCM    | 0.426    | -0.253 | 0.151  | 0.200  | 0.397 | 1.000 |       |        |
| LogCF    | 0.240    | -0.097 | -0.006 | 0.267  | 0.718 | 0.190 | 1.000 |        |
| LogSBC   | −0.117   | 0.003  | -0.382 | 0.188  | 0.283 | 0.021 | 0.416 | 1.000  |

*Source* Author's calculations

does not exist as only some of the factors have a moderate association among themselves. The natural logarithm assets are seen to have positive associations with NPL's, equity to liabilities, logBS, logCM, and logCF, while adverse associations exist between net interest margin and logSBC. Whereas, other correlations are presented in detail in Table 2.

Empirical analysis requires a flow logic before commenting on the results, applying some diagnostic tests on the appropriateness of the applied approach. Therefore, the diagnostics of the approach was initially done through $R^2$, wherein our case is 63.7 percent of the independent variables explain the dependent variables, also for multicollinearity testing was applied VIF, where the mean value of the variables is 9.62 which proves that the data do not have multicollinearity problem. Furthermore, in terms of heteroskedasticity, our test proves that the applied data do not have such problems, and finally, the Wald chi (2) for the endogeneity and constancy of the instruments proves that the applied approach is adequate since its value is 126.66 (Table 3).

Findings results argue that financial features have an influence on the preservation of banks' assets, and what is considered crucial that the two components of corporate governance board size and independent committees have resulted in the confirmation of hypotheses that they have an impact on preserving and increasing of the value of banks. The results on gender diversity have shown insignificant results that are expectedly based on the hypotheses presented. The initial outcomes on financial indicators show that adequate management of credit risk management, respectively non-performing loans, directly affects the increase of net interest margin, and an increase in net interest margin affects the increase of the bank's productivity. On the other hand, inadequate management of liquid assets affects the reduction of banks'

**Table 3** Empirical results of 2SLS regression

| *Instrumental variables (2SLS) regression* | | | | | |
|---|---|---|---|---|---|
| | Coefficient | Std. Err | Z | P > [z] | VIF |
| NIM | 59.59506 | 7.345415 | 8.11 | 0.000 | 0.845529 |
| NPL's | 4.99476 | 2.576601 | 1.94 | 0.003 | 0.802854 |
| ELR | −1.405229 | 0.6358486 | −2.21 | 0.027 | 0.812879 |
| LogBS | 8.191799 | 4.197526 | 1.95 | 0.011 | 0.030106 |
| LogCM | −3.028255 | 2.819797 | −1.07 | 0.283 | 0.059954 |
| LogCF | −1.532275 | 1.696478 | −0.90 | 0.366 | 0.081734 |
| LogSBC | −1.176367 | 0.6812793 | −1.73 | 0.004 | 0.657489 |
| _cons | 5.502088 | 1.322457 | 4.16 | 0.000 | **Mean 9.62** |
| $R^2$ | | | | | 0.6376 |
| Wald chi2 | | | | 126.66 | 0.0000 |
| Breusch-Pagan/Cook-Weisberg test ($\mathcal{X}^2$ heteroscedascity) | | | | 13.32 | 0.0003 |

*Note* (***), (**), (*) significant, respectively, at 1, 5, and 10 %
*Source* Author's calculations

assets, which is also found in our confirmed hypothesis where P = 0.027. Board size as the most important component of corporate governance is confirmed at the significance level of 1%. And this gives us indications that the proper functioning of the board of directors has a positive impact on the preservation of bank assets. These results are in the spirit of revisions steered by [13] arguing that an optimal equilibrium of board of directors and their monitoring activities should provide a higher value for shareholders. While the other equally important component is the independent committees where it has turned out to hurt maintaining the bank's assets at the level of 1 percent significance. This argues that their lack of establishment and adequate functioning reduces the wealth of financial institutions. This finding is in line with many studies conducted by [22], and [15].

# 5 Conclusion

This study examined the association of mechanisms of corporate governance, some of the crucial financial indicator's vis-a-vis (NIM, NPL's, ELR, log board size, log male composition, log females' compositions, and log sub board committees), and the protection of shareholders value. Surprisingly, not such an investigation has been made to analyze the impact of corporate governance on efficiency. Moreover, in Kosovo, among 10 licensed banks by the Central Bank of Kosovo, nine banks show productivity, and what are the reasons behind the results are still unexplored. Therefore, research aiming to explore the reason behind such limitations of the banking sector carries huge significance. Such research needs to shape the aim of this research that attempts to examine the determinants of a bank's efficiency by giving special heed to ownership structures and board diversification.

In the framework of this research, panel data for banks in Kosovo were used, applying a dynamic approach through instrumental regression analysis, respectively 2SLS. The conclusions reached under this approach argue that the corporate governance components at the significance level of 1 percent confirm the hypotheses that the size of the board of directors and independent committees are considered the main protectors of shareholder capital, and the bank as a whole. Moreover, in Kosovo, the maximum number of board sizes reached 11, while the smallest number is 5. This number is considered an optimal size to govern professionally and without any other complications.

Another significant finding in this study is that the regulatory body, namely the CBK, has approved the directives derived as an obligation from the European Central Bank and the Basel Committee. As a result, the financial industry is required to establish CG bodies and strictly enforce the laws' directives. According to the study, all banks operating in Kosovo have met all of the requirements for the application of sound CG principles. Finally, we must emphasize that this research is of particular importance in the field of corporate governance as a whole, but a crucial significance for financial institutions in Kosovo. In future studies, to further enrich the sustainability and importance of corporate governance, it is suggested to apply other more specific corporate governance variables to analyze in more detail the effect of corporate governance components to protect shareholder value.

# References

1. Almoneef A, Samontaray DP (2019) Corporate governance and firm performance in the Saudi banking industry. Banks Bank Syst 14(1):147–158. https://doi.org/10.21511/bbs.14(1).2019.13
2. AlHares A, Ntim C (2017) A cross-country study of the effect of institutional ownership on credit ratings. Int J Bus Manag 12:80–99
3. Báez AB, Báez-García AJ, Flores-Muñoz F, Gutiérrez-Barros J (2018) Gender diversity, corporate governance and firm behavior: the challenge of emotional management. Euro Res Manag Bus Econ 24(3):121–129, ISSN 2444-8834, https://doi.org/10.1016/j.iedeen.2018.07.001
4. Aman H, Nguyen P (2007) Do stock prices reflect the corporate governance quality of Japanese firms? Journal of the Japanese and International Economies 22:647–662
5. Bauer R, Gunster N, Otten R (2004) Empirical evidence on corporate governance in Europe: The effect on stock returns, firm value and performance. J Asset Manag 5:91–104
6. Benvenuto M, Avram RL, Avram A, Viola C (2021) Assessing the Impact of Corporate Governance Index on Financial Performance in the Romanian and Italian Banking Systems. Sustainability 2021(13):5535. https://doi.org/10.3390/su13105535
7. Berger PG, Ofek E, Yermack DL (1997) Managerial Entrenchment and Capital Structure Decisions. Journal of Finance 52(4):1411–1438
8. Bezawada, B, and Adavelli, SR (2020). Corporate governance, board characteristics and performance of Indian Banks: An Empirical Study, International Journal of Economics and Financial Issues. Volume 10 (3), 83–87. https://doi.org/10.32479/ijefi.9536.
9. Bhagat S, Bolton B (2008) Corporate governance and firm performance. J Corp Finan 14:257–273
10. Bistrowa J, Lace N (2012) Corporate governance best practice and stock performance: Case of CEE Companies. Systemics, Cybernetics and Informatics 10:63–69
11. José García C, Herrero B, Morillas F (2021) Corporate board and default risk of financial firms. Economic Research-Ekonomska Istraživanja. https://doi.org/10.1080/1331677X.2021.1909490
12. Caprio, G., & Levine, R. (2002). Corporate governance of banks: Concepts and international observations. Paper presented in the Global Corporate Governance Forum Research Network Meeting, April 5, 2002.
13. Chen Q, Goldstein I, Jiang W (2008) Directors' ownership in the US mutual fund industry. J Financ 63(6):2629–2677
14. Craig Doidge G, Karolyi A, Stulz RM (2007) Why do countries matter so much for corporate governance? Journal of Financial Economics, 86(1), 2007. ISSN 1–39:0304-405X
15. Dalton CM, Dalton DR (2005) Boards of directors: Utilizing empirical evidence in developing practical prescriptions. Br J Manag 16:S91–S97
16. Drobetz W, Shillhofer A, Zimmermann H (2003) Corporate governance and expected stock returns: Evidence from Germany. Eur Financ Manag 10:267–293
17. Gompers P, Joy I, Metrick A (2003) Corporate governance and equity prices. Quarterly Journal of Financial Economics 118:107–155
18. Goncharov I, Werner J, Zimmermann J (2006) Does compliance with the German corporate governance code have an impact on stock valuation? An empirical analysis. Corp Gov 14:432–445
19. Kamalnath, Akshaya, (2018). The Corporate Governance Case for Board Gender Diversity: Evidence from Delaware Cases. Albany Law Review, 2018, Available at SSRN: https://ssrn.com/abstract=3128272.

20. Okhmatovskiy I (2017) Self-regulation of corporate governance in Russian firms: Translating the national standard into internal policies. J Manage Governance 21:499–532
21. Renders A, Gaeremynck A, Sercu P (2010) Corporate-governance ratings and company performance: A cross-European study. Corporate Governance: An International Review 18:87–106
22. Romano, G., Ferretti, P., and Quirici, M.C. (2012). Corporate Governance and efficiency of Italian Bank Holding companies during the financial crisis: an empirical analysis, 102–133.
23. Rose C (2016) Firm performance and comply or explain disclosure in corporate governance. Eur Manag J 34:202–222
24. Roy A, Pay AM (2017) Corporate governance compliance, governance structures, and firm performance. Indian Accounting Review 21:31–50
25. Seidl D, Sanderson P, Roberts J (2013) Applying the 'comply-or-explain' principle: Discursive legitimacy tactics with regards to codes of governance. J Manage Governance 17:791–826
26. Shrives PJ, Brennan NM (2015) A typology for exploring the quality of explanations for non-compliance with UK corporate governance regulations. Br Account Rev 47:85–99
27. Stiglbauer M, Velte P (2014) Impact of soft law regulation by corporate governance code on firm valuation: The case of Germany. Corp Gov 14:395–406
28. Wahid AS (2019) The Effects and the Mechanisms of Board Gender Diversity: Evidence from Financial Manipulation. J Bus Ethics 159:705–725. https://doi.org/10.1007/s10551-018-3785-6

# Cryptocurrencies in Turkey; Facts, Figures and Trends

**Ahmet Salih İkiz**

**Abstract** A decade has passed since the global financial system introduced a new phenomenon, cryptocurrencies and blockchain technology. That is evolving in Turkey where there is strong involvement with world finance capital. In this paper, we first investigate main dynamics of cryptocurrencies. Their characteristics, usage, functions and role in globe briefly. The remainder of the chapter devoted solely Turkish case. How the legal system fits into those new instruments. We also searched the drives of Turkish citizen's demand for these new assets and their ability to cope with this investment opportunity. In the end, we do share our humble opinions for the future trends of cryptocurrency for Turkish economy.

**Keywords** Cryptocurrency · Blockchain · e-commerce · e-money · Turkey · Turkish economy · Turkish financial system

## 1 Introduction

Cryptocurrencies are becoming quite popular in Turkey like other parts of the world. As an emerging G20 country, Turkey has strong financial links to global markets from different perspectives. In first part of our paper, the emergence and dimensions of cryptocurrency will discussed in general context. Following this chapter, the use of crypto currency in Turkey is going to be discussed with the volume of money market and crypto world. We will emphasize the scale of the virtual currency market. With recent developments in legal system. The research paper then explore macroeconomic indicators such as inflation and monetary risks for Turkish economy and their impacts on demand for cryptocurrencies. We do aim to provide insight for these new phenomena for Turkey.

A. S. İkiz (✉)
Faculty of Economics and Administrative Studies, Muğla Sıtkı Koçman University, Muğla, Turkey
e-mail: ahmet@mu.edu.tr

© The Author(s), under exclusive license to Springer Nature Switzerland AG 2022
S. G. Yaseen (ed.), *Digital Economy, Business Analytics, and Big Data Analytics Applications*, Studies in Computational Intelligence 1010,
https://doi.org/10.1007/978-3-031-05258-3_45

## 2   Cryptocurrencies How They Emerge

Classical functions of money can be classified are saving, means of transaction and value determination. Globalization in financial markets dominated by strong currencies such as USD and others. Following an increase in e-commerce volume there are new technologies appearing in payment and saving for capital owners, such as Bitcoin.

The idea of creating of cryptocurrency reaches as far back as 1983 by a man: cryptographer David Chaum [8]. He developed a transaction system called eCash, later with DigiCash. The foundational element of those are that transactions were anonymous and conducted over a public network.

Satoshi Nakamoto (till today we do not know his real name and country of origin) published a brief note on the potential of CC (cryptocurrency) and conducted the first transaction with the most well-known cryptocurrency to this day: BitCoin. Satoshi Nakamoto also invented the blockchain database whereby BitCoin and a majority of the other cryptocurrency platforms conduct business and house information about the actors within the platforms. Users create that new way of payment system not official authorities. This opened a new era in payment systems where money supposed was issued by independent national agencies such as central banks.

The important part is the classical functions of money cannot fit new crypto currencies because they are very volatile for saving; they cannot be used for transactions and cannot determine value of products. Even if some cases where you can use those new crypto currencies for transaction it is still not too common. However, this very speculative asset attracts many investors. There is always a confusion about whether electronic money and CC are same. E-money is like a shadow for hard currency, which is issued by central banks. The main backbone of CC is official authorities of nation state i.e. central banks do not issue them. They are issued and produced in different sources in computer-associated platforms. From the perspective of political science, this way of monetary activities plays a revolutionary role in globe. For the first time money, which is the main element for all transactions, is not issued by a central bank and other official means. Blockchain process is very independent from nation state authorities and even diminish their sovereignty power sized from their economic hegemony. Meanwhile political distortions emerging from government's intervention to money supply would be eliminated which is experienced by both developed and developing countries. In developing countries, inflationary boom risk would be eliminated with these new phenomena as CC supply increase. In developed world quantitative easing policies in recent years creates collateral effects in other parts of the world by currency attacks and upsurge. In this view, CC would eliminate those risks if it has broader usage.

After the launching of Bitcoin there were, many cryptocurrencies emerged with similar technical bases. They are called altcoin due to that they are alternative to bitcoin. Every altcoin has similarities and differences to Bitcoin. Due to open source code of Bitcoin it can easily reached and transferred to other coins electronic infrastructure. There are thousands of different altcoins in world. There are many crypto currencies in world and the market is dominated largely by Bitcoin. The volume of all transactions reached more than 2 trillion dollars globally. There are millions of users of this currency and the main risk element is internet security and hacking. According to ING Bank research findings, 1/3 of all crypto user prefer Bitcoin and 1/3 of all users thought that cryptocurrencies will be main investment assets in coming years. Even though there are some hesitations about its possible illegal usage, taxation issues, cases about secret transactions, nonexistence of possible government monitoring were questioned, experts in financial and banking sectors advocate a revolutionary role for them in near future due to the distributed record management system. On the other side, some research done by WTO argues the possible destructive impacts of Ripple (another cryptocurrency) in financial sector. IMF President Christine Lagarde mentioned fast and low cost payment advantages of crypto currencies and showed positive attitude in 2018 [2].

Another aspect is creating digital central bank money. Because cash usage declining in some countries and the rise of private digital tokens (like bitcoin). Thus creating a Central Bank Digital Currency (CBDC) means if also interest bearing could be an attractive asset to hold. It may also compete with other money-like assets, such as T-Bills, reverse repos, CPs and commercial bank deposits [7]. In following graph, Schrimpf gave potential place of CBDC in monetary system as money flower.

## The money flower

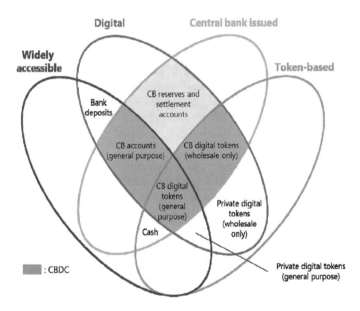

Source: [7]

In case of regulations, there is not a unique successive regulation set in different countries. Since it is quite a new technology governments approach cautious to that. The main criticism of crypto currencies is their possible role in money laundering, which is the main financial crime in many countries.

## 3 Crypto Currencies in Turkey

Turkey has sustainable strong growth performance even after COVID-19 as G20 country compared to most members of OECD. Since 1980's its economic model was based on export promotion and liberal economic policies that strengthened its ties with global finance capital. The economy is open to financial flows and financial instrument like CC like most parts of the developed world.

CC has important aspect of Turkish financial investment market. Interesting point is its legal framework mainly related with electronic money which is totally different from CC. Decree number 6493 for e-money hardly covers CC transactions and there is limited legal instruction for CC in Turkish Trade Law.

Turkish citizens aim to use Bitcoin as a strong shield against inflation and the lira's unparalleled depreciation against the dollar due to the harsh currency crisis after March 2021. Meantime foreign residents in Turkey also fueled crypto currency demand. To clarify, crypto trading volumes in Turkey hit 218 billion liras ($27bn) from early February to 24 March, up from just over 7 billion liras in the same period a year earlier, according to data analyzed by Reuters in 2021 [6]. Turkish traditional investment asset gold replaced mainly Bitcoin and its derivatives after 2010 in order to avoid ongoing financial crisis. The demand for CC is highly correlated with the volatility of USD/TR parity and BIST100 index of BorsaIstanbul. Research reveals that there is tendency to invest CC when there is low level in gains for those assets USD and BIST100 [1]. Expectations for negative trends in national currency would hamper demand for money and encourage CC demand in near future according to some economists. This is simply shows famous Gresham law that good money overthrows bad money. That all depends upon to having a shield on potential financial crisis in Turkish economy.

Many economic agents prefer shadow economic activities to avoid government regulations. The CC is very useful since they are easy to carry and non-traceable [3]. Since most of the banking transactions are recorded and monitored no one cannot hide monetary transfers and thus CC become quite attractive financial instrument for shadow economic transactions. CC are then can be considered as very versatile financial asset that cannot be classified yet whether that is a speculative boom or trustable investment platform in Turkey [5].

The very first time when cryptocurrency has entered Turkey in 2009 in a way that caught the attention of a good number of investors and entrepreneurs instantly. In 2013, The Banking and Regulation and Supervision Agency (BDDK), the watchdog body that safeguards the rights and benefits of depositors ensures the financial stability in the market, released some decrees about cryptocurrencies. The few regulation on CC created maximum freedom for market players with almost none responsibilities for their harms against third parties. Also there are number of coin exchanges funded both in Turkish Republic of Cyprus and Mainland. Last couple of decades Turkey manage to use noncash payment alternatives in every aspects of life. However, it seems that CC technology is not too much common rather used as speculative asset. We have to bear in mind that the penetration of CC usage in Turkey is still quite low compared to western countries. Financial Market Infrastructure is the key point for well-functioning CC. That means payment systems, central clearing institutions, data storage organizations. Turkish financial infrastructure is well performing due to the capital market operations in BorsaIstanbul with its own historical tradition. That provides solid background for CC exchange [9].

Most radical regulatory made in 16th of April, CB of Turkey released a public statement outlawing the widespread usage of cryptocurrencies and crypto assets for purchases, citing irreparable harm, transaction risks, excessive volatility and possible flaws allowing illegal activities. In the Official Gazette where legislations are published, the CB said cryptocurrencies and other types of digital assets predicated upon the distributed ledger technology could not be used directly or indirectly in exchange for goods and services. The impact of this decision on Turkey's cryptocurrency market was much more severe in comparison with international markets simply because the crypto market has gained momentum in recent months with investors seeking to hedge themselves against Turkish Lira's depreciation and inflation that reached a high in last month. Citizens interest in cryptocurrencies will not appear to last to hedge themselves against long-standing inflation problem.

Historical cornerstone of Bitcoin and others start with pandemic, which means high risk and uncertainty. The global demand for crypto currencies may change for investment but focal point is increasing electronic trade with high isolation would flourish those payment methods. At the beginning, cryptocurrencies behave like the traditional assets, but they start to become a hedge as the effect of COVID-19 deepens. As the number of reported cases and deaths rise, governments impose additional restrictions and those restrictions are likely to increase the demand for non-traditional assets. Bitcoin and Blockchain technology are theoretically capable of mitigating some of the issues that come with the new realities that the pandemic has brought. Investors should consider including cryptocurrencies in their portfolios depending on the COVID-19 phases. Cryptocurrencies will not only provide benefits in terms of hedge against the pandemic, but they can also be used as a payment and money transfer instrument [4].

# 4 Conclusion

Turkey introduced with CC simultaneously with developed world. Even though the volume of CC trade still remarkably low compared to western world Turkish people are eager to use CC. Inflationary expectations and financial risks made CC for an alternative investment opportunity in Turkey, which is dominated by real estate and gold investments. There was too much regulation for CC usage in Turkey until the mid-2021. Thus CC used in both transactions and investments. Latest developments regulated and banned CC in daily business transactions such as buying and selling goods and services with CC. The implications so far not too much harmful for CC exchanges where there is still ongoing demand for CC in different platforms. The future trends of CC demand and volume in CC exchanges mainly will depend upon the volatility and fragility of Turkish economy where many investors prefer to have long position due to the fact that they are easy to carry and hard to trace for shadow economic gains.

# References

1. Akdağ, M. (2019). Ph.D. Dissertation. Crypto monetization and an empirical study for Turkish economy. Erzurum, Atatürk University Sosyal Bilimler Enstitüsü
2. Alnıaçık B (2019) An analysis of current status of crypto currencies over the World and Turkey. Res Stud Anatolia J, 21–30
3. Alpago H (2018) From Bitcoin to Selfcoin: Cryptocurrencies. IBAD, 411–428
4. Demir EB (2020) The relationship between cryptocurrencies and COVID-19 pandemic. Eurasian Econ Rev, 349–360
5. Dilek Ş (2018) Blokchain Teknolojisi ve Bitcoin. 7 22, 2021 tarihinde https://setav.org/assets/uploads/2018/02/231.-Bitcoin1.pdf adresinden alındı
6. Gezginci HK (2021) Academia AHBVÜ. 2021 tarihinde https://d1wqtxts1xzle7.cloudfront.net/66622020/Factors_Affecting_Prices_of_Cryptocurrencies.pdf?1619200462=&response-content-disposition=inline%3B+filename%3DFACTORS_DETERMINING_THE_PRICE_OF_CRYPTOC.pdf&Expires=1625427896&Signature=Po6pxq8EhUBF6npJgt5J7~d adresinden alındı
7. Schrimpf A (2018) Central Bank of Turkey. 2021 tarihinde https://www.tcmb.gov.tr/wps/wcm/connect/8604eb70-ff34-4527-9a35-f48057b4f78a/schrimpf.pdf?MOD=AJPERES adresinden alındı
8. Software Guild (2021) A brief history of cryptocurrency. 7 2021 tarihinde https://www.thesoftwareguild.com/blog/a-brief-history-of-cryptocurrency/ adresinden alındı
9. Üzer B (2017) Sanal Para Birimleri. 2021 tarihinde https://www.tcmb.gov.tr/wps/wcm/connect/f4b2db90-7729-4d94-8202-031e98972d0f/Sanal+Para+Birimleri.pdf?MOD=AJPERES&CACHEID=ROOTWORKSPACE-f4b2db90-7729-4d94-8202-031e98972d0f-m3f Bagn adresinden alındı

# Effect of FDI on Carbon Emissions in Tunisia

Lamia Jamel

**Abstract** This article investigates the impact of FDI (foreign direct investment) inflows on $CO_2$ emissions in Tunisia, with the aim of studying the authenticity of pollution paradise hypothesis, using the cointegration methodology with time series data during the period from 1960 to 2019. From the bivariate cointegration assessment, we conclude that the used indicators are cointegrated. Furthermore, the empirical results of the DOLS (Dynamic Ordinary Least Square) and FMOLS (Fully Modified Ordinary Least Square) estimators show that FDI as a positive long-term link with $CO_2$ emissions. However, corresponding to the findings of Engle-Granger causality test, foreign direct investment inflows decrease carbon dioxide emissions in the brief run but behave inversely in the lengthy run. So, we assume that dirty industries tend to meet environmental regulations and standards at first, but in the long run they also become polluting. It is suggested to maintain to draw these applications whilst placing in position processes and mechanisms to decrease $CO_2$ emissions within the context of thorough environmental procedures.

**Keywords** FDI · $CO_2$ emissions · Cointegration assessment · Tunisia

## 1 Introduction

Over the previous quarter of a period, FDI flows must increase significantly around the world, registering an increase of \$1.297 billion between 1990 and 2018, or more than 530%. This increase is mainly due to the important opportunity represented by FDI, particularly for developing countries, to materialize their economic development strategies, given that these flows contribute positively and strongly to the economic growth of host countries [6, 14].

Furthermore, to its function as a growing component in developing nations [19], foreign direct investment is seen as a means of transferring new technologies between

L. Jamel (✉)
Department of Finance and Economic, College of Business Administration, Taibah University, Al Madinah, Saudi Arabia
e-mail: lajamel@yahoo.fr

© The Author(s), under exclusive license to Springer Nature Switzerland AG 2022 573
S. G. Yaseen (ed.), *Digital Economy, Business Analytics, and Big Data Analytics Applications*, Studies in Computational Intelligence 1010,
https://doi.org/10.1007/978-3-031-05258-3_46

countries, especially from developed to developing countries [12] and can participate positively in reducing unemployment by creating new employment opportunities [22].

Corresponding to the 2019 Report of the United Nations Conference on Trade and Development (UNCTAD) on World Investment, FDI flows to Africa recorded $46 billion, an expansion of 11% from matched to the preceding year. Despite the weakness of these flows compared to the global level, around 3.5% of global movements, they continue commonly durable and escaping the decline recorded in the previous 3 years. In addition, the ratio of foreign direct investment to African Gross Domestic Product (GDP) achieves 5.1%, the highest in the world, thus confirming the significance of these flows for the financial and economic growth of the region.

Likewise, foreign direct investment flows to Tunisia have remarkably increased by 25%, reaching $2.8 billion, making it the fourth country in North Africa, in terms of volume, after Morocco, Egypt, and Algeria thanks to investments in finance and the automotive sector.

However, FDI flows might be harmful to the natural environment of present countries, usually developing countries. This problem has been the subject of vigorous discussion amongst economic expert, researchers, and political groups, giving rise mainly to two controversial hypotheses. The first states that as part of trade liberalization, companies producing contaminating goods will transfer from developed countries with ecological requirements to developing countries with relatively low ecological directives. Hence, developing nations would turn into contamination sanctuaries for polluting businesses in developed nations. This assumption was first stated in 1994 and referred to as the Pollution Harbor Hypothesis (HHP) [5]. Alternatively, Porter's Hypothesis (HP) contends that stringent ecological procedures in countries promote enterprises to participate more in sanitary and effective machineries. These clean machineries lessen low expenditure and improve production, getting companies additional competitive [15].

In this framework, Tunisia would not be eliminated from this discussion. While its $CO_2$ emissions represented only 0.14% of worldwide emissions in 2018, the last fourth century has seen an overall increase of 166%, from 19,213 megatonnes in 1991 to 57,326 megatonnes in 2018 [16]. Thus, to predict the country's environmental future and establish the required recommendations for its environmental advancement, it is essential to understand and analyze the connection amongst foreign direct investment flows and $CO_2$ emissions. In our paper, we intend to examine the authenticity of the pollution haven hypothesis for the Tunisian case, by assessing the causality amongst foreign direct investment flows and carbon dioxide emissions.

Our study will be structured as follows: Sect. 2 presents a review of the literature on the connection amongst the progression of foreign direct investment and the confirmation of stated assumption. Section 3 describes the used dataset and econometric methodology in this paper. Section 4 recapitulates the main empirical findings. Finally, Sect. 5 concludes.

# 2   Literature Review

Over the past decades, the debate on the environmental impact of economic development and the increase in industrial activity has continued to increase among different researchers and academics, thus generating many controversial hypotheses on this subject. One of these most studied hypotheses and that of the Environmental Kuznets Curve (EKC) which states that a national pollution concentration increases with development and industrialization to a point where it decreases again when the country uses its increased affluence to further reduce pollution concentrations [9]. The way to reduce the concentration of pollution leaves room for different possibilities; the first assumes that developed countries adopt cleaner technologies in production processes, the second states that developed countries will specialize in the production of clean goods while polluting industries will be transferred to developing countries, in the form of l'IDE, where there are fewer environmental restrictions.

Research on pollution harbors is generally based on different estimation methods, various data samples, models and changing variables, which impact their results. This generates opposing arguments, which often can support or deny the validity of the hypothesis and therefore the existence of the impact of FDI on the environment in host countries.

Starting from a trade perspective, A panel data study for manufacturing industries for the case of Turkey, between 1994 and 1997, showed that exports increase as much as polluting industries, thus confirming the existence of the hypothesis [2]. In addition, for the same country, examining the relationships between FDI inflows, $CO_2$ emissions and economic growth during the period 1987Q1–2009Q4, based on the Granger causality test generated from the error correction model (ECM), also demonstrated the existence of a causal relationship between the variables, thus supporting the pollution haven hypothesis [13].

In the case of Ghana, many variables are used as determinants of pollution for the period 1980–2012. Their study focused on the Gross Domestic Product (GDP), the square of GDP, energy consumption, renewable energy consumption, fossil energy consumption, FDI flows, the quality of institutions, urbanization, and commercial opening. Its results confirm the existence of a long-term relationship between the variables, as well as a positive impact on $CO_2$ emissions, which confirms the validity of the hypothesis [20].

Similarly, Riti investigated the relationship between FDI inflows, manufactured exports, and the environment in Nigeria between 1980 and 2013 to test the validity of the pollution haven hypothesis. Using an ARDL model to estimate the short- and long-term parameters, he showed that these fluxes significantly influence the variation in $CO_2$ emissions, while the parameter of exports of manufactured products was found to be positive although not significant. The Granger causality result indicates the existence of a unidirectional causality from FDI to $CO_2$ emissions [17]. Furthermore, the statistical results of Spearman's correlation analysis, used to examine the validity of the hypothesis for the case of Indonesia, between carbon emissions and FDI flows, gross domestic product, and population size, indicate that $CO_2$ emissions have a significant relationship, negative and positive respectively, with the FDI and the GDP.

Real gross domestic product (GDP) and the size of the population in the Indonesian economy. However, there is a weak and insignificant relationship between $CO_2$ emissions and FDI inflows over this period, indicating weak support for the hypothesis [18].

To examine how environmental regulation shapes the FDI model and thereby assess the pollution haven hypothesis, a study [4] analyzed the structure of South Korean FDI over the period. 2000–2007, a period during which Korean companies relied on old production technologies despite the rapid strengthening of environmental standards. The results obtained reveal strong evidence that polluting industries tend to invest more in countries where environmental regulations are laxer, both in terms of the number of investments (intensive margin) and the number of new foreign subsidiaries (extensive margin). A similar result is obtained when analyzing imports.

For the Chinese case, the challenge of climate change seems increasingly difficult to meet. However, there is yet no consensus on whether the influx of FDI should be responsible for the domestic situation of rising $CO_2$ emissions. One study [21] aim to examine the validity of the pollution haven hypothesis using annual time series data for the period 1980–2012. By adopting the ARDL approach, the study integrated gross domestic product (GDP), square of GDP, energy use, foreign direct investment (FDI), economic freedom, urbanization, financial development as well as trade openness as the main factors of $CO_2$ emissions. The results of the limit tests show that there is a stable long-term relationship between the variables chosen in each model. $CO_2$ emissions would increase by 0.058% with a 1% increase in inward FDI flows (% of GDP), indicating that the pollution haven hypothesis does indeed exist in China [21].

## 3   Data and Methodology

In view of the above, we find that the issue of the effect of FDI flows on carbon emissions remains controversial, both validated by empirical studies and sometimes dismissed. However, for the Tunisian case, this field of research remains poorly explored and the problem of the environment little exploited, hence the obligation to initiate studies in this direction.

A first step would be to deal with the issue of foreign direct investment flows, which represents an important lever for Tunisian economic growth, from an environmental point of view and to verify the existence of the hypothesis of the pollution haven in Tunisia by identifying the impact of these flows on $CO_2$ emissions (Carbon emissions) both in the brief term and in the lengthy term. Therefore, our central question would be to know to what extent inward FDI impacts carbon emissions in Tunisia, and if this impact varies over time or remains the same if over the cost or the long term.

To answer this problem, we use the appropriate empirical methods and optimize the choice of the variables concerned. The approach and the model to be implemented, the phases of experimental assessment and the explanation of the dataset should additionally be debated.

## 3.1 Stylish Facts

It is first necessary to present the findings in the form of stylized facts before proceeding to an experimental demonstration of the supposed correlation amongst inflows of FDI and $CO_2$ emissions per capita. Then, Fig. 1 demonstrates the evolution of FDI in terms of flow in billions of current dollars and carbon emissions in tons per capita in Tunisia as presented in Fig. 1.

Between 1960 and 2012, carbon emissions per capita experienced a positive trend almost linearly with an average annual growth rate of 3.88%. On the other hand, net inward FDI flows evolved weakly and little volatile between 1970 and 1987. But from the 1990s until 2012, outward FDI stocks increased exponentially; thanks to several factors set out below; and a very volatile variation because of the exogenous shocks of the world economy such as the world financial crisis of 2007. Such an evolution can be explained by the integration of Tunisia into the international economy through commercial and political treaties investment.

Also, the two variables have experienced a downward trend since 2013–2015 with an average annual decrease rate of 3.7% for carbon emissions per capita and 12.99% for FDI inflows to Tunisia.

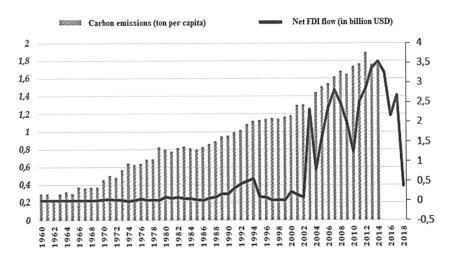

**Fig. 1** The evolution of net FDI inflows and $CO_2$ emissions in ton per capita during the period 1960–2018 in Tunisia. *Source* World Bank (2019)

It turns out that said FDI and carbon emissions per capita in Tunisia are positively associated as an upward trend in one indicates the other is following the same pace, and vice versa. However, correlation does not mean causation, an econometric demonstration is needed to verify such an impact.

## 3.2 The Econometric Tests

The literature review provides an opportunity to identify the different channels through which foreign direct investment can impact carbon emissions in the host country, namely economic growth, industrialization, the rate of urbanization, open rate, etc. However, our objective would not be to identify the role of the transmission channels between FDI and carbon emissions but to capture the overall effect without going through the said auxiliary variables. The cointegration approach offers a most important reason in preference of the bivariate methodology, while avoiding estimation problems such as serial correlation, heteroskedasticity, and multicollinearity. In addition, this approach permits us to identify the influence of foreign direct investment on short-term and long-term carbon emissions using the dynamic Angle-Granger methodology.

In accordance with current practice in studies employing the bivariate cointegration approach [1, 3, 8, 10, 11], we propose a model bivariate presented below.

## 3.3 The Methodology

Drawing inspiration from studies employing the dynamic bivariate cointegration methodology, two empirical models are presented as follows:

Model (1) presents the impact of *IDE* on *EC* in the long term:

$$CE_t = \alpha + \beta \text{FDI}_t + \varepsilon_t \tag{1}$$

With $CE_t$ carbon emissions in Tunisia during the year $t$ (in metric tons per capita) and FDI and net inflows of FDI in Tunisia during the same year $t$ (in current US dollars).

Model (2) presents the dynamic impact of FDI on short/long term CEs by assessing the DECM (Dynamic Error Correction Model) proposed by Engle-Granger as follows:

$$\Delta CE_t = \alpha + \lambda \varepsilon_{t-1} + \sum_q \theta_q \Delta CE_{t-q} + \sum_q \beta_q \Delta \text{FDI}_{t-q} + \mu_t \tag{2}$$

In this equation, $q$ is the optimal offset, $\theta$ is the quickness of correction to the lengthy-term equilibrium path, $\lambda$ is the lengthy-term influence of on CE, $\beta$ is the brief-term impact of the IDE on the CEs and $\mu$ is the white noise error term.

## 3.4  Data

To confirm the effect of inward FDI flows on $CO_2$ emissions in Tunisia, the empirical analysis is based on unbalanced data between 1960 and 2019, i.e., 60 years. The dataset for the 2 indicators is obtained from the World Bank database which defines foreign direct investment as any net investment income allowing the acquisition of a lasting participation (10% or more of the stocks with voting rights) in a business functioning in an economy supplementary than that of the stockholder. It is the quantity of equity, invested revenues, other lengthy-term capital, and short-term capital recorded in the balance of payments. Also, the World Bank defines carbon emissions as those emanating from the combustion of remnant fuels and the manufacturing of cement. Then, they incorporate $CO_2$ emissions from the expenditure of consistent, liquefied, and gaseous fuels and from burning.

## 4  Empirical Results

### 4.1  Unit Root Tests

Before testing cointegration and identifying the impact of FDI on long-term CEs, it should be checked whether the variables are integrated in first order I (Table 1).

First, we use four tests: Dickey-Fuller Augmented (ADF), Dickey-Fuller GLS (DF GLS-ERS), Philipe-Perron (PP) and Kwiatkowski–Phillips–Schmidt–Shin (KPSS). According to the four-unit root tests, the two variables are not stationary in level since the null assumption of a unit root is not refused for the ADF, DF GLS and PP tests, and the null assumption of stationarity of the variables is eliminated at the 1% significance level for the KPSS test. Moreover, the two variables become stationary at the first difference since the null hypothesis of a unit root is denied at the significance

**Table 1**  Unit root test

| Variables | ADF | DF GLS (ERS) | PP | KPSS |
| --- | --- | --- | --- | --- |
| FDI | I(1)** | I(1)** | I(1)** | I(1)** |
| CE | I(1)* | I(1)** | I(1)** | I(1)** |

*Notes* ***, **, * denote significance at the 1%, 5% and 10% level, respectively. Also, the choice of the maximum postponement is automated by the software employing the SIC as a character reference

**Table 2** Fisher cointegration assessment (Johansen combined)

| Null hypothesis | Max-Eigen value | Trace |
|---|---|---|
| None cointegration | 13.52674 (0.0265)** | 13.65384 (0.0379)** |
| There is at most a cointegration relation | 0.04256 (0.8695)* | 0.04256 (0.8704)* |

Notes ***, **, * denote significance at the 1%, 5% and 10% level, respectively. Also, the choice of the maximum postponement is automated by the software employing the SIC as a character reference. The trend hypothesis: no deterministic trend

level of 1% for the tests of ADF, DF GLS and PP, and since the null assumption of the stationarity of the variables is not rejected for the KPSS test. We then conclude that the two indicators are integrated in the first order I (1).

## 4.2 Cointegration Results

After having confirmed that the two variables are integrated in the same order I (1), it is advisable to test the cointegration amongst said variables using Fisher's tests as demonstrated in Table 2.

The Fisher, Trace and Max-Eigen assessments, founded on Johansen's method, allocate examination for cointegration amongst these 2 indicators. First, the 2 Fisher tests refuse the null assumption of non-cointegration among FDI and $CO_2$ emissions at the 5% level and acknowledge the alternative assumption that the indicators are cointegrated. Furthermore, the 2 tests cannot refuse the null assumption, that there is at greatest one cointegrating connection. To this end, one could assume, corresponding to the 2 Fisher tests, that the indicators are cointegrated, specifically the presence of a lengthy-term relationship.

## 4.3 Estimation of the Lengthy-Term Impact Coefficient

Founded on the beyond outcomes, we estimate the lengthy-term parameter of the model (1) employing 2 cointegration regression techniques of Dynamic Ordinary Least Squares (DOLS) and Fully Modified Ordinary Least Squares (FMOLS). The DOLS regression is utilized to prevent estimation difficulties associated to autocorrelation, heteroscedasticity, and multi-collinearity, while the FMOLS regression is employed to prevent difficulties of normality of residuals connected to the sample, insignificant, endogeneity of the independent variable (Table 3).

The two cointegration regression methods show that net FDI inflows to Tunisia have a positive influence on $CO_2$ emissions with a significant level of 1%. In the case of the DOLS technique, an expansion of US $1 million in FDI could consequence

**Table 3** Estimation of the long-term coefficient

|        | Coefficient | P-value |
|--------|-------------|---------|
| DOLS   | $1.25 \times 10^{-10}$ | 0.0002*** |
| FMOLS  | $6.29 \times 10^{-10}$ | 0.0001*** |

*Notes* ***, **, * denote significance at the 1%, 5% and 10% level, respectively. Also, the choice of the maximum postponement is automated by the software employing the SIC as a character reference. The trend hypothesis: no deterministic trend

**Table 4** Results of causality in the sense of Angle-Granger

| Dependent variable | Independent variables | |
|--------------------|-----------------------|---|
|                    | $\Delta$FDI (Short-term impact) | $\varepsilon_{t-1}$ (long-term impact) |
| $\Delta$CE         | $-4.82 \times 10{-}11$ (0.0273)** | $-0.418654$ (0.0081)*** |

*Notes* ***, **, * denote significance at the 1%, 5% and 10% level, respectively. Also, the choice of the maximum postponement is automated by the software employing the SIC as a character reference. The trend hypothesis: no deterministic trend

from 0.00125 metric tons of $CO_2$ per capita while for the FMOLS method an increase of US $1 million in FDI would be followed by an expansion of 0.00629 metric tons of $CO_2$ per capita.

## 4.4  Granger-Engle Causality Estimation

The Engle-Granger test permits us to dynamically investigate the causality, formerly assumed, amongst the 2 variables, to distinguish amongst the brief term and the lengthy term. Taking $CO_2$ emissions as the dependent variable and EDI as an explicative variable, it transforms out that the long-term causality of FDI to $CO_2$ emissions is validated by the Engle Granger test since the value of u ($-1$) is significant at the 1% level. However, in the short term, FDI has a negative impact on per capita carbon dioxide emissions at the significant 5% level of which an increase of US $1 million in FDI could consequence in a reduction of 0.000482 metric tons of $CO_2$ per capita (Table 4).

## 5  Conclusion

In this paper, we use unscrambled time series data, over the period 1960–2019, to investigate the association amongst foreign direct investment inflows and $CO_2$ emissions to examine the authenticity of the pollution haven assumption for Tunisia. First, we employ the unit root test, which demonstrated that two indicators are integrated in the first order I, then the cointegration analysis asserts the existence of a lengthy-term causality ranging from FDI to carbon dioxide emissions.

Based on the above results, dynamic Ordinary Least Squares and Fully Modified Ordinary Least Squares cointegration regression techniques were assessed, and their outcomes suggest that the foreign direct investment inflows have a positive influence on $CO_2$ emissions per capita with a significant level of 1%. However, the Granger-Engle causality test, which utilized to investigate short-term and long-term causality amongst used variables, demonstrates that per capita flows, and $CO_2$ emissions have a negative short-term causative connection. in other words, FDI flows contribute to the reduction of $CO_2$ emissions, whereas in the lengthy run, they contribute to the increase of these emissions. In the long term, the results of this study confirm the results of our previous study on the legality of the pollution haven assumption for African states throughout the identical period of study [7].

Indeed, we can justify the involvement of per capita to the decrease of $CO_2$ emissions by the statement that investors, when entering Tunisia, have a tendency to appreciate environmental procedures and worldwide requirements. However, in the long term, it is possible that these firms will become less and less respectful of these standards and therefore their contributions remain harmful to the environment. For these reasons, it is recommended that Tunisia strengthen its environmental policy, by increasing processes and mechanisms for decreasing $CO_2$ emissions, such as ecological taxes and carbon secure and storage area, as well as monitoring. Activity of polluting industries especially after their long-term installations in Tunisia.

**Acknowledgements** The author would to think the editor and the reviewers for their supportive and important remarks to improve the quality of this paper.

# References

1. Ajaga E, Nunnenkamp P (2008) Inward FDI, value added and employment in US States: a panel cointegration approach. Kiel Institute for the World Economy, Kiel Working Papers, No. 1420
2. Akbostanci E, Tunç Gİ, Türüt-Aşik S (2007) Pollution haven hypothesis and the role of dirty industries in Turkey's exports. Environ Dev Econ 12(2):297–322. https://doi.org/10.1017/S13 55770X06003512
3. Chakraborty C, Nunnenkamp P (2008) Economic reforms, FDI, and economic growth in India: a sector level analysis. World Dev 36(7):1192–1212. https://doi.org/10.1016/j.worlddev.2007. 06.014
4. Chung S (2014) Environmental regulation and foreign direct investment: evidence from South Korea. J Dev Econ 108:222–236. https://doi.org/10.1016/j.jdeveco.2014.01.003
5. Copeland BR, Taylor MS (1994) North-South trade and the environment. Q J Econ 109(3):755–787. https://doi.org/10.2307/2118421
6. Fadhil MA, Almsafir MK (2015) The role of FDI inflows in economic growth in Malaysia (Time series: 1975–2010). Procedia Econ Finan 23:1558–1566. https://doi.org/10.1016/S2212-5671(15)00498-0

7. Gharnit S, Bouzahzah M, Soussane JA (2019) Foreign direct investment and pollution havens: evidence from African countries. Arch Bus Res 7(12):244-252. https://doi.org/10.14738/abr.712.7531
8. Herzer D, Nunnenkamp P (2012) The effect of foreign aid on income inequality: evidence from panel cointegration. Struct Chang Econ Dyn 23(3):245–255. https://doi.org/10.1016/j.strueco.2012.04.002
9. Jbara BW (2007) Exploring the causality between the pollution haven hypothesis and the environmental Kuznets curve. Honors Projects. https://digitalcommons.iwu.edu/econ_honproj/21
10. Khan MA, Khan SA (2011) Foreign direct investment and economic growth in Pakistan: a sectoral analysis. Pakistan Institute of Development Economics, PIDE-Working Papers, No. 2011/67.
11. Mehmood B, Siddiqui W (2013) What causes what? Panel cointegration approach on investment in telecommunication and economic growth: case of Asian countries
12. Melnyk L, Kubatko O, Pysarenko S (2014) The impact of foreign direct investment on economic growth: case of post communism transition economies. Probl Perspect Manag 12:17–24
13. Mutafoglu TH (2012) Foreign direct investment, pollution, and economic growth: evidence from Turkey. J Developing Soc 28(3): 281-297. https://doi.org/10.1177/2F0169796X12453780
14. Nistor P (2014) FDI and economic growth, the case of Romania. Procedia Econ Finan 15:577–582. https://doi.org/10.1016/S2212-5671(14)00514-0
15. Porter ME, Linde CVD (1995) Toward a new conception of the environment-competitiveness relationship. J Econ Perspect 9(4): 97–118. https://doi.org/10.1257/jep.9.4.97
16. Publications Office of the European Union (2018) Fossil $CO_2$ emissions of all world countries—2018 Report. EU Science Hub, European Commission
17. Riti JS, Sentanu IGEPS, Cai A, Sheikh S (2016) Foreign direct investment, manufacturing export and the environment in Nigeria: a test of pollution haven hypothesis. NIDA Dev J 56(2): 73–98. https://so04.tci-thaijo.org/index.php/NDJ/article/view/57944
18. Shofwan S, Fong M (2011) Foreign direct investment and the pollution haven hypothesis in Indonesia. J Law Gov 6(2): 27–35. https://doi.org/10.15209/jbsge.v6i2.202
19. Sokang K (2018) The impact of foreign direct investment on the economic growth in Cambodia: empirical evidence. Int J Innov Econ Dev 4(5):31–38. http://dx.doi.org/https://doi.org/10.18775/ijied.1849-7551-7020.2015.45.2003
20. Solarin SA, Al-Mulali U, Musah I, Ozturk I (2017) Investigating the pollution haven hypothesis in Ghana: an empirical investigation. Energy 124:706–719. https://doi.org/10.1016/j.energy.2017.02.089
21. Sun C, Zhang F, Xu M (2017) Investigation of pollution haven hypothesis for China: an ARDL approach with breakpoint unit root tests. J Cleaner Prod 161:153–164. https://doi.org/10.1016/j.jclepro.2017.05.119
22. Zeb N, Qiang F, Sharif S (2014) Foreign direct investment and unemployment reduction in Pakistan. Int J Econ Res 5(2):10–17

# Factors Affecting the Intention to Use Cloud Accounting in SMEs: Evidence from Vietnam

Malik Abu Afifa, Hien Vo Van, and Trang Le Hoang Van

**Abstract** Cloud accounting and its application in business activities are relatively new research topics in Vietnam. This study investigates the factors affecting the intention to use cloud accounting according to managerial and accountant perspective in small and medium-sized enterprises (SMEs). It provides empirical evidence from Vietnam. Data is collected quantitatively utilizing an online-survey questionnaire. It was sent by email to managers and accountants in 300 SMEs. The results show that perceived usefulness and availability to embrace technology have a positive impact on the intention to use cloud accounting. A rather interesting result is that perceived ease of use technology has also a positive effect on the intention to use cloud accounting. These results provide reliable evidence for managers and accountants to strategically accelerate the cloud accounting application process and improve financial and business efficiency. These results are also a reference for software manufacturers providing products with more suitable features.

**Keywords** Cloud accounting · Intent to use · Perceived usefulness · Availability to embrace technology · SME · TAM model

## 1 Introduction

Using new technology in small and medium-sized enterprises (SMEs) provides several benefits and functionality, including decrease efforts and costs, improve effective services, as well as improve managerial capabilities and fewer errors [1, 2]. A cloud technology such as cloud accounting (CA) is a modern idea for processing accounting data through the use of a collection of information distribution systems and applications within the idea of cloud information without requiring users to be

M. Abu Afifa (✉)
Faculty of Business, Accounting Department, Al-Zaytoonah University of Jordan, Amman, Jordan
e-mail: M.abuafifa@zuj.edu.jo

H. V. Van · T. Le Hoang Van
Tien Giang University, My Tho, Tien Giang Province, Vietnam

© The Author(s), under exclusive license to Springer Nature Switzerland AG 2022
S. G. Yaseen (ed.), *Digital Economy, Business Analytics, and Big Data Analytics Applications*, Studies in Computational Intelligence 1010,
https://doi.org/10.1007/978-3-031-05258-3_47

aware of physical location and system structure [3, 4]. CA, often known as online accounting, serves the same purpose as accounting software that is installed on the client's computer. However, it is run on the server of the cloud computing service provider [5, 6].

According to previous literature, CA is the most effective tool for assisting businesses in achieving their financial goals. CA not only transforms the way companies operate swiftly and effectively, but it also saves time and costs to provide better financial performance [7–9]. Accounting systems based on cloud computing can assist businesses in swiftly disclosing information and improving production efficiency [10]. This encourages the widespread adoption of cloud computing, not just in companies that provide services, but also in companies that use the services.

Vietnam is one of the countries that has established a set of criteria and technical specifications for cloud computing infrastructure. According to the report of ANA Technology Service Trade Joint Stock Company (ANA) in 2020, in the context of global competition, utilizing an accounting system based on cloud computing applications will assist Vietnamese companies in maintaining good management and a steady financial source [11]. As well, an accounting system based on cloud computing applications is a must for firms to establish a firm foothold in the market. Therefore, CA provides more benefits than traditional accounting, and its use in company activities is an unavoidable trend.

Previous studies (such as [12]) documented that the intention to use CA in SMEs is substantially higher than in large-sized enterprises. However, the use of CA in Vietnam is still in its infancy [13]. Hence this study aims to investigate the factors affecting the intention to use CA in SMEs. It provides empirical evidence from Vietnam.

The contributions of this study can present as follows. Firstly, this study provides a reliable scientific basis for practitioners who can make decisions relevant to using CA in enterprises to improve financial performance, business efficiency and competitiveness. Secondly, it provides empirical evidence from Vietnam as a developing country, where there have been fewer attempts to search in this context in the developing countries compared to developed countries. Thirdly, the results of this study can show the main factors affecting the intention to use CA in SMEs. These results help managers, policymakers, and other interested parties to develop policies and procedures that will increase the levels of cloud accounting applications and so achieve their benefits, especially in Vietnamese SMEs.

This paper's rest is organized as follows. Section 2 shows the theoretical framework. Section 3 presents data and methodology. Section 4 shows data analysis and results. Section 5 provides a discussion. The end section shows the conclusion.

## 2 Theoretical Framework

### 2.1 Cloud Accounting

The purpose of cloud computing is to deliver computer hardware and software applications over the Internet. It enables users to save data and access applications across multiple devices situated in various locations [14]. After being used in various types of businesses, cloud computing has made its way into the accounting profession. Accounting is inextricably linked to the business. It must be an integral part of the enterprise and play an important role in its operation. To accomplish this, the accounting model should be connected with other aspects of the business. This can increase the value of both the financial aspects and the enterprise itself [15, 16].

CA emerged as a result of the convergence of cloud computing and enterprises that provide accounting software [17]. It provides accounting services using cloud computing solutions [18, 6]. In reality, though a formal definition of CA has yet to be defined, it will be better articulated through its benefits and functions. Its key advantage is that it allows you to access accounting services without installing any software, lowering your investment expenses in computer infrastructure. Applications are accessed through web browsers [7]. Customer data is safely stored and processed on the server of the cloud service provider [9]. Additionally, Table 1 shows the comparison between CA and traditional accounting.

A cloud-based accounting solution offers the capacity to perform complicated activities and operations through an integrated online system since enterprises that adopt CA will reduce the number of accountants [19, 8, 20]. In truth, the provision of CA services is diversified, and it always meets the needs of the customer.

**Table 1** Comparison between CA and traditional accounting

| Criteria | CA | Traditional accounting |
| --- | --- | --- |
| Hardware and software investment costs | The cost for hardware and software is quite low | The high cost of hardware and maintenance costs after that |
| Human costs | Quite low | Quite high |
| Build system | Hire an outside cloud service provider | Build accounting system itself |
| Data saving | Distributed storage technique of many independent devices, large data storage space | Centralized storage technology, low data recovery ability if the machine is broken |
| Update | Service update is not interrupted Ensuring work progress | System must pause to update |
| Accessible ability | The system can be accessed anytime, anywhere as long as there is internet | The accounting system can only be used in local networks |

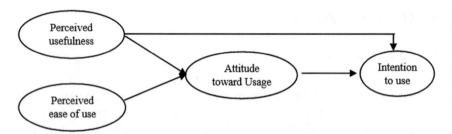

**Fig. 1** The technology acceptance model (TAM)

## 2.2 Background Theory

In 1989, Davis built the Technology Acceptance Model (TAM) with the concepts of usefulness and ease of use to explain the factors that affect an individual's intention when they accept receive certain technologies [5]. After that, TAM was used extensively in the assessment of user adoption. TAM also explains user behavior by assessing the impact of perception on attitude and intention. Davis's study developed and confirmed new scales for two specific variables, namely the perceived usefulness and perceived ease of use. These are two mechanical factors that decide the user's intention and acceptable behavior [21].

In terms of Fig. 1, the perceived usefulness factor is the degree to which a person believes that implementing information technology can improve performance. Individuals' performance can be shown in the fact that they are not very interested in information technology but believe that adopting it will give them prospects for progress [5, 22]. Besides, perceived ease of use is described as a person's perception of using information technology without exerting significant effort. In the other words, individuals who use technology can do tasks easily. Furthermore, TAM demonstrates that the usage of certain technology boosts productivity since perceived usefulness is closely related to behavioral intention [23].

Previous studies (such as [24, 25, 26, 6], and [13]) have used the TAM model to investigate the intention to adopt and use information technology in enterprises. Therefore, this study uses the TAM model to investigate the factors affecting the intention to use CA in SMEs.

## 2.3 Research Framework and Hypotheses Development

Prior studies based on TAM showed that perceived usefulness and perceived ease of use have a strong impact on intention to use information technology in enterprises [26, 6]. Other studies also indicated that the factor of attitude toward usage has an insignificant influence on intention to use it [25, 22, 21, 13, 27]. Besides, the

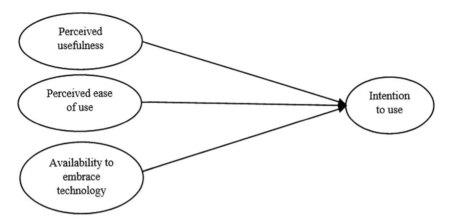

**Fig. 2**  Research framework

availability to embrace technology is regarded as a relatively novel factor that is highly related to the intention to use new technology [22].

Finally, the above discussions motivate us to propose a model to evaluate the factors that affect accountants' and managers' intention to use CA in Vietnamese SMEs (see Fig. 2).

### 2.3.1  Perceived Usefulness, Perceived Ease of Use, and Intention to Use

Davis [5] confirmed that the intention to use information technology is affected by perceived usefulness and perceived ease of use. However, the impact of perceived usefulness on the intention to use technology is higher than perceived ease of use. Recently, Ardiansah et al. [24] investigated the intention to purchase through e-commerce channels, and the results showed that perceived usefulness has a stronger influence on the intention to purchase than perceived ease of use. A study by Iqbal and Bhatti [23] pointed out that both perceived usefulness and perceived ease of use affect the intention to learn via mobile devices.

Rogers [28] studied the intention to use the computer accounting system in small enterprises in central Ohio, USA. The result showed that both perceived usefulness and perceived ease of use have positive and strong influence on the intention to use computer accounting system in small enterprises. Lanlan et al. [26] also indicated that perceived usefulness and perceived ease of use have a positive correlation with an intention to use the computer accounting system in small enterprises in Shaanxi, China. Besides, Chen [29] showed that perceived usefulness and perceived ease of use have a positive and strong effect on the intention to use cloud computing software in Taiwanese enterprises. Pramuka and Pinasti [6] also documented similar results when investigating factors affecting the intention to use cloud computing software in small enterprises in Purwokerto, Indonesia.

Continuously, the current study aims to investigate the effect of perceived usefulness and perceived ease of use on the intention to use CA in Vietnamese SMEs, and thus the current study's first and second hypotheses are as follows:

*H1*: *Perceived usefulness positively affects the intention to use CA.*

*H2*: *Perceived ease of use positively affects the intention to use CA.*

### 2.3.2 Availability to Embrace Technology and Intention to Use

Based on prior research, the availability to embrace technology has been integrated into the TAM model in the context of customers adopting electronic service systems. For example, Lin et al. [30] documented that availability to embrace technology affects the intention to use the stock trading system. Habiba et al. [22] noted that the application of computerized accounting information systems in SMEs affected by the availability to embrace technology. As well, Damerji and Salimi [31] argued that the intention to use new technology is based on the availability to embrace technology in enterprises. To employ new technology, it is necessary to share knowledge and obtain support from specialists.

Therefore, the third hypothesis of this study can be structured as follows:

*H3*: *Availability to embrace technology positively affects the intention to use CA.*

## 3 Data and Methodology

### 3.1 Sampling and Data Collection Method

This study's population consists of all listed SMEs in Vietnam's North, Central, and South. The kind of SMEs is determined by the number of employees in Vietnam (less than 200 employees). This study uses the convenient sampling method, with a random sample of 300 SMEs. The primary data is then gathered via an online-survey questionnaire using the Google Docs forms. It was distributed by email to managers and accountants in the selected sample. Three emails were sent as reminders to boost the response rate and encourage these individuals to respond. However, only 77 surveys were returned, representing a response rate of 25.67%.

### 3.2 Measurement of Variables

The model of this study includes three independent variables, namely the perceived usefulness, perceived ease of use, and availability to embrace technology, and one dependent variable (intention to use CA). Table 2 outlines how these variables were measured and where they came from.

**Table 2** Measurement of variables

| Factor | Symbol | Variable type | Measurement | Sources |
|---|---|---|---|---|
| Perceived usefulness | PU | Independent | 4 items using a 5- point likert scale | [31, 5, 13] |
| Perceived ease of use | PE | Independent | 3 items using a 5- point likert scale | |
| Availability to embrace technology | AE | Independent | 4 items using a 5-point likert scale | [22, 30] |
| Intention to use cloud accounting | IU | Dependent | 5 items using a 5- point likert scale | [5, 26, 13, 30] |

## 3.3   Data Analysis Method

Descriptive statistics are used in this study to emphasize the features of the study sample and to clarify the independent and dependent variables. Cronbach's alpha coefficient is also used in the study to assess measurement reliability. Following that, the Exploratory Factor Analysis (EFA) test is used to reconfigure scale groups, examine variable group convergence and differentiation, eliminate junk observations, finally, improve study outcomes [32, 33]. The regression analysis model is then utilized to investigate the study hypotheses.

## 4   Data Analysis and Results

## 4.1   Profile of Respondents

Table 3 displays the descriptive statistics of the respondents' profiles. The results indicate that the male and female proportions of respondents account for 49.4% and 50.6% respectively. The rate of graduating from college, bachelor and postgraduate (Masters or Ph.D.) is 22.1%, 70.1% and 7.8% respectively. Respondents over 40 years old account for 35%, followed by 30–40 years old at 33.8%, and those under 30 years old at 31.2%. Besides, it's worth noting that the proportions of managers and accountants are identical (46.8%), as well as the percentage of respondents who are working in other jobs (6.4%). The majority of those surveyed have 5–10 years of experience (36.3%). Respondents with more than 10 years of experience account for 32.5% of those surveyed, while those with less than 5 years of experience account for 31.2%.

**Table 3** Profile of respondents

| Sample characteristics | | Quantity | Percent |
|---|---|---|---|
| Gender | Male | 38 | 49.4 |
| | Female | 39 | 50.6 |
| Academic level | College | 17 | 22.1 |
| | Bachelor | 54 | 70.1 |
| | Postgraduate (Masters or Ph.D.) | 6 | 7.8 |
| Age | Under 30 | 24 | 31.2 |
| | From 30 to 40 | 26 | 33.8 |
| | Over 40 | 27 | 35.0 |
| Job position | Manager | 36 | 46.8 |
| | Accountant | 36 | 46.8 |
| | Other job position | 5 | 6.4 |
| Experience | Less than 5 years | 24 | 31.2 |
| | From 5 to 10 years | 28 | 36.3 |
| | Over 10 years | 25 | 32.5 |
| Total | | 77 | 100 |

## 4.2 Descriptive Statistics of Variables

Based on the results in Table 4, the variables PU and IU have quite similar indicators. The average value of PU and IU is 3.756 and 3.740 respectively, as well as the minimum value of these two variables is 3.649 and 3.636 respectively, while their maximum values are 3.844 and 3.831 respectively. The variable PE has an average value of 4.013, the minimum value is 3.974, and the maximum value 4.065. The variable AE has the maximum value of 3.078, the minimum value of 3.000, and the average value of 3.029.

**Table 4** Descriptive statistics of variables and reliability

| Variables | Mean | Minimum | Maximum | Std. deviation | Cronbach's Alpha |
|---|---|---|---|---|---|
| PU | 3.756 | 3.649 | 3.844 | 0.792 | 0.784 |
| PE | 4.013 | 3.974 | 4.065 | 0.750 | 0.813 |
| AE | 3.029 | 3.000 | 3.078 | 0.781 | 0.878 |
| IU | 3.740 | 3.636 | 3.831 | 0.770 | 0.762 |

## 4.3  Goodness of Measures

To check the data validity and reliability (or consistency), Cronbach's alpha coefficient is used in the study. The results in Table 4 show that the Cronbach's alpha coefficients for all variables in the study model are more than 0.7, which means that the strength of that consistency is satisfactory for analysis [34, 35]. Additionally, the Exploratory Factor Analysis (EFA) test is used to determine the core configuration of a large number of variables. The results of the EFA test indicate that the KMO coefficient is 0.746, and the significance of the Bartlett test is 0.000 (< 0.05) (see Table 5). As well, the factor loadings for all items are more than 0.7, which means that the factor takes enough variance from that variable (see Table 6). Hence we categorize the factors as follows: The first factor is AE (AE1, AE2, AE3, and AE4); the second factor is PU (PU1, PU2, PU3, and PU4); and the third factor is PE (PE1, PE2, and PE3).

**Table 5** Test of KMO and Bartlett

| Kaiser–Meyer–Olkin measure of sampling adequacy | | 0.746 |
|---|---|---|
| Bartlett's test of sphericity | Approx. Chi-Square | 430.129 |
| | Df | 55 |
| | Sig | 0.000 |

**Table 6** Matrix of rotation factors

| Factors | 1 | 2 | 3 |
|---|---|---|---|
| AE1 | 0.863 | | |
| AE2 | 0.863 | | |
| AE3 | 0.846 | | |
| AE4 | 0.783 | | |
| PU4 | | 0.796 | |
| PU1 | | 0.763 | |
| PU2 | | 0.739 | |
| PU3 | | 0.727 | |
| PE2 | | | 0.879 |
| PE1 | | | 0.826 |
| PE3 | | | 0.825 |
| Initialization Eigenvalues | 3.886 | 2.562 | 1.360 |
| % variance | 35.325 | 23.287 | 12.363 |
| % accumulate | 35.325 | 58.612 | 70.975 |

**Table 7**  Results of the regression analysis model

| Model | Unstandardized coefficients | | Standardized coefficients (Beta) | t value | Sig | VIF | Hypothesis testing |
|---|---|---|---|---|---|---|---|
| | (B) | Std. error | | | | | |
| Constant | 0.830 | 0.345 | | 2.406 | 0.019 | | |
| PU | 0.571 | 0.072 | 0.640 | 5.902 | 0.000 | 1.260 | Supported—H1 |
| PE | 0.026 | 0.064 | 0.031 | 1.412 | 0.041 | 1.067 | Supported—H2 |
| AE | 0.217 | 0.066 | 0.264 | 3.313 | 0.001 | 1.220 | Supported—H3 |
| R2 adjusted | | | 0.605 | | | | |
| Durbin-Watson | | | 1.921 | | | | |
| F | | | 39.734 | | | | |
| Sig. (ANOVA) | | | 0.000 | | | | |
| N | | | 77 | | | | |

## 4.4  Results of Regression Analysis Model

Table 7 shows the results of the regression analysis model. The results indicate that the variables PU (Beta = 0.640), PE (Beta = 0.031), and AE (Beta = 0.264) have Sig. < 0.05, indicating that their effects on IU are statistically and positively significant. VIF coefficients are all less than 5; therefore, there is no multicollinearity problem [32, 33]. Additionally, Sig. value of ANOVA equals 0.000 (<0.05) and the adjusted $R^2$ is 0.605, so this model can be extrapolated and applied to the whole. Finally, Hypotheses H1, H2, and H3 are accepted.

## 5  Discussion

Cloud-based accounting is still quite new in Vietnam, so determining the decisive factors in the intention to use this tool is essential. The current study aims to investigate the factors affecting the intention to use CA in Vietnamese SMEs using the TAM model. The results of this study showed that the factors PU, PE, and AE affect IU. These results are strongly supported by previous studies that use information technology behaviors [36, 29, 5, 37]. They stated that implementing CA will raise productivity and efficiency in SMEs by reducing costs such as human resources and investment in computer technology. Besides, practitioner awareness and promotion of the remarkable benefits of the CA above traditional accounting raise the intention to use it.

Managers and accountants appreciate PU when they believe that CA will provide tools to make accounting work easier [13, 6]. In this study, it is quite interesting that PE has a negligible positive effect on IU. It can be explained that if a new technology is simple and convenient for human operations but does not provide many predicted benefits, managers and accountants will find it difficult to use. These results are also fully supported by previous studies (such as [24, 5, 6]). On the other hand, these results are not supported by Mohd Sam et al. [38], when they noted that the PE factor has a high negative association with the implementation of a computerized accounting system in SMEs. Furthermore, several earlier studies on information technology usage behavior do not even include the PE element in the measurement model, such as the intention to purchase e-commerce channels [25], the intention to use the Internet for learning [27], and the intention to use E-Payment System (EPS) [36].

CA is consistently used by SMEs that have the technical infrastructures, such as computer hardware and software, and a staff of technology specialists. This positive correlation is supported by the behavioral studies using the online stock trading system [30], and a computerized accounting system [22]. Damerji and Salimi [31] also documented the similar result when studying the intention to use artificial intelligence in accounting and auditing fields of accounting. The majority of these studies have also documented that, through innovative initiatives, SMEs have pushed their staff to learn and use this new accounting tool.

# 6  Conclusion

This study aims to determine the factors affecting the intention to use CA in SMEs. It provides empirical evidence from Vietnam. The convenient sampling method was used, with a random sample of 300 SMEs in North, Central and South Vietnam. Then, primary data was collected through an online-survey questionnaire using the Google docs forms. The online questionnaire was sent by email to managers and accountants. However, only 77 surveys were returned, representing a response rate of 25.67%.

The results showed that IU is positively affected by PU, PE and AE. The PU has the strongest influence and PE has the lowest influence. The results provide a reliable scientific basis for managers and accountants in SMEs that can accelerate the process of using CA technology. Enterprises that provide accounting software based on cloud platforms can also refer to make more suitable products. Furthermore, this study is based on Davis's TAM model for the intention to use information technology, hence it has added some empirical evidence to the body of the knowledge in this context.

Although this is a relatively new study that has made significant contributions, there are some limitations. While we apply the quantitative analysis method, the sample size in the study is rather small. Because of the recent global crisis (the covid-19 pandemic), the study only obtained data from 77 managers and accountants. More data may have an impact on the outcomes. However, due to the random sampling procedure, the sample does not include all Vietnamese SMEs. Further studies should scale up the sample size to ensure the representativeness of the study. Another limitation related to the enterprise characteristics (such as size, growth, profitability index, so on), where the model of this study does not include such characteristics. These characteristics as control variables may have an impact on the outcomes. Thus, further studies should use the enterprise characteristics variables as control variables to have more outcomes.

# Appendix: Research Instrument

I    **Profile of the individuals**

**1. Gender**
☐ Male                                          ☐ Female

**2. Working experience (years)**
☐ Less than 5                                   ☐ From 5 to 10
☐ Over 10

**3. Academic level**
☐ Bachelor                                      ☐ Postgraduate (Masters or PhD)
☐ College

**4. Job position**
☐ Manager                                       ☐ Accountant
☐ Other

**5. Age (years)**
☐ Under 30                                      ☐ From 30 to 40
☐ Over 40

II    **The Part of Respondents to Be Surveyed**

The following questions were answered based on perceptions with 5 levels: Totally disagree (1); Disagree (2); Neutral (3); Agree (4) and Totally agree (5).

| No. | Content |
|-----|---------|
| **1. Perceived usefulness (PU)** | |
| 1 | Cloud accounting is very useful for the job |
| 2 | Cloud accounting can increase productivity |
| 3 | Cloud accounting improves efficiency |
| 4 | Cloud accounting provides tools to help make accounting work easier |
| **2. Perceived ease of use (PE)** | |
| 5 | Interaction with cloud accounting is clear and easy to understand |

(continued)

(continued)

| No. | Content |
|-----|---------|
| 6 | It is easy to learn how to use cloud accounting |
| 7 | Existing accounting software easily connects data with cloud accounting |
| **3. Availability to embrace technology (AE)** | |
| 8 | Our firm are willing to invest in technical infrastructure to use cloud accounting |
| 9 | Our firm is willing to hire experts to use cloud accounting |
| 10 | Our firm has programs to encourage employees to learn and use cloud accounting |
| 11 | Our firm really wants to use a new accounting tool |
| **4. Intention to use cloud accounting (IU)** | |
| 12 | I have intend to use cloud accounting in the future |
| 13 | I intend to use cloud accounting frequently |
| 14 | Cloud accounting is convenient since it is easily integrated with devices connected to the internet; therefore, I recommend firms to use it |
| 15 | Cloud accounting is very useful; therefore, I recommend firms to use it |
| 16 | Cloud accounting is easy to use; therefore, I recommend firms to use it |

# References

1. Lutfi A (2020) Investigating the moderating effect of environment uncertainty on the relationship between institutional factors and ERP adoption among Jordanian SMEs. J Open Innov Technol Market Complex 6(3):91–110
2. Lutfi AA, Idris KM, Mohamad R (2016) The influence of technological, organizational and environmental factors on accounting information system usage among Jordanian small and medium-sized enterprises. Int J Econ Finan Issues 6(7S):240–248
3. Mihalache AS (2011) Cloud Accounting. Ovidius Univ Ann Econ Sci Ser 11(2):782–787
4. Prasad A, Green P (2015) Governing cloud computing services: reconsideration of IT governance structures. Int J Acc Inf Syst 19(1):45–58
5. Davis FD (1989) Perceived usefulness, perceived ease of use, and user acceptance of information technology. MIS Q 13(3):319–340
6. Pramuka AB, Pinasti M (2020) Does cloud-based accounting information system harmonize the small business needs? J Inf Organ Sci 44(1):141–156
7. Khanom T (2017) Cloud accounting: a theoretical overview. IOSR J Bus Manag 19(6):31–38
8. Prichici C, Ionescu B (2015) Cloud accounting—a new paradigm of accounting policies. SEA–Pract Appl Sci 1(7): 489–496
9. Wicaksono A, Kartikasary M, Salma N (2020) Analyze cloud accounting software implementation and security system for accounting in MSMEs and cloud accounting software developer. In: 2020 international conference on information management and technology, IEEE, Bandung, pp 538–543
10. Chang BY, Hai PH, Seo DW et al (2013) The determinant of adoption in cloud computing in Vietnam. In: 2013 international conference on computing, management and telecommunications, IEEE, Vietnam, pp 407–409
11. ANA Technology Service Trade Joint Stock Company (2020) Online accounting software and how is effective it for businesses? ANA, Vietnam

12. Van den Bergh K, Kloppers SR (2019) The absorption and usage of cloud accounting technology by accounting firms in Cape Town for services provided to their clients. Afr J Sci Technol Innov Dev 11(2):161–180

13. Le O, Cao Q (2020) Examining the technology acceptance model using cloud-based accounting software of Vietnamese enterprises. Manage Sci Lett 10(12):2781–2788

14. Buyya R, Yeo CS, Venugopal S (2008) Market-oriented cloud computing: vision, hype, and reality for delivering it services as computing utilities. In: 2008 10th IEEE international conference on high performance computing and communications, Shanghai, pp 5–13

15. Oliveira T, Thomas M, Espadanal M (2014) Assessing the determinants of cloud computing adoption: an analysis of the manufacturing and services sectors. Inf Manage 51(5):497–510

16. Senarathna I, Wilkin C, Warren M et al (2018) Factors that influence adoption of cloud computing: an empirical study of Australian SMEs. Australas J Inf Syst 22(1):1–31

17. Lewis SX, Burks EJ, King EW et al (2012) Cloud computing: items professional firms consider in selecting data storage firms. J Case Res Bus Econ 4(1):1–12

18. Dimitriu O, Matei M (2014) A new paradigm for accounting through cloud computing. Procedia Econ Finan 15(1):840–846

19. Arsenie-Samoil MD (2011) Cloud accounting. Ovidius Univ Ann Econ Sci Ser 9(2):782–787

20. Zhang L, Gu W (2013) The simple analysis of impact on financial outsourcing because of the rising of cloud accounting. Asian J Bus Manage 5(1):140–143

21. Lai P (2017) The literature review of technology adoption models and theories for the novelty technology. J Inf Syst Technol Manag 14(1):21–38

22. Habiba Y, Azhar MN, Annuar BMN et al (2019) Computerized accounting information system adoption among small and medium enterprises in Addis Ababa, Ethiopia. Int J Account Finan Bus (IJAFB) 4(19):44–60

23. Iqbal S, Bhatti AZ (2015) An investigation of university student readiness towards m-learning using technology acceptance model. Int Rev Res Open Distrib Learn 16(4):83–103

24. Ardiansah M, Chariri A, Rahardja S et al (2020) The effect of electronic payments security on e-commerce consumer perception: an extended model of technology acceptance. Manage Sci Lett 10(7):1473–1480

25. Arora S, Sahney S (2018) Antecedents to consumers' show rooming behavior: an integrated TAM-TPB framework. J Consum Mark 35(4):438–450

26. Lanlan Z, Ahmi A, Popoola OMJ (2019) Perceived ease of use, perceived usefulness and the usage of computerized accounting systems: a performance of micro and small enterprises (MSEs) in China. Int J Recent Technol Eng 8(2):324–331

27. Mallya J, Lakshminarayanan S (2017) Factors influencing usage of internet for academic purposes using technology acceptance model. DESIDOC J Libr Inf Technol 37(2):119–124

28. Rogers AD (2016) Examining small business adoption of computerized accounting systems using the technology acceptance model. Ph.D. thesis, Walden University, Washington

29. Chen LY (2015) Determinants of Software-as-a-Service adoption and intention to use for enterprise applications. Int J Innov Appl Stud 10(1):138–148

30. Lin CH, Shih HY, Sher PJ (2007) Integrating technology readiness into technology acceptance: the TRAM model. Psychol Mark 24(7):641–657

31. Damerji H, Salimi A (2021) Mediating effect of use perceptions on technology readiness and adoption of artificial intelligence in accounting. Account Educ, 1–24

32. Hair JF, Black WC, Babin BJ et al (2006) Multivariate data analysis, 6th edn. Prentice-Hall, Uppersaddle River

33. Hair JF, Black W, Babin B et al (2009) Multivariate data analysis, 7th ed. Pearson Prentice Hall, Hoboken

34. Nguyen DT (2012) Scientific research method in business. Labor—Social Publishing House, Hanoi

35. Nunnally JC, Bernstein IH (1994) Psychometric theory, 3rd edn. McGraw-Hill, New York

36. Barkhordari M, Nourollah Z, Mashayekhi H et al (2017) Factors influencing adoption of e-payment systems: an empirical study on Iranian customers. IseB 15(1):89–116

37. Suki NM, Suki NM (2011) Exploring the relationship between perceived usefulness, perceived ease of use, perceived enjoyment, attitude and subscribers' intention towards using 3G mobile services. J Inf Technol Manage 22(1):1–7
38. Mohd Sam MF, Hoshino Y, Tahir MNH (2012) The adoption of computerized accounting system in small medium enterprises in Melaka, Malaysia. Int J Bus Manage 7(18):12–25

# Drivers and Inhibitors to the Use of the E-Disclosure System by Jordanian Listed Companies: A Conceptual Model for Future Research

**Moayyad Tahtamouni, Esraa Alkhatib, Shatha Abd Khaliq, and Maha Ayoush**

**Abstract** The purpose of this study is to develop a conceptual model for identifying the drivers and inhibitors to the use of the e-disclosure system by listed companies to the ASE and JSC in Jordan. The factors that incorporated in our conceptual model: (1) expected benefits, (2) adoption costs, and (3) network effects. We also suggest that the association between the network effects and the use of the e-disclosure system is mediated by the expected benefits. In doing so, it adds to the emerging literature by extending our understanding of the costs and benefits of adopting the e-disclosure system. It also improves to the theory by developing a conceptual model of the factors driving and inhibiting the use of e-disclosure via XBRL. Our study will be of interest to listed companies and their accountants, auditors, government agencies in Jordan, and other jurisdictions looking for e-disclosure initiatives or planning to reduce the financial reporting burden on listed companies.

**Keywords** DOI theory · E-disclosure · Jordan · Listed companies · TOE framework · XBRL

## 1 Introduction

The goal of this research is to develop a conceptual model to identify the facilitators and barriers to the use of the e-disclosure system by listed companies to the Amman

M. Tahtamouni
Qatar Stock Exchange, Doha, Qatar
e-mail: Moayyad.tahatamouni@qe.qa

E. Alkhatib (✉) · S. A. Khaliq · M. Ayoush
AL-Zaytoonah University of Jordan, Amman, Jordan
e-mail: E.Alkhatib@zuj.edu.jo

S. A. Khaliq
e-mail: Shatha@zuj.edu.jo

M. Ayoush
e-mail: Maha.Ayoush@zuj.edu.jo

© The Author(s), under exclusive license to Springer Nature Switzerland AG 2022
S. G. Yaseen (ed.), *Digital Economy, Business Analytics, and Big Data Analytics Applications*, Studies in Computational Intelligence 1010,
https://doi.org/10.1007/978-3-031-05258-3_48

601

Stock Exchange (ASE) and the Jordan Securities Commission (JSC) in Jordan. In 2016, the ASE with JSC has adopted the e-disclosure initiative. This was in a response to national and international policies aimed at reaching further standardization and harmonization of international business reporting standards [30].

The term 'E-disclosure refers to 'the electronic system that enables listed companies to prepare and disseminate their financial and non-financial disclosures in structured and standardized formats for the electronic exchange' [3, p. 6]. It is established on international open standards, such as eXtensible Business Reporting Language (XBRL) [17], which allows business information to be created, distributed, analyzed, and reused through the internet and facilitates the transmission of structured data between enterprises and governmental institutions [17]. This drives out duplicated data and unnecessary descriptions, thus reducing the burden of reporting [5, 7].

The e-disclosure project needs the creation of a national taxonomy that serves as a "data dictionary," allowing each piece of information to be categorized in a systematic and structured form [17]. As each country has its specific corporate law, and Generally Accepted Accounting Principles (GAAP), the XBRL taxonomy in Jordan is developed following the International Financial Reporting Standards (IFRS). In this context, this system in Jordan is similar to the digital financial reporting and standard business reporting (SBR) initiatives which aim to reduce the reporting burden on businesses in the UK, USA, Finland, Australia, and the Netherlands, and other countries [14, 16]. These initiatives facilitate the joint filing of the financial and business reports, and other documents through a common governmental gateway [17]. This gateway helps in removing duplicated information by enabling companies to file one set of the required business information to multiple government agencies concurrently [16].

Both the ASE and JSC in Jordan are working together to prepare special forms that enable companies to prepare and file their financial statements in structured and standard formats [33]. These forms are suitable to fit filing the information by different sectors of the listed companies (financial, industry, and services). Since June 2020, all listed companies have been required to use the e-disclosure system via XBRL for filing and submitting their financial and non-financial information to the JSC [33]. Recently, more than 90% of Jordanian listed companies successfully filed and submitted their quarterly digital reports to ASE using the e-disclosure system via XBRL for the period ending 31 March 2021 [32]. This implies that listed companies show high compliance with the laws and regulations in terms of transparency and disclosure principles in Jordan.

Several studies have attempted to extend the understanding of the factors that drive or inhibit similar initiatives to the e-disclosure system via XBRL technology. However, most of these studies have focused on governmental agencies and listed companies in the developed countries such as the USA [21], and the UK [10]. Having reviewed the relevant literature, we found that the research and theory in this area are still very limited in developing countries. This study argues that research in the context of industrialized economies needs to be re-examined in the context of developing countries for two reasons. First, developing countries may have very different economic and regulatory environments, which may lead to different results in terms of factors influencing the adoption of the e-disclosure system via XBRL. Second, each country has its own country-specific XBRL taxonomy, thus its implementation will vary across countries as suggested by Ojala et al. [17], Troshani and Rao [29].

Although the few studies conducted on Jordan provided valuable insights about the development of the e-disclosure system via XBRL, they have been conducted before mandating the innovation. A recent study by Alkhatib et al. [2] argues that the actual benefits of XBRL-based reporting to filers and users are still in their infancy stage in the research. To our knowledge, this is the first study on the e-disclosure system after it became mandatory by JSC for listed companies in Jordan in 2020. Motivated by these literature gaps, our study addresses the following research question:

- What are the drivers and inhibitors to the use of the e-disclosure system by Jordanian listed companies?

The remainder of the study is structured as follows. We review the literature in the next section, develop the hypotheses, and propose a conceptual model. We then describe the proposed methodology for future research. The study closes by concluding and providing some implications to practice and suggesting avenues for further research.

## 2   Literature Review and Theoretical Background

Having reviewed IS literature, the diffusion of innovation (DOI) theory by Rogers [22] and the technology-organization-environment (TOE) framework by Tornatzky and Fleischer [27] have been widely employed to study the factors influencing the decision of the technology adoption at an organizational level. In 1995, the DOI theory was proposed [22], and it is considered as the classic theory to study the characteristics of the innovation that affect the adoption of the technology [34]. Diffusion means "the process in which an innovation is communicated through certain channels over time among the members of a social system" [23, p. 5]. The innovation characteristics include relative advantage, compatibility, complexity, trialability, and observability [20]. A meta-analysis of 75 information systems studies by Tornatzky and Klein [28] revealed that relative advantage, compatibility, complexity, and the cost of technology are the main factors influencing the adoption of innovation. This theory focuses on the economic perspective that considers the costs and benefits of adopting innovation [35].

In 1990, the TOE framework was proposed and considered as an important theoretical perspective for identifying specific contextual factors at an organizational level [27]. The contextual factors identified by the TOE framework represent three aspects [34]. They may influence organizational adoption of technology innovations that are technology, organization, and environment contexts [27].

The technological context of the TOE framework refers to the characteristics of both internal and external technologies relevant to an organization [1, 27]. Internal technologies are those currently adopted by the organization and external technologies represent technologies that are existing in the marketplace but are not used by the organization [1]. Technological factors in IT studies are based on and drawn from the diffusion of innovations (DOI) theory developed by Rogers in 1983 [19]. According to the DOI theory, the technological characteristics of the innovation are: "relative advantage, compatibility, complexity, observability, and trialability" [22, p. 211]. Several IT studies found that combing the DOI theory with the TOE framework is very useful for understanding the costs and benefits of the technology [8].

The organizational context of the TOE framework describes the characteristics and resources of an organization that determines the adoption of innovative technology [18]. The organizational factors that determine the adoption of innovative technology are firm size, top management support, organization structure and culture, and the availability of human and financial resources [18]. The environmental context of the TOE framework represents the external arena in which a company conducts its business, its industry, competitors, and dealings with the government [27]. The external relationships with those parties may influence adoption decisions [34].

The DOI theory and TOE framework were tested by a large number of IS studies and have consistent empirical support [34]. There are also several studies on XBRL-reporting based on the DOI theory and the TOE framework that examined the factors affecting the take-up of this innovation in Australia [9], the USA, the UK [2], and other countries [12]. For example, [2] conducted 11 interviews with some participants who were involved in the implementation of XBRL in Australia to investigate the factors that facilitate and inhibit XBRL adoption. They confirmed that these theories are applicable and useful to understand the adoption of new technologies such as XBRL. They provided empirical evidence that the education and training, mandating XBRL, limited software tools, instability of XBRL specifications, and the complexity of XBRL are important factors that influence the companies' decision to adopt this technology in Australia.

Another evidence by Henderson et al. [12] also employed the DOI and institutional theories to develop the theoretical model to investigate internal and inter-organizational XBRL adoption. Their model was tested by 65 international surveys and statistically analyzed. The results revealed that the drivers of internal adoption differ from those of inter-organizational adoption. The multivariate analysis showed also that the factors that are driving the XBRL adoption are: relative advantage, compatibility, complexity, technological expertise, learning from external sources. They suggested that testing the TOE framework in both an internal and inter-organizational context is important to enhance our understanding of the diffusion of XBRL.

A recent study by Alkhatib et al. [2] examined the factors that influence the voluntary adoption of digital reporting via XBRL by small private companies in the UK. Their model was developed and based on both the DOI and the TOE theories. A survey data from 343 accountants from the ACCA database were collected and statistically analyzed. They found that three factors positively influence the voluntary adoption of digital reporting: the relative advantage, the technical competence, and support from top management. They also showed that the technology costs and the complexity of the company's accounting system negatively influence this voluntary adoption.

These studies confirm the validity and usefulness of both theories in improving the understanding of the diffusion of the complex are innovations. The model by Thong [26] that was based on them also can be extended to add further new factors which may enhance our understanding of technology adoption [26]. Drawing on these reasons, this study proposes a theoretical framework based on the DOI theory and TOE framework to identify the factors that drive and inhibit the use of e-disclosure systems via XBRL by Jordanian listed companies.

## 3    The Conceptual Model and Hypotheses Development

Based on the above discussion, we propose the conceptual model as shown in Fig. 1. We specify three driving and inhibiting factors of using the e-disclosure system by listed companies in Jordan (H1–H3): adoption costs, expected benefits, and network effects. These independent factors are derived from previous studies, the DOI theory, and the TOE framework.

As mentioned above, we also examine the mediation effect of the expected benefits on the linkage between network effects and using of the e-disclosure system (H4). Following the literature [35], we incorporate industry dummies and firm size as control variables to the model as our study represents several types of industry. We

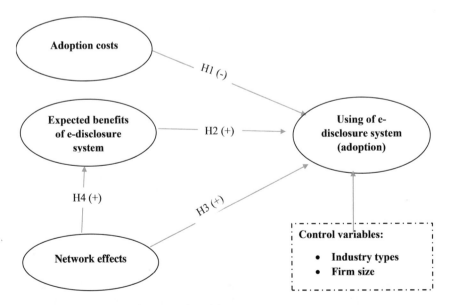

**Fig. 1** Presents the proposed conceptual model

adopted the operational definitions and the measurement of these variables based on previous IS studies [35] and the XBRL literature in particular [2, 12], with making minor changes to suit e-disclosure system context as shown in Appendix.

### Adoption Costs

We propose the adoption costs to be an important hindering factor that influences the decision to use e-disclosure system via XBRL. We have conceptualized three dimensions of adoption costs, following some studies on similar initiatives like open-standards IOS [35]: The financial costs, management complexity, and transactional hazards that come with adopting new technology are all factors to consider. The financial costs refer to the required financial resources that should be invested to enable the company to use the e-disclosure system. Managerial complexity is defined as "the extent to which an innovation is perceived as a relatively difficult process to understand and use" [23, p. 242). Several studies provide evidence that XBRL –based reporting is a complex filing task due to the required tagging process of the financial reports in Australia [9] and the USA [21]. In contrast, [2] found that digital reporting via XBRL is not complex for small private companies in the UK. Transactional risk is defined as risk and security concerns about transactions conducted over the Internet platform [9]. These three factors represent key barriers to the adoption of the e-disclosure system via XBRL. They are specified in our model as a second-order construct, adoption costs. Hence, our first hypothesis:

H1: *The adoption costs of XBRL negatively influence the decision to use the e-disclosure system by listed companies.*

### Expected Benefits

We consider the expected benefits to be an important driving factor that influences the decision to use e-disclosure system via XBRL. This factor has been supported by DOI theory in which relative advantage of new technology (or perceived benefits) is an important factor of new technology adoption [22]. Expected benefits refer to "the degree to which an innovation is perceived as being better than the idea it supersedes" [22, p. 213]. XBRL technology, according to various studies, provides a number of advantages to filers and users of financial data. For example, [21] found that XBRL-based reporting is faster than paper reporting since all calculations are performed automatically. Dunne et al. [10] provided evidence that XBRL is convenient for filing information to multiple governmental agencies as it is available 24/7.

Other evidence indicated that because XBRL-based reporting eliminates human interaction in the process, it eliminates manual errors. Roohani and Zheng [24], the digital financial information will be more accurate and reliable [10, 31]. A study by Baldwin et al. [4] confirmed that XBRL reporting ensures better compliance with statutory requirements than paper filing because it facilitates real-time preparation of financial reports. Evidence by Henderson et al. [12] examined a large number of expected benefits from adopting XBRL technology. They found that the expected benefits of XBRL have a stronger positive influence on internal adoption than on inter-organizational adoption. Alkhatib et al. [2] confirmed that the relative advantage from standardization benefits positively influences the voluntary adoption of digital reporting via XBRL by small private companies in the UK. This study takes a coherent approach by following previous studies in which a range of benefits of XBRL will be examined in the context of the e-disclosure system. Based on the above discussion we postulate the second hypothesis.

H2: *The expected benefits of XBRL positively influence the decision to use the e-disclosure system by listed companies.*

### Network Effects

We expect the network effects to be an important driving factor that influences the decision to use e-disclosure system via XBRL. The term network effects can be defined as "the value of membership to one user [which] is positively affected when another user joins and enlarges the network" [13, p. 94]. Once a critical mass of users has been reached, the external benefit of technology adoption emerges, attracting other users to join, resulting in diffusion of innovation [15].

Previous studies on the open-standard IOS identified two factors that contribute to network effects [25, 35]: First, vertical trading partners (or trading community) who can be upstream or downstream in the supply chain. Second, horizontal peers who are at the same level in the supply chain. Therefore, the size of the technology network grows when two of these trading partners and peers join it and support the adoption of the same technology. Since network effects have not been previously examined in the context of XBRL-based reporting, we based on a study by Zhu et al. [35] on similar open standard adoption to define these factors to fit our research context. The trading community influence refers to "the extent to which a firm's customers, suppliers, and other vertical partners in its trading community are willing to use or support the use of the e-disclosure system. Peer influence refers to "the extent of e-disclosure system diffusion among horizontal peers in the same industry". Following previous studies, we specified network effects as a second-order construct and operationalized by two sub-constructs: trading-community influence and peer influence. Thus, we postulate our third hypothesis:

H3: *Network effects of XBRL positively influence the decision to use the e-disclosure system by listed companies.*

Most importantly, this study suggests that the above hypothesis is mediated by expected benefits. Based on the network effect theory by Katz and Shapiro [13], the benefits that adopters derive from a network technology are positively associated with the size of the network [13]. The empirical evidence by Zhu et al. [35] confirmed the theoretical importance of network effects. They showed that network effects are posited as another independent variable that can lead indirectly through the expected benefits of open standards IOS adoption. Therefore, this leads to the view that there is an indirect mediation effect that can be postulated in the following hypotheses:

H4: *Expected benefits of XBRL positively mediate the positive association between network effects and the decision to use the e-disclosure system by listed companies.*

# 4　Methodology

The data was collected in two stages between 2020 and 2021. The first stage of the study included some interviews with a focus group to preliminary test and pilot our model. The group comprised of six people who had participated in the implementation of the e-disclosure system in Jordan since 2016. We used focus group in our study because: (1) it combines interviewing and observation and can be used to develop knowledge of a new phenomenon, (2) enables the researchers to generate propositions from the issues that emerge, and assists in the development of survey questions based on the outcomes of research in which the focus group members actively participated [6].

The interviews were semi-structured, face to face, and conducted during January 2020. Approximately, each interview lasted 20 min, and all interviews were recorded. The participants provided us with useful insights into the costs and benefits of the e-disclosure system via XBRL at the ASE and the JSC. The model was subjected to several checks and revisions by this focus group to ensure content and face validity and to fit the context of our study. As suggested by Flick [11], The list of open-ended questions is distributed to participants one week ahead of time to give them time to prepare and allow them to reflect on their experiences as fully as possible. We also collected data from online documents obtained from governmental agencies' reports websites, XBRL consortium website. The data focuses on the e-disclosure system via XBRL in Jordan.

The second stage involved two online survey questionnaires that were developed and designed to collect the data from the filers and users of digital information by e-disclosure system. The first survey was for those using digital information by the e-disclosure system at the ASE and the JSC. The second survey was for those filing information using the e-disclosure system in Jordanian listed companies via XBRL. The data is collected during 2021. Then, we will use the partial least squares structural equation modeling (PLS-SEM) technique to validate our conceptual model.

# 5　Conclusions

Having reviewed previous studies on XBRL- reporting, a considerable amount of literature has focused on the factors that affect XBRL adoption by large listed companies in developed countries. Little is known about the factors that drive and inhibit the use of this innovation in developing countries, particularly in Jordan. This study seeks to narrow this gap by proposing a conceptual model based on the DOI theory and the TOE framework and developed 4 hypotheses. The model considers that the expected benefits and network effects are driving factors of using the e-disclosure system whereas the adoption costs are inhibiting factor of using this system via XBRL by listed companies in Jordan. It also suggests that the association between the network effects and the use of the e-disclosure system is mediated by expected benefits.

This study's main theoretical contribution to the literature is identifying the drivers and inhibitors of using the e-disclosure system. By doing so, it enriches our understanding of how the digitization of business reporting to listed companies in Jordan. The results will provide significant implications for practice. First, all stakeholders (e.g. the ASE, the JSC, and other government agencies, XBRL Consortium, accountancy professions, and auditing firms) should cooperate to implement joint filing that is an essential aspect of the e-disclosure system and similar to other initiatives implemented in other countries such the UK, Finland, the Netherlands, and Australia. Companies will be able to file a single set of information rather than filing it many times in different forms to different governmental authorities for various purposes because of the joint filing facility. Thus, it will offer a single template or 'one-stop shop' for simultaneously fulfilling their statutory filing obligations to two governmental agencies [2]. Second, the ASE and the JSC should collaborate with other governmental agencies to increase the current scope of e-disclosure in Jordan to facilitate sharing of the companies' information.

The major limitation of our study is that caution should be considered when comparing the factors with other studies in other countries. However, this limitation would be a fruitful area for further research. The study also should be repeated by focusing on the factors that determine the adoption of e-disclosure in-house or outsourced with a third-party adoption. The second limitation is that our conceptual model currently has not yet been validated. Further research is required to be conducted to test and validate our proposed model which allows the results to be generalized to the whole population.

## 6  Appendix

See Table 1.

**Table 1** The measurements of variables

| Constructs | Indicators | References |
|---|---|---|
| **Relative advantage of e-disclosure system** (*5-point Likert scale*) | Using e-disclosure system facilitates continuous auditing | [2, 12] |
| | Using e-disclosure system reduces financial statements audit costs | |
| | Using e-disclosure system leads to improvements in internal controls | |
| | Using e-disclosure system enhances the consistency of information across companies | |
| | Using e-disclosure system enhances the comparability of information across companies | |
| | Using e-disclosure system reduces the company's cost of capital | |
| | Using e-disclosure system improves decision making | |
| | Using e-disclosure system improves the efficiency of the financial reporting process | |
| | Using e-disclosure system facilitates continuous reporting | |
| | Using e-disclosure system allows data from disparate AIS to be reconciled more efficiency | |
| | Using e-disclosure system provides more reliable information due to fewer filing errors | |
| | Using e-disclosure system speeds up the accessibility of information | |
| | Using e-disclosure system increases the relevance of information | |
| | Using e-disclosure system enhances the accuracy of information | |
| | Using e-disclosure system improves the analysis of information | |
| | Using e-disclosure system makes information more transparent | |
| | Using e-disclosure system ensures better compliance with regulatory authorities | |
| | Using e-disclosure system is more convenient because it is available 24/7 | |

(continued)

**Table 1** (continued)

| Constructs | Indicators | References |
|---|---|---|
| | Using e-disclosure system is faster than paper filing system | |
| **Financial costs** (*5-point Likert scale*) | Costs of using e-disclosure system are high | [2, 12, 35] |
| | Cost of the commercial software to produce electronic reports is high | |
| | Training cost to use e-disclosure system is high | |
| **Managerial complexity** (*5-point Likert scale*) | Using e-disclosure system is difficult to understand | [2, 12, 35] |
| | Using e-disclosure system is difficult to use | |
| | It is complex task to find staff with expertise of using e-disclosure system | |
| **Transactional risks** (*5-point Likert scale*) | Companies have concerns about data security when using e-disclosure system | [2, 12, 35] |
| | Using e-disclosure system is not sufficiently protected by laws | |
| **Peer influence** (*5-point Likert scale*) | The extent that peer companies used e-disclosure system | [2, 12, 35] |
| **Trading community influence** (*5-point Likert scale*) | The extent that the company's decision is influenced by suppliers | [2, 12, 35] |
| | The extent that the company's decision to use e-disclosure system is influenced by customers | |
| | The extent that the company's decision to use e-disclosure system is influenced by accountancy firms and auditors | |
| | The extent that the company's decision to use e-disclosure system is influenced by government | |
| **The use of the e-disclosure (adoption)** (*5-point Likert scale*) | Your organization is using e-disclosure system to convert data into electronic forms and use it for communicating with trading partners | [2, 35] |
| | Your organization is using e-disclosure system to convert information into electronic forms and use it for communicating with ASE and JSC | |

# References

1. Al-Hujran O, Al-Lozi EM, Al-Debei MM, Maqableh M (2018) Challenges of cloud computing adoption from the TOE framework perspective. Int J e-Bus Res 14(3):77–94
2. Alkhatib EA, Ojala H, Collis J (2019) Determinants of the voluntary adoption of digital reporting by small private companies to companies house: evidence from the UK. Int J Acc Inf Syst 34:100421
3. ASE (2018) https://w ww.ase.com.jo/sites/default/files/2018-06/March%202018.pdf
4. Baldwin AA, Brown CE, Trinkle BS (2006) XBRL: an impacts framework and research challenge. J Emerg Technol Acc 3(1):97–116
5. BIS (2009) Better regulation simplification plan 2009: delivering a better business environment. Department for Business Innovation & Skills. Retrieved 9 Oct 2020, from https://webarchive. nationalarchives.gov.uk/20100104173841/, http://www.berr.gov.uk/files/file53978.pdf
6. Collis J, Hussey R (2013) Business research: a practical guide for undergraduate and postgraduate students. Macmillan International Higher Education
7. Collis J, Alkhatib E, de Cesare S (2018) Costs and benefits to small businesses of digital reporting. Association of chartered certified accountants. Retrieved 18 Oct 2020, from https://www.accaglobal.com/gb/en/technical-activities/technical-resources-search/2018/ january/-costs-and-benefits-to-small-companies-of-digital-reporting.html
8. Daoud H (2019) Information technology investments and productivity changes in the Jordanian banking industry between 1993–2015. Jordan J Econ Sci 6(1):99–118
9. Doolin B, Troshani I (2007) Organizational adoption of XBRL. Electron Mark 17(3):199–209
10. Dunne T, Helliar C, Lymer A, Mousa R (2013) Stakeholder engagement in internet financial reporting: the diffusion of XBRL in the UK. Br Acc 45(3):167–182
11. Flick U (2002) Qualitative research-state of the art. Soc Sci Inf 41(1):5–24
12. Henderson D, Sheetz SD, Trinkle BS (2012) The determinants of inter-organizational and internal in-house adoption of XBRL: a structural equation model. Int J Acc Inf Syst 13(2):109–140
13. Katz ML, Shapiro C (1994) Systems competition and network effects. J Econ Perspect 8(2):93–115
14. Lim N, Perrin B (2014) Standard business reporting in Australia: past, present, and future. Australas J Inf Syst 18(3)
15. Lin CP, Bhattacherjee A (2008) Elucidating individual intention to use interactive information technologies: the role of network externalities. Int J Electron Commer 13(1):85–108
16. OECD (2020) Forum on tax administration: taxpayer services sub-group, guidance note standard business reporting. Centre for tax policy and administration. Retrieved 10 Oct 2020, from http://www.oecd.org/tax/administration/43384923.pdf
17. Ojala H, Penttinen E, Collis J, Virtanen TH (2018) Design principles for standard business reporting (SBR) taxonomy development: evidence from Finland. Nord Bus 67(1):4–26
18. Oliveira T, Martins MF (2011) Literature review of information technology adoption models at firm level. Electron. J Inf Syst Eval 14(1):110
19. Oliveira T, Thomas M, Espadanal M (2014) Assessing the determinants of cloud computing adoption: an analysis of the manufacturing and services sectors. Inf Manage 51(5):497–510
20. Picoto WN, Bélanger F, Palma-dos-Reis A (2014) An organizational perspective on m-business: usage factors and value determination. Eur J Inf Syst 23(5):571–592
21. Pinsker R, Li S (2008) Costs and benefits of XBRL adoption: early evidence. Commun ACM 51(3):47–50
22. Rogers EM (1995) Diffusion of innovations, 4th edn. Free Press, New York
23. Rogers EM (2003) Diffusion of innovations, 5th edn. Free Press, New York
24. Roohani SJ, Zheng X (2013) Determinants of the deficiency of XBRL mandatory filings. Int Res J Appl Financ 4(3):502–518
25. Teo HH, Wei KK, Benbasat I (2003) Predicting intention to adopt interorganizational linkages: an institutional perspective. MIS Q, pp 19–49

26. Thong JY (1995) An integrated model of information systems adoption in small businesses. J Manag Inf Syst 15(4):187–214
27. Tornatzky LG, Fleischer M (1990) The process of technology innovation. Lexington Books, Lexington
28. Tornatzky LG, Klein KJ (1982) Innovation characteristics and innovation adoption-implementation: a meta-analysis of findings. IEEE Trans Eng Manage 1:28–45
29. Troshani I, Rao S (2007) Drivers and inhibitors to XBRL adoption: a qualitative approach to building a theory in under-researched areas. Int J e-Bus Res 3(4):98–111
30. Valentinetti D, Rea MA (2012) IFRS Taxonomy and financial reporting practices: the case of Italian listed companies. Int J Acc Inf Syst 13(2):163–180
31. Vasarhelyi MA, Chan DY, Krahel JP (2012) Consequences of XBRL standardization on financial statement data. J Inf Syst 26(1):155–167
32. XBRL (2020) https://www.xbrl.org/news/jordan-securities-commission-mandates-xbrl-disclosures/
33. XBRL (2021) https://www.xbrl.org/news/over-90-of-companies-in-jordan-file-xbrl-reports/#
34. Zhu K, Dong S, Xu SX, Kraemer KL (2006) Innovation diffusion in global contexts: determinants of the post-adoption digital transformation of European companies. Eur J Inf Syst 15(6):601–616
35. Zhu K, Kraemer KL, Gurbaxani V, Xu SX (2005) Migration to open-standard interorganizational systems: network effects, switching costs, and path dependency. MIS Q 515–539

# Does Perceived Organizational Support Have a Mediating Role in Directing the Relationship Between E-Banking and Corporate Digital Responsibility?

Madher Ebrahim Hamdallah, Anan F. Srouji, and Orman Ahmad Al-Ibbini

**Abstract** The aim of this research paper is to investigate E-Banking impact on Corporate Digital Responsibility in the Jordanian banking service sector. Meanwhile, then this work also focuses to propose a mediating role of Perceived Organizational Perspective (POS) to measure the association influences between exogenous and endogenous variables. Modeling tests based on hypotheses was performed through e-surveys, totaling 88 respondents from the financial department gathered from 16 Jordanian banks. Results based on the respondents' opinions reveal determination that E-Banking is a direct element in the dimensional model that assists in indicating a positive relationship Social Corporate Digital Responsibility (SCDR), Economic Corporate Digital Responsibility (ECDR), and Technological Corporate Digital Responsibility (TechCDR). POS has a partial positive mediating effect in the relationship between E-Banking and ECDR, and between E-Banking and TechCDR, with no mediating effect in the relationship between E-Banking and SCDR.

**Keywords** E-Banking · Corporate digital responsibility · Social corporate digital responsibility · Economic corporate digital responsibility · Environmental corporate digital responsibility · Technological corporate digital responsibility

M. E. Hamdallah (✉) · O. A. Al-Ibbini
Accounting Department, Business Faculty, Al-Zaytoonah University, Airport Road, Amman, Jordan
e-mail: M.Hamdella@zuj.edu.jo

O. A. Al-Ibbini
e-mail: o.alibbini@zuj.edu.jo

A. F. Srouji
Accounting Department, King Talal School of Business Technology, Princess Sumaya University for Technology, Amman, Jordan
e-mail: a.srouji@psut.edu.jo

© The Author(s), under exclusive license to Springer Nature Switzerland AG 2022
S. G. Yaseen (ed.), *Digital Economy, Business Analytics, and Big Data Analytics Applications*, Studies in Computational Intelligence 1010,
https://doi.org/10.1007/978-3-031-05258-3_49

# 1 Introduction

Improving organizations' performance is a worldwide endless issue, which develops and changes depend on all-inclusive events and circumstances creating a new basis of persistence defies; as the effects of the COVID-19 disease and community separation; which is known as a global health crisis that lie in earth's civilization corruption [1]. As it is the major challenge faced since the World War two, pushing organizations to respond quickly to any changes in businesses to endorse and operate in effective and efficient manner [2]. Rafindadi and Ilhan [3] stated that operative business procedures are anticipated to develop companies' financial performance in a positive manner. Depending mainly on technology utilization as a major pillar to follow up with such required changes [4]. Business models compete and are essential for predictions. Different practices may create real valued decisions in explaining the relationships [5].

Interest in corporate social responsibility (CSR) has gained through the past eras; where the trend of having CSR practices in banks is inevitable [6]; as CSR is exemplifies as a competitive advantage that augments the performance of institutions and social supportive technique [7]. Digital CSR have the potential to create immense value for both consumers and brands and [8] and even services provided by the banks. Digital transformation and the changes in the forms of information technology empower businesses to implement their social responsibilities toward clients; in addition to its effectiveness in decreasing costs [9], where digital technologies and social networking allows fundamental accepted services provided by banks. So looking into the past and the future to develop digital responsibility learnings is a must for businesses. From the ancestries of many increases, it determines potential prospects of business transformation [10]. Two main questions guide the following discussion: How is E-Banking effecting the corporate digital responsibility (CDR) in Jordan. How can Perceived Organizational Support (POS) break the path and explain the relationship between E-Banking and CDR?

# 2 Literature Review and Hypotheses Development

Based on the new pandemic requirements, and in order to satisfy customers' needs CDR is an essential necessity. However, [11] mentioned in The World Economic Forum previously that the main purpose of digital transformation is saving our planet, by Shaping the Future of Digital Economy and Society. Practicing CDR based on the increase in E-Banking practice is a new implication concerning Jordanian Banks based on the authors point of view, for that more studies are compulsory to indorse the new techniques and measurements, literature review and hypotheses building underpin this paper to help answer some questions related to the relationship between E-Banking and its effect on the CDR in Jordan, and then test the mediating role of POS and its influence on this relation.

## 2.1  E-Banking

As Banking businesses are essential industries that support the development and economy of countries [12] exertions are required to develop literacy programs that directly involve financial institutions; on continuous basis. Such continuous improvement is directed through CSR [13]. Enchanting internal bank management assessment, financial reporting is the convenience of evidence disclosed to affluence decision making [14]. This knowledge gives the management ability to better understand how investments are required for long-term sustainability, and it will assist users preserve, inflate, or reduction the size of their bank funds, investments and etc.

E-Banking is a new important new system required by clients. Attractiveness of E-Banking amplified along with global warming and collaborative awareness by using new methods [15]. Where E-Banking amenities necessitate bank investment in technological and personnel mannerisms, all require improvement in the banking services [16].

## 2.2  Corporate Digital Responsibility (CDR)

CDR was developed to achieve widespread acceptance, desirability, and efficiency [17]. After the global digital transformation CSR had to be indicated and designated through the financial reports, which was inducted afterwards as digital responsibility [18]. Even though digital transformation changed people's natural life and comportments, banks comprehend the importance of undertaking digital podiums concerning social responsibility [9].

Through online platforms and podiums business may benefit widely via retrieving broader markets; either national or international. Digital markets help clients' save time and cost [7]. Digital social responsibility (DSR) enterprises have a positive significant influence on the clients, in addition to playing a mediating role in social business perspectives [7, 9].

Based on [19] study, which indicated that digital corporate responsibility technology augments significant and dependable of accounting measurements. Lobschat et al. [20; p.879] stated that "Designing data models for consumer data and models to analyze and predict should be guided by CDR norms, which then can help data scientists determine which data they can collect ethically and the conditions in which they can process that data," supported by the Stakeholders' theory. However limited studies of impacts made by CDR initiatives are published [21, 22] not only in emerging markets only. This study came to focus on these relationships and impacts, hoping to enrich the theoretical and literature library. Depending on [23] statement that there are 4 Categories of CDR, as in Fig. 1; to hold the responsibility to promote ethical and sustainable businesses; as they are the pillars of competitive differentiation in a new digital world. For that this research came to build the hypotheses testing

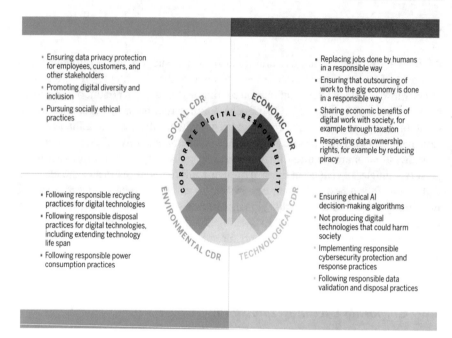

- Ensuring data privacy protection for employees, customers, and other stakeholders
- Promoting digital diversity and inclusion
- Pursuing socially ethical practices

- Replacing jobs done by humans in a responsible way
- Ensuring that outsourcing of work to the gig economy is done in a responsible way
- Sharing economic benefits of digital work with society, for example through taxation
- Respecting data ownership rights, for example by reducing piracy

- Following responsible recycling practices for digital technologies
- Following responsible disposal practices for digital technologies, including extending technology life span
- Following responsible power consumption practices

- Ensuring ethical AI decision-making algorithms
- Not producing digital technologies that could harm society
- Implementing responsible cybersecurity protection and response practices
- Following responsible data validation and disposal practices

**Fig. 1** Corporate digital responsibility categories [23]

on this theoretical framework, and through the evidence of the other literatures, and how it may be influenced by E-Banking.

### 2.2.1 Social Corporate Digital Responsibility (SCDR)

Economic crises and the cause of financial shocks afford inadequate social reporting coverage to country citizens [24], as declared by Lobschat et al. [20], Galbreath [25] society prerequisite an influencing role in a firm's orientation towards the responsible treatment of stakeholders. In current eons, an amplified use of digital responsibility is perceived, where conjoint reporting practices are needed regarding the type of data disclosed [18], based on International Financial Reporting Standards (IFRS) and countries or markets corporate governance [26].

Since COVID-19 pandemic has been promptly and inclusively effecting the societies [27] CDR outlooks as the innovative bearing, blending ethical deliberations, separable and social level, confirming the importance of CDR [11]. Vollero et al. [28] findings highlight that several differences are found in CSR communication on corporate websites, covering the social requirements based on technological modifications and developments. Srouji et al. [6] claimed that the relationship between Banks with clients and managers behavior differ. Where the covering of CDR to serve a function of social responsibility should be effectively identified by the managers based on their new e-services providing's. Therefore, against the previous backdrop we hypothesize that:

H1: There is a positive relationship between E-Banking and SCDR.

### 2.2.2 Economic Corporate Digital Responsibility (ECDR)

Demanding the world-wide lockdown during the COVID-19 pandemic, personal visiting and dealing with each other and via businesses has been proscribed. Client losses pretentious the income of industry transversely the nations. Using digital businesses or at least including digitals in businesses, are agreed upon to be indispensable to people's lives and underpinning their economy. Therefore, governments should work on supporting the lifecycle of businesses in their societies, through engaging and encouraging representatives to aiding, funding, and supporting [29]. The deteriorations of economic outcomes and the world-wide recession caused fewer firms responsibility, related to the decrease of revenues and declined employee work spans; providing initial indications needed to upsurge the economic activities of the businesses [30]. Authenticating the aims of this study; the following hypothesis is derived:

H2: There is a positive relationship between E-Banking and ECDR.

### 2.2.3 Environmental Corporate Digital Responsibility

Identifying new prospects for environmental enhancement is a significant part of an institute that aims to minimize total costs, improve the financial performance of micro, small, and medium business processes [31]. E-business includes imitating the ability to sense environmental vicissitudes using web-based systems and electronic networks [32] and cybersecurity, which is a must in the e-banking industry.

Beneficiaries of the banking amenities are to be related to the entire society. Financial efficiency and stability with social commitment is essential [33] including environmental fortification and responsible management. As environmental responsibility may include illusive evidence related to corporate social responsibility [34], environmental responsibility and its reporting disclosures has to be renewed and improved into a mutual business indicator around the world [35]. In the case of ambiguity anticipation of innovative techniques in the field of CSR, structured environmental techniques are desired [28], where e-banking proved to be an effective up-to-date method; the research came with the succeeding hypothesis:

H3: There is a positive relationship between E-Banking and EnvCDR.

### 2.2.4 Technological Corporate Digital Responsibility

"New technologies are compulsory techniques used these days to save the earth, banks will progressively offer green products and services, which became a new customary in enterprises and the banking operations and services [15]. After the occurrence of Covid-19 it was indicated that good business processes were required to provision the organization's financial performance by evolving new techniques to survive crises [31]. However [36] mentioned that digital transformation should be used in the most ethical use of new technologies in any type of business.

In a technological eon interruption nowadays, digital technology is playing as a critical part in businesses [7]. E-Banking has become easier to implement with the support of technology, where banks are irreplaceable in the epoch of life evolution and entrepreneurship. Regardless of technologies benefits disadvantages should not be neglected [15]. Banks and Financial institutions can implement 'go-green' policy [36]. Based on the previous discussions, it directed us to the following hypothesis:

H4: There is a positive relationship between E-Banking and TechCDR".

## 2.3 Perceived Organizational Support

"POS is related to an incentive idea that increases the consummation of psychological requirements [37], which also reflects an institutions significance on its assets [38]. Employees are the best to evaluate POS; hence, it is essential to devour the institutional support and its employee prosperity [39]. POS supports employees' requirements and responsibilities toward the institution in order to positively change [40] and POS values the degree of employees' contribution and concern via management [41]. POS can be achieved through effective leadership, proper job conditions, fair treatment, and human resource practices. Throughout POS valuation, management rely on their employees to pay more attention towards the practices of the institution preference, and its implementation of corporate governance [42].

POS of sophisticated intensities among employees would also lead to commitment feelings and the efforts would be directed toward facilitating the institutions achievement of its goals and intentions [43]. In addition to the idea that POS also fulfils socioemotional needs [44] POS has been associated with increases in many positive employee outcomes [37]. For that this research came to evaluate the mediating role of POS and its effect on the relationship between E-Banking services provided by banks and corporate digital responsibility toward its clients. To our knowledge, an exhaustive examination of such relationship is understudied, thus validating the purpose of this study; the following hypotheses are derived:

H5: There is a mediating effect of POS in the relationship between E-Banking and SCDR.

H6: There is a mediating effect of POS in the relationship between E-Banking and ECDR.

H7: There is a mediating effect of POS in the relationship between E-Banking and EnvCDR.

H8: There is a mediating effect of POS in the relationship between E-Banking and TechCDR."

# 3 Study Design and Methodology

Descriptive and analytical research design is tested through an e-survey, which were carried out from April to June 2021; via emails. After distributing the e-survey to individuals working in the financial departments in Jordanian banks, 88 were valid for the analysis, all after confirming the confidentiality of the individuals and banks names; and insuring them that their responses will be used for research purposes only. Partial least square (PLS) was used for a confirmatory factor analysis and model measurement to test the hypotheses. Path Model Approach is a diagram that tests the influences between the variables by Structural Equational Modeling (SEM) to test the hypotheses [45]. Since financial institutions are directly related to technological developments, economic, environmental, and the society, this study came to test the banking sector as a sample.

In this research model, E-Banking is the exogenous variable; while SCDR, ECDR, EnvCDR, and TechCDR are endogenous; In addition to testing the mediating role of POS. Survey questions have been gathered from [2, 7, 9, 23, 41, 46, 47], as indicated in Table 1.

"The measurement model is made up of 30 items via a five-point Likert-type scale. The questions were of nominal scale character, grouped into six constellations. The structural model is based on the following linear regression equation

$$LVj = \beta 0 \sum_{i \to j} \beta ji LVi + error\ i$$

where, $LVj$ is the endogenous variables
$LVi$ is the exogenous variables
$\beta 0$ is the intercept
$\beta ji$ is the path coefficients
and the error $j$ is for the residuals.
"Just like in the inner model, the outer model relationships are also deliberated to be linear" [48, p. 37], via:

$$X\ jk = \lambda jk + \lambda jk\ LVj + error\ jk$$

where, $\lambda jk$ = coefficients loadings;
$\lambda_0$ = intercept,
and $jk$ is the residual errors.
To test the mediating effect of POS, Structural Equational Modeling is developed, which implicates multiple regression analysis, path and confirmatory factor [49]."

**Table 1** Construct and measurement items

| "Construct | Code | Measurement items |
|---|---|---|
| Social corporate Digital responsibility | To what extent do the following statements apply to your bank | |
| | SCDR1 | Ensuring data privacy protection for employees, customers, and other stakeholders |
| | SCDR2 | Promoting digital diversity and inclusion |
| | SCDR3 | Pursuing socially ethical practices |
| Economic corporate digital responsibility | To what extent do the following statements apply to your bank | |
| | ECDR1 | Replacing jobs done by humans in a responsible way |
| | ECDR2 | Ensuring that outsourcing of work to the gig economy is done in a responsible way |
| | ECDR3 | Sharing economic benefits of digital work with society, for example through taxation |
| | ECDR4 | Respecting data ownership rights, for example by reducing piracy |
| Environment corporate digital responsibility | To what extent do the following statements apply to your bank | |
| | EnvCDR1 | Following responsible recycling practices for digital technologies |
| | EnvCDR2 | Following responsible disposal practices for digital technologies, including extending technology life span |
| | EnvCDR3 | Following responsible power consumption practices |
| Technological corporate digital responsibility | To what extent do the following statements apply to your bank | |
| | TechCDR1 | Ensuring ethical AI decision-making algorithms |
| | TechCDR2 | Not producing digital technologies that could harm society |
| | TechCDR3 | Implementing responsible cybersecurity protection and response practices |
| | TechCDR4 | Following responsible data validation and disposal practices |
| E-Banking | To what extent do you agree with the following statements | |
| | E-Banking1 | Our bank has invested heavily in internet banking |
| | E-Banking2 | All our corporate clients use internet banking |

(continued)

**Table 1** (continued)

| "Construct | Code | Measurement items |
| --- | --- | --- |
| | E-Banking3 | All our retail clients use internet banking |
| | E-Banking4 | Internet banking has improved the image of our bank among its customers |
| | E-Banking5 | Our bank always ensures security of data and information that is operated on the internet banking platform |
| | E-Banking6 | Customers fear internet banking due to fear of hacking of their accounts by web hackers |
| | E-Banking7 | Customers are provided with encrypted passwords in order to protect their information and transactions |
| | E-Banking8 | Internet service is operated in a restricted and controlled environment in order to safe guard customer information |
| Perceived organizational support | To what extent do you agree with the following statements | |
| | POS1 | The bank cares about my opinions |
| | POS2 | The bank cares about my well-being |
| | POS3 | The bank appreciates any extra effort from me |
| | POS4 | The bank would ignore any complaint from me |
| | POS5 | Even if I did the best job possible, my organization would fail to notice |
| | POS6 | The bank cares about my general satisfaction at work |
| | POS7 | The bank shows very little concern for me |
| | POS8 | The bank takes pride in my accomplishments at work" |

# 4 Data Analysis and Findings

## 4.1 Measurement Model Evaluation

"The model was evaluated via validity and reliability indexes; as Cronbach's α, Average Variance Extracted, and factor loading. Construct reliability outcomes are stated in Table 2, the internal consistency by using Cronbach's α is accepted at levels above 0.7; where the validity of all the survey constructs are accepted based on [50] measurement. Where the AVE values measure the level of inconsistency in the hypothesis in contradiction with the level due to magnitude error; very good is deliberated for values exceeding 0.7; however, 0.5 is satisfactory. Meanwhile the composite

**Table 2** Measurements model results

| "Constructs | Items | Cronbach's alpha | Composite reliability | Average variance extracted | Factor loading |
|---|---|---|---|---|---|
| E-Banking | EB1 | 0.853 | 0.931 | 0.753 | 0.754 |
| | EB2 | | | | 0.843 |
| | EB3 | | | | 0.929 |
| | EB4 | | | | 0.729 |
| | EB5 | | | | −0.775 |
| | EB6 | | | | 0.775 |
| | EB7 | | | | 0.785 |
| | EB8 | | | | 0.844 |
| SCDR | SCDR1 | 0.812 | 0.876 | 0.652 | 0.760 |
| | SCDR2 | | | | 0.820 |
| | SCDR3 | | | | 0.710 |
| ECDR | ECDR1 | 0.754 | 0.901 | 0.727 | 0.908 |
| | ECDR2 | | | | 0.752 |
| | ECDR3 | | | | 0.766 |
| | ECDR4 | | | | −0.771 |
| ENVCDR | ENVCDR1 | 0.702 | 0.635 | 0.604 | −0.837 |
| | ENVCDR2 | | | | 0.749 |
| | ENVCDR3 | | | | 0.799 |
| TECHCDR | TECHCDR1 | 0.716 | 0.788 | 0.659 | −0.763 |
| | TECHCDR2 | | | | 0.934 |
| | TECHCDR3 | | | | 0.859 |
| | TECHCDR4 | | | | 0.750" |

reliability test amounts above 0.7 is predictable [51]. Factor analysis results are indicated in Table 2; indicating that the item loading factor are above the minimum least amount 0.707 [52]."

## 4.2 Assessing the Structural Model

To envisage the hypothesized interactions between the variables path coefficient explains the association among the exogenous factors [53] and the coefficient of determination and their weights is indicated by bootstrapping of 500 re-samples [46]; As stated in Table 3. Making up the structural model adopts a significant relationship between different models by means of SEM [54]. It stated in Table 3 three of four hypotheses are positively significantly accepted with a high β for the relationships

**Table 3** Hypotheses path coefficient

| "Hypo | "Relationship | Std. beta | Std. error | t-value | P-value" |
|---|---|---|---|---|---|
| H1 | E-Banking → SCDR | 0.957 | 0.257 | 3.373 | 0.001*** |
| H2 | E-Banking → EcoCDR | 0.899 | 0.465 | 1.411 | 0.015** |
| H3 | E-Banking → EnvCDR | 0.417 | 1.666 | 0.575 | 0.566 |
| H4 | E-Banking → TechCDR | 0.880 | 0.457 | 1.830 | 0.068*" |

Significant at $p*** \leq 0.01$
Significant at $p** \leq 0.05$
Significant at $p* \leq 0.1$

between e-banking and SCDR, EcoCDR and TechCDR, at $p \leq 0.01$, 0.05, and 0.1 respectively.

$R^2$ analysis determines the "proportion of variance in the endogenous variable explained by the exogenous variable" [6, p. 11]. Where $R^2$ has a strong effective size when higher than 0.7, mediate between 0.5 and 0.7, weak between 0.3 and 0.5, and none or very weak when less than 0.3 [55, p.138]. The research SEM results are illustrated in Fig. 2.

Table 4 indicates the association direction among the endogenous and exogenous variables; for $H_1$ is elucidated by approximately 92%, the second hypothesis by 81%; and the fourth by 77.5% all indicating a robust outcome of elucidations erected by ($R^2$) as in [55]. However and based on the structural model results neither the path coefficient nor the $R^2$ indicated any relationship between e-banking and EnvCDR.

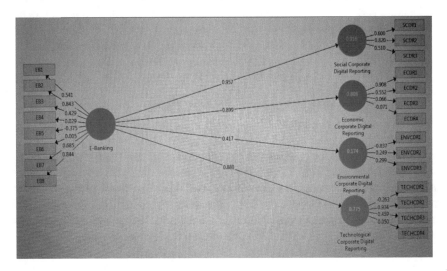

**Fig. 2** Structural model results (see online version for colors)

**Table 4** $R^2$ of the endogenous variables

| Constructs | $R^2$ | Result |
|------------|-------|--------|
| SCDR | 0.916 | Strong |
| ECDR | 0.808 | Strong |
| ENVCDR | 0.174 | Very weak |
| TECHCDR | 0.775 | Strong |

## 4.3 Mediator Analysis of POS Using PLS-SEM

A mediator variable give a whole or partial explanations concerning a predictor and outcome association in the PLS path model and its direction [45]. Thus, mediators divulge the 'true' association between the variables. Later, the mediator was included in the research in the PLS path model to test its significance; if results indicated a significant relationship then confidence interval is observed.

"After adding the mediator and bootstrapping the total indirect effect between E-banking and POS positively significant, where ($\beta = 0.99$, $t\text{-value} = 232.8$, $p\text{-value} = 0.00$). Path analysis exposed that all the associations concerning the exogenous variables and all the endogenous variables after adding the mediator has a positive significance relationship, except for EnvCDR which demonstrated no significant relationship, for that no further tests will be held forward in relation to the mediating variables and rejecting H3. Analysis results indicated that the relationship between E-Banking and SCDR as ($\beta = 0.94$, $t\text{-value} = 50.51$, $p\text{-value} = 0.000$) and between E-Banking and ECDR as ($\beta = 0.279$, $t\text{-value} = 28.1$, $p\text{-value} = 0.001$) and were E-Banking and TechCDR between ($\beta = 0.46$, $t\text{-value} = 20.73$, $p\text{-value} = 0.000$). The mediator was also associated to the endogenous variables, excluding SCDR indicating no significant value ($\beta = 0.017$, $t\text{-value} = 0.054$, $p\text{-value} = 0.957$) POS and ECDR ($\beta = 0.61$, $t\text{-value} = 12.059$, $p\text{-value} = 0.043$) and POS and TechCDR ($\beta = 0.408$, $t\text{-value} = 19.49$, $p\text{-value} = 0.023$); as exemplified in Table 5 Part one and Fig. 3."

Bootstrapping confidence interval of the lower and upper level should be done after testing the indirect effect of the mediator. As the preceding associations are only indicators, deprived of giving clarifications, and whether it's a mediator or not. Ensuring such results are based on intervals that do not intercept by 0 [56]. Bootstrapping confidence interval in Table 5 Part two confirmed that POS actually does not play a mediating role between E-Banking and SCDR as the UL and LL rate intercept at zero. Meanwhile, POS as the mediator essentially absorbs and elucidates the association between E-Banking with both ECDR and TechCDR, supporting Hypotheses 7 and 8 demonstrating an influence of the mediating variables in model 2 and 3.

**Table 5** Bootstrapping analyses

| Relationship | Std. beta | Std. error | T-value | p-value | Decision |
|---|---|---|---|---|---|
| *Analysis: part one (Bootstrapping results of the total indirect effect)* | | | | | |
| E-Banking → SCDR | 0.938 | 0.019 | 50.512 | 0.000*** | Supported |
| E-Banking → EcoCDR | 0.279 | 0.031 | 28.104 | 0.001*** | Supported |
| E-Banking → EnvCDR | 0.424 | 0.439 | 0.966 | 0.335 | Not supported |
| E-Banking → TechCDR | 0.462 | 0.042 | 20.725 | 0.000*** | Supported |
| E-Banking → POS | 0.990 | 0.004 | 232.814 | 0.000*** | Supported |
| POS → SCDR | 0.017 | 0.320 | 0.054 | 0.957 | Not supported |
| POS → EcoCDR | 0.610 | 0.053 | 12.059 | 0.0453** | Supported |
| POS → TechCDR | 0.408 | 0.038 | 19.492 | 0.023** | Supported |

| *Analysis: part two (Bootstrapping confidence interval)* | | | | | |
|---|---|---|---|---|---|
| Hypothesis models | | | 95% LL | 95% UL | Decision |
| Model 1 = E-Banking → POS → SCDR | | | −0.364 | 0.614 | Not supported |
| Model 2 = E-Banking → POS → ECDR | | | 0.088 | 1.996 | Supported |
| Model 3 = E-Banking → POS → TechCDR | | | 0.469 | 1.386 | Supported |

Significant at $p*** \leq 0.01$
Significant at $p** \leq 0.05$
Significant at $p* \leq 0.1$

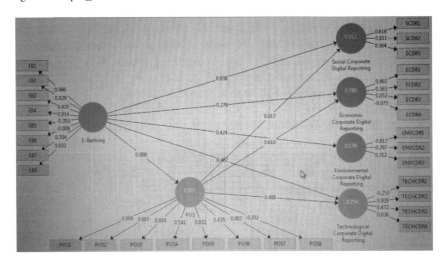

**Fig. 3** Structural results via mediating variable

# 5  Discussion

This research paper meant to investigate E-Banking influence on CDR in the Jordanian banking service sector. Results based on the respondents' opinions reveal determination that E-Banking is a direct element in the dimensional model that assists in indicating a positive relationship (SCDR), which supports [7] study indicating the need to start emphasizing on CDR strategic preparation and application. Which is an additional burden of bank managers to convey the idea of E-Banking to clients and at the same time balance this idea with digital responsibility in and efficient and effective manner. Srouji et al. [6] claimed that the accomplishments vary from developed and developing countries and their regions for the reason of different economies, demanding of broader studies and research on social responsibility in general and digital in specific.

Therefore, it is important for businesses to consider and identify the different mediation effects on the direct relationship between CDR and preferred services, for that institutions need to pay attention to such changes [9] and observe any required changes based on economic changes or crisis. Economic Corporate Digital Responsibility (ECDR), and the Corona crisis has left individuals frantic for money, and a prompt necessity of the government to; supporting and in alignment with [9] study results. Regarding the stakeholders both financial and non-financial information concerning any technological improvements and changes have to be disclosed to satisfy their needs [18], as [57] added that the satisfaction of clients is the main concern of technological based services, which are directly related to TechCDR.

Once mediating the role of Perceived Organizational Perspective (POS), results indicated that POS indicated a partial mediating influence on relationship between E-Banking—ECDR, and between E-Banking—TechCDR in a positive manner, where the convention was associated with enjoyment and acceptance of intrinsically motivated by mediating the role of POS, with no mediating role between E-Banking—SCDR, which does not compromise with [58] as their study stated that POS works as a interactive mechanism in relation to socialization tactics.

The results of this study stated no effect of E-Banking on CDR, which may be caused by the absence of satisfactory environmental policy [15] and were the least to be disclosed in [59] study that took place in Lebanon banks. Miah et al. [36] concluded that eco-projects can be applied meritoriously in banks, which is still an unexploited opportunity. Meanwhile [7] stated that clients behaviour has toward digital services and having an environmental impact.

# 6  Conclusion and Recommendations

After testing the hypothesis and explaining the results of digital responsibility and how it is affected by E-Banking and the mediating role of POS, further testing is needed. Results are based on Jordanian Banks listed in Amman Stock Exchange only, where taking different samples and including other sectors and even expanding the sample to include other countries either in the Middle East or the MENA region. Not to forget mentioning that the increased concern over E-Banking and digital responsibility and the shortage of research relative to this field of developing countries in general [36] as in Jordan. Results are influenced by the period of investigation as the country is still suffering the COVID-19 crisis and effects; so changing the time span post COVID-19 may diverge the results. Therefore, generalising the outcomes of this study is inadequate. Thus, the research findings recommend that further studies may focus on different sectors or even expand the research sample."

# References

1. Hopper T (2020) Swimming in a sea of uncertainty–business, governance and the coronavirus (COVID-19) pandemic. J Account Organ Change
2. Al-Omoush KS, Simón-Moya V, Sendra-García J (2020) The impact of social capital and collaborative knowledge creation on e-business proactiveness and organizational agility in responding to the COVID-19 crisis. J Innov Knowl 5(4):279–288
3. Rafindadi AA, Ilhan O (2017) Dynamic effects of financial development, trade openness and economic growth on energy consumption: evidence from South Africa 7(3):74–85
4. AlShamsi A, Mohaidat J, Hinai NA, Samy A (2020) Instructional and business continuity amid and beyond COVID-19 outbreak: a case study from the higher colleges of technology. Int J High Educ 9(6):118–135
5. Andries P, Debackere K, Van Looy B (2020) Simultaneous experimentation as a learning strategy: business model development under uncertainty—relevance in times of COVID-19 and beyond 556–559
6. Srouji AF, Abed SR, Hamdallah ME (2019) Banks performance and customers' satisfaction in relation to corporate social responsibility: mediating customer trust and spiritual leadership: what counts! Int J Bus Innov Res 19(3):358–384
7. Puwirat W, Tripopsakul S (2019) The impact of digital social responsibility on customer trust and brand equity: an evidence from social commerce in Thailand
8. Okazaki S, Plangger K, West D, Menéndez HD (2020) Exploring digital corporate social responsibility communications on Twitter. J Bus Res 117:675–682
9. Puriwat W, Tripopsakul S (2021) Customer engagement with digital social responsibility in social media: a case study of COVID-19 situation in Thailand. J Asian Fin Econ Bus 8(2):475–483
10. Schaltegger S (2020) Sustainability learnings from the COVID-19 crisis. Opportunities for resilient industry and business development. Sustain Account Manage Policy J
11. Newman D (2017) How digital transformation aligns with corporate social responsibility. Interactive: https://www.forbes.com/sites/danielnewman/2017/11/21/how-digital-transformation-aligns-withcorporate-social-responsibility
12. Bukhari SA, Hashim F, Amran A (2020) The journey of Pakistan's banking industry towards green banking adoption. South Asian J Bus Manage Cases 9(2):208–18
13. Krisnawati A, Yuliana S (2019) Does corporate social responsibility disclosure affect stock return? (An empirical study in Indonesia crude petroleum and natural gas industry). Sustainable collaboration in business, technology, information and innovation (SCBTII) Jan 15
14. Hamdallah ME, Srouji AF, Abed SR (2021) The nexus between reducing audit report lags and divining integrated financial report governance disclosures: should ASE directives be more conspicuous? Afro-Asian J Fin Account 11(1):81–103
15. Ziolo M, Pawlaczyk M, Sawicki P (2019) Sustainable development versus green banking: where is the link? Financing sustainable development. Palgrave Macmillan, Cham, pp 53–81

16. Bose S, Khan HZ, Monem RM (2021) Does green banking performance pay off? Evidence from a unique regulatory setting in Bangladesh. Corp Gov Int Rev 29(2):162–187
17. Orbik Z, Zozuľaková V (2019) Corporate social and digital responsibility. Manage Syst Prod Eng 27(2):79–83
18. Coluccia D, Fontana S, Solimene S (2016) Disclosure of corporate social responsibility: a comparison between traditional and digital reporting. An empirical analysis on Italian listed companies. Int J Manag Finan Account 8(3–4):230–246
19. Shan YG, Troshani I (2020) Digital corporate reporting and value relevance: evidence from the US and Japan. Int J Manag Finance 17(2): 256–281
20. Lobschat L, Mueller B, Eggers F, Brandimarte L, Diefenbach S, Kroschke M, Wirtz J (2021) Corporate digital responsibility. J Bus Res 122:875–888
21. Chu SC, Chen HT, Gan C (2020) Consumers' engagement with corporate social responsibility (CSR) communication in social media: evidence from China and the United States. J Bus Res 110:260–271
22. Fatma M, Khan I, Rahman Z, Pérez A (2020) The sharing economy: the influence of perceived corporate social responsibility on brand commitment. J Prod Brand Manag 28(4):941–956
23. Wade M (2020) Corporate responsibility in the digital era, MITSloan Managmenet Review. https://sloanreview.mit.edu/article/corporate-responsibility-in-the-digital-era/. Accessed 15 Mar 2021
24. Yap OF (2020) A new normal or business-as-usual? Lessons for COVID-19 from financial crises in East and Southeast Asia. Eur J Dev Res 32(5):1504–1534
25. Galbreath J (2010) Drivers of corporate social responsibility: the role of formal strategic planning and firm culture. Br J Manage 21(2):511–525
26. Hamdallah ME (2012) Corporate governance and credibility gap: empirical evidence from Jordan. Int Bus Res 5(11):178
27. Zhang Z, Shen Y, Yu J (2020) Combating COVID-19 together: China's collaborative response and the role of business associations. Nonprofit Voluntary Sector Q 49(6):1161–1172
28. Vollero A, Siano A, Palazzo M, Amabile S (2020) Hoftsede's cultural dimensions and corporate social responsibility in online communication: are they independent constructs? Corp Soc Responsib Environ Manag 27(1):53–64
29. Kyung A, Whitney SA (2020) Study on the financial and entrepreneurial risks of small business owners amidst covid-19. In2020 IEEE international IOT, electronics and mechatronics conference (IEMTRONICS) Sep 9. IEEE, pp 1–4
30. Li M (2021) Did the small business administration's COVID-19 assistance go to the hard hit firms and bring the desired relief? J Econ Bus 115:105969
31. Widarti W, Desfitrina D, Zulfadhli Z (2020) Business process life cycle affects company financial performance: micro, small, and medium business enterprises during the Covid-19 Period. Int J Econ Financ Issues 10(5):211
32. Oh LB, Teo HH (2006) The impacts of information technology and managerial proactiveness in building net-enabled organizational resilience. In: IFIP international working conference on the transfer and diffusion of information technology for organizational resilience. Springer, Boston, pp 33–50
33. Zhu N, Stjepcevic J, Baležentis T, Yu Z, Wang B (2017) How does corporate social responsibility impact banking efficiency: a case in China. Econ Manag 20(4):70–87
34. Delmas MA, Burbano VC (2011) The drivers of greenwashing. Calif Manag Rev 54(1):64–87
35. El Ghoul S, Guedhami O, Nash R, Patel A (2019) New evidence on the role of the media in corporate social responsibility. J Bus Ethics 154(4):1051–1079
36. Miah MD, Rahman SM, Mamoon M (2021) Green banking: the case of commercial banking sector in Oman. Environ Dev Sustain 23(2):2681–2697
37. Mitchell JI, Gagné M, Beaudry A, Dyer L (2012) The role of perceived organizational support, distributive justice and motivation in reactions to new information technology. Comput Hum Behav 28(2):729–738
38. Sun L (2019) Perceived organizational support: a literature review. Int J Hum Resour Stud 9(3):155–175

39. Andriyanti NP, Supartha IW (2021) Effect of perceived organizational support on organizational citizenship behavior with job satisfaction as mediating variables

40. Maan AT, Abid G, Butt TH, Ashfaq F, Ahmed S (2020) Perceived organizational support and job satisfaction: a moderated mediation model of proactive personality and psychological empowerment. Future Bus J 6(1):1–2.https://doi.org/10.1186/s43093-020-00027-8

41. Chen H, Eyoun K (2021) Do mindfulness and perceived organizational support work? Fear of COVID-19 on restaurant frontline employees' job insecurity and emotional exhaustion. Int J Hospitality Manage 94:102850

42. Anwar S, Chandio JA, Ashraf M, Bhutto SA (2021) Does transformational leadership effect employees' commitment? A mediation analysis of perceived organizational support using VB-SEM.

43. Na-Nan KH, Joungtrakul J, Dhienhirun A (2018) The influence of perceived organizational support and work adjustment on the employee performance of expatriate teachers in Thailand. Mod Appl Sci 12(3):105

44. Kurtessis JN, Eisenberger R, Ford MT, Buffardi LC, Stewart KA, Adis CS (2017) Perceived organizational support: a meta-analytic evaluation of organizational support theory. J Manage 43(6):1854–1884

45. Hair Jr JF, Hult GT, Ringle CM, Sarstedt M (2021) A primer on partial least squares structural equation modeling (PLS-SEM). Sage Publications Jul 20

46. Mateka M, Gogo J, Omagwa J (2016) Effects of internet banking on financial performance of listed commercial banks in Kenya. https://repository.maseno.ac.ke/handle/123456789/3114. Accessed 16 May 2021

47. Kartadjumena E, Rodgers W (2019) Executive compensation, sustainability, climate, environmental concerns, and company financial performance: evidence from Indonesian commercial banks. Sustainability 11(6):1673

48. Sanchez G (2013) PLS path modeling with R. Berkeley: Trowchez Editions, 383

49. Hussey DM, Eagan PD (2007) Using structural equation modeling to test environmental performance in small and medium-sized manufacturers: can SEM help SMEs? J Cleaner Prod. 15(4):303–312

50. Maditinos D, Mitsinis N, Sotiriadou D (2008) Measuring user satisfaction with respect to websites. Zagreb Int Rev Econ Bus 1(1):81–97. https://doi.org/10.18052/www.scipress.com/ILSHS.21.9. Accessed 19 June 2021

51. Fornell C, David FL (1981) Evaluating structural equation models with unobservable variables and measurement error. J Mark Res 18(1):39–50

52. Chin WW (1981) The partial least squares approach to structural equation modeling. Mod Methods Bus Res 295(2):295–336

53. Wixom BH, Watson HJ (2001) An empirical investigation of the factors affecting data warehousing success. MIS Q 1:17–41

54. Martínez-Torres MR (2006) A procedure to design a structural and measurement model of intellectual capital: an exploratory study. Inf Manage 43(5):617–626

55. Moore DS, Notz W, Fligner MA (2003) The basic practice of statistics. WH Freeman, New York

56. Hayes AF (2017) Introduction to mediation, moderation, and conditional process analysis: a regression-based approach. Guilford Publications Dec 13

57. Ejigu SN (2017) E-Banking service quality and its impact on customer satisfaction in state owned banks in East Gojjam Zone; Ethiopia. Global J Manage Bus Res

58. Allen DG, Shanock LR (2013) Perceived organizational support and embeddedness as key mechanisms connecting socialization tactics to commitment and turnover among new employees. J Organ Behav 34(3):350–369

59. Khalil S, O'sullivan P (2017) Corporate social responsibility: Internet social and environmental reporting by banks. Meditari Accountancy Res 25(3):414–446

# COVID-19 and the Economic Activity of Jordanian Companies: The Mediating Role of the Community Response

Najm A. Najm, Jasser A. Al-nsour, A. S. H. Yousif, and Abdulazez B. Al-nadawy

**Abstract** This study aimed at determining the impact of Covid-19 on the economic activity of Jordanian companies through three dimensions of the pandemic (damage and risks, working under the pandemic, and the need for treatment) as factors affecting economic activity during two stages: the recession and recovery. The study also sought to identify the impact of the societal response as a mediating variable on the causal relationship between the dimensions of COVID-19 and the economic activity of Jordanian companies. The results of this study have contributed considerably to understanding the impact of the pandemic on Jordanian companies. On the other hand, the outcomes of the study have confirmed that Nigeria damages and risks had the greatest effect in creating the economic recession. The results of the study also, have confirmed that the need for medical treatment had stronger impact on economic stagnation than its impact on recovery.

**Keywords** Covid-19 · Breakout · Economic activity · Recession · Recovery · Harms and risks · Treatment · Community response

**JEL Classification** M16

## 1 Introduction

The post-World War II era has witnessed the birth of the United Nations. This world eminent event was followed by the creation of UN specialized organizations, including, financial organizations of the (World Bank and, the International Monetary Fund), world health organization (WHO) which, is concerned with fighting diseases and pandemics, and (Food and Agriculture Organization) (FAO) that is strive hunger. This is similar to the concepts and mechanisms used to protect the state of welfare in

N. A. Najm (✉) · J. A. Al-nsour · A. S. H. Yousif · A. B. Al-nadawy
Al Zaytoonah University of Jordan, Amman, Jordan
e-mail: najimnajim@zuj.edu.jo

© The Author(s), under exclusive license to Springer Nature Switzerland AG 2022
S. G. Yaseen (ed.), *Digital Economy, Business Analytics, and Big Data Analytics Applications*, Studies in Computational Intelligence 1010,
https://doi.org/10.1007/978-3-031-05258-3_50

developed countries and confront potential economic crises on the basis of economic and humanitarian cooperation.

Covid-19 is a catastrophe of a special kind, unparalleled in our contemporary world except in World Wars 1 and 2, which for a long time affected all areas of life and different sectors of the economies of the entire world. It also represents a kind of force majeure event in imposing harsh measures [4]. COVID-19 is different in nature from natural catastrophes, such as, earthquakes, floods, and hurricanes, [29] and [32]. This pandemic has resulted in a highest global disruption in economic activities and social life accompanied by unprecedented fears of pandemic risks in a highly interconnected world. Destructive earthquakes, for instant, usually strike on a specific area in a country and lead to casualties and damage with the halt of production and economic activities in that area. As the earthquake stops (which, usually, lasts for seconds or minutes) life returns to normal immediately, within unaffected areas and, perhaps after days or weeks in the affected areas. In the case of Covid-19, it is totally different as the turbulent and instable conditions are still going on in the whole area of a country and round the world. As an overall assessment, COVID-19 pandemic has led to worsen instability state, continued economic downturn, job losses, and disrupted supply chains. Some developing countries were hardly hit by the pandemic, as they are experiencing financial and humanitarian crises, which will increase the risks of increasing sudden waves of migration, governments collapse or internal conflicts [31]

In light of the foregoing, this study has focused on the impact of Covid-19 pandemic, in its three adopted dimensions (damage and risks, working under the pandemic, and the need for treatment) on economic activities of Jordanian companies in the stagnation and recovery states. This study has, also, tested the impact of local community response as a mediating variable on the causal relationship between the pandemic and the economic activities of Jordanian companies.

## 2 Covid-19 and Economic Activity

The emergence of negative effects of Covid-19 have made a clear picture for our strongly interconnected world. This type of world interconnection has facilitated the fast deployment of the epidemic. According to this concept, the starting point was in Wuhan (the largest city in Hubei Province in China). The first turn was in February 2020,which witnesses the deployment of the virus throughout 54 countries within four world infection transmission complexes: China (centered in Hubei), East Asia (based in South Korea and Japan), Middle East (based in Iran) and Western Europe (based in Italy) [17]. During this period, it was found, that the highly hit countries by the epidemic accounted for about 40% of the global economy. In April 2020 was the occurrence of the second turning point, as the virus spread in the vast majority of the world countries. At this phase most the world governments have recognized that COVID-19 is a true pandemic representing a very serious and dangerous challenge to all societies and economies in the world. Despite the remarkable medical achievements and major investments in the health sector in our world, the virus has spread at a speed that exceeds the abilities and capacities of the world countries responses to these type of risks. According to a published statistical data by the World Health Organization, the number of infections by the Covid-19 virus met about 53.7 million confirmed cases, plus 1.3 million deaths up to November 15, 2020 [41]. In November 9, 2020, the (WHO) organization announced the development of the first vaccine for Covid-19, that was Pfizer Vaccine [42].

The COVID-19 pandemic has been characterized by its multi-waves and mutations that, have complicated health organizations challenges and increasing risks they are facing. At first, researchers have paid a considerable attention about the first and second waves of COVID-19. By the wave that started in March 2020 (it hit Italy and Spain) a home-confinement accompanied by a strict closures were imposed by the two governments. After Spain government was easing these measures in June 2020, in October 2020, the second wave was began in Italy [13, 37]. As the pandemic has hit all world countries, creating an increasing kinds risks, then researchers have turned out from focusing on COVID-19 waves to concentrating on Covid-19 variants and the mutated strains of the virus such as The Kent variant, or B117, the B.1.617 variant in the UK and The Delta Plus, or Delta-AY.1 variant.

COVID-19 pandemic has been characterized by many waves and mutations causing an expansion in health challenges and risks. These new variants brought about the out breaking of the virus and the emergence of new symptoms, such as, a fast fulmination, and greater risks. This development was behind the confusion of health systems, between their need to know how to confront Covid-19 variants, and the strong feeling that the virus still carries new surprises and risks in other countries. The Confrontation with this pandemic was seeking two main goals: protecting citizens from risk of infection and death, and urgent need to return back to normal situation of businesses [1]. Due to the quarantine policy, closure and social distancing, Covid-19 pandemic has severely impacted economic activity in various countries of the world, whether at macro or micro level.

At macro level, world national economies have entered into a wide recession that comprised most economic sectors that were facing either a reduction or a stoppage. The fight against Covid-19 imposed many severe measures of closure, partial or total quarantine, and social distancing worldwide. These measures have caused a decline in investments with a rise in unemployment to unprecedented levels. This situation necessitated the development of national contingency plans by world countries to aid those workers who lost their jobs obtaining the basic financial benefits to meet their normal needs. In the United States, for example, the 2021–2022 rescue plan was approved, as a package of measures to stimulate its economy assigning an amount of $1.9 trillion [15]. On the other side the European Union adopted a European rescue plan of 700 million pounds to overcome the economic effects of the pandemic [38]. At the micro level, the decline in demand for goods and services was a general feature for the majority of companies within the physical economic sector. By contrast, the digital economy achieved significant increases in the demand for its services with significant gains and advantages [15]. In many countries of the world, the

Fourth Industrial Revolution (the application of digitization, Artificial Intelligence, analytics, to all phases of economic activity, from design to production) contributed to the re-increase in productivity that was greatly affected by the pandemic [36]. Despite the fact that Covid-19 had a direct impact on companies in different sectors, this impact was, widely, varied from one sector to another. For instance, the travel and tourism sector was the most badly hit, but, then it has shown a partial recovered, despite of the massive vaccination campaigns that were conducted by many countries round the world. The banking sector has, also, witnessed an economic stagnation due to a decline in the demand for loans, and that caused a drab in interest. In term of health care services, hospitals have witnessed an increase in the demand for medical services, generally, related to the pandemic, associated with a decrease in the demand for general health services due to the reluctance of individuals to visit hospitals in commitment to distance and fear from pandemic infection. The measures of total or partial closure, spacing at the workplace, and declining in market demand, have made companies suffering from an inability for either, using their full production capacity, or completely stop operating. This situation necessitated the intervention of governments to provide an urgent aid for companies and their employees.

The impact of covid-19 pandemic on Jordanian national economy and companies. After one year in the pandemic, it was realized that Jordanian enterprises were badly hit by Covid-19 pandemic. In a collaborated survey conducted by Fafo Institute for Labour and Social Research, ILO, and UN included 2000 Jordanian companies [27]. This survey focused on the negative impacts of the pandemic on Jordanian companies. It has come up with the following outcomes: 98% (1960 companies) of the surveyed enterprises confirmed that they were negatively impacted in one way or another, 89% (1780 companies) reported lower demand for their products and services with declined revenue, 50% (1000 companies) were impacted by the mandatory closures, 33% (660 companies) were impacted by closed marketplaces, 90% of start-ups (1800 companies) stated that they were heavily impacted, 40% (800 companies) confirmed that they had laid off one or more employees since the outbreak of the COVID-19 pandemic, the largest percentage of lay-offs 48% was in the tourism sector, followed by construction with 45% and manufacturing with 34%, 20% (400 companies) of the interviewed enterprises confirmed that their debt grew, 25% (500 companies) had bank loans, 32% (640 companies) had either supplier credit or informal credit from family or friends, 20% (400 companies) enterprises were extremely pessimistic and replied that they would never bounce back to pre-pandemic levels of operation. According to the study, these results led to a decline in the economic growth rate in Jordan to reach $-2\%$ with very high unemployment level of 88.5. A set of field studies and opinion surveys conducted in Jordan regarding the effects of the Covid-19 pandemic on various economic activities in Jordan, have come up with the following outcomes:

- In UNDP survey it was found that 58.6% of employed people have lost their income, 17.1% their income has become "much lower, 9.4% a "slightly lower" income, with only 11.3% unaffected by the crisis [46].

- At governorate level, there was a significant variation in term of losing income from 69% in Zarqa (similar to Irbid at 65% and Amman at 62%) to 32% in Tafileh, 39% in Ma'an and 40% in Ajloun. Younger age groups have been affected more [46]
- The impact of COVID-19's new case was high and negative over the stock market's performance [5, 45]
- It was reported that the travel and tourism sector of services economy individually comprises about 13.8% of the total GDP [38].
- In a CSIS interviewed with the Jordanian Finance Minister about the economic impacts of COVID-19, has announced through it, that Jordan has been suffering for more than a decade of severity and shocks that accounted for about 45% of its entire GDP. During COVID-19 emergence, the per capita GDP of the country had declined to 14% [19]. According to Singh [44], the economy of Jordan is now predicted to go down with a negative 3.5% of the net GDP in the latter year. The overall economic crisis of Jordan is a challenging facet for the government because approximately 97% of the GDP is additionally strained into the public debt comprising about $42 billion [41].
- Alharbi survey of Jordanian universities has reveal that 32.7% of the respondents strongly agree that the funding of the universities will be reduced, 53.8% strongly agree that universities should form an emergence team to deal with consequences of the pandemic, and 44.2% of the respondents strongly agree that covid-19 pandemic means a huge downturn in universities economic activities [3, 4].
- A survey conducted by Saraireh [42] has revealed that 66% of respondents reported that their financial resources will last less than one week, 20.3% said that they have enough to last between 1 and 2 weeks, 7.7% can last between 2 and 4 weeks with their current resources, and less than 6% of respondents who can last over a month in their current situation (38).

In short, these outcomes are, in fact, the main factors behind the recession that hit the Jordanian economy since the spread of the Corona pandemic.

## 3   Two Stages: Recession and Recovery

The recession caused by the pandemic was reflected in many overall economic indicators, such as a decrease in the gross domestic product (GDP), the unemployment rate, decrease in capital investments, and an increase in public debt. However, this stagnation at macro level has hit all sectors and companies, in various degrees. While some companies were suffering from complete closure, others were experiencing partial shutdown. BY contrast, it was found that some other companies took advantage from the pandemic conditions by achieving better business results. For example, digital companies have reached a considerable growth in their services due to the closure other companies and the replacement of physical work by digital activities, as what was happened at schools and universities. The reliance upon digital activities did not materialize with the SARS pandemic, which hit the world two decades ago [5]. This disparity of the recession impact on companies was, also reflected by the inability of companies to recover after the pandemic recedes. For example, tourism companies will face greater difficulties in restoring the recovery situation they were in before the pandemic. Banks may witness a significant increase in demand on loans and other banking services, as they were postponed due to the pandemic.

Economic activity is, usually, influenced by the business cycle consists of the expansions of several economic activities, that often followed each period of a general recessions [5]. Economic crises accelerate the emergence of recessionary periods, while a financial crisis limits confidence and new investments as an engine of economic activity, a pandemic leads to a complete or partial cessation of economic activities. This situation, often, leads to a halt in the generation of companies and individuals incomes, causing a slowdown in demand, with a decline in economic performance. By adopting some stimulus measures such as rescue plans and aiding programs for companies to revitalize the economy, the gradual or a leaping transition to the stage of recovery will accrue. During crises eras, such as, COVID-19 pandemic, economic activity at both macro and micro levels goes through two successive.

**Recession** The International Monetary Fund (IMF) defines a global recession as an economical growth below 3 percent, because this level is far too weak to keep up with the demands for jobs in the emerging markets that witness growing population [3]. The pandemic negatively affected most economic sectors, and the greatest impact was on the informal sector [36]. Perhaps the exception is the digital business sector [25, 37]. During all past crises that our modern world has experienced, the economic stagnation has been lasting for specific periods of times ranging from 13 months (the economic crisis of 1937–1938) to 18 months (the financial crisis of 2007–2009). The most important indicators of those economic crises were a declining in gross domestic product, deterioration in industrial production, and an increase in unemployment rate. As for as COVID-19 pandemic is concerned (from April 2020 to July 2021), it continues interacting, after the passing of 18 months so far (see Table 1).

**Recovery** The overcoming of an economic crisis caused by a pandemic should come cross two stages: recovery (i.e. returning the economy back to its level before the crisis), and growth to ensure a subsequent economic development. Recovery is an approach for overcoming the recession and its worrying indicators. Any economic recession is, usually, followed by an economic recovery, however, this recovery can come quickly or be very slow. During the Great Depression period (1929–1933) the recovery was late due to the non-intervention policy adopted by government at that time, although many researchers supported that policy as best decision taken by the state. It might reasonable to say that, the exacerbation of problems at that time was, strongly contributed to the development of a New Keynesianism economic theory. This theory, supports a state intervention by steps opposite to the recession. The New Deal policy came in 1933 to begin the recovery phase in the American economy [7]. This approach was repeated when there was a need for a global New Deal in the 2008–2009 crisis [11].

**Table 1** World recessions between 1929 and 2020

| Recession | Duration | Indicators | Drivers |
|---|---|---|---|
| 1929–1933 | | – 15.0% decline in real GDP<br>– The Dow Jones industrial average dropped from 381 to 198<br>– 20.0% of unprecedented unemployment rate | – The 1929 stock market crash<br>– Business cycle<br>– Delayed government intervention (new deal: 1933–1936) |
| 1937–1938 | 13 months | – 18.2% decline in real GDP<br>– 32.4% decline in industrial production<br>– 20.0% of unemployment rate | – Business cycle |
| 1973–1975 | 16 months | – 4.9% decline in real GDP<br>– 15.3% decline in industrial production<br>– 9.0% of unemployment rate | – The increase in oil prices |
| 1981–1982 | 16 months | – 3.0% decline in real GDP<br>– 12.3% decline in industrial production<br>– 10.0.3% of unemployment rate | – Sharp interest-rate increases by the American Federal Reserve and big round of layoffs |
| 2007–2009 | 18 months | – 3.8% decline in real GDP<br>– 16.9% decline in industrial production<br>– 9.0.5% of unemployment rate | – The bursting of the technology bubble and the economic repercussions of the 9/11 attacks<br>– The housing bubble in the US |
| 2020–? | 18 months to date | – A 0.3- to 0.7 percentage-point reduction in global GDP [45]<br>– 58.6% of employed people have lost their income (Jordan) [46]<br>– 33% of companies have laid off the worlds due to closed markets [27]<br>– 48% layoffs in the tourism sector [27]<br>– The informal sector, which represents 70% of the workforce, is the most affected (Nigeria) [38] | – The risks and threats of Covid-19 comprised all countries round the world |

*Sources* [6, 10, 11, 18, 38, 46]

Many studies based on previous experiences of pandemics operation, have presented different scenarios for the evolution of the pandemic behavior and its waves, according to the mitigation and strictness measures that were taken by each country. A United States, study conducted by the Center for Infectious Diseases Research and Policy has relied on the best comparative model, for the current Pandemic, that is the influenza pandemic. Since the early 1700s, at least eight global influenza pandemics have occurred, four of which hit the world between 1918 and 2009–10. That study has identified three time-stages scenarios of the pandemic outbreak. In its first wave, which began in the spring of 2020, increasing of infections cases and then started decreasing according to two different patterns, while the third scenario develops without a specific wave pattern [30].

The International Institute for Strategic Studies (IISS) has developed a comprehensive scenario for the world in the face of Covid-19, basing on 40 drivers, encompasses three main areas: politics (comprising two clusters of political and geopolitical governance), the economy (embracing two clusters of economic reordering, and economic consequences), the military (involving clusters armed conflict and military situation). The scenario covered all three domains by type $2 \times 2$ matrices representing the best and worst expected outcomes. The politics domain matrix, included the geopolitical conflict or cooperation axis/the axis of strengthening or disintegration of governance, the economy domain matrix represented the axis of collapse or economic recovery axis/the modified globalization and Fragmentation axis, and finally, the army domain matrix: it represents the stalling of Evolution or accelerating of transformation axis /the axis of pandemic peace or the outbreak of conflicts.

The study of Sharma et al. [43], introduced four possible scenarios, which were described in letter forms: V–, U–, L– or W–shaped economic recovery after Covid-19. A V-shape refers to a quick and sustained recovery after a sharp economic decline, U-shape refers to a sharply decline and a continuous depression before recovering again, L-shape refers to a slow rate of recovery, and the W-shape refers to an economic recovery (it might needs more time to recover as, the economy often, drops twice before it recovers). There is also the K-shape recovery introduced by JP Morgan which introduced different sectors of an economy that get back at starkly different rates [19]. As there are different sectors affected differently by the pandemic, there was another scenario described by another letter: the K-shape scenario. There is also the K-shape recovery introduced by JP Morgan in which different sectors of the economy recover at starkly different rates [19] and as a pattern to widen the gap between winners and losers in the stock recovery [22]. Therefore, it would be objective to suggest that, the figure-K scenario is the most realistic one because it states that sectors and companies do not suffer from the pandemic in a similar way. This study covered five business sectors (tourism, hospitals, universities, banks, and malls) that were affected differently by current pandemic.

In all financial and economic crises, there was a next stage, represents the economic recovery stage. But, the question in this context is, what about the economic recession caused by Covid-19 pandemic? The answer to this question can be determined in the light of three main aspects:

- It is certain that recovery cannot be achieved under the implementation of a continuous businesses closure policy and quarantining individuals inside their cities and counties in response to alarming health risks. The ongoing of the pandemic out breaking will impose the continuation of closure procedures wholly or partially. Therefore, recovery cannot be achieved within these circumstances.
- The variation in the levels of outbreaks between countries has resulted in an unequal recovery by them. This will negatively affect the supply chains between those countries which are handling the pandemic by implementing a closing policy. On the other hand some others are adopting a policy of openness and a gradual return to normal life. According to the magazine, "The uneven recovery of global economy creates new challenges [28], as there are crowded ports while others are closed due to the pandemic."
- Variation in the capabilities of some countries in adopting policies and programs of rescue and economic stimulus in order to return-back to the pre-pandemic economic situation. This will allow these attaining growth rates higher than that of the pre-pandemic.

## 4   Covid-19 Pandemic

According to WHO, Coronavirus disease (COVID-19) is an infectious disease caused by a recently discovered coronavirus [42]. The Covid-19 pandemic is the most global dangerous challenge facing the whole world in this century. So far there are 141 million infectious cases and 4 million deaths. Developed countries, despite their huge financial resources and advanced health care systems, have experienced an outbreak and increase in the number of deaths that were no less dangerous than that of developing countries. Furthermore according to a recently updated data by the World Health Organization (updated on July 12, 2021), confirm that the two countries which have witnessed the first outbreak of the pandemic were Italy with (4.3 million cases), Spain with (3.9 million cases), and the United States with (33.5 million cases plus 601 thousand deaths) [42]. In term of world interconnection by flight lines, a journey between the farthest two points destinations on our planet can take one day or so under normal conditions, but it will be not so with Covid-19 pandemic. The fact is, that all world countries are live under the threat of this pandemic. Relying on the huge amount of published data and information, it could be concluded that, there is a strong linear relationship between globalization activities and the speedily separation of the pandemic. This conclusion means that, our whole world nations are sailing in the same boat and facing together the same dangers of this pandemic. The economic effects of Covid-19 pandemic are evident, under the case of stagnation at the level of the national economy level of each country. This result, often takes many different forms and levels related to companies global economic activities. The process of overcoming the pandemic negative effects is, usually, accompanied by an economic recovery with gradual return to normal life of the national economy, companies, and whole society.

This study sought to determine the impact of three dimensions of COVID-19 on the economic activity of companies in two stages. These dimensions: damages and risk (such as infection, additional expenses due to business closure, and loss of sales), working under the pandemic conditions (fear of working with others, feelings of anxiety and depression at work, relationships tension and quarrels between employees), the need for treatment (the need for frequent examinations, the need to provide vaccine, the company's actions to protect its employees). In the confrontation with this pandemic, the different patterns of community response can accelerate or slow down the spread of the pandemic. Therefore, the societal response represented an intermediate variable that could affect the relationship between COVID-19 and economic activities of companies. Based on these potential effects, the study hypotheses were formulated as follows:

H1: The COVID-19 pandemic, with its three dimensions (damages and risks, working under the pandemic, and the need for treatment), affects the recession of economic activity in Jordanian companies.

H2: The COVID-19 pandemic affects the recovery of economic activity in Jordanian companies.

H3: The community response as a mediating variable affects the causal relationship between the COVID-19 pandemic and economic activity.

## 5 Methods

**Measures** This study, has used the questionnaire as a tool to collect the required data that consist of five sections: "demographic data, imposed restrictions associated with Covid-19, the three dimensions of the pandemic (damage and risks, working under the pandemic, and the need for treatment), economic activities related to the cases of (recession and recovery), and finally, the community response as a mediating variable. Each, dimension of COVID-19 pandemic was measured by a representative set of elements. The dimension of damages and risks (DAR) was measured by seven elements (A1–A7): "level of risk (A1), lack of resources due to the pandemic (A2)", "additional expenditures due to protection measures (A3)", "working at partial capacity (A4)", "fears of contagion (A5)", "sales decrease due to pandemic outbreak (A6)", "the risks of closure and none full stoppage of work (A7)".

To measure the dimension of "work under the pandemic (WUP)" six elements six elements were used (B1-B6): "fear of working with others (B1)", "risk of injury at work (B2)", "increased negative feelings, such as, anxiety and depression at work (B3)", "working fewer hours (B4)", "stressful and conflict relationships (B5)", "working according to the rule of: less work, less risk (B6)". The dimension of the need for treatment (NFT) was measured by seven elements comprising (C1–C7): "the need for frequent examination (C1)", "the need to take the vaccine (C2)", "and the request for providing the vaccine by the company (C3)", " demanding of employees for increasing protection measures (C4)", "the complaints of employees against the company due to the pandemic (C5)", "the increasing number of employees asking to visit a doctor (C6)", "ensuring a continuous effectiveness of the protective measures adopted by the company' (C7)". Regarding economic activity, companies recession (RES) was measured by five questionnaire items (X1–X5): "working stoppage at the company (X1)", lack in utilizing production or service capacity (X2)", "layoffs of employees (X3)", " Sales decrease (X4)", and inability to pay financial obligations (X5).

While the recovery dimension (RCO) is represented by six questionnaire items (Y1–Y6): "increasing company's sales (Y1)", "extra investments for expansion (Y2)", "introducing new products and services (Y3)", "adopting future plans for expansion (Y4)", "introducing new technology for confronting the pandemic (Y5)", "partnerships and alliances to enter new markets (Y6)". Finally, the community response dimension (COM) as a mediating variable was represented by five elements (M1–M4): "good community awareness of about social circumstances imposing by covid-19 pandemic (M1)"," individual and collective social solidarity (M2)", "high commitment to health instructions (M3)", and "acting responsibly for overcoming the risks of the pandemic: accepting vaccination (M4)". The IBM SPSS AMOS, Version 23 was used to represent the model of the study and determine the direct effect of the three dimensions of COVID-19 (independent variables) on economic activity under the recession and recovery stages (dependent variables) as well as the effect of the societal response (median variable) (see Fig. 1).

**The Sample** In response to the pandemic circumstances, a random sample was used by this study, where 105 respondents were withdrawn from five sectors: "tourism, hospitals, banks, malls and universities". Each sector was represented by 20 respondents, except universities sector, which was represented by (25) respondents. The main characteristics of the current study sample shows that, males represented about (69%). The age category over 30 years old represented (63%.) Half of the respondents (51%) hold a bachelor's degree (see Table 2).

# 6   Economic Restrictions of Covid-19

Covid-19 has been accompanied by an unprecedented way, with widely varied restrictions. These restrictions have led to an inability of companies practicing many of their normal activities, in term of using their human, financial, productive and service resources efficiently. These constraints have, also, played a negative role in limiting companies' expansion plans by introducing, new products and services, new technology or entering new markets. Therefore, companies were facing a problem of lack of resources that could cause a severe layoffs, suspension of training and social responsibility programs at the company. The questionnaire encompasses paragraphs concern 15 economic and financial constraints that companies were suffering from,

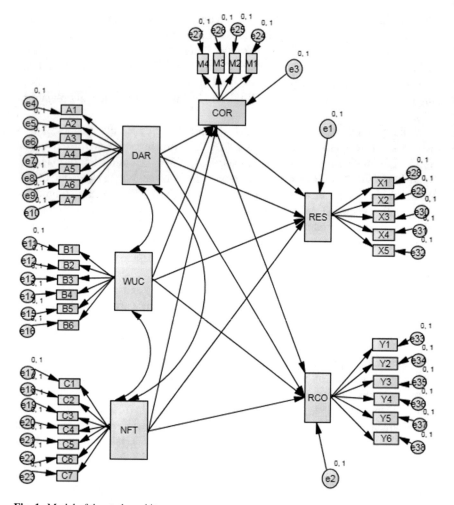

**Fig. 1** Model of the study and its measures

under the pandemic conditions. The results of the study have confirmed that Jordanian companies have suffered totally or partially from these restrictions (see Table 3).

**Table 2** The characteristics of sample

| Characteristics | | Frequency | % |
|---|---|---|---|
| Gender | Male | 72 | 68.6 |
| | Female | 33 | 31.4 |
| | Total | 105 | 100.0 |
| Age | < 30 | 37 | 35.2 |
| | 30–39 | 32 | 30.5 |
| | 40–49 | 18 | 17.1 |
| | 50–59 | 15 | 14.3 |
| | ≤ 60 | 3 | 2.9 |
| | Total | 105 | 100.0 |
| Marital status | Single | 38 | 36.2 |
| | Married | 67 | 83.8 |
| | Total | 105 | 100.0 |
| Education | Secondary | 14 | 13.3 |
| | Diploma | 12 | 11.4 |
| | Bachelor | 53 | 50.5 |
| | Master | 11 | 10.5 |
| | Ph.D. | 15 | 14.3 |
| | Total | 105 | 100.0 |
| Job title | Manager | 27 | 25.7 |
| | Officer | 67 | 63.8 |
| | Technician | 3 | 2.9 |
| | Other | 8 | 7.6 |
| | Total | 105 | 100.0 |
| Working experience (years) | < 5 | 30 | 28.5 |
| | 5–9 | 32 | 30.5 |
| | 10–14 | 24 | 22.9 |
| | ≥ 15 | 19 | 18.1 |
| | Total | 105 | 100.0 |

# 7 Tests

## 7.1 Reliability and Validity

To ensure an internal consistency between the questionnaire items used to measure each variable, Cornbrash's alpha was used. The values of the multi-item variables ranged between (798–0.844). These values indicate that the questionnaire items that used to measure each variable are characterized by an internal consistency. Constructed validity was used to assess the extent to which the questionnaire's items measures what it seeks to measure. Factor analysis loadings were used to examine all the study variables. According to Hair et al. [20, p. 151], factor loadings values within a range of ± 0.30 to ± 0.40 are considered meeting the minimal level for interpretation of structure. The results confirmed, that these loadings ranged from 0.440 to 0.885 greater than the accepted minimum value. Table 4, also, presents, that, the KMO values for the three dimensions of COVID-19 are greater than 0.50. The factor

**Table 3** Restrictions on economic activity in companies due to COVID-19

| Restrictions | Yes (%) | Some (%) | No (%) |
|---|---|---|---|
| 1. Restrictions of productive or service capacity | 53.3 | 38.1 | 8.6 |
| 2. Restrictions on overtime | 58.1 | 26.7 | 16.2 |
| 3. Interruptions in the supply chain | 49.5 | 33.3 | 17.2 |
| 4. Layoffs restrictions | 51.4 | 31.4 | 17.2 |
| 5. Restrictions on company investments | 43.8 | 41.9 | 14.3 |
| 6. Restrictions of obtaining loans | 43.8 | 33.3 | 22.9 |
| 7. Constraints leading to spending cuts | 59.0 | 25.7 | 15.2 |
| 8. Restrictions leading to reduction of social programs | 50.5 | 29.5 | 20.0 |
| 9. Restrictions on introducing new products and services | 47.6 | 37.1 | 15.2 |
| 10. Restrictions leading to reduction of training programs | 47.6 | 31.4 | 21.0 |
| 11. Restrictions leading to company closing | 56.2 | 23.8 | 20.0 |
| 12. Restrictions on company development activities | 41.9 | 41.9 | 16.2 |
| 13. The company's failure to enter new markets | 47.6 | 37.1 | 15.2 |
| 14. Salary Improvement Constraints | 66.7 | 0.26.7 | 7.6 |
| 15. Travel restrictions for work | 60.0 | 25.8 | 14.2 |

**Table 4** Descriptive data, inter-correlations, and CR

| Variables | M | SD | CR | DAR | WUC | NFT | RES | RCO | COR |
|---|---|---|---|---|---|---|---|---|---|
| DAR | 3.479 | 0.850 | 0.854 | 1 | | | | | |
| WUC | 3.217 | 0.887 | 0.896 | 0.434 | 1 | | | | |
| NFT | 3.324 | 0.795 | 0.867 | 0.344 | 0.578 | 1 | | | |
| RES | 3.556 | 0.877 | 0.871 | 0.585 | 0.190 | 0.234 | 1 | | |
| RCO | 2.739 | 0.819 | 0.847 | 0.281 | 0.017 | 0.295 | −0.233 | 1 | |
| COR | 3.235 | 0.872 | 0.879 | 0.204 | -0.168 | 0.044 | 0.083 | 0.092 | 1 |

loadings and KMO values indicate a strong relationship between the questionnaire's items used to measure each variable, which, represent a good indicator, that, the data of this study is adequate for any advanced statistical analysis. The inter-correlations, composite reliability (CR) tests were conducted. The results are portrayed by Table 4, which indicate that there are positive and negative correlations between variables, as, the values of the inter-correlation ranged between (−0.233 and 0.585). These results confirm that there is a problem of multicollinearity between the variables. Also, Table 5, also, shows that the CR values for all variables were greater than "70%", not above "95%" [18].

**Table 5** Effect of Covid-19 on recession (H1) and recovery (H2)

| Hypotheses | Variables | Estimate | SE | CR | P-value | Support |
|---|---|---|---|---|---|---|
| *Covid-19 >>>> Economic recession* | | | | | | |
| H1 | DAR >>> RES | **0.516** | **0.088** | **5.246** | **0.000** | Yes |
| | WUC >>> RES | **−0.161** | **0.110** | **−1.462** | **0.030** | Yes |
| | NFT >>> RES | **0.174** | **0.115** | **1.506** | **0.000** | Yes |
| *Covid-19 >>>> Economic recovery* | | | | | | |
| H2 | **DAR >>> RCO** | **0.091** | **0.104** | **0.880** | **0.379** | No |
| | WUC >>> RCO | 0.101 | 0.116 | −0.056 | 0.002 | Yes |
| | NFT >>> RCO | 0.064 | 0.121 | 2.010 | 0.044 | Yes |

DAR: damages and risks, WUP: work under the pandemic, NFT: need for treatment, RES: companies recession, RCO: company recovery

## 7.2 Goodness-of-Fit Indices

To determine the direct effect of COVID-19 variables on economic activity using the IBM SPSS AMOS, the estimates that represent this effect are usually, associated with a set of model-fit indices. According to Kline [31], there are four groups of model appropriateness indicators (absolute, incremental, parsimonious, and predictive) [29]. All these fit-indices are acceptable at a specified cut-off point or threshold [2] and [20]. Using more than one indicator use to be the best option that should be considered to determine the fit of the model. While, those indices are affected by the sample size, others indicators are independent from the sample size, and therefore they can be used, when the sample size is not large. A large sample size smooth out the use these indicators, but as the study sample is not large, the researchers, had to adopt some of these indicators. According to the testing results of this study, the indicators of fit of the model, pointed out, that Chi-square ($\chi 2 = 9.787$) and its relative Chi-square/degree of freedom or (CMIN/DF) at a significance level ($p = 0.02$) are valid. The normed Fit Index: NFI was (0.917) which is larger than the cutting point ($> 0.90$), the incremental Fit Index: IFI ($0.925 > 0.90$) and the comparative Fit Index: CFI ($0.915 > 0.90$). These results approve that the indicators of model fit of this study were fairly accepted and appropriate for testing the study hypotheses.

## 8 Causal Model Results

Regarding the first hypothesis (H1), the results of the causal model illustrated in Table 5 and Fig. 2 have confirmed that DAR (0.516) and NFT (0.174) had a positive impact, at significance level of ($p < 0.05$), on economic recession of Jordanian companies. While working under Covid-19 (WUC) had a negative effect ($−0.161$) at a significant level of ($p < 0.05$) on the recession. In Jordan as it was the case worldwide, during

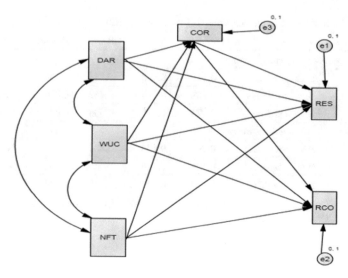

**Fig. 2** Model of the study

the first wave of the pandemic (from March 2020 to the end of the year) the need for treatment (NFT) (as, no vaccine was available, where, the first announcement, of the Pfizer vaccine was on November 17, 2020), that has greatly, contributed to an increasing economic recession due to fears of lack of access to get the treatment against this pandemic. The negative impact of working under the pandemic (WUC) reveals that companies dislike closure, but on the contrary, they were encouraging the return of employees to work under a preventive measures of protection.

These results can also be interpreted according to the hard preventive measures that, have been adopted by Jordanian authorities during the first period. These measures included complete closure, complete or partial cessation of work, and curfew. This situation has, clearly, contributed to the increase of burdens and economic difficulties on companies. But, with the beginning of 2021, where, Jordan has obtained vaccination doses of multiple vaccines types, these measures were eased to ensure a gradually return to work of Jordanian companies. In testing hypothesis (H2), the results of this test have confirmed that there is a positive effect of WUC (0.10) and NFT (0.24) at a significant level of ($p < 0.05$) on, economic recovery of Jordanian companies. By contrast DAR had insignificant effect ($p = 0.379$) on recovery (see Fig. 3). These results could be interpreted, in the light of the fact that, the acquisition of vaccines and the expansion of vaccination gave companies, as well as, individuals the feeling, that a health recovery can be begin to contribute to an economic recovery and return- back to work.

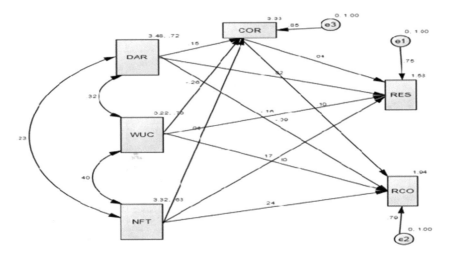

**Fig. 3** The results of the causal model

## 9 Mediation Analysis

In order to determine the indirect effect, the mediation analysis was implemented through the three-step model. According to Kenny et al. [28], Baron and Kenny [9], Judd and Kenny [26] and Wu et al. [53], the model for testing mediating variable includes four steps, which can be reduced to three steps as was the in this study. The three steps are: the independent variable (X) affects the dependent variable (Y), the mediator variable (M) affects the dependent variable (Y), and the independent and the mediator variables (X-M) affect the dependent variable (Y). The differences between the values of the first and third steps at an acceptable level of significance, indicates that, the mediating variable has an effect on the causal relationship between the independent (X) and dependent (Y) variables. Table 6 presents the results of mediation analysis (H3).

Figure 4a shows the direct effect of COVID-19, by its three, components (DAR, WUC and NFT) on recession and recovery without the impact of mediating variable (the societal response). Figure 4b displays the indirect effect of the mediating variable, on the relationship between the study variables. This indirect effect on the causal relationship between COVID-19 and economic activity indicates, that, the mediating variable had a significant effect ($p < 0.05$) on DAR, as the effect value of DAR on the recession has decreased from a (0.52), without the mediator variable, to a (0.50), with it. There is also a positive effect of the mediating variable on the causal relationship between WUE and NFT and economic recession. The results also confirmed that there was no significant effect of the mediating variable on the relationship between DAR and economic recovery. In return, the mediating variable had a significant positive effect ($p < 0.05$) on the relationship between each of the two variables (WUE and NFT) and economic recovery.

**Table 6** The three steps of mediation analysis (H3)

| Steps | Variables | Estimate | SE | P | Support |
|---|---|---|---|---|---|
| *Recession (RES)* | | | | | |
| One | DAR >>> RES | 0.516 | 0.093 | 0.000 | Yes |
| | WUC >>> RES | −0.161 | 0.110 | 0.030 | Yes |
| | NFT >>> RES | 0.174 | 0.115 | 0.000 | Yes |
| Second | DAR. >>> COR | 0.061 | 0.104 | 0.030 | – |
| | WUC >>> COR | −0.170 | 0.116 | 0.056 | – |
| | NFT >>> COR | 0.062 | 0.121 | 0.044 | – |
| Third | DAR >>> COR ≫ RES | 0.503 | .OSS | 0.000 | Yes |
| | WUC ≫ COR RES | 0.13 7 | 0.097 | 0.006 | Yes |
| | NFT ≫ COR ≫ RES | 0.263 | 0.105 | 0.012 | Yes |
| *Recovery (RCO)* | | | | | |
| One | DAR >>> RCO | 0.091 | 0.101 | 0.3 79 | No |
| | WUC >>> RCO | 0.101 | 0.124 | 0.002 | Yes |
| | NFT >>> RCO | 0.064 | 0.122 | 0.044 | Yes |
| Second | DAR >>> COR | 0.064 | 0.110 | 0.161 | – |
| | WUC >>> C OR | 0.070 | 0.120 | 0.030 | – |
| | NFT >>> COR | 0.154 | 0.132 | 0.019 | – |
| Third | DAR ≫ COR ≫ RCO | −0.021 | 0.100 | 0.533 | No |
| | WUC ≫ COR ≫ RCO | 0.032 | 0.091 | 0.036 | Yes |
| | NFT ≫ COR ≫ RCO | 0.206 | 0.099 | 0.037 | Yes |

## 10  Discussion

Some researchers believe that the disruptive effects of Covid-19 leading to a wide economic recession are not certain, and that indicators such as GDP and inflation, as well as the search for safe havens for investment and safe assets for funds vary between countries and sectors [8, 30]. The different paths of economic recovery (as in L, U, V, W-shaped recovery) also indicate the uncertainty related to the post-recession period [40]. On the other hand, some other researchers believe that business concerns go beyond acknowledging the recession, and went farther, by discussing the expected scenarios for getting out of the recession to enter into an economic recovery phase [1, 19, 35]. At the macro level, according to McKibbin and Fernando [33], COVID-19 will lead to a significant drop in GDP. The cost of the economic repercussions of this pandemic has become estimated in trillions. Unemployment is also expected to be doubled in 2022, rising from 140 to 245 million [24]. These repercussions have appeared very clearly throughout economic activities in the companies. Business companies are affected by the economic crises in the country, but in the conditions of the Covid-19, they were the most affected due to the closure procedures and the cessation of economic activities in various fields. Therefore, the economic crisis

## a. Recession with nd without mediating variable

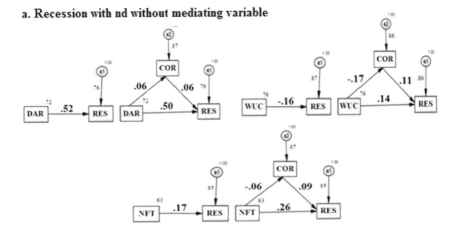

## b. recovery: with and without mediating variable

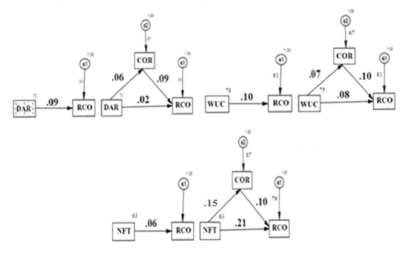

**Fig. 4** With and without mediating variable

caused by the pandemic appears first in the immediate and short-term decisions and practices adopted by companies. This study, has recognized (15) restrictions associated with Covid-19 pandemic that contributed to limiting economic activities of Jordanian companies. These findings were consistent with that of other studies in Jordan [27, 38, 46] and in other countries emphasized, job cuts and increased unemployment [22], company closure restrictions [21], disrupted supply chain restrictions [31], restrictions on social programs and salary improvements by companies [26].

The results of the study confirmed that the damages and risks of COVID-19 and the need for treatment (NFT) have a direct impact on the economic recession that Jordanian companies were suffering from. Companies that confronting risks of injuries and inability to operate with full production and service capacity, have experienced a great difficulty, due to their deficiency to fulfill obligations, as well as, inability to expand their economic activities (increasing their production and sales, introducing new products and services and entering new markets). The need for medical treatment, when there was no such treatment available during the first period of the pandemic, has imposed some alternatives of preventive measures such as stopping work, working for fewer hours, and social distancing. These preventive measures were reflected in damages and economic risks, such as a decrease and interruption of income [27, 46]. When such measures continue for a long time, they will defiantly, lead to a decline in economic activity of companies. Within this context, many studies in Jordan and [4]. Al-Saeed and El Khalil [5], OECD [38] other countries [10, 12], have confirmed that Financial crises can arise from a pandemic. These crises often lead to job destruction, high unemployment rate, work with less capacity, and an increase in the debt ratio of private sector companies greater than that of public sector organizations ratios. Working under the pandemic conditions in Jordan during 2020 was very hard for companies, due to a suspension of work, total or partial closure of companies and, a fragile and disrupted supply chain, which had worsening the negative impact of economic recession.

The results of the current study have, also, confirmed that the damages and risks of COVID-19 (DAR) do not have a significant impact on the economic recovery of companies. This result can be explicated by the fact that the damages and risks of Covid-19 remain a major obstacle to economic recovery, as, companies seek to adopt safe measures and hesitating in going ahead in their new investments whenever, these damages and risks increase. On the other hand, the results have confirmed that the work under Covid-19 (WUC) and the need for treatment (NFT) both have a positive and significant effect on economic recovery. The general indicators in the cases of epidemics emphasize that, providing a treatment or a vaccine represents the first step for overcoming the danger of the epidemic, with a gradual return to normal life. This return encourages the company's return to its normal activities, and its expansion programs that, will lead to an accumulation of opportunities for economic recovery.

The mediation analysis, has indicated that, the community response as a mediating variable had an indirect effect on the relationship between HIV-19 and economic recession. The effect of the need for treatment has increased with the adoption of the mediating variable compared with the direct effect, when the mediating variable was absent. This indirect impact has decreased during working under COVID-19 in Comparison with the direct impact. This result can be explained in that, the community response to return to work was not efficient enough to increase this impact to achieve economic recovery. The societal response did not have a significant impact on the relationship between Covid-19 and both economic recession and recovery by Jordanian companies. This result can be expounded throughout the fact that, damages and risks of Covid-19 on the economic activity of companies are greater than the community response to reduce this effect.

## 11    Conclusions and Implications

The first important result was that damages and risks had the greatest effect on economic recession (estimate: 0.52). This result can be explained by the fact that companies have transformed from working with their full available capacity to shut down and its repercussions on closing markets, work stoppage, and other negative results such as layoffs. The effect of these damages and risks were diminished during the economic recovery stage (estimate: 0.09) as companies returning back to work with optimism after complete closure, which helps companies to accept damages and risks they were experiencing. This result also indicates that, the conditions of economic recovery were characterized by a strong influence on the company and its employees to overcome fears from the returning of the pandemic and its various variants (The last one is the Indian Delta plus variant). The outcomes of the study also confirmed that the need for medical treatment had a strong impact on economic recession (estimate; 0.17) in comparison with the stage of recovery (estimate: 0.06). This result reveals that the fear of infection by the virus was greater in the beginning (recession stage), but this effect decreased with the return to work after the closure period, and perhaps due to the increasing number of vaccinations among employees in companies.

Covid-19 pandemic might represent a truly unique phenomenon as, it differs from all previous crises that it lasted for a relatively long period and spread very quickly and widely throughout all continents, regions and countries. Therefore, studying of this dangerous phenomenon by researchers round the world is a national and humanitarian responsibility. The limited sample size of the study would not make the generalization of its outcomes possible, to overcome this limitation, samples of multiple companies within five different economic sectors, were taken to ease the generalization of our study's outcomes. The reasonability of these results was based on the fact that all types of Jordanian companies share and face the same challenges and repercussions. In practical implications, companies urgently need to develop their own appropriate methods in order to reduce damage and risk and develop flexibility in their working environment and times (including the expansion and development of their electronic services).

Our study sought to determine the impact of Covid-19 at the micro level (sectors and companies). So that studying the impact of the pandemic at macro-level (such as investment, national productivity, levels of employment and unemployment) will contribute significantly to clarifying and interpreting other important dimensions of this pandemic. In addition to that, comparative studies of the pandemic effects between countries and regions will help achieving an understanding of the reciprocal relations and the model of transmission of these effects.

# 12 Limitations

The current study, has some limitations. It was limited to only five economic sectors, so its results cannot be generalized to cover all other economic sectors, except when conducting an evaluation of the similarities between the adopted sectors by this study and those which are not. This study was concerned with the cases of recession and recovery in light of Covid-19, where all countries and economic sectors suffered from it, so the study of financial and economic crises in particular country or region cannot match this study. Finally, Covid-19 pandemic was studied through three variables (damage and risks, working under the pandemic, and the need for treatment), however, it is possible to study this pandemic through other variables that may be more important in other economic conditions.

# References

1. Aleta A, Martin-Corral D, Piontti AP, Ajelli M, Litvinova M, Chinazzi M, Dean NEM, Halloran E, Longini Jr IM, Merler S, Pentland A, Vespignani AA, Moro E, Moreno Y (2020) Modelling the impact of testing, contact tracing and household quarantine on second waves of COVID-19, 4:964–971
2. Alexandre JS, Morin AJS, Marsh HW, Nagengast B (2013) Exploratory structural equation modeling. In Hancock GR, Mueller RO (eds) Structural equation modeling: a second course. Information Age Publishing, Charlotte, pp 395–438
3. Alharbi M (2020) The economic effect of coronavirus (COVID-19) on higher education in Jordan: an analytical survey. Int J Adv Sci Technol 29(11):78–86
4. Alharbi M (2020) The economic effect of coronavirus (COVID-19) on higher education in Jordan: an analytical survey. Int J Econ Bus Adm 8(2):521–532
5. Al-Saeed A, El Khalil Z (2020) Jordan: COVID-19 pandemic weighs heavily on the economy, as it does on the region. Retrieved 27 Aug 2020, from: https://www.worldbank.org/en/news/press-release/2020/07/14/jordan-covid-19-pandemic-weighs-heavily-on-the-economy-as-it-does-on-the-region
6. Altug SG (2010) Business cycles: fact, fallacy, and fantasy. World Scientific Publishing Co, New Jersey
7. Andrew J, Baker M, Guthrie J, Martin-Sardesai A (2020) Australia's COVID-19 public budgeting response: the straitjacket of neoliberalism. J Public Budg Acc Financ Manag 32(5):759–770
8. Baldwin R, di Mauro BW (2020) Introduction. In: Economics in the time of COVID-19. CEPR Press, London, pp 1–30
9. Baron RM, and DA Kenny (1986) The moderator-mediator variable distinction in social psychological research: Conceptual, strategic and statistical considerations. J Pers Soc Psychol 51(6):1173–1182
10. Basco S (2018) Housing bubbles: origins and consequences. Palgrave Macmillan, Cham
11. Bosen EA (2005) Roosevelt, the great depression, and the economics of recovery. University of Virginia Press, Virginia
12. Carlsson-Szlezak P, Reeves M, Swarts P (2020) What coronavirus could mean for the global economy. Harvard Business Review. Available at: https://hbr.org, retrieved 25 June 2021
13. Carneiro A, Portugal P, Varejão J (2014) Catastrophic job destruction during the Portuguese economic crisis. J Macroecon 39:444–457
14. Chowdhury A, Islam I (2018) Rethinking macroeconomics for employment and development. Nova Science Publishers Inc., New York

15. Constâncio, V. (2014).The European crisis and the role of the financial system. J Macroecon 39, Part B, 250–259
16. Chinco F, Nucara G, Szerpak L (2021) COVID-19 mortality in Italy: the first wave was more severe and deadly, but only in Lombardy region. J Infect. Letter to the editor, Published May 31, 2021, Available at: https://www.journalofinfection.com/
17. Congress of the US of America (2021) American Rescue Plan Act of 2021, https://www.congress.gov
18. Craven M, Liu L, Mysore M, Wilson M (2020) Covid-19: implications for business, McKinney & Company, Available at: https://www.mckinsey.com. Retrieved 10 Jan 2021
19. CSIS (2020) Jordan's economy during Covid-19. Center for Strategic and International Studies. Retrieved from https://www.csis.org/analysis/jordans-economy-during-covid-19
20. Gates L (2020) Coronavirus versus a broken economy. Gelbstein Media, Berlin
21. Hair JF Jr, Black WC, Badin BJ, Anderson RE (2019) Multivariate data analysis. Cengage Learning, Australia
22. Henderson R, Platt E (2020) K-shaped' stock recovery widens gap between winners and losers. Financial Times, 22 August. Available: https://www.ft.com. Retrieved 20 May 20, 2021
23. Hooper D, Coughlan J, Mullen MR (2008) Structural equation modeling: guidelines for determining model fit, the electronic. J Bus Res Methods 6(1):53–60, Available at: https://www.ejbrm.com
24. ILO (2020) ILO monitor: COVID-19 and the world of work. Geneva. Available at: https://www.ilo.org
25. Jiang X (2020) Digital economy in the post-pandemic era. J Chin Econ Bus Stud 18(4):333–339
26. Judd CM, Kenny DA (1981) Process analysis: estimating mediation in treatment evaluation. Eval Rev 5:602–619
27. Kebede TA, Stave SE, Kattaa M, Prokop M (2020) Impact of the COVID-19 pandemic on enterprises in Jordan. Fafo Institute for Labour and Social Research, Oslo
28. Kenny DA, Korchmaros JD, Bolger N (2003) Lower level mediation in multilevel models. Psychol Methods 8:115–128
29. Lee HH, Park D (2021) Post-COVID Asia: Deglobalization, fourth industrial revolution, and sustainable development. World Scientific Publishing Co, New Jersey
30. Li T, Liu Y, Li M, Qian X, Dai SY (2020) Mask or no mask for COVID-19: a public health and market study. PloSOne 15(8):1–17
31. Kline RB (2016) Principles and practice of structural equation modeling. Guilford Press, New York
32. MacDonald SB (2021) Congested ports, economic problems and challenged supply chains. The National Interests. Available at: https://nationalinterest.or. Retrieved 2 Jul 2021
33. McKibbin W, Fernando R (2020) The global macroeconomic impacts of COVID-19: Seven scenarios, centre for applied macroeconomic analysis: CAMA, Working paper, Available at: https://papers.ssrn.com
34. Mooijaart A, Montfort K (2004) Statistical power in path models for small sample sizes. In: Montfort et al (eds) Recent developments on structural equation modeling. Kluwer Academic Publishers, New York, pp 1–11
35. Moore KA, DPhil ML, Barry JM, Osterholm MT (2020) The future of the COVID-19 pandemic: lessons learned from pandemic influenza. Center for Infectious Disease Research and Policy (CIDRAP), University of Minnesota, Minnesota
36. ODNI: Office of the Director of National Intelligence (2021) Annual threat assessment of the US intelligence community. Washington, D.C., Available at: https://www.dni.gov. Retrieved 20 May 2021
37. OECD (2020) Keeping the internet up and running in times of crisis. OECD Publishing, Paris
38. Omobowale AO, Oyelade OK, Omobowale MO, Falase OS (2020) Contextual reflections on COVID-19 and informal workers in Nigeria. Int J Sociol Soc Policy 40(9/10):1041–1057
39. Osterholm MT, Olshaker M (2020) Chronicle of a pandemic foretold learning from the Covid-19 failure before the next outbreak arrives. Foreign Affairs, July/August 2020, https://www.foreignaffairs.com Retrieved 24 Nov 2020

40. Pembroke M (2020) America in retreat: the decline of US leadership from WW2 to Covid-19. Oneworld Publications, London
41. Santucci E (2020) What lies ahead as Jordan faces the fallout of COVID-19? Atlantic Council.org Retrieved from: https://www.atlanticcouncil.org/blogs/menasource/what-lies
42. Saraireh SAM (2021) The economic effects of coronavirus (COVID-19) on Jordan. J Contemp Issues Bus Gov 27(2):1183–1197
43. Sharma D, Bouchaud J-P, Gualdi S, Tarzia M, Zamponi F (2021) V–, U–, L– or W–shaped economic recovery after Covid-19: insights from an agent based model. PLoS ONE 16(3). Available at: https://www.ncbi.nlm.nih.govRetrieved 20 May 2021
44. Singh M (2020) Jordan after COVID-19: from crisis adjustment to crisis management. Washington Institute, Washington DC
45. Sneader K, Singhai S (2021) Trends that will define 2021 and beyond: six months on. McKinsey & Company
46. Soriano SV, Ganado-Pinillaa P, Sanchez-Santos M, Gómez-Gallego F, Barreiro PC, de Mendoza C, Corral O (2021) Main differences between the first and second waves of COVID-19 in Madrid. Int J Infect Dis 105:374–376
47. The International Institute for Strategic Studies (2020) The COVID-19 pandemic: scenarios to understand the international impact. London, October 2020
48. The World Bank (2020) Jordan: COVID-19 pandemic weighs heavily on the economy, as it does on the region. Retrieved from: https://www.worldbank.org/en/news/press,release/2020
49. UNDP: United Nations Development Program in Jordan (2020) COVID-19 impact on households in Jordan: a rapid assessment, case study publications, Jordan, UNDP/English
50. WHO (2021) Covid-19 Research and innovation achievement. Geneva
51. WHOa (2020) Coronavirus disease 2019 (COVID-19): situation report 100. Geneva
52. WHOb (2020) Weekly epidemiological update 17 Nov 2020. https://www.who.in, Retrieved 7 June 2021
53. Wu AD, Zumbo BD (2008) Understanding and using mediators and moderators. Soc Indic Res 87:367–392

# Big Data and Big Data Analytics in Audit Brainstorming Sessions: A Canadian Qualitative Research

**Yahya Marei, Malik Abu Afifa, Ahmad Abdallah, Maha Ayoush, and Arwa Amoush**

**Abstract** This exploratory research aims to ascertain participants' perspectives on the use of Big Data and Big Data Analytics methods during audit brainstorming sessions at Canadian audit firms, and whether such methods aid in the risk assessment process to fraud detection. A Canadian qualitative research method is applied in this study to provide an overview of the results from audit industry interviews. The complete sample included twenty-two external auditor participants who attended an office interview in Canada during the third and fourth quarters of 2019; twelve auditors from the Big-4 and ten from mid-size audit firms. The research discovered that, on average, brainstorming sessions interpret the impact of Big Data and Big Data Analytics favourably. Additionally, our study findings show that utilising Big Data Analytics during brainstorming sessions improves the efficiency and effectiveness of fraud risk evaluations substantially. Finally, the paper discussed how Big Data had altered auditors' positions and professional decisions. Our analysis and results have a broader effect on the purpose of fraud detection brainstorming sessions and the quantification of fraud risk in an audit context.

**Keywords** Big data · Big data analytics · Brainstorming · Audit · Canadian audit firms

## 1 Introduction

Accounting literature has a significant amount of academic research centred on financial fraud, particularly in the last decade [31]. According to Hogan et al. [30], the outcome of financial fraud was more than 900 billion dollars between 1997 and 2004. In their reports, the Association of Certified Fraud Examiners [3] estimated the losses

Y. Marei (✉)
Seneca College of Applied Arts and Technology, Toronto, Canada
e-mail: yahya.marei@senecacollege.ca

M. Abu Afifa · A. Abdallah · M. Ayoush · A. Amoush
Faculty of Business, Accounting Department, Al-Zaytoonah University of Jordan, Amman, Jordan

© The Author(s), under exclusive license to Springer Nature Switzerland AG 2022
S. G. Yaseen (ed.), *Digital Economy, Business Analytics, and Big Data Analytics Applications*, Studies in Computational Intelligence 1010,
https://doi.org/10.1007/978-3-031-05258-3_51

due to financial fraud in typical organisations at around 5%. Recently, Gepp et al. [26] indicated that global losses due to fraud are almost 4 trillion dollars. Therefore, The Canadian Public Accountability Board (CPAB) has contained the fraud subject on their top main concern.

Canadian audit standards, Generally Accepted Auditing Standards (GAAS), established by the Auditing and Assurance Standards Board require the auditor to plan for the engagement and identify risks related to fraud; one of the auditors' tools during the auditing plan is brainstorming. The brainstorming sessions' mission is to collect intellectual perspectives from the engagement team and promote new suggestions and the best approach to the audit engagement [47]. The brainstorming session, which capitalises on mutual knowledge and encourages idea creation, provides an advantageous atmosphere for such discussion [47]. One major issue of brainstorms does not link to the effectiveness of the brainstorming sessions but rather the execution, which causes complexity in conducting the session. Besides, one of the reasons for the complication of the brainstorming sessions is due to the inefficient usage of unstructured data. Researchers have highlighted that the existing audit procedure for identifying fraud risk factors needs improvement [44]. Our motivation in this research is to determine whether Canadian audit firms face the same problem during brainstorming sessions and whether Big Data Analytical tools enable additional assistance to come over the complexity during the brainstorming sessions.

The current audit standards encourage auditors to utilise their professional judgement [34, 43, 44]. Therefore, different auditors will apply different perspectives based on their interpretations. Auditors will also not provide a complete picture of fraud due to a shortage of information such as executives meetings and reports [34]. Hence, the significant loss of revenue promotes auditors to utilise Data Analytical (DA) in detecting fraud. For instance, Chang et al. [13] provided a DA model to examine the banks' system, which can handle millions of transactions interactivity and, thus, detect fraud. Another critical study by Debreceny and Gray [20] applied data mining of journal entries to identify the fraud and assist auditors in developing and selecting audit methodologies appropriate to the journal entries' characteristics. Therefore, they focused on the financial part and ignored the non-financial part, such as email, videos, and the quality of employees who prepared the financial reports. Therefore, the second aim of this research is to attempt to cover the flaws in Debreceny and Gray's [20] study and examine the use of Big Data during brainstorming in detecting fraud factors in Canadian audit firms.

We documented Canadian audit firms overcome the challenges by utilising BDA during the brainstorming sessions. Furthermore, to produce a reliable result during such sessions, we found auditors in the engagement team utilise Big Data to compare the client industry data over periods to effectively identify abnormalities, which significantly enhances audit engagement quality by analysing data through duration and industries. Our result aligns with prior studies [17], who stated that auditors could quickly spot anomalies by performing descriptive procedures.

Further, auditors demonstrate that Big Data Analytics (BDA) in brainstorming sessions permits the engagement team to utilise the unstructured data and investigate fraud factors closely related to the fraud triangle. We uncovered the effectiveness and efficiency of using BDA in recognising fraud's risks. Our results are supported by prior research such as [43] and [47], who indicated that auditors, through analysing business dynamics, use cutting-edge techniques to track and stop illegal activity. Minor deviations can be recorded, tracked automatically and flagged as potential fraud.

We also documented the superiority of electronic brainstorming sessions in Canadian audit firms. It permits the use of a large set of information and increases discussion quality while reducing the production blocking. Prior studies (e.g., [24, 44, 55]) purpose the advantage of electronic brainstorming sessions; auditors conduct real-time brainstorming sessions and collaborate to use BDA tools in the process, including data analysis, data aggregation, fraud signal identification, community meetings, conclusions, and monitoring. We captured the auditors' response to the task of archiving information acquired by collaboration, while the brainstorming sessions aided in the process.

Our contribution to this research can be summarised as follows: First and foremost, it aims to describe the process by which Canadian auditors perform brainstorming sessions. Second, it seeks to provide evidence for the importance of BDA in the audit function, including the relationship between auditors' roles and BDA, the ways in which BDA alters auditors' roles, and the role of BDA in financial fraud prevention through brainstorming sessions. Thirdly, current and prospective auditors and other official parties may use the findings of this study to enhance fraud risk assessment and detection in companies, thus increasing taxpayers, creditors, lenders, and other stakeholders' trust in financial reports audited by qualified audit firms and enabling them to make sound, related decisions. Researchers may wish to equate Canadian audit firms to those located in other countries.

# 2 Literature Review and Research Directions

The following sections discuss the Big Data and BDA, and brainstorming sessions in Canadian Audit Standards and financial fraud. Additionally, financial fraud, auditors' function and Big Data.

## 2.1 Big Data and Big Data Analytics

The concept of "Big Data" was coined in computer science in 2005 by Roger Magoulas of O'Reilly Media to describe a large amount of data that conventional data management methods cannot handle or process due to its complexity and size [29, 45]. Nowadays, the Big Data concept is approached from various angles, with implications in a wide range of fields. Big Data is characterised by its size, which consists of a massive, complex, and autonomous collection of data sets, each of which has the potential to interact [48]. Additionally, due to the inconsistency and unpredictable nature of the potential configurations, Big Data cannot be managed using conventional data processing techniques.

The primary significance of Big Data is the potential to enhance productivity in the context of using a huge volume of data of various types. Suppose Big Data is specified correctly and used appropriately, companies will gain a better understanding of their business, leading to increased productivity in areas such as sales, product improvement, and so on [53]. Therefore, Big Data can be used efficiently in a variety of fields, including information technology to increase security and troubleshooting by examining patterns in existing logs; customer service by using information from call centres to obtain customer patterns and, thus, improve customer satisfaction by customising services; enhancing services and goods by using social media content. Various studies have assessed the efficacy of big data, such as [17] and [46], who described that a business which knows prospective customers' preferences allows it to adapt its product in order to reach a broader group of individuals.

Consequently, since the world we live in today is based on data [6]; today's lives are influenced by an organisations' ability to dispose of, interrogate, and handle data [4]. The development of technology infrastructure is tailored to assist in generating data, allowing all offered services to be upgraded as they are used. So, Big Data management (BDM) is required, wherein its foundation is the collection and organ-isation of related data [52, 54]. Data analytics seeks to comprehend what occurred, why it occurred, and predict what will occur in the future. Deeper analytics imply new analytical tools for gaining deeper insights [38]. BDA is quickly becoming the preferred solution to the business and technological developments disrupting the conventional data management and processing landscape. Enterprises that are early adopters of BDA will achieve a competitive advantage. Despite the fact that BDA can be technically difficult, businesses should not put it off [41].

The BDA initiative should be a collaborative effort between Information Tech-nology (I.T.) and business [38]. I.T. should be in charge of deploying the appropriate BDA techniques and putting in place sound data management practices. Both groups must recognise that success will be measured by the value added by the initiative's business enhancements [15]. Thus, the purpose of this research is to provide some evidence regarding the role of BDA in the Canadian audit function.

## 2.2 Brainstorming Sessions in Canadian Audit Standards and Financial Fraud

Auditing standards entail the engagement partner and other key engagement group members to review the client's financial reports' vulnerability to material misstate-ment [8]. The conversation provides an opportunity for the audit team to share their insights about the business and its environment, including their understanding of internal controls. Chief auditors are extensively involved in all performance audit teams in brainstorming sessions, not only for the knowledge but also for discussing common concerns and challenges. A brainstorming session is a method for creating as many suggestions or solutions to a situation or topic as possible. However, it is not a method for deciding the most appropriate solution to a problem [23].

According to the Canadian Institute of Chartered Accountants (CPA Canada), audit groups need to brainstorm and evaluate material misstatement risks relating to the fraud. Particularly, Canadian audit standards, mainly CAS 315 and 240, require that discussion among engagement audit members consider the client's financial reports' susceptibility to fraud and material misstatement due to errors. CAS 315 also requires a discussion among the engagement team members and a determination by the engagement partner as to which matters are to be communicated to those team members not involved in the discussion. This discussion shall place particular emphasis on how and where the entity's financial statements may be susceptible to material misstatement due to fraud, including how fraud might occur. Additionally, CAS 240 indicates that, throughout the investigation, the auditor must retain professional scepticism, acknowledging the risk of misstatements due to fraud. Hence, a discussion among the engagement team members supports professional scepticism, and contributes to better understanding the risk of misstatements due to fraud. For instance, Carpenter [12] examined the auditor's judgement during the brainstorming sessions and the result suggests that auditors who participate in brainstorming sessions significantly have a higher qualified opinion on fraud assessment.

Nevertheless, Trompeter et al. [49] emphasise the role of the auditors in deducting fraud and discussing the matter during the brainstorming sessions. One of the critical studies on brainstorming activities indicated that brainstorming discussion allows auditors to review the engagement and assess fraud activities' likelihood [7]. However, they noted that auditors did not share their opinion during the brainstorm, because participants may fear criticism from their team leader and/or peer judgment. This may lead to a misunderstanding of the fraud risk evaluation. DeZoort and Harrison [23] assessed auditor responsibility for fraud detection by interviewing 878 auditors on the role of brainstorming. The study documented the impact of practical brainstorming sessions on audit quality, where practical brainstorming sessions contribute to better audit quality and, thus, detecting fraud.

As indicated previously, most of the studies examined the effectiveness of brainstorming sessions during the engagement. However, Lynch et al. [32] took a different direction and examined the effectiveness of electronic brainstorming sessions. As expected, these were more effective in assessing the entity fraud risk. The study suggests that the electronic brainstorming sessions identified more risk factors than face-to-face sessions, which is also consistent with other studies [2, 22]. These studies pointed out that, in the face-to-face brainstorming session, participants must wait for approval to speak and must not be interrupted by other members. This may create production blocking, whereas, in electronic sessions, members can simultaneously engage and mitigate productivity blocking. The superiority of electronic sessions comes from interactive brainstorming as members can input their ideas in real-time to all members to discuss and not rely on memories during face-to-face sessions. On the other hand, Cockrell and Stone [18] documented that face-to-face discussions are more productive and better to identify risks. There is also less narcissism. More-over, Brazel et al. [11] have a different view, whereby they pointed out that auditors in face-to-face sessions have more accountability than electronic preparing for the sessions.

## 2.3 Financial Fraud, Auditors Function and Big Data

Financial statements are considered the most essential reports companies publish during the fiscal years. The information provided in the reports is helpful for stake-holders, lenders, creditors and other interested parties [6, 38]. For instance, investors can decide to invest more or withdraw their investment merely based on these reports. Governments and banks utilise these reports to provide funds and grants and deter-mine the tax implications. Therefore, earnings management is utilised by executives to alter financial statements [35]. Some executives take it a step further; for instance, Enron and WorldCom are examples of such executives who committed fraud to misrepresent their financial reports. Maka et al. [33] classify financial fraud into financial statement fraud and transaction fraud; financial statement fraud has a signif-icant adverse valuation consequence on companies. Abu Afifa et al. [1] documented a significant negative valuation on companies that committed frauds.

Prior works of literature discuss the fraud triangle in-depth since its existence. Clinard and Cressey [16] argued that fraudulent activities are committed by whether an individual or organisation adopts the three components of the fraud triangle (pres-sure, opportunity and rationalisation). Boyle et al. [9] stated the prior model is missing the capability to commit the act, while other researchers included different factors that contribute to fraud. They documented that conflicts of interest [36], knowledge of accounting position [21], compensations [25], morality [37] and absence or weak-ness of an internal control system [5] are important factors that contribute to fraud. Power [40] also classified and distinguished between fraud risks and fraud activ-ities and mentioned that the misconception of the business environment is due to the world of auditing and position of the fraud, and, thus, it contributes to increase earnings management and fraud. Therefore, these factors and others add challenge to the auditors and require highly competent auditors and robust technology.

Davis and Pesch [19] documented that individuals who committed fraud or pro fraud exist in mechanism environments that help spread the fraud, hence auditors could have a challenge to assess the risk and detect fraud; therefore, auditors should have a uniform and modern approach. Similarly, Peecher et al. [39] and Trotman and Wright [50] indicated that auditors should know the company's business envi-ronment to assess the level of fraud risk. Also, auditors may require acquiring to evaluate external pieces of evidence based on the business model. Therefore, compe-tent auditors who utilise analytical procedures is essential. Thus, auditors require an understanding of Big Data and analytical technologies to assess the risk and fraud deduction; some of these technologies utilised by auditors are data mining and DA [28], outlier techniques [51], algorithms and support vector machine [14]. Here, this research aims to provide more explanations based on Canadian auditors' perceptions about the following main questions:

*Q1: What is the role of Big Data Management and Big Data Analytics in financial fraud detection through brainstorming sessions?*

*Q2: How have Big Data and Big Data Analytics changed the auditors' roles?*

# 3　Sample and Research Method

The complete sample included twenty-two external auditor's participants who attended an office interview in Canada during the third and fourth quarters of 2019; twelve auditors from the Big-4 and ten from mid-size audit firms. As a result of initial interviews, we identified and worked with individuals with expertise in BDA and business process automation tools, with an eye to building a network of contacts to be used during an audit. We conducted semi-structured face-to-face interviews in four major Canadian cities (Toronto, Montreal, Niagara Falls, and Mississauga). The participants all had at least five years of relevant experience. On average, the participants' experience is nearly nine years (see Table 1).

Interviews provided us with an opportunity to probe deeper into the issues at hand, providing the data with a 'richness' that was impossible to achieve through other methods. We audio-recorded each office interview, and the conversations typically lasted for about one hour. These answers to the research questions above enabled us to identify the audit firm's approach to Canada's engagements and understand whether the audit firm employs the Big Data Analytics process during the brainstorming sessions.

In the field of thematic analysis and based on Braun and Clarke [10], the following steps are also involved. Familiarising with the data, generating initial codes, searching for themes, reviewing themes, describing and naming them, as well as drafting a report of the outcomes. Conclusions are drawn by using a thematic analysis method. Prior to the initiation of interviews, general areas were identified as the primary focus of research and these codes were revised as the interviews progressed. As a consequence of this approach, we could begin to code interviews early and increase the reliability of the analysis. Coding the qualitative data involves identifying causal links between themes and describing and arranging the interview observations systematically. We followed well-established procedure from a variety of studies, such as Glaser and Strauss [27] in processing the data. Data collection was accomplished using a combination of coding, memo-writing, and diagramming techniques simultaneously to identify textual patterns from the interview transcriptions.

**Table 1** List of interview participants

| Code | Participant | Audit firm | Position (s) | Experience (s) |
|------|-------------|-----------|--------------|----------------|
| P1 | Partner—audit methodology | Big four | Audit manager | 10 years |
| P2 | Partner—audit methodology | Big four | Audit manager | 15 years |
| P3 | Partner—audit methodology | Mid-tier audit firm | Audit manager | 13 years |
| P4 | Partner—audit assurance | Big four | Auditor | 16 years |
| P5 | Partner—audit assurance | Big four | Auditor | 10 years |
| P6 | Partner—audit assurance | Mid-tier audit firm | Auditor | 7 years |
| P7 | Partner—audit assurance | Mid-tier audit firm | Auditor | 10 years |
| P8 | Partner—audit risk analytics | Mid-tier audit firm | Auditor | 8 years |
| P9 | Partner—audit risk analytics | Big four | Audit manager | 7 years |
| P10 | Partner—audit risk analytics | Mid-tier audit firm | Audit manager | 5 years |
| P11 | Partner—data assurance | Big four | Auditor | 11 years |
| P12 | Partner—data assurance | Big four | Auditor | 13 years |
| P13 | Partner—data assurance | Big four | Audit manager | 7 years |
| P14 | Partner—data assurance | Mid-tier audit firm | Audit manager | 5 years |
| P15 | Partner—data assurance | Mid-tier audit firm | Audit manager | 5 years |
| P16 | Partner—data assurance | Mid-tier audit firm | Audit manager | 8 years |
| P17 | Data analytics auditor | Big four | Auditor | 10 years |
| P18 | Data analytics auditor | Big four | Audit manager | 12 years |
| P19 | Partner—audit assurance | Big four | Auditor | 7 years |
| P20 | Partner—audit assurance | Big four | Audit manager | 7.5 years |
| P21 | Data analytics auditor | Mid-tier audit firm | Audit manager | 8 years |
| P22 | Data analytics auditor | Mid-tier audit firm | Audit manager | 9 years |

## 4   Findings and Discussions

The following subsections present twenty-two Canadian auditors' perceptions about the research questions, and discussions of them based on the previous literature.

### 4.1   Intensive Brainstorming and the Usefulness of Big Data

Embedding BDA into brainstorming sessions becomes a way to improve their efficiency [45, 48]. To begin, Big Data has the potential to expand the depth of knowledge. We documented that in utilising BDA during the brainstorming time, by integrating or summarising various forms of data using Big Data software, Canadian auditors gain access to the records containing both financial and non-financial information about their clients. Also, we found that Big Data aids in brainstorming sessions by exchanging information.

Using electronic brainstorming as a foundation, Big Data reintroduces computerised programmes to monitor fraud consultations during the brainstorming time, documents the chain of thought, and maintains potential options. Canadian auditors were able to submit longer feedback that includes images, flowcharts, and videos to emphasise their thoughts. Audit managers indicated that using Big Data during the brainstorming engagement session of an audit partnership results in the generation of more high-quality fraud insights than without its use. Additionally, the results indicate that audit teams' fraud risk evaluations after the brainstorming session are considerably higher than previous assessments without Big Data analysis.

One of the findings that emerges from these interviews is auditors' managers prefer face-to-face brainstorming in some engagements as it increases transparency and provides better-quality judgement on the part of preparers as compared to electronic review. However, all agreed, face-to-face brainstorming seems to suffer more from process failure, especially production blocking, while electronic brainstorming is used in the majority of engagement brainstorming as it provides high adequate dialogue and monitors the feedback, and auditors can revisit the sessions for greater clarification. Our results align with Carpenter [12] who identified the effectiveness of brainstorming on identifying fraud risks. The study documented that, despite the fact that the total number of ideas generated during the brainstorming session is limited, brainstorming audit firms produce more high-quality fraud insights than independent auditors. The comments below from participant auditors documented that the brainstorming sessions increase the efficiency of audit fraud risk assessment higher than those made prior to the brainstorming using the BDA.

> Group brainstorming, when conducted correctly, will encourage strategic thinking, and get a team together, as well as assist you in coming up with the best idea. [P1]

> ….since the team is working for a shared purpose in a creative and welcoming setting, brainstorming as a group increases group morale.[….] brainstorming allows teammates to share face-to-face time together, which is beneficial to team unity. [P5]

> Group brainstorming will seem like a victory when the group leaves feeling energised, fulfilled, and ready about the next move. This can be accomplished by the use of effective brainstorming tactics. When a brainstorming session fails, whether it is unproductive, tedious, or pessimistic, the team may feel uninspired collectively. [P2]

> The aim of brainstorming is to collect intelligent feedback from various participates and stimulate fresh thoughts, this format seems to be an efficient means of manipulating unstructured data and collecting anomalies. […] We agree that brainstorming sessions can improve overall fraud assessment; however, implementation can be difficult due to the difficulty of conducting discussions with a high quality and the constraints of the brainstorming format (e.g. production blocking). [P8]

## 4.2   The Functions of Auditors in the Era of Big Data

A considerable amount of literature has been published on BDA. These studies indicated that the adoption of business decision analysis in the audit and assurance services market has just begun to occur at the innovation stage [42]. The twenty-two participants indicated, based on the results of this research, that this technology is already influencing the way auditors work and the way they manage their audit in Canada. The majority of participants were pleased with the BDA, but three participants were not so certain. These few auditors indicated that they are struggling to identify patterns in Big Data visualisations, and it has fewer advantages when presented before auditors review more traditional audit facts. Our interpretation of this group is that they may lack the necessary technical expertise and are inexperienced with data processing methods. Another explanation could be the manner in which data is gathered and recorded, in which business uses data, and in which reporting items are analysed and assembled; especially for inexperienced auditors, it may create challenges.

Nevertheless, the majority group, specifically, risk analysts and audit managers, pointed out that Big Data has the potential to revolutionise auditors' roles and audit. We documented that auditor's performance improved due to BDA's propensity to move away from minor sample size procedures into areas where they are now able to provide a higher level of assurance for a large or complete set of transactions. This outcome is contrary to that of Rose et al. [43], who found that auditors do not get many positive BDA results. This research result may be explained by the fact that auditors are being freed from tedious and repetitive manual tasks, and they use their time more efficiently to focus their minds and skills in areas where they are most needed, such as more critical evaluation. Almost all the participants in the research stated that their use had enabled auditors to think outside the box and focus on making critical judgements. Auditors' managers insist on the fundamental idea of auditors' roles within a Big Data environment, since auditors must be capable of reproducing the data lifecycle or transaction route, which is, in most of the cases, achievable in an electronic system with adequate validity.

Some of the comments below from the participant auditors documented that they have already benefited from the efficient use of BDA technologies and the added value of their efforts add to audits and audit clients [38].

> As a result of digitalisation, audit firms would be able to expand their offerings by offering new services. It would also increase audit accuracy by analysing all consumer data. Finally, as a result of digitalisation, a new auditor persona emerges, allowing audit firms to foster an innovative culture. [P5]

> Big Data is the latest wave that is sweeping through business processes. Businesses who can use the scale, range, and pace of Dig Data can make smarter decisions, cut operating costs, and keep up with changing consumer demands. [P10]

> Big Data is used in a wide range of industries. The ability of companies to collect and interpret data in real-time is an influential weapon that fuels innovation in industries ranging from retail to banking and supply chain management. [P8]

In the field of Big Data Analytics in auditing. Big Data's function and implementations in auditing can be difficult to imagine. Big Data has many applications in auditing. Predictive and prescriptive analytics may be used to measure the probability of possible results by gathering knowledge from historical events. Furthermore, auditors would be more confident in assessing the effectiveness of real business processes. [P7]

## 4.3　Practical Use of Big Data and Identity Theft Fraud Detection

The appearance of red flags alerts auditors to the possibility of misconduct, thus encouraging early notice to businesses. As a result, the probability of not finding wrongdoing decreases if auditors recognise warning flags and exercise specialist "scepticism". Our finding in terms of whether the BDA is capable of identifying fraud risk factors demonstrates that auditor expertise and previous performance in detecting fraud are reliably important predictors of fraud detection for individual audit cycles and cumulative period forecasts. Combining with Big Data Analytical, BDA enables the auditor to adjust the fraud risk assessment based on new information provided by BDA. Inability to update the change in fraud risk assessments when BDA demonstrates otherwise can lead to audit failure [4, 6]. Auditors' manager and auditors "Risk Analyst" highlighted that the emerging Big Data is assessed in terms of its effect on the sufficiency, competence, and reliability of audit evidence. The evidence usually derived from the external context is more probabilistic in nature and must be weighed in light of information's characteristics.

We documented that Big Data enables auditors to review the whole population; the change of perspective is most likely due to the auditor's timely access to appropriate data and auditors use of multiple data analytic methods to assess and view the information in a useful and efficient manner. As well as the automating data extraction and utilisation by structured measure of the effect in a significantly higher degree of efficiency than manual processes which marks more valuable fraud risks assessment. However, auditors pointed out that measuring the entire population could eliminate the risk associated with specific area; subsequently, the auditor's task is to determine how to extract meaning from the enhanced volume of knowledge to which they are subject and to ensure that audit risks assessments are based on accurate and reliable evidence. BDA assistance is used with more advanced audit methods that automate the processing and analysis of critical audit priorities and procedures. As a result of these processes, auditors are able to produce professional judgement driven by algorithms.

The comments below from the participant auditors documented that auditors' skills and computational resources are essential to prevent fraud using data analytics. In the same vein, Yoon et al. [54] highlighted the issue related to the development of Big Data, digital evidence, and electronic traces, and that the concept and essence of audit evidence are evolving. In our paper, we show a dramatic increase in the amount of evidence, and analytics offering auditors help to summarise and clarify its significance. The analytics used in Canadian audit firms differ among audit firms depending on the context, but are generally heavily reliant on automatic data that are constantly monitored for effectiveness in deducing risks factors.

> Big Data is now assisting financial auditors in streamlining monitoring and detecting fraud. This expert can detect market risks faster and perform more appropriate and reliable audits. [...] Before you can use analytics during auditing, you must first have appropriate data collection and aggregation systems in place. All data sets reviewed for auditing must first be tested for consistency, timeliness, and capability. As a result, the downstream decisions made by auditors are focused on accurate and high-quality evidence. [P12]

> Big Data also enables you to simplify various aspects of the auditing process. Human negligence is a frequent cause in companies falling out of compliance or spending too much money on audit-related standards. [...] Auditors will set up different controls in advance to track how well an organisation adheres to existing standards by automating manual and routine activities. [P1]

> ....The advantage of having a consistent approach in place is that analytics can be strategically implemented into each phase of the operation. You may use artificial intelligence to make better strategic choices by automating mundane activities. [P5]

## 5    Conclusion

This exploratory research aimed to ascertain participants' perspectives on the use of Big Data and Big Data Analytics methods during audit brainstorming sessions at Canadian audit firms, and whether such methods aid in the risk assessment process to fraud detection. This research is a Canadian qualitative research and based on twenty-two interviews with external auditors during the third and fourth quarters of 2019; twelve auditors from the Big-4 and ten from mid-size audit firms.

We addressed current problems in fraud detection systems, with a particular emphasis on the difficulty of analysing unstructured data. We documented an improvement in auditor productivity as a result of BDA's proclivity for turning away from procedures that need a limited sample size and towards procedures that need a greater depth of comprehension for a large or complete series of transactions. Auditors now have more available means to collect the information necessary for their audit's engagements and judgement on the risk levels and factors. Auditors also have access to a wealth of knowledge, both financial and non-financial, and the Big Data Analytics significantly boost the audit performance. In addition, it enables auditors in using statistical data analysis to classify fraud and points out outlier and abnormal transactions in financial statements, all of which are manageable during the brainstorming sessions.

We solicited feedback from auditors about the risk evaluation conducted during the brainstorming session and its efficacy. Overall, participants indicated that integrating BDA into brainstorming sessions improved their effectiveness when it came to identifying and tracking fraud risk factors. Perhaps most significantly, auditors acquire insight about the regulatory environment, which allows them to provide superior services. Auditors agreed that integrating BDA into brainstorming techniques, such as those demonstrated, has the ability to provide substantial evidence and validation of their subsequent professional judgement. In addition, once the risk assessment has been completed, the auditors have gained greater knowledge of the company and examined further evidence so they can apply additional analytical methods to risk evaluation during the brainstorming sessions. Lastly, auditors collectively agreed that automated brainstorms are more effective and produce better results than face-to-face. However, a few audit managers pointed out that some of the engagement requires face-to-face brainstorming, especially when the engagement has high risks factors.

# 6　Implications

The primary contribution in this research is to try to explain the audit technique used in the brainstorming sessions, as well as the procedure that is more effective at detecting fraud. However, only mid-sized and Big-4 audit firms were interviewed. Additionally, this research leads to and establishes the critical link between auditors' positions and Big Data. This research can assess potential researchers to make comparisons between various countries. Finally, this research is beneficial for auditors in terms of gaining a better understanding of the value of Big Data through brainstorming sessions. It is also beneficial for clients in terms of increasing their trust in the audit process and risk management.

We propose additional queries that can be explored for future research in regard to brainstorming sessions, including the following: comparing analytical systems that make use of methods for analysing large amounts of data during brainstorm and the long-term consequences of using Big Data during brainstorming sessions.

# References

1. Abu Afifa M, Alsufy F, Abdallah A (2020) Direct and mediated associations among audit quality, earnings quality, and share price: the case of Jordan. Int J Econ Bus Adm 8(3):500–516
2. Anson R, Bostrom R, Wynne B (1995) An experiment assessing group support system and facilitator effects on meeting outcomes. Manage Sci 41(2):189–208
3. Association of Certified Fraud Examiners (2016) Report to the Nations on occupational fraud and abuse. Global Fraud Study. ACFE, U.S.
4. Augustine FK Jr, Woodside J, Mendoza M et al (2020) Analytics, accounting and big data: enhancing accounting education. J Manage Eng Integr 13(1):1–8
5. Baker CR, Cohanier B, Leo NJ (2017) Breakdowns in internal controls in bank trading information systems: the case of the fraud at Société Générale. Int J Acc Inf Syst 26:20–31
6. Balios D (2021) The impact of big data on accounting and auditing. Int J Corp Finan Acc (IJCFA) 8(1):1–14
7. Bellovary JL, Johnstone KM (2007) Descriptive evidence from audit practice on SAS No. 99 brainstorming activities. Curr Issues Auditing 1(1):1–11
8. Blay AD, Notbohm M, Schelleman C et al (2014) Audit quality effects of an individual audit engagement partner signature mandate. Int J Audit 18(3):172–192

9. Boyle DM, DeZoort FT, Hermanson DR (2015) The effect of alternative fraud model use on auditors' fraud risk judgments. J Acc Public Policy 34(6):578–596

10. Braun V, Clarke V (2020) One size fits all? What counts as quality practice in (reflexive) thematic analysis? Qual Res Psychol, 1–25.

11. Brazel JF, Carpenter TD, Jenkins JG (2010) Auditors' use of brainstorming in the consideration of fraud: reports from the field. Acc Rev 85(4):1273–1301

12. Carpenter TD (2007) Audit team brainstorming, fraud risk identification, and fraud risk assessment: implications of SAS No. 99. Acc Rev 82(5):1119–1140

13. Chang R, Lee A, Ghoniem M et al (2008) Scalable and interactive visual analysis of financial wire transactions for fraud detection. Inf Visualisation 7(1):63–76

14. Chen YJ, Wu CH, Chen YM et al (2017) Enhancement of fraud detection for narratives in annual reports. Int J Acc Inf Syst 26:32–45

15. Cho Y, Mun J, Park Y et al (2020) Data processing method in a context-aware system to provide intelligent robot services based on big-data. In: IOP conference series: materials science and engineering, vol 715(1). IOP Publishing, U.K., p 012091

16. Clinard MB, Cressey DR (1954) Other people's money: a study in the social psychology of embezzlement. Am Sociol Rev 19(3):362–363

17. Cockcroft S, Russell M (2018) Big data opportunities for accounting and finance practice and research. Aust Acc Rev 28(3):323–333

18. Cockrell C, Stone DN (2011) Team discourse explains media richness and anonymity effects in audit fraud cue brainstorming. Int J Acc Inf Syst 12(3):225–242

19. Davis JS, Pesch HL (2013) Fraud dynamics and controls in organisations. Acc Organ Soc 38(6–7):469–483

20. Debreceny RS, Gray GL (2010) Data mining journal entries for fraud detection: an exploratory study. Int J Acc Inf Syst 11(3):157–181

21. Dellaportas S (2013) Conversations with inmate accountants: motivation, opportunity and the fraud triangle. Acc Fórum 37(1):29–39

22. Dennis AR, Wixom BH (2002) Investigating the moderators of the group support systems use with meta-analysis. J Manag Inf Syst 18(3):235–257

23. DeZoort FT, Harrison PD (2018) Understanding auditors' sense of responsibility for detecting fraud within organisations. J Bus Ethics 149(4):857–874

24. Earley CE (2015) Data analytics in auditing: opportunities and challenges. Bus Horiz 58(5):493–500

25. Fung SYK, Raman KK, Sun L et al (2015) Insider sales and the effectiveness of clawback adoptions in mitigating fraud risk. J Acc Public Policy 34(4):417–436

26. Gepp A, Linnenluecke MK, O'Neill TJ et al (2018) Big data techniques in auditing research and practice: current trends and future opportunities. J Acc Lit 40:102–115

27. Glaser BG, Strauss AL (2017) Discovery of grounded theory: strategies for qualitative research, discovery of grounded theory: strategies for qualitative research. Routledge, New York

28. Gray GL, Debreceny RS (2014) A taxonomy to guide research on the application of data mining to fraud detection in financial statement audits. Int J Acc Inf Syst 15(4):357–380

29. Halevi G, Moed HF (2012) The evolution of big data as a research and scientific topic: overview of the literature. Res Trends 30(1):3–6

30. Hogan CE, Rezaee Z, Riley Jr RA et al (2008) Financial statement fraud: Insights from the academic literature. Auditing J Pract Theory 27(2):231–252

31. Karpoff JM, Lee DS, Martin GS (2008) The cost to firms of cooking the books. J Financ Quant Anal 43(3):581–611

32. Lynch AL, Murthy US, Engle TJ (2009) Fraud brainstorming using computer-mediated communication: the effects of brainstorming technique and facilitation. Acc Rev 84(4):1209–1232

33. Maka K, Pazhanirajan S, Mallapur S (2020) Selection of most significant variables to detect fraud in financial statements. Mater Today Proc (in press)

34. Manita R, Elommal N, Baudier P et al (2020) The digital transformation of external audit and its impact on corporate governance. Technol Forecast Soc Chang 150:119751

35. Marei Y (2020) The impact of earnings management on innovation strategies in developed, developing, and transition economies. Ph.D. thesis, University of Salford, UK
36. Mehran H, Stulz RM (2007) The economics of conflicts of interest in financial institutions. J Financ Econ 85(2):267–296
37. Morales J, Gendron Y, Guénin-Paracini H (2014) The construction of the risky individual and vigilant organisation: a genealogy of the fraud triangle. Acc Organ Soc 39(3):170–194
38. Odia JO, Akpata OT (2021) Role of data science and data analytics in forensic accounting and fraud detection. In: Handbook of research on engineering, business, and healthcare applications of data science and analytics. IGI Global, USA, pp 203–227
39. Peecher ME, Schwartz R, Solomon I (2007) It's all about audit quality: perspectives on strategic-systems auditing. Acc Organ Soc 32(4–5):463–485
40. Power M (2013) The apparatus of fraud risk. Acc Organ Soc 38(6–7):525–543
41. Qasim A, Kharbat FF (2020) Blockchain technology, business data analytics, and artificial intelligence: use in the accounting profession and ideas for inclusion into the accounting curriculum. J Emerg Technol Acc 17(1):107–117
42. Rogers EM, Singhal A, Quinlan MM (2019) Diffusion of innovations. In: An integrated approach to communication theory and research, 3rd edn. Routledge, New York, pp 418–434
43. Rose AM, Rose JM, Sanderson KA et al (2017) When should audit firms introduce analyses of big data into the audit process? J Inf Syst 31(3):81–99
44. Salijeni G, Samsonova-Taddei A, Turley S (2019) Big data and changes in audit technology: contemplating a research agenda. Acc Bus Res 49(1):95–119
45. Skyt J, Jensen CS, Mark L (2003) A foundation for vacuuming temporal databases. Data Knowl Eng 44(1):1–29
46. Spiess J, T'Joens Y, Dragnea R et al (2014) Using big data to improve customer experience and business performance. Bell Labs Tech J 18(4):3–17
47. Tang J, Karim KE (2019) Financial fraud detection and big data analytics–implications on auditors' use of fraud brainstorming session. Manag Audit J 34(4):324–337
48. Tang JJ, Karim K (2017) Big data in business analytics: implications for the audit profession. CPA J 87(6):34–39
49. Trompeter GM, Carpenter TD, Desai N et al (2013) A synthesis of fraud-related research. Auditing J Pract 32(1):287–321
50. Trotman KT, Wright WF (2012) Triangulation of audit evidence in fraud risk assessments. Acc Organ Soc 37(1):41–53
51. Van Capelleveen G, Poel M, Mueller RM et al (2016) Outlier detection in healthcare fraud: a case study in the Medicaid dental domain. Int J Acc Inf Syst 21:18–31
52. Vasarhelyi MA, Kogan A, Tuttle BM (2015) Big data in accounting: an overview. Acc Horiz 29(2):381–396
53. Vitari C, Raguseo E (2020) Big data analytics business value and firm performance: linking with environmental context. Int J Prod Res 58(18):5456–5476
54. Yoon K, Hoogduin L, Zhang L (2015) Big data as complementary audit evidence. Acc Horiz 29(2):431–438
55. Zhang J, Yang X, Appelbaum D (2015) Toward effective big data analysis in continuous auditing. Acc Horiz 29(2):469–476

# Factors Affecting Palestinian Consumer Behavioral Intentions Toward Online Shopping During COVID-19: The Moderating Role of Perceived Trust

**Mohammed Salem and Samir Baidoun**

**Abstract** The aim of this study is to identify the factors that influence the Palestinian consumer behavioral intentions towards online shopping during COVID-19 pandemic and examine the moderating role of perceived trust. Primary data was collected by a survey questionnaire to test the research hypotheses. In total 1 173 MBA graduates and currently enrolled students from two major Palestinian universities received the questionnaire. The findings provide evidence that the perceived usefulness, perceived ease of use, and perceived behavioral control affect the consumer's intentions towards online shopping based on several reasons discussed thoroughly in this paper. Additionally, perceived trust strengthens the effect of (a) perceived usefulness, (b) perceived ease of use, and (c) perceived behavioral control on online shopping.

**Keywords** Perceived usefulness · Perceived ease of use · Perceived behavioral control · Perceived trust · Online shopping

## 1 Introduction

Recently online shopping has become one of the widely used shopping methods as a result of technology and media advancement [1]. The revenues of online shopping and the number of online shopping consumers witnessed a continuous and significant increase over the years [2]. Online shopping for many scholars has more advantages than traditional shopping including convenience where shopping is available anywhere and anytime [3]; time saving [4]; a wide range of products to compare and choose [5], and cost saving by purchasing lower prices products [6].

M. Salem (✉)
University College of Applied Sciences, Gaza, Palestine
e-mail: Palestinemrdd_salem@hotmail.com

S. Baidoun
Department of Business Administration and Marketing, Birzeit University, Birzeit, Palestine

© The Author(s), under exclusive license to Springer Nature Switzerland AG 2022
S. G. Yaseen (ed.), *Digital Economy, Business Analytics, and Big Data Analytics Applications*, Studies in Computational Intelligence 1010,
https://doi.org/10.1007/978-3-031-05258-3_52

Currently, all over the world, countries, businesses, and individuals face dramatic changes due to the COVID-19 pandemic [7]. The spread of COVID-19 world-wide has increased uncertainty level regarding the consumption and investment among involved stakeholders, including suppliers, trading partners, investors and most importantly consumers [8]. According to Hashem [9], the pandemic succeeded to change consumer attitude, intentions, and behavior towards online shopping as a daily life choice. Furthermore, Pham et al., [10] attributed this change to the fact that the COVID-19 pandemic requires social distancing where most people, could not go to work or shopping to avoid direct contact.

Scholars reported several factors that affect online shopping including: perceived usefulness [11, 12], perceived ease of use [13, 14], and perceived behavioral controls [15, 16]. Singh & Srivastava [17] reported that these factors have a significant influence on purchase intention. Based on the Technology Acceptance Model-TAM [18] other researchers provided evidence that these factors represent the major benefits of online shopping and have important and positive effect on consumers' intention [19]. In addition, the perception of usefulness and the ease of use also influences the consumers' online shopping intention [1, 20]. Other studies used the Theory of Planned Behavior—TPB [21] to investigate shopping intention of online consumers. The results reveal that attitude, subjective norms and perceived behavioral control affect consumers' shopping intention.

According to Slade et al., [22], most previous studies were conducted in the developed countries with highly developed technology infrastructure. Positive atti-tude, high level of control behavior and low perceived risks for online shopping were reported [23]. However, many contradicting results were reported by previous studies which require further investigation in specific contexts especially with the many evidences reporting differences in the factors affecting consumers' intentions among developed and developing countries [24].

For Ajzen [21] shopping intention is a decisive factor that influence consumers' shopping behaviors. Therefore, online sellers need to determine the hindering and the motivating factors that affect consumers' online shopping intention in order to influence their intentions [1]. To have better understanding of factors that influence Palestinian consumers' shopping intention, the study targeted the MBA alumni who graduated in the past three years and the currently enrolled MBA students at two major Palestinian universities as university students constitute the majority of internet users.

One of the study's contributions is providing an empirical of why consumers decide to use online shopping in Palestine. This research also investigates previous studies reported inconsistent relationships. More precisely, this paper aims to provide empirical evidence from one of the developing countries to the body of knowledge by defining the relationship between the independent variables of the "perceived useful-ness, perceived ease of use, and perceived behavioral control" and the dependent variable "the consumer's intentions towards online shopping" in light of COVID-19 in Palestine as the dependent variable while examining the role of perceived trust as a moderating variable.

The remaining sections of the paper are organized as follows: Sect. 2 presents theoretical background and hypotheses development. In Sect. 3, research methods are explained. The analysis of results is provided in Sect. 4. Finally, discussion of findings, limitations and future research are presented in Sect. 5.

## 2 Theoretical Background and Hypotheses Development

### 2.1 Technology Adoption Model (TAM)

Organizations in the Arab World are commonly considered to be late adopters of the Internet and its applications [26]. Despite a rise in Internet usage, recent IT adoption studies show that Arabs are still hesitant to utilize and embrace technology for a variety of economic and cultural reasons [27–29]. It is essential to consider the elements that influence IT adoption in the Arab world. As a result, a number of theories and models have been created to explain the link between user beliefs, attitudes, and technological behavior intentions [27, 30]. Theory of reasoned action (TRA), theory of planned behavior (TPB), the innovation diffusion theory (IDT), model of perceived credibility utilization model (PCUM), motivational model (MM), social cognitive theory (SCT), and technology acceptance models (TAM) are just a few of these theories [27].

Davis' [18] TAM model (Perceived Usefulness, Perceived Ease of Use, and User Acceptance of Information Technology) is an information system theory that aims to investigate users' acceptance of certain workplace systems. This approach has been widely utilized to analyze customers' attitudes regarding online shopping since e-commerce is technology-related [31]. TAM is an extension of the Theory of Reasoned Action (TRA), a popular theory that states that a person's behavior is governed by his intention to execute the behavior [32]. Davis [18] modified TRA by including two key beliefs, perceived usefulness and perceived ease of use, that particularly account for technology adoption. TAM demonstrated an intention to affect the connection between perceived usefulness, perceived ease of use, and usage behavior by acting as a mediator [18].

### 2.2 Online Behavioral Intention

One of the elements that influence consumers' buying behavior is their shopping intention [32]. Intention is a driving element that encourages one individual to process an action [21] thus it is a criterion for determining the likelihood of future behavior [33]. Meanwhile, Akbar et al. [34] suggest that intention is a specific goal consumers have in mind when they analyze one or more actions. Consumers may have a variety of intentions of which is shopping intention. Shopping intention is a plan to decide where customers would buy products, for that, Delafrooz et al. [5] define online shopping intention as the degree to which a consumer intends to engage in a certain purchase activity over the Internet.

Customer online behavioral intention has sparked a lot of studies [35, 36]. Behavioral intention is a component of cognitive behavior that describes how a buyer intends to purchase a certain brand. In the theories of reasoned action [37] and planned behavior [38], intention is a crucial factor in making a purchasing choice. Intentional measurements may be more successful than behavioral measures in capturing customers' minds since they may have made purchases owing to limitations rather than true preferences [39]. As a result, the strength of a customers' intention to engage in a certain purchase behavior over the internet is determined by their online behavioral intention [40]. Online behavioral intention, as defined by Pavlou [41], is the context in which a consumer is willing and intends to engage in an online transaction. Online transactions are activities that involve the retrieval, transmission, and purchase of information and products through the internet.

## 2.3   Factors Affecting Consumer Behavioral Intentions Toward Online Shopping

### 2.3.1   Perceived Usefulness

Consumers' behavioral intentions are influenced by their perception of usefulness Davis [18]. The perceived usefulness is defined as the degree to which a person believes that adopting a specific system would improve his or her performance [18]. In the context of online shopping, perceived usefulness represents the degree to which a consumer believes that online shopping will improve the efficiency of his/her purchase [1]. There is evidence that perceived usefulness has a substantial impact on online shopping intention [19]. Therefore, the proposed hypothesis is:

H1: Perceived usefulness has positive effect on online shopping intention.

### 2.3.2   Perceived Ease of Use

Perceived ease of use is defined as the customer's view of the ease of interaction on e-commerce web sites and the ability to finish online purchasing without much effort [42]. When customers find a website easy to use and navigate, they will consider it a new option to try which increases their likelihood of making an online purchase [43]. Despite the fact that there are differences in how people use the internet and how they shop online, perceived ease of use has an impact on the overall information system [44]. The behavioral intention in adopting information systems is highly connected to perceived ease of use [18]. Furthermore, online retailers must provide comfort, efficiency, and convenience to their consumers in terms of perceived ease of use in order to build a competitive edge for their website [45]. Consumers always look for an easier method to buy products online. Consumers value the readily accessible products, simple access to websites, the ability to compare products and prices, and the ability to understand online purchasing, when they have the intention to buy online [25, 41]. Thus, we set the following hypothesis:

H2: Perceived ease of use has positive effect on online shopping intention.

### 2.3.3 Perceived Behavior Control

Ajzen [21] introduced the notion of an individual's self-belief and how it affects their intentions. This notion was later known as "perceived behavioral control." It refers to a person's own belief in his or her own ability to display behavior [46]. Perceived behavioral control refers to an individual's capacity to manage his/her actual behavior [47]. Furthermore, the perceived behavioral control supports individual engagement in finding appropriate information and whether he or she has the required capabilities, resources, and a sense of control over the decision to perform or not [48]. Since intention is influenced by perceived behavioral control [49], we propose the following hypothesis:

H3: Perceived behavioral control has positive effect on online shopping intention.

## 2.4 The Moderating Role of Perceived Trust

Perceived trust as a moderating variable, is defined as a factor that influences the direction and strength of the connection between independent and dependent variables [50]. Online shopping activities are driven by the nature and dynamics of perceived trust. Thus, online shopping firms profit from a high level of confidence in website information, quality, credibility, security, and protection. According to Ganguly et al., [51], trust aspects offer customers confidence since they may obtain greater advantages with less risk while shopping online.

Trust is the expectation that others with whom one interacts would not take unfair advantage of one's reliance on them. This is the idea that all parties involved will act in an ethical, reliable, and socially suitable manner and will keep their promises [52]. McKnight et al. [53] think that trust in the context of online shopping is the willingness to accept the risks (vulnerability) from online shopping websites after collecting information about them. Similarly, trust may be defined as the willingness to tolerate vulnerability risks (vulnerability) in order to make a transaction with an online shop [54]. Trust is a critical component of exchange relationships [53] and an influential factor to customer behavior in purchasing methods (both online and offline) [19]. When it comes to online shopping, trust is especially crucial since the consumers' sense of transaction risk is higher when they don't have direct connection with the vendor or the products they intend to buy [55]. Consumer trust in a particular retail website has been shown in previous research to have a significant influence on buying intentions [19, 41, 52]. Consumers are reluctant to purchase online for a

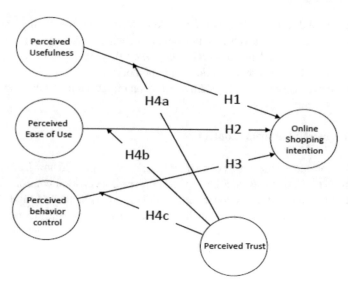

**Fig. 1** The research framework

variety of reasons, one of which is a lack of trust. Therefore, the fourth hypothesis H4:

Perceived trust strengthens the effect of (a) perceived usefulness, (b) perceived ease of use, and (c) perceived behavioral control on online shopping intention.

The research framework of the study is presented in Fig. 1. The model presents the relationship between the independent variables (i.e., perceived usefulness, perceived ease of use, and perceived behavioral control), and the dependent variable (i.e., the consumer's intentions towards online shopping). In addition, the moderation influence practiced by perceived trust is also shown.

## 3  Methodology

### 3.1  Participants

In total 362 respondents completed and returned the questionnaire. Figure 2 shows that 221 of the respondents are males, while 141 are females. In total, 245 of the respondents are current MBA students while the MBA graduates were 117 respondents. Finally, the majority of respondents (309) are between 25 and 35 years old.

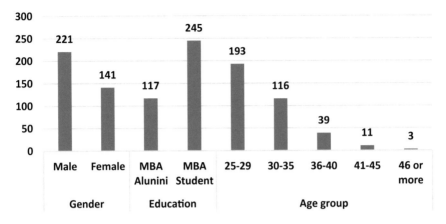

**Fig. 2** Respondents' demographics

## 3.2 Procedure

Palestine with a population of about 5.2 million of which 22% are between the age of 18–29 years old, has 3.65 million internet users with 3.1 million social media users. The social media platform has become a major channel for global marketing communication, especially for younger generation [25]. The "success" in the digital life is therefore of paramount importance to the young. Therefore, a university student sample is more suitable for the purpose of this study, as they constitute the majority of internet users. For this reason, the researchers approached the MBA students and alumni (who graduated in the past 3 years) of 2 major Palestinian universities via their emails and social media groups with a link to the questionnaire. The questionnaire was developed and converted to Google Forms to collect primary data needed to test the hypotheses. Respondents were asked whether they have previous online shopping experience or not in the first question. If the answer was "No", then the respondents were denied access to the rest of the survey questions. Therefore, the population of this study represents all the MBA alumni who graduated in the past three years and the currently enrolled MBA students at Birzeit and Alquds University. In total 1173 respondents were reached out by email with the link for the questionnaire. 362 responses were received and analyzed for this study.

## 3.3 Measures

A 5-point Likert scale questionnaire (5 = strongly agree, 4 = agree, 3 = no opinion/neutral, 2 = disagree, and 1 = strongly disagree) was developed as survey tool. It has five parts, the part 1 examines the perceived usefulness by using 4 items based on the work of Rahman, et al. [56], and Davis et al., [18]. Part 2 measures the perceived ease of use using 4 items and was adapted in reference to Chin & Goh [57], and Rahman, et al. [56]. Part 3 examines the perceived behavioral controls through adopting 4 items stemmed from Moshrefjavadi et al., [58]. Part 4 addresses the perceived trust by using 5 items adapted from Constantinides et al., [59]. Lastly, part 5 uses 3 items to assess the consumer purchase intention toward online shopping and was adapted from Thananuraksakul [60].

# 4   Results

## 4.1   Measurement Model Assessment

The data was analyzed using the PLS-SEM model fitting method. The evaluation of structural equation models was done in two steps [61]. The first is concerned with the validity and reliability of the model, whereas the second is concerned with the structural model. Convergent and discriminant validity tests were used to calculate the (Outer) model in terms of validity and reliability. Furthermore, construct reliability steps such as internal consistency and construct reliability were taken into account. To assess discriminating validity, the square root of Average Variance Extracted (AVE) was compared to the relationship between latent constructs. The Structural (Inner) model and hypotheses, on the other hand, were examined. To anticipate the relationships of the study model's hypotheses, the Coefficient of Determination ($R^2$) was computed.

The evaluation method in this study follows the criteria proposed by Nunnally and Bernstein [62], which suggests good reliability since the Cronbach alpha (used to measure construct reliability) values are greater than (0.7), as shown in Table 1. Similarly, the AVE values of all constructs are larger than the 0.5 threshold when using another assessment criterion suggested by Henseler et al. [63] to test convergent validity, proving the measuring instrument's convergence validity.

**Table 1**  Reliability of the constructs

| Construct | Composite reliability "CR" | Cronbach's alpha "CA" | Average variance extracted "AVE" |
|---|---|---|---|
| Perceived Usefulness | 0.872 | 0.743 | 0.721 |
| Perceived ease of use | 0.846 | 0.754 | 0.743 |
| Perceived behavioral control | 0.832 | 0.756 | 0.654 |
| Perceived trust | 0.826 | 0.776 | 0.653 |
| Intention toward Online shopping | 0.824 | 0.746 | 0.673 |

**Table 2** Discriminant validity of the constructs

|     | PU    | PEU   | PBC   | PT    | IOS   | ME1   | ME2   | ME3   |
|-----|-------|-------|-------|-------|-------|-------|-------|-------|
| PU  | 0.884 |       |       |       |       |       |       |       |
| PEU | 0.605 | 0.864 |       |       |       |       |       |       |
| PBC | 0.365 | 0.685 | 0.887 |       |       |       |       |       |
| PT  | 0.638 | 0.691 | 0.448 | 0.838 |       |       |       |       |
| IOS | 0.519 | 0.681 | 0.362 | 0.597 | 0.812 |       |       |       |
| ME1 | 0.321 | 0.076 | 0.149 | 0.128 | 0.135 | 1.000 |       |       |
| ME2 | 0.140 | 0.098 | 0.133 | 0.164 | 0.141 | 0.598 | 1.000 |       |
| ME3 | 0.053 | 0.035 | 0.128 | 0.035 | 0.053 | 0.239 | 0.462 | 1.000 |

*Note PU* Perceived usefulness; *PEU* Perceived ease of use; *PBC* Perceived behavioral control; *PT* Perceived trust; *IOS* The consumer's intentions towards online shopping; *ME1* Moderating Effect 1; *ME2* Moderating Effect 2; *ME3* Moderating Effect 3

The discriminant validity was assessed using the Fornell-Larker criterion [64]. As shown in Table 2, the criteria requires the AVE to be larger than the highest squared correlation among all other constructs for each latent construct.

## 4.2 Structural Model Assessment

The results of the structural model assessment that examined the relationship between the various variables in the recommended model, are presented in Table 3; for each hypothesized path, path coefficients were tested. The sign, magnitude, and meaning of the path coefficients are all taken into account. To assess the significance of each estimated path, bootstrapping using 5000 re-samples drawn with substitution was used. The coefficient of determination $R^2$ is used to assess the structural model's explanatory capacity and evaluate the model's ability to anticipate endogenous constructs.

The findings indicate that there is a positive relationship between the consumer's intentions towards online shopping and the perceived usefulness ($\beta = 0.121$, t = 2.148, p < 0.038); the perceived ease of use ($\beta = 0.158$, t = 2.154, p < 0.024); and the perceived behavioral control ($\beta = 0.162$, t = 3.748, p < 0.000). These findings also provide evidence to support the hypotheses H1, H2, and H3. The findings also support the moderating role of perceived trust in the relationships between perceived usefulness ($\beta = 0.126$, t = 2.162, p < 0.035), perceived ease of use ($\beta = 0.172$, t = 2.348, p < 0.000), and perceived behavioral control ($\beta = 0.124$, t = 2.075, p < 0.000), providing support to H4.

Finally, the fit model in PLS was assessed by employing the Stone-Geisser $Q^2$ (predictive relevance) [65] and the standardized root mean square residual (SRMR). $Q^2$ determines how well the model and its anticipated parameters match the observed values. A $Q^2$ value greater than 0 indicates predictive relevance. The composite model

**Table 3** Results of structural equation modeling

| Dependent variable: intentions towards online shopping | Model (1) | | Model (2) | |
|---|---|---|---|---|
| Path model (*n*, model fit indices) | Coef. | *t*-value | Coef. | *t*-value |
| (1) Base model (SRMR = 0.072, d_ULS = 0.862, d_G = 0.384, NFI, 0.908) | | | | |
| PU | 0.121 | 2.148* | 0.146 | 2.458* |
| PEU | 0.158 | 2.154* | 0.162 | 2.264* |
| PBC | 0.162 | 3.748*** | 0.178 | 3.974*** |
| (2) Perceived trust (SRMR = 0.079, d_ULS = 1.254, d_G = 0.537, NFI, 0.916) | | | | |
| PT | | | 0.136 | 3.045** |
| PU × PT | | | 0.126 | 2.162* |
| PEU × PT | | | 0.172 | 2.348** |
| PBC × PT | | | 0.124 | 2.075* |

*Note PU* Perceived usefulness; *PEU* Perceived ease of use; *PBC* Perceived behavioral control; *PT* Perceived trust; *IOS* The consumer's intentions towards online shopping; *ME1* Moderating Effect 1; *ME2* Moderating Effect 2; *ME3* Moderating Effect 3
* $p < 0.05$
** $p < 0.01$
*** $p < 0.001$

SRMR value of independent and dependent variables was 0.072, which is lower than Hu and Bentler's [66] recommended value of 0.08, indicating a strong model fit. The adjusted $R^2$ value was 0.527, this means that the independent factors are responsible for 52.7 percent of the variation in consumer intentions toward online shopping. In the existence of the moderating variable (perceived trust), the composite model SRMR value was 0.079, which is lower than Hu and Bentler's [66] recommended value of 0.08, revealing a good model fit. The adjusted $R^2$ value was 0.586, suggesting that the independent variables, when the moderating effect of perceived trust is taken into account, explain 58.6 percent of the variation in consumer intentions towards online shopping.

# 5   Discussion

## 5.1   Discussion of Findings

This research investigates the moderating role of perceived trust in the relationships between (perceived usefulness, perceived ease of use, and perceived behavioral control), and the consumer's intentions towards online shopping. The findings show that perceived usefulness has a favorable impact on the intention to shop online. Time savings, lower pricing, easier product comparison, and the removal of geographical restrictions are all advantages of online shopping. Customers' shopping intentions are also higher if they can see the benefits that shopping online may provide. This finding is comparable to that of earlier research [1, 42, 52, 57].

Furthermore, the findings show a favorable link between perceived ease of use and customer intentions to purchase online. In terms of content and functionality. A well-designed website has a substantial impact on online buying intention. According to Mansori, et al. [67], a website with rich content enhances the chance of an online purchase since it aids customers in their decision-making process by offering more information, therefore increasing their confidence to purchase on the website. This is also in line with prior research [31].

The findings also reveal that consumers' perceived behavioral control have a strong positive association with their intentions to purchase online. Customers' purchasing intentions improve when they believe they have all of the essential conditions for online shopping, and vice versa. This finding is consistent with prior studies [1, 20].

Finally, the findings suggest that perceived trust, as a moderating variable, strengthens the relationships between the independent variables (i.e., perceived usefulness, perceived ease of use, and perceived behavioral control) on the one hand, and the consumers' intentions toward online shopping on the other. In trade interactions, trust is a crucial component that is characterized by uncertainty and vulnerability [53]. Because the consumer's perception of transaction risk in an online environment is higher when the customer does not have direct connection with the seller or the products they want to buy, trust is especially essential in the online shopping context. As a result, one of the primary factors preventing customers from purchasing online is a lack of trust. As a result, once trust has been created, shopping intention will increase, and vice versa. This finding is comparable to that of Ha et al., [1], Rehman et al., [46], and Pavlou [41].

## 5.2 Theoretical and Practical Implications

Despite the widespread use of information technologies in a variety of industries, only a few studies have lately used the TAM and its expanded variants to assess technology adoption in Arab countries (e.g. [26, 27, 30]). As a result, the use of technological acceptance models at universities is still in its infancy, particularly in Arab countries. The majority of these studies are concerned with the essential elements that influence user acceptability in many industries, but not exclusively in universities. Furthermore, none of these researches looked at the elements that influence Palestinian consumers' online shopping intentions in light of COVID-19: The moderating impact of perceived trust.

The findings provide the online retailers (vendors) with helpful information about the elements that may influence the acceptability of e-commerce in the Arab countries (including Palestine). This study also looks at the inconsistencies in the results reported by prior studies. More specifically, this paper adds an empirical evidence from one of the developing countries to the open literature by defining an integrated model of the four factors "perceived usefulness, perceived ease of use, and perceived behavioral control" as independent variables and the dependent variable "the consumer's intentions towards online shopping" during the COVID-19 pandemic in Palestine, as well as examining the role of perceived trust as a moderating variable. As a result, retailers, website programmers and designers should pay attention to these variables. It is suggested that websites should be easy to use, interactive, and suitable for the users in order to assist them in understanding and finding what they are looking for. Additionally, the terminology and instructions for surfing the shopping platforms should be simple to comprehend [68, 69]. Finally, user-friendliness of e-commerce websites is explicitly necessary to enhance users' adoption.

## 5.3   Limitations and Further Research

The study's empirical findings have contextual and application limitations. The first limitation stems from the study's geographical area, which is Palestine. Even though the findings apply to similar economies, they may not be applicable to places that do not share the Palestinian national culture. Second, the study was limited to MBA students and alumni (those who graduated within the last three years), limiting the results' potential to be generalized to other population groups in Palestine.

In terms of future research, we recommend that scholars replicate the study in other environments or cultural context, utilizing additional moderating factors other than perceived trust, to better understand the linkages and processes of the issue at hand.

## 6   Conclusion

Despite the fact that numerous prior studies have looked into IT adoption and use, there has been little research done in the Arab World. In light of COVID-19, the aim of the study is to identify the elements that influence Palestinian consumer behavioral intentions toward online shopping and to investigate the moderating impact of perceived trust. To evaluate the hypotheses, a survey questionnaire was designed and utilized to gather primary data. A total of 1173 MBA alumni and current students from two prominent Palestinian universities received the questionnaire. The findings show that, for a variety of reasons detailed in this study, perceived usefulness, perceived ease of use, and perceived behavioral control influence consumers' intentions toward online shopping. Furthermore, the effects of (a) perceived usefulness, (b) perceived ease of use, and (c) perceived behavioral control on online shopping are strengthened by perceived trust. As a consequence, more qualitative and quantitative research are needed to develop a better understanding of the complexities of technology adoption in the Arab virtual business environment. Researchers need to create new scales for technology adoption in the Arab World that take cultural values into account. Future research in this area should probably focus on comparative studies between Arab and other business environments.

# References

1. Ha NT, Nguyen TLH, Pham TV, Nguyen THT (2021) Factors influencing online shopping intention: an empirical study in Vietnam. J Asian Financ Econ Bus 8(3):1257–1266
2. Ozen H, Engizek N (2014) Shopping online without thinking: being emotional or rational? Asia Pac J Mark Logist 26(1):78–93
3. Lester DH, Forman AM, Loyd D (2005) Internet shopping and buying behavior of college students. Serv Mark Q 27(2):123–138
4. Abbad M, Abbad R, Saleh M (2011) Limitations of e-commerce in developing countries: Jordan case. Educ Bus Soc Contemp Middle East Issues 4(4):280–291
5. Delafrooz N, Paim LH, Khatibi A (2011) a research modeling to understand online shopping intention. Aust J Basic Appl Sci 5(5):70–77
6. Choi J, Park J (2006) Multichannel retailing in Korea: Effects of shopping orientations and information seeking patterns on channel choice behavior. Int J Retail Distrib Manag 34(8):577–596
7. Prentice C, Chen J, Stantic B (2020) Timed intervention in COVID-19 and panic buying. J Retail Consum Serv 57:1–11
8. Donthu N, Gustafsson A (2020) Effects of COVID-19 on business and research. J Bus Res 117:284–289
9. Hashem N (2020) Examining the influence of COVID 19 pandemic in changing customers' orientation towards E-shopping. Mod Appl Sci 14(8):59–76
10. Pham VK, Nguyen TL, Do TTH, Tang MH, Thu Hoai HL (2020) A study on switching behavior toward online shopping of Vietnamese consumer during the Covid-19 time. Social Sciences and Humanities Open, Under Review, pp 1–27
11. Harrigan M, Feddema K, Wang S, Harrigan P, Diot E (2021) How trust leads to online purchase intention founded in perceived usefulness and peer communication. J Consum Behav 20(5)
12. Ventre I, Kolbe D (2020) The impact of perceived usefulness of online reviews, trust and perceived risk on online purchase intention in emerging markets: a Mexican perspective. J Int Consum Mark 32(4):287–299
13. Lui TK, Zainuldin MH, Yii KJ, Lau LS, Go YH (2021) Consumer adoption of Alipay in Malaysia: the mediation effect of perceived ease of use and perceived usefulness. Pertanika J Soc Sci Humanities 29(1):389–418
14. Wicaksono A, Maharani A (2020) the effect of perceived usefulness and perceived ease of use on the technology acceptance model to use online travel agency. J Bus Manag Rev 1(5):313–328
15. Rogus S, Guthrie JF, Niculescu M, Mancino L (2020) Online grocery shopping knowledge, attitudes, and behaviors among SNAP participants. J Nutr Educ Behav 52(5):539–545
16. Peña-García N, Gil-Saura I, Rodríguez-Orejuela A, Siqueira-Junior JR (2020) Purchase intention and purchase behavior online: a cross-cultural approach. Heliyon 6(6):e04284
17. Singh S, Srivastava S (2018) Moderating effect of product type on online shopping behaviour and purchase intention: an Indian perspective. Cogent Arts Humanit 5(1):1–27
18. Davis FD (1989) perceived usefulness, perceived ease of use, and user acceptance of information technology. MIS Q 13(3):319–340
19. Ha NT, Nguyen TLH, Nguyen TPL, Nguyen TD (2019) The effect of trust on consumers' online purchase intention: an integration of TAM and TPB. Manag Sci Lett 9(9):1451–1460
20. Lin HF (2007) Predicting consumer intentions to shop online: an empirical test of competing theories. Electron Commer Res Appl 6(4):433–442

21. Ajzen I (1991) The theory of planned behavior. Organ Behav Hum Decis Process 50(2):179–211
22. Slade EL, Dwivedi YK, Piercy NC, Williams MD (2015) Modeling consumers' adoption intentions of remote mobile payments in the United Kingdom: extending UTAUT with innovativeness, risk, and trust. Psychol Mark 32(8):860–873
23. Hanafizadeh P, Behboudi M, Koshksaray AA, Tabar MJS (2014) Mobile-banking adoption by Iranian bank clients. Telematics Inform 31(1):62–78
24. Malaquias RF, Hwang Y (2016) An empirical study on trust in mobile banking: a developing country perspective. Comput Hum Behav 54:453–461
25. Abeza G, O'Reilly N, Finch D, Séguin B, Nadeau J (2020) The role of social media in the co-creation of value in relationship marketing: a multi-domain study. J Strateg Mark 28(6):472–493
26. Dajani D, Yaseen SG (2016) The applicability of technology acceptance models in the Arab business setting. J Bus Retail Manag Res 10(3):46–56
27. Salem MZ, Baidoun S, Walsh G (2019) Factors affecting Palestinian customers' use of online banking services. Int J Bank Mark 37(2):426–451
28. Shanab EAA (2005) Internet banking and customers' acceptance in Jordan: the unified model's perspective. Southern Illinois University at Carbondale
29. Akour I, Alshare K, Miller D, Dwairi M (2006) An exploratory analysis of culture, perceived ease of use, perceived usefulness, and internet acceptance: the case of Jordan. J Internet Commer 5(3):83–108
30. Dajani D (2016) Using the unified theory of acceptance and use of technology to explain e-commerce acceptance by Jordanian travel agencies. J Comp Int Manag 19(1):99–118
31. Chong TY (2019) Factors affecting online shopping intention in Malaysia. Editorial Rev Board Members 51
32. Lim YJ, Osman A, Salahuddin SN, Romle AR, Abdullah S (2016) Factors influencing online shopping behavior: the mediating role of purchase intention. Procedia Econ Finance 35:401–410
33. Blackwell RD, Miniard PW, Engel JF (2001) Consumer behavior, 9th edn. Dryden, New York
34. Akbar W, Hassan S, Khurshid S, Niaz M, Rizwan M (2014) Antecedents affecting customer's purchase intentions towards green products. J Sociol Res 5(1):273–289
35. Sethuraman P, Thanigan J (2019) An empirical study on consumer attitude and intention towards online shopping. Int J Bus Innov Res 18(2):145–166
36. Musleh JSA, Marthandan G, Aziz N (2015) An extension of UTAUT model for Palestine e-commerce. Int J Electron Bus 12(1):95–114
37. Fishbein M, Ajzen I (1975) Belief, attitude intention, and behavior: an introduction to theory and research. Addison-Wesley, Reading, MA
38. Ajzen I (1985) From intentions to actions: a theory of planned behaviour. In Kuhl J, Beckman J (eds) Action control: from cognition to behavior. Springer, New York
39. Day G (1969) A two dimensional concept of brand loyalty. J Advert Res 93(3):29–35
40. Salisbury WD, Pearson RA, Pearson AW, Miller DW (2001) Perceived security and World Wide Web purchase intention. Ind Manag Data Syst 10(4):165–176
41. Pavlou PA (2003) Consumer acceptance of electronic commerce: integrating trust and risk with the technology acceptance model. Int J Electron Commer 7(3):101–134
42. Wen C, Prybutok VR, Xu C (2011) An integrated model for customer online repurchase intention. J Comput Inf Syst 52(1):14-23
43. Juniwati J (2014) Influence of perceived usefulness, ease of use, risk on attitude and intention to shop online. Eur J Bus Manag 6(27):218–228
44. Chiu C-M, Chang C-C, Cheng H-L, Fang Y-H (2009) Determinants of customer repurchase intention in online shopping. Online Inf Rev 33(4):761–784
45. Cha J (2011) Exploring the Internet as a unique shopping channel to sell both real and virtual items a comparison of factors affecting purchase intention and consumer characteristics. J Electron Commer Res 12(2):115–132
46. Rehman SU, Bhatti A, Mohamed R, Ayoup H (2019) The moderating role of trust and commitment between consumer purchase intention and online shopping behavior in the context of Pakistan. J Glob Entrep Res 9(1):1–25

47. Francis J, Eccles MP, Johnston M, Walker A, Grimshaw JM, Foy R, Kaner EF, Smith L, Bonetti D (2004) Constructing questionnaires based on the theory of planned behaviour: a manual for health services researchers. Centre for Health Services Research, University of Newcastle upon Tyne, United Kingdom

48. Gao L, Bai X (2014) A unified perspective on the factors influencing consumer acceptance of internet of things technology. Asia Pac J Mark Logist 26(2):211–231

49. Casaló LV, Flavián C, Guinalíu M (2010) Determinants of the intention to participate in firm-hosted online travel communities and effects on consumer behavioral intentions. Tour Manage 31(6):898–911

50. Baron RM, Kenny D (1986) The moderator–mediator variable distinction in social the moderator-mediator variable distinction in social psychological research: conceptual, strategic, and statistical considerations. J Pers Soc Psychol 51(6):1173–1182

51. Ganguly B, Dash SB, Cyr D, Head M (2010) The effects of website design on purchase intention in online shopping: the mediating role of trust and the moderating role of culture. Int J Electron Bus 8(4/5):302–330

52. Gefen D, Karahanna E, Straub DW (2003) Inexperience and experience with online stores: the importance of TAM and trust. IEEE Trans Eng Manag 50(3):307–321

53. McKnight DH, Choudhury V, Kacmar C (2002) Developing and validating trust measures for e-commerce: an integrative typology. Inf Syst Res 13(3):334–359

54. Lee MK, Turban E (2001) A trust model for consumer internet shopping. Int J Electron Commer 6(1):75–91

55. Verhagen T, Meents S, Tan YH (2006) Perceived risk and trust associated with purchasing at electronic marketplaces. Eur J Inf Syst 15(6):542–555

56. Rahman MK, Jalil MA, Mamun A-A, Robel S (2014) Factors influencing Malaysian consumers' intention towards E-shopping. J Appl Sci 14(18):2119–2128

57. Chin S-L, Goh Y-N (2017) Consumer purchase intention toward online grocery shopping: view from Malaysia. Glob Bus Manag Res 9(4):221–238

58. Moshrefjavadi MH, Dolatabadi HR, Nourbakhsh M, Poursaeedi A, Asadollahi A (2012) An analysis of factors affecting on online shopping behavior of consumers. Int J Mark Stud 4(5):81

59. Constantinides E, Lorenzo-Romero C, Gómez MA (2010) Effects of web experience on consumer choice: a multicultural approach. Internet Res 20(2):188–209

60. Thananuraksakul S (2007) Factors influencing online shopping behavior intention: a study of Thai consumers. AU J Manag 5(1):41–46

61. Smid SC, Rosseel Y (2020) SEM with small samples: two-step modeling and factor score regression versus Bayesian estimation with informative priors. In: Small sample size solutions. Routledge, pp 239–254

62. Nunnally J, Bernstein I (1994) Psychometric theory, 3rd edn. McGraw-Hill, New York, NY

63. Henseler J, Ringle CM, Sinkovics RR (2009) The use of partial least squares path modeling in international marketing. Adv Int Mark 20(1):277–319

64. Fornell C, Larcker D (1981) Evaluating structural equation models with unobservable variables and measurement error. J Mark Res 18(1):39–50

65. Geisser S (1975) The predictive sample reuse method with applications. J Am Stat Assoc 70(350):320–328

66. Hu L, Bentler P (1999) Cutoff criteria for fit indexes in covariance structure analysis: conventional criteria versus new alternatives. Struct Equ Modeling 6(1):1–55

67. Mansori S, Cheng BL, Lee HS (2012) A study Of E-shopping intention in Malaysia: the influence of generation X & Y. Aust J Basic Appl Sci 6(8):28–35

68. Im I, Kim Y, Han HJ (2008) The effects of perceived risk and technology type on users' acceptance of technologies. Inf Manag 45(1):1–9

69. Huh HJ, Kim TT, Law R (2009) A comparison of competing theoretical models for understanding acceptance behavior of information systems in upscale hotels. Int J Hosp Manag 28(1):121–134

# Study Concerning the Taxpayers' Perception on the Phenomenon of Tax Evasion from Romania

Maria Grosu, Nicoleta Ionașcu, and Camelia Cătălina Mihalciuc

**Abstract** Tax evasion is an acute global problem, being a complex economic and social phenomenon. In countries with emerging economies, such as Romania, tax evasion is a major problem that slows down the achievement of economic, social and political goals. Eradicating tax evasion is difficult to achieve in practice, with efforts focusing mainly on limiting it. The purpose of this study aims, first of all, to identify the factors that most influence the perception of Romanian taxpayers to tax evasion and to determine the general level of tolerance of citizens working in various fields towards tax evasion. Secondly, the research conducted highlights the taxpayers' perception about the causes of tax evasion and what would be the best methods to limit this phenomenon, from the perspective of taxpayers in Romania. The research results highlight the fact that Romanian taxpayers do not tolerate evasionist behavior, and this aspect has positive connotations for society.

**Keywords** Tax evasion · Taxpayers · Fiscal pressure · Relevance of accounting information · Causes · Combating methods

## 1 Introduction

Tax evasion can be characterized as a complex and extremely important economic and social phenomenon that states face today, regardless of their tax system and level of economic development. In practice, it is only tried to limit tax evasion, because

M. Grosu · N. Ionașcu
Faculty of Economics and Business Administration, "Alexandru Ioan Cuza" University of Iasi, Iasi, Romania
e-mail: maria.grosu@uaic.ro

N. Ionașcu
e-mail: ionascu.nicoleta@feaa.uaic.ro

C. C. Mihalciuc (✉)
Faculty of Economics and Public Administration, "Stefan cel Mare" University of Suceava, Suceava, Romania
e-mail: camelia.mihalciuc@usm.ro

© The Author(s), under exclusive license to Springer Nature Switzerland AG 2022
S. G. Yaseen (ed.), *Digital Economy, Business Analytics, and Big Data Analytics Applications*, Studies in Computational Intelligence 1010,
https://doi.org/10.1007/978-3-031-05258-3_53
689

the eradication of this phenomenon is difficult to achieve. The analyzed topic is also important and topical, given the negative effects of the phenomenon of tax evasion on the state budget, as well as on society as a whole. Specifically, it reduces the investment needed to be made in important sectors such as health, transport, education. The main objectives of this study are: identifying the economists taxpayers' perception on fiscal policy and the existence of legislative gaps, identifying the relevance of accounting information to prevent and combat tax evasion, identifying factors generators of tax evasion, the incentives for non-compliance tax, and to identify methods to which can be used to commit tax evasion; identifying the perception of other taxpayers on evasive behavior and their willingness to comply with tax law, as well as identifying the factors that influence their tax compliance; identifying the causes of tax evasion and prevention and control measures from the perspective of Romanian taxpayers.

## 2   Literature Review and Formulating Research Hypotheses

The spirit of tax evasion is born of the simple game of interest. Some who would never have the slightest idea of their neighbor's property will evade their tax debts without hesitation [1]. The modernization of economic life has led to a rethinking of crime and its adaptation to everyday life. Classical crimes have remained a thing of the past, now shaping a much more complex economic crime, here finding its place and tax evasion [2]. The extension of tax evasion to all types of companies and to all social classes determines its universality, and the age of the phenomenon is linked to the emergence of taxes and fees [3]. Some authors argue that taxation also played an important role in the development of writing, and the mandatory nature of taxation undoubtedly creates a great deal of public hostility [4].

From ancient times, in ancient Rome, rich Roman citizens made false declarations to avoid wealth taxes, and these actions had detrimental consequences on the economy, accentuating even the subsequent social crises [5]. Over time, the meaning of the phrase "tax evasion" has changed, and is now frequently associated with wealthy individuals hiding offshore assets and corporations who deliberately manipulate their businesses to reduce their tax liability [6]. From the taxpayer's perspective, taxation is an opportunity. Therefore, studies on taxation usually aim to determine the most appropriate methods of limiting its negative effects on the taxpayer and on his tax behavior [7]. The definition of the concept of tax evasion in Romania, by referring to concepts related to this phenomenon, such as tax fraud, tax avoidance, abuse of rights, informal economy, along with the presentation of ways of manifestation, causes and methods to combat it has been done in different studies [8–17]. There are many explanations for the expression: tax evasion, but, succinctly, it can be considered that tax evasion consists in the intention of a person to evade the payment of taxes, based on gaps and other weaknesses of the law in order to hide reality [18]. When tax evasion is illegal, flagrantly violating the law, by evading taxes, we are dealing with tax fraud [19]. Tax evasion could be eradicated only in the utopian hypothesis that a state could exist and function efficiently without taxes and duties. This state seems impossible and we get confirmation in this regard from Benjamin Franklin, who emphasized in his letter to Jean-Baptiste Leroy that "nothing in this world is safer than death and taxes." [20].

At European level, there is a continuing concern for the identification of *methods to prevent and combat the phenomenon of evasion*, due to the fact that it has become increasingly widespread. Paradoxically, element that favored the increase is one of the basic principles of the European Union, namely the free movement of goods, services and capital, including human [21]. As a multifaceted problem, tax evasion requires the establishment of appropriate control measures, which are the responsibility of each EU Member State. Member of the European Union, Romania has one of the lowest levels of institutional trust, along with one of the highest levels of perception of corruption, which can increase the tax evasion of citizens [22]. The level of tax evasion is considered to be proportional to the level of corruption [3]. Tax evasion can also have detrimental consequences on the purchasing power of the national currency and even on the stability of the national economy [23].

In the specialized literature there are several forms that tax evasion can take: traditional, legal, accounting, by evaluation, national, international, artisanal, industrial, long-term, short-term, committed by a natural or legal person [24, 25]. The most used *methods of achieving legal tax evasion* are: *investing a part of the profit in the acquisition of machines and technical equipment for which the state regulates reductions of the profit tax; involvement in philanthropic actions; deduction from taxable income of advertising or protocol expenses* [3]. Fraudulent tax evasion is more common than legal evasion and is done in violation of legal provisions. This *can manifest itself in many forms: transfer of unreal figures in the accounting registers; drawing up false declarations and fictitious payment documents; reducing turnover by not recognizing income; reporting errors; non-declaration of taxable matter; keeping double-entry records; sales without supporting documents; falsification of the balance sheet* [26]. In practice, achieving a boundary between licit and the illicit tax evasion is difficult, because between them there is a natural continuation—taxpayer efforts to exploit legislative gaps lead him often to fraud [27]. *The factors that generate the phenomenon of tax evasion*, both in its legal form and in its illicit form can be: *psycho-social, legislative and administrative and economic* [23]. The trend towards tax evasion is found in any individual whose income is taken away by someone [28]. Specialists wrote 80 years ago that the education of the Romanian taxpayer is so poorly trained that he has only one concern: to pay as little or not at all [29]. The needs of the state can create a high fiscal pressure on taxpayers. Under these conditions, in order to survive, taxpayers are forced to look for ways to avoid paying taxes [30].

In order to combat tax evasion, it is necessary to build a unitary and effective strategy, which requires, first of all, the identification *of the real causes of tax evasion*, followed by the elaboration and application *of coherent measures to combat it. The main causes of tax evasion* are: *fiscal pressure, imperfections in tax legislation, inadequate fiscal control and inadequate tax education of taxpayers* [31]. *The fiscal pressure* at national level represents the ratio between the revenues collected from taxes, fees and social contributions and the gross domestic product [32]. *Fiscal control* is the central element in the management of any socio-economic system and a direct way to prevent tax evasion [33]. The inefficiency of the Romanian tax system is often determined by political involvement in public administration [34]. Regarding the imperfections of the fiscal legislation, it was highlighted the fact that the Romanian fiscal legislative system is incomplete, it presents great gaps, inaccuracies and even ambiguities, making it possible evasion have a wide margin of maneuver in trying to evade the payment of tax obligations [35]. According to some authors [36], tax evasion is also caused by the fact that the tax authorities do not understand the reasons behind taxpayers' beliefs that they can evade the fulfillment of tax obligations. In addition, the authorities need to highlight the interdependence between fiscal and budgetary policy so that taxpayers understand how the money is spent [37]. Some taxpayers are trying to get rid of the national tax regime, looking for more favorable regimes in other parts of the world. The phenomenon of international tax evasion is stimulated by the presence of tax-free zones, considered tax havens. Currently, there are over 40 regions in the world considered tax havens [38].Revenues attracted to the state budget ensure an economic balance at the national level, because they become the instrument through which the state intervenes in the economy and implements economic adjustment measures [13]. Taking into account the harmful effects of tax evasion, preventing and combating tax evasion must be a priority for Romania as well. The first law, which aimed to monitor the realization of public revenues of the state, dates from March 21, 1877 and stated that it was subject to prosecution, the taxpayer who did not fulfill his duty until the 15th day of the second month of each quarter [39]. Later (1933), tax evasion offenses were fined four times the tax difference found to have been stolen [40]. After 2013, the possibility of applying the fine was eliminated from the legislation, the prison sentence being the only sanction for the acts of tax evasion, a fact also criticized in the specialized works [41]. At the level of companies, accounting information provides the necessary support to identify the evasion phenomenon. Accounting information contributing to the detection and quantification of production tax evasion because: *ensures access to the possibility of identifying tax evasion, ensures the direct link with the supporting documents attesting the production or intention to produce tax evasion, ensures the verification of compliance with accounting principles, ensures the quantification of tax evasion produced* [8]. For most countries in the world, preventing and combating tax evasion is a priority, and in order to succeed in limiting the spread of this phenomenon, it is necessary to apply methods, among which we mention: *increasing the penalties, changing the taxpayer's mentality, tightening the conditions for setting up companies, improving the fiscal apparatus, relaxing the fiscal policy, elaborating complete and clear normative acts* [3]. According to the literature, tax evasion decreases at the same rate as the probability of detecting cases of tax evasion increases and increases the sanctions against such acts [42]. Some authors believe that the increase in penalties should be large enough to eliminate criminal behavior [31]. The degradation of fiscal discipline must be combated. The methods can be original or can be borrowed from other economies of the world. For example, in the United States, the authorities are combating tax evasion by using as methods: increasing penalties and large-scale controls, on the one hand, or offering tax amnesty, on the other [43]. Through the tax amnesty, taxpayers are exempted from paying penalties and/or interest on debts and, most importantly, are not penalized for their actions [44]. Other authors consider that other ways to prevent tax evasion can be: tightening the conditions for setting up and authorizing economic entities, and the authorization will be made exclusively for those legal entities that prove that they have no debts to the state budget [45] or exemptions or reductions from payment of taxes and fees for a period of time; in China, for example, a regulation has been adopted stating that legal entities do not pay taxes in the first five years of economic activity [46]. Internationally, a series of articles have been published that have made a visible delimitation between tax evasion and terms close to this concept, presenting by theoretical justifications ways to estimate the volume of tax evasion and its determinants, as well as methodological aspects related to measuring tax evasion and combating it [47–56]. Numerous scientific studies have resorted to the conceptualization of econometric models, tax evasion being either a dependent variable or an independent variable, following the link between tax evasion and corporate debt and other determinants factors of it [57–60]. Tax evasion regarding the profit tax is one of the topics treated in the literature, the authors of the studies trying to show the relationship between tax avoidance and company performance [61–65].

Considering the results identified at the level of the consulted specialized literature, in the present study the following research hypotheses are proposed for testing and validation: (1) *From the economists point of view, the legislative factors, the fiscal pressure, but also the fiscal facilities favor the tax evasion;* (2) *The relevance of accounting information helps to prevent and combat tax evasion, from the economists point of view;* (3) *The income obtained by other taxpayers differentiates the factors regarding their fiscal compliance: civic debt and personal integrity;* (4) *The imperfections of the legislation and the dissatisfaction of the Romanian taxpayers require the elaboration of complete and clear normative acts.*

## 3 Research Methodology: Population, Sample, Variables, Data Source, Data Analysis Methods

In order to identify the perception of Romanian taxpayers regarding: *factors that favor tax evasion, the role of accounting information in preventing and combating tax evasion, factors related to tax compliance, causes and measures to prevent and combat tax evasion,* the survey method was used based on a questionnaire, as *a method of data collection.* The questionnaire was structured in two sections, one for *economists* and one for *other taxpayers.* This option has been used, as economists have in-depth knowledge of taxation and can answer more accurately the more difficult questions characteristic of the tax field. The questionnaire was distributed online, using a survey management software—Google Forms. The questions asked in the questionnaire were mostly quantitative, with closed questions using the 5-grade Likert scale and qualitative (open-ended and multiple-choice questions).

The questions addressed to the respondents were divided into three categories: common, addressed only to economists and addressed to other taxpayers. *The common questions* considered demographic and educational and occupational data, as well as issues related to the causes of tax evasion and measures to prevent and combat it. *The questions addressed only to economists* focused on their perception of fiscal policy and the existence of legislative gaps, the relevance of accounting information to prevent and combat tax evasion, the factors generating tax evasion, the factors that promote tax non-compliance, and the methods that can be used to commits tax evasion. *Questions to other taxpayers* focused on their perceptions of evasive behavior and their willingness to comply with tax law, as well as the factors that influence taxpayers' tax compliance. The questions asked also represent *the variables* for which the data were collected following the answers received. There were 71 respondents among the taxpayers. In order to determine the sample size, an analysis of *the sample* size from similar studies, both national and international, was carried out and it was established that an appropriate sample should be between 60–250 participants. *The data analysis methods* take into account the quantitative analysis, systematization, comparison, but also the use of the multivariate data analysis method [66], respectively the Factorial Analysis of Multiple Correspondences (FAMC), and the software used is SPSS 23.0.

## 4   Results and Discussions

In the first section of the questionnaire, for common questions, all respondents had to indicate demographic data and related to their education and profession, such as *gender, age, level of education and occupation*. The results of the descriptive statistics can be found in Table 1.

From Table 1 it can be seen that of the 71 taxpayers who participated in the survey, 45.1% are economists and 54.9% have other professions. The majority of respondents were women (83.1%), aged 18–34 (56.3%).

To test Hypothesis 1: *From the point of view of economists, the legislative factors, the fiscal pressure, but also the fiscal facilities favor the tax evasion*, the identified variables and the data processing are presented in Table 2.

According to the processed data, reflected in Table 2, it can be seen that the main factors that generate tax evasion are *legislative and administrative factors* (50%). Regarding *the economists' perception of the factors that favor fiscal non-compliance*, respondents were asked to express their opinion using a scale with options between 1 (To a very small extent) and 5 (To a very large extent), and the average value of the results shows that the fiscal pressure is the main factor (average value-4.03). Regarding *the existence of gaps in financial-accounting legislation*, the majority, 81.3%, answered in the affirmative. In order to determine the frequency of tax evasion in practice, respondents working in the economic field were asked to rate on a scale

**Table 1** Descriptive statistics on the common variables analyzed

| Variable | Number | Percent |
|---|---|---|
| (I) **Gender** | **71** | **100** |
| Female | 59 | 83.1 |
| Male | 12 | 16.9 |
| (II) **Level of education** | **71** | **100** |
| Undergraduated | 35 | 49.3 |
| Master degree | 36 | 50.7 |
| (III) **Age** | **71** | **100** |
| 18–24 years | 40 | 56.3 |
| 25–34 years | 16 | 22.5 |
| 35–49 years | 9 | 12.7 |
| 50 + | 6 | 8.5 |
| (IV).**Profession** | **71** | **100** |
| Economist | 32 | 45.1 |
| Other professions | 39 | 54.9 |

*Source* Own processing in SPSS 23.0

between options 1 (Never) and 5 (Very often), how often *methods of evasion are used from the payment of taxes and fees.* It is observed that the most common are *the legal methods* of tax evasion (average value 3.63), as well as the decrease in taxable income of advertising or protocol expenses (average value 3.59). *Illegal ways* of evasion are considered to be less common: non-declaration of taxable matter with an average of 3.28, preparation of false statements (average values: 2.97 and 2.72).

To test Hypothesis 2: *The relevance of accounting information contributes to preventing and combating tax evasion, from the point of view of economists,* the variables identified and data processing are presented in Table 3.

Regarding *the role of accounting information in preventing and combating tax evasion,* the answers given show that the accounting information is relevant for the activity of preventing and combating tax evasion, registering an average value of 4.63. At the same time, it can be seen that the economists participating in the survey admit that the authorities are insufficiently involved in the activity of preventing and combating tax evasion (average value 3.41), and the methods, techniques and tools to combat and prevent this phenomenon are not fully adapted to the economic reality (average value 3.44). In addition, it is found that tax evasion contributes to the increase of inequality between taxpayers, the average value for this statement being high, of 4.38.

To test Hypothesis 3: *The income obtained by other taxpayers differentiates the factors regarding their fiscal compliance: civic debt and personal integrity,* identified variables and data processing are presented in Table 4.

The questions addressed to other taxpayers whose answers are summarized in Table 4 are questions of tax ethics designed to highlight *the attitude of taxpayers*

**Table 2** Variables for Hypothesis 1, description, occurrences (%) or average

| Variable | Description | Frequency (%) | Mean |
|---|---|---|---|
| *Economists' perception of the factors generating tax evasion* | | | |
| Psy-Soc_Fact | Psycho-social factors | 6.2% | – |
| Econ_Fact | Economic factors | 43.8% | – |
| Legis_and_Adm_Fact | Legislative and administrative factors | 50% | – |
| *Economists' perception of the factors that favor fiscal non-compliance* | | | |
| Avoid_Tax | The desire to avoid tax burdens | – | 3.66 |
| Distr_State | Distrust of the state and the purpose of the tax system | – | 3.66 |
| Limited_Resources | Limited financial and material resources | – | 3.66 |
| Unstable_Frame | Unstable and ambiguous legislative framework | – | 3.75 |
| Unfair_Comp | Unfair competition from evasionist firms | – | 3.53 |
| Comp_Acc | Complex accounting and tax legislation | – | 3.44 |
| Exist_Under | The existence of the underground economy | – | 3.72 |
| Fiscal_Press | Fiscal pressure | – | 4.03 |
| Gaps_Fin_Acc | Are there gaps in financial-accounting legislation that would encourage economic entities to commit tax evasion | Yes 81.3% | Not 18.7% |
| *Methods of tax evasion* | | | |
| Tax_Facilities | Tax facilities in case of reinvestment of profit in equipment | – | 3.63 |
| Phil_Act | Philanthropic actions | – | 3.09 |
| Ded_Prot_Exe | Deduction from taxable income of protocol expenses | – | 3.59 |
| Offshore_Comp | Establishment of offshore companies | – | 3.03 |
| Rec_Depr | Recording of an overvalued depreciation from an economic point of view | – | 3.25 |
| Unr_Amounts | Unrealistic amounts in the accounting records | – | 3.16 |
| False_State | Making false statements | – | 2.97 |
| Fict_Pay | Preparation of fictitious payment documents | – | 3.06 |
| Non_Decl_Tax | Non-declaration of taxable matter | – | 3.28 |
| Double_Acc | Double accounting records | – | 2.91 |
| Sales_Without_Doc | Sales without supporting documents | – | 3.47 |
| Fals_Bal_Sheet | Falsification of the balance sheet | – | 2.72 |

*Source* Own processing in SPSS 23.0

**Table 3** Variables for Hypothesis 2, description, occurrences (%) or average

| Variable | Description | Frequency (%) | Mean |
|---|---|---|---|
| *Acc_Inf_Relev* | The accounting information is relevant for the activity of preventing and combating tax evasion | – | 4.63 |
| *Auth_Suff_Involved* | The state authorities are sufficiently involved in the activity of preventing and combating tax evasion | – | 3.41 |
| *Meth_Tech* | The methods, techniques and tools to combat and prevent tax evasion are adapted to reality | – | 3.44 |
| *Prof_Acc* | Professional accountants and financial auditors have an important role to play in preventing and combating evasion | – | 4.28 |
| *Ineq_People* | Tax evasion contributes to increasing inequality between people | – | 4.38 |
| *Possib_Ident_Tax_Ev* | Offers the possibility to identify tax evasion | 28.1% | – |
| *Ens_Verif_Acc_Princ* | Ensures the verification of the observance of the accounting principles | 12.5% | – |
| *Ens_Doc_Tax_Ev* | Ensures the direct connection with the primary documents attesting the tax evasion | 50% | – |
| *Ens_Warn_tax_Ev* | Ensures early warning about the creation of conditions for future tax evasion | 9.4% | – |

*Source* Own processing in SPSS 23.0

*who practice other professions towards tax evasion and whether they condemn this fact or not.* In addition, from the answers provided, it can be deduced the extent to which *respondents are willing to comply with tax rules and whether they would resort to tax evasion in practice.* Respondents were asked to express their agreement or disagreement on 15 statements by selecting an option on a Likert scale between 1 (Total Disagree) and 5 (Total Agreement) for each answer. Extremely low average values of respondents' consent received the first 6 statements, which indicates that evasionist behavior is not tolerated. However, it can be seen that the respondents agreed to exploit the legislative gaps (average value-2.18). Statements 7–12 and 15 have moderate average values (ranging from 2.33 to 3.74), which reflects that, although many condemn the tax evasion action and consider that it is the civic duty of each to be subject to tax obligations (average-4.21), most would not report tax evasion. Regarding the involvement of the state in preventing and combating tax evasion, most respondents believe that the state does not do everything possible in this regard and offers unequal opportunities for different categories of taxpayers to reduce their tax burden (average value—3.74). However, most respondents perceive tax evasion as a reprehensible act, supported by the high average level (3.95) of the statement: all evaders should be punished. Regarding *the factors that exert a considerable influence on the taxpayers' behavior in terms of tax compliance,* the

**Table 4**  Variables for Hypothesis 3, description, frequency (%) and average

| Variable | Description | Frequency (%) | |
|---|---|---|---|
| *Month_Income* | Monthly income | < 460 Euro | 38.5% |
| | | 460–600 Euro | 35.9% |
| | | > 600 Euro | 25.6% |
| *Taxpayers' perception of evasionist behavior and their willingness to comply with tax law* | | Mean | |
| *Not_Decl_Income* | It is not necessary to declare all income | 2.08 | |
| *Not_Mand_Compl* | In tax matters, it is not mandatory to comply with all laws | 1.46 | |
| *Not_Get_Caught* | The most important thing is not to get caught | 1.51 | |
| *Legis_expl* | Legislative loopholes are allowed to be exploited | 2.18 | |
| *Tax_Ev_Not_Crime* | Tax evasion is not a serious crime | 1.49 | |
| *Inf_Ill_Act* | Information about illegal activities should not be disclosed | 2.10 | |
| *Eng_Tax_Ev* | A person should report anyone engaging in tax evasion | 2.79 | |
| *Non_Decl_Inc* | A person should be fined, but not punished with imprisonment for total non-declaration of income | 2.64 | |
| *Midlle_Income* | Middle-income taxpayers should not be subject to all tax obligations | 2.44 | |
| *Legis_looph* | A person who has not paid any income tax cannot be charged if he has made use of legislative loopholes | 2.44 | |
| *State_Opport* | The state offers unequal opportunities for different categories of taxpayers to reduce their tax burden | 3.74 | |
| *Prevent_Combat* | The state is doing everything possible to prevent and combat tax evasion | 2.46 | |
| *Civic_duty* | It is the civic duty of every citizen to pay their taxes and duties | 4.21 | |
| *Evas_Acc* | All evasionists should be held accountable | 3.95 | |
| *Fair_Amount* | Taxpayers should only pay what they consider to be a fair amount | 2.33 | |

(continued)

**Table 4** (continued)

| Variable | Description | Frequency (%) |
|---|---|---|
| *Factors influencing taxpayers' tax compliance* | Mean | |
| *Pers_Integr* | Personal integrity | 2.85 |
| *Third_Parties* | Third parties who report your income to ANAF | 2.59 |
| *Audit* | Fear of an audit | 2.51 |
| Report_Sincerely | The belief that your friends and associates report sincerely and pay all obligations | 2.36 |

*Source* Own processing in SPSS 23.0

respondents had to express their opinion, using a scale of 4 items: from 1 (No influence) to 4 (High influence), and the result was that personal integrity has a moderate and high influence (average value-2.85) in terms of tax compliance.

Given the results of the processing-*civic spirit and personal integrity are factors that influence the tax compliance of taxpayers*-and taking into account the income obtained by other taxpayers (without economists), according to the data in Table 4, it can be established an association between the three variables using the multivariate data analysis method, FAMC. Figure 1 shows the results of statistical processing.

From Fig. 1, it can be seen that taxpayers with lower monthly incomes (<460 Euro) do not express an opinion about civic spirit and personal integrity, as factors

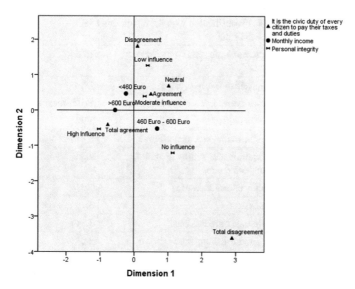

**Fig. 1** The association between civic spirit, personal integrity and the income of other taxpayers. *Source* Own processing in SPSS 23.0, using FAMC

that influence compliance or not with regard to tax evasion. Taxpayers with average incomes taken into account in the analyzed sample (460–600 Euro) consider that personal integrity would not have any influence on tax evasion. Instead, taxpayers with higher incomes (>600 Euro) consider that personal integrity has a high influence on the tax compliance of citizens and agree that it is the duty of each taxpayer to pay their taxes. In order to test Hypothesis 4: *Imperfections of legislation and dissatisfaction of Romanian taxpayers requires the development of complete and clear regulations*, the identified variables and data processing are presented in Table 5.

Regarding the causes of tax evasion, it can be seen from Table 5 that both economists and other taxpayers considered, in proportion of over 50%, that the imperfections of tax legislation and taxpayers' dissatisfaction with the conditions offered by the state cause tax evasion. For the effectiveness of measures to combat tax evasion, respondents had to express their opinion using a scale between 1 (Not effective) and 5 (Extremely effective), and the opinion of respondents often coincides, placing in the foreground the elaboration complete and clear normative acts to

**Table 5** Variables for Hypothesis 4, description, occurrences (%) or average

| Variable | Description | | |
|---|---|---|---|
| | *Causes of tax evasion* | Econ (%) | Other (%) |
| Fiscal_Press | Fiscal pressure | 43.8 | 28.2 |
| Ineff_Fisc_Ctrl | Inefficiency of fiscal control | 46.9 | 46.2 |
| Imperf_Tax_Legis | Imperfections in tax legislation | 65.6 | 67.7 |
| Insuff_Fisc_Ed | Insufficiency of fiscal education | 37.5 | 51.3 |
| Tax_Diss | Taxpayers' dissatisfaction with the conditions offered by the state | 56.3 | 66.7 |
| Emerg_Econ | Romania's economic status—emerging economy | 15.6 | 10.3 |
| Prom_Selfi | Promoting selfishness as a value | 18.8 | 30.8 |
| *Assessing the effectiveness of measures to combat tax evasion* | | Econ (mean) | Other (mean) |
| Incr_Penalt | Increasing penalties | 3.22 | 3.10 |
| Chang_Ment | Changing the taxpayer's mentality | 3.81 | 3.38 |
| Sever_Cond_Auth | Severe conditions for setting up and authorizing companies | 2.84 | 2.82 |
| Fiscal_Pol | Fiscal policy easing measures | 2.87 | 3.00 |
| Fiscal_Amn | Fiscal amnesty | 3.19 | 3.10 |
| Tax_Exem | Tax exemptions or reductions | 3.81 | 3.31 |
| Improv_Fisc | Improving the fiscal apparatus | 3.94 | 3.72 |
| Adm_Coop | Administrative cooperation | 3.78 | 3.87 |
| Elab_Norm | Elaboration of complete and clear normative acts | 4.22 | 4.15 |

*Source* Own processing in SPSS 23.0

remove the possibility for some taxpayers to use their ingenuity and find loopholes in the tax legislation. High average values are also recorded for the improvement of the fiscal apparatus (3.94-Economists, 3.72-Other taxpayers), administrative cooperation (3.78-Economists, 3.87-Other taxpayers), changing the mentality of taxpayers, but also granting some exemptions from taxes and fees, from the point of view of economists (average value-3.81). As demonstrated in other cited studies, most respondents do not tolerate evasionist behavior, and this has positive connotations for society.

# 5 Conclusions

The modernization of economic life has led to a reinvention of crime and its adaptation to everyday life. A phenomenon that brings great damage to a state budget is tax evasion. The research conducted highlights the perception of Romanian taxpayers, economists and other professions, on the phenomenon of tax evasion. It was found that both categories of taxpayers admit that tax evasion contributes to increasing inequality between taxpayers. Most respondents condemn evasive behavior and some respondents would tolerate tax evasion, but only in its legal form. Although it is considered that it is the civic duty of every citizen to be subject to tax obligations, a large part of the respondents consider that they are not obliged to denounce tax evaders. More than half of the respondents showed that personal integrity has a moderate and high influence when it comes to tax compliance. Most respondents comply with tax laws and would not resort to evasion to ease their tax burden. In the opinion of economists, the methods of tax evasion most commonly encountered in practice are the legal ways of tax evasion. Regarding the main causes and factors that lead to tax evasion, it is found that both categories of taxpayers participating in the study are convinced of the existence of gaps in financial-accounting legislation, and the factors that affect the tax compliance behavior of taxpayers are, first and foremost, legislative and administrative. According to the taxpayers surveyed, tax evasion is caused primarily by the imperfections of tax legislation and taxpayers' dissatisfaction with the conditions offered by the state, and the main factor that determines taxpayers to commit tax evasion is tax pressure. In terms of the accounting information, it was found that it is relevant in the process of preventing and combating tax evasion. Regarding the measures to combat tax evasion, the most effective are, in the opinion of all respondents, the elaboration of complete and clear normative acts, then the administrative cooperation and the improvement of the fiscal apparatus. In summary, this study was aimed to identify the general level of tolerance of citizens working in various fields towards tax evasion. It is considered that the objectives have been achieved, and the results are satisfactory: most respondents do not tolerate evasionist behavior, although many have expressed distrust in the effectiveness of fiscal policy and are discouraged by corruption in Romania. The small sample size is the main limitation of the study. In addition, most respondents are between 18 and 34 years old. The study was attended only by citizens with higher education, therefore it is necessary to extend the sample in case of further research. In conclusion, although, apparently, tax evasion only affects the state budget, we must not forget that the harmful effects of this phenomenon are felt by all citizens, both evasionists and honest ones. Evasionist behavior should not be encouraged in any way. The more it expands, the greater the social inequity and the distortion of market mechanisms. Therefore, in order to limit this phenomenon as much as possible, a relationship of cooperation and mutual respect must be established between the state and the taxpayers, because the state institutions are the mirror of society and the state is the citizens.

# References

1. Şaguna DD (2003) Treaty of financial and fiscal law/Tratat de drept financiar și fiscal, All Beck, Bucharest
2. Jurj-Tudoran R, Şaguna DD (2018) Money laundering. Judicial theory and practice/Spălarea banilor. Teorie și practică judiciară, C.H. Beck, Bucharest
3. Aniței N-C, Lazăr RE (2016) Tax evasion between legality and crime/Evaziunea fiscală între legalitate și infracțiune. The Legal Universe/Universul Juridic, Bucharest
4. Smith S (2015) Taxation: a very short introduction. Editura Oxford University Press, Oxford
5. Pană I (2019) Crime control and prevention/Controlul și prevenirea infracțiunilor. Bull Natl Defense Univ "Carol I" 112--120
6. McGill RK, Haye CA, Lipo S (2017) G.A.T.C.A.A practical guide to global anti-tax evasion frameworks. Polgrave Macmillan, Londra
7. Pop O (2003) Tax evasion/Evaziunea fiscală, Mirton, Timișoara
8. Dinga E (2008) Theoretical considerations on tax evasion vs. tax fraud/Considerații teoretice privind evaziunea fiscală vs frauda fiscală, Financial Studies-Theoretical approaches and modeling, pp 43–47
9. David D, Pojar D (2011) Approaches regarding the tax evasion in Romania. Annals of Faculty of Economics, University of Oradea, Faculty of Economics, vol 1, issue 1, pp 376--381
10. Manea AC (2011) Means and methods of preventing and combating tax evasion adopted by Romania and Moldova. Bull Transilvania Univ Brașov Ser VII Soc Sci Law 4(53):127–132
11. Constantin SB (2014) A study on tax evasion in Romania. In: ECOTREND conference proceedings, Academica Brâncuşi
12. Popescu AM, Cochințu I (2014) Legal regulation of the fight against and prevention of tax evasion by ways of legislative delegation. Constitutional limits. AGORA Int J Juridical Sci 1:132–140
13. Manea CA (2015) Tax evasion in Romania—a national security issue. Bull Transilvania Univ Brașov Ser VII Soc Sci Law 8(57):163–172
14. Perpelea SG, Beldiman D (2016) Tax evasion in Romania: new approaches. J Financ Challenges Future Universitaria 16(18):109--117
15. Ungureanu MA, Calugareanu M, Caraus M, Bartalis AM (2016) Conceptual approaches to tax evasion in Romania. Rom Econ Bus Rev 11(3):7–14
16. Comândaru AM, Stănescu SG, Păduraru A (2018) The phenomenon of tax evasion and the need to combat tax evasion. Contemp Econ J 3(3):124–133
17. Socoliuc M, Mihalciuc C, Cosmulese G (2018) Tax evasion in Romania-between past and present. In: Năstase C (ed) The 14th economic international conference: strategies and development policies of territories. Stefan cel Mare University of Suceava, Romania, pp 406–412
18. Popescu CI (2015) Tax evasion/Evaziune fiscală, Aius, Craiova

19. Georoceanu AM (2015) Drept fiscal român, Editura Pro Universitaria, Bucuresti
20. Le Borgne E, Baer K (2008) Tax amnesties: theory, trends, and some alternatives. International Monetary Fund, IMF Special Issues
21. Rusu IE, Gornig G (2009) Dreptul Uniunii Europene, Editura C.H. Beck, Bucuresti
22. Vâlsan C, Druică E, Ianole-Călin R (2020) State capacity and tolerance towards tax evasion: first evidence from Romania, Administrative Sciences, p 14
23. Vîrjan B (2016) Infracțiunile de evaziune fiscală, Editura C.H. Beck, Bucuresti
24. Văcărel I, Bistriceanu GD, Anghelache G, Bodnar M, Bercea F, Moșteanu T, Georgescu F (2006) Finanțe publice, Editura Didactica si Pedagogica Bucuresti
25. Hoanță N (2010) Evaziunea fiscala, Editura C.H. Beck, Bucuresti
26. Șaguna DD, Șova D (2009) Drept fiscal. Editura C.H. Beck, Bucuresti
27. Mrejeru T, Florescu D, Safta D, Safta M (2000) Evaziunea fiscală. Practică judiciară. Legislație aplicabilă, Revista Tribuna Economica, Bucuresti
28. Balaban C (2003) Evaziunea fiscală. Aspecte controversate de teorie și practică judiciară, Editura Rosetti, Bucuresti
29. Tăutu CN (1940) Evoluția și tehnica impozitelor directe din România, Bucuresti
30. Sasu H, Pătroi D, Cuciureanu F (2009) Practici de sustragere de la înregistrarea și plata impozitelor. La limita fraudei și dincolo de ea, Editura C.H. Beck, Bucuresti
31. Șaguna DD, Marin AA (2020) Evaziunea fiscală. Prevenire și combatere, Editura Universul Juridic, Iasi
32. Dobrotă G, Chirculescu MF (2011) Fiscal pressure in EU Member States/Presiunea fiscală la nivelul statelor membre UE, Annals of the "Constantin Brâncuși" University, Târgu Jiu, Economic Series, nr.1, p. 157
33. Voinea GM, Ștefura G, Boariu A, Soroceanu M (2002) Impozite, taxe și contribuții, Editura Junimea, Iași
34. Luca I (2006) Elements of current economic and financial crime/Elemente ale criminalității economico-financiare actuale. Dunărea de Jos, Galați (2006)
35. Șaguna DD, Radu DI (2018) Drept fiscal. Editura C.H. Beck, Bucuresti
36. Comaniciu C (2010) The possible causes of tax evasion in Romania. In: Proceedings of the 5th WSEAS international conference on economy and management transformation, vol II, Timișoara
37. Devos K (2008) Tax evasion behavior and demographic factors: an exploratory study in Australia. Revenue Law J 18:5–6
38. Antonescu M, Buziernescu R, Ciora IL (2004) Current trends in some forms of tax evasion internationally, Public Finance and Accounting no. 1
39. Pantea M, Șanta O (2012) Tax evasion, Journal of Crime Investigation/Evaziunea fiscală, Crime Investigation Journal, 1, Legal Universe, Bucharest
40. Ene-Corbeanu E (2020) Evaziunea fiscala, Editura Hamangiu, Bucuresti
41. Costaș CF (2016) Drept financiar, Editura Universul Juridic, București
42. Zagler M (Editor) (2010) International tax coordination: an interdiciplinarity perspective on virtues and pitfalls, Milton Park, Abingdon, Oxon; 1st edn. Routledge, New York
43. Tresch R (2008) Public sector economics. Palgrave Macmillan, New York
44. Baer K, Borgne E (2008) Tax amnesties: theory, trends, and some alternatives. Int Monetary Fund, IMF Special Issues
45. Durdureanu C, Soroceanu M (2010) The display of evasion and tax fraud phenomena in the context of current economic and financial crisis. Yearbook of Petre Andrei University of Iași, Issue 5, Lumen, Iași, p 127
46. Phyllis LLM (2003) Tax avoidance and anti-avoidance measures in major developing economies. Greenwood Publishing Group
47. Freedman J (2004) Defining taxpayer responsibility: in support of a general anti avoidance principle. Br Tax Rev 4:332–357
48. Richardson G, Lanis R (2007) Determinants of the variability in corporate effective tax rates and tax reform, evidence from Australia. J Account Public Policy 26(6):689–704. https://doi.org/10.1016/j.jaccpubpol.2007.10.003

49. Lipatov V (2012) Corporate tax evasion. The case for specialists. J Econ Behavior Organ 81(1):185–206. https://doi.org/10.1016/j.jebo.2011.09.015
50. Khlif H, Achek I (2015) The determinants of tax evasion: a literature review. Int J Law Manag 57(5):486—497. https://doi.org/10.1108/IJLMA-03-2014-0027. (Publisher Emerald Group Publishing Limited, 2015)
51. Zidkova H, Tepperova J, Heiman K (2016) How do opinions on tax evasion relate to shadow economy and VAT gap? In: Sedmihradska L (ed) Theoretical and practical aspects of public finance. Proceedings paper 21st international conference on theoretical and practical aspects of public finance 2016, University of Economics, Department of Public Finance, Prague, pp 119–124
52. Allen A, Francis BB, Wu Q, Zhao Y (2016) Analyst coverage and corporate tax avoidance. J Bank Financ 73:84–98
53. Gebhart MS (2017) Measuring corporate tax avoidance—an analysis of different measures. Junior Manag Sci 3:43–60
54. Christians A (2018) Distinguishing tax avoidance and tax evasion. In: Hashimzade N, Epifantseva Y (eds) The routledge companion to tax avoidance research. Routledge, Abingdon, UK/New York, US, pp 417–429
55. Păunescu RA, Vintilă G (2018) Study of influence factors of effective corporate tax rate from the Baltic Countries. J Acc Auditing Res Pract 1–20
56. Mocanu M, Constantin SB, Răileanu V (2021) Determinants of tax avoidance—evidence on profit tax-paying companies in Romania. Econ Res-EkonomskaIstraživanja. https://doi.org/10.1080/1331677X.2020.1860794
57. Chen S, Chen X, Cheng Q, Shevlin T (2010)Are family firms more tax aggressive than non-family firms? J Financ Econ 95(1):41–61. https://doi.org/10.1016/j.jfineco.2009.02.003
58. Armstrong C, Blouin J, Larcker D (2012) The incentives for tax planning. J Account Econ 53(1–2):391–411. https://doi.org/10.1016/j.jacceco.2011.04.001
59. Steijvers T, Niskanen M (2014) Tax aggressiveness in private family firms: an agency perspective. J Family Bus Strategy 5(4):347–357. https://doi.org/10.1016/j.jfbs.2014.06.001
60. Richardson G, Lanis R, Leung SCM (2014) Corporate tax aggressiveness, outside directors, and debt policy. An empirical analysis. J Corp Finan 25:107–121. https://doi.org/10.1016/j.jcorpfin.2013.11.010
61. Inger KK (2013) Relative valuation of alternative methods of tax avoidance. J Am Taxation Assoc 36(1):27--55
62. Yorke SM, Amidu M, Agyemin-Boateng C (2016) The effects of earnings management and corporate tax avoidance on firm performance. Int J Manag Pract 9(2):112–131
63. Chen KS, Tsai H (2018) Taxing the rich policy, evasion behavior, and portfolio choice: a sustainability perspective. Cogent Bus Manag 5(1):1–25
64. Lazăr S, Istrate C (2018) Corporate tax-mix and firm performance. A comprehensive assessment for Romanian listed companies. Econ Res-EkonomskaIstraživanja 31(1):1258–1272
65. Zhu N, Mbroh N, Monney A, Bonsu MO (2019) Corporate tax avoidance and firm profitability. Eur Sci J 15(7):61–70
66. Pintilescu C (2007) Multivariate Statistical Analysis/Analiză statistică multivariată, "Alexandru Ioan Cuza" University Publishing House, Iasi

# Impact of Pro-cyclicality Fluctuations, Financial Assets' Fair Value, Equity Ratio on Banks' Financial Position Statement

Osama Samih Shaban

**Abstract** This research aims to study two different financial performance situations for the impact of Pro-cyclicality on financial assets fair value and on equity ratios. The study also compared this relationship across two different years, one is an ordinary year 2019, and the second year is the Covid-19 year 2020. The study population is formed out from the Jordanian Commercial Banks. Moreover, one sample T-test was performed to assess the difference between Pro-cyclicality disclosure, Financial Assets' Fair Value disclosure, and equity ratio performance of the year 2019 and the year 2020. The study concluded that, Jordanian banks were having an inverse relationship between pro-cyclical fluctuations and the fair value of financial assets, also most of these banks were having positive relationship between pro-cyclical fluctuations and equity ratio. Also, Banks equity ratios, Pro-cyclicality disclosure, and Financial Assets' Fair Value disclosure in the ordinary year 2019 is significantly different in the pandemic year 2020.

**Keywords** Covid-19 · Pro-cyclicality · Financial position statement · Fair value · Equity ratio

## 1 Introduction

Pro-cyclicality, or cyclical changes, is one of the downsides of fair value accounting assessment. The amplification of natural cyclical economic movements, both in booms and busts, is known as pro-cyclicality, and it creates the conditions for increased financial system instability and fragility. Fair value accounting, according to accounting standards, can cause exaggerated cycles and exacerbate cyclical fluctuations in asset and liability values by recording profits or losses that follow economic cycles [1].

Fair value accounting and its impact on banks' financial statements have been extensively examined by researchers all around the world over the last decade. The

O. S. Shaban (✉)
Department of Accounting, Al_Zaytoonah University of Jordan, Amman, Jordan
e-mail: drosama@zuj.edu.jo

© The Author(s), under exclusive license to Springer Nature Switzerland AG 2022
S. G. Yaseen (ed.), *Digital Economy, Business Analytics, and Big Data Analytics Applications*, Studies in Computational Intelligence 1010,
https://doi.org/10.1007/978-3-031-05258-3_54

pro-cyclical and its effect on the valuation of fair value is the current trend among researchers during the current period, particularly during the Covid-19 pandemic, as it is considered a complex issue in the financial markets, raising questions about the use of market prices below "theoretical valuation" and the credibility of "false sales. "Financial products were assessed at fair value notwithstanding concerns that current market prices did not adequately reflect the product's underlying cash flows or the price at which it could be sold in the future. Selling decisions based on fair value pricing resulted in further price declines in a bad market with already low prices, reflecting the market's illiquidity premium [2].

It's worth noting that there are counter-arguments to the claim that fair value accounting is pro-cyclical and so unsuitable for use under US Generally Accepted Accounting Principles (GAAP) [1]. While proponents acknowledge that it is pro-cyclical they say that this isn't always a bad thing and that valuing assets and debts at fair value during periods of economic recovery can be advantageous. This can be explained by the fact that fair value measures might reveal "early warning signals of an impending crisis" that can be used to lessen the severity of the crisis and price decline a feature that the argument for periodic fair value accounting often overlooks. Furthermore while cyclical fluctuations can cause some market volatility financial position statement volatility is generally caused by asset owners' risk management framework and investment decision protocols rather than the fair value accounting system itself [3].

The current study derived its importance from the expected benefits that will affect accounting department of commercial banks, researchers, and traders at financial markets.

The primary goal of this study is to study two different financial performance situations for the impact of Pro-cyclicality on financial assets fair value and on equity ratios, the research population will include Jordanian Commercial Banks. The current study looks at the principles and applications of fair value accounting (FVA), as well as the ramifications of its features and how they will affect banks' financial statements. It provides empirical support for public discussions of the pro-cyclical FVA in banks' financial status statements using a simple model. It explores the application of fair value through banks' financial position statements during an average business cycle as well as during extreme shocks, such as the recent pandemic, using representative financial position statements of Jordanian commercial banks. The study went over the findings and then suggested some limitations.

## 2   Literature Review

Companies using fair value accounting experienced significant fluctuations in the value of their financial assets during recessions or depressions. Many corporations write down the value of such assets to new discount market values, causing financial market prices to fall even lower. This occurs because financial institutions may attempt to counteract a write-down of their financial assets by selling securities in illiquid markets, even though the securities would have held them until maturity [4]. Keeping securities till maturity means that the firm or the investor intends to keep the securities until the issuing firm repays the securities, and selling them before maturity will result in a lower exit price than anticipated [1]. Other firms utilize these forced sales as observable inputs to evaluate their assets at fair value. Other firms must devalue their assets more than they would otherwise in order to reverse the lower exit price that would be paid in the current market if they use such information from forced sales. Furthermore non-performing corporations may opt to wait until the market price recovers from the dip caused by the forced sales before selling those securities in this market [5].

This waiting or deferral of sales may prevent the market price from reverting to a value that accurately reflects the securities' future gains, so prolonging the period of illiquidity and recession. Forced sales can also lead the price of an asset to be determined by market liquidity rather than the strength of the asset's future revenues, because if the firm has no alternative but to sell the asset, the market price will be lower if the market is less liquid than usual [3].

It's also worth noting that unrealized gains recorded when the market improves can be deceiving because the market will eventually return to normal, offsetting the recorded profits if the firm holds the assets until the market returns to normal. Furthermore, such profits provide ineffective income, which is not the type of income preferred by investors. Furthermore, ineffective income is generated by changes in the environment and the market in which these enterprises operate, as well as the accompanying increases in the value of these firms' investments [1]. Investors prefer active income which is gained via ongoing work and management efforts [6], because management has no control over the revenue created by market movements and it is difficult to replicate. Fair value accounting critics argue that embedding ineffective income with active income earned by a firm through the evaluation of its financial statement can be misleading to investors, particularly those who may not realize that a portion of the firm's income is a result of uncontrollable market fluctuations that the firm has approved [7]. Furthermore, if the firm keeps part of its assets or obligations, it may result in unrealized gains and losses being reversed in certain instances. It occurs when market prices are excessively high. These prices are regarded as unrealistic because they are inflated by market optimism and abundant liquidity and do not reflect the fundamental values of assets and liabilities [1].

The purpose of broadening the extent of disclosure necessary for fair value accounting is to bring more transparency to financial statements and make them less confusing for intended users, as well as to make it easier to value fair value using additional measurements such historical cost [8, 9].

Certain disclosures of things measured at fair value under statement SFAS 159, as well as other extra disclosures, are required by SFAS 159. The firm must report the fair value of assets and liabilities for which the fair value option is selected separately from the carrying values of equivalent items measured by another measurement feature in the financial situation statement. It can also be done by putting them in two columns in the financial position statement, one for fair value items and the other for amounts listed without their fair value, or by grouping them together under one item with the total amounts of all similar assets or liabilities and putting the fair book value in parentheses. The SFAS 159 disclosure is intended to aid users of financial statements in making comparisons between firms using different measurement attributes for similar assets and liabilities, as well as between the assets and liabilities in a firm using different measurement attributes for similar assets and liabilities [10].

When determining the fair value of an asset or obligation, the standard assumes that the fair value is the price of exchange in a structured transaction between market participants to sell the asset or transfer the liability at the measurement date. Structured transaction means that the supposed transaction was offered in the market for a period previous to the measuring date, allowing the regular and familiar marketing activities of transactions involving assets or liabilities to take place without being rushed or coerced. To put it another way, it's not a coercive transaction (e.g. forced liquidation or distressed sale) [10].

Furthermore, on the measurement date, the transaction of selling an asset or transferring an obligation is a hypothetical transaction. Furthermore, it is regarded in terms of market participation, as the goal of fair value assessment is to establish the price that will be received to sell an asset or paid to transfer an obligation on the measurement date (the exit price). The standard also implies that when calculating fair value, a transaction to sell an asset or transfer an obligation takes place in a principal market, and that if there isn't one, an asset or liability should be evaluated in the most favorable market. With the biggest volume and degree of activity in the market, the financial reporting entity sells an asset or transfers an obligation to it. The most advantageous market is one in which the financial reporting entity sells an asset or transfers a liability at a price that is higher than the amount that could be received for an asset or lower than the amount that will be paid to transfer a liability, after transaction costs in the relevant market are taken into account [11]. In either instance, from the standpoint of the financial reporting entity, the major (or most favorable) market and market players should be considered, taking into account differences across organizations with various operations. If an asset or liability has a primary market, fair value measurement must reflect the price in that market (whether the price is directly observable or determined in another way using a valuation technique). Furthermore, even if the market price is different, using the historical cost in the assessment and compilation of financial statements may be more effective [10].

The standard also states that the price in the major (or most beneficial) market used to calculate an asset's or liability's fair value is not modified for transaction costs. The added direct expenses of selling the asset or transferring the liability in the major (or most beneficial) market for the asset or liability are also included in the cost of the business transaction. The costs of the business process are not an asset or a liability; they are unique to the business process and will change depending on how the financial reporting entity is treated. The standard also specifies that the price in the major (or most beneficial) market used to calculate an asset's or liability's fair value is not modified for transaction costs. The added direct expenses of selling an asset or transferring an obligation in the primary (or most beneficial) market for the asset or liability are also included in the cost of the business transaction. The costs of the business process are not an asset or a liability; they are specific to the business process and will change based on the financial reporting entity's approach [8].

International Financial Reporting Standards IFRS 13 adopts fair value approach in measuring assets and liabilities instead of historical approach. The amount received for selling an asset or paid to transfer a liability between a seller and a buyer at a synchronized moment is defined by the International Financial Reporting Standards (IFRS). One of the IFRS 13 measurement methodologies is the market method. An asset or a liability is measured using appropriate pricing and information in this manner [12]. Market transactions involving same or equivalent assets create prices and other relevant information, which are used in the market approach. The market approach is represented through a variety of ways. The data sources used in the most commonly used market approach methodology for valuing unquoted equity securities are based on the data sources used (for example, listed prices of public companies or prices from mergers and acquisitions transactions) [13].

The cost method is the second approach. The amount paid to replace the original topic of measurement, referred to as the current re-placement cost, is considered fair value under this method. The price of a trading transaction (i.e. the entrance price) may be an appropriate starting point for assessing the fair value of an unquoted equity instrument on the measurement date when an investor buys in an instrument identical to an unquoted equity instrument that is being valued at fair value. According to IFRS 13, it represents the instrument's fair value at the time of initial recognition. However, from the date of initial recognition until the measurement date, the investor must use all information about investment performance and activities that becomes reasonably available to the investor. Because this information may affect the fair value of the investee's unlisted equity instrument at the measurement date, cost is only a credible estimate of the fair value at the measurement date in very restricted circumstances. [12, 14].

The revenue approach is the third option. It reflects current market expectations of future flows by converting future cash flows into a single quantity of money. To put it another way, the income technique reduces future numbers (such as cash flows or income and expenses) to a single current (or decreased) quantity. The discounted cash flow approach is commonly used to do this, and it is applied to an organization's cash flows or, less frequently, to equity cash flows. Investors must estimate the investment's expected cash flows when using the discounted cash flow approach. In practice, most models estimate cash flows for a discrete period and then use either a constant growth model (such as the Gordon growth model), in which a capitalization rate is applied to the cash flow immediately after the end of the period discrete value, or the exit multiple to estimate the final value when an investor expects an indefinite lifetime of cash flows [12, 13].

The purpose of calculating fair value, according to IFRS 13, is to identify the price at which a market participant could sell an asset or transfer a liability in an orderly transaction at the measurement date under current market conditions (i.e. estimating the exit price). Measurement of provisions in accordance with IAS 37 for Provisions, Liabilities, and Contingent Assets is an example of when a corporation is needed to generate other financial reporting purposes. The timing and/or amounts of future cash flows, as well as other aspects, will frequently be uncertain or certain in financial reporting measurement. This standard illustrates how to use a range of valuation methodologies to determine the fair value of unquoted equity securities. It contains judgment not just in the implementation of the assessment approach, but also in its selection. Part of this procedure include examining the information available to the investor. An investor is more likely to focus on a comparable firm valuation multiples technique when there is a counterparty to the firm that can be correctly compared, or when the background or specifics of the firm can be compared [12].

Regarding off-Financial Position Statement Entities and Pro-cyclicality, Alicia et al. [2] stated that recent market volatility has enhanced public awareness of banks' and financial institutions' extensive use of off-Financial Position Statement Entities. Regarding off-Financial Position Statement Entities and Pro-cyclicality, Alicia et al. 2 stated that recent market volatility has enhanced public awareness of banks' and financial institutions' extensive use of off-Financial Position Statement Entities. Any stake in securitized financial assets must be reported on the financial position statement and valued at fair market value, which is commonly done in the trading book. Off-financial-position-statement entities are rarely subject to mandatory reporting. Their absence may have contributed to market uncertainty and pro-cyclical behavior by causing the market to believe that banks were behind their off-balance-sheet firms [2].

## 3  Method

Pro-cyclicality fluctuations, equity ratios, and the fair value of financial instruments. were derived from Jordanian commercial banks' annual reports. Furthermore, the equity ratio was determined by dividing total equity by total assets, and the equity ratio included all assets and equity indicated on the financial status statement. The higher the equity-to-assets ratio, the less leverage the company has, implying that the company and its investors hold a larger share of its assets. While 100 percent is desirable, a lesser proportion is not necessarily cause for alarm.

In addition, Pro-cyclicality reserves were extracted from the financial position statements of the Jordanian commercial banks. Pro-cyclicality reserves were disclosed in these statements under the name of pro-cyclicality reserves, or under the name of fair value reserves, because as we mentioned earlier that pro-cyclicality is one of the weaknesses in the measurement of fair value accounting so some of these banks named it fair value reserves. The period of these figures will cover two subsequent years 2019 and 2020. Year 2019 represents an ordinary financial performance year, and year 2020 represents a recession or the pandemic year of Covid-19. The study sample will cover seven commercial banks out of 13 banks. The aim is to study two different financial performance situations for the impact of Pro-cyclicality on financial assets fair value. Moreover, one sample T-test will be performed to assess the difference between Pro-cyclicality disclosure of the year 2019 and the year 2020. And also assessing the difference between Financial Assets' Fair Value disclosure of the year 2019 and the year 2020. And finally assessing the difference between equity ratio performance of the year 2019 and the year 2020.

## 3.1 Development of Hypotheses

**H01**: There is no statistical significance difference between Pro-cyclicality disclosure of the year 2019 and the year 2020.

**H02**: There is no statistical significance difference between Financial Assets' Fair Value disclosure of the year 2019 and the year 2020.

**H03**: There is no statistical significance difference between equity ratio performance of the year 2019 and the year 2020.

Table 1, shows the Pro-cyclicality reserves disclosed in the financial position statements of the Jordanian commercial banks, and the percentage of these reserves to total assets for the years 2019 and 2020.

Table 2, shows the fair value (FV) of the financial assets disclosed in the financial position statements of the Jordanian commercial banks, and the percentage of these figures to total assets for the years 2019 and 2020. Table 2 shows also the equity ratios calculated for the years 2019 and 2020 in order to illustrate the relationship between Pro-cyclicality and equity ratio.

## 4 Statistical Analysis

Percentages of pro-cyclicality reserves were calculated by dividing the Pro-cyclicality reserves by the total assets for the years 2019 and 2020. It noticed that

**Table 1** Pro-cyclicality reserves (total and percentage of total assets)

| Bank name | 2019 JD | 2020 JD | 2019 % total assets | 2020 % total Assets |
|---|---|---|---|---|
| Jordan Ahli Bank | 3,678,559 | 3,678,559 | 0.133 | 0.129 |
| Arab Bank | 298,403,000 | 295,797,000 | 0.580 | 0.540 |
| Housing Bank | 5,400,864 | 9,654,188 | 0.063 | 0.116 |
| Bank of Jordan | 24,954,157 | 6,092,218 | 0.921 | 0.022 |
| Jordan Kuwait Bank | 2,296,466 | 4,571,425 | 0.083 | 0.162 |
| Arab Jordan Investment Bank | 2,067,878 | 3,079,877 | 0.097 | 0.139 |
| Jordan Commercial Bank | 2,211,406 | 2,020,984 | 0.159 | 0.149 |

**Table 2** Financial assets' fair value and equity ratio

| Bank name | 2019 JD | 2020 JD | 2019 % total assets | 2020 % total assets | 2019 equity ratio | 2020 equity ratio |
|---|---|---|---|---|---|---|
| Jordan Ahli Bank | 25,014,042 | 25,744,834 | 0.9 | 0.9 | 7.3 | 3.05 |
| Arab Bank | 519,053,000 | 304,054,000 | 1.01 | 0.56 | 14.4 | 14.2 |
| Housing Bank | 388,454,051 | 371,882,600 | 4.60 | 4.47 | 13.31 | 13.98 |
| Bank of Jordan | 80,865,636 | 84,526,410 | 2.98 | 3.12 | 15.51 | 17.04 |
| Jordan Kuwait Bank | 96,124,306 | 85,867,274 | 3.49 | 3.06 | 16.67 | 16.24 |
| Arab Jordan Investment Bank | 20,059,841 | 17,198,214 | 0.94 | 0.78 | 10.12 | 10.21 |
| Jordan Commercial Bank | 11,105,937 | 29,053,113 | 0.80 | 2.15 | 10.04 | 10.32 |

these percentages were unstable during the subsequent years. Housing bank almost doubled its Pro-cyclicality reserves, 0.063% in the year 2019 to 0.116% in the year 2020. At the contrary Bank of Jordan witnessed a dramatic drop of its reserves between the years 2019 and 2020, Table 3 illustrates these results.

Percentages of financial assets fair value were also calculated by dividing financial assets fair value by total assets for the years 2019 and 2020. It noticed that these percentages were stable during the subsequent years except Arab Bank and Jordan

**Table 3** Pro-cyclicality percentages, equity ratios & FV percentages

| Bank name | 2019 pro-cy-% total assets | 2020 pro-cy-% total assets | 2019 FV % total assets | 2020 FV % total assets | 2019 equity ratio | 2020 equity ratio |
|---|---|---|---|---|---|---|
| Jordan Ahli Bank | 0.133 | 0.129 | 0.9 | 0.9 | 7.3 | 3.05 |
| Arab Bank | 0.580 | 0.540 | 1.01 | 0.56 | 14.4 | 14.2 |
| Housing bank | 0.063 | 0.116 | 4.60 | 4.47 | 13.31 | 13.98 |
| Bank of Jordan | 0.921 | 0.022 | 2.98 | 3.12 | 15.51 | 17.04 |
| Jordan Kuwait Bank | 0.083 | 0.162 | 3.49 | 3.06 | 16.67 | 16.24 |
| Arab Jordan Investment Bank | 0.097 | 0.139 | 0.94 | 0.78 | 10.12 | 10.21 |
| Jordan Commercial Bank | 0.159 | 0.149 | 0.80 | 2.15 | 10.04 | 10.32 |

Commercial Bank. Arab Bank financial assets fair value dropped by 44% due to Covid-19 pandemic conservatism regulations adopted by the bank. As for Jordan Commercial Bank its financial assets fair value increased by 135% between the years 2019 and 2020, the notes of the specified bank explained the cause of such increase, and mentioned that, the financial assets fair value reflects the shares value quoted in an active markets at time of preparing financial statements. Table 3 illustrates these results.

Equity ratios were also calculated by dividing total equity by total assets for the years 2019 and 2020. The cause of including equity ratios is due to the relationship between pro-cyclicality and fair value. Alicia et al. 2 mentioned that in the normal cycle, a fair evaluation of the financial position statement results in moderate volatility compared to the economic boom scenarios. The high influence of fair value appraisal of liabilities has a counter-cyclical effect on bank equity. The value of the bank's liabilities drops with poorer economic activity and a higher chance of default under fully fair value (FFV), which mitigates the decline in equity. This effect emerges as a result of the structure of assets, or liabilities, on investment banks' financial position statements, which includes a substantial share of financial liabilities assessed at fair value. Liabilities in the FFV can provide cyclicality by functioning as an implied stabilizing hedge for the fair valuation of assets, as certain US investment banks do. The study sample (Jordan Ahli Bank, Arab Bank, Housing Bank, Arab Jordan Investment Bank) has a favorable association between pro-cyclical fluctuations and equity ratios, according to the findings. Pro-cyclical variations and equity ratios have an inverse relationship at Bank of Jordan, Jordan Kuwait Bank, and Jordan Commercial Bank. Table 3 illustrates these results.

Also, the analysis shows that, the majority of the study sample (Housing bank, Bank of Jordan, Jordan Kuwait Bank, Arab Jordan Investment Bank, Jordan Commercial Bank) is having an inverse relationship between pro-cyclical fluctuations and the fair value of financial assets. Arab Bank has a positive relationship between pro-cyclical fluctuations and the fair value of financial assets. Jordan Ahli Bank has no signs of change in its financial assets' fair value in both years, year 2019 which represents normal cycle, and year 2020 which represents the recession cycle or the pandemic year of Covid-19. Table 3, illustrates these results.

## 4.1 Descriptive Statistics

The Table 4, shows the mean and standard deviations of the study variables, Table 4, indicates that the mean of the independent variable Pro-cyclicality for the year 2019 is around 29%, and therefore it reflects a better average than Pro-cyclicality in the pandemic year 2020, which scored 17.9%. Equity ratio for the year 2019 and 2020 had the highest mean value, indicating that commercial banks managements highest priority is towards equity ratio whether it is an ordinary year or a pandemic year. Mean and standard deviation of Fair value for the ordinary performance year 2019,

**Table 4** Mean and standard deviations of pro-cyclicality, FV, and equity ratio

| One sample statistics | N | Mean | Std. deviation |
|---|---|---|---|
| Pro-cyclicality for 2019 | 7 | 0.2908571 | 0.33057850 |
| Fair value for 2019 | 7 | 2.1028571 | 1.56099022 |
| Equity ratio for 2019 | 7 | 12.4785714 | 3.40264323 |
| Pro-cyclicality for 2020 | 7 | 0.1795714 | 0.16544169 |
| Fair value for 2020 | 7 | 2.1485714 | 1.47867540 |
| Equity ratio for 2020 | 7 | 12.1485714 | 4.79781349 |

and for the pandemic year 2020 were almost the same, indicating that commercial bank' managements had kept the same strategy towards fair value.

## 4.2 Testing Hypotheses

**H01**: There is no statistical significance difference between Pro-cyclicality disclosure of the year 2019 and the year 2020.

One sample T-Test was performed to assess if Pro-cyclicality for the year 2019 (ordinary year) differ significantly in disclosure from the year 2020 (Pandemic year). The results reveal a significant difference in the disclosure of the ordinary year 2019, and pandemic year 2021, $t = 2.328$, $p < 0.059$. Hence average banks Pro-cyclicality disclosure in the ordinary year is significantly different from the average banks Pro-cyclicality disclosure in the pandemic year 2020. Hence we reject the null hypothesis.

**H02**: There is no statistical significance difference between Financial Assets' Fair Value disclosure of the year 2019 and the year 2020.

One sample T-Test was performed to assess if Financial Assets' Fair Value for the year 2019 (ordinary year) differ significantly in disclosure from the year 2020 (Pandemic year). The results reveal a significant difference in the disclosure of the ordinary year 2019, and pandemic year 2021, $t = 3.564$, $p < 0.012$. Hence average banks Financial Assets' Fair Value disclosure in the ordinary year is significantly different from the average banks Financial Assets' Fair Value disclosure in the pandemic year 2020. Hence we reject the null hypothesis.

**H03**: There is no statistical significance difference between equity ratio performance of the year 2019 and the year 2020.

One sample T-Test was performed to assess if equity ratio performance for the year 2019 (ordinary year) differ significantly in disclosure from the year 2020 (Pandemic year). The results reveal a significant difference in the disclosure of the ordinary year 2019, and pandemic year 2021, $t = 3.564$, $p < 0.012$. Hence average banks equity ratio performance in the ordinary year is significantly different from the average banks equity ratio performance in the pandemic year 2020. Hence we reject the null hypothesis. Table 5 illustrate the hypotheses testing.

**Table 5** One-sample test for pro-cyclicality, FV, and equity ratio

| One-sample test | | | | | |
|---|---|---|---|---|---|
| | Test value = 0 | | | | |
| | t | df | Sig. (2-tailed) | 95% confidence Interval of the difference | |
| | | | | Lower | Upper |
| Pro-cyclicality for 2019 | 2.328 | 6 | 0.059 | − 0.0148770 | 0.5965913 |
| Fair value for 2019 | 3.564 | 6 | 0.012 | 0.6591820 | 3.5465323 |
| Equity ratio for 2019 | 9.703 | 6 | 0.000 | 9.3316513 | 15.6254916 |
| Pro-cyclicality for 2020 | 2.872 | 6 | 0.028 | 0.0265634 | 0.3325795 |
| Fair value for 2020 | 3.844 | 6 | 0.009 | 0.7810248 | 3.5161181 |
| Equity ratio for 2020 | 6.699 | 6 | 0.001 | 7.7113340 | 16.5858088 |

# 5 Conclusions

It was noticed that most of the study sample were having an inverse relationship between pro-cyclical fluctuations and the fair value of financial assets. This means, whenever pro-cyclicality increases, the fair value of financial assets decreases. Moreover, It was noticed also that most of the study sample were having positive relationship between pro-cyclical fluctuations and equity ratio. This means whenever pro-cyclicality increases, equity ratio also increases and vice versa. Pro-cyclical fluctuations in recession cycle mostly have a diverse effect on the results obtained. This means, whenever pro-cyclicality decreases, the cause of such decrease is the abnormal event of the pandemic year of Covid-19.

Banks Pro-cyclicality disclosure, and Financial Assets' Fair Value disclosure, in the ordinary year 2019 is significantly different in the pandemic year 2020. Moreover, banks equity ratio performance in the ordinary year is significantly different from the average banks equity ratio performance in the pandemic year 2020.

# 6 Research Limitations and Future Studies

While performing this research paper, the following few limitations could be considered:

This research paper concentrated on banks industry. Thus, further studies may include other industries, such as manufacturing, trading etc. Moreover, The study sample is narrow and that because of the difficulty of having actual figures from the financial statement of the banks.

Because the study is limited to the commercial banking sector, the results gained have limits in terms of generality.

Future studies should concentrate on a broader range of foreign and Islamic banks.

**Contribution of the Study** This research paper is an original research paper, it contributes to the current literature of fair value measurement by reviewing the basic ideas of fair value application, its characteristics, and how pro-cyclicality influences fair value and, ultimately, banks' financial position statements.

# References

1. Ryan S (2008) Fair value accounting: understanding issues raised by the credit crunch. http://www.uic.edu
2. Alicia N, Jodi S, Juan S et al (2009) Pro-cyclicality and fair value accounting. IMF Working Paper Monetary and Capital Markets Department
3. Lefebvre R, Simonova E, Mihaela S et al (2009) Fair value accounting: the road to be most travelled. http://www.cga-canada.org
4. Shaban O, Alqotaish A, Alqatawneh A (2020) The Impact of fair value accounting on earnings predictability: evidence from Jordan. Asian Econ Financ Rev 10(12):1466–1479
5. Laux C, Leuz C. (2009) The crisis of fair value accounting: making sense of the recent debate. Booth School of Business, The University of Chicago, Chicago, IL, Working Paper No. 33. 2021. http://papers.ssrn.com
6. Saleh I, Abu Afifa M (2020) The effect of credit risk, liquidity risk and bank capital on bank profitability: evidence from an emerging market. Cogent Econ Financ 8(1):1814509
7. Muller I, Riedl E, Sellhorn T et al (2008) Consequences of voluntary and mandatory fair value accounting: evidence surrounding IFRS adoption in the EU real estate industry. Working Papers. Harvard Business School Division of Research, 1–40
8. Financial Accounting Standards Board [FASB] (2018) Statement of financial accounting standards no. 157: fair value measurements (SFAS No. 157). http://www.fasb.org. Accessed 15 June 2021
9. Financial Accounting Standards Board [FASB] (2007) Statement of financial accounting standards no. 159: the fair value option for financial assets and financial liabilities (SFAS No. 159). http://www.fasb.org. Accessed 15 June 2021
10. AICPA (2009) Fair value measures. www.aicpa.org Accessed 20 June 2021
11. Yassin MM, Al-Khatib E (2019) Internet financial reporting and expected stock return. J Account Financ Manag Strategy 14(1)
12. IFRS (2018) International financial reporting standards' annual report, statement of financial position. Accessed on 29 June 2021. Retrieved from http://www.ifrs.org
13. Riahi O, Khoufi W (2016) Effect of fair value accounting on the company's reputation. Int J Account Econ Stud 4(1):36–45
14. Penman S (2007) Financial reporting quality: is fair value a plus or a minus? Accounting Bus Res 33–44

# Legal Protection of Private Electronic Life: Problems and Solutions

**Ali Awad Al-Jabra, Muneer Mohammed Shahada Al-Afaishat, and Sarah Mahmoud Al-Arasi**

**Abstract** As the concept of individual's private electronic life has broadened, it became a real challenge for humanity which is subject to infringement, violations, and penetration on one hand and difficult to technologically and legally control, being a human right, on the other. Thus, methods for controlling and protecting it have become more difficult. Consequently, the difficulty to control methods of committing the crime of penetrating private electronic life and to include that in strict legal provisions assist to achieve a successful protection for it. Undoubtedly, the absence of legislative provisions for some forms of crimes endangers the future of individual's private electronic life, hence fails in to punish penetrators and parasites, due to the legislative provision which stipulates that no crime and no punishment can be inflected without a provision. The current study sheds light on private electronic life of individuals through presenting problems and challenges that it encounters, besides the preventive remedial possible solutions for this phenomenon which disturbs whoever deals with information technology.

**Keyword** Individuals digital data · Privacy penetration · Private electronic life

## 1 Introduction

In a report presented to the American congress in 1967, the American jurist, Arthur Miller wrote. "The avid computer that is never saturated with information, and the reputation of being infallible, besides its memory from which the saved can never be forgotten or deleted, might become the neural center for a control system that changes the society into a transparent world where homes, financial transactions, meetings, mental and corporal situations lie naked to any passing scene.

Though more than half a century elapsed on the aforementioned statement, humanity started to feel the size of danger and the amount and impact of change that information technology and computers would leave ion individual's private life.

A. A. Al-Jabra · M. M. S. Al-Afaishat · S. M. Al-Arasi (✉)
Al-Zaytoonah University of Jordan, Amman, Jordan
e-mail: dr.arasi@yahoo.com

© The Author(s), under exclusive license to Springer Nature Switzerland AG 2022
S. G. Yaseen (ed.), *Digital Economy, Business Analytics, and Big Data Analytics Applications*, Studies in Computational Intelligence 1010,
https://doi.org/10.1007/978-3-031-05258-3_55

The wicked and sinful practices of some pushed them to trespass the boundaries and encroach upon individual's privacy within the framework of information technology. Such a thing broadened the circle of practices which made of it a real threat and a nightmare for some so that obligates enacting regulatory legislations to face it and to protect individual's private electronic life. From the start, the legislations enacted called for respect of private life, being a human right whose protection it caters for. The change of the traditional concept of private life to an electronic one became a big challenge for legislators to protect life, due to difficulty of control as exemplified in Jordanian legislation. Therefore, this study tackles criminal confrontation of private electronic life in Jordanian legislation and to what extent it succeeded in protecting that life through presenting problems and their solutions.

The problem of the study stems from the broad use of information technology witnessed these days in all fields of life, specifically in individual's privacy that switched from traditional to electronic. Such a change pushed the researchers to pose the following pivotal questions for which the current study attempts to find answers.

The questions are: To what extent did the Jordanian legislator succeed in enacting regulation and legal provisions to protect private electronic life? What are the scientific and empirical problems and challenges that face adopting such a protection? What are the best solutions for that?

The study also endeavors to answer a set of subsidiary questions such as:

- What is meant by private electronic life?
- What are the foremost problems and challenges that face private electronic life?
- What is the legal vision to secure legal protection for private electronic life?

The study's significance lies in its focus on legal provisions that tackle criminal protection of private life and lack of legislation, if any, besides methods of treatment. In addition, the study is also theoretically significant as the topic was not sufficiently discussed before, due to its novelty and to the problems it arouses. From the scientific side, its importance lies in correlating the subject to the essence of legislation and to what concerns researchers in legal studies, courts, lawyers, in addition to legislative institutions in Jordan.

To achieve objectives of the study, the researchers adopted the descriptive analytical approach by which they presented the case first, then discussed the legal provisions it faces, if any.

The study afterwards presents pro-and con-attitudes, critiques, then proposes solutions relevant to the protection of private electronic life.

To shed light on the foremost axes and problems, the researchers divided their plan into the following sections.

## 1.1 First Topic: Problems of Criminal Protection for Private Electronic Life

This was subdivided into the following:
First requisite: Idiomatic crisis of private electronic life.
Second requisite: Difficulties met in securing criminal protection for private electronic life.
Second topic: challenges facing criminal protection of private life and solutions.

This was divided into the following subdivisions:
First requisite: Adequacy of criminalization in Jordanian legislating with regard to protecting private electronic life.
Second requisite: A vision relevant to securing active protection of private electronic life.

## 2 Discussion and Analysis

First topic: problems of criminal protection for private electronic life.

Due to the rapid development of information technology in several fields of life: social, educational, financial or economical, this affected various human practices related to personal electronic and social life, in addition to penetration, hacking electronic websites, robbing data and information, spreading fake news, encroaching upon privacy, impersonation, blackmailing, promoting rumors, person's assassination through electronic sexual blackmailing sometimes. Such practices diverted the legal correct role information technology plays to an illegal one, where it tampered with the course and requirement of private electronic personal life via illegal and unfamiliar acts. Such a thing prompted legislators to look into those practices that diverted the track for that information technology was created, as that negatively affects personal and human security. Thus, it was necessary to use criminal law to secure an active protection for information technology against what might affect it from such illegal practices. This issue created problems for private electronic life which will be addressed by the two following requisites:

**First: Idiomatic Crisis of Private Electronic Life**

**Second: Difficulty of Securing Criminal Protection for Private Electronic Life**

As for the first requisite, one can note that though humans are identical, yet that doesn't hamper diversification with regard to their ideology, nature, beliefs, humor, and lifestyle which is echoed in their personal life. That obligates retaining personal secrets, a natural life phenomenon. Everyone has the right to keep his own personal secrets that shouldn't be exposed to everyone or be a topic of promulgations. "Everybody has the right to be left alone to enjoy a peaceful life away from lights and publicity" [2].

There is no doubt that the right to private life is one of the inseparable concepts attached to man as long as he exists. Yet, it is one of the changing renewable concepts for which it is difficult to put down criteria and restraints. Most of jurisdiction identifies it with right to privacy, the other face of the coin of private life. Pursuant to that, the right to private life was important and of great concern for ancient civilizations [10], in her dissertation noted that during the Roman era, law included criminal provisions that criminalize harming or assaulting on persons. In addition, every home has its sanctity and every aggrieved whose home was forcefully broken into has the right to litigate the aggressor. Most of heavenly religions showed a great interest in respecting the sanctity of private life. For example, Islamic Shariah (jurisdiction) prohibited spying on others, sneaking looks and hearing and disclosing secrets because of the damage that incur private life, promulgating its secrets as stated in chapters of the holy Quran [9], (chapter 49 (Al-Hujurat), verse 12):

> O ye who believe! Avoid suspicion as much as possible: for suspicion in some cases is a sin: And spy not on each other, Nor speak ill of each other behind their backs. Would any of you like to eat the flesh of his dead brother? Nay, he would abhore it [1, 9]

And Chap. 24 (Al-Noor), [9], verses 27, 28.[1][2]

> O ye who believe! Enter not houses other than your own until ye have permission and salute those in them: that is best for you [9].

Private life was the focus of interest by international conventions. The International declaration of Human Rights tackled the issue of private human life and how to protect it from any arbitrary interference. Article (12) of 1948 declaration stipulated that "Nobody should be susceptible to arbitrary interference in his personal life, family affairs, home, correspondences or any campaign of defamation. Every person has the right to be protected by law from such interferences or campaigns" [6]. In addition to the International declaration, other agreements such as: European International Agreement on Human on Civil and Political Rights, 1969, all reiterated the same stance. In the aftermath of 1968 United Nations Conference held in Tehran, the first legislation on the issue of dangers of information technology on human rights regarding privacy emerged. The legislation was subjected to successive modifications in 80 s and 90 s. Among the major objectives of the conference and what the participating states came up to where: to respect private human life, to care for individual's security in light of technological development through an ideal use of technology, to balance accelerating development of technology with human rights [1]. In 1968 Montreal convention was held. Among some of its important recommendations were: to pay attention to what threatens private life from technology like the internet, to steer governmental and non-governmental bodies to eliminate threats resulting from electronic media and other agreements like that of Budapest 1995 which showed concern with global criminology [11].

---

[1] Quran, Holy. Ali, Yusuf (trans) 1983.
  Ibid.

As for the Jordanian legislator, he coped with such advanced legislations. The Jordanian constitution of 1952 and its amendments for item (22), article (7) stipulated that: any assault on public rights and freedoms or private life of Jordanians is considered a crime punishable by law. Article (45) of the Egyptian constitution of 1971 also stipulated that private life of citizens has its own sanctity protected by law. The Iraqi constitution of 2005 in article (17), item (1) also indicated that every citizen has the right to enjoy personal privacy, pending that it doesn't conflict with rights of others and public morals.

Most of penalty legislations including the Jordanian, Egyptian, and Iraqi avoided defining what private life means. That might be because it is difficult to determine what is private and what is public in human's life as that broadens and narrows in accordance with people's and individual's conditions and cultures. If all agree that private life has to be legally protected, this obligates keeping it confidential [8]. pointed out that the definition of private life is controversial in legal studies.

The researchers see that the difficulty arises from the term itself because the concept of private electronic life differs from the traditional one, for being relative and loose and can't be accurately specified. The normal person cares for keeping a part of his life private away from people and publicity. In fact, there is no specific criterion which can be enacted by jurists to measure the internal and external issues that relate to individual's private life, due to the changing nature of the concept in accordance with time and place changes, in addition to the influences of political, economic, social, and health circumstances on that type of life.

Therefore, the legislations that were enacted on the issue couldn't provide a specific definition for the term. Despite that, individual's private life aspects can be determined though.

On the other hand, one finds that judicially, privacy according to (Suroor, Fathi, 1979) means "the invaluable piece of human entity that can't be detached from the individual because then he becomes a solid tool devoid of any creative ability" [11]. Every individual has his own secrets and feelings which nobody else can enjoy except himself. Such a private life needs to be protected through legislation the way freedom is.

Jurisprudentially speaking, some consider private life to be represented in the right of the individual to keep some information related to his personal life confidential. This confidentiality strongly correlates with the concept of private life.

Despite the strong correlation between the concept of private life and private secret, yet the researchers see that the idea of private electronic life, as a legal concept, is broader and more resilient than the idea of confidentiality. It concerns the individual, and the others have nothing to do with it. Whatever the secret is, it relates to private life. Yet, the opposite is not correct as the idea of private life includes private secrets and all that is associated with personal affairs which others are not legally allowed to pursue.

The researchers point out that such secrets might involve information about crime or about rights of others. Thus, the legislator might consider enacting legal provisions in this respect an infringement of the sanctity of private life which is legally considered first priority [10].

Judge Douglas defined the right to privacy to be the individual's right to choose his own personal behavior and actions whenever he socializes with others through free expression of his ideas, interests, and personal taste. Individual's freedom involves having children and bringing them up, in addition to enjoying dignity, being free from coercion and oppression [3].

Based on what preceded and despite the judicial definition of private life, the researchers didn't find a comprehensive definition for private electronic life. The absence of such a definition might be because the concept of electronic privacy varies from one society to another in relevance to place, environment, traditions, and norms. It is eventually a situation, not disclosed to the public that the person loves to keep as a secret in electronic space.

The concept of private electronic life is so resilient to include all issues related to individua's emotional, familial, matrimonial affairs, talks, political and religious beliefs, homeland, domicile, correspondences, images, taboo pictures, professional life, leisure time activities, etc.

## 2.1 Second Requisite: Difficulty of Securing Criminal Protection for Private Electronic Life

Securing protection for private electronic life at present is difficult due to openness of information systems and computerized programs in all fields of life. Thus, information systems became indispensable for every individual to process his data, save them and exchange them with others. Such things created an electronic space that contains banks of private information on normal and corporate people. The internet, electronic exchange of data, spread of electronic applications, and social media sites such as Facebook, Instagram, and what's app all boosted saving data in such banks and made people more inclined to share their private lives with a limited number of persons. Thus, it is the individual's rights to decide when, how, and to what extent they share their private information with others. Despite the strict supervision over such information systems by the designers, yet some criminal minds succeeded in deciphering them and in exposing date of individual's private life that anybody anywhere could have access to. This created a widespread criminal phenomenon manifested in parasitism, penetration, and encroaching upon the sanctity of individual's private electronic life. As a result, transparency and privacy of humans lay naked to the global information system. Such phenomena obligate enacting rules and legislations to protect this kind of life and to deter parasites and criminals.

Due to the variety of technological methods used to encroach upon private life saved on websites, it is very difficult to secure a comprehensive type of protection for all forms of encroachment such as piracy, impersonation and control over private secret numbers associated with individual's private lives.

The difficulty can also be noted in the inability to identify pirates and criminals who use pseudonyms to avoid litigation, but spread malware using viruses that destroy information systems and websites. Piracy can also be noted in what is known as electronic forgery for digital and personal data which can be tampered with by the intruder who can change program language that results in a change of rights incorporated in the data, in addition to intercepting messages or redirecting them. such a kind of piracy disrupts information and date, particularly those of banking or stock exchange [7]. pointed out that the criminals (in this case the intruders) can easily promulgate rumors and defame anybody through modern technology that doesn't need any effort.

# 3 Criminal Protection Challenges and Solutions to Private Life

There is no doubt that technological crimes that affect private life are the greatest challenge that encounter countries at the present time, due to the social, economic, and political changes that occur. Therefore, there is a need to look into the illegal practices and to identify their features in order to repulse their dangers and to achieve a comprehensive security for the country and its citizens. The researchers will here down present challenges and solutions in the following requisites:

**First: Adequacy of Criminology in Jordanian Legislation for Protecting Private Electronic Life**

**Second: A Vision Regarding the Protection of Private Electronic Life**

With regard to the first requisite, the Jordanian legislator followed the traces of most comparative legislatures by enacting a special regulation for electronic crimes number 27 for 2015. The law listed the foremost criminal features related to information technology and networks; it didn't directly refer to private electronic life, but included general provisions such as: place of crime, and means of execution. Because the legislator realized that it was difficult to identify the criminal features committed through technological means in that law, article (15) ascertained that "whoever commits any crime punishable in compliance with any legislation in force using the network, any other information system, or any website, participate or incite he should be punished according to the provision stipulated on in that legislation". (Article 3 of electronic crimes, 2015). The legislator also indirectly elaborated on the distinct forms of encroachment on private electronic life as represented by: privacy crimes, non-permissible entry to networks and websites.

The legislator later added what is now known as programs of spreading malware that destroy system, programs and networks. Article (4) of electronic crimes law, No. 27, 2015 didn't stipulate that information technology should be protected by any of the anti-virus programs. This is classified as crimes of result which imply that if the doer fails to achieve his objects even though he tried, yet he can't be punished. It is considered a kind of non-punishable felony that has is no law provision for it.

The legislator's attention was also drawn to what is known as the illegal interception of messages sent through network or information system. It is the closest provision to the protection of private electronic life, though it doesn't have a direct provision. According to article (76) of Jordanian communication law, it is considered a crime regardless date or information sent in any form via website or information system. The Jordanian legislator through the traditional penal law maintained protecting sound and image sneaking [4]. There are several provisions in penal legislations which might be utilized for the protection of private electronic life according to article (15). These are: central bank law, illegal competition law, general statistics law, social security law, and finally customs law.

By analyzing the approach of Jordanian legislator for protecting individual's private electronic law, one finds that it is not sufficient to include some criminal forms. Still, he neglected electronic sexual blackmailing, artificial intelligence crimes, crimes of encroaching upon intellectual property, author's copyrights, patent rights, crimes of spreading electronic rumors, in addition to personalities assassination. Despite that, the protection of private electronic life needs to be reviewed and a legal regulation to be enacted to complement aspects of protection in order to achieve an active protection for private electronic life.

## 3.1 Second Requisite: A Vision Regarding the Criminal Protection for Private Electronic Life

Despite the spread of the phenomenon of penetrating private electronic life with its criminal, political economic and social impacts, the Jordanian legislator didn't add any provision to the electronic criminal law to protect private life, in contrast to what was stipulated on in the penal law [4]. More than that, he didn't even mention the term of individual's private life in any of provisions of that law [4]. Mentioned that the Jordanian legislator didn't even look into the crimes related to protecting the previously mentioned private life, as they were all considered felonies because such crimes are never penalized except through provision.

The researchers add that the electronic criminal law by using the criminal method of alternating behavior to determine the criminal one in such crimes weakens the legislative provision, from their perspective. Such a use is odd in light of regulating a crime linked to developed and changeable electronic means. It was better for the legislator to add a statement, for example, on encroachment upon environments and information to give a chance for Jurisdiction and judiciary to interpret the meaning and to apply it in light of time and place variables. Although article (348) included a provision which was intended to be comprehensive for protecting private life, yet the legislator failed in distinguishing the criminal behavior considered to be a penetration. It was better for him to be satisfied with any means that copes with aims of the provision.

Hence, the researchers conclude thereof that the Jordanian legislator succeeded to an extent in legal regulations for protecting private electronic life to deter such a crime. No one can imagine that any penetrator or parasite on private electronic life of others could escape in light of the previously mentioned provisions. The only thing the researchers have against the legislator is absence of penalty to be inflicted on initiators of such crimes. In addition, they feel that legal codification of legislative provisions need to be improved. They also see that a set of certain provisions need to be added to Jordanian penal law or to a new law under the name of law of protecting private electronic life whose aim is to contain all forms of potential criminology that might possibly occur [5].

## 4 Conclusion

The researchers came up to the following findings:

1. Right of private life is a right that all international agreements, constitutions and legislation confirm.
2. Private electronic life is a loose and resilient term that narrows and broadens in accordance with place and time. Therefore, it is difficult to put down a comprehensive definition that engulfs all meanings relevant to the concept of the term.
3. The Jordanian legislator didn't directly and accurately identify the law of electronic crimes.
4. Sentimental conviction of the judge of crimes is what determines the implication of private electronic life in every criminal lawsuit.
5. Thematically, the legislator succeeded to a great extent in protecting private life, but failed in codifying such provisions.
6. The punishments contained in electronic crimes law are not proportional to the criminal dangers they represent.
7. The Jordanian legislator succeeded in adding a provision to article (15) of electronic crimes act since it can be applied to any electronic encroachment on private life.

## 5 Recommendations

The study recommends the following:

1. To enact an independent law called "Law of encroachment upon individual private life", or to add a set of provisions to the law of electronic crimes to frankly criminalize all forms of encroachment.
2. To re-codify legislative provisions concerned with criminalizing private life penetration and to stop using the style of crime alternating behavior so as to allow jurists and judges to apply the developing and changing method of electronic crimes.

3. To assign a competent court to investigate crimes of encroachment upon private electronic life, to put down procedural provisions to deal with procedures, and to establish deterrent punishment to stop the spread of such crimes on all social media means.

4. To increase international cooperation through conferences, courses and seminars to put down common provisions which help in protecting private life.

5. To modernize and to develop as electronic crimes unit, and to train and qualify cadres specialized in electronic crimes.

6. To increase community awareness about the need for preventive protection to stop electronic penetration and piracy committed against internet users.

# References

1. Abdul Badee (2002) Right to private life and the extent criminal law secures protection. (Unpublished dissertation), University of Cairo, Faculty of Law, Egypt, p 109
2. Ayoub P (2019) Legal protection of personal life in the field of information. AL-Halabi Legal Circulars, Lebanon, p 40
3. Douglas (1978) The privacy opinion of justice. Yale Law J 168
4. Jordanian Penal Law (1961) No. 16. Jordanian law of electronic crimes. Jordan
5. Majali N (2010) Interpretation of penal law. Dar Al-Thaqafa for Publishing and Distribution, Amman, p 26
6. Ministry of Justice (2016) A study on merging regional and international laws in teaching. Judicial Institute, Amman, p. 42
7. Ostaz S (2013) Violation of private life through the internet. (Unpublished M.A Thesis). J Damascus Univ Law Econ Sci (3):431 (Damascus)
8. Qayed O (1994) Criminal protection of private life and information banks. Dar An-Nahda Al-Arabiyeh, 3rd edn. Cairo, Egypt, p 9
9. Quran, Holy. Ali Y (trans) (1983) Amana Corp. Maryland, pp 903, 406
10. Shammat K (2005) Right to private life. (Unpublished Dissertation), Damascus University, Damascus
11. Sroor A (1979) The mediocre in criminal procedures law. Dar An-Nahda Al-Arabiyeh, Cairo

# Challenges of Taxation and Blockchain Technology

## A Legal Study

**Qabas Hasan Awad Albadrani**

**Abstract** There are constants laid down the legal foundations and principles for conducting financial transactions among individuals that cannot be violated according to what laws have been established and what people have come to know about and according to public order too, and these foundations are important to ensure security, confidence and protection for the parties, moreover giving legal status to the disposition or the event achieved in the sense of granting it legal recognition official and general acceptance of people and official bodies beside being the basis for government authorities to adopt when they intervene to impose taxes, but the frequency of using modern technologies, Blockchain especially would be a great challenge of imposing and collecting taxes under a system does not recognize the legal constants that depend on the paper and official proof provided, and the material tangible in the money or the place of the contract.

**Keywords** Financial law · Blockchain · Tax · Financial sovereignty

## 1 Introduction

This paper addresses modern technical variables, In particular, blockchain relating to the electronic circulation of goods and services, and its effectiveness to legal concepts both national and internationally applicable in the sense of tax law system, highlighting the most important concepts that will inevitably be affected by these technical variables, how states in both their domestic administrations or the work of the international legal system in briefing and attempting to regulate such activities legally.

Q. H. A. Albadrani (✉)
University of Mosul-College of Law, Mosul, Iraq
e-mail: Qabas.hasan@uomosul.edu.iq

© The Author(s), under exclusive license to Springer Nature Switzerland AG 2022  727
S. G. Yaseen (ed.), *Digital Economy, Business Analytics, and Big Data Analytics Applications*, Studies in Computational Intelligence 1010,
https://doi.org/10.1007/978-3-031-05258-3_56

# 2 Research Hypothesis

1. What Is considered to be a variable at a moment in time may be viable, and with viability and survival through application and experimentation until it takes place among the rest of the constants.
2. Despite the nature of financial law in absorbing societal variables internally and externally, Due process remains the basis for any solutions, for that, we find the international legal system provides solutions to control modern technologies based on traditional and relatively consistent concepts of financial law.

## 2.1 Manifestations of Blockchain Technology

All behaviors that take place in the Blockchain System are characterized by a set of features and aspects resulting from the increasing use of modern technology, Blockchain, which clearly differs and completely contradicts the usual legal formal principles prevailing in legal dealings that we all know, it includes:

1. The absence of a certification body, there is no official authority empowered by a law issued by a state to practice this documentation for the circulation of funds, services, and goods. There is no third party, but only two parties, especially in light of the countries 'endeavor to recognize this technology and issue legislation that establishes legal infrastructure to contain this global commercial movement the range [1].
2. There are no documents, paper documents, or electronic files that contain a scan of a signed electronic copy, and thus the need for a third party to keep the paper and electronic copies are eliminated.
3. The two factors of guarantee and trust in the circulation of funds, goods, and services are achieved based on the idea of collective knowledge from the users of this technology, which created the guarantee directly. The distribution of information and data among Blockchain users will provide high security from any hacking or attempted theft if successful, this means that those behind this movement controlled 51% of the nodes or blocks and layers of the Blockchain system, and this matter according to the technicians of the Blockchain System is impossible. So, any modification or change to any layer or block and node will not change the rest of the nodes and other blocks in which the data also resides [2].
4. The principle of confidentiality is available in this technology through the use of a coding system or encryption for the owners of blocks or nodes in the chain, and this means that the circulation of data on the process of transferring ownership of a property will be traded among the owners of the blocks without knowing the identity of the property owner to whom the ownership was transferred, as he is given a symbol or code Especially in it, on the other hand, this technology provides the idea of publicity and displaying information and data

in front of technology users without concealing the information concerning the confidentiality of the names of the owners of the parties to the commercial or civil relationship.

The legal formalism in its traditional form is completely inconsistent with the rules and foundations of trading within the blockchain technology, which is a conflict between two concepts each insist on preserving its constants and foundations, and the challenge is prominent in the hands of lawmakers, how they will provide solutions to this conflict and surround all the actions and facts achieved within this virtual environment. Since it is the result of transactions for the transfer of property and money between natural and moral persons, meaning thinking of applicable solutions and not the trend towards prohibiting dealing with this technology and similar ones or recommending that it violates legal and societal constants and may even reach the possibility of it contradicting religious constants in some religions, and therefore the legal legislator is not obligated nor The legal authorities are to fall into the problem of not keeping pace with the movement of money and goods and the emergence of new markets outside the scope of general legal regulation, which means losing profit first, in addition to achieving negative effects on the stability of societies. The normal environment within the Blockchain environment and similar ones, as it is said, while this technology has its law, so there is no world with There is no law or regulations [3]. So that for individuals to trust and guarantee protection for their money, this must be characterized by clarity, understanding, and full knowledge of all the details of this circulation and what is permitted and not permitted regardless of the presence of the official third party in this technical financial deal, which raises an obligatory question: Is that The virtual environment for commercial and financial transactions will create a general system parallel to the traditional and physical public system, or will it also depend on what is prevalent in normal and non-electronic legal relations?

## 2.2 Challenge of Imposing and Collecting Taxes in the Electronic Environment

Taxes remain one of the most important tools that express the legal sovereignty of the state overall funds and people in its territory, this fixed idea has been facing a continuous challenge for years, especially since the widespread use of technologies and the electronic environment in various economic transactions. The extent of imposing taxes on companies that deal in electronic commerce over the Internet and have their clear and information addresses, but this controversy no longer has an effect in the face of a virtual environment that relies on numbers and symbols, and there are no addresses known to the authorities for those who deal with Blockchain technology and similar ones. The profit was achieved or the money was produced in it + A specific amount of money + A person who pays the tax + A time during which the event that established the right of the tax authorities to claim the tax debt, These constants are governed by an important rule, which is that The Tax law is an Internal law whose application limits are clear within the spatial scope of state sovereignty It is not for it to exceed its application to the borders of another country except in cases that are mostly agreed upon bilaterally [4], and this means that the philosophy of imposing a tax is strong on the existence of tangible physical manifestations, and because of that the tax administrations faced problems in applying traditional legal rules to electronic commerce in all its forms, and when countries began to set unified standards for tax dealing with these electronic activities, a hypothetical system appeared more complex and far from the material, and it is known that trade is characterized by the development in its patterns It is not stopped by geographic or political borders, but paper and electronic documents remain approved by the tax authorities of any country, as they need documented information. Therefore, a network of administrative procedures regulated by law has been created that provides a tax database on which the tax authority relies when taxing money, and here the conflict appears clear between Blockchain technology and who deals with it, and what the tax authorities want through:

1.  The tax authorities cannot obtain the data verified in The Blockchain system, which means that these trades will not impose any tax on them.
2.  If The tax authorities were to impose and obtain information related to the circulation of funds in The Blockchain system, which tax law shall be applied and what are the limits of its application, especially since The Blockchain system does not contain countries or national or international laws, it is a system that represents interconnected blocks with each other in an encrypted manner. Distribution of information, circulation of funds, and the transfer of ownership among them.
3.  If any tax authority was able to restrict the place in which the transfer of money was achieved, who is the person who owned this money and achieved the profit, which one will pay the tax, one of the characteristics of the Blockchain system is that it is a record of digital facts and actions with symbols and not just a normal documentation record but rather includes Smart transactions, "smart contracts", in the form of encrypted codes, without any risk or likelihood of being stopped or censored [5].

## 2.3  Procedures of the Organization for Economic Cooperation and Development

Perhaps there are some international attempts to create a major structure through which to find keys to deal with these virtual technologies, despite the trends of some countries towards not imposing local taxes and fees on these hypothetical transactions. Therefore, we find that some US states support the hypothesis of not imposing local taxes and fees on the use of these technologies, as in the states of Nevada and Vermont in the Americas [6] [However, we find that the Organization for Economic Cooperation and Development directed in its report issued by its conference held in October 2019 a report in which it clarified the necessity of dealing with these digital technologies [Blockchain and cryptocurrencies[—Blockchain and virtual money as in Bitcoin. And similar ones—as it considered that the digital or knowledge economy has communicated tax rules that are considered to be installed and approved as rules of universal application—and this matter was the responsibility of the Organization for International Economic Cooperation and Development—in the face of a real danger that should not be ignored by countries. The members are represented by [7]:

1. Individual countries dealing with these digital challenges by issuing legislation and recognizing some of these technologies within some forms of circulation of movable and immovable funds, as is the case in South Korea.
2. Some countries do not apply internationally agreed standards for taxation.

   Therefore, the report presented several solutions to confront digital technologies in the scope of global trade by adopting the idea of electronic residence, and it is known that one of the foundations of taxation and considering persons subject to tax fulfills the fact of residence for a specific period, and therefore the idea of approved residence in most tax laws in countries can be applied to activities. Digital without requiring the idea of physical or actual residence, as is the case in traditional economic activities, and despite the relevance of this opinion, the matter needs studies and maturation of the idea, which can be said despite the relevance of taking it, but that residence in itself is a material idea that depends on physical and physical existence for a period of time most of the laws The tax takes a period of no less than 4 to 6 months in the country of residence, and if we take the idea of electronic residence, then this means searching for elements for this residence that depends on the Internet law through electronic use systems and the controls it takes to install accounts, and even under the Blockchain system, we are in front of blocks or Nodes distributed in the names and encrypted accounts containing economic trading data, and therefore some US states, as previously mentioned, tend to recognize eagerness The electronic mail and its formula adopted in the Blockchain system is understandable and is a necessary step if the legislator decides not to ignore this technology and similar ones, finally we must have a new legal vision towards blockchain system and its subject, Recognition of this smart legal person the same as physical and legal person with respect to the elements of Blockchain system.

## 3 Conclusions

1. National tax systems must be dealt with seriously about trading and actions that take place through Blockchain technology and take advantage of the capabilities available therein, and therefore there must be a set of procedures that we consider as a legal basis regulating the imposition of taxes on this type of commercial behavior.
2. There must be internal legal recognition of these behaviors used for Blockchain and similar ones as a preliminary step for their inclusion in the national legal system.

3. Coming up with realistic solutions and benefiting from some normal legal standards and The possibility of applying them to the actions achieved in The Blockchain system, as in the idea of electronic residence, which is a new legal idea of an old principle, which is a physical residency that needs maturation and deepening.

4. Searching for new legal standards and trying to develop definitions of traditional legal concepts, but with a modern vision and dimensions that were not previously known, as in the term Sovereignty—Procedures—Documentation—Legal Formalism.

# References

1. Iansiti M, Lakhani KR (2017) The truth about block chain. Harvard Business Review 4
2. Dajani D, Yaseen SG (2016) The applicability of technology acceptance models in the Arab business setting. J Bus Retail Manag Res 3–10
3. Viswanathan M (2017) Tax compliance in a decentralizing economy. Ga St UL Rev 34:283
4. Nashid DSA (2008) The phenomenon of International tax evasion and its effects on the economics of developing countries in Arabic. Al-Halabi Legal Publications, Beirut
5. Cao S et al (2017) A review of researches on the Blockchain, Wuhan International Conference on e-Business. Association for Information Systems, Wuhan-China 2017
6. OECD Secretary-general Tax report G20, October 2019, Finance ministers and Central Bank governors OECD, 4, s.l.: s.n. 2019
7. The Next Web 2017. https://thenextweb.com/news/7-interesting-laws-blockchain-apps-will-force-to-change
8. Ashawai M (2017) A look at state and law in Arabic, 1st edn. Al-Thakera for distribution and publication, Baghdad

# On the Boundary Between Rest Time and Working Hours in a Digital Environment

**Andriyana Andreeva, Galina Yolova, and Diana Dimitrova**

**Abstract** This paper examines some current issues related to the boundary between the institutes of rest and working hours in a digital environment. The focus is on some contemporary aspects of these concepts, conditioned by the digital environment for labor performance. Through a legal analysis of the applicable national labor law in Bulgaria and reference to ongoing processes at European level and the relevant legal acts the authors seek to conclude on some ongoing transformations. Some summaries and concrete proposals for improvement of the regulation are formulated in the paper's conclusion.

**Keywords** Right to rest · Working hours · Remote work

## 1 Introduction

In labor law, the issues of working hours and rest have been the subject of interest on the part of many authors during the various stages of development of the legal branch. They are linked to the evolution of the industry and are the driving force behind the protection of employees' rights.

At present, the challenges facing these concepts are determined by several factors. *On the one hand*, digitalization leads to a transformation that affects all spheres of life, from the personal to the professional, and concomitantly changes occur in the norms of communication: social, legal, ethical, moral. Given the aggressive pace of digitalization, it is almost impossible to make sufficiently reliable forecasts for the

A. Andreeva (✉) · G. Yolova · D. Dimitrova
Legal Sciences Department, University of Economics—Varna, Knyaz Boris I blvd., 9000 Varna, Bulgaria
e-mail: a.andreeva@ue-varna.bg

G. Yolova
e-mail: ina_yolova@ue-varna.bg

D. Dimitrova
e-mail: dianadim@ue-varna.bg

© The Author(s), under exclusive license to Springer Nature Switzerland AG 2022    733
S. G. Yaseen (ed.), *Digital Economy, Business Analytics, and Big Data Analytics Applications*, Studies in Computational Intelligence 1010,
https://doi.org/10.1007/978-3-031-05258-3_57

future state of the labor market and the organization of people's working lives. [1] In this regard, the process of digitalization requires a rethinking of concepts established in labor law, their legal updating in order to achieve the goal of adequate regulation of public relations. Along with the need to update the institutes within the framework of the national legislations, the topic also raises a number of issues at the doctrinal level related to the improvement of the institutes in view of the new social realities related to digitalization. The concept of working time is a measure of the due service in terms of time provided by the employee. At the same time, it limits the employer's interference in the employee's private life.

*Secondly*, the pandemic caused by COVID-19 has necessitated urgent measures to protect human health and limit the presence work related to traditional jobs. In view of this, a large number of employees have switched to remote work. This has further exacerbated the need to rethink the traditional and interconnected concepts of working hours and rest.

The correlation work-rest acquires a new meaning, when placed on the plane of the emergency epidemiological measures and work in a remote environment. The world is on the one hand in the period of the fourth industrial revolution, and on the other hand in a pandemic situation. These two factors have a relevance to all societies and are refracted through the prism and specifics of social development. In such a situation, it is not possible for the legislative mechanism to respond in a timely and adequate manner, given the dynamics of the processes and the unpredictability of the COVID-19 pandemic.

*Thirdly*, the digitalization of public relations has imposed a new approach to the regulation of classical employment contracts, introducing distance work contracts into their system and creating a new culture of Internet nomads, in which the concepts of working hours and rest time are reshaped by the circumstances of the new digital reality in labor performance.

At the same time, the issue has not only a legal significance, it also puts ethical norms to the test and causes the need to activate moral pillars related to responsibility and tolerance. Responsibility on the part of the employee who assumes new commitments traditionally typical for the employer, as well as of his employer's authority, i.e. control over the work process, carried out in a home environment, which accidentally replaces the traditional workplace. The employee is also assigned the atypical task of self-control over labor discipline. The employer, on the other hand, is also in an atypical situation, characterized by the lack of direct control over the labor process and compensation with digital methods of exercising part of their employer power.

This implies greater tolerance and increased confidence in the observance of labor discipline by the employee.

This is what provokes the authors' interest in studying the problem of the boundary between rest and working hours, the boundary between legal and ethical norms in the conditions of digitalization and pandemic. The role of science is to be a catalyst for progress, and that of the legal science—a pillar of justice in the regulation of social relations.

*The relevance of the studied issues* is complex and can be presented as follows: *First*, in its interrelationship the issue of the boundary between rest and work in modern labor law considered in the context of digitalization and the pandemic has not been the subject of theoretical research. To date, there are no theoretical works analyzing this correlation, and practice shows the need to clarify these dependencies in order to change the conditions in which work is performed and the need for adequacy on the part of employers. *Secondly*, the relevance stems from the need for legislative changes that would provide additional guarantees for the observance of the individual labor rights of employees, but also the interests of employers. The achievement of the status quo of equality between the parties in the employment relationship is put to the test. *Thirdly*, it is quite clear that reforms are needed in the classical doctrinal formulations for the institutes of working time and rest, in the light of their understanding as dynamic legal constructions, whose change in the digital environment is yet to come.

It is the above-mentioned complexity that outlines the importance of the researched issues, which have mainly been the subject of independent legal research on working hours and rest, on employer power and control, on responsibility and self-discipline.

*The aim of the present study* is to examine, through a complex normative analysis, some current issues related to the institutes of working hours and rest, both in the Bulgarian legislation and in view of the tendencies of the common European policy. Emphasis is placed on some contemporary aspects of these concepts, conditioned by the modern digital environment of work performance. The authors seek to conclude about some ongoing transformations through legal analysis of the current labor law and reference to the ongoing processes. In conclusion, summaries have been formulated and specific proposals have been made for improving the regulation.

To achieve this goal, the following *research tasks* have been implemented: (1) analysis of the institute of working hours with its current manifestations in the modern digital environment; (2) clarification of the interdependence of working hours and rest in the labor process; (3) formulation of summaries, conclusions, specific proposals for improvement of the legislation.

The *research methodology* includes the complex application of the following methods: legal dogmatic, normative, induction and deduction. The material complies with the legislation as of April 30, 2021.

## 2   On Some Issues of Working Hours and Rest—Contemporary Aspects

European legislation contains a number of rules relating to the question of the boundary between working hours and rest. Here are some of them that bear importance in the context of the present study:

- Art. 2 of the European Social Charter of the Council of Europe of 3 May 1996, regulating the right to fair working conditions, including of reasonable working hours and rest.
- The report of the European Foundation for the Improvement of Living and Working Conditions (Eurofound) of 31 July 2019, entitled "The right to switch off".
- The case law of the Court of Justice of the European Union on the criteria for determining working hours, including on-duty time and on-call time, the meaning of rest time, the requirement to measure working hours and the criteria for determining worker status.[1]
- The European Pillar of Social Rights, and in particular Principles № 5, 7, 8, 9 and 10.

There is currently no specific legislation in the EU regulating the boundaries between rest and working hours and the related right of workers to switch off the digital tools used for carrying out the work.

Digitization and digital tools are undoubtedly useful for both parties to the employment relationship: economically beneficial for the employer and providing flexibility and increased autonomy for employees. Flexible forms of employment have not only been regulated in the legislation of many countries, including Bulgaria, but they are also an important component in the struggle against the COVID-19 pandemic.

At the same time, however, many questions arise for which the legislation does not contain an answer, and even less so in the state sources norms. Ethical problems are combined with legal ones in the regulation of working hours: increase of the assigned work, extension of working hours. This unregulated process crosses many boundaries: between personal and professional life, between working hours and rest. In this situation, the responsibility is blurred. On the one hand, the lack of explicit normative regulation presupposes a shift in the layers of engagement of the parties. Traditionally, working hours are determined in state sources within a certain range, limiting its maximum duration, but allowing freedom to the parties to provide for more favorable rules in non-state acts /CLA, internal acts, etc./. However, the digital environment and the use of digital tools are changing this reality. Employers and workers are already connected not only within the established working hours, but also outside them, and this is a lasting and deepening trend. This "permanently secured line of interaction for official purposes" transforms the essence of the employment relationship and shifts the layers of commitments in favor of the employer. Indeed, for the employer, as the organizer of the labor process, it is more favorable to have an employee available at all times for the performance of the respective labor functions. At the same time, however, this violates the basic individual rights of the worker or

---

[1] See the judgments of the Court of Justice of the EU of 5 October 2004, Pfeiffer et al., C-397/01 to C-403/01, ECLI:EU:C:2004:584, item 93; of September 7, 2006, Commission/UK, C-484/04, ECLI:EU:C:2006:526, item 36; of November 17, 2016, Betriebsrat der Ruhrlandklinik, C-216/15, ECLI:EU:C:2016:883, item 27; of February 21, 2018, Matzak, ECLI:EU:C:2018:82, C-518/15, item 66; and of May 14, 2019, Federación de Servicios de Comisiones Obreras (CCOO), C-55/18, ECLI:EU:C:2019:402, item 60.

employee. [2] On the one hand, there are the typical and traditional labor rights—the right to rest, to a leave of absence, to remuneration (given the overtime work performed), to safe and healthy working conditions. On the other hand, rights related to privacy, invading the family sphere, etc. are also affected.

The responsibility for this avalanche-like process of digital instruments entering labor relations and the adverse effects we have mentioned cannot be defined unequivocally as being imposed solely on the employer. Indeed, the employer is usually the party that implements digital tools and this corresponds to his employer's authority. However, it is not uncommon for employees to "violate" their rights by transferring work to non-working hours. The reasons for this are many and can be related to the lack of an explicit regulation for switching off from work, as well as to the goal of personal professional realization and growth in the corporate hierarchy. These issues are only part of the overall problem, which requires interdisciplinary research and proposing complex solutions. Account should be taken of the fact that the workforce is inseparable from the personality of the worker or employee. This is what should be kept in mind when preparing complex regulatory mechanisms for the use of digital instruments. Their overuse can have a negative impact on a person's personal development and the effects of this are yet to be felt (according to the WHO more than 300 million people worldwide suffer from depression and common work-related mental disorders, and 38.2% of the EU population suffer from mental disorders each year) [3].

In the period before the pandemic, digitalisation was studied from the point of view of a high level of benefits and the desire for regulation with a view to expansion and application in more areas and professions. To date, the pandemic has changed this reality—remote work and distance work have increased during the COVID-19 crisis and are expected to remain at a higher level than before the virus crisis. Employees work from home without this being provided as an opportunity upon establishing their employment relationship. Under the threat of the spread of the infection and with a view to protecting public health, employers have the extended power to determine the regime of work at their discretion. In these cases, workers are usually in a new function of "self-control" over the observance of working hours. At the same time, there are often hypotheses where employees themselves disrupt their own rest in an effort to be at the right level and in order to cope with their new tasks, transformed or modified when working remotely, and through the mandatory use of digital tools.

Working time is one of the components of the work discipline. It is exactly this element that is carefully monitored by employers through various means of control, including through various technological means of Internet supervision.

Employer surveillance and supervision have expanded without being limited to the workplace and working hours. The monitoring systems detect operation and activity in the network, respectively strict reading of the switch-off time. This is indirectly a complete control over the distribution of professional and personal time of the individual and unregulated intrusion into his privacy. Article 8 of the European Convention on Human Rights (ECHR) provides that "everyone has the right to the protection of personal data". This should be linked to the privacy of employees in the context of employment relationships. In this regard, employers should make their employees aware both of the monitoring and control measures introduced and the justification for the need for them. The participation of trade unions in the introduction of control mechanisms by employers should be expanded. The internal regulations no longer comprise the workplace only, their perimeter of action should be expanded also in view of the restriction of the employer's intervention in distance or remote work.

All this creates the urgent need, both at the international and national level, to regulate with imperative norms the right to switch off from work. The European legislator responds to this need through specific procedures for drafting the Directive on detachment from the working environment. The idea is to set out minimum regulatory requirements introduced and established in national labor laws, respectively at the level of collective bargaining or internal departmental rules, without prejudice to the right of Member States to introduce or maintain more favorable provisions. The act also contains a definition of "detachment from the work environment", namely not to carry out, directly or indirectly, activities or communications related to work through the use of digital tools outside working hours.

This is the basis which the member states should use to adapt their national legislation, respectively to further develop in the non-state sources this new modern law, which will provide a boundary between working hours and rest, between personal and professional life.

Undoubtedly, this step is necessary because the achievements in the development of labor law must not only be defended and guaranteed, but they must be further developed in accordance with the time in which the labor is provided.

# 3   Conclusion

The question of the boundary between working time and rest is a legal, as well as an ethical issue. This traditional question has been important in the evolution of labor law and has reflected the degree of development of the respective society. At the same time, it is both a controversial and a delicate issue in the doctrinal and practical terms of labor law. Its new forms in the digital environment increasingly pose questions of a different nature and character, on which national legislation is currently unable to find an unambiguous and easy-to-regulate answer.

*Based on the research, the authors can formulate the following conclusions and suggestions concerning the relationship between working time and rest in the contemporary conditions of work performance*:

1.   Frameworks should be established urgently within the framework of acts of a different nature governing working hours, which are adequate to the new needs, both concerning the use of working hours and for guaranteeing the right to rest.
2.   To rethink the traditional understanding of the employer's competence to monitor and sanction labor discipline related to the violation of working hours by employees.

3. To rethink the issue of responsibility in the labor law, and along with its traditional types to expand the scope of the employer's responsibility, providing for the violation of the right to rest and intrusion into the private life of the employee.

4. To limit the legal means for control over the observance of working hours and especially those representing infringement of sensitive data within the meaning of the General Regulation concerning the protection of personal data (Regulation (EU) 2016/679).

5. The boundaries between personal and professional life, respectively between working hours and rest should be regulated, particularly at the level of non-state sources (internal acts of the employer, collective bargaining and tripartite cooperation).

6. To apply the rules for the right to detach from the working environment in cases that lead to continuous connectivity, overtime work and violation of the right to rest and leave of absence.

In the course of the research the authors have posed some traditional questions concerning the legal doctrine, and at the same time a number of ethical problems have become evident, caused by the new working conditions related to the entry of digitalization in all spheres of work, and the use of AI. It is the intertwining of legal and ethical norms in the modern digital age that presupposes a decisive transformation of the legal mechanism. Solutions should not only be sought in the field of legislative adjustments, they should combine the achievements of the legal doctrine with ethical and moral norms that focus on the individual and give the latter priority over the widespread technology and AI.

Not all questions that arise can be answered unequivocally and this is not the purpose of this paper. The authors have sought to make the public aware of some emerging needs, to identify trends and to suggest some possible solutions. However, the overall solution should be the result of the work of the international community and the unification of law and ethics in one for fair and dignified work in the new digital reality.

# References

1. Blagoycheva H (2019) Employers' social responsibility in the context of digitization. In: Bulgarian–Chinese forum International cluster policies: a collection of papers from an international conference 2018. Science and Economics, University of Economics–Varna, Varna, pp 94–111

2. Andreeva A, Yolova G, Blagoycheva H, Aleksandrov A, Banov H, Yordanov Z (2020) Protection of individual labor rights (of the worker and employee). Science and Economics, Varna (in Bulgarian)

3. European Parliament, https://www.europarl.europa.eu/doceo/document/TA-9-2021-0021_BG. html; WHO, https://www.euro.who.int/en/home

Printed in the United States
by Baker & Taylor Publisher Services